Advanced Linear Algebra with Applications

Mohammad Ashraf · Vincenzo De Filippis ·
Mohammad Aslam Siddeeque

Advanced Linear Algebra
with Applications

Mohammad Ashraf
Department of Mathematics
Aligarh Muslim University
Aligarh, Uttar Pradesh, India

Vincenzo De Filippis
Department of Engineering
University of Messina
Messina, Italy

Mohammad Aslam Siddeeque
Department of Mathematics
Aligarh Muslim University
Aligarh, India

ISBN 978-981-16-2169-7   ISBN 978-981-16-2167-3   (eBook)
https://doi.org/10.1007/978-981-16-2167-3

Mathematics Subject Classification: 15A03, 15A04, 15A18, 15A20, 15A21, 15A23, 15A42, 15A63

This Springer imprint is published by the registered company Springer Nature Singapore Pte Ltd.
The registered company address is: 152 Beach Road, #21-01/04 Gateway East, Singapore 189721,
Singapore

Preface

Linear Algebra is a fundamental branch of mathematics and is regarded as the most powerful mathematical tool which has acquired enormous applications in diverse fields of studies viz. Basic Sciences, Engineering, Computer Science, Economics, Ecology, Demography, Genetics etc. The present book has been designed for advanced undergraduate students of linear algebra. The subject matter has been developed in such a manner that it is accessible to the students who have very little background of linear algebra. A rich collection of examples and exercises have been added at the end of each section just to help the readers for conceptual understanding. Basic knowledge of various notions e.g., sets, relations, mappings have been pre-assumed.

This book comprises twelve chapters. It is expected that students would already be familiar with most of the material presented in the first chapter which starts with a quick review of basic literature on groups, rings and fields which are essential components for defining a vector space over a field. Further, it goes into the details of the matrices and determinants. Besides discussing elementary row operations and rank of matrices, method for finding solution of system of linear equations have been given. Chapter 2 opens with the introduction of vector space and its various examples. Apart from discussing linearly independent sets, linearly dependent sets, subspaces and their algebras, linear span, quotient space, basis and dimension, several exercises have been included on these topics. After a brief geometrical interpretation in case of vector spaces over the real field, change of basis has been included. Chapter 3 introduces linear transformation on vector spaces, kernel and range of a linear transformation, nonsingular, singular and invertible linear transformations. Further, basic isomorphism theorems, algebra of linear transformations, matrix of a linear transformation and vice-versa, effect of change of basis on matrix representation of a linear transformation are also discussed in this chapter. Chapter 4 treats linear functional and dual spaces. Some results related to annihilator of a subset of a vector space and hyperspaces (or hyperplanes) have been included. Finally, dual (or transpose) of a linear transformation and its matrix representation relative to ordered dual basis have been discussed.

v

Chapter 5 is based on inner product spaces. Orthogonality and orthonormality of a subset of a vector space are discussed. It also contains applications of Gram-Schmidt orthonormalization process in obtaining orthonormal basis, orthogonal complements, Riesz representation theorem in finite dimensional inner product spaces. Finally, operators viz. adjoint operators, self adjoint operators, isometries on inner product spaces have been studied. Chapter 6 is devoted to the study of canonical form of an operator. This chapter opens with eigenvalues and eigenvectors of a linear transformation. Further, triangularizable, diagonalizable operators are highlighted. We conclude this chapter with a study of Jordan canonical form of an operator, Cayley-Hamilton theorem, minimal polynomial and normal operators on inner product spaces. Chapter 7 deals with the study of bilinear and quadratic forms on Euclidean spaces. Proofs of the main theorems regarding the symmetric, skew-symmetric and quadratic forms have been included. Chapter 8 is devoted to the study of sesquilinear and Hermitian forms of an operator. Besides the matrix of sesquilinear forms the effect of change of basis has also been included. Chapter 9, which is novel in a book of this kind, illustrates a more advanced view of tensor product of vector spaces and tensor product of linear transformations. A unified procedure is presented to establish existence and uniqueness of tensor product of two vector spaces. Furthermore, we provide a fairly detailed introduction of tensor algebra. Final section comes up with an introduction of exterior Algebra or Grassmann Algebra.

The last three chapters focus on empowering readers to pursue interdisciplinary applications of linear algebra in numerical methods, analytical geometry and in solving linear system of differential equations. In detail, Chap. 10 initially provides an introduction to LU and PLU decompositions of matrices and their applications in solving the system of linear equations. Furthermore, we discuss briefly dominant eigenvalue and the corresponding dominant eigenvector of a square matrix and use the application of power method that gives an approximation to the eigenvalue of the greatest absolute value and a corresponding eigenvector. This chapter ends with singular value decomposition which is a powerful tool in machine learning, detaining, telecommunications, digital image processing, spectral decomposition, discrete optimization etc. We have demonstrated an idea how SVD compression works on images. Chapter 11 is dedicated to estimate distances between points, lines and planes, as well as angles between lines and planes in Euclidean spaces by using methods involved in the theory of inner product spaces. The chapter concludes with application of the progress of eigenvalues, eigenvectors and canonical forms of linear operators and quadratic form to study and classify the conic curves and quadric surfaces in the affine, Euclidean and projective spaces. Chapter 12 is devoted to study methods for solving linear systems of ordinary differential equations, by using techniques associated with the calculation of eigenvalues, eigenvectors and generalized eigenvectors of matrices. Further, the method to represent a system of differential equations in a matrix formulation and then using the Jordan canonical form and, whenever possible, the diagonal canonical form of matrices, a process aimed at solving linear systems of differential equations in a very efficient way has been described. Finally, it is shown that the method linked with the solution of systems also presents a way of dealing with the problem of solving differential equations of order n.

This book is self contained. Throughout the book stress is laid on the under-standing of the subject. We hope that this book will help the readers to gain enough mathematical maturity to understand and pursue any advance course in linear algebra with greater ease and understanding. We welcome comments and suggestions from the readers.

The authors gratefully acknowledge their deep gratitude to all those, whose text books on linear algebra have been used by them in learning the subject and writing this text book. We are also thankful to our colleagues and research scholars for making valuable comments and suggestions.

Aligarh, India   Mohammad Ashraf
Messina, Italy   Vincenzo De Filippis
Aligarh, India   Mohammad Aslam Siddeeque

Contents

About the Authors

Mohammad Ashraf is Professor at the Department of Mathematics, Aligarh Muslim University, India. He completed his Ph.D. in Mathematics from Aligarh Muslim University, India, in the year 1986. After completing his Ph.D., he started his teaching career as Lecturer at the Department of Mathematics, Aligarh Muslim University, elevated to the post of Reader in 1987 and then became Professor in 2005. He also served as Associate Professor at the Department of Mathematics, King Abdulaziz University, KSA, from 1998 to 2004.

His research interests include ring theory/commutativity and structure of rings and near-rings, derivations on rings, near-rings & Banach algebras, differential identities in rings and algebras, applied linear algebra, algebraic coding theory and cryptography. With a teaching experience of around 35 years, Prof. Ashraf has supervised the Ph.D. thesis of 13 students and is currently guiding 6 more. He has published around 225 research articles in international journals and conference proceedings of repute. He received the Young Scientist's Award from Indian Science Congress Association in the year 1988 and the I.M.S. Prize from Indian Mathematical Society for the year 1995. He has completed many major research projects from the UGC, DST and NBHM. He is also Editor/ Managing Editor of many reputed international mathematical journals.

Vincenzo De Filippis is Associate Professor of Algebra at the University of Messina, Italy. He completed his Ph.D. in Mathematics from the University of Messina, Italy, in 1999. He is the member of the Italian Mathematical Society (UMI) and National Society of Algebraic and Geometric Structures and their Applications (GNSAGA). He has published around 100 research articles in reputed journals and conference proceedings.

Mohammad Aslam Siddeeque is Associate Professor at the Department of Mathematics, Aligarh Muslim University, India. He completed his Ph.D. in Mathematics from Aligarh Muslim University, India, in 2014 with the thesis entitled "On derivations and related mappings in rings and near-rings". His research interest lies in derivations and its various generalizations on rings and near-rings, on which he has published articles in reputed journals.

Symbols

$\mathbb{N}$	The set of natural numbers
$\mathbb{Z}$	The set of integers
$\mathbb{Q}$	The set of rational numbers
$\mathbb{R}$	The set of real numbers
$\mathbb{C}$	The set of complex numbers
$a \in S$	a is an element of a set S
$a \notin S$	a is not an element of a set S
$S \subseteq T, T \supseteq S$	S is a subset of a set T
$S = T$	Sets S and T are equal (have same elements)
$\varnothing$	The empty set
$A \cup B$	Union of sets A and B
$A \cap B$	Intersection of sets A and B
$A \backslash B$	Difference of sets A and B
A'	Complement of A
(a, b)	Ordered pair consisting of a, b
$A \times B$	Cartesian product of A and B
$f : S \to T$	Function f from a set S to a set T
$f(s)$	Image of an element s under a function f
$f \circ g, fg$	Composition or product of functions f and g
S_n	Symmetric group of degree n
$m\|n$	m divides n
$m \nmid n$	m does not divide n
$\|G\|$ or $o(G)$	Order of a group G
$Z(G)$	Center of a group G
$a \sim b$	a is equivalent to b in a specified sense
$a \equiv b \pmod{n}$	a is congruent to b modulo n
Ker ϕ	Kernel of a homomorphism ϕ
G/N	Quotient of a group G by a subgroup N
$\mathbb{F}[x]$	Polynomial ring over a field $\mathbb{F}$
deg $p(x)$	Degree of a polynomial $p(x)$
$g(x)\|f(x)$	Polynomial $g(x)$ divides $f(x)$

$R[x]$	Polynomial ring over a ring R
$\mathbb{F}_n[x]$	Set of all polynomials of at most degree n over a field $\mathbb{F}$
$\mathscr{C}[a,b]$	Set of all real valued continuous functions defined on $[a,b]$
$v \in V$	Vector v in a vector space V
αv	Scalar α times a vector v
$\alpha_1 v_1 + \cdots + \alpha_n v_n$	Linear combination of vectors $v_1, \ldots, v_n$
$\langle v_1, v_2, \ldots, v_n \rangle$	Subspace spanned by $v_1, v_2, \ldots, v_n$
$L(S)$	Linear span of a set S of vectors
$V \oplus W$	Direct sum of vector spaces V, W
$dim(V)$	Dimension of a vector space V
$U + W$	Sum of subspaces U, W of V
$R(T)$	Range of a linear transformation T
$N(T)$ or $Ker\, T$	Kernel of a linear transformation T
$r(T)$	Rank of a linear transformation T
$n(T)$	Nullity of a linear transformation T
A^t	Transpose of a matrix A
$r(A)$	Rank of a matrix A
$f'(x)$	Formal derivative of a polynomial $f(x)$
$m(T)$	Matrix associated with a linear transformation T
$V \cong W$	Isomorphism of two vector spaces V and W
$\widehat{V}$	Dual of a vector space V
$\widehat{\widehat{V}}$	Second dual or bidual of a vector space V
$S^{\perp}$	Orthogonal complement of a subset S of an inner product space V
S°	Annihilator of a subset S of an inner product space V
$det(A)$ or $|A|$	Determinant of a matrix A
$\text{trace}(A)$ or $\text{tr}(A)$	Trace of a matrix A
$d(a,b)$	Distance between two vectors a and b
T^*	Adjoint of a linear transformation T
$\|v\|$	Norm (or length) of a vector v
$L(U,V)$	Set of all linear transformations from a vector space U to V
$\mathscr{A}(V)$	Set of all linear transformations from a vector space V to itself
$M_{m \times n}(\mathbb{F})$	Set of all $m \times n$ matrices over a field $\mathbb{F}$
$M_n(\mathbb{F})$	Set of all $n \times n$ matrices over a field $\mathbb{F}$
$m(T)_{(B,B')}$ or $[T]_B^{B'}$	Matrix of a linear transformation T relative to ordered bases B and B'
$[v]_B$	Coordinate vector of v relative to an ordered basis B
$\widehat{T}$	Dual of a linear transformation T
$W_1 \odot W_2 \odot \cdots \odot W_n$	Orthogonal direct sum of subspaces $W_1, W_2, \ldots, W_n$
$\otimes$	Tensor product
$\wedge$	Exterior product
$\oplus^{\text{ext}}$	External direct sum

Chapter 1
Algebraic Structures and Matrices

The present chapter is aimed at providing background material in order to make the book as self-contained as possible. However, the basic information about set, relation, mapping etc. have been pre-assumed. Further, an appropriate training on the basics of matrix theory is certainly the right approach in studying linear algebra. Everybody knows that essentially, handling any problem relating to linear transformations, eigenvalues, normal form of operators, quadratic forms and applications to geometry, sooner or later, the focal point will be solving a linear system of equations. That is why matrix theory is so important. Matrices give us both the opportunity to represent data of problems and methods to solve any associated linear system.

1.1 Groups

Definition 1.1 Let S be a nonempty set. A mapping $* : S \times S \rightarrow S$ is called a binary operation on S.

Remark 1.2 (i) In the above definition $*(a, b)$ is denoted by $a * b$. It can be easily seen that usual addition and multiplication are binary operations on the set of natural numbers $\mathbb{N}$, the set of integers $\mathbb{Z}$, the set of rational numbers $\mathbb{Q}$, the set of real numbers $\mathbb{R}$ and on the set of complex numbers $\mathbb{C}$.

(ii) If we consider the set of vectors V in a plane, vector product is a binary operation on V while the scalar product on the set V is not a binary operation.

(iii) Let S be a nonempty set equipped with a binary operation $*$, i.e., $* : S \times S \rightarrow S$ is a mapping, then $(S, *)$ is called groupoid. For example, $(\mathbb{Z}, +), (\mathbb{Z}, -)$ are groupoids while $(\mathbb{N}, -)$ is not a groupoid.

(iv) A nonempty set G equipped with a binary operation $*$, say $(G, *)$, is said to be a semigroup if $*$ is an associative binary operation, i.e., an associative groupoid is called a semigroup. For example, $(\mathbb{Z}, +)$, $(\mathbb{Q}, +)$, $(\mathbb{R}, +)$, $(\mathbb{C}, +)$, $(\mathbb{N}, \cdot)$, $(\mathbb{Z}, \cdot)$, $(\mathbb{Q}, \cdot)$, $(\mathbb{R}, \cdot)$, $(\mathbb{C}, \cdot)$, are semigroups but $(\mathbb{Z}, -)$, $(\mathbb{R}^* = \mathbb{R}\backslash\{0\}, \div)$ are not semigroups.

Definition 1.3 A nonempty set G equipped with a binary operation $*$, say $(G, *)$, is said to be a group if

(1) For every $a, b, c \in G$, $a * (b * c) = (a * b) * c$, i.e., $*$ is an associative binary operation on G.
(2) For every $a \in G$, there exists an element e in G such that $a * e = e * a = a$. Such an element e in G is called the identity element in G.
(3) For every $a \in G$, there exists an element $b \in G$ such that $a * b = b * a = e$. Such an element $b \in G$ is said to be the inverse of a in G.

Remark 1.4 (i) In the above definition, G is a group under the binary operation $*$. Throughout this chapter by a group G means a group under multiplication unless otherwise mentioned. For the sake of convenience the product between any two elements a, b of a multiplicative group G will be denoted by ab instead of $a \cdot b$.

(ii) It can be easily seen that the identity element e in a group G is unique. For if e, e' are two identities in a group G, then $e = ee' = e'e = e'$.

(iii) The inverse of each element in a group G is unique. If a' and a'' are two inverses of an element a in a group G, then $aa' = a'a = e$ and $aa'' = a''a = e$, where e is the identity element in G. Then it can be easily seen that $a'' = a''e = a''(aa') = (a''a)a' = ea' = a'$.

(iv) If a^{-1} denotes the inverse of a in G, then in view of axiom (3), one can write $a = b^{-1}$ and $b = a^{-1}$. This is also easy to see that $(ab)^{-1} = b^{-1}a^{-1}$ for all $a, b \in G$.

(v) A group G is said to be abelian (or commutative) if $ab = ba$ holds for all $a, b \in G$. Otherwise, G is said to be a nonabelian group.

(vi) If a group G contains finite number of elements, then G is said to be a finite group. Otherwise group G is said to be infinite. The number of elements in a finite group G is called the order of G and is generally denoted by $o(G)$ or $|G|$.

(vii) For any $a \in G$ and n a positive integer, $a^n = \underbrace{a \cdot a \cdots a}_{n-\text{ times}}$, $a^0 = e$, the identity of the group G, $a^{-n} = (a^{-1})^n$ and hence it is straightforward to see that $a^n a^m = a^{n+m}$, $(a^n)^m = a^{nm}$.

Example 1.5 (1) It can be easily seen that the groupoids $(\mathbb{Z}, +)$, $(\mathbb{Q}, +)$, $(\mathbb{R}, +)$ and $(\mathbb{C}, +)$ form abelian groups under addition, while $(\mathbb{Q}^* = \mathbb{Q}\backslash\{0\}, \cdot)$, $(\mathbb{R}^* = \mathbb{R}\backslash\{0\}, \cdot)$ and $(\mathbb{C}^* = \mathbb{C}\backslash\{0\}, \cdot)$ form abelian groups under multiplication.

(2) The set $G = \{1, -1\}$ forms an abelian group under multiplication.

(3) Consider the set $G = \{1, -1, i, -i\}$ the set of fourth roots of unity. This is an abelian group of order 4 under multiplication.

(4) The set of all positive rational numbers $\mathbb{Q}^+$ forms an abelian group under the binary operation $'*'$ defined as $a * b = \frac{ab}{2}$. Note that $e = 2$ is the identity element in $\mathbb{Q}^+$ while for any $a \in \mathbb{Q}^+$, $\frac{4}{a}$ is the inverse of a in $\mathbb{Q}^+$.

(5) The set $\mathbb{Z}$ of all integers forms an abelian group with respect to the binary operation $'*'$ defined by $a * b = a + b + 1$, for all $a, b \in \mathbb{Z}$. Note that $e = -1$ is the identity element in $\mathbb{Z}$, while any $a \in \mathbb{Z}$ has the inverse $-2 - a$.

(6) The set $\{1, \omega, \omega^2\}$, where ω is the cube root of unity, forms an abelian group of order 3 under multiplication of complex numbers.

(7) The set of matrices

$$M = \left\{ \begin{bmatrix} 1 & 0 \\ 0 & 1 \end{bmatrix}, \begin{bmatrix} -1 & 0 \\ 0 & 1 \end{bmatrix}, \begin{bmatrix} 1 & 0 \\ 0 & -1 \end{bmatrix}, \begin{bmatrix} -1 & 0 \\ 0 & -1 \end{bmatrix} \right\}$$

forms an abelian group of order 4 under matrix multiplication.

(8) The set of quaternions $\{\pm 1, \pm i, \pm j, \pm k\}$ forms a nonabelian group of order 8 under multiplication $i^2 = j^2 = k^2 = -1$, $ij = -ji = k$, $jk = -kj = i$, $ki = -ik = j$.

(9) The set of matrices $M = \left\{ \begin{bmatrix} a & a \\ a & a \end{bmatrix} \mid 0 \neq a \in \mathbb{R} \right\}$ forms an abelian group under matrix multiplication. The identity of the group M is $\begin{bmatrix} \frac{1}{2} & \frac{1}{2} \\ \frac{1}{2} & \frac{1}{2} \end{bmatrix}$, and for any nonzero $a \in \mathbb{R}$ the inverse of $\begin{bmatrix} a & a \\ a & a \end{bmatrix} \in M$ is $\begin{bmatrix} \frac{1}{4a} & \frac{1}{4a} \\ \frac{1}{4a} & \frac{1}{4a} \end{bmatrix}$.

(10) The set $K = \{e, a, b, c\}$ forms an abelian group under the binary operation defined on K as $ab = ba = c, bc = cb = a, ca = ac = b, a^2 = b^2 = c^2 = e$. This group is known as Klein 4-group.

(11) The set of all permutations defined on n symbols ($n \geq 3$) forms a nonabelian group of order $n!$ and is denoted by S_n. For example, $S_3 = \{I, (12), (13), (23), (123), (132)\}$ is a nonabelian group of order 6.

(12) The set of all $n \times n$ invertible matrices over the set $\mathbb{R}$ (or $\mathbb{C}$) forms a group under matrix multiplication generally known as general linear group, denoted by $GL_n(\mathbb{R})$ (or $GL_n(\mathbb{C})$). Similarly, the set of all $n \times n$ real matrices with determinant one is a group under multiplication, denoted as $SL(n, \mathbb{R})$ and known special linear group. The set of all $n \times n$ orthogonal matrices over reals forms a group and is called the orthogonal group, denoted as $O(n, \mathbb{R})$.

Definition 1.6 If G is a group, then the order of an element $a \in G$ is the least positive integer n such that $a^n = e$, the identity of group G. If no such positive integer exists we say that a is of infinite order. We use the notation $o(a)$ for the order of a.

Proposition 1.7 *If G is a group, then the following hold:*

(i) For any $a, g \in G$, $o(a) = o(gag^{-1})$.

(ii) For any $a, b \in G$, $o(ab) = o(ba)$.

(iii) If $o(a)$ is finite, then for any integer m, $a^m = e$ implies that $o(a)$ divides m.

Proof (i) For any positive integer n, $a^n = e$ if and only if $(gag^{-1})^n = ga^n g^{-1} = e$ for all $g \in G$. Hence $o(gag^{-1}) = o(a)$.

(ii) Since $ba = b(ab)b^{-1}$, we find that $o(ab) = o(ba)$, by (i).

(iii) Suppose that $o(a) = n$ and m be a positive integer such that $a^m = e$. Application of Euclid's algorithm yields that $m = nq + r$ for some integer q and r with $0 \le r < n$. Hence, in this case $e = a^m = (a^n)^q a^r = a^r$. But since n is the least positive integer such that $a^n = e$, we arrive at $r = 0$, and hence $o(a)|m$.

Proposition 1.8 *Let G be a group and n be a positive integer. Then the following hold:*

(i) *G is abelian if and only if $(ab)^2 = a^2 b^2$, for all $a, b \in G$.*
(ii) *G is abelian if and only if $(ab)^k = a^k b^k$, for all $a, b \in G$, where $k = n, n + 1, n + 2$.*

Proof (i) Suppose that G is abelian. Then

$$(ab)^2 = (ab)(ab) = a(b(ab)) = a((ba)b) = a(ab)b = a^2 b^2.$$

Conversely, if $(ab)^2 = a^2 b^2$, then $a^{-1}((ab)(ab)) = a^{-1}(a^2 b^2)$, and this implies that $b(ab) = ab^2$. Similarly, operating by b^{-1} from the right side in the latter equation, we find that $ba = ab$, and hence G is abelian.

(ii) We have $(ab)^n = a^n b^n$, $(ab)^{n+1} = a^{n+1} b^{n+1}$ and $(ab)^{n+2} = a^{n+2} b^{n+2}$. Using the first two conditions, we get $(a^n b^n)(ab) = (ab)^{n+1} = a^{n+1} b^{n+1}$. Now using cancellation laws we arrive at $b^n a = ab^n$. Similarly, using the latter two conditions we find that $(a^{n+1} b^{n+1})(ab) = (ab)^{n+2} = a^{n+2} b^{n+2}$, and hence $b^{n+1} a = ab^{n+1}$. But in this case $ab^{n+1} = b^{n+1} a = b(b^n a) = b(ab^n)$ and using cancellation laws we find that G is abelian.

Definition 1.9 A nonempty subset H of a group $(G, *)$ is said to be a subgroup of G if

(i) H is closed under the binary operation $'*'$,
(ii) H is a group under the same binary operation $'*'$.

Example 1.10 (1) For any group G, G is a subgroup of G. Similarly, the trivial group $\{e\}$ is a subgroup of G. Any subgroup of G which is different from G and $\{e\}$ is said to be a proper (nontrivial) subgroup of G.

(2) The additive group $(\mathbb{Z}, +)$ is a subgroup of additive group $(\mathbb{Q}, +)$, the additive group $(\mathbb{Q}, +)$ is a subgroup of $(\mathbb{R}, +)$ and the additive group $(\mathbb{R}, +)$ is a subgroup of the additive group $(\mathbb{C}, +)$.

(3) The multiplicative group $\{1, -1, i, -i\}$, $i^2 = -1$, is a subgroup of all complex numbers under multiplication. It is straightforward to see that $SL(n, \mathbb{R})$ and $O(n, \mathbb{R})$ are subgroups of $GL(n, \mathbb{R})$.

(4) The subsets $\{e, a\}$, $\{e, b\}$ and $\{e, c\}$ are subgroups of Klein 4-group $K = \{e, a, b, c\}$.

(5) Let G be the multiplicative group of all nonsingular 2×2 matrices over the set of complex numbers. Consider the subset

$$H = \left\{ \pm \begin{bmatrix} 1 & 0 \\ 0 & 1 \end{bmatrix}, \pm \begin{bmatrix} i & 0 \\ 0 & -i \end{bmatrix}, \pm \begin{bmatrix} 0 & i \\ i & 0 \end{bmatrix}, \pm \begin{bmatrix} 0 & 1 \\ -1 & 0 \end{bmatrix} \right\}$$

of the group G. It can be easily seen that H is a subgroup of G.

(6) Consider the subset $H' = \left\{ \pm \begin{bmatrix} 1 & 0 \\ 0 & 1 \end{bmatrix}, \pm \begin{bmatrix} i & 0 \\ 0 & -i \end{bmatrix} \right\}$ of H in the above example. It can be seen that H' is a subgroup of H.

Proposition 1.11 *A nonempty subset H of a group G is a subgroup of G if and only if $ab^{-1} \in H$ for any $a, b \in H$.*

Proof If H is a subgroup of G, then for any $a, b \in H$, $b^{-1} \in H$ and consequently $ab^{-1} \in H$.

Conversely, assume that for any $a, b \in H$, $ab^{-1} \in H$. For any $a \in H$, $e = aa^{-1} \in H$, and H contains identity. Now for any $a \in H$, and $e \in H$, $a^{-1} = ea^{-1} \in H$, i.e., every element in H has its inverse in H. Also for $a, b \in H$, $b^{-1} \in H$, and hence $ab = a(b^{-1})^{-1} \in H$. Consequently, H is closed. Since binary composition on G is associative, the induced binary composition is also associative on H. Thus H is a subgroup of G.

Definition 1.12 Let G be a group and H a subgroup of G. For any $a \in G$, the set $Ha = \{ha \mid h \in H\}$ (resp. $aH = \{ah \mid h \in H\}$) is called a right (resp. left) coset of H determined by a in G.

Example 1.13 Let $S_3 = \{I, \varphi, \xi, \xi^2, \varphi\xi, \varphi\xi^2\}$, where $\varphi = (23)$, $\xi = (123)$, $\varphi^2 = I$, $\xi^3 = I$, be symmetric group defined on three symbols. Consider the subset $H = \{I, \varphi\}$ which is a subgroup of S_3. The left coset $\xi H = \{\xi, \xi\varphi\}$ and the right coset $H\xi = \{\xi, \varphi\xi\}$ of H in S_3 are distinct.

Remark 1.14 (i) If the binary composition in G is addition and H a subgroup of G, then the left (resp. right) coset of H in G determined by a is defined as $a + H = \{a + h \mid h \in H\}$ (resp. $H + a = \{h + a \mid h \in H\}$).

(ii) If a group G is abelian and H is a subgroup of G, then every left coset of H in G is always the same as the right coset of H in G.

(iii) If H is a subgroup of a group G, then it can be seen that there is one-to-one correspondence between any two left cosets of H in G. If aH, bH are any two left cosets of H in G, define a map $\theta : aH \rightarrow bH$ such that $\theta(ah) = bh$. If $\theta(ah_1) = \theta(ah_2)$, then $bh_1 = bh_2$ implies that $h_1 = h_2$ and hence $ah_1 = ah_2$, i.e., θ is one-to-one. Clearly, θ is onto and hence there is one-to-one correspondence between aH and bH.

Definition 1.15 A subgroup H of G is said to be a normal subgroup of G if $aH = Ha$ for all $a \in G$.

Proposition 1.16 *A subgroup H of a group G is normal in G if and only if* $ghg^{-1} \in$ *H for any* $g \in G, h \in H$.

Proof Since $gH = Hg$, $gh \in gH = Hg$, and hence $gh = h_1 g$ for some $h_1 \in H$. Thus, $ghg^{-1} = h_1 \in H$, for any $g \in G, h \in H$.

Conversely, assume that $ghg^{-1} \in H$ for any $g \in G, h \in H$. Let a be an arbitrary element in G. Then $aha^{-1} \in H$ for any $h \in H$. Hence $ha = a(a^{-1}h(a^{-1})^{-1}) \in aH$, and consequently, $Ha \subseteq aH$. Similarly, if $aha^{-1} \in H$, then it can be seen that $ah = aha^{-1}a \in Ha$, i.e., $aH \subseteq Ha$, which shows that $aH = Ha$.

Proposition 1.17 *If H is a normal subgroup of G, then the set of left (right) cosets of H in G forms a group under the binary operation* $aHbH = abH$.

Proof First, we show that the given binary operation is well-defined, i.e., for any $a_1, a_2, b_1, b_2 \in G$, $a_1 H = a_2 H$, $b_1 H = b_2 H$ implies that $a_1 b_1 H = a_2 b_2 H$. If $a_1 H = a_2 H$, $b_1 H = b_2 H$, then $a_1^{-1} a_2 \in H$ and $b_1^{-1} b_2 \in H$. This yields that $(a_1 b_1)^{-1} a_2 b_2 = b_1^{-1} a_1^{-1} a_2 b_2 = b_1^{-1} (a_1^{-1} a_2) b_1 (b_1^{-1} b_2)$. Since H is normal in G, $b_1^{-1} (a_1^{-1} a_2) b_1 \in H$. But $b_1^{-1} b_2 \in H$ yields that $(a_1 b_1)^{-1} a_2 b_2 \in H$, and hence $a_1 b_1 H = a_2 b_2 H$. This operation is also associative. In fact, for any $a, b, c \in G$, $(aHbH)cH = (ab)HcH = (ab)cH = a(bc)H = aH(bcH) = aH(bHcH)$. Now if e is the identity element in G, then $eH = H$ acts as the identity element, i.e., $eHaH = (ea)H = aH$. For any $a \in G$, $a^{-1}H$ is the inverse of aH. Hence the set of cosets of H in G forms a group.

Remark 1.18 If H is a normal subgroup of G, then the set of left(right) cosets of H in G forms a group which is known as the quotient group or factor group with respect to H and generally denoted as G/H.

Lemma 1.19 *Let G be a group and H a subgroup of G. Then G is the union of all left cosets of H in G and any two distinct left cosets of H in G are disjoint.*

Proof Obviously, union of all left cosets of H in G is a subset of G. Conversely, for any $a \in G$, $a = ae \in aH$, and hence G is the union of all left cosets of H in G. Now assume that aH, bH are any two distinct cosets of H in G which are not disjoint, i.e., $aH \cap bH \neq \emptyset$. If $x \in aH \cap bH$, then $x = ah_1 = bh_2$ for some $h_1, h_2 \in H$. Let y be an arbitrary element in aH so that $y = ah, h \in H$. Then

$$y = ah = (xh_1^{-1})h = x(h_1^{-1}h) = bh_2(h_1^{-1}h) = b(h_2 h_1^{-1} h),$$

where $h_2 h_1^{-1} h \in H$. Hence $y \in bH$ and $aH \subseteq bH$. Similarly, it can be shown that $bH \subseteq aH$. Consequently, $aH = bH$, i.e., aH and bH are not distinct. This leads to a contradiction.

Theorem 1.20 (Lagrange's Theorem) *Let G be a finite group and H a subgroup of G, then* o(H) *divides the* o(G).

Proof Since G is a finite group, the number of left cosets of H in G is finite. If the distinct left cosets of H in G are denoted by $H = a_1 H, a_2 H, \ldots, a_t H$, then by Lemma 1.19, G is the union of its all left cosets of H in G, i.e., $G = a_1 H \cup a_2 H \cup \cdots \cup a_t H$. But each left coset has the same number of elements which is equal to the number of elements in $H = a_1 H$, $o(G) = t \circ (H)$. Hence $o(H)$ divides $o(G)$.

Definition 1.21 Let G and G' be any two groups. A mapping $\theta : G \to G'$ is said to be a homomorphism if $\theta(ab) = \theta(a)\theta(b)$ holds for all $a, b \in G$.

Remark 1.22 (i) Note that if binary operations in groups G and G' are different, say $*$ and o, respectively, then θ satisfies the property $\theta(a * b) = \theta(a)o\theta(b)$ for all $a, b \in G$. For the sake of convenience, it has been assumed that both G and G' are multiplicative groups.

 (ii) A homomorphism of a group which is also onto is called an epimorphism. A homomorphism of a group which is also one-to-one is called monomorphism. Further, a homomorphism of a group G into itself is called an endomorphism.

(iii) A group homomorphism which is one-to-one and onto is said to be an isomorphism. An isomorphism of a group G onto itself is called an automorphism of G.

(iv) If $\theta : G \to G'$ is a homomorphism, then $\theta(G)$ is a subgroup of G'.

 (v) Let $\theta : G \to G'$ be a group homomorphism of G onto G'. If G is abelian, then G' is also abelian.

(vi) If $\theta : G \to G'$ is a group homomorphism of G onto G', where e and e' are the identities of G and G', respectively, then $\theta(e) = e'$ and $\theta(a^{-1}) = (\theta(a))^{-1}$ for all $a \in G$.

Example 1.23 (1) For any group G, the identity mapping $I_G : G \to G$ is a group homomorphism.

(2) For any two groups G and G', the mapping $\theta : G \to G'$ such that $\theta(a) = e'$, the identity of G' is a group homomorphism and is called trivial homomorphism.

(3) The mapping $\theta : \mathbb{R}^* \to \mathbb{R}^*$ defined on multiplicative group $\mathbb{R}^* = \mathbb{R}\backslash\{0\}$ such that $\theta(a) = |a|$ is a homomorphism, but $\theta : \mathbb{R} \to \mathbb{R}$ such that $\theta(a) = |a|$ is not homomorphism on additive group $\mathbb{R}$.

(4) The map $\theta : \mathbb{R}^2 \to \mathbb{R}^2$ such that $\theta(a, b) = a$ is a homomorphism.

(5) If $\theta : \mathbb{R}^2 \to \mathbb{C}$ such that $\theta(a, b) = a + ib$, then θ is an isomorphism.

(6) Let $a \in G$ be a fixed element of G and $\theta : G \to G$ such that $\theta_a(g) = aga^{-1}$, for all $a \in G$. Then it can be seen that θ is an automorphism. In fact, for any $x, y \in G$, $\theta(xy) = a(xy)a^{-1} = axa^{-1}aya^{-1} = \theta_a(x)\theta_a(y)$, i.e., θ is a homomorphism. Further, $\theta_a(x) = \theta_a(y)$ implies that $axa^{-1} = aya^{-1}$, i.e., $x = y$ and hence θ is one-to-one. For any $x \in G$, there exists $a^{-1}xa \in G$ such that $\theta_a(a^{-1}xa) = a(a^{-1}xa)a^{-1} = x$ and hence θ is onto. Thus θ is an automorphism of G.

Definition 1.24 Let $\theta : G \to G'$ be a homomorphism of a group G to G'. Then the set $\{a \in G \mid \theta(a) = e'$, the identity of $G'\}$ is called the kernel of θ and is denoted by $Ker\theta$.

Proposition 1.25 *Let $\theta : G \to G'$ be a homomorphism of a group G to G'. Then*

(i) $Ker\theta$ is a subgroup of G,

(ii) θ is one-to-one mapping if and only if $Ker\theta = \{e\}$, where e is the identity of G.

Proof (i) Since $\theta(e) = e'$, where e' is the identity of G', hence $e \in Ker\theta$, i.e., $Ker\theta \neq \emptyset$. Now let $a, b \in Ker\theta$. Then $\theta(a) = e', \theta(b) = e'$. Now

$$\theta(ab^{-1}) = \theta(a)\theta(b^{-1}) = \theta(a)(\theta(b))^{-1} = e'e'^{(-1)} = e'e' = e',$$

and hence $ab^{-1} \in Ker\theta$, i.e., $Ker\theta$ is a subgroup of G.

(ii) Suppose that θ is one-to-one mapping and $a \in Ker\theta$. Then $\theta(a) = e' = \theta(e)$, which yields that $a = e$, and hence $Ker\theta = \{e\}$.

Conversely, assume that $Ker\theta = \{e\}$ and $\theta(a) = \theta(b)$. Then $e' = \theta(a)(\theta(b))^{-1} = \theta(a)\theta(b^{-1}) = \theta(ab^{-1})$, and hence $ab^{-1} \in Ker\theta = \{e\}$. Consequently, $ab^{-1} = e$ and hence $a = b$.

Exercises

1. Show that the set of all nth complex roots of unity forms a group with respect to ordinary multiplication.

2. If G is a group, then show that the set $Z(G) = \{a \in G \mid ab = ba, \text{ for all } b \in G\}$ (is said to be center of G) is a normal subgroup of G.

3. If a is a fixed element of a group G, then show that the set $N(a) = \{x \in G \mid ax = xa\}$ (is said to be normalizer of a in G) is a subgroup of G.

4. If every element in a group G is self-inverse, i.e., $a^{-1} = a$ for all $a \in G$, then show that G is abelian.

5. In a group G, for all $a, b \in G$ show that $(aba^{-1})^n = aba^{-1}$ if and only if $b = b^n$, where n is any integer.

6. Let G be a group and m, n be two relatively prime positive integers such that $(ab)^m = a^m b^m$ and $(ab)^n = a^n b^n$ hold for all $a, b \in G$. Then show that G is abelian.

7. Let G be a group such that $(ab)^2 = (ba)^2$ for all $a, b \in G$. Suppose G also has the property that $c^2 = e$ implies that $c = e$, $c \in G$. Then show that G is abelian.

8. Show that a group G in which $a^m b^m = b^m a^m$ and $a^n b^n = b^n a^n$ hold for all $a, b \in G$, where m, n are any two relatively prime positive integers, is abelian.

9. Show that intersection of two subgroups of a group G is a subgroup of G. More generally, show that intersection of any arbitrary family of subgroups of G is a subgroup of G.

10. If H is a finite subset of a group G such that $ab \in H$ for all $a, b \in H$, then show that H is a subgroup of G.

11. Show that a group cannot be written as a set theoretic union of two of its proper subgroups.

12. If K is a subgroup of H and H is a subgroup of G, then show that K is a subgroup of G.
13. If a is an element of a group G, and n is any nonzero integer, then show that $o(a) = o(a^n)$ if and only if n is relatively prime to $o(a)$.

1.2 Rings

This section is devoted to the study of algebraic structure equipped with two binary operations, namely rings. Several basic properties of rings and subrings are given.

Definition 1.26 A nonempty set R equipped with two binary operations, say addition $'+'$ and multiplication $'\cdot'$, is said to be a ring if it satisfies the following axioms:

(1) $(R, +)$ is an abelian group.
(2) $(R, \cdot)$ is a semigroup.
(3) For any $a, b, c \in R$, $a \cdot (b + c) = a \cdot b + a \cdot c$, $(b + c) \cdot a = b \cdot a + c \cdot a$.

Remark 1.27 (i) The binary operations $'+'$ and $'\cdot'$ are not necessarily usual addition and multiplication. Moreover, these are only symbols used to represent both binary operations of R. For the convenience we write ab instead of $a \cdot b$. A ring R is said to be commutative if $ab = ba$ holds for all $a, b \in R$. Otherwise, R is said to be noncommutative. If a ring R contains an element e such that $ae = ea = a$ for all $a \in R$, we say that R is a ring with identity, and the identity e in a ring R is usually denoted by 1. In general, a ring R may or may not have identity. But if the ring R has identity 1, then it is unique.
 (ii) It can be easily seen that $a0 = 0a = 0$ for all $a \in R$.
 (iii) For any $a, b \in R$, $a(-b) = (-a)b = -ab$, $(-a)(-b) = ab$.
 (iv) In a ring R, for a fixed positive integer n, $na = \underbrace{a + a + \cdots + a}_{n-\text{ times}}$ and $a^n = \underbrace{a \cdot a \cdots a}_{n-\text{ times}}$.

Example 1.28 (1) $(\mathbb{Z}, +, \cdot)$, $(\mathbb{Q}, +, \cdot)$, $(\mathbb{R}, +, \cdot)$ and $(\mathbb{C}, +, \cdot)$ are examples of commutative rings.
 (2) Every additive abelian group G is a ring if we define multiplication in G as $ab = 0$ for all $a, b \in G$, where 0 is the additive identity. This ring is called zero ring.
 (3) The set $M_{n \times n}(\mathbb{Z})$ of all $n \times n$ matrices over integers forms a ring under the binary operations of matrix addition and matrix multiplication, which is an example of a noncommutative ring.
 (4) Let $Q = \{a + bi + cj + dk \mid a, b, c, d \in \mathbb{R}\}$. Define addition in Q as $(a_1 + b_1 i + c_1 j + d_1 k) + (a_2 + b_2 i + c_2 j + d_2 k) = (a_1 + a_2) + (b_1 + b_2)i + (c_1 + c_2)j + (d_1 + d_2)k$ under which Q forms an abelian group. Now multiply any two members of Q as multiplication of polynomials by using the rules

$ij = -ji = k,\ jk = -kj = i,\ ki = -ik = j$. Then Q forms a ring known as ring of real quaternions which is noncommutative.

(5) $\mathbb{R}^n = \{(a_1, a_2, \ldots, a_n) \mid a_1, a_2, \ldots, a_n \in \mathbb{R}\}$ forms a ring under the addition and multiplication defined as follows:

$$(a_1, a_2, \ldots, a_n) + (b_1, b_2, \ldots, b_n) = (a_1 + b_1, a_2 + b_2, \ldots, a_n + b_n)$$

$$(a_1, a_2, \ldots, a_n)(b_1, b_2, \ldots, b_n) = (a_1 b_1, a_2 b_2, \ldots, a_n b_n).$$

This is a commutative ring.

(6) Let R be any ring. Consider the set $R[x]$ of all symbols $a_0 + a_1 x + \cdots + a_n x^n$ where n is any nonnegative integer and the coefficients $a_0, a_1, \ldots, a_n \in R$. If $f(x) = a_0 + a_1 x + \cdots + a_n x^n$, $g(x) = b_0 + b_1 x + \cdots + b_m x^m$, then $f(x) = g(x)$ if and only if $a_i = b_i$, $i \geq 0$. Define addition and multiplication in $R[x]$ as follows:

$$f(x) + g(x) = (a_0 + b_0) + (a_1 + b_1)x + \cdots + (a_k + b_k)x^k$$

$$f(x)g(x) = c_0 + c_1 x + \cdots + c_k x^k,$$

where $c_k = a_0 b_k + a_1 b_{k-1} + \cdots + a_k b_0$. It can be seen that $R[x]$ is a ring under the above operation of addition and multiplication, known as polynomial ring in indeterminate x.

(7) The set $M = \left\{ \begin{bmatrix} a & b \\ 0 & 0 \end{bmatrix} \mid a, b \in \mathbb{Z}_2 \right\}$ forms a noncommutative ring with four elements.

(8) The set $\mathbb{Z}[i] = \{a + ib \mid a, b \in \mathbb{Z}\}$ is a ring under usual addition and multiplication, which is known as the ring of Gaussian integers.

(9) The set $\{a + b\sqrt{2} \mid a, b \in \mathbb{Z}\}$ forms a ring under usual addition and multiplication.

(10) Let S be any set. Consider $\mathscr{P}(S)$, the power set of S and define binary operations in $\mathscr{P}(S)$ as $X + Y = (X \cup Y)\backslash(X \cap Y)$, $\quad XY = X \cap Y$. Then $\mathscr{P}(S)$ is a commutative ring.

(11) For any fixed positive number $n \geq 2$, let $\mathbb{Z}_n = \{0, 1, 2, \ldots, n - 1\}$ and define addition modulo n, denoted as $\oplus_n$, as $a \oplus_n b = r$, where r is the remainder obtained by dividing the ordinary sum of a and b by n, where $a, b \in \mathbb{Z}_n$. Similarly, the multiplication modulo n is defined as $a \otimes_n b = r$, r is the remainder obtained by dividing the ordinary product of a and b by n. It can be seen that the set $(\mathbb{Z}_n, \oplus_n, \otimes_n)$ forms a commutative ring with identity 1 under the addition and multiplication modulo n.

(12) The set $\mathbb{R}[x]$ of all polynomials with real coefficients under addition and multiplication of polynomials forms a commutative ring with identity.

In the previous section, we have defined the notion of subgroup H of a group G, i.e., a nonempty subset H of G which is itself a group under the same binary operation. Analogously, the notion of a subring has been introduced in the case of a ring also.

Definition 1.29 A nonempty subset S of a ring $(R, +, \cdot)$ is said to be a subring of $(R, +, \cdot)$, if S is itself a ring under the same binary operations.

Example 1.30 (1) $(\mathbb{Z}, +, \cdot)$ is a subring of $(\mathbb{Q}, +, \cdot)$, $(\mathbb{Q}, +, \cdot)$ is a subring of $(\mathbb{R}, +, \cdot)$ and $(\mathbb{R}, +, \cdot)$ is a subring of $(\mathbb{C}, +, \cdot)$.

(2) $S = \left\{ \begin{bmatrix} a & b \\ 0 & 0 \end{bmatrix} \mid a, b \in \mathbb{Z} \right\}$ is a subring of $M_2(\mathbb{Z}) = \left\{ \begin{bmatrix} a & b \\ c & d \end{bmatrix} \mid a, b, c, d \in \mathbb{Z} \right\}$.

(3) $S = \left\{ \begin{bmatrix} a & a \\ a & a \end{bmatrix} \mid a \in \mathbb{R} \right\}$ is a subring of $M_2(\mathbb{R}) = \left\{ \begin{bmatrix} a & b \\ c & d \end{bmatrix} \mid a, b, c, d \in \mathbb{R} \right\}$.

(4) $(n\mathbb{Z}, +, \cdot)$ is a subring of the ring $(\mathbb{Z}, +, \cdot)$ for any fixed integer $n \in \mathbb{Z}$.

Proposition 1.31 *A nonempty subset S of a ring R is a subring of R if and only if for any $a, b \in S, a - b \in S$ and $ab \in S$.*

Proof Let S be a subring of R and $a, b \in S$. Then by definition $(S, +, \cdot)$ is a ring and for any $a, b \in S, a, -b \in S$ and hence $a - b \in S$ and $ab \in S$. Conversely, assume that S is a nonempty subset of R such that for any $a, b \in S, a - b \in S$ and $ab \in S$. If $a - b \in S$ for any $a, b \in S$, then it can be seen that $(S, +)$ is an additive subgroup of $(R, +)$. Also since S is closed under multiplication, $(S, \cdot)$ is a sub-semigroup of $(R, \cdot)$. Further, since the laws of distributivity hold in R, these also hold in S. Hence $(S, +, \cdot)$ is a ring under the induced operation of addition and multiplication. Hence S is a subring of R.

Remark 1.32 (i) A subring need not contain the identity of a ring. For example, $(\mathbb{Z}, +, \cdot)$ has the identity 1 while its subring $(2\mathbb{Z}, +, \cdot)$ has no identity. $(\mathbb{Z}_6, \oplus_6, \otimes_6)$ has the identity 1, but its subring $(\{0, 3\}, \oplus_6, \otimes_6)$ has the identity 3. It is also to remark that a ring has no identity element but it has a subring which contains the identity element. For example, $\mathbb{Z} \times 3\mathbb{Z}$ has no identity while its subring $\mathbb{Z} \times \{0\}$ has the identity $(1, 0)$.

(ii) Trivially, subring of a commutative ring is commutative. But a noncommutative ring may have a commutative subring. For example,

$$M_2(\mathbb{Z}) = \left\{ \begin{bmatrix} a & b \\ c & d \end{bmatrix} \mid a, b, c, d \in \mathbb{Z} \right\}$$

is a noncommutative ring while its subring

$$S = \left\{ \begin{bmatrix} a & 0 \\ 0 & d \end{bmatrix} \mid a, d \in \mathbb{Z} \right\}$$

is commutative.

(iii) Let $R = \left\{ \begin{bmatrix} a & b \\ 0 & q \end{bmatrix} \mid a, b \in \mathbb{R}, q \in \mathbb{Q} \right\}$. It can be seen that R is a ring with

identity $\begin{bmatrix} 1 & 0 \\ 0 & 1 \end{bmatrix}$. But it has a subring $S = \left\{ \begin{bmatrix} a & b \\ 0 & 2m \end{bmatrix} \mid a, b \in \mathbb{R}, m \in \mathbb{Z} \right\}$ without

identity.

Definition 1.33 A nonempty subset I of a ring $(R, +, \cdot)$ is said to be a left (resp. right) ideal of R if for all $a, b \in I, r \in R, a - b \in I, ra \in I$ (resp. $a - b \in I, ar \in I$). If I is both a left as well as a right ideal of R, I is said to be an ideal of R.

Example 1.34 In the ring $M_2(\mathbb{Z})$ of all 2×2 matrices over $\mathbb{Z}$, the ring of integers consider the subsets $A = \left\{ \begin{bmatrix} a & b \\ 0 & 0 \end{bmatrix} \mid a, b \in \mathbb{Z} \right\}$ and $B = \left\{ \begin{bmatrix} c & 0 \\ d & 0 \end{bmatrix} \mid c, d \in \mathbb{Z} \right\}$ of $M_2(\mathbb{Z})$. It can be easily seen that A is a right ideal of $M_2(\mathbb{Z})$ which is not a left ideal of $M_2(\mathbb{Z})$ while B is a left ideal of $M_2(\mathbb{Z})$ which is not a right ideal of $M_2(\mathbb{Z})$. If we consider the subset $C = \left\{ \begin{bmatrix} a & 0 \\ b & c \end{bmatrix} \mid a, b, c \in \mathbb{Z} \right\}$ of $M_2(\mathbb{Z})$, then it can be seen that C is a subring of $M_2(\mathbb{Z})$, but neither a left nor a right ideal of $M_2(\mathbb{Z})$.

Definition 1.35 Let $(R, +, \cdot)$ be a ring and I be an ideal of R. Then the set $R/I = \{a + I \mid a \in R\}$ forms a ring under the operations: $(a + I) + (b + I) = (a + b) + I$ and $(a + I)(b + I) = ab + I$. This is known as quotient ring.

Definition 1.36 Let R_1, R_2 be any two rings. A mapping $\theta : R_1 \to R_2$ is said to be a homomorphism if $\theta(a + b) = \theta(a) + \theta(b)$ and $\theta(ab) = \theta(a)\theta(b)$ hold for all $a, b \in R_1$.
A ring homomorphism which is also one-to-one and onto is called a ring isomorphism. Moreover, if R_1, R_2 have the identity elements e_1 and e_2, respectively, then $\theta(e_1) = e_2$.

Example 1.37 (1) A map $\theta : \mathbb{C} \to \mathbb{C}$ such that $\theta(z) = \bar{z}$, the complex conjugate of z, is a ring homomorphism which is both one-to-one and onto.

(2) Let $\theta : \mathbb{C} \to R$, where $R = \left\{ \begin{bmatrix} a & -b \\ b & a \end{bmatrix} \mid a, b \in \mathbb{R} \right\}$ is a ring under matrix addition and multiplication. For any $z = x + iy \in \mathbb{C}$, define $\theta(z) = \begin{bmatrix} x & -y \\ y & x \end{bmatrix}$. It can be easily verified that θ is a homomorphism which is both one-to-one and onto.

Definition 1.38 Let $\theta : R_1 \to R_2$ be a homomorphism of a ring R_1 to a ring R_2. Then the kernel of θ, denoted as $ker\theta$, is the collection of all those elements of R_1 which map to the zero of R_2, i.e., $Ker\theta = \{a \in R_1 \mid \theta(a) = 0\}$.

Remark 1.39 (*i*) It can be easily seen that $Ker\theta$ is an ideal of R_1. In fact, $\theta(0) = 0, 0 \in Ker\theta$ and hence $Ker\theta \neq \emptyset$. If $a, b \in Ker\theta$, then $\theta(a) = 0, \theta(b) = 0$. Hence $\theta(a - b) = \theta(a) - \theta(b) = 0$, i.e., $a - b \in Ker\theta$. Again if $a \in Ker\theta$ and $r \in R_1$, $\theta(ar) = \theta(a)\theta(r) = 0$, and $\theta(ra) = \theta(r)\theta(a) = 0$. Hence ar and $ra \in Ker\theta$.

This shows that $ker\theta$ is an ideal of R_1.

(ii) If $\theta : R_1 \to R_2$ is a homomorphism and I is an ideal of R_1, then $\theta(I)$ is an ideal of $\theta(R_1)$. Note that $\theta(I) = \{x \in R_2 \mid$ there exists $a \in I$ such that $\theta(a) = x\}$. Since $0 = \theta(0), 0 \in \theta(I)$, i.e., $\theta(I) \neq \emptyset$. If $x, y \in \theta(I)$, then there exist $a, b \in I$ such that $\theta(a) = x, \theta(b) = y$. Hence $x - y = \theta(a) - \theta(b) = \theta(a - b)$. Since $a - b \in I$, $x - y \in \theta(I)$. If $r_1 \in \theta(R_1)$, then there exists $r \in R_1$ such that $r_1 = \theta(r)$. Hence $r_1 x = \theta(r)\theta(a) = \theta(ra)$. Since $ra \in I$, $r_1 x \in \theta(I)$. In a similar way, it can be seen that $xr_1 \in \theta(I)$, and hence $\theta(I)$ is an ideal of $\theta(R_1)$.

Proposition 1.40 *Let $\theta : R_1 \to R_2$ be a ring homomorphism. Then θ is one-to-one if and only if $Ker\theta = \{0\}$.*

Proof Suppose that $Ker\theta = \{0\}$, and $\theta(a) = \theta(b)$, $a, b \in R_1$. Then $\theta(a - b) = \theta(a) - \theta(b) = 0$, and hence $a - b \in Ker\theta = \{0\}$, i.e., $a - b = 0$ and θ is one-to-one.
Conversely, assume that θ is one-to-one. If $a \in Ker\theta$, then $\theta(a) = 0 = \theta(0)$, and since θ is one-to-one, we find that $a = 0$, i.e., $Ker\theta = \{0\}$.

Definition 1.41 A nonzero element $a \in R$ is called a zero divisor if there exists a nonzero element $b \in R$ such that $ab = 0$ and $ba = 0$. A commutative ring R is said to be an integral domain if R contains no zero divisors.
It is obvious to see that $\mathbb{Z}$, the ring of integers, is free from zero divisor and hence is an integral domain.

Definition 1.42 A ring R with identity 1, in which each nonzero element has multiplicative inverse, i.e., if $a \in R$ is a nonzero element, then there exists $b \in R$ such that $ab = ba = 1$, is called a division ring.

Definition 1.43 The characteristic of a ring R, denoted as $char(R)$, is the smallest positive integer n such that $na = 0$ for all $a \in R$. If no such integer exists, the characteristic of R is said to be zero.

Proposition 1.44 *If R is an integral domain, then $char(R) = 0$ or $char(R) = p$, where p is a prime.*

Proof Suppose that $char(R) = n \neq 0$, and n is not prime. Then $n = n_1 n_2$, where n_1 and n_2 are proper divisors of n. For any $0 \neq a \in R$, $0 = na^2 = n_1 n_2 a^2 = (n_1 a)(n_2 a)$. Since R is an integral domain, $n_1 a = 0$ or $n_2 a = 0$. If $n_1 a = 0$, then it can be seen that $n_1 r = 0$ for any $r \in R$. In fact, $(n_1 r)a = r(n_1 a) = 0$, and since $a \neq 0$, we arrive at $n_1 r = 0$ for all $r \in R$, where $n_1 < n$. This contradicts the minimality of n, and hence n is a prime.

Definition 1.45 A ring is called a simple ring if it has no proper ideals.

Example 1.46 (1) The ring $R = \left\{ \begin{bmatrix} a & b \\ c & d \end{bmatrix} \mid a, b, c, d \in \mathbb{Q} \right\}$ is a simple ring.

(2) Every division ring is a simple ring.

Polynomial Ring Over a Ring

Let S be a ring and R be a subring of S. Then S is called a ring extension of R. An element $\alpha \in S$ is called an algebraic element over R if there exists a nonzero polynomial $f(x) = \alpha_0 + \alpha_1 x + \alpha_2 x^2 + \cdots + \alpha_n x^n$ over R such that $f(\alpha) = 0$, where we define $f(\alpha) = \alpha_0 + \alpha_1 \alpha + \alpha_2 \alpha^2 + \cdots + \alpha_n \alpha^n$. Clearly $\mathbb{R}$ and $\mathbb{Q}$ are extensions of $\mathbb{Z}$. $\sqrt{3} \in \mathbb{R}$ is algebraic over $\mathbb{Z}$ because it satisfies the polynomial $x^2 - 3$ over $\mathbb{Z}$. It is to be noted that $e, \pi \in \mathbb{R}$ are transcendental over $\mathbb{Z}$.

Let S be an extension of R. Let $x \in S$ be a transcendental element over R. Further, suppose that x commutes with all the elements of R. Then it can be easily verified that $T = \{\alpha_0 + \alpha_1 x + \alpha_2 x^2 + \cdots + \alpha_n x^n \mid \alpha_0, \alpha_1, \alpha_2 \cdots \alpha_n \in \mathbb{R}, n$ any nonnegative integer$\}$ is a subring of R. This ring is known as a polynomial ring in x over the ring R. Each element of T is known as a polynomial in x over R. It is usually denoted by $R[x]$. Obviously, $R \subseteq T = R[x]$ but $x \notin R[x]$. It can be easily verified that each element of $R[x]$ has its unique representation. Moreover, if R is commutative, then $R[x]$ is also commutative and if R is an integral domain, then $R[x]$ is also an integral domain.

In particular, if the ring S and its subring R have the same identity 1, then it can be also shown that $R[x] = [R \cup \{x\}]$, the smallest subring of S containing $R \cup \{x\}$. Moreover, in this case $R[x]$ is a ring with the identity 1 and $x \in R[x]$. Let $f(x) = \alpha_0 + \alpha_1 x + \alpha_2 x^2 + \cdots + \alpha_n x^n \in R[x]$. Here $\alpha_0, \alpha_1, \alpha_2, \ldots, \alpha_n$ are known as coefficients of $f(x)$. Here n is called the degree of $f(x)$ provided $\alpha_n \neq 0$. If $f(x) = 0 + 0x + 0x^2 + \cdots + 0x^n$, then it is called the zero polynomial. Its degree is undefined.

Remark 1.47 It can be easily seen that the definition of a polynomial ring given above is equivalent to the definition of a polynomial ring via Example 1.28(6).

Exercises

1. Show that a ring R with identity is commutative if and only if $(ab)^2 = a^2 b^2$ for all $a, b \in R$.
2. Justify the existence of the identity in the above result.
3. Prove that a ring R is commutative if it satisfies the property $a^2 = a$ for all $a \in R$.
4. Show that a ring R is commutative if and only if $(a + b)^2 = a^2 + 2ab + b^2$ for all $a, b \in R$.
5. If $(R, +, \cdot)$ is a system satisfying all the axioms of a ring with unity except $a + b = b + a$, for all $a, b \in R$, then show that $(R, +, \cdot)$ is a ring.
6. Let R be a ring (may be without unity 1) satisfying any one of the following identities:

(a) $(xy)^2 = xy$, for all $x, y \in R$,
(b) $(xy)^2 = yx$, for all $x, y \in R$,
(c) $(xy - yx)^2 = xy - yx$, for all $x, y \in R$.

Then show that R is commutative.
7. Find all rings up to isomorphism with two elements (three elements).
8. Show that it is impossible to find a homomorphism of $\mathbb{Z}_4$ onto $\mathbb{Z}_5$.
9. Show that the rings $\mathbb{Z}$ and $2\mathbb{Z}$ are not isomorphic.
10. Prove or disprove that homomorphic image of an integral domain is an integral domain.
11. If $char(R) = p$, p a prime, then show that $(a + b)^p = a^p + b^p$, $a, b \in R$.
12. If R is a ring satisfying $a^2 = a$ for all $a \in R$, then show that $char(R) = 2$.
13. Show that every ring containing six elements is necessarily commutative.
14. Show that it is impossible to get an integral domain with six elements.
15. Let R be a ring (may be without unity 1) satisfying any one of the following identities:

(a) $(xy)^n = xy$, for all $x, y \in R$, where $n > 1$ is a fixed positive integer,
(b) $(xy)^n = yx$, for all $x, y \in R$, where $n \geq 1$ is a fixed positive integer.

Then show that R is commutative.
16. Let $n \geq 1$, be a fixed positive integer and R a ring with the identity 1 satisfying $(ab)^k = a^k b^k$, for all $a, b \in R$, where $k = n, n + 1, n + 2$. Prove that R is commutative. Moreover, justify the existence of the identity 1 in this result.

1.3   Fields with Basic Properties

This section is devoted to study field, its various examples and some basic properties which have been used freely in the subsequent chapters.

Definition 1.48 A commutative division ring is called a field.
The set $\mathbb{Q}$, $\mathbb{R}$, and $\mathbb{C}$ under usual addition and multiplication are known examples of fields.

Example 1.49 (1) $(\mathbb{Z}_6, \oplus_6, \otimes_6)$ is neither an integral domain nor a field.
(2) $(\mathbb{Z}_p, \oplus_p, \otimes_p)$, p a prime, is a field.
(3) Consider the ring of real quaternions Q (see Example 1.23(4)), which is non-commutative and with the identity 1. For any nonzero element $x = a + bi + cj + dk \in Q$, it can be seen that there exists an element $y = \frac{a - bi - cj - dk}{a^2 + b^2 + c^2 + d^2} \in Q$ such that $xy = yx = 1$, i.e., every nonzero element has its multiplicative inverse. This is an example of a division ring which is not a field.
(4) Define addition and multiplication in $\mathbb{R}^2 = \{(a, b) \mid a, b \in \mathbb{R}\}$ as $(a, b) + (c, d) = (a + c, b + d)$, $(a, b)(c, d) = (ac - bd, ad + bc)$. It can be seen that $(\mathbb{R}^2, +, \cdot)$ is a field. In fact, for any nonzero $(a, b) \in \mathbb{R}^2$, there exists $(a, b)^{-1} = (\frac{a}{a^2 + b^2}, \frac{-b}{a^2 + b^2})$.

(5) Consider the set $M = \left\{ \begin{bmatrix} a & b \\ -\bar{b} & \bar{a} \end{bmatrix} \mid a, b \in \mathbb{C} \right\}$. Then M is a ring under matrix

addition and matrix multiplication with identity $\begin{bmatrix} 1 & 0 \\ 0 & 1 \end{bmatrix}$. For any nonzero matrix

$X = \begin{bmatrix} x+iy & a+ib \\ -a+ib & x-iy \end{bmatrix} \in M$, there exists $Y = \begin{bmatrix} \frac{x-iy}{x^2+y^2+a^2+b^2} & \frac{-(a+ib)}{x^2+y^2+a^2+b^2} \\ \frac{a-ib}{x^2+y^2+a^2+b^2} & \frac{x+iy}{x^2+y^2+a^2+b^2} \end{bmatrix} \in$

M such that $XY = YX = \begin{bmatrix} 1 & 0 \\ 0 & 1 \end{bmatrix}$. Hence M is a division ring which is not a

field.

Proposition 1.50 *Every field is an integral domain.*

Proof Suppose that R is a field, and there exist $a \neq 0, b \neq 0$ in R such that $ab = 0$. Since $b \neq 0$ is an element of the field R, there exists $c \in R$ such that $bc = cb = 1$. Now $0 = (ab)c = a(bc) = a1 = a$, a contradiction. Hence $ab \neq 0$ and R is an integral domain.

Remark 1.51 Every integral domain need not be a field. For example, $\mathbb{Z}$, the ring of integers, is an integral domain which is not a field. But in case of a finite integral domain, the converse of the above result is true.

Proposition 1.52 *Every finite integral domain is a field.*

Proof Let $R = \{a_1, a_2, \ldots, a_n\}$ be a finite integral domain consisting of distinct elements $a_1, a_2, \ldots, a_n$, and let a be a nonzero element of R. Then $aa_1, aa_2, \ldots, aa_n$ are all distinct and belong to R. In fact, if $aa_i = aa_j$, $1 \leq i, j \leq n$, then $a(a_i - a_j) = 0$ which yields that $a_i = a_j$, $1 \leq i, j \leq n$, a contradiction. Hence the set $\{aa_1, aa_2, \ldots, aa_n\}$ coincides with R. In particular, $aa_k = a$ for some $1 \leq k \leq n$. It can be seen that a_k is the multiplicative identity of R. If a_ℓ is an arbitrary element of R, then $a_\ell = aa_i$ for some $1 \leq i \leq n$. But $a_\ell a_k = a_k a_\ell = a_k(aa_i) = (a_k a)a_i = aa_i = a_\ell$ shows that a_k is the identity of R, denote it by 1. Hence if $0 \neq a \in R$, we find that $aa_m = a_m a = 1$, for some $1 \leq m \leq n$. Thus every nonzero element in R has its multiplicative inverse, R is a field.

Remark 1.53 (i) We have earlier shown that the characteristic of an integral domain is either 0 or p, where p is a positive prime integer. But we have also earlier proved that every field is an integral domain. Thus we conclude that characteristic of a field is either 0 or p, where p is a positive prime integer.
(ii) The identity of a subfield is the same as the identity of the field.

Proposition 1.54 *A field has no proper ideals. In particular, every field is a simple ring.*

Proof Let $\mathbb{F}$ be a field. Next, suppose that I is an ideal of $\mathbb{F}$. If $I = \{0\}$, then nothing to do. If not, there exists a nonzero element $a \in I$. As $\mathbb{F}$ is a field, there exists the element $a^{-1} \in \mathbb{F}$ such that $a^{-1}a = 1$. Since I is an ideal of $\mathbb{F}$, it follows that $a^{-1}a = 1 \in I$. As a result, $I = \mathbb{F}$. Hence $\mathbb{F}$ has no proper ideal. Finally, it can be said that every field $\mathbb{F}$ is a simple ring.

Proposition 1.55 *Let* $f : \mathbb{F} \longrightarrow R$ *be any ring homomorphism, where* R *is any arbitrary ring. Then* f *is either an injective ring homomorphism or the zero homomorphism.*

Proof We know that $Ker f$ is an ideal of $\mathbb{F}$. But as proved above a field has no proper ideals. Thus we have only two possibilities here for $Ker f$. Either $Ker f = \{0\}$, or $Ker f = \mathbb{F}$. If the first case happens then f becomes an injective ring homomorphism, and if the second case happens then f becomes the zero homomorphism.

Polynomial Ring Over a Field

As we described a polynomial ring over a ring R in the previous section, in the present section, we will focus only on polynomial ring over a field $\mathbb{F}$ and its some important properties, which have been used in different chapters of this book. First, we define some basic terms analogously as we did in the previous section.

Let $\mathbb{F}$ be a subfield of K. Then K is called a field extension of $\mathbb{F}$. An element $\alpha \in K$ is called an algebraic element over $\mathbb{F}$ if there exists a nonzero polynomial $f(x) = \alpha_0 + \alpha_1 x + \alpha_2 x^2 + \cdots + \alpha_n x^n \in \mathbb{F}[x]$ such that $f(\alpha) = 0$, where we define $f(\alpha) = \alpha_0 + \alpha_1 \alpha + \alpha_2 \alpha^2 + \cdots + \alpha_n \alpha^n$. Otherwise, $\alpha \in K$ is known as a transcendental element over $\mathbb{F}$. Clearly, $\mathbb{C}$ is an extension of $\mathbb{R}$ and $\mathbb{R}$ is an extension of $\mathbb{Q}$. Obviously, $\sqrt{3} \in \mathbb{R}$ is algebraic over $\mathbb{Q}$ because it is a root of $x^2 - 3$ over $\mathbb{Q}$. It is to be noted that $e, \pi \in \mathbb{R}$ are transcendental over $\mathbb{Q}$. An extension K of a field $\mathbb{F}$ is called an algebraic extension of $\mathbb{F}$ if its each element is algebraic over $\mathbb{F}$. Observe that $\mathbb{C}$ is an algebraic extension of $\mathbb{R}$ because any $\alpha + i\beta \in \mathbb{C}$ is a root of the polynomial $(x - \alpha)^2 + \beta^2 \in \mathbb{R}[x]$. It is also clear that $\mathbb{R}$ is not an algebraic extension of $\mathbb{Q}$.

Let K be an extension of $\mathbb{F}$. Let $x \in K$ be a transcendental element over $\mathbb{F}$. Similarly, as discussed in preceding section, one concludes that the polynomial ring $\mathbb{F}[x]$ is given as $\mathbb{F}[x] = \{\alpha_0 + \alpha_1 x + \alpha_2 x^2 + \cdots + \alpha_n x^n | \alpha_0, \alpha_1, \alpha_2 \ldots \alpha_n \in \mathbb{F}, n$ any nonnegative integer$\}$. This ring $\mathbb{F}[x]$ is an integral domain with identity. The polynomial ring $\mathbb{F}[x]$ has many important properties. Division algorithm, Euclidean algorithm and Factor theorem are some of the important and well-known results in the ring $\mathbb{F}[x]$. Here, we prove only Division algorithm, which has been used in proving so many results of this book.

Proposition 1.56 (Division Algorithm) *Let* $f(x), 0 \neq g(x) \in \mathbb{F}[x]$. *There exists unique polynomials* $q(x), r(x) \in \mathbb{F}[x]$ *such that* $f(x) = g(x)q(x) + r(x)$, *where either* $r(x) = 0$ *or* $\deg r(x) < \deg g(x)$. *Here* $q(x)$ *and* $r(x)$ *are called quotient and remainder, respectively, when* $f(x)$ *is divided by* $g(x)$.

Proof First, we prove the existence of such polynomials $q(x), r(x) \in \mathbb{F}[x]$. If $f(x) = 0$ or a constant polynomial then it is obvious. On the other hand, if $g(x)$ is a constant polynomial then existence is also trivial. Now consider the case when $\deg f(x) = m \geq 1$ and $\deg g(x) = n \geq 1$. Here if $m \leq n$ then it is easy to get existence of such polynomials. Finally, we tackle the case when $m > n$. In this case proof follows by transfinite induction applying on $\deg f(x)$. Result is obvious when $m = 1$. Let us suppose induction hypothesis, i.e., the result holds for all polynomials belonging to $\mathbb{F}[x]$ whose degree is less than the degree of $f(x)$. Let $f(x) = \alpha_0 + \alpha_1 x +$

$\alpha_2 x^2 + \cdots + \alpha_m x^m$ and $g(x) = \beta_0 + \beta_1 x + \beta_2 x^2 + \cdots + \beta_n x^n$. Now construct the polynomial $h(x) = f(x) - \alpha_m \beta_n^{-1} x^{m-n} g(x)$. Clearly, $h(x) \in \mathbb{F}[x]$ and deg $h(x) <$ deg $f(x)$. Hence by induction hypothesis there exist polynomials $q_1(x), r_1(x) \in \mathbb{F}[x]$ such that $h(x) = g(x) q_1(x) + r_1(x)$. This implies that $f(x) - \alpha_m \beta_n^{-1} x^{m-n} g(x) = g(x) q_1(x) + r_1(x)$, i.e., $f(x) = g(x)(\alpha_m \beta_n^{-1} x^{m-n} + q_1(x)) + r_1(x)$. Thus existence of polynomials for $f(x)$ stands proved. Thus result holds for any $f(x), 0 \neq g(x) \in \mathbb{F}[x]$.

Now, we prove the uniqueness of this existence of polynomials. If possible suppose that there exist polynomials $q_1(x) \neq q_2(x); r_1(x) \neq r_2(x) \in \mathbb{F}[x]$ such that $f(x) = g(x) q_1(x) + r_1(x)$, where either $r_1(x) = 0$ or deg $r_1(x) <$ deg $g(x)$ and $f(x) = g(x) q_2(x) + r_2(x)$, where either $r_2(x) = 0$ or deg $r_2(x) <$ deg $g(x)$. Equating two values for $f(x)$, we obtain that $g(x)(q_1(x) - q_2(x)) = r_1(x) - r_2(x)$. As $\mathbb{F}[x]$ is an integral domain and $q_1(x) - q_2(x) \neq 0$, hence we conclude that $r_1(x) - r_2(x) \neq 0$. Thus deg $(r_1(x) - r_2(x)) < n$ because deg $r_1(x) < n$ and deg $r_2(x) < n$. On the other hand, deg $(r_1(x) - r_2(x)) =$ deg $[g(x)(q_1(x) - q_2(x))] =$ deg $g(x) +$ deg $(q_1(x) - q_2(x)) \geq n$, which leads to a contradiction. This completes the proof.

Remark 1.57 (i) Let $f(x) \in \mathbb{F}[x]$. Let $\alpha \in \mathbb{F}$ be a root of $f(x)$. Divide $f(x)$ by $(x - \alpha)$ according to Division algorithm, there exist unique polynomials $q(x)$ and $r(x)$ such that $f(x) = (x - \alpha) q(x) + r(x)$, either $r(x) = 0$ or deg $r(x) <$ deg $(x - \alpha) = 1$. This implies that $r(x) = 0$ or λ, a nonzero constant polynomial. If $r(x) = \lambda$, then we get $f(x) = (x - \alpha) q(x) + \lambda$. Using the fact that α is a root of $f(x)$, we arrive at $\lambda = 0$, i.e., $f(x) = (x - \alpha) q(x)$. This shows that $(x - \alpha)$ is a factor of $f(x)$. It is known as factor theorem.

(ii) Let $f(x), g(x) \in \mathbb{F}[x]$. We say that a nonzero polynomial $f(x)$ divides a polynomial $g(x)$ in $\mathbb{F}[x]$, symbolically written as $f(x)/g(x)$, if there exists $h(x) \in \mathbb{F}[x]$ such that $g(x) = f(x) h(x)$. Here $f(x)$ and $h(x)$ are called as factors of $g(x)$. In particular, if deg $f(x) <$ deg $g(x)$ and deg $h(x) <$ deg $g(x)$, then $f(x)$ and $h(x)$ are known as proper or nontrivial factors of $g(x)$. Otherwise, $f(x)$ and $h(x)$ are known as improper or trivial factors of $g(x)$.

(iii) Let $f(x) \in \mathbb{F}[x]$ be any nonconstant polynomial. If $f(x)$ has no proper factors then $f(x)$ is known as irreducible polynomial. Otherwise, it is known as reducible polynomial.

(iv) Let $f(x) \in \mathbb{F}[x]$ be any nonconstant polynomial. Then $f(x) = f_1(x) f_2(x) \cdots f_n(x)$, where $f_1(x), f_2(x), \ldots, f_n(x)$ are some irreducible polynomials over $\mathbb{F}$.

(v) Let K be an extension of $\mathbb{F}$. An element $\alpha \in K$ is called a root of $f(x) = \alpha_0 + \alpha_1 x + \alpha_2 x^2 + \cdots + \alpha_n x^n \in \mathbb{F}[x]$ if $f(\alpha) = 0$, where $f(\alpha) = \alpha_0 + \alpha_1 \alpha + \alpha_2 \alpha^2 + \cdots + \alpha_n \alpha^n$.

(vi) Let $f(x)$ and $g(x)$ be any two nonzero polynomials over $\mathbb{F}$. Then deg $(f(x) + g(x)) \leq$ max $($ deg $f(x),$ deg $g(x))$, deg $(f(x) g(x)) =$ deg $f(x) +$ deg $g(x)$.

(vii) Let $f(x) \in \mathbb{F}[x]$ be any polynomial of degree $n \geq 1$. Then total number of roots of $f(x) = n$.

(viii) Let K be an extension of $\mathbb{F}$. Let $\alpha \in K$ be algebraic over $\mathbb{F}$. Let $T = [\mathbb{F} \cup \{\alpha\}]$
be the subfield of K generated by $\mathbb{F} \cup \{\alpha\}$. Then T is an algebraic extension
of $\mathbb{F}$.

Definition 1.58 Let $\mathbb{F}$ be a field. A subset S of $\mathbb{F}$ with minimum 2 elements is called
a subfield of $\mathbb{F}$ if it forms a field with regard to the operations of $\mathbb{F}$.

Example 1.59 (1) $\mathbb{Q}$ is a subfield of $\mathbb{R}$.
(2) $\mathbb{Q}$ and $\mathbb{R}$ are subfields of $\mathbb{C}$.
(3) Every field is a subfield of itself, known as trivial subfield.
(4) We know that the set $X = \{a + b\sqrt{2} \mid a, b \in \mathbb{Q}\}$ consisting of special type of
real numbers is a field with regard to usual addition and usual multiplication. It
is obvious to observe that $\mathbb{Q}$ is a subfield of X.

Proposition 1.60 *Let $\mathbb{F}$ be a field. A subset S of $\mathbb{F}$, with minimum two elements is a
subfield of $\mathbb{F}$ if and only if*

(i) $a, b \in S \Rightarrow a - b \in S$,
(ii) $a, 0 \neq b \in S \Rightarrow ab^{-1} \in S$.

Proof Suppose that $\mathbb{F}$ is a field. And S is a subfield of $\mathbb{F}$. This implies that $(S, +)$ is
a subgroup of $(\mathbb{F}, +)$. Hence $a, b \in S \Rightarrow a + (-b) = a - b \in S$. Similarly, $(S^* =
S \setminus \{0\}, \cdot)$ is a subgroup of $(\mathbb{F}^* = \mathbb{F} \setminus \{0\}, \cdot)$. This implies that $a, b \in S^* \Rightarrow ab^{-1} \in S^*$.
In turn, we have $ab^{-1} \in S$. On the other hand, if $a = 0$ and $b \in S^*$, then $ab^{-1} = 0 \in
S$. Now including the previous two statements, we have shown that $a, 0 \neq b \in S \Rightarrow
ab^{-1} \in S$.

Conversely, suppose that a subset S of $\mathbb{F}$, with minimum two elements satisfies
$(i)\ a, b \in S \Rightarrow a - b \in S$; $(ii)\ a, 0 \neq b \in S \Rightarrow ab^{-1} \in S$. To show that $(S, +, \cdot)$ is
a field, we prove that $(S, +, \cdot)$ is a commutative ring with identity in which each
nonzero element is a unit. As $|S| \geq 2$, there exists an element $0 \neq x \in S$. Now using
(i), we have $x - x = 0 \in S$ and using (ii) we have $xx^{-1} = 1 \in S$. Next suppose
that $a, b \in S$. If at least one of them is zero then $ab = 0 \in S$. Now suppose the case
when a and b both are nonzero. Using (ii), we have $1b^{-1} = b^{-1} \in S$. Obviously
$b^{-1} \neq 0$. Now using (ii) again, we have $a(b^{-1})^{-1} = ab \in S$. Thus, we have shown
that $a, b \in S \Rightarrow ab \in S$. Now utilizing (i) and the previous conclusion, we have
shown that S is a subring of $\mathbb{F}$ with identity. As $\mathbb{F}$ is a commutative ring, we are
bound to conclude that S is a commutative ring with the identity 1 in which each
nonzero element is a unit. Thus S is a field but it is contained in $\mathbb{F}$. Hence S is a
subfield of $\mathbb{F}$.

Definition 1.61 A field $\mathbb{F}$ is called a prime field if it has no proper subfield. This
says that only subfield of $\mathbb{F}$ is $\mathbb{F}$ itself.

Proposition 1.62 *The fields $\mathbb{Q}$ and $\mathbb{Z}_p$, where p is a positive prime integer, are prime
fields.*

Proof Let S be a subfield of $\mathbb{Q}$. It is obvious that $1 \in S$. Let $q \in Q$. If $q = 0$, then it obviously belongs to S. On the other hand, if $q \neq 0$, then it will look like $q = \frac{a}{b}$, where $a \neq 0, b \neq 0 \in \mathbb{Z}$. Without loss of generality, we can assume that $b > 0$. As $1 \in S$, $\underbrace{1 + 1 + \cdots + 1}_{b-\text{times}} = b \in S$. But characteristic of $\mathbb{Q}$ is 0. As a result, $b \neq 0$, but

S is a field. Hence $b^{-1} \in S$. If $a > 0$, then like the previous arguments $a \in S$. On the other hand, if $a < 0$, then $-a > 0$ and by previous arguments $-a \in S$. But as S is a field, hence $-(-a) = a \in S$. At this stage, we have proved that $a, b^{-1} \in S$. Using the fact that S is a field, we conclude that $ab^{-1} = \frac{a}{b} \in S$. Now, it follows that $\mathbb{Q} \subseteq S$. Finally, we have shown that $S = \mathbb{Q}$. As a result, $\mathbb{Q}$ has no proper subfield. Thus $\mathbb{Q}$ is a prime field.

Consider $\mathbb{Z}_p = \{\bar{0}, \bar{1}, \ldots, \overline{(p-1)}\}$, the field of residue classes modulo p, where p is a positive prime integer. Let S be a subfield of $\mathbb{Z}_p$. Clearly, $\bar{1} \in S$. Let $\bar{x} \in \mathbb{Z}_p$. If $\bar{x} = \bar{0}$, then obviously $\bar{x} \in \mathbb{Z}_p$. If $\bar{x} \neq \bar{0}$, then clearly x is an integer such that $1 \leq x \leq p$. We have $\bar{x} = \underbrace{\bar{1} + \bar{1} + \cdots + \bar{1}}_{x-\text{times}}$. We obtain that $\bar{x} \in S$ due to the fact that

S is a field. Thus we have shown that $S = \mathbb{Z}_p$. Hence $\mathbb{Z}_p$ is a prime field.

Proposition 1.63 *Intersection of all subfields of a field $\mathbb{F}$ is a prime field.*

Proof Let $\{S_i\}_{i \in \Lambda}$ be the family of all subfields of the field $\mathbb{F}$, where Λ is an indexing set. We have to prove that $\bigcap \{S_i\}_{i \in \Lambda}$ is a prime field. We know that intersection of a family of subfields is a subfield. Thus $\bigcap \{S_i\}_{i \in \Lambda}$ is a subfield of $\mathbb{F}$. Next we prove that $\bigcap \{S_i\}_{i \in \Lambda}$ has no proper subfield. For this let S be a subfield of $\bigcap \{S_i\}_{i \in \Lambda}$. This implies that S is obviously a subfield of $\mathbb{F}$. But $\{S_i\}_{i \in \Lambda}$ be the family of all subfields of the field $\mathbb{F}$, thus there exists $i_0 \in \Lambda$ such that $S = S_{i_0}$. Obviously, $\bigcap \{S_i\}_{i \in \Lambda} \subseteq S_{i_0} = S$. Thus we conclude that $S = \bigcap \{S_i\}_{i \in \Lambda}$. Therefore $\bigcap \{S_i\}_{i \in \Lambda}$ has no proper subfield. Hence $\bigcap \{S_i\}_{i \in \Lambda}$ is a prime field.

Proposition 1.64 *Let $\mathbb{F}$ be a field. If char $(\mathbb{F}) = 0$, then there exists a subfield S_1 of $\mathbb{F}$ such that $S_1 \cong \mathbb{Q}$. Further, if char $(\mathbb{F}) = p$, where p is a positive prime integer, then there exists a subfield S_2 of $\mathbb{F}$ such that $S_2 \cong \mathbb{Z}_p$.*

Proof Let us suppose that char $(\mathbb{F}) = 0$. Define a map $f : \mathbb{Q} \longrightarrow \mathbb{F}$ such that $f(\frac{m}{n}) = (m1)(n1)^{-1}$, where 1 is the identity of the field $\mathbb{F}$. As char $(\mathbb{F}) = 0$, so $(n1) \neq 0$ for any nonnegative integer n. Therefore $(n1)^{-1}$ exists. It can be easily verified that f is a well-defined map and also it is an injective ring homomorphism. Thus $\mathbb{Q} \approx f(\mathbb{Q})$. But we know that an isomorphic image of a field is a field. This shows that $f(\mathbb{Q})$ is a field. Now our required subfield S_1 of $\mathbb{F}$ is $f(\mathbb{Q})$, i.e., $S_1 = f(\mathbb{Q}) \subseteq \mathbb{F}$.

Now let us suppose that char $(\mathbb{F}) = p$. Since $\mathbb{Z}_p = \{\bar{0}, \bar{1}, \ldots, \overline{(p-1)}\}$. Now define a map $f : \mathbb{Z}_p \longrightarrow \mathbb{F}$ such that $f(\bar{x}) = x1$, where 1 is the identity of the field $\mathbb{F}$ and x is an integer such that $0 \leq x \leq p - 1$. It is easy to verify that f is a well-defined map and also it is an injective ring homomorphism. Hence $\mathbb{Z}_p \approx f(\mathbb{Z}_p)$. But $f(\mathbb{Z}_p)$ is a field because isomorphic image of a field is a field. Thus our desired subfield S_2 of $\mathbb{F}$ is $f(\mathbb{Z}_p)$, i.e., $S_2 = f(\mathbb{Z}_p) \subseteq \mathbb{F}$.

Definition 1.65 A field containing a finite number of elements is called a finite field or a Galois field.

Example 1.66 $\mathbb{Z}_p$ is a finite field.

Remark 1.67 (i) The characteristic of a finite field $\mathbb{F}$ cannot be 0. For otherwise, i.e., if char($\mathbb{F}$) = 0, then $1 + 1, 1 + 1 + 1, 1 + 1 + 1 + 1, \ldots$ will belong to $\mathbb{F}$ as $1 \in \mathbb{F}$ and also all these elements will be distinct. But under such a situation $\mathbb{F}$ will contain infinite number of elements. This leads to a contradiction. Thus the characteristic of a finite field $\mathbb{F}$ will be a positive prime integer p.

Here it is to be noted that converse of this statement is not true. There exists infinite field of characteristic p. We are constructing an example of such a field. Let $\mathbb{Z}_p$ denotes the field of residue classes modulo p. Let x be a transcendental element over the field $\mathbb{Z}_p$. Let $\mathbb{Z}_p[x]$ denotes the polynomial ring over the field $\mathbb{Z}_p$. Now define a set given by $T = \{\frac{f(x)}{g(x)} | f(x), g(x) \in \mathbb{Z}_p[x]\}$. It can be easily verified that T is a field with regard to addition and multiplication of rational functions. It is also obvious to observe that T is an infinite field but char $(T) = p$.

(ii) A field $\mathbb{F}$ is finite if and only if its multiplicative group is cyclic.

(iii) The number of elements in a finite field will be of the form p^n, where p is a positive prime integer, n is any positive integer and char $(\mathbb{F}) = p$. A finite field or a Galois field containing p^n elements is usually denoted by $GF(p^n)$.

(iv) Given a positive prime p and a positive integer n, there exists a finite field containing p^n elements.

(v) Any two finite fields having the same number of elements are isomorphic.

Definition 1.68 A field $\mathbb{F}$ is called an algebraically closed field if every polynomial over $\mathbb{F}$ has a root in $\mathbb{F}$.

Example 1.69 (1) The field of complex numbers $\mathbb{C}$ is algebraically closed. Since by Fundamental Theorem of Algebra, we know that every polynomial over field of complex numbers has all roots in $\mathbb{C}$.

(2) The field of real numbers $\mathbb{R}$ is not algebraically closed. Since $x^2 + 1$ is a polynomial over field of real numbers which has roots $\pm i$ but these roots do not lie in $\mathbb{R}$.

(3) The field of rational numbers $\mathbb{Q}$ is not algebraically closed. Since $x^2 - 3$ is a polynomial over field of rational numbers which has roots $\pm \sqrt{3}$ but these roots do not lie in $\mathbb{Q}$.

(4) Any finite field $\mathbb{F}$ cannot be algebraically closed. Let $\mathbb{F} = \{a_1, a_2, \ldots, a_m\}$. Now construct a polynomial $P(x) = 1 + (x - a_1)(x - a_2) \cdots (x - a_m)$ over $\mathbb{F}$. It is obvious to observe that no element of $\mathbb{F}$ is a root of $P(x)$. Thus $\mathbb{F}$ is not algebraically closed.

Proposition 1.70 *Let $\mathbb{F}$ be a field. Then following statements are equivalent for $\mathbb{F}$:*

(i) $\mathbb{F}$ is algebraically closed.

(ii) $\mathbb{F}$ has no proper algebraic extension.

(iii) Every irreducible polynomial over $\mathbb{F}$ *is of degree* 1.
(iv) Every polynomial over $\mathbb{F}$ *has a root in* $\mathbb{F}$.
(v) Every polynomial over $\mathbb{F}$ *has all its roots in* $\mathbb{F}$.
(vi) Every polynomial over $\mathbb{F}$ *breaks into linear factors over* $\mathbb{F}$.

Proof $(i) \Rightarrow (ii)$ Let $\mathbb{F}$ he algebraically closed. We have to prove that $\mathbb{F}$ has no proper algebraic extension. Suppose on contrary, i.e., there exists a proper algebraic extension K of $\mathbb{F}$. This implies that there exists $\alpha \in K$ such that $\alpha \notin \mathbb{F}$ and α is a root of a polynomial $f(x)$ over $\mathbb{F}$. Let $f(x) = \alpha_0 + \alpha_1 x + \alpha_2 x^2 + \cdots + \alpha_n x^n$, where $n \geq 1$ be a polynomial over the field $\mathbb{F}$. As $\mathbb{F}$ is algebraically closed, hence $f(x)$ will have a root in F. Let this root be $\beta_1 \in \mathbb{F}$. Hence by factor theorem $f(x) = (x - \beta_1)g(x)$, where $g(x) \in \mathbb{F}[x]$. Using hypothesis again $g(x)$ will have a root $\beta_2 \in \mathbb{F}$. Now using again the factor theorem, we arrive at $f(x) = (x - \beta_1)(x - \beta_2)g_1(x)$, where $g_1(x) \in \mathbb{F}[x]$. Now proceeding inductively, after finite number of steps and using the fact that any polynomial over $\mathbb{F}$ of degree n has n roots, we obtain that $f(x) = \lambda(x - \beta_1)(x - \beta_2) \cdots (x - \beta_n)$, where $\lambda, \beta_n \in \mathbb{F}$. Now we conclude that $\beta_1, \beta_2, \ldots, \beta_n$ are precisely the roots of $f(x)$. But α is also a root of $f(x)$. Thus, we get that $\alpha = \beta_i$ for some $i : 1 \leq i \leq n$. This implies that $\alpha \in \mathbb{F}$, which leads to a contradiction. Thus $\mathbb{F}$ has no proper algebraic extension.

$(ii) \Rightarrow (iii)$ Let $\mathbb{F}$ has no proper algebraic extension. We have to show that every irreducible polynomial over $\mathbb{F}$ is of degree 1. Suppose on contrary. Hence there exists an irreducible polynomial $f(x)$ over $\mathbb{F}$ of degree $n \geq 2$. It is obvious to observe that no root of $f(x)$ will lie in $\mathbb{F}$. Let α be a root of $f(x)$ lying outside $\mathbb{F}$. Clearly α is algebraic over $\mathbb{F}$. Let K= $[\mathbb{F} \cup \{\alpha\}]$ be the subfield generated by $\mathbb{F} \cup \{\alpha\}$. As α is algebraic over $\mathbb{F}$ and $K \neq \mathbb{F}$. Thus K is a proper algebraic extension of $\mathbb{F}$. This leads to a contradiction. Hence every irreducible polynomial over $\mathbb{F}$ is of degree 1.

$(iii) \Rightarrow (iv)$ Let every irreducible polynomial over $\mathbb{F}$ is of degree 1. Suppose that $f(x)$ is any polynomial over $\mathbb{F}$. We know that $f(x) = f_1(x)f_2(x) \cdots f_n(x)$, where $f_1(x), f_2(x), \ldots, f_n(x)$ are irreducible polynomials over $\mathbb{F}$. But according to hypothesis, we get $f_1(x) = a_1 x + b_1, f_2(x) = a_2 x + b_2, \ldots, f_n(x) = a_n x + b_n$. This shows that $f_1(x), f_2(x), \ldots, f_n(x)$ have their roots in $\mathbb{F}$. As a result, $f(x)$ has all its roots in $\mathbb{F}$. This completes our proof.

$(iv) \Rightarrow (v)$ Suppose that every polynomial over $\mathbb{F}$ has a root in $\mathbb{F}$. Let $f(x) \in \mathbb{F}[x]$ of degree $n \geq 1$. Hence $f(x)$ will have a root in F. Let this root be $\beta_1 \in \mathbb{F}$. Hence by factor theorem $f(x) = (x - \beta_1)g(x)$, where $g(x) \in \mathbb{F}[x]$. Using hypothesis again $g(x)$ will have a root $\beta_2 \in \mathbb{F}$. Now using again the factor theorem, we arrive at $f(x) = (x - \beta_1)(x - \beta_2)g_1(x)$, where $g_1(x) \in \mathbb{F}[x]$. Now proceeding inductively, after finitely number of steps and using the fact that degree of $f(x) = n$, we obtain that $f(x) = \lambda(x - \beta_1)(x - \beta_2) \cdots (x - \beta_n)$, where $\lambda, \beta_n \in \mathbb{F}$. Now we conclude that $\beta_1, \beta_2, \ldots, \beta_n$ are precisely the roots of $f(x)$ and these lie in $\mathbb{F}$. We are done.

$(v) \Rightarrow (vi)$ Let every polynomial over $\mathbb{F}$ have all its roots in $\mathbb{F}$. Suppose that $f(x) \in \mathbb{F}[x]$ of degree $n \geq 1$. By hypothesis $f(x)$ will have all roots in $\mathbb{F}$. But we know that the total number of all roots of $f(x)$ is equal to n, the degree of the polynomial. Thus, let these roots be $\alpha_1, \alpha_2, \ldots, \alpha_n \in \mathbb{F}$. Thus, by factor theorem $(x - \alpha_1), (x - \alpha_2), \ldots, (x - \alpha_n)$ will be factors of $f(x)$. Thus, we can write $f(x) = (x - \alpha_1)(x - \alpha_2) \cdots (x - \alpha_n)g(x)$, where $g(x) \in \mathbb{F}[x]$. Now comparing the degrees of both sides in the previous relation, we conclude that $g(x) = \lambda \in \mathbb{F}$. Hence $f(x) = \lambda(x - \alpha_1)(x - \alpha_2) \cdots (x - \alpha_n)$. Now proof is completed.

$(vi) \Rightarrow (i)$ Suppose that every polynomial over $\mathbb{F}$ breaks into linear factors over $\mathbb{F}$ and let $f(x) = \lambda(x - \alpha_1)(x - \alpha_2) \cdots (x - \alpha_n)$. Clearly α_1 is a root of $f(x)$ which lies in $\mathbb{F}$. This shows that $\mathbb{F}$ is algebraically closed.

Exercises

1. Prove that the only isomorphism of $\mathbb{Q}$ (resp. $\mathbb{R}$) onto $\mathbb{Q}$ (resp. $\mathbb{R}$) is the identity mapping I_Q (resp. I_R).
2. Show that the only isomorphism of $\mathbb{C}$ onto itself which maps reals to reals is the identity mapping $I_{\mathbb{C}}$ or the conjugation mapping.
3. Let R be a commutative ring with unity. Prove that R is a field if R has no proper ideals.
4. Show that a finite field cannot be of characteristic zero.
5. Prove that for a fixed prime p, $\mathbb{Q}(\sqrt{p}) = \{a + b\sqrt{p} \mid a, b \in \mathbb{Q}\}$ forms a field under usual addition and usual multiplication.
6. Let $\mathbb{F}$ be a field of characteristic $p > 0$ and $a \in \mathbb{F}$ be such that there exists no $b \in \mathbb{F}$ with $b^p = a$ (i.e., $\sqrt[p]{a} \notin \mathbb{F}$). Then show that $x^p - a$ is irreducible over $\mathbb{F}$.
7. If f is a function from a field $\mathbb{F}$ to itself such that $f(x) = x^{-1}$ if $x \neq 0$ and $f(0) = 0$, then show that f is an automorphism if and only if $\mathbb{F}$ has either 2, 3, 4 elements.
8. Find all roots of $x^5 + \bar{3}x^3 + x^2 + \bar{2}x \in \mathbb{Z}_5[x]$ in $\mathbb{Z}_5$.
9. Show that a factor ring of a field is either the trivial ring of one element or is isomorphic to the field.
10. Show that for a field $\mathbb{F}$, the set of all matrices of the form
    $$\left\{ \begin{bmatrix} a & b \\ 0 & 0 \end{bmatrix} \mid \text{ for all } a, b \in \mathbb{F} \right\} \text{ is a right ideal of } M_2(\mathbb{F}) \text{ but not a left ideal of}$$
    $M_2(\mathbb{F})$. Moreover, find a subset of $M_2(\mathbb{F})$ which is a left ideal of $M_2(\mathbb{F})$, but not a right ideal of $M_2(\mathbb{F})$.
11. Let R be a ring with identity. If char $(R) = 0$, then show that there exists a subring S_1 of R such that $S_1 \cong \mathbb{Z}$, the ring of integers. Further, if char $(R) = n$, where n is a positive integer, then prove that there exists a subring S_2 of R such that $S_2 \cong \mathbb{Z}_n$, the ring of residue classes (mod n).

1.4   Matrices

Let us start by introducing the basic definitions:

Let $\mathbb{F}$ be a field. Most part of the results we are going to discuss hold in the case $\mathbb{F}$ is an arbitrary field, in all that follows we always assume that the characteristic of $\mathbb{F}$ is different from 2.

A *matrix* over $\mathbb{F}$ of size $m \times n$ is a rectangular array with m rows and n columns of the form

$$A = \begin{bmatrix} a_{11} & a_{12} & \dots & a_{1n} \\ a_{21} & a_{22} & \dots & a_{2n} \\ \dots & \dots & \dots & \dots \\ a_{m1} & a_{m2} & \dots & a_{mn} \end{bmatrix}$$

where $a_{ij} \in \mathbb{F}$, for $i = 1, \dots, m$ and $j = 1, \dots, n$. The indices i, j corresponding to any a_{ij} indicate that the element appears in the ith row and jth column of the matrix. A matrix is said to be *square* of order n, if it has the same number n of rows and columns. Any matrix A can be represented in the compact form $A = (a_{ij})$, where the number of rows and columns must clearly be intended. Sometimes it will be useful to look at a matrix A as the sequence of its rows or a sequence of its columns. In those cases, denoting by

$$R_i = [a_{i1} \quad a_{i2} \quad \dots \quad a_{in}] \quad \text{and} \quad C_j = \begin{bmatrix} a_{1j} \\ a_{2j} \\ \vdots \\ a_{mj} \end{bmatrix}$$

respectively, the ith row and the jth column of A, we may represent A just listing its rows or columns, that is as

$$A = [R_1, R_2, \dots, R_m] \quad \text{or} \quad A = [C_1, C_2, \dots, C_n].$$

Definition 1.71   Let A be a matrix over $\mathbb{F}$ of order $m \times n$. A submatrix of the matrix A is a rectangular array with $s \leq m$ rows and $t \leq n$ columns obtained from A by deleting some of its $m - s$ rows and $n - t$ columns. This is equivalent to say that, if we pick s rows $R_{i_1}, \dots, R_{i_s}$ and t columns $C_{j_1}, \dots, C_{j_t}$ of A, we obtain a submatrix B of A by collecting the elements belonging to those rows and columns

$$B = \begin{bmatrix} a_{i_1 j_1} & a_{i_1 j_2} & \dots & a_{i_1 j_t} \\ a_{i_2 j_1} & a_{i_2 j_2} & \dots & a_{i_2 j_t} \\ \dots & \dots & \dots & \dots \\ a_{i_s j_1} & a_{i_s j_2} & \dots & a_{i_s j_t} \end{bmatrix}.$$

A principal submatrix is a square submatrix obtained by picking the same number of rows and columns in A, i.e., $s = t$ and $i_1 = j_1, \ldots, i_s = j_s$. The leading principal submatrix of size $k \times k$ is a principal submatrix which arises by picking precisely the first k rows and the first k columns of A.

Two matrices are said to be equal if they have the same orders and the corresponding entries are all equal to each other. The $m \times n$ *zero matrix* is the matrix whose all entries a_{ij} are 0, for any $i = 1, \ldots, m$ and $j = 1, \ldots, n$. We denote such a matrix by 0, or $0_{m \times n}$ in case its order needs to be recalled.

In literature, the set of all $m \times n$ matrices over a field $\mathbb{F}$ is denoted by $M_{mn}(\mathbb{F})$ and in particular, for $m = n$, the set of all square matrices of order n over $\mathbb{F}$ is usually denoted by $M_n(\mathbb{F})$.

In case of a square matrix $A = (a_{ij}) \in M_n(\mathbb{F})$, we give the following additional definitions:

(1) The set consisting of all elements with the same indices a_{ii}, for $i = 1, \ldots, n$ is said to be the *main diagonal* of the square matrix.
(2) A is said to be a *diagonal matrix* if all its off-diagonal terms are 0.
(3) A is called *identity matrix of size n* if all its diagonal terms are equal to 1, while other remaining entries are 0, i.e., $a_{ii} = 1$ and $a_{ij} = 0$ for $i \neq j$. We denote such a matrix by I, or I_n if it is important to keep track of its size.
(4) A is said to be a *scalar matrix* if it is of the form αI_n, for some $\alpha \in \mathbb{F}$. Then it is a diagonal matrix whose all diagonal terms are equal to α.
(5) A is called an *upper triangular* if every element below the main diagonal is 0, that is $a_{ij} = 0$ for any $i > j$.
(6) A is called a *lower triangular* if every element above the main diagonal is 0, that is $a_{ij} = 0$ for any $j > i$.

Matrix Operations

Let $M_{mn}(\mathbb{F})$ be the set of all $m \times n$ matrices over $\mathbb{F}$. If $A, B \in M_{mn}(\mathbb{F})$, then the *sum* $A + B$ is the matrix in $M_{mn}(\mathbb{F})$ obtained by adding together the corresponding entries in the two matrices. In other words, if $A = (a_{ij})$ and $B = (b_{ij})$ then $A + B = C \in M_{mn}(\mathbb{F})$ is the matrix $C = (c_{ij})$ such that $c_{ij} = a_{ij} + b_{ij}$, for any $i = 1, \ldots, m$ and $j = 1, \ldots, n$:

$$C = \begin{bmatrix} a_{11} + b_{11} & a_{12} + b_{12} & \cdots & a_{1n} + b_{1n} \\ a_{21} + b_{21} & a_{22} + b_{22} & \cdots & a_{2n} + b_{2n} \\ \cdots & \cdots & \cdots & \cdots \\ a_{m1} + b_{m1} & a_{m2} + b_{m2} & \cdots & a_{mn} + b_{mn} \end{bmatrix}.$$

Let now $\alpha \in \mathbb{F}$ and $A = (a_{ij}) \in M_{mn}(\mathbb{F})$. The *scalar multiplication* of A by α is the matrix in $M_{mn}(\mathbb{F})$ obtained by multiplying each entry of A by the scalar α, that is, the matrix $\alpha A = (\alpha a_{ij})$:

$$\alpha A = \begin{bmatrix} \alpha a_{11} & \alpha a_{12} & \cdots & \alpha a_{1n} \\ \alpha a_{21} & \alpha a_{22} & \cdots & \alpha a_{2n} \\ \cdots & \cdots & \cdots & \cdots \\ \alpha a_{m1} & \alpha a_{m2} & \cdots & \alpha a_{mn} \end{bmatrix}.$$

Appealing to the field axioms and focusing the attention to an entry-wise, it is easy to see that addition and scalar multiplication satisfy the following properties:

(1) For any $A, B \in M_{mn}(\mathbb{F})$, $A + B = B + A$.
(2) For any $A, B, C \in M_{mn}(\mathbb{F})$, $A + (B + C) = (A + B) + C$.
(3) For any $A \in M_{mn}(\mathbb{F})$, $A + 0_{m \times n} = 0_{m \times n} + A = A$.
(4) For any $A = (a_{ij}) \in M_{mn}(\mathbb{F})$, there exists the matrix $B = (-a_{ij}) \in M_{mn}(\mathbb{F})$ such that $A + B = 0_{m \times n}$. Usually such a matrix B is denoted by $-A$.
(5) For any $\alpha \in \mathbb{F}$ and $A, B \in M_{mn}(\mathbb{F})$, $\alpha(A + B) = \alpha A + \alpha B$.
(6) For any $\alpha, \beta \in \mathbb{F}$ and $A \in M_{mn}(\mathbb{F})$, $(\alpha + \beta)A = \alpha A + \beta A$.
(7) If $1_{\mathbb{F}}$ is the identity element of $\mathbb{F}$, then $1_{\mathbb{F}}A = A$, for any $A \in M_{mn}(\mathbb{F})$.
(8) For any $\alpha, \beta \in \mathbb{F}$ and $A \in M_{mn}(\mathbb{F})$, $\alpha(\beta A) = (\alpha \beta)A$.

In particular, note that $M_{mn}(\mathbb{F})$ is a commutative group with respect to the addition between matrices. Moreover, it is an example of a vector space over $\mathbb{F}$. The concept of vector space is discussed in the next chapter.

Definition 1.72 Let $A \in M_n(\mathbb{F})$. The sum of all entries on the main diagonal of A is called the *trace* of A and denoted by $tr(A)$.

The following properties satisfied by the trace are easy consequences of its definition.

(1) If $A, B \in M_n(\mathbb{F})$, then $tr(A + B) = tr(A) + tr(B)$.
(2) If $A \in M_n(\mathbb{F})$ and $\alpha \in \mathbb{F}$, then $tr(\alpha A) = \alpha tr(A)$.

Definition 1.73 Let $A \in M_{mn}(\mathbb{F})$. The *transpose* of A, usually denoted by A^t, is the $n \times m$ matrix obtained by interchanging rows and columns of A in such a way that the ordered elements in the ith row (column) of A are exactly the ordered elements of the ith column (resp. row) of A^t. In other words, if $\alpha \in \mathbb{F}$ appears in the (r, s)-entry of A, then it appears in the (s, r)-entry of A^t.

Here we also list a number of easy properties, whose proofs depend only on the definition of the transpose.

(1) For any $A \in M_{mn}(\mathbb{F})$, $(A^t)^t = A$.
(2) For any $A, B \in M_{mn}(\mathbb{F})$, $(A + B)^t = A^t + B^t$.
(3) For any $A \in M_n(\mathbb{F})$ and $\alpha \in \mathbb{F}$, $(\alpha A)^t = \alpha A^t$.
(4) For any $A \in M_n(\mathbb{F})$, $tr(A^t) = tr(A)$.

A particular class of matrices, highly significant in further discussions, consists of those which satisfy the condition of coinciding with their corresponding transposes.

Definition 1.74 If a square matrix A satisfies the condition $A = A^t$, (resp. $A = -A^t$) we say that it is a *symmetric* (resp. *skew-symmetric*) matrix.

Generalizing the previous concepts, we introduce the following:

Definition 1.75 The *conjugate* of a complex matrix $A = (a_{ij}) \in M_{mn}(\mathbb{C})$ is the matrix $\overline{A} = (a'_{ij}) \in M_{mn}(\mathbb{C})$ such that $a'_{ij} = \overline{a_{ij}}$ for any $i = 1, \ldots, m$ and $j = 1, \ldots, n$. The *conjugate transpose* of A is the matrix $A^* = (b_{ij}) \in M_{nm}(\mathbb{C})$ such that $b_{ij} = \overline{a_{ji}}$ for any $i = 1, \ldots, n$ and $j = 1, \ldots, m$. A square matrix that is equal to its conjugate transpose is called *hermitian*. If $A = -A^*$, then A is said to be *skew-hermitian*.

If A is real, then its conjugate transpose is same as its transpose, and hence hermitian is same as symmetric.

Now we describe matrix multiplication as follows: Let $A = (a_{ij}) \in M_{mn}(\mathbb{F})$ and $B = (b_{ij}) \in M_{nq}(\mathbb{F})$. The matrix product AB is defined as the matrix $C = (c_{ij}) \in M_{mq}(\mathbb{F})$ such that the entry c_{rs} of C is computed by

$$c_{rs} = \sum_{k=1}^{n} a_{rk}b_{ks} = a_{r1}b_{1s} + a_{r2}b_{2s} + a_{r3}b_{3s} + \ldots + a_{rn}b_{ns}.$$

It is clear that the product AB does not make sense if $A \in M_{mn}(\mathbb{F})$ and $B \in M_{tq}(\mathbb{F})$ with $n \neq t$. Hence, the products AB and BA are simultaneously possible only if A and B are matrices of the orders $m \times n$ and $n \times m$, respectively. Nevertheless, even when possible, it is not generally true that $AB = BA$. Thus the matrix product is not commutative. Associativity of matrix multiplication and distributive properties are given as below:

Proposition 1.76 (i) If $A = (a_{ij}) \in M_{mn}(\mathbb{F})$, $B = (b_{ij}) \in M_{nt}(\mathbb{F})$, $C = (c_{ij}) \in M_{tq}(\mathbb{F})$ then $(AB)C = A(BC)$, that is, the matrix product is associative.
(ii) For any $A \in M_{mn}(\mathbb{F})$, $B \in M_{nt}(\mathbb{F})$, $C \in M_{nt}(\mathbb{F})$, we have $A(B + C) = AB + AC$, i.e., the matrix product is distributive over the matrix addition.

Note 1.77 The earlier mentioned identity matrix of order n is the identity element for the matrix product in $M_n(\mathbb{F})$. All the properties we have discussed lead us to the conclusion that, for any $n \geq 1$, the set $M_n(\mathbb{F})$, equipped by matrix addition and multiplication is a (noncommutative) ring having unity.

Two relevant additional properties for the matrix product are the following:

(i) If $A = (a_{ij}) \in M_{mn}(\mathbb{F})$ and $B = (b_{ij}) \in M_{nr}(\mathbb{F})$, then $(AB)^t = B^t A^t$.
(ii) If $A = (a_{ij}) \in M_n(\mathbb{F})$ and $B = (b_{ij}) \in M_n(\mathbb{F})$, then $tr(AB) = tr(BA)$.

The Determinant of a Square Matrix

Definition 1.78 Let $A = (a_{ij}) \in M_n(\mathbb{F})$. Then the *determinant* of A, written as det A or $|A|$ is the element $\sum_{\sigma \in S_n} \text{Sign}(\sigma) a_{1\sigma(1)} a_{2\sigma(2)} \cdots a_{n\sigma(n)} \in \mathbb{F}$, where Sign $(\sigma) = 1$ if σ is an even permutation and sign $(\sigma) = -1$ if σ is an odd permutation and S_n is the permutation group on n symbols, i.e., $1, 2, \ldots, n$. We will also use the notation

$$\begin{vmatrix} a_{11} \cdots a_{1n} \\ \vdots \quad \vdots \quad \vdots \\ a_{n1} \cdots a_{nn} \end{vmatrix}$$

for the determinant of the matrix

$$\begin{bmatrix} a_{11} \cdots a_{1n} \\ \vdots \quad \vdots \quad \vdots \\ a_{n1} \cdots a_{nn} \end{bmatrix}.$$

Remark 1.79 (i) For $n = 1$, i.e., if $A = (a)$ is a 1×1 matrix, then we have $|A| = a$.

(ii) For $n = 2$, i.e., if $A = \begin{bmatrix} a_{11} \ a_{12} \\ a_{21} \ a_{22} \end{bmatrix}$ is a 2×2 matrix, then we have $|A| = a_{11}a_{22} - a_{12}a_{21}$.

(iii) If A is a diagonal matrix with diagonal entries $\lambda_1, \lambda_2, \ldots, \lambda_n$, then $|A| = \lambda_1\lambda_2 \cdots \lambda_n$.

Definition 1.80 Any square matrix having determinant equal to zero (respectively different from zero) is said to be *singular* (resp. *nonsingular*).

Determinants have the following well-known basic properties:

Theorem 1.81 *Let $A \in M_n(\mathbb{F})$. Then*

(i) *interchanging two rows of A changes the sign of det A,*

(ii) *det A= det A^t,*

(iii) *for any $B \in M_n(\mathbb{F})$, det (AB) =det A det B,*

(iv) *the determinant of an upper triangular or lower triangular matrix is the product of the entries on its main diagonal,*

(v) *if A has the block diagonal form, i.e., $A = \begin{pmatrix} B_1 & 0 & \cdots & 0 \\ 0 & \ddots & \cdots & \vdots \\ \vdots & \cdots & \ddots & \vdots \\ 0 & \cdots & 0 & B_m \end{pmatrix}$, where B_1,*

$B_2, \ldots, B_m$ *are square matrices, then det A=det B_1 det $B_2 \cdots$ det B_m.*

Elementary Array Operations in $\mathbb{F}^n$

Let us consider a set of arrays $R_1, \ldots, R_k \in \mathbb{F}^n$. For any $\alpha_1, \ldots, \alpha_k \in \mathbb{F}$, the element

$$\alpha_1 R_1 + \alpha_2 R_2 + \cdots + \alpha_k R_k \in \mathbb{F}^n \tag{1.1}$$

is said to be a *linear combination* of $R_1, \ldots, R_k$.

Definition 1.82 Let $R_1, \ldots, R_k \in \mathbb{F}^n$ and $\mathbf{0}$ be the zero element in $\mathbb{F}^n$. It seems clear that any R_i can be regarded as a matrix with exactly one row and n columns, that is $R_i \in M_{1n}(\mathbb{F})$. Therefore, addition $R_i + R_j$ and scalar product αR_i, by an element $\alpha \in \mathbb{F}$, are just particular cases of the more general ones related to addition of $m \times n$ matrices and scalar product of a $m \times n$ matrix by an element $\alpha \in \mathbb{F}$.

We say that $R_1, \ldots, R_k$ are *linearly dependent* if there exist nontrivial choice of scalars $\alpha_1, \ldots, \alpha_k \in \mathbb{F}$ such that

$$\alpha_1 R_1 + \alpha_2 R_2 + \cdots + \alpha_k R_k = 0.$$

Conversely, we say that $R_1, \ldots, R_k$ are *linearly independent* in the case

$$\alpha_1 R_1 + \alpha_2 R_2 + \cdots + \alpha_k R_k = 0 \quad \text{if and only if} \quad \alpha_1 = \alpha_2 = \cdots = \alpha_k = 0.$$

Remark 1.83 Suppose that $R_1, \ldots, R_k \in \mathbb{F}^n$ are linearly dependent and let $\alpha_1, \ldots, \alpha_k \in \mathbb{F}$ such that

$$\alpha_1 R_1 + \alpha_2 R_2 + \cdots + \alpha_k R_k = 0,$$

where $\alpha_i \neq 0$, for some $i = 1, \ldots, k$. Then

$$\alpha_i R_i = -\alpha_1 R_1 - \alpha_2 R_2 - \cdots - \alpha_{i-1} R_{i-1} - \alpha_{i+1} R_{i+1} - \cdots - \alpha_k R_k$$

so that

$$R_i = -\frac{\alpha_1}{\alpha_i} R_1 - \frac{\alpha_2}{\alpha_i} R_2 - \cdots - \frac{\alpha_{i-1}}{\alpha_i} R_{i-1} - \frac{\alpha_{i+1}}{\alpha_i} R_{i+1} - \cdots - \frac{\alpha_k}{\alpha_i} R_k,$$

where $\frac{1}{\alpha_i} = \alpha_i^{-1}$ is the inverse of α_i as an element of the field $\mathbb{F}$, implying that R_i can be expressed as a linear combination of $\{R_1, \ldots, R_{i-1}, R_{i+1}, \ldots, R_k\}$.

Remark 1.84 Let $\{R_1, \ldots, R_k\}$ be a set of elements in $\mathbb{F}^n$. If one of them is the zero element $0 \in \mathbb{F}^n$, then $R_1, \ldots, R_k$ are linearly dependent.

For the sake of clarity, if $R_1, \ldots, R_k$ are linearly dependent (or independent) arrays, we sometimes say that the set $\{R_1, \ldots, R_k\}$ is a *linearly dependent set* (resp. *linearly independent set*).

Given a set S of elements of $\mathbb{F}^n$, one of the most important questions in linear algebra, if not the most important one, is how to recognize the (largest) number of linearly independent arrays in S. To answer this question we need to permit some remarks.

We first assume that the set

$$\{R_1, \ldots, R_k\} \subset \mathbb{F}^n \tag{1.2}$$

is linearly dependent and there are $\alpha_1, \ldots, \alpha_k \in \mathbb{F}$ such that

$$\alpha_1 R_1 + \alpha_2 R_2 + \ldots + \alpha_k R_k = 0 \tag{1.3}$$

for some $\alpha_i \neq 0$.

A first and obvious remark is that the order of comparison of the arrays does not affect the fact that the set is linearly dependent. Thus, any other set obtained by permuting $R_1, \ldots, R_k$, is yet linearly dependent.

Let now $0 \neq \beta \in \mathbb{F}$ and consider the following set of elements in $\mathbb{F}^n$:

$$\{R_1, \ldots, R_{i-1}, \beta R_i, R_{i+1}, \ldots, R_k\} \subset \mathbb{F}^n.$$

From the fact that

$$\gamma_1 R_1 + \gamma_2 R_2 + \cdots + \gamma_i(\beta R_i) + \cdots + \gamma_k R_k = \mathbf{0},$$

where $\gamma_h = \beta \alpha_h$, for any $h \neq i$, and $\gamma_i = \alpha_i$, it follows that

$$\{R_1, \ldots, R_{i-1}, \beta R_i, R_{i+1}, \ldots, R_k\}$$

is linearly dependent. Hence, scalar product of an array by an element from $\mathbb{F}$ does not affect the fact that the new set of elements in $\mathbb{F}^n$ is linearly dependent, as the original set was.

We now fix $0 \neq \beta \in \mathbb{F}$ and two distinct arrays $R_i, R_j \in \{R_1, \ldots, R_k\}$. Consider the following set of elements in $\mathbb{F}^n$:

$$\{R_1, \ldots, R_{i-1}, R_i + \beta R_j, R_{i+1}, \ldots, R_k\}. \tag{1.4}$$

Notice that

$$\gamma_1 R_1 + \gamma_2 R_2 + \cdots + \gamma_i(R_i + \beta R_j) + \cdots + \gamma_k R_k = \mathbf{0},$$

where $\gamma_h = \alpha_h$, for any $h \neq j$, and $\gamma_j = \alpha_j - \beta \alpha_i$. We then conclude that the set described in (1.4) is linearly dependent.

Suppose now that the set at (1.2) is linearly independent. Of course, any other set obtained by permuting (1.2) is yet linearly independent.

Moreover, if we assume there exists some $0 \neq \beta \in \mathbb{F}$ such that

$$\{R_1, \ldots, R_{i-1}, \beta R_i, R_{i+1}, \ldots, R_k\}$$

is linearly dependent, we would conclude that

$$\gamma_1 R_1 + \gamma_2 R_2 + \cdots + \gamma_i(\beta R_i) + \cdots + \gamma_k R_k = \mathbf{0}$$

for some $0 \neq \gamma_i \in \mathbb{F}$, contradicting the fact that (1.2) is linearly independent. Also in this case, the scalar product by a nonzero scalar preserves the linear independence between elements of the new set.

Finally, for $0 \neq \beta \in \mathbb{F}$ and two distinct rows $R_i, R_j \in \{R_1, \ldots, R_k\}$, assuming that

$$\{R_1, \ldots, R_{i-1}, R_i + \beta R_j, R_{i+1}, \ldots, R_k\} \tag{1.5}$$

is linearly dependent, it should follow that

$$\gamma_1 R_1 + \gamma_2 R_2 + \cdots + \gamma_i R_i + \cdots + (\gamma_j + \gamma_i \beta) R_j + \cdots + \gamma_k R_k = 0,$$

where $0 \neq \gamma_h \in \mathbb{F}$ for some $1 \leq h \leq k$, once again a contradiction. We then conclude that the set described in (1.5) is linearly independent.

In summary, starting from a given set of array (1.2), we have obtained three different sets of elements in $\mathbb{F}^n$, by performing the following three operations on the elements of (1.2):

(1) Interchanging two arrays.
(2) Multiplying an array R_i by a nonzero scalar α in such a way that αR_i replaces R_i in the new set of arrays.
(3) Adding a constant multiple of an array R_j (namely αR_j) to another array R_i, in such a way that $R_i + \alpha R_j$ replaces R_i in the new set of arrays.

The listed operations are called *elementary operations*. We have in particular proved that, even if elementary operations generate subsets that are different from the starting one, they preserve the property of linear dependence (or independence) between the elements constituting those sets. Hence, when we determine the largest number of linearly independent arrays in a given set, it could be useful to perform some elementary operations to obtain a new set of arrays, for which the answer to the question can be more easily achieved.

Definition 1.85 Two sets of arrays $S = \{R_1, \ldots, R_k\} \subset \mathbb{F}^n$ and $S' = \{R'_1, \ldots, R'_k\} \subset \mathbb{F}^n$ are said to be *equivalent* if S can be transformed into S' by performing a finite sequence of elementary operations on it.

Consider then the set $S = \{R_1, \ldots, R_k\} \subset \mathbb{F}^n$ and assume that it is linearly dependent. Without loss of generality, we may assume that R_k is linearly depending from $R_1, \ldots, R_{k-1}$ (if not, it would be enough to permute the order of arrays). By following the argument in Remark 1.83, R_k being a linear combination of $R_1, \ldots, R_{k-1}$, there exist $\alpha_1, \ldots, \alpha_k \in \mathbb{F}$ such that

$$R_k = -\frac{\alpha_1}{\alpha_k} R_1 - \frac{\alpha_2}{\alpha_k} R_2 - \cdots - \frac{\alpha_{k-1}}{\alpha_k} R_{k-1}.$$

At this point, replacing R_k by

$$R_k + \frac{\alpha_1}{\alpha_k} R_1 + \frac{\alpha_2}{\alpha_k} R_2 + \cdots + \frac{\alpha_{k-1}}{\alpha_k} R_{k-1}$$

we obtain a set S' of arrays which is equivalent to S. In particular, the kth array of S' is now precisely 0:

$$S' = \{R_1, \ldots, R_{k-1}, 0\}.$$

Starting from this, we consider the subset $\{R_1, \ldots, R_{k-1}\}$. If it is linearly independent, we have no chance to replace any array by $\mathbf{0}$. But, in case it is linearly dependent, by repeating the process, we are able to replace, for instance, R_{k-1} by $\mathbf{0}$ and obtain again a new set of arrays S'' which is equivalent to S' :

$$S'' = \{R_1, \ldots, R_{k-2}, \mathbf{0}, \mathbf{0}\}.$$

Therefore, if we suppose that, after t steps, the set of arrays is

$$S^{(t)} = \{R_1, \ldots, R_{k-t}, \underbrace{\mathbf{0}, \ldots, \mathbf{0}}_{t-times}\}$$

and the subset $\{R_1, \ldots, R_{k-t}\}$ is linearly independent, then we stop the process. Having in mind the above, it becomes clear that the largest number $k - t$ of linearly independent arrays in $S^{(t)}$ is equal to the largest number of linearly independent arrays in the starting set S. Moreover, if the original set S consists of all linearly independent elements, then there is no sequence of elementary operations that can transform it into a set S' having some zero element.

Equivalent Matrices

We have previously remarked how any matrix $A = (a_{ij}) \in M_{mn}(\mathbb{F})$ can be represented just listing its rows $R_1, \ldots, R_m$.

In light of what we said above, k rows $R_1, \ldots, R_k$ extracted from the matrix are linearly dependent (or independent) if they are dependent (or independent) as elements of $\mathbb{F}^n$. In particular, to recognize the largest number of linearly independent rows in the set $S = \{R_1, \ldots, R_k\}$, we may perform some appropriate elementary operations on S and obtain an equivalent set S' of arrays in $\mathbb{F}^n$. The elements of S' are not necessarily rows from A; nevertheless the relationship of linear dependence (or independence) between elements of S' has not changed since the relationship between elements of S.

Note 1.86 We introduce the following notations for elementary row operations on a matrix A :

(I)   Interchanging two rows R_i and R_j will be denoted $R_i \leftrightarrow R_j$.

(II) Multiplying a row R_i by a nonzero scalar α, in such a way that αR_i replaces R_i in the new set of arrays, will be denoted by $R_i \to \alpha R_i$.

(III) Adding a constant multiple of a row R_j (namely αR_j) to another row R_i, in such a way that $R_i + \alpha R_j$ replaces R_i in the new set of arrays, will be denoted by $R_i \to R_i + \alpha R_j$.

Definition 1.87 Two matrices $A, A' \in M_{mn}(\mathbb{F})$ are called *equivalent* if A can be transformed into A' by performing a finite sequence of elementary row operations on it.

Remark 1.88 Let $A, A' \in M_n(\mathbb{F})$ be equivalent matrices. Thus A' is obtained from A by performing a finite sequence of elementary row operations. This means that, any submatrix B' of A' is obtained from a suitable submatrix B of A, by performing the same sequence of row operations. In particular, if B' and B are equivalent square submatrices respectively of A' and A, then B' is singular (resp. nonsingular) if and only if B is.

Reduced Row Form of a Matrix

Let $A = (a_{ij}) \in M_{mn}(\mathbb{F})$ and $A = [R_1, \ldots, R_m]$ be its compact representation, where any $R_i \in \mathbb{F}^n$ is a row of A.

Definition 1.89 An element $a_{ij} \in R_i$ is said to be the *leading entry* of ith row if it is the first nonzero entry of the row.

Definition 1.90 The matrix $A = (a_{ij}) \in M_{mn}(\mathbb{F})$ is said to be in *reduced row form* or *echelon form* if it satisfies the following conditions:

(1) All the rows that consist entirely of zeros are below any row that contains nonzero entries, that is the nonzero rows of A precede the zero rows.
(2) The column numbers $c_1, \ldots, c_k$ of the leading entries of the nonzero rows $R_1, \ldots, R_k$ form an increasing sequence of numbers, that is $c_1 < c_2 < \cdots < c_k$. In other words, the leading element of each row is in a column to the right of the leading elements of preceding rows.

Hence, we may, in general, say that the reduced row form of a matrix is the following:

$$
\begin{bmatrix}
0 \cdot 0\, a_{1j_1} \cdot \cdot & \cdot & \cdot & \cdots & a_{1n} \\
0 \cdot \cdot \ 0 & \cdot 0\, a_{2j_2} \cdot \cdot & \cdot & \cdots & a_{2n} \\
0 \cdot 0\ 0\ \ 0 \cdot & 0 \ \cdot 0\, a_{3j_3} \cdots & \cdot & \cdots & a_{3n} \\
\cdot \cdot \cdot \cdot & \cdot \cdot \cdot & \cdot & \cdots & \cdot \\
\cdot \cdot \cdot \cdot & \cdot \cdot \cdot & \cdot & \cdots & \cdot \\
\cdot \cdot \cdot \cdot & \cdot \cdot \cdot & \cdot & \cdots & \cdot \\
0\ 0 \cdot & \cdot \cdot & 0 \cdot \ 0 & \cdots 0\, a_{mj_m} & \cdots a_{mn}
\end{bmatrix}
$$

in case there is no row consisting entirely of zeros, or also

$$
\begin{bmatrix}
0 \cdot 0\, a_{1j_1} & \cdots & & \cdot & a_{1n} \\
0 \cdot \cdot \ 0 & \cdots 0\, a_{2j_2} & & \cdot & a_{2n} \\
0 \cdot 0\ 0\ \ 0 & \cdots \ 0 & \cdots 0\, a_{3j_3} \cdots & \cdot & a_{3n} \\
\cdot \cdot \cdot & & & & \cdot \\
0\ 0 \cdot & \cdots & 0 \cdots \ 0 & \cdots 0\, a_{kj_k} & \cdot\ a_{kn} \\
0\ 0 \cdot & & & \cdot \cdot & \cdot\ 0\ 0 \\
\cdot \cdot \cdot & & & & \\
0\ 0 \cdot & & & \cdot & 0\ 0
\end{bmatrix}
$$

if there are precisely $m - k$ zero rows in the matrix.

Theorem 1.91 (Gaussian elimination method) *By performing a finite sequence of elementary row operations, any matrix $A \in M_{mn}(\mathbb{F})$ can be transformed in a matrix having a reduced row form.*

Definition 1.92 A leading entry of a matrix used to zero out entries below it by means of elementary row operations is called a *pivot*.

We have the following important property when a matrix is transformed into a row-reduced echelon form:

Proposition 1.93 *If $A \in M_{mn}(\mathbb{F})$ has the reduced row form, then the $m - r$ nonzero rows $R_1, \ldots, R_{m-r}$ are linearly independent.*

The following results are important in explaining the relationships between linearly independent rows and linearly independent columns in matrices:

Theorem 1.94 *(i) Let $A \in M_n(\mathbb{F})$. The set of all rows $\{R_1, \ldots, R_n\}$ from A is linearly independent if and only if A is nonsingular.*

(ii) Let $A \in M_{mn}(\mathbb{F})$. The largest number of linearly independent rows from A is equal to the largest number of linearly independent columns from A.

The Rank of a Matrix

Definition 1.95 Let $A \in M_{mn}(\mathbb{F})$. The *rank* of A, denoted by $r(A)$, is the largest order of any nonsingular square submatrix in A.

According to Theorem 1.94, the above definition of rank of a matrix can be reworded as follows:

Definition 1.96 Let $A \in M_{mn}(\mathbb{F})$. The *rank* of A is the largest number of linearly independent rows (or columns) in A.

Theorem 1.97 *Let $A, B \in M_{mn}(\mathbb{F})$ be two distinct matrices. If A, B are equivalent then $r(A) = r(B)$.*

Proof We proved earlier that row operations do not affect the linear relationships among rows. Also, since the rank is the largest number of linearly independent rows, we may assert that two equivalent matrices have the same rank.

Therefore, in order to compute the rank of a matrix, we may change it into a reduced row form by appropriate elementary operations such that we can directly see how large its rank is: the count for the number of nonzero rows in its reduced row form is precisely the rank of the matrix.

Definition 1.98 A square matrix $A \in M_n(\mathbb{F})$ is said to be *invertible* if there exists a square matrix $B \in M_n(\mathbb{F})$, called the *inverse* of A, such that $AB = BA = I_n$.

Since matrix multiplication is not commutative, one can give the notions of both right and left inverse for a right invertible and left invertible matrix, respectively. We have the following important elementary properties of invertible matrices:

Proposition 1.99 *(i)* $A \in M_n(\mathbb{F})$ *is invertible if and only if* $\det A \neq 0$.
(ii) Also, $\det A^{-1} = (\det A)^{-1}$.

Elementary Matrices

Definition 1.100 A square matrix of order n is said to be an *elementary matrix* if it is obtained by performing one of the three types of elementary operations on the identity matrix of order n.

Hence we may obtain three different types of elementary matrices E having order n :

(1) Interchanging two rows R_i and R_j in the identity matrix I_n. In this case, $|E| = -1$.
(2) Multiplying a row R_i of I_n by a nonzero scalar α. In this case $|E| = \alpha$.
(3) Adding a constant multiple of a row R_j from I_n to another row R_i of I_n. In this case $|E| = 1$.

Notice that since any elementary matrix of order n is equivalent to the identity matrix I_n, its rank is precisely equal to n; therefore it is not singular and so invertible. In particular, easy computations show that:

(1) If E is the elementary matrix obtained from I_n by interchanging two different rows, then $E^{-1} = E$.
(2) If E is the elementary matrix obtained from I_n by multiplying a row R_i by a nonzero scalar α, then E^{-1} is the elementary matrix obtained from I_n by multiplying the same row R_i by the nonzero scalar α^{-1}.
(3) If E is the elementary matrix obtained from I_n by adding a constant multiple αR_j of a row R_j to another row R_i, then E^{-1} is the elementary matrix obtained from I_n by adding the constant multiple $-\alpha R_j$ of the row R_j to the row R_i.

As a simple exercise in matrix multiplication, one may verify that if $A \in M_{mn}(\mathbb{F})$, then the matrix resulting by applying one of the three types of elementary row operations to A can also be obtained by left multiplying the matrix A by the corresponding same type of elementary matrix of order m. In fact, if E is the elementary matrix obtained from I_m by interchanging rows R_i and R_j, then EA is the matrix obtained from A by interchanging the same rows i and j. If E is the elementary matrix obtained from I_m by multiplying a row R_i by a nonzero scalar α, then EA is the matrix obtained from A by multiplying the same row R_i by $\alpha \neq 0$. Finally, if E is the elementary matrix obtained from I_m adding the constant multiple αR_j to the row R_i, then EA is the matrix obtained from A by adding its jth row multiplied by α to its ith row.

At this point it seems clear that if $E_1, E_2, \ldots, E_k$ are elementary matrices, then $(E_1 E_2 \ldots E_k)A$ is a matrix obtained by performing, one by one, all the row operations represented by $E_1, E_2, \ldots, E_k$, according to the order of matrix multiplication:

$$A \rightarrow E_k A \longrightarrow E_{k-1} E_k A \longrightarrow \cdots \longrightarrow E_1 E_2 \cdots E_k A.$$

Remark 1.101 (i) If the elementary matrix E results from performing a row operation on I_m and if A is an $m \times n$ matrix, then the product EA is the matrix that results when the same row operation is performed on A.

(ii) Every nonsingular square matrix of order n is equivalent to the identity matrix I_n and there exists elementary matrices $E_1, \ldots, E_t$ such that

$$I_n = E_t E_{t-1} \cdots E_1 A.$$

Moreover

$$A = (E_t E_{t-1} \cdots E_1)^{-1} = E_1^{-1} E_2^{-1} \cdots E_t^{-1}.$$

Recalling that the inverse of any elementary matrix is again an elementary matrix, we also may assert that any nonsingular square matrix can be expressed by the product of a finite number of elementary matrices.

Exercises

1. For what values of $k \in \mathbb{R}$, is the rank of the following matrix equal to 1, 2 or 3?

$$\begin{bmatrix} 1 & 2 & 1 & 1 \\ k & 1 & k & 1 \\ k & 1 & k & k \end{bmatrix}.$$

2. Determine the rank of the following matrix

$$\begin{bmatrix} 3 & 1 & 2 & 1 \\ 1 & 2 & 1 & 2 \\ 1 & 1 & 1 & 1 \\ 4 & 4 & 3 & 4 \\ 1 & -2 & 0 & -2 \end{bmatrix}.$$

3. Determine a reduced row form of the following matrix

$$\begin{bmatrix} 1 & 1 & 1 & 0 & 1 \\ 2 & 1 & 3 & 1 & 2 \\ 2 & 2 & 2 & 1 & 1 \\ 3 & 2 & 4 & 2 & 2 \end{bmatrix}.$$

4. Determine a reduced row form of the following matrix

$$\begin{bmatrix} k-1 & 2 & 1 & 1 \\ 1 & 3 & 2 & k \\ 1 & 1 & 1 & 0 \end{bmatrix}$$

where $k \in \mathbb{R}$.

5. Let $A \in M_n(\mathbb{C})$ be a Hermitian matrix. Show that its determinant is a real number.
6. Let $A \in M_n(\mathbb{R})$ be such that $A^t = -A$. Show that, if the order n is odd, then $|A| = 0$.
7. Compute the determinant of the following matrix

$$\begin{bmatrix} 1 & 1 & 1 & \cdots & 1 \\ 1 & \alpha & \alpha^2 & \cdots & \alpha^{n-1} \\ 1 & \alpha^2 & \alpha^4 & \cdots & \alpha^{2n-2} \\ \cdots & \cdots & \cdots & \cdots & \cdots \\ 1 & \alpha^{n-1} & \alpha^{2(n-1)} & \cdots & \alpha^{(n-1)^2} \end{bmatrix},$$

where $\alpha = \cos \frac{2\pi}{n} + i \sin \frac{2\pi}{n}$.

8. Let $\mathbb{Z}_7 = \{\bar{0}, \bar{1}, \bar{2}, \bar{3}, \bar{4}, \bar{5}, \bar{6}\}$ be the field of residue classes (mod 7). Using elementary row operations, find the inverse of the matrix $A \in M_3(\mathbb{Z}_7)$, where

$$A = \begin{bmatrix} \bar{2} & \bar{3} & \bar{1} \\ \bar{0} & \bar{4} & \bar{5} \\ \bar{6} & \bar{3} & \bar{4} \end{bmatrix}.$$

1.5 System of Linear Equations

One of the most important problems in mathematics is that of solving a system of linear equations. Actually, solving a linear equation can be considered the first step for studying problems in mathematics. In fact, linear equations occur, at some stage, in the process of solution for the great part of mathematical problems encountered in any area of scientific research.

A *linear equation* in n unknowns is an equation that can be written in the form

$$a_1 x_1 + a_2 x_2 + \cdots + a_n x_n = b \tag{1.6}$$

where $a_1, \ldots, a_n, b$ are known scalars and $x_1, \ldots, x_n$ denote the unknowns. Usually the scalars $a_1, \ldots, a_n$ are called the *coefficients* and b is called the *constant* of the equation. A *solution* to Eq. (1.6) is any sequence of scalars $c_1, \ldots, c_n$ such that

$$a_1 c_1 + a_2 c_2 + \cdots + a_n c_n = b$$

that is such that the substitution $x_1 = c_1, x_2 = c_2, \ldots, x_n = c_n$ satisfies Eq. (1.6).

An $m \times n$ *system of linear equations* is a set of m equations of the form (1.6). So, it can be put in the form

$$a_{11}x_1 + a_{12}x_2 + \cdots + a_{1n}x_n = b_1$$
$$a_{21}x_1 + a_{22}x_2 + \cdots + a_{2n}x_n = b_2$$
$$\cdots\cdots\cdots$$
$$a_{m1}x_1 + a_{m2}x_2 + \cdots + a_{mn}x_n = b_m,$$

$$(1.7)$$

where $a_{i1}, \ldots, a_{in}$ are the coefficients and b_i is the constant term of the ith equation in the system. A *solution* to the system (1.7) is a sequence of scalars $c_1, \ldots, c_n$ that is simultaneously a solution to any equation in the system, that is

$$a_{11}c_1 + a_{12}c_2 + \cdots + a_{1n}c_n = b_1$$
$$a_{21}c_1 + a_{22}c_2 + \cdots + a_{2n}c_n = b_2$$
$$\cdots\cdots\cdots$$
$$a_{m1}c_1 + a_{m2}c_2 + \cdots + a_{mn}c_n = b_m.$$

A system of linear equations may have either infinitely many solutions, no solution or a unique solution. The system is called *consistent* if it has at least one solution, and it is called *inconsistent* if it has no solution. Among consistent systems, the one having infinitely many solutions are said to be *indefinite*, the one having a unique solution are said to be *definite*.

Definition 1.102 Two system of linear equations involving the same variables $x_1, \ldots, x_n$ are said to be *equivalent* if they have the same set of solutions.

To obtain the solutions of a linear system, we may then decide to study an equivalent system, having the same set of solutions as the original one. The goal is to simplify the given system in order to obtain an equivalent one, whose solutions are easier to get. In this regard, we will emphasize the role played by three types of operations that can be used on a system to obtain an equivalent system.

(1) If we interchange the order in which two equations of a system occur, this will have no effect on the solution set.
(2) If one equation of a system is multiplied by a nonzero scalar, this will have no effect on the solution set.
(3) By adding one equation to a multiple of another one, we create a new equation certainly satisfied by any solution of the original equations. Thus, the system consisting of both the original equations and the resulting new one is equivalent to the original system.

In short, it seems natural that the main questions we must ask ourselves are

(1) how to recognize if a system is consistent;
(2) if it is consistent, how to describe the set of its solutions, by simply relying on an equivalent system.

To answer to these questions, we will refer to the matrix theory previously developed. In doing so, the first step is to collect all coefficients from the equations of the system (1.7), in order to present them in tabular form

$$A = \begin{bmatrix} a_{11} & a_{12} & \ldots & a_{1n} \\ a_{21} & a_{22} & \ldots & a_{2n} \\ \ldots & \ldots & \ldots & \ldots \\ a_{m1} & a_{m2} & \ldots & a_{mn} \end{bmatrix}.$$

The matrix A is called the *coefficient matrix* for the system (1.7). Moreover, if we augment the coefficient matrix with the extra column

$$B = \begin{bmatrix} b_1 \\ b_2 \\ \ldots \\ \ldots \\ b_m \end{bmatrix}.$$

whose entries are the constants of equations from the system, we obtain a $m \times n + 1$ matrix

$$C = \begin{bmatrix} a_{11} & a_{12} & \ldots & a_{1n} & | & b_1 \\ a_{21} & a_{22} & \ldots & a_{2n} & | & b_2 \\ \ldots & \ldots & \ldots & \ldots & | & \ldots \\ a_{m1} & a_{m2} & \ldots & a_{mn} & | & b_m \end{bmatrix}$$

that is called the *augmented matrix* for the system (1.7). The augmented matrix C is usually denoted by $[A|B]$ or $A^\star$. Hence, if we display the coefficients and constants for the system in matrix form, and by introducing the array

$$X = \begin{bmatrix} x_1 \\ x_2 \\ \ldots \\ \ldots \\ x_n \end{bmatrix}$$

whose entries are the n unknowns used in the equations, we may express compactly the system (1.7) as follows:

$$AX = B. \tag{1.8}$$

From this point of view, it is clear that the elementary row operations for the augmented matrix representing a system of linear equations exactly duplicate the above described operations for equations in the system. This means that, if $C = [A|B] \in M_{mn+1}(\mathbb{F})$ is the augmented matrix for the original system (1.8) and $C' = [A'|B'] \in M_{mn+1}(\mathbb{F})$ is row equivalent to C, then C' is the augmented matrix for a system that is equivalent to (1.8). Therefore, to reduce (1.8) to an equivalent and easier system, we just perform elementary row operations on the augmented matrix of the system, in order to obtain its reduced row form. Then consider the system associated with this last reduced matrix to recognize if it is consistent and, in case of positive answer, to describe its set of solutions. At this point we need to fix a method that states unequivocally when a system is consistent or not.

Theorem 1.103 *A $m \times n$ system of linear equations $AX = B$ is consistent if and only if $rank(A) = rank([A|B])$.*

Therefore we conclude that, in case of consistent systems having rank r lesser than the number n of unknowns, the general solution of the system can be found as follows:

(i) Recognize the r leading unknowns: the ones whose coefficients are precisely the pivots of augmented matrix in its reduced row form.
(ii) Assign arbitrary values to $n - r$ free unknowns.
(iii) For any given set of values for the free unknowns, the values of the leading ones are determined uniquely from the equations in the system, by using the back substitution.

Hence the system admits infinitely many solutions depending on $n - r$ free parameters.

Homogeneous Systems Let us focus our attention on an important special case of systems of linear equations, more precisely when the constant terms are all zero, i.e., $b_i = 0$ for any $i = 1, \ldots, m$ in (1.7). Such a system is called *homogeneous* and, of course, admits always at least the trivial solution $x_i = 0$, for all $i = 1, \ldots, n$. In fact, there is no doubt that homogeneous systems are consistent. The augmented matrix is obtained from the coefficient one by adjoining a zero column, and this would thus not change the rank of the matrix. Hence, the only real question is whether a homogeneous system admits other solutions besides the trivial one. To answer this question it is sufficient to recall the arguments presented as a consequence of Theorem 1.103 so that we may assert the following:

Theorem 1.104 *A $m \times n$ homogeneous system of linear equations admits only the trivial solution if and only if the rank of its coefficient matrix equals the number of unknowns. Moreover, in case the rank r of coefficient matrix is lesser than the number n of unknowns, the system admits infinitely many solutions (besides the trivial one) depending on $n - r$ free parameters.*

Exercises

1. Consider the system

$$2x_1 + kx_2 = 2 - k$$
$$kx_1 + 2x_2 = k$$
$$kx_2 + kx_3 = k - 1$$

in three unknowns and one real parameter k. Determine, in case there exist, the values of the real parameter k for which the system is consistent. Then, in those cases, determine all the solutions of the system.

2. Determine, in case there exist, the solutions of system

$$(3 - k)x_1 - 2x_2 + (k - 2)x_3 = 4$$
$$2x_1 - 6x_2 - 3x_3 \quad\quad = 0$$
$$kx_1 + 4x_2 + 2x_3 \quad\quad = 7$$

in three unknowns and one real parameter k.

3. Determine, in case there exist, the nontrivial solutions of the following homogeneous system:

$$x_1 + 2x_2 - x_3 + x_4 = 0$$
$$2x_1 + 4x_2 - 2x_3 - x_4 = 0$$
$$x_1 - x_2 + 2x_4 = 0$$
$$2x_1 + x_2 - x_3 = 0.$$

4. Let $\mathbb{C}$ be the field of complex numbers. Are the following two systems of linear equations equivalent? If so, express each equation in each system as a linear combination of the equations in the other system.

$$2x_1 + (-1 + i)x_2 + x_4 = 0$$
$$3x_2 - 2ix_3 + 5x_4 = 0$$

and

$$(1 + \tfrac{i}{2})x_1 + 8x_2 - ix_3 - x_4 = 0$$
$$\tfrac{2}{3}x_1 - \tfrac{1}{2}x_2 + x_3 + 7x_4 = 0.$$

5. Determine, in case there exist, the solutions of system

$$x_1 + 2x_2 - x_3 + kx_4 = 1$$
$$2x_1 + x_2 + x_4 = 1$$
$$3x_1 + 3x_2 - x_3 + (3 - k)x_4 = -1$$

in four unknowns and one real parameter k.

6. Determine, in case there exist, the solutions of system

$$x_1 + x_2 + 2x_4 = 2$$
$$-x_1 - x_2 + x_3 + x_4 = k^2$$
$$3x_1 + 2x_3 + 3x_4 = k - 2$$

in three unknowns and one real parameter k.

7. Determine, in case there exist, the solutions of system

$$x_1 + 2x_2 - kx_3 = k$$
$$x_1 + x_2 - kx_3 = 1$$
$$3x_1 + (2 + k)x_2 - (2 + k)x_3 = 0$$

in three unknowns and one real parameter k.

8. Find all solutions to the system of equations

$$(1 - i)x_1 - ix_2 = 0$$
$$2x_1 + (1 - i)x_2 = 0.$$

Chapter 2
Vector Spaces

If we consider the set V of all vectors in a plane (or in a 3-dimensional Euclidean space) it can be easily seen that the sum of two vectors is a vector again and under the binary operation of vector addition $'+'$, V forms an additive abelian group. One can also multiply a vector with a scalar (real numbers) and it is also straightforward to see that the product of a scalar with a vector is again a vector in the plane (or in a 3-dimensional Euclidean space); that is, there exists an external binary operation (or scalar multiplication) $\mathbb{R} \times V \to V$ satisfying certain properties. Motivated by these two basic operations on V, the notion of an algebraic structure, viz., vector space was introduced. It is a very important notion not only in Algebra but also has a significant role in the study of various notions in analysis. Throughout this chapter $\mathbb{F}$ will denote a field.

2.1 Definitions and Examples

Definition 2.1 A nonempty set V equipped with a binary operation say $'+'$ and an external binary operation $\mathbb{F} \times V \to V$ such that $(\alpha, v) \mapsto \alpha.v$ is said to be a vector space over the field $\mathbb{F}$ if it satisfies the following axioms:

(1) $(V, +)$ is an abelian group
(2) $\alpha.(v+w) = \alpha.v + \alpha.w$
(3) $(\alpha + \beta).v = \alpha.v + \beta.v$
(4) $(\alpha\beta).v = \alpha.(\beta v)$
(5) $1.v = v$ for all $v \in V$, where 1 is the identity of $\mathbb{F}$
    for all $\alpha, \beta \in \mathbb{F}, v, w \in V$

M. Ashraf et al., *Advanced Linear Algebra with Applications*,
https://doi.org/10.1007/978-981-16-2167-3_2

Remark 2.2 (i) The elements of V are called vectors while the elements of $\mathbb{F}$ are said to be scalars. Throughout, the product between scalar α and vector v will be denoted as αv instead of $\alpha.v$. In axiom (3) the sum $\alpha + \beta$ is defined between two scalars, the elements of the field $\mathbb{F}$ while the sum $\alpha v + \beta v$ is defined between two vectors, the elements of V. There should be no confusion, the context will clear the intension. For the sake of convenience the symbol $+$ will stand for the addition of two vectors as well as scalars.

(ii) If $\mathbb{F} = \mathbb{R}$ then V is said to be the real vector space while if $\mathbb{F} = \mathbb{C}$, then V is said to be a complex vector space. For any $\alpha, \beta \in \mathbb{F}$ the difference $\alpha - \beta$ represents $\alpha + (-\beta)$, where $(-\beta)$ is the additive inverse of $\beta \in \mathbb{F}$, considered as an element of additive group $(\mathbb{F}, +)$, while for any $u, v \in V$ the difference $u - v$ represents the vector $u + (-v)$, where $-v$ is the additive inverse of $v \in V$ in the additive group $(V, +)$.

The following lemma is a direct consequence of the axioms of a vector space:

Lemma 2.3 *Let V be a vector space over a field $\mathbb{F}$. Then for any $\alpha, \beta \in \mathbb{F}$ and $v, w \in V$*

(i) $\alpha \mathbf{0} = \mathbf{0}$.
(ii) $0v = \mathbf{0}$.
(iii) $-(\alpha v) = \alpha(-v) = (-\alpha)v$.
(iv) $\alpha v = \mathbf{0}$ *if and only if* $\alpha = 0$ *or* $v = \mathbf{0}$.
(v) $(\alpha - \beta)v = \alpha v - \beta v$.
(vi) $\alpha(v - w) = \alpha v - \alpha w$.

There should be no confusion in 0 and $\mathbf{0}$. In the above result 0 is the additive identity of the field $\mathbb{F}$ while $\mathbf{0}$ denotes the additive identity of the additive group V. Henceforth, we shall also denote the additive identity $\mathbf{0}$ by 0. In use it can be easily understood whether 0 denotes the additive identity of a field or the additive identity of the group V.

Example 2.4 (1) Every field $\mathbb{F}$ is a vector space over its subfield. If $\mathbb{E}$ is a subfield of $\mathbb{F}$, then $\mathbb{F}$ is a vector space over $\mathbb{E}$, under the usual addition of $\mathbb{F}$ and the scalar multiplication αv (or the multiplication of $\mathbb{F}$), for any $\alpha \in \mathbb{E}, v \in \mathbb{F}$. In particular, every field $\mathbb{F}$ is a vector space over itself, the field of complex numbers $\mathbb{C}$ is a vector space over the field of reals $\mathbb{R}$ and finally the field $\mathbb{R}$ of reals is a vector space over the field $\mathbb{Q}$ of rational numbers.

By (1), every field of characteristic 0 can be regarded as a vector space over the field $\mathbb{Q}$ of rational numbers because each field of characteristic 0 contains a subfield isomorphic to the field $\mathbb{Q}$ of rational numbers. Similarly every field of characteristic p, where p is a positive prime integer, can be regarded as a vector space over the field $\mathbb{Z}_p$ of the field of residue classes modulo p because each field of characteristic p contains a subfield isomorphic to the field $\mathbb{Z}_p$ of residue classes modulo p.

(2) The set

$$\left\{ \begin{bmatrix} a_{11} & a_{12} & \cdots & a_{1n} \\ a_{21} & a_{22} & \cdots & a_{2n} \\ \cdots & \cdots & \cdots & \cdots \\ a_{m1} & a_{m2} & \cdots & a_{mn} \end{bmatrix} \mid a_{ij} \in \mathbb{F}, i = 1, 2, \ldots, m, j = 1, 2, \ldots, n \right\}$$

of all $M_{m \times n}$ matrices over a field $\mathbb{F}$ is a vector space over $\mathbb{F}$ under the matrix addition and scalar multiplication.

(3) If we consider the set $\mathbb{F}^n = \{(a_1, a_2, \ldots, a_n) \mid a_i \in \mathbb{F}, i = 1, 2, \ldots, n\}$. Then any two elements $x = (x_1, x_2, \ldots, x_n)$, $y = (y_1, y_2, \ldots, y_n) \in \mathbb{F}^n$ are said to be equal if and only if $x_i = y_i$ for each $i = 1, 2, \ldots, n$. Now for any $\alpha \in \mathbb{R}$, define addition and scalar multiplication in $\mathbb{F}^n$ as follows:

$$x + y = (x_1 + y_1, x_2 + y_2, \ldots, x_n + y_n),$$

$$\alpha x = (\alpha x_1, \alpha x_2, \ldots, \alpha x_n).$$

It can be easily seen that $\mathbb{F}^n$ is a vector space over $\mathbb{F}$. For $\mathbb{F} = \mathbb{R}$ and $n = 2$, there is a correspondence between any vector in the plane to a unique ordered pair of real numbers and conversely. Under this correspondence, coordinate-wise addition and coordinate-wise scalar multiplication correspond, respectively, to vector addition and scalar multiplication, and hence $\mathbb{R}^2$ represents the space of all vectors in a plane. Note also that the real vector space $\mathbb{R}^2$ has special geometric significance and is called an Euclidean plane.

(4) Let A be a nonempty set, $\mathbb{F}$ be a field and let $V = \{f \mid f : A \to \mathbb{F}\}$. For any $f, g \in V$ and $\alpha \in \mathbb{F}$ define $f + g, \alpha f : A \to \mathbb{F}$ such that for any $x \in A$

$$(f + g)(x) = f(x) + g(x),$$

$$(\alpha f)(x) = \alpha f(x).$$

It can be easily verified that the set V of all functions from A into $\mathbb{F}$ is a vector space over $\mathbb{F}$.

(5) Let $\mathbb{F}[x]$ be the polynomial ring in indeterminate x over the field $\mathbb{F}$. In the abelian group $(\mathbb{F}[x], +)$ define scalar multiplication as follows:

$$\alpha(\alpha_0 + \alpha_1 x + \cdots + \alpha_n x^n) = \alpha \alpha_0 + \alpha \alpha_1 x + \cdots + \alpha \alpha_n x^n.$$

It can be easily seen that $\mathbb{F}[x]$ is a vector space over $\mathbb{F}$.

(6) In the above example, consider the set $\mathbb{F}_n[x]$ of all polynomial of degree less than or equal to n in $\mathbb{F}[x]$. Using the natural addition and scalar multiplication of polynomials, as defined above, it can be seen that V is a vector space over the field $\mathbb{F}$.

(7) Consider the set of all real-valued continuous functions $\mathscr{C}[0, 1]$ defined on the interval $[0, 1]$, i.e., $f : [0, 1] \to \mathbb{R}$. Define addition and scalar multiplication in $\mathscr{C}[0, 1]$ as follows:

$$(f + g)(x) = f(x) + g(x),$$

$$(\alpha f)(x) = \alpha f(x)$$

for any $\alpha \in \mathbb{R}$, $f, g \in \mathscr{C}[0, 1]$, $x \in [0, 1]$. Then $\mathscr{C}[0, 1]$ is a vector space over $\mathbb{R}$.

(8) Let $\mathbb{F}[[x]]$ be the set of all formal power series in indeterminate x over a field $\mathbb{F}$, that is, a collection of the expression of the form $f(x) = \sum_{i=0}^{\infty} a_i x^i$, where $a_i \in \mathbb{F}$. Any two power series $f(x) = \sum_{i=0}^{\infty} a_i x^i$ and $g(x) = \sum_{i=0}^{\infty} b_i x^i$ are equal if and only if $a_i = b_i$ for all i. For any $\alpha \in \mathbb{F}$ define

$$f(x) + g(x) = \sum_{i=0}^{\infty} (a_i + b_i) x^i,$$

$$\alpha f(x) = \sum_{i=0}^{\infty} (\alpha a_i) x^i.$$

It can be seen that $\mathbb{F}[[x]]$ is a vector space over the field $\mathbb{F}$.

Definition 2.5 A nonempty subset W of a vector space V over a field $\mathbb{F}$ is said to be a subspace of V if W is itself a vector space over the same field $\mathbb{F}$ with regard to induced binary operations.

Lemma 2.6 *A nonempty subset W of a vector space V over a field $\mathbb{F}$ is a subspace of V if and only if for any $w, w_1, w_2 \in W$ and $\alpha \in \mathbb{F}$*

(i) $w_1 + w_2 \in W$,
(ii) $\alpha w \in W$.

Proof If W is a subspace of V, then W is itself a vector space over the same field $\mathbb{F}$ and hence for any $\alpha \in \mathbb{F}$ and $w \in W$, $\alpha w \in W$, as W is closed with regard to scalar multiplication. Being an additive group W is nonempty and closed under the operation of addition, that is, for any $w_1, w_2 \in W$, $w_1 + w_2 \in W$.

Conversely, if the conditions (i) and (ii) hold, then by (i) W is closed under the addition and by (ii) W is closed under the scalar multiplication. Since W is nonempty, there exists $w \in W$. Now by condition (ii), $0w = 0 \in W$ and also for any $w \in W$, $-w = (-1)w \in W$. The operation of addition in V being associative and commutative is also associative and commutative in W, and thus $(W, +)$ is an abelian group. The axioms (2)–(5) in the Definition 2.1 hold in W, as they hold in V. Hence W is itself a vector space over the field $\mathbb{F}$ with regard to induced binary operations and therefore W is a subspace of V.

Remark 2.7 Let V be any vector space. Then $\{0\}$ and V are always subspaces of V. These two subspaces are called trivial or improper subspaces of V. Any subspace W of V other than $\{0\}$ and V is called nontrivial or proper subspace of V.

Example 2.8 (1) In the vector space $\mathbb{R}^3$, consider the subset $W = \{(x, y, z) \in \mathbb{R}^3 \mid \alpha x + \beta y + \gamma z = 0\}$, where $\alpha, \beta, \gamma \in \mathbb{R}$. Since $(0, 0, 0) \in W$, $W \neq \emptyset$. Let $(x_1, y_1, z_1), (x_2, y_2, z_2) \in W$. Then

$$\alpha(x_1 + x_2) + \beta(y_1 + y_2) + \gamma(z_1 + z_2) = (\alpha x_1 + \beta y_1 + \gamma z_1) + (\alpha x_2 + \beta y_2 + \gamma z_2) = 0$$

and hence $(x_1, y_1, z_1) + (x_2, y_2, z_2) \in W$. Also since for any $\delta \in \mathbb{R}$ and $(x_1, y_1, z_1) \in W$, $\alpha(\delta x_1) + \beta(\delta y_1) + \gamma(\delta z_1) = \delta(\alpha x_1 + \beta y_1 + \gamma z_1) = \delta 0 = 0$ implies that $\delta(x_1, y_1, z_1) \in W$. Hence W is a subspace of $\mathbb{R}^3$.

(2) Consider the subsets $W_1 = \{(x, 0) \mid x \in \mathbb{R}\}$ and $W_2 = \{(0, y) \mid y \in \mathbb{R}\}$ of $\mathbb{R}^2$. It can be easily seen that W_1 and W_2 are subspaces of the vector space $\mathbb{R}^2$ over the field $\mathbb{R}$. Note that W_1 and W_2 are the X and Y axes, respectively. Similarly, if we consider the XY-plane $W_3 = \{(x, y, 0) \mid x, y \in \mathbb{R}\}$ in $\mathbb{R}^3$, it can be easily verified that W_3 is a subspace of the vector space $\mathbb{R}^3$ over $\mathbb{R}$.

(3) In Example 2.4(6), the subset $\mathbb{F}_n[x]$ of $\mathbb{F}[x]$ is a subspace of $\mathbb{F}[x]$ over the field $\mathbb{F}$.

(4) Let $V = \mathscr{C}[0, 1]$ be the vector space of all real-valued continuous functions on $[0, 1]$. Then W, the subset of V, consisting of all differentiable functions is a subspace of V.

(5) Let $M_n(\mathbb{R})$ denote the vector space of all $n \times n$ matrices with real entries over the field of real numbers. The subsets W_1 and W_2 of V, consisting of all symmetric and skew symmetric matrices respectively, are subspaces of the vector space $M_n(\mathbb{R})$.

Lemma 2.9 *A nonempty subset W of a vector space V over a field $\mathbb{F}$ is a subspace of V if and only if for any $w_1, w_2 \in W$ and $\alpha, \beta \in \mathbb{F}$, $\alpha w_1 + \beta w_2 \in W$.*

Proof If W is a subspace of V, then for any $w_1, w_2 \in W$ and $\alpha, \beta \in \mathbb{F}$, $\alpha w_1, \beta w_2 \in W$ by Lemma 2.6, $\alpha w_1 + \beta w_2 \in W$. Conversely, let W be a nonempty subset of V such that for any $w_1, w_2 \in W$ and $\alpha, \beta \in \mathbb{F}$, $\alpha w_1 + \beta w_2 \in W$. Since $1 \in \mathbb{F}$, $w_1 + w_2 = 1 w_1 + 1 w_2 \in W$. Also since $0 \in \mathbb{F}$, $\alpha w_1 = \alpha w_1 + 0 w_2 \in W$, for any $\alpha \in \mathbb{F}$ and $w_1 \in W$. Hence by Lemma 2.6, W is a subspace of V.

Definition 2.10 Let $W_1, W_2, \ldots, W_n$ be subspaces of a vector space V over $\mathbb{F}$. Then the sum of $W_1, W_2, \ldots, W_n$ is defined as $\sum_{i=1}^{n} W_i = W_1 + W_2 + \cdots + W_n = \{w_1 + w_2 + \cdots + w_n \mid w_i \in W_i, i = 1, 2, \ldots, n\}$.

Remark 2.11 It can be easily seen that $\sum_{i=1}^{n} W_i$ is a subspace of V. In fact, $0 \in \sum_{i=1}^{n} W_i$, $\sum_{i=1}^{n} W_i \neq \emptyset$ and if $x, y \in \sum_{i=1}^{n} W_i$, then $x = w_1 + w_2 + \cdots + w_n$, $y = w'_1 + w'_2 + \cdots + w'_n$, where $w_i, w'_i \in W_i$ for each $i = 1, 2, \ldots, n$. Since each $W_i's$ is a subspace we find that $x + y = (w_1 + w'_1) + (w_2 + w'_2) + \cdots + (w_n + w'_n) \in \sum_{i=1}^{n} W_i$ and $\alpha \in \mathbb{F}$, $\alpha x = \alpha w_1 + \alpha w_2 + \cdots + \alpha w_n \in \sum_{i=1}^{n} W_i$. Hence by Lemma 2.6, we find that $\sum_{i=1}^{n} W_i$ is a subspace of V.

Lemma 2.12 *If V is a vector space over $\mathbb{F}$ and $\{W_i\}_{i \in I}$, where I is an index set, is a collection of subspaces of V, then $W = \bigcap_{i \in I} W_i$ is a subspace of V.*

Proof Obviously $0 \in W$, $W \neq \emptyset$. Let $w, w' \in W$. Then $w, w' \in W_i$ for each $i \in I$. Since each W_i is a subspace, for any $\alpha \in \mathbb{F}$ both $w + w' \in W$ and $\alpha w \in W$ and hence W is a subspace of V.

The above result shows that the intersection of two subspaces over a field $\mathbb{F}$ is a subspace over $\mathbb{F}$, but it is to be noted that the union of two subspaces need not be a subspace.

Example 2.13 (1) $\mathbb{R}^2$ is a vector space over $\mathbb{R}$. Consider two subspaces of $\mathbb{R}^2$, namely, $W_1 = \{(x, 0) \mid x \in \mathbb{R}\}$ and $W_2 = \{(0, y) \mid y \in \mathbb{R}\}$. If we take $(1, 0) \in W_1$ and $(0, 1) \in W_2$, then $(1, 0), (0, 1) \in W_1 \cup W_2$ but their sum $(1, 0) + (0, 1) = (1, 1) \notin W_1 \cup W_2$ and hence $W_1 \cup W_2$ is not a subspace of $\mathbb{R}^2$.

(2) In the vector space $\mathbb{R}^3$, $W_1 = \{(x, y, 0) \mid x, y \in \mathbb{R}\}$, i.e., XY-plane and $W_2 = \{(0, y, z) \mid y, z \in \mathbb{R}\}$, i.e., YZ-plane are subspaces of $\mathbb{R}^3$. Then it can be easily seen that $\mathbb{R}^3 = W_1 + W_2$. In fact, for any $(x, y, z) \in \mathbb{R}^3$, $(x, y, z) = (x, y, 0) + (0, 0, z) \in W_1 + W_2$, as $(x, y, 0) \in W_1$, $(0, 0, z) \in W_2$. This shows that $\mathbb{R}^3 \subseteq W_1 + W_2$. Also $W_1 + W_2$, being a subspace of $\mathbb{R}^3$, is a subset of $\mathbb{R}^3$ and hence $\mathbb{R}^3 = W_1 + W_2$.

Definition 2.14 Let W_1, W_2 be two subspaces of a vector space V over $\mathbb{F}$. Then V is said to be a direct sum of W_1 and W_2 if $V = W_1 + W_2$ and $W_1 \cap W_2 = \{0\}$, denoted as $V = W_1 \oplus W_2$. Each of $W_i's$ are called direct summand of V. For each subspace W_2 of V, W_2 is also known as complement of W_1 and W_1 is a complement of W_2.

Lemma 2.15 *Let V be a vector space over $\mathbb{F}$, and W_1, W_2 be two subspaces of V. Then $V = W_1 \oplus W_2$ if and only if every $v \in V$ can be uniquely written as $w_1 + w_2$, $w_1 \in W_1$, $w_2 \in W_2$.*

Proof Assume that $V = W_1 \oplus W_2$ and $v \in V$. Then $v = w_1 + w_2$, $w_1 \in W_1$, $w_2 \in W_2$. If v has another representation say $v = w_1' + w_2'$, $w_1' \in W_1$, $w_2' \in W_2$, then $w_1 - w_1' = w_2' - w_2 \in W_1 \cap W_2 = \{0\}$. This implies that $w_1 = w_1'$ and $w_2 = w_2'$ and hence every element $v \in V$ can be uniquely written as $v = w_1 + w_2$, where $w_1 \in W_1$, $w_2 \in W_2$.

Conversely, assume that $v \in V$ can be uniquely written as $v = w_1 + w_2$, where $w_1 \in W_1$, $w_2 \in W_2$. Then obviously $V = W_1 + W_2$. Now let $w \in W_1 \cap W_2$. Then $w \in W_1$ and $w \in W_2$. Hence $w = w + 0$, $w \in W_1$, $0 \in W_2$ and also $w = 0 + w$, $0 \in W_1$, $w \in W_2$. By the uniqueness of the expression for w, we find that $w = 0$, and hence $W_1 \cap W_2 = \{0\}$ i.e., $V = W_1 \oplus W_2$.

In view of the above result one can generalize the definition of direct sum to a finite number of subspaces as follows:

Definition 2.16 Let V be a vector space and $W_1, W_2, \ldots, W_n$ be subspaces of V. Then V is said to be a direct sum of $W_1, W_2, \ldots, W_n$ if every $v \in V$ can be uniquely expressed as $v = w_1 + w_2 + \cdots + w_n$, where $w_i \in W_i$ for each $i = 1, 2, \ldots, n$. Note that the above direct sum is in fact internal direct sum of $W_1, W_2, \ldots, W_n$.

Consider any finite number of vector spaces $V_1, V_2, \ldots, V_n$ over the same field $\mathbb{F}$ and let $V = \{(v_1, v_2, \ldots, v_n) \mid v_i \in V_i, i = 1, 2, \ldots, n\}$. Any two elements $(v_1, v_2, \ldots, v_n)$, $(v'_1, v'_2, \ldots, v'_n) \in V$ are said to be equal if and only if $v_i = v'_i$ for each $i = 1, 2, \ldots, n$. For any $(v_1, v_2, \ldots, v_n), (v'_1, v'_2, \ldots, v'_n) \in V$ and $\alpha \in \mathbb{F}$ define addition and scalar multiplication in V as follows:

$$(v_1, v_2, \ldots, v_n) + (v'_1, v'_2, \ldots, v'_n) = (v_1 + v'_1, v_2 + v'_2, \ldots, v_n + v'_n)$$

$$\alpha(v_1, v_2, \ldots, v_n) = (\alpha v_1, \alpha v_2, \ldots, \alpha v_n).$$

It can be easily seen that V is a vector space over the field $\mathbb{F}$ with regard to the above operations. We call V as the external direct sum of $V_1, V_2, \ldots, V_n$, denoted as $V = V_1 \oplus V_2 \oplus \cdots \oplus V_n$.

Lemma 2.17 *Let $W_1, W_2, \ldots, W_n$ be subspaces of a vector space over $\mathbb{F}$. Then V is an internal direct sum of $W_1, W_2, \ldots, W_n$, i.e., $V = W_1 \oplus W_2 \oplus \cdots \oplus W_n$ if and only if $V = W_1 + W_2 + \cdots + W_n$ and $W_i \cap (W_1 + W_2 + \cdots + W_{i-1} + W_{i+1} + \cdots + W_n) = \{0\}$ for every $i = 1, 2, 3, \ldots, n$.*

Proof Suppose that $V = W_1 \oplus W_2 \oplus \cdots \oplus W_n$. Then it is obvious to observe that $V = W_1 + W_2 + \cdots + W_n$. Let $w \in W_i \cap (W_1 + W_2 + \cdots + W_{i-1} + W_{i+1} + \cdots + W_n)$. This implies that $w \in W_i$ and $w = w_1 + w_2 + \cdots + w_{i-1} + w_{i+1} + \cdots + w_n$ for some $w_1 \in W_1, w_2 \in W_2, \ldots, w_{i-1} \in W_{i-1}, w_{i+1} \in W_{i+1}, \ldots, w_n \in W_n$. Here $w \in V$ has two representations as: $w = 0 + 0 + \cdots + 0 + w + 0 + \cdots + 0$ and $w = w_1 + w_2 + \cdots + w_{i-1} + 0 + w_{i+1} + \cdots + w_n$. But as each element of V has unique representation, we are forced to conclude that $w = 0$. Hence it follows that $W_i \cap (W_1 + W_2 + \cdots + W_{i-1} + W_{i+1} + \cdots + W_n) = \{0\}$ for every $i = 1, 2, 3, \ldots, n$.

Conversely, suppose that $V = W_1 + W_2 + \cdots + W_n$ and $W_i \cap (W_1 + W_2 + \cdots + W_{i-1} + W_{i+1} \cdots + W_n) = \{0\}$ for every $i = 1, 2, 3, \ldots, n$. It is clear that if $v \in V$, then there exists $w_1 \in W_1, w_2 \in W_2, \ldots, w_n \in W_n$ such that $v = w_1 + w_2 + \cdots + w_{i-1} + w_i + w_{i+1} + \cdots + w_n$. To prove the uniqueness of this representation, let us suppose that $v = w'_1 + w'_2 + \cdots + w'_{i-1} + w'_i + w'_{i+1} + \cdots + w'_n$ also, where $w'_1 \in W_1, w'_2 \in W_2, \ldots, w'_n \in W_n$. This implies that $w'_i - w_i = (w_1 - w'_1) + (w_2 - w'_2) + \cdots + (w_{i-1} - w'_{i-1}) + (w_{i+1} - w'_{i+1}) + \cdots + (w_n - w'_n) \in W_i \cap (W_1 + W_2 + \cdots + W_{i-1} + W_{i+1} \cdots + W_n)$. Using our hypothesis we are forced to conclude that $w_i = w'_i$, for each $i = 1, 2, \ldots, n$. This shows that representation is unique and therefore V is an internal direct sum of $W_1, W_2, \ldots, W_n$.

Theorem 2.18 *Let $W_1, W_2, \ldots, W_n$ be any $n \geq 1$ subspaces of vector space V over a field $\mathbb{F}$. Then $V = W_1 \oplus W_2 \oplus \cdots \oplus W_n$ if and only if $V = W_1 + W_2 + \cdots + W_n$*

and for any $v_i \in W_i$; $1 \leq i \leq n$, $\quad v_1 + v_2 + \cdots + v_n = 0$ *implies that* $v_1 = v_2 = \cdots = v_n = 0$.

Proof For $n = 1$, $V = W_1$ and the result is obvious, and hence assume that $n \geq 2$. Suppose that $V = W_1 \oplus W_2 \oplus \cdots \oplus W_n$. Hence by Lemma 2.17, $V = W_1 + W_2 + \cdots + W_n$ and $W_i \cap (W_1 + W_2 + \cdots + W_{i-1} + W_{i+1} + \cdots + W_n) = \{0\}$. Now for some $v_j \in W_j$, let $v_1 + v_2 + \cdots + v_n = 0$. Then $v_i = -\sum_{j=1, j \neq i}^{n} v_j \in W_i \cap (W_1 + W_2 + \cdots + W_{i-1} + W_{i+1} + \cdots + W_n)$. This yields that $v_i = 0$ for every i.
Conversely, assume that $V = W_1 + W_2 + \cdots + W_n$ and for any $v_i \in W_i$; $1 \leq i \leq n$, $v_1 + v_2 = \cdots + v_n = 0$ implies that $v_1 = v_2 = \cdots = v_n = 0$. For some i let $v \in W_i \cap (W_1 + W_2 + \cdots + W_{i-1} + W_{i+1} + \cdots + W_n)$. Then $v = w_i \in W_i$ and $v \in W_1 + W_2 + \cdots + W_{i-1} + W_{i+1} + \cdots + W_n$ implies that $v = v_1 + v_2 + \cdots + v_{i-1} + v_{i+1} + \cdots + v_n$ and consequently $w_i = v_1 + v_2 + \cdots + v_{i-1} + v_{i+1} + \cdots + v_n$. This implies that $\sum_{j=1}^{n} v_j = 0$, where $v_i = -w_i$. By our hypothesis $v_j = 0$ and in particular $v_i = 0$ and hence $v = -v_i = 0$. Therefore $W_i \cap (W_1 + W_2 + \cdots + W_{i-1} + W_{i+1} + \cdots + W_n) = \{0\}$ and hence $V = W_1 \oplus W_2 \oplus \cdots \oplus W_n$.

Definition 2.19 Let V be a vector space over a field $\mathbb{F}$ and $v_1, v_2, \ldots, v_n \in V$. Then any element of the form $\alpha_1 v_1 + \alpha_2 v_2 + \cdots + \alpha_n v_n$, where $\alpha_i \in \mathbb{F}$, $i = 1, 2, \ldots, n$, is called a linear combination of $v_1, v_2, \ldots, v_n$ over $\mathbb{F}$.

Definition 2.20 Let S be a non-empty subset of a vector space V over a field $\mathbb{F}$, then the linear span of S, denoted as $L(S)$ is the set of all linear combinations of finite sets of elements of S, i.e.,

$$L(S) = \left\{ \sum_{i=1}^{k} \alpha_i v_i \mid \alpha_i \in \mathbb{F}, v_i \in S, k \in \mathbb{N} \right\}.$$

Note that in the above definition α_i, v_i, k are chosen from their respective domains. Since $\alpha 0 = 0$ for all $\alpha \in \mathbb{F}$, for $S = \{0\}$, $L(S) = \{0\}$. Moreover, if $S = \{v\}$ for some $v \in V$, then $L(S) = \{\alpha v \mid \alpha \in \mathbb{F}\}$.

Theorem 2.21 *Let* S, T *be nonempty subsets of a vector space* V *over a field* $\mathbb{F}$, *then*

(i) $L(S)$ *is the smallest subspace of* V *containing* S,
(ii) *If* $S \subseteq T$, *then* $L(S) \subseteq L(T)$,
(iii) $L(S \cup T) = L(S) + L(T)$,
(iv) $L(L(S)) = L(S)$,
(v) $L(S) = S$ *if and only if* S *is a subspace of* V.

Proof (i) First we shall show that $L(S)$ is a subspace of V. Since $0 = 10$, $0 \in L(S)$ and hence $L(S) \neq \emptyset$. Let $v, v' \in L(S)$. Then $v = \sum_{i=1}^{k} \alpha_i v_i, \alpha_i \in \mathbb{F}$, $v_i \in S$ and $v' = \sum_{i=1}^{\ell} \beta_i v_i', \beta_i \in \mathbb{F}$, $v_i' \in S$, and hence $v + v' = \sum_{i=1}^{k} \alpha_i v_i + \sum_{i=1}^{\ell} \beta_i v_i' \in L(S)$ and for any $\gamma \in \mathbb{F}$, $\gamma v = \sum_{i=1}^{k} (\gamma \alpha) v_i \in L(S)$. This shows that $L(S)$ is a subspace of V. For any $v \in S$, $v = 1v$, $v \in L(S)$, and hence $L(S)$ is a subspace of V containing

S. To show that $L(S)$ is the smallest subspace of V containing S, assume that W is a subspace of V which contains S. Then for any $v \in L(S)$, $v = \sum_{i=1}^{k} \alpha_i v_i$, $v_i \in S$. Since S is contained in W, we find that each of $v_i \in W$ and W being a subspace of V contains v and hence $L(S) \subseteq W$ for any subspace containing S. This proves our assertion.

(ii) Assume that $S \subseteq T$. Then by the above case we find that $S \subseteq T \subseteq L(T)$ and hence $S \subseteq L(T)$. But by (i), $L(S)$ is the smallest subspace of V containing S, $L(S) \subseteq L(T)$.

(iii) Consider any two subsets S, T of V and assume that $W = L(S) + L(T)$. But since $S \subseteq L(S) \subseteq W$ and $T \subseteq L(T) \subseteq W$, we find that $S, T \subseteq W$ and hence $S \cup T \subseteq W$. Therefore using (i) $L(S \cup T) \subseteq W$. Now assume that $w \in W$ and hence $w = w_1 + w_2$ such that $w_1 \in L(S)$, $w_2 \in L(T)$. Since $S, T \subseteq S \cup T$, $L(S), L(T) \subseteq L(S \cup T)$. This implies that $w = w_1 + w_2 \in L(S \cup T)$ and hence $W \subseteq L(S \cup T)$. Combining the results so obtained we find that $L(S \cup T) = L(S) + L(T)$.

(iv) Obviously $L(S) \subseteq L(L(S))$. Now let $v \in L(L(S))$. Then $v = \sum_{i=1}^{k} \alpha_i v_i$, $\alpha_i \in \mathbb{F}$, $v_i \in L(S)$. Now each of v_i in $L(S)$ can be written as $v_i = \sum_{j=1}^{m} \beta_{i_j} v_{i_j}$, $\beta_{i_j} \in \mathbb{F}$, $v_{i_j} \in S$. This yields that $v = \sum_{i=1}^{k} \sum_{j=1}^{m} \alpha_i \beta_{i_j} v_{i_j}$, where $\alpha_i, \beta_{i_j} \in \mathbb{F}$, $v_{i_j} \in S$. This shows that $L(L(S)) \subseteq L(S)$. Thus result stands proved.

(v) Assume that $L(S) = S$. By (i), we know that $L(S)$ is a subspace of V. As a result, S is a subspace of V. Conversely, suppose that S is a subspace of V. As by (i), $L(S)$ is the smallest subspace of V containing S, i.e., $S \subseteq L(S)$. But S is also a subspace of V containing S as a subset, thus we have $L(S) \subseteq S$. Finally, we have proved that $L(S) = S$.

In view of Theorem 2.21(iii) and (v), the following result follows directly:

Corollary 2.22 *If W_1, W_2 are any two subspaces of a vector space V over $\mathbb{F}$, then $W_1 + W_2$ is a subspace of V spanned by $W_1 \cup W_2$.*

Definition 2.23 Let X be a subset of a vector space V over a field $\mathbb{F}$. A subspace W of V is said to be generated by X, denoted as $\langle X \rangle$, if $X \subseteq W$ and for any subspace W' of V, $X \subseteq W'$ implies that $W \subseteq W'$.

Theorem 2.24 *Let S be a nonempty subset of a vector space V over a field $\mathbb{F}$, then*

(i) $\langle S \rangle = L(S)$,
(ii) $\langle S \rangle = \bigcap_{i \in I} W_i$, *where I is an index set and W_i is a subspace of V containing S as a subset.*

Proof (i) By definition, it is clear that $\langle S \rangle$ is a subspace of V, containing S as a subset. But $L(S)$ is the smallest subspace of V, containing S as a subset. Thus we conclude that $L(S) \subseteq \langle S \rangle$. Again by definition of $\langle S \rangle$, it is obvious that $\langle S \rangle \subseteq L(S)$. Finally, it follows that $\langle S \rangle = L(S)$.

(ii) For each $i \in I$, W_i is a subspace of V containing S as a subset. This shows that $\bigcap_{i \in I} W_i$ is also a subspace of V containing S as a subset. By the definition of $\langle S \rangle$, we obtain that $\langle S \rangle \subseteq \bigcap_{i \in I} W_i$. Let $w \in \bigcap_{i \in I} W_i$. This shows that $w \in W_i$, for each $i \in I$. As $\langle S \rangle$ is also a subspace of V containing S as a subset, in particular we find that $\langle S \rangle = W_{i_0}$ for some $i_0 \in I$. Now we conclude that $w \in W_{i_0} = \langle S \rangle$. This gives us $\bigcap_{i \in I} W_i \subseteq \langle S \rangle$. Hence, we get $\langle S \rangle = \bigcap_{i \in I} W_i$.

Remark 2.25 If S is a subset of a vector space V, where $S = \emptyset$, then we can also talk about $\langle S \rangle$ and $L(S)$. Actually here $\langle S \rangle = \bigcap_{i \in I} W_i$, where W_i is a subspace of V containing $S = \emptyset$ as a subset. But each subspace of V contains $\emptyset$ as its subset. This shows that W_i's, $i \in I$ are precisely subspaces of V. In particular, $W_i = \{0\}$ for some $i \in I$. Thus we conclude that in this case $\langle S \rangle = \bigcap_{i \in I} W_i = \{0\}$. On the other hand, $L(S) = \{0\}$ is vacuously true. Finally, we can say that the above theorem holds if S is any arbitrary subset of V.

Theorem 2.26 *Let $\{v_1, v_2, \ldots, v_n\}$ be a set of vectors which spans a subspace W of a vector space V. If for some j; $1 \leq j \leq n$, v_j is a linear combination of the remaining $n - 1$ vectors, then $\{v_1, v_2, \ldots, v_{j-1}, v_{j+1}, \ldots, v_n\}$ also spans W.*

Proof Since v_j is a linear combination of $v_1, v_2, \ldots, v_{j-1}, v_{j+1}, \ldots, v_n$, there exist scalars α_i such that $v_j = \sum_{i=1}^{j-1} \alpha_i v_i + \sum_{i=j+1}^{n} \alpha_i v_i$. Now let v be an arbitrary vector in W, then

$$
\begin{aligned}
v &= \sum_{\ell=1}^{n} \beta_\ell v_\ell \\
&= \sum_{\ell=1, \ell \neq j}^{n} \beta_\ell v_\ell + \beta_j v_j \\
&= \sum_{\ell=1, \ell \neq j}^{n} \beta_\ell v_\ell + \beta_j \left\{ \sum_{i=1}^{j-1} \alpha_i v_i + \sum_{i=j+1}^{n} \alpha_i v_i \right\}.
\end{aligned}
$$

This shows that $\{v_1, v_2, \ldots, v_{j-1}, v_{j+1}, \ldots, v_n\}$ spans W.

Definition 2.27 Let V be a vector space over a field $\mathbb{F}$ and W a subspace of V. Let

$$
V/W = \{v + W \mid v \in V\}.
$$

Define addition and multiplication in V/W as follows:

(i) $(v_1 + W) + (v_2 + W) = v_1 + v_2 + W$, for any $v_1 + W$, $v_2 + W \in V/W$.
(ii) $\alpha(v_1 + W) = \alpha v_1 + W$ for any $\alpha \in \mathbb{F}$, $v_1 + W \in V/W$.

It can be easily seen that addition and scalar multiplication both are well-defined and

(1) $(V/W, +)$ is an abelian group.
(2) For any $\alpha, \beta \in \mathbb{F}$ and $v_1 + W$, $v_2 + W \in V/W$
    (i) $\alpha\{(v_1 + W) + (v_2 + W)\} = \alpha(v_1 + W) + \alpha(v_2 + W)$
    (ii) $(\alpha + \beta)(v_1 + W) = \alpha(v_1 + W) + \beta(v_1 + W)$
    (iii) $\alpha\beta(v_1 + W) = (\alpha\beta)v_1 + W = \alpha(\beta v_1) + W = \alpha(\beta v_1 + W)$
    (iv) $1(v_1 + W) = 1v_1 + W = v_1 + W$.

V/W is a vector space over $\mathbb{F}$ and is called quotient space of V relative to W.

Remark 2.28 Since $(V/W, +)$ is the quotient group of $(V, +)$ with regard to its normal subgroup $(W, +)$, it is obvious to observe the following for all $v, v_1, v_2 \in V$:

(i) $v + W \neq \emptyset$.

(ii) $v + W = W$ if and only if $v \in W$.

(iii) $v_1 + W = v_2 + W$ if and only if $(v_1 - v_2) \in W$.

(iv) Any two elements of V/W are either equal or mutually disjoint.

(v) Union of all the elements of V/W equals V. Thus the quotient space V/W gives a partition of the vector space V.

Example 2.29 Let V be the vector space of all polynomials of degree less than or equal to n over the field $\mathbb{R}$ of real numbers, where $n \geq 2$ is a fixed integer. Assume that W is a subspace of V, consisting of all polynomials of degree less than or equal to $(n - 2)$. Hence

$$
\begin{aligned}
V/W &= \{(\alpha_0 + \alpha_1 x + \alpha_2 x^2 + \cdots + \alpha_{n-2} x^{n-2} + \alpha_{n-1} x^{n-1} + \alpha_n x^n) + W | \\
&\quad \alpha_0, \alpha_1, \alpha_2, \ldots, \alpha_{n-2}, \alpha_{n-1}, \alpha_n \in \mathbb{R}\} \\
&= \{((\alpha_0 + \alpha_1 x + \alpha_2 x^2 + \cdots + \alpha_{n-2} x^{n-2}) + W \\
&\quad + (\alpha_{n-1} x^{n-1} + \alpha_n x^n) + W) | \alpha_0, \alpha_1, \alpha_2, \ldots, \alpha_{n-2}, \alpha_{n-1}, \alpha_n \in \mathbb{R}\} \\
&= \{(\alpha_{n-1} x^{n-1} + \alpha_n x^n) + W | \alpha_{n-1}, \alpha_n \in \mathbb{R}\}
\end{aligned}
$$

It is easy to notice that $V/W = \{\alpha_{n-1}(x^{n-1} + W) + \alpha_n(x^n + W) | \alpha_{n-1}, \alpha_n \in \mathbb{R}\}$. This implies that $V/W = L(S)$, where $S = \{(x^{n-1} + W), (x^n + W)\} \subseteq V/W$.

Exercises

1. Let U and W be vector spaces over a field $\mathbb{F}$. Let V be the set of ordered pairs (u, w) where $u \in U$ and $w \in W$. Show that V is a vector space over $\mathbb{F}$ with regard to addition in V and scalar multiplication on V defined by $(u, w) + (u', w') = (u + u', w + w')$ and $\alpha(u, v) = (\alpha u, \alpha v)$, where $\alpha \in \mathbb{F}$.

2. Let $AX = B$ be a nonhomogeneous system of linear equations in n unknowns, that is $B \neq 0$. Show that the solution set is not a subspace of F^n.

3. Let V be the vector space of all functions from the real field $\mathbb{R}$ into $\mathbb{R}$. Prove that W is a subspace of V if W consists of all bounded functions.

4. Suppose U, W_1, W_2 are subspaces of a vector space V over a field $\mathbb{F}$. Show that $(U \cap W_1) + (U \cap W_2) \subseteq U \cap (W_1 + W_2)$.

5. Give examples of three subspaces U, W_1, W_2 of a vector space V such that $(U \cap W_1) + (U \cap W_2) \neq U \cap (W_1 + W_2)$.

6. Let $S = \{(x_n) \mid x_n \in \mathbb{R}\}$ be the set of all real sequences. Then S is a vector space over $\mathbb{R}$ under the following operations:

$$(x_n) + (y_n) = (x_n + y_n),$$

$$\alpha(x_n) = (\alpha x_n), \quad \alpha \in \mathbb{R}.$$

Let $\mathscr{C}$ be the set of all convergent sequences and $\mathscr{C}_0$ be the set of all null sequences. Then show that $\mathscr{C}$ and $\mathscr{C}_0$ are vector subspaces of S.

7. Why does a vector space V over $\mathbb{F}(=\mathbb{C}, \mathbb{R}$ or $\mathbb{Q})$ contain either one element or infinitely many elements? Given $v \in V$, is it possible to have two distinct vectors $u, w \in V$ such that $u + v = 0$ and $w + v = 0$?

8. Let $\mathbb{H}$ be the collection of all complex 2×2 matrices of the form $\begin{bmatrix} a & b \\ -\bar{b} & \bar{a} \end{bmatrix}$. Show that $\mathbb{H}$ is a vector space under the usual matrix addition and scalar multiplication over $\mathbb{R}$. Is $\mathbb{H}$ also a vector space over $\mathbb{C}$?

9. Show that if W_1 is a subspace of a vector space V, and if there is a unique subspace W_2 such that $V = W_1 \oplus W_2$, then $W_1 = V$.

10. Let $U = \{(a, b, c) \mid a = b = c, a, b, c \in \mathbb{R}\}$ and $W = \{(0, b, c) \mid b, c \in \mathbb{R}\}$ be subspaces of $\mathbb{R}^3$. Show that $\mathbb{R}^3 = U \oplus W$.

11. Let $U_1 = \{(a, b, c) \mid a = c, a, b, c \in \mathbb{R}\}, U_2 = \{(a, b, c) \mid a + b + c = 0, a, b, c \in \mathbb{R}\}$ and $U_3 = \{(0, 0, c) \mid c \in \mathbb{R}\}$ be subspaces of $\mathbb{R}^3$. Show that

(a) $\mathbb{R}^3 = U_1 + U_2$,
(b) $\mathbb{R}^3 = U_2 + U_3$,
(c) $\mathbb{R}^3 = U_1 + U_3$.

When is the sum a direct sum?

12. Let W_1, W_2 and W_3 be subspaces of a vector space V. Show that $W_1 + W_2 + W_3$ is not necessarily a direct sum even though

$$W_1 \cap W_2 = W_1 \cap W_3 = W_2 \cap W_3 = \{0\}.$$

13. Show that $M_2(\mathbb{R}) = W_1 \oplus W_2$, where

$$W_1 = \left\{ \begin{bmatrix} a & b \\ -b & a \end{bmatrix} \mid a, b \in \mathbb{R} \right\},$$

$$W_2 = \left\{ \begin{bmatrix} c & d \\ d & -c \end{bmatrix} \mid c, d \in \mathbb{R} \right\}.$$

14. Let $C(\mathbb{R})$ be the vector space of all real-valued functions over $\mathbb{R}$, and let W_1 and W_2 be the collections of even and odd continuous functions on $\mathbb{R}$, respectively. Show that W_1 and W_2 are subspaces of $C(\mathbb{R})$. Show further that $C(\mathbb{R}) = W_1 \oplus W_2$.

15. Give an example of a vector space V having any three different nonzero subspaces W_1, W_2, W_3 such that $V = W_1 \oplus W_2 = W_2 \oplus W_3 = W_3 \oplus W_1$.

2.2 Linear Dependence, Independence and Basis

Definition 2.30 Let $S = \{v_1, v_2, \ldots, v_n\}$ be a finite set, having n vectors in a vector space V over a field $\mathbb{F}$. Then S is said to be linearly independent if for any $\alpha_1, \alpha_2, \ldots, \alpha_n \in \mathbb{F}$, $\sum_{i=1}^{n} \alpha_i v_i = 0$ implies that $\alpha_i = 0$ for each $i = 1, 2, \ldots, n$. If S is not linearly independent then S is said to be linearly dependent, that is, if there exist scalars $\alpha_1, \alpha_2, \ldots, \alpha_n \in \mathbb{F}$, not all zero, such that $\sum_{i=1}^{n} \alpha_i v_i = 0$.

On the other hand, an infinite subset S of V is said to be linearly independent if every finite subset of S is linearly independent. In this case also if S is not linearly independent then S is said to be linearly dependent, i.e., if there exists a finite subset of S, which is linearly dependent.

Example 2.31 (1) Let $V = \{(a, b, c) \mid a, b, c \in \mathbb{F}\}$. This is a vector space over $\mathbb{F}$ in which vectors $e_1 = (1, 0, 0)$, $e_2 = (0, 1, 0)$, $e_3 = (0, 0, 1)$ are linearly independent. In fact, if there exist scalars $\alpha_1, \alpha_2, \alpha_3 \in \mathbb{F}$ such that $\alpha_1(1, 0, 0) + \alpha_2(0, 1, 0) + \alpha_3(0, 0, 1) = (\alpha_1, \alpha_2, \alpha_3) = (0, 0, 0)$ implies that $\alpha_1 = \alpha_2 = \alpha_3 = 0$.

(2) Let V denote the vector space of all polynomials in x over the field $\mathbb{R}$ of real numbers, i.e., $V = \{\alpha_0 + \alpha_1 x + \alpha_2 x^2 + \cdots + \alpha_n x^n \mid \alpha_0, \alpha_1, \alpha_2, \ldots, \alpha_n \in \mathbb{R}, n \in \mathbb{N} \cup \{0\}\}$. Assume S is an infinite subset of V, where $S = \{1, x, x^2, x^3, x^4, \ldots, x^n, \ldots\}$. The set S is a linearly independent subset of V. This is due to the fact that if we take any finite subset of S, then it will be of the form $\{x^{i_1}, x^{i_2}, x^{i_3}, \ldots, x^{i_m}\}$, where $i_1, i_2, i_3, \ldots, i_m$ are some nonnegative integers and if $\sum_{k=1}^{m} \lambda_{i_k} x^{i_k} = 0$, this shows that $\lambda_{i_k} = 0$ for each $k = 1, 2, \ldots, m$ and hence $\{x^{i_1}, x^{i_2}, x^{i_3}, \ldots, x^{i_m}\}$ is a linearly independent subset of V.

Remark 2.32 (i) If $0 \in S$, then S is linearly dependent.

(ii) $\{v\}$, where $v \in V$, is linearly independent if and only if $v \neq 0$.

(ii) If any two vectors u, v are linearly independent in V, then $u + v$ and $u - v$ are also linearly independent.

(iii) Any two nonzero vectors are linearly dependent in V if and only if one is a scalar multiple of the other.

(iv) Every superset of a linearly dependent subset of V is linearly dependent.

(v) Every subset of a linearly independent subset of V is linearly independent.

(vi) The empty subset $\emptyset$ of V is linearly independent, as the condition of linear independence holds vacuously for $\emptyset$.

Theorem 2.33 *Let V be a vector space over a field $\mathbb{F}$. A subset $S = \{v_1, v_2, \ldots, v_n\}$ of nonzero vectors of V is linearly dependent if and only if $v_i = \sum_{j=1, j \neq i}^{n} \alpha_j v_j, \alpha_j \in \mathbb{F}$, for some i.*

Proof If S is linearly dependent, then there exist scalars $\beta_i \in \mathbb{F}$ not all zero such that $\sum_{i=1}^{n} \beta_i v_i = 0$. Suppose that $\beta_i \neq 0$ for some i. Then the above expression can be written as $v_i = \sum_{j=1, j \neq i}^{n} (-\beta_i^{-1} \beta_j) v_j$, i.e., $v_i = \sum_{j=1, j \neq i}^{n} \alpha_j v_j$, where $\alpha_j = -\beta_i^{-1} \beta_j \in \mathbb{F}$.

Conversely, if for some i, v_i can be expressed as a linear combination of v_j, $j \neq i$, i.e., $v_i = \sum_{j=1, j \neq i}^n \alpha_j v_j$, where $\alpha_j \in \mathbb{F}$, then this yields that $\alpha_1 v_1 + \alpha_2 v_2 + \cdots + (-1)v_i + \alpha_{i+1}v_{i+1} + \cdots + \alpha_n v_n = 0$. This shows that there exist scalars $\alpha_1, \alpha_2, \ldots, \alpha_n$ with $\alpha_i = -1 \neq 0$ such that $\sum_{i=1}^n \alpha_i v_i = 0$, and hence S is linearly dependent.

Theorem 2.34 *Let V be a vector space over a field $\mathbb{F}$. An ordered subset $S = \{v_1, v_2, \ldots, v_n\}$ of nonzero vectors of V is linearly dependent if and only if there exists some vector v_k, $2 \leq k \leq n$, which is a linear combination of preceding ones.*

Proof Assume that ordered set S is linearly dependent. Consider the set $S_1 = \{v_1\}$, which is obviously a linearly independent set. Now consider the set $S_2 = \{v_1, v_2\}$. If S_2 is a linearly dependent set, then there exist scalars $\alpha_1, \alpha_2 \in \mathbb{F}$, not both zero such that $\alpha_1 v_1 + \alpha_2 v_2 = 0$. Here we claim that $\alpha_2 \neq 0$. For otherwise S_1 becomes a linearly dependent set, leading to a contradiction. Now multiplying both sides of the previous relation by α_2^{-1}, we arrive at $v_2 = \lambda v_1$, where $\lambda = -\alpha_2^{-1}\alpha_1$, i.e., v_k is a linear combination of preceding vector, where $k = 2$. If S_2 is a linearly independent set, then we consider the set $S_3 = \{v_1, v_2, v_3\}$. If S_3 is a linearly dependent set, then there exist scalars $\beta_1, \beta_2, \beta_3 \in \mathbb{F}$, not all zero such that $\beta_1 v_1 + \beta_2 v_2 + \beta_3 v_3 = 0$. Using the similar argument as above, it can be easily shown that $\beta_3 \neq 0$ and $v_3 = \epsilon v_1 + \delta v_2$, for some $\epsilon, \delta \in \mathbb{F}$, i.e., v_k is a linear combination of preceding vectors, where $k = 3$. Let m be the least positive integer such that the set $S_m = \{v_1, v_2, v_3, \ldots, v_m\}$ is linearly dependent. Using similar argument as above, it can be shown that $v_m = \delta_1 v_1 + \delta_2 v_2 + \cdots + \delta_{m-1} v_{m-1}$, i.e., v_k is a linear combination of preceding vectors, where $k = m$. In the worst case $k = n$, will do our job. It is obvious that $2 \leq k \leq n$. Conversely, suppose that v_k, $2 \leq k \leq n$ is a linear combination of preceding vectors, i.e., there exist scalars $\epsilon_1, \epsilon_2, \ldots, \epsilon_{k-1}$ such that $v_k = \epsilon_1 v_1 + \epsilon_2 v_2 + \cdots + \epsilon_{k-1} v_{k-1}$. This shows that $\epsilon_1 v_1 + \epsilon_2 v_2 + \cdots + \epsilon_{k-1} v_{k-1} + (-1)v_k + 0v_{k+1} + 0v_{k+2} + \cdots + 0v_n = 0$ and thus the ordered set $S = \{v_1, v_2, \ldots, v_n\}$ is linearly dependent.

Definition 2.35 A subset B of a vector space V over $\mathbb{F}$ is called a basis of V if

(i) B is linearly independent.
(ii) B spans V, i.e., $L(B) = V$.

Remark 2.36 (i) Basis of a vector space need not be unique.
(ii) Every field $\mathbb{F}$ is a vector space over itself. If $0 \neq \alpha \in \mathbb{F}$, then $\{\alpha\}$ is a basis of $\mathbb{F}$.
(iii) Let $B = \{v_1, v_2, \ldots, v_n\}$ be a subset of a vector space V over $\mathbb{F}$. Then B is a basis of V if and only if every vector $v \in V$ can be expressed uniquely as $v = \sum_{i=1}^n \alpha_i v_i$ where $\alpha_i \in \mathbb{F}$, i.e., if $v = \sum_{i=1}^n \alpha_i v_i = \sum_{i=1}^n \beta_i v_i$, $\alpha_i, \beta_i \in \mathbb{F}$, then $\alpha_i = \beta_i$ for each $i = 1, 2, \ldots, n$.

Example 2.37 (1) Consider the vector space $V = \mathbb{F}^n$ over a field $\mathbb{F}$, where n is a positive integer. Then the set of n-tuples $e_1 = (1, 0, 0, \ldots, 0)$, $e_2 = (0, 1, 0, \ldots, 0)$, $\ldots, e_n = (0, 0, 0, \ldots, 1) \in V$. For any $v = (v_1, v_2, \ldots, v_n) \in V$ can be written as $v = e_1 v_1 + e_2 v_2 + \cdots + e_n v_n$, and hence the set $\{e_1, e_2, \ldots, e_n\}$ spans V. Moreover, $\{e_1, e_2, \ldots, e_n\}$ is linearly independent subset of V. Hence, $\{e_1, e_2, \ldots, e_n\}$ is a basis of V, which is called the standard basis of $V = \mathbb{F}^n$.

(2) Let $V = \mathbb{F}_n[x]$, the set of all polynomials in indeterminate x of degree less than or equal to n. V is a vector space over the field $\mathbb{F}$. Now let $B = \{1, x, x^2, \ldots, x^n\} \subseteq V$. It can be easily seen that any $f(x) \in V$ can be written as $f(x) = \alpha_0 1 + \alpha_1 x + \cdots + \alpha_n x^n$ with $\alpha_i \in \mathbb{F}$. This shows that B spans V. Moreover, $\alpha_0 1 + \alpha_1 x + \cdots + \alpha_n x^n = 0$ yields that $\alpha_0 = \alpha_1 = \cdots = \alpha_n = 0$ and hence B is linearly independent. Thus $B = \{1, x, x^2, \ldots, x^n\}$ is a basis of V, which is called the standard basis of V.

(3) The set M_2 of all 2×2 matrices over $\mathbb{R}$ forms a vector space over $\mathbb{R}$. The set
$$B = \left\{ \begin{bmatrix} 1 & 0 \\ 0 & 0 \end{bmatrix}, \begin{bmatrix} 0 & 1 \\ 0 & 0 \end{bmatrix}, \begin{bmatrix} 0 & 0 \\ 1 & 0 \end{bmatrix}, \begin{bmatrix} 0 & 0 \\ 0 & 1 \end{bmatrix} \right\} \text{ is a standard basis of } M_2.$$

Definition 2.38 Let V be a vector space over a field $\mathbb{F}$. A subset S of V is called a maximal linearly independent set if

(i) S is linearly independent.
(ii) $S \subset S'$, where $S' \subseteq V$, implies that S' is a linearly dependent set.

Definition 2.39 Let V be a vector space over a field $\mathbb{F}$. A subset S of V is called a minimal set of generators if

(i) $\langle S \rangle = V$.
(ii) $S' \subset S$, where $S' \subseteq V$, implies that $\langle S' \rangle \neq V$.

Theorem 2.40 *Let V be a vector space over a field $\mathbb{F}$ and let $S \subseteq V$. Then following statements are equivalent.*

(i) S is a maximal linearly independent set of V.
(ii) S is a minimal set of generators of V.
(iii) S is a basis of V.
(iv) Every element of V can be uniquely written as a linear combination of finitely number of elements of S, i.e., for any $v \in V$ if $v = \alpha_1 v_1 + \alpha_2 v_2 + \cdots + \alpha_n v_n$ and $v = \beta_1 v_1 + \beta_2 v_2 + \cdots + \beta_n v_n$, where $\alpha_i, \beta_i \in \mathbb{F}$ and $v_i \in S, i = 1, 2, 3, \ldots, n$ then $\alpha_i = \beta_i$ for all $i = 1, 2, 3, \ldots, n$.

Proof $(i) \Rightarrow (ii)$ We know that $\langle S \rangle = L(S)$. First we prove that $L(S) = V$. $L(S) \subseteq V$ holds obviously. Let $v \in V$. If $v \in S$, then obviously $v = 1v$, which shows that $v \in L(S)$. On the other hand, if $v \in V \setminus S$, then $S \cup \{v\}$ is a linearly dependent set since S is a maximal linearly independent set. There exists a finite subset T of $S \cup \{v\}$ containing the element v of S, which is a linearly dependent set. Let $T = \{v_1, v_2, \ldots, v_n, v\}$. There exist scalars $\alpha_1, \alpha_2, \ldots, \alpha_n, \alpha$, not all zero such that $\alpha_1 v_1 + \alpha_2 v_2 + \cdots + \alpha_n v_n + \alpha v = 0$. Here we claim that $\alpha \neq 0$, for otherwise the set $\{v_1, v_2, \ldots, v_n\}$ becomes a linearly dependent set, leading to a contradiction. Now we get $v = (-(\alpha^{-1}\alpha_1))v_1 + (-(\alpha^{-1}\alpha_2))v_2 + \cdots + ((-\alpha^{-1}\alpha_n))v_n$. This implies that $v \in L(S)$ and thus $V \subseteq L(S)$. Finally we have proved that $L(S) = V$.

Now we prove that S is a minimal set such that $\langle S \rangle = V$. Suppose on contrary, i.e., there exists $P \subsetneq S$ such that $\langle P \rangle = V$, i.e., $L(P) = V$. Let $w \in S - P$. There exist scalars $\beta_1, \beta_2, \ldots, \beta_m$ such that $w = \beta_1 w_1 + \beta_2 w_2 + \cdots + \beta_m w_m$ for some

$w_1, w_2, \ldots, w_m \in P$. This shows that the set $\{w_1, w_2, \ldots, w_m, w\} \subseteq S$ is a linearly dependent set which is a contradiction. Including the above arguments, we have shown that S is a minimal set of generators.

$(ii) \Rightarrow (iii)$ Suppose that S is a minimal set of generators. Then we have to prove that S is a basis of V. By hypothesis $\langle S \rangle = V$, i.e., $L(S) = V$. We have to only prove that S is a linearly independent set. Suppose on contrary, i.e., S is a linearly dependent set. Then there exist a set $\{v_1, v_2, \ldots, v_n\} \subseteq S$, which is a linearly dependent set. Hence there exist scalars $\alpha_1, \alpha_2, \ldots, \alpha_n$, not all zero such that $\alpha_1 v_1 + \alpha_2 v_2 + \cdots + \alpha_i v_i + \cdots + \alpha_n v_n = 0$. Let us suppose that $\alpha_i \neq 0$. Now preceding relation can be written as $v_i = (-(\alpha_i^{-1}\alpha_1))v_1 + (-(\alpha_i^{-1}\alpha_2))v_2 + \cdots + (-(\alpha_i^{-1}\alpha_{i-1})v_{i-1} + (-(\alpha_i^{-1}\alpha_{i+1})v_{i+1} + \cdots + (-(\alpha_i^{-1}\alpha_n)v_n$. This shows that $< S - \{v_i\} >= V$, which is a contradiction. Hence S is a linearly independent set. Thus S is a basis of V.

$(iii) \Rightarrow (iv)$ Let S be a basis of V. This implies that $L(S) = V$. For any $v \in V$ if $v = \alpha_1 v_1 + \alpha_2 v_2 + \cdots + \alpha_n v_n$ and $v = \beta_1 v_1 + \beta_2 v_2 + \cdots + \beta_n v_n$, where $\alpha_i, \beta_i \in \mathbb{F}$ and $v_i \in S, i = 1, 2, 3, \ldots, n$. This implies that $(\alpha_1 - \beta_1)v_1 + (\alpha_2 - \beta_2)v_2 + \cdots + (\alpha_n - \beta_n)v_n = 0$. As S is a linearly independent set, we are forced to conclude that $(\alpha_i - \beta_i) = 0$ for all $i = 1, 2, 3, \ldots, n$, i.e., $\alpha_i = \beta_i$ for all $i = 1, 2, 3, \ldots, n$. Hence each element of V is uniquely expressible as a linear combination of finitely number of elements of S.

$(iv) \Rightarrow (i)$ First we prove that S is a linearly independent set. Let $\{v_1, v_2, \ldots, v_n\} \subseteq S$ and $\alpha_1 v_1 + \alpha_2 v_2 + \cdots + \alpha_n v_n = 0$ for some scalars $\alpha_1, \alpha_2, \ldots, \alpha_n$. But we also have $0 = 0v_1 + 0v_2 + \cdots + 0v_n$. Using hypothesis we are forced to conclude that $\alpha_i = 0$ for all $i = 1, 2, 3, \ldots, n$. This implies that S is a linearly independent set.

Next we show that S is a maximal linearly independent set. Let $S \subsetneq T \subseteq V$. Suppose that $v \in T \setminus S$. Hence there exist unique scalars $\beta_1, \beta_2, \ldots, \beta_m$ such that $v = \beta_1 v_1 + \beta_2 v_2 + \cdots + \beta_m v_m$ where $v_1, v_2, \ldots, v_m \in S$. This shows that the set $\{v, v_1, v_2, \ldots, v_m\}$ is a linearly dependent set. Hence T is a linearly dependent set. Thus we have proved that S is a maximal linearly independent set.

Lemma 2.41 *Let S be a linearly independent subset of a vector space V and let $S = S_1 \cup S_2$ with $S_1 \cap S_2 = \emptyset$. If W, W_1, W_2 are subspaces of V spanned by S, S_1, S_2 respectively, then $W = W_1 + W_2$ and $W_1 \cap W_2 = \{0\}$.*

Proof In fact, $L(S_1 \cup S_2) = L(S_1) + L(S_2)$ i.e., $W = W_1 + W_2$. Moreover, if $0 \neq v \in W_1 \cap W_2$, then $v = \sum_{i=1}^{m} \alpha_i v_i = \sum_{j=1}^{n} \beta_j v_j'; v_i \in S_1, v_j' \in S_2$ for some $m \geq 1, n \geq 1$. This yields that $\sum_{i=1}^{m} \alpha_i v_i - \sum_{j=1}^{n} \beta_j v_j' = 0$. Since $v \neq 0$, there exist some $\alpha_i \neq 0$ and some $\beta_i \neq 0$. As $S_1 \cap S_2 = \emptyset$, $B = \{v_1, v_2, \ldots, v_m, v_1', v_2', \ldots, v_n'\}$ is a subset of S with $m + n$ vectors. But the latter relation shows that B is linearly dependent. This contradicts the fact that S is linearly independent, and hence $W_1 \cap W_2 = \{0\}$.

Example 2.42 Let V be a vector space over a field $\mathbb{F}$ with $dim V = k \geq 2$, and let $B = \{v_1, v_2, \ldots, v_k\}$ be a basis of V. For each $1 \leq i \leq k$, let $W_i = \langle v_i \rangle = \{\alpha v_i \mid \alpha \in \mathbb{F}\}$, then it can be seen that V is internal direct sum of k subspaces $W_1, W_2, \ldots, W_k$ each of dimension one. Consider $v \in V$. Then there exist scalars $\alpha_1, \alpha_2, \ldots, \alpha_k \in \mathbb{F}$

such that $v = \alpha_1 v_1 + \alpha_2 v_2 + \cdots + \alpha_k v_k$ i.e., $v = \beta_1 + \beta_2 + \cdots + \beta_k$; $\beta_i = \alpha_i v_i \in W_i$. Suppose also that $v = \beta_1' + \beta_2' + \cdots + \beta_k'$; $\beta_i' = \alpha_i' v_i \in W_i$. Hence $v = \alpha_1' v_1 + \alpha_2' v_2 + \cdots + \alpha_k' v_k$, which implies that $\sum_{i=1}^{k} \alpha_i v_i = \sum_{i=1}^{k} \alpha_i' v_i$ i.e., $\sum_{i=1}^{k} (\alpha_i - \alpha_i') v_i = 0$. But since B is linearly independent, we find that $\alpha_i = \alpha_i'$ and therefore $\beta_i = \alpha_i v_i = \alpha_i' v_i = \beta_i'$ and v has a unique representation. Hence $V = W_1 \oplus W_2 \oplus \cdots W_k$.

Theorem 2.43 *If a vector space V over a field $\mathbb{F}$ has a basis containing m vectors, where m is a positive integer, then any set containing n vectors $n > m$ in V is linearly dependent.*

Proof Suppose that $\{v_1, v_2, \ldots, v_m\}$ is a basis of V and let $\{w_1, w_2, \ldots, w_n\}$ be any subset of V containing n vectors $(n > m)$. Since $\{v_1, v_2, \ldots, v_m\}$ spans V, there exist scalars $\alpha_{ij} \in \mathbb{F}$ such that

$$w_j = \sum_{i=1}^{m} \alpha_{ij} v_i,$$

for each j such that $1 \le j \le n$.

In order to show that $\{w_1, w_2, \ldots, w_n\}$ is linearly dependent, we have to find $\beta_1, \beta_2, \ldots, \beta_n \in \mathbb{F}$ not all zero such that

$$\sum_{j=1}^{n} \beta_j w_j = 0.$$

This yields that $\sum_{j=1}^{n} \beta_j (\sum_{i=1}^{m} \alpha_{ij} v_i) = 0$, i.e., $\sum_{i=1}^{m} (\sum_{j=1}^{n} \alpha_{ij} \beta_j) v_i = 0$.

But since $\{v_1, v_2, \ldots, v_m\}$ is a basis of V, $0 = 0v_1 + 0v_2 + \cdots + 0v_m$ is the unique representation for 0 vector. Hence the above expression yields that $\sum_{j=1}^{n} \alpha_{ij} \beta_j = 0$, for each i such that $1 \le i \le m$. This is a system of m homogeneous linear equations in n unknowns. Thus by Theorem 1.104, there exists nontrivial solution say $\beta_1, \beta_2, \ldots, \beta_n$. This ensures that there exist scalars $\beta_1, \beta_2, \ldots, \beta_n$ not all zero such that $\sum_{j=1}^{n} \beta_j w_j = 0$ and hence the set $\{w_1, w_2, \ldots, w_n\}$ is linearly dependent.

Definition 2.44 A vector space V over a field $\mathbb{F}$ is said to be finite dimensional (resp. infinite dimensional) if there exists a finite (resp. infinite) subset S of V which spans V, i.e., $L(S) = V$.

Remark 2.45 If a vector space V has a basis with a finite (resp. an infinite) number of vectors, then it is finite (resp. infinite) dimensional. The number of vectors of a basis of V is called the dimension of V denoted as $dim V$. If $V = \{0\}$, then its dimension is taken to be zero.

Theorem 2.46 *Let V be a vector space over a field $\mathbb{F}$. If it has a finite basis, then any two bases of V have the same number of elements.*

Proof Let $B = \{v_1, v_2, \ldots, v_n\}$ and $B' = \{w_1, w_2, \ldots, w_m\}$ be two bases of V. As B is a basis and the set B' is a linearly independent set of V, by Theorem 2.43, we arrive at $m \leq n$. Similarly, as B' is a basis of V and the set B is a linearly independent set of V, again by Theorem 2.43, we conclude that $n \leq m$. This yields that $m = n$.

Theorem 2.47 *Let V be an n-dimensional vector space over a field $\mathbb{F}$. Then any linearly independent subset of V consisting n elements is a basis of V.*

Proof Let $S = \{v_1, v_2, \ldots, v_n\}$ be a linearly independent subset of V containing n vectors. Then in view of Theorem 2.43, for any $v \in V$, the set $\{v, v_1, v_2, \ldots, v_n\}$ consisting of $n + 1$ vectors is linearly dependent. Therefore, there exist scalars $\alpha, \alpha_1, \alpha_2, \ldots, \alpha_n \in \mathbb{F}$ not all zero such that $\alpha v + \alpha_1 v_1 + \alpha_2 v_2 + \cdots + \alpha_n v_n = 0$. Our claim is that $\alpha \neq 0$. If $\alpha = 0$, then this contradicts the fact that S is linearly independent. Hence $v = (-(\alpha^{-1}\alpha_1))v_1 + (-(\alpha^{-1}\alpha_2))v_2 + \cdots + (-(\alpha^{-1}\alpha_n))v_n$. This shows that S spans V and hence a basis of V.

Theorem 2.48 *Let V be an n-dimensional vector space over a field $\mathbb{F}$. Then any linearly independent subset $\{v_1, v_2, \ldots, v_m\}$, $m \leq n$, of V can be extended to a basis V.*

Proof Given that $\{v_1, v_2, \ldots, v_m\}$ is a linearly independent subset of V. Let W be the subspace of V generated by $\{v_1, v_2, \ldots, v_m\}$. If $W = V$, then $\{v_1, v_2, \ldots, v_m\}$ is a basis of V. If $W \subset V$, then there exists $v_{m+1} \in V$ such that $v_{m+1} \notin W$, that is $v_{m+1} \in V \setminus W$. Then we claim that the set $\{v_1, v_2, \ldots, v_m, v_{m+1}\}$ is linearly independent. If it is not, then there exist scalars $\alpha_1, \alpha_2, \ldots, \alpha_m, \alpha_{m+1}$ not all zero such that $\alpha_1 v_1 + \alpha_2 v_2 + \cdots + \alpha_m v_m + \alpha_{m+1} v_{m+1} = 0$. For if $\alpha_{m+1} \neq 0$, then $v_{m+1} = \sum_{i=1}^{m}(-(\alpha_{m+1}^{-1}\alpha_i))v_i \in W$, a contradiction to the choice of v_{m+1}. Now let W' be the subspace generated by $\{v_1, v_2, \ldots, v_m, v_{m+1}\}$. If $W' = V$, then $\{v_1, v_2, \ldots, v_m, v_{m+1}\}$ is a basis of V. Otherwise choose $v_{m+2} \in V \setminus W'$ and continue this process. Since V is finite dimensional, a linearly independent subset in V has utmost n elements. After a finite number of $n - m$ steps, we arrive at a linearly independent subset $\{v_1, v_2, \ldots, v_m, v_{m+1}, \ldots, v_n\}$ of V such that the subspace generated by $\{v_1, v_2, \ldots, v_m, v_{m+1}, \ldots, v_n\}$ coincides with V. Hence there exist vectors $v_{m+1}, v_{m+2}, \ldots, v_n \in V$ such that $\{v_1, v_2, \ldots, v_m, v_{m+1}, \ldots, v_n\}$ is a basis of V.

Theorem 2.49 *Let W be a subspace of a finite dimensional vector space V. Then there exists a subspace W' of V such that $V = W \oplus W'$.*

Proof Let $dim V = n$ and $dim W = m$. If $W = \{0\}$, then $W' = V$ and, on the other hand, if $W = V$, then $W' = \{0\}$. Hence assume that neither $W = \{0\}$ nor $W = V$. In this case $1 \leq m < n$. Let $B_1 = \{v_1, v_2, \ldots, v_m\}$ be a basis of W. Since B is linearly independent, by the above theorem B can be extended to a basis of V say $B = \{v_1, v_2, \ldots, v_m, v_{m+1}, \ldots, v_n\}$. Now $B = B_1 \cup B_2$ with $B_2 = \{v_{m+1}, v_{m+2}, \ldots, v_n\}$ and $B_1 \cap B_2 = \emptyset$. Let W' be spanned by B_2. Since V and W are spanned by B and B_1, respectively, by Lemma 2.41, $V = W + W'$ and $W \cap W' = \{0\}$, i.e., $V = W \oplus W'$.

Theorem 2.50 *Let V be a finite dimensional vector space over a field $\mathbb{F}$ and U a subspace of V. Then $dim U \leq dim V$. Equality holds only when $U = V$.*

Proof Let $B = \{v_1, v_2, \ldots, v_n\}$ be a basis of the vector space V. Let B_1 be a basis of subspace U. This implies that B_1 is linearly independent set in U and hence B_1 is also linearly independent set in V. Using Theorem 2.43, we arrive at the number of elements in the set $B_1 \leq n$, i.e., $dim U \leq dim V$.

When $dim U = dim V$. B_1, a basis of U is also a linearly independent subset of V. But $dim V = n$ forces us to conclude that B_1 is a basis of V also. Now we conclude that $L(B_1) = U = V$. This implies that $U = V$.

Remark 2.51 It is to be noted that if V is an infinite dimensional vector space then its subspaces are of both finite and infinite dimensions. For justification, let V denote the vector space of all polynomials in x with real coefficients over the field of real numbers. It is easy to observe that V is an infinite dimensional vector space. One of its basis is $\{1, x, x^2, x^3, \ldots, x^n, \ldots\}$. It is easy to observe that $W_1 = \langle\{1\}\rangle$, $W_2 = \langle\{1, x\}\rangle$, $W_3 = \langle\{1, x, x^2\}\rangle, \ldots$ are subspaces of V of dimensions $1, 2, 3, \ldots$, respectively. If we take $W = \langle\{x^4, x^5, x^6, x^7, \ldots\}\rangle$, then it is easy to observe that W is a subspace of V of infinite dimension.

Theorem 2.52 *Let V be a finite dimensional vector space over a field $\mathbb{F}$ and W_1, W_2 be any two subspaces of V, then $dim(W_1 + W_2) = dim W_1 + dim W_2 - dim(W_1 \cap W_2)$.*

Proof Since $W_1 \cap W_2$ is a subspace of V, which is a finite dimensional vector space. Therefore using Theorem 2.50, we have $dim W_1 \cap W_2 \leq dim V$. Now assume that $\{v_1, v_2, \ldots, v_k\}$ is a basis of $W_1 \cap W_2$. But as $W_1 \cap W_2$ is a subset of both W_1 and W_2, the set $\{v_1, v_2, \ldots, v_k\}$ is a subset of both W_1 and W_2. It is obvious to see that $\{v_1, v_2, \ldots, v_k\}$ is a linearly independent subset of both W_1 and W_2. Thus the set $\{v_1, v_2, \ldots, v_k\}$ can be extended to a basis $\{v_1, v_2, \ldots, v_k, w_1, w_2, \ldots, w_r\}$ of W_1 and a basis $\{v_1, v_2, \ldots, v_k, u_1, u_2, \ldots, u_s\}$ of W_2. Now our claim is that the set $B = \{v_1, v_2, \ldots, v_k, w_1, w_2, \ldots, w_r, u_1, u_2, \ldots, u_s\}$ spans $W_1 + W_2$. Let $x \in W_1 + W_2$. Then $x = y_1 + y_2$ where $y_1 \in W_1$, $y_2 \in W_2$. Hence $y_1 = \sum_{i=1}^{k} \alpha_i v_i + \sum_{j=1}^{r} \beta_j w_j$ and $y_2 = \sum_{i=1}^{k} \gamma_i v_i + \sum_{\ell=1}^{s} \delta_\ell u_\ell$, for some $\alpha_i, \beta_j, \gamma_i, \delta_l \in \mathbb{F}, i = 1, 2, \ldots, k; j = 1, 2, \ldots, r; \ell = 1, 2, \ldots, s$. This implies that $x = y_1 + y_2 = \sum_{i=1}^{k}(\alpha_i + \gamma_i)v_i + \sum_{j=1}^{r} \beta_j w_j + \sum_{\ell=1}^{s} \delta_\ell u_\ell$. This shows that the set B spans $W_1 + W_2$. Now to show that B is linearly independent, let

$$\sum_{i=1}^{k} \alpha_i v_i + \sum_{j=1}^{r} \beta_j w_j + \sum_{\ell=1}^{s} \delta_\ell u_\ell = 0.$$

This implies that

$$\sum_{i=1}^{k} \alpha_i v_i + \sum_{j=1}^{r} \beta_j w_j = -\sum_{\ell=1}^{s} \delta_\ell u_\ell.$$

The expression in the right side is in W_2 and in the left side is in W_1. Therefore, $-\sum_{\ell=1}^{s} \delta_\ell u_\ell \in W_1 \cap W_2$, and hence can be written as $-\sum_{\ell=1}^{s} \delta_\ell u_\ell = \sum_{j=1}^{k} \gamma_j v_j$ for some $\gamma_j \in F$, $j = 1, 2, \ldots, k$. But since $\{v_1, v_2, \ldots, v_k, u_1, u_2, \ldots, u_s\}$ is a basis of W_2, we find that $\delta_\ell = 0$, $\gamma_j = 0$ for all ℓ and j. Hence in particular,

$$\sum_{i=1}^{k} \alpha_i v_i + \sum_{j=1}^{r} \beta_j w_j = -\sum_{\ell=1}^{s} \delta_\ell u_\ell = 0.$$

Since the set $\{v_1, v_2, \ldots, v_k, w_1, w_2, \ldots, w_r\}$ being a basis of W_1 is linearly independent, the latter relation yields that $\alpha_i = 0$, $\beta_j = 0$ for all i, j. Thus $\alpha_i = 0$, $\beta_j = 0$, $\delta_\ell = 0$ for all i, j, ℓ and the set B is linearly independent. Thus B forms a basis for the subspace $W_1 + W_2$. Hence $dim(W_1 + W_2) = k + r + s = (k + r) + (k + s) - k = dim W_1 + dim W_2 - dim(W_1 \cap W_2)$.

If V is a finite dimensional vector space over a field $\mathbb{F}$ and W_1, W_2, W_3 are any three subspaces of V, then by replacing W_1 with $W_1 + W_2$ and W_2 with W_3 in the above theorem and using the fact that $W_1 \cap (W_1 + W_2) \supseteq (W_1 \cap W_2) + (W_1 \cap W_3)$, we arrive at the following corollary:

Corollary 2.53 *Let V be a finite dimensional vector space over a field $\mathbb{F}$, and W_1, W_2, W_3 be any three subspaces of V. Then $dim(W_1 + W_2 + W_3) \leq dim W_1 + dim W_2 + dim W_3 - dim(W_1 \cap W_2) - dim(W_2 \cap W_3) - dim(W_3 \cap W_1) + dim(W_1 \cap W_2 \cap W_3)$.*

Remark 2.54 If V is a direct sum of subspaces W_1 and W_2, i.e., $V = W_1 \oplus W_2$, then $dim V = dim W_1 + dim W_2$.

Theorem 2.55 *Let W be a subspace of a finite dimensional vector space V over a field $\mathbb{F}$. Then $dim(V/W) = dim V - dim W$.*

Proof Let $dim V = n$ and $dim W = m$, and let $\{v_1, v_2, \ldots, v_m\}$ be a basis of W. Since $\{v_1, v_2, \ldots, v_m\}$ is a linearly independent subset of V, the set $\{v_1, v_2, \ldots, v_m\}$ can be extended to a basis of V, say

$$\{v_1, v_2, \ldots, v_m, v_{m+1}, \ldots, v_n\}.$$

Now consider the $n - m$ vectors $v_{m+1} + W, v_{m+2} + W, \ldots, v_n + W$ in the quotient space V/W. Our claim is that the set $S = \{v_{m+1} + W, v_{m+2} + W, \ldots, v_n + W\}$ is a basis of V/W and then $dim(V/W) = n - m = dim V - dim W$. First we show that S spans V/W. Let $v + W \in V/W$. Then $v \in V$ and there exist scalars $\alpha_1, \alpha_2, \ldots, \alpha_n \in \mathbb{F}$ such that $v = \alpha_1 v_1 + \alpha_2 v_2 + \cdots + \alpha_n v_n$. Therefore

$$
\begin{aligned}
v + W &= (\alpha_1 v_1 + \alpha_2 v_2 + \cdots + \alpha_m v_m) + W + (\alpha_{m+1} v_{m+1} + \alpha_{m+2} v_{m+2} + \\
&\quad \cdots + \alpha_n v_n) + W \\
&= (\alpha_{m+1} v_{m+1} + \alpha_{m+2} v_{m+2} + \cdots + \alpha_n v_n) + W \\
&= \alpha_{m+1}(v_{m+1} + W) + \alpha_{m+2}(v_{m+2} + W) + \cdots + \alpha_n(v_n + W)
\end{aligned}
$$

This shows that S spans V/W. Further, we show that S is linearly independent. Let $\beta_{m+1}, \beta_{m+2}, \ldots, \beta_n \in \mathbb{F}$ such that

$$\beta_{m+1}(v_{m+1} + W) + \beta_{m+2}(v_{m+2} + W) + \cdots + \beta_n(v_n + W) = W$$

$$(\beta_{m+1}v_{m+1} + \beta_{m+2}v_{m+2} + \cdots + \beta_n v_n) + W = W.$$

This implies that $\beta_{m+1}v_{m+1} + \beta_{m+2}v_{m+2} + \cdots + \beta_n v_n \in W$. Therefore there exist $\delta_1, \delta_2, \ldots, \delta_n \in \mathbb{F}$ such that $\beta_{m+1}v_{m+1} + \beta_{m+2}v_{m+2} + \cdots + \beta_n v_n = \delta_1 v_1 + \delta_2 v_2 + \cdots + \delta_m v_m$, i.e.,

$$\delta_1 v_1 + \delta_2 v_2 + \cdots + \delta_m v_m + (-\beta_{m+1})v_{m+1} + (-\beta_{m+2})v_{m+2} + \cdots + (-\beta_n)v_n = 0.$$

Since $\{v_1, v_2, \ldots, v_m, v_{m+1}, \ldots, v_n\}$ is a basis of V, we find that $\delta_1 = \delta_2 = \cdots = \delta_m = -\beta_{m+1} = -\beta_{m+2} = \cdots = -\beta_n = 0$ and hence in particular $\beta_{m+1} = \beta_{m+2} = \cdots = \beta_n = 0$ and S is linearly independent. This completes the proof of our theorem.

Exercises

1. Let K be a subfield of a field L and L be a subfield of a field M. Suppose M is of dimension n over L and L is of dimension m over K. Prove that M is of dimension mn over K.
2. Suppose that $S_1, S_2, \ldots$ are linearly independent sets of vectors and $S_1 \subseteq S_2 \subseteq \cdots$. Show that the union $S = S_1 \cup S_2 \cup \cdots$ is also linearly independent.
3. Let $\mathbb{C}$, $\mathbb{R}$ and $\mathbb{Q}$ denote the field of complex numbers, real numbers and rational numbers, respectively. Show that

   (a) $\mathbb{C}$ is an infinite dimensional vector over $\mathbb{Q}$.
   (b) $\mathbb{R}$ is an infinite dimensional vector space over $\mathbb{Q}$.
   (c) The set $\{\alpha + i\beta, \gamma + i\delta\}$ is a basis of $\mathbb{C}$ over $\mathbb{R}$ if and only if $(\alpha\delta - \beta\gamma) \neq 0$, hence $\mathbb{C}$ is a vector space of dimension 2 over $\mathbb{R}$.

4. Construct an example of an infinite dimensional vector space V and it's subspace W such that there exists a subspace W_1 of V such that $V = W \oplus W_1$.
5. Construct an example of an infinite dimensional vector space V over $\mathbb{F}$ with a subspace W such that V/W is a finite dimensional vector space.
6. Let $U = Span\{(1, 2, 3), (0, 1, 2), (3, 2, 1)\}$ and $W = Span\{(1, -2, 3), (-1, 1, -2), (1, -3, 4)\}$ be two subspaces of $\mathbb{R}^3$. Determine the dimension and a basis for $U + W$, and $U \cap W$.
7. Consider the following subspaces of $\mathbb{R}^5$:

$$U = Span\{(1, 3, -2, 2, 3), (1, 4, -3, 4, 2), (2, 3, -1, -2, 9)\},$$

$$W = Span\{(1, 3, 0, 2, 1), (1, 5, -6, 6, 3), (2, 5, 3, 2, 1)\}.$$

Find a basis and the dimension of $U + W$, and $U \cap W$.

8. Let $U = \{(x_1, x_2, x_3, x_4) \in \mathbb{R}^4 \mid x_2 + x_3 + x_4 = 0\}$ and $W = \{(x_1, x_2, x_3, x_4) \in \mathbb{R}^4 \mid x_1 + x_2 = 0, x_3 = 2x_4\}$ be subspaces of $\mathbb{R}^4$. Find bases and dimensions for U, W, $U \cap W$ and $U + W$.

9. Let U and W be two distinct $(n - 1)$-dimensional subspaces of an n-dimensional vector space V. Then prove that $dim(U \cap W) = n - 2$.

10 Suppose U and W are distinct 4-dimensional subspaces of a vector space V, where $dim\, V = 6$. Find the possible dimensions of $U \cap W$.

11. Let W_1 be a subspace of $M_{n \times n}(\mathbb{R})$ consisting of all $n \times n$ symmetric matrices. Find a basis and the dimension of W_1. Further, find a subspace W_2 of $M_{n \times n}(\mathbb{R})$ such that $M_{n \times n}(\mathbb{R}) = W_1 \oplus W_2$.

12. Let $W = \{(x, y) \in \mathbb{R}^2 \mid ax + by = 0\}$ for a fixed $(a, b) \in \mathbb{R}^2 \setminus (0, 0)$. Show that W is a 1-dimensional subspace of $V = \mathbb{R}^2$ and that the cosets of W in V are lines parallel to the line $ax + by = c$ for $c \in \mathbb{R}$.

13. Let $\mathscr{C}(\mathbb{R})$ be the vector space of all real-valued continuous functions over $\mathbb{R}$ with addition $(f + g)(x) = f(x) + g(x)$ and scalar multiplication $(\alpha f) = \alpha f(x)$, $\alpha \in \mathbb{R}$. Show that $sin\, x$ and $cos\, x$ are linearly independent and the vector space generated by $sin\, x$ and $cos\, x$, i.e., $Span\{sin\, x, cos\, x\} = \{a sin\, x + b cos\, x \mid a, b \in \mathbb{R}\}$ is contained in the solution set to the differential equation $y'' + y = 0$. Are $sin^2 x$ and $cos^2 x$ linearly independent? What about 1, $sin^2 x$ and $cos^2 x$? Find $\mathbb{R} \cap Span\{sin\, x, cos\, x\}$ and $\mathbb{R} \cap Span\{sin^2 x, cos^2 x\}$.

14. Let $V = M_3(\mathbb{R})$, W be set of symmetric matrices, W' be the set of skew symmetric matrices. Prove that W and W' are subspaces of V and find the $dim\, W$ and $dim\, W'$. Moreover, prove that $V = W \oplus W'$.

15. Let $V = W_1 + W_2$ for some finite dimensional subspaces W_1 and W_2 of V. If $dim\, V = dim\, W_1 + dim\, W_2$, show that $V = W_1 \oplus W_2$.

16. Let $u \in \mathbb{R}$ be a transcendental number. Let W be the set of real numbers which are of the type $\alpha_0 + \alpha_1 u + \cdots + \alpha_k u^k$; $\alpha_i \in \mathbb{Q}$, $k \geq 0$. Prove that W is an infinite dimensional subspace of $\mathbb{R}$ over $\mathbb{Q}$.

17. Prove that the vector space $\mathscr{C}$ of all continuous functions from $\mathbb{R}$ to $\mathbb{R}$ is infinite dimensional.

18. If $W_1, W_2, \ldots, W_n$ are finite dimensional subspaces of a vector space V, then show that $W_1 + W_2 + \cdots + W_n$ is finite dimensional and $dim(W_1 + W_2 + \cdots + W_n) \leq dim\, W_1 + dim\, W_2 + \cdots + dim\, W_n$.

19. Suppose that $W_1, W_2, \ldots, W_n$ are finite dimensional subspaces of a vector space V such that $W_1 + W_2 + \cdots + W_n$ is a directsum. Then show that $W_1 \oplus W_2 \oplus \cdots \oplus W_n$ is finite dimensional and $dim(W_1 \oplus W_2 \oplus \cdots \oplus W_n) = dim\, W_1 + dim\, W_2 + \cdots + dim\, W_n$.

2.3   Geometrical Interpretations

Vector spaces and notions involved there can be interpreted geometrically in some cases. We have chosen the vector spaces $\mathbb{R}^2$ and $\mathbb{R}^3$ over the field $\mathbb{R}$ of real numbers for geometrical interpretation.

Geometrically $\mathbb{R}^2$ is the Euclidean plane or 2-dimensional cartesian plane. $\mathbb{R}^2$ is a vector space of dimension 2. $\{(0, 0)\}$ and $\mathbb{R}^2$ are trivial subspaces of $\mathbb{R}^2$ of dimensions 0 and 2, respectively. Let W be a nontrivial subspace of $\mathbb{R}^2$. Then dim $W = 1$. Let $\{v\}$ be a basis of W. It is obvious that $v \neq 0$, choose $v = (\alpha, \beta) \neq (0, 0) \in \mathbb{R}^2$. Therefore $W = L(\{v\}) = \{\lambda(\alpha, \beta) \mid \lambda \in \mathbb{R}\}$. If $(x, y) \in W$, then it implies that $x = \lambda\alpha$ and $y = \lambda\beta$, i.e., $\frac{x-0}{\alpha} = \frac{y-0}{\beta}$, which is obviously an equation of a straight line passing through origin and having slope $\frac{\beta}{\alpha}$. This shows that each nontrivial subspace of $\mathbb{R}^2$ represents a straight line passing through origin. Conversely, suppose a straight line passing through origin be given by $\frac{x-0}{\delta} = \frac{y-0}{\eta}$, where $\frac{\eta}{\delta}$ is the slope of the line. If W' denotes the set of all the points lying on this line, then $W' = \{(\mu\delta, \mu\eta) \mid \mu \in \mathbb{R}\}$ or $W' = \{\mu(\delta, \eta) \mid \mu \in \mathbb{R}\} \subset \mathbb{R}^2$. It can be easily verified that W' is a nontrivial subspace of $\mathbb{R}^2$. Thus all the straight lines passing through origin are precisely nontrivial subspaces of $\mathbb{R}^2$.

Let W be a nontrivial subspace of $\mathbb{R}^2$. Then clearly $W = \{\lambda(\alpha, \beta) \mid \lambda \in \mathbb{R}\}$, which is geometrically a straight line passing through origin and having slope $\frac{\beta}{\alpha}$. Let $\frac{\mathbb{R}^2}{W}$ denote the quotient space of $\mathbb{R}^2$ with regard to the subspace W and is given by $\frac{\mathbb{R}^2}{W} = \{v + W \mid v \in \mathbb{R}^2\}$. Consider a coset $(v + W) \in \frac{\mathbb{R}^2}{W}$, where $v = (a, b)$. Here $(v + W) = \{(a + \lambda\alpha, b + \lambda\beta) \mid \lambda \in \mathbb{R}\}$. If (x, y) be any arbitrary element of $(v + W)$, then we get $\frac{x-a}{\alpha} = \frac{y-b}{\beta}$. Geometrically the latter equation gives a straight line passing through (a, b) and parallel to the straight line represented by W. This shows that geometrically each coset $(v + W)$ is a straight line passing through point v and parallel to the line represented by the subspace W. Thus geometrically the quotient space $\frac{\mathbb{R}^2}{W}$ is the collection of all the straight lines, which are parallel to the straight line represented by the subspace W.

Consider $\mathbb{R}^3$, the vector space over field of real numbers. It is a 3-dimensional vector space. Geometrically $\mathbb{R}^3$ is the 3-dimensional Euclidean space. Its nontrivial subspaces will be either of dimension 1 or of dimension 2. Let W_1 be a subspace of $\mathbb{R}^3$ of dimension 1. Let $\{v\}$ be a basis of W_1. It is obvious that $v \neq 0$, choose $v = (\alpha, \beta, \gamma) \neq (0, 0, 0) \in \mathbb{R}^3$. Therefore $W_1 = L(\{v\}) = \{\lambda(\alpha, \beta, \gamma) \mid \lambda \in \mathbb{R}\}$. If $(x, y, z) \in W_1$, then it implies that $x = \lambda\alpha$, $y = \lambda\beta$ and $z = \lambda\gamma$, i.e., $\frac{x-0}{\alpha} = \frac{y-0}{\beta} = \frac{z-0}{\gamma}$, which is obviously an equation of a straight line passing through origin and having direction ratios α, β and γ. This shows that the nontrivial subspace W_1 of $\mathbb{R}^3$ represents a straight line passing through origin. On the other hand, let a straight line passing through origin be given by: $\frac{x-0}{\delta} = \frac{y-0}{\eta} = \frac{z-0}{\psi}$, where δ, η, ψ are direction ratios of the line. If W' denotes the set of all the points lying on this line, then $W' = \{(\mu\delta, \mu\eta, \mu\psi) \mid \mu \in \mathbb{R}\}$ or $W' = \{\mu(\delta, \eta, \psi) \mid \mu \in \mathbb{R}\} \subset \mathbb{R}^3$. It can be easily seen that W' is a nontrivial 1-dimensional subspace of $\mathbb{R}^3$. Thus all the straight lines passing through origin are precisely 1-dimensional subspaces of $\mathbb{R}^3$.

Let W_2 be a subspace of $\mathbb{R}^3$ of dimension 2. Let $\{v_1, v_2\}$ be a basis of W_2. Let $v_1 = (\alpha_1, \beta_1, \gamma_1)$, $v_2 = (\alpha_2, \beta_2, \gamma_2)$. Therefore $W_2 = L(\{v_1, v_2\}) = \{\lambda(\alpha_1, \beta_1, \gamma_1) + \mu(\alpha_2, \beta_2, \gamma_2) \mid \lambda, \mu \in \mathbb{R}\}$. If $(x, y, z) \in W_2$, then it implies that $x = \lambda\alpha_1 + \mu\alpha_2$, $y = \lambda\beta_1 + \mu\beta_2$ and $z = \lambda\gamma_1 + \lambda\gamma_2$. Eliminating λ, μ from the latter relations we get $(\beta_1\gamma_2 - \beta_2\gamma_1)x + (\gamma_1\alpha_2 - \gamma_2\alpha_1)y + (\alpha_1\beta_2 - \alpha_2\beta_1)z = 0$, which is an equation of a plane passing through origin, v_1 and v_2. This shows that geometrically W_2 represents a plane passing through origin. Conversely, suppose a plane passing through origin be given by: $px + qy + rz = 0$, where $(0, 0, 0) \neq (p, q, r) \in \mathbb{R}^3$. If W'' denotes the set of all the points lying on this plane, then $W'' = \{(\frac{q\lambda - r\mu}{p}, \lambda, \mu) \mid \lambda, \mu \in \mathbb{R}\}$ or $W'' = \{\lambda(\frac{q}{p}, 1, 0) + \mu(\frac{-r}{p}, 0, 1) \mid \lambda, \mu \in \mathbb{R}\} \subset \mathbb{R}^3$. It can be easily verified that W'' is a subspace of $\mathbb{R}^3$ of dimension 2. Thus all the planes passing through origin are precisely 2-dimensional subspaces of $\mathbb{R}^3$. Hence we conclude that lines passing through origin and planes passing through origin are precisely subspaces of $\mathbb{R}^3$.

Consider the quotient spaces $\frac{\mathbb{R}^3}{W'}, \frac{\mathbb{R}^3}{W''}$. Then it can be easily seen that geometrically $\frac{\mathbb{R}^3}{W'}$ represents all the straight lines, which are parallel to the line represented by the subspace W', whereas the quotient space $\frac{\mathbb{R}^3}{W''}$ represents all the planes, which are parallel to the plane represented by the subspace W''.

Exercises

1. Let $\mathbb{R}^2$ be the vector space over the field $\mathbb{R}$ of real numbers. Let S be any subset of $\mathbb{R}^2$ as given below. Then find the subspace generated by S, i.e., $\langle S \rangle$. Also find out the equation of the curve represented by this subspace.

   (a) $S = \{(3, 5)\}$.
   (b) $S = \{(2, -3), (4, -6)\}$.
   (c) $S = \{(-3, -8), (3, 8), (3\sqrt{5}, 8\sqrt{5})\}$.

2. Let $\mathbb{R}^3$ be the vector space over the field $\mathbb{R}$ of real numbers. Let S be any subset of $\mathbb{R}^3$ as given below. Then find the subspace generated by S, i.e., $\langle S \rangle$. Also find out the equations of the curves or surfaces represented by these subspaces.

   (a) $S = \{(-5, 11, 3)\}$.
   (b) $S = \{(5, -3, 17), (3, -8, -11)\}$.
   (c) $S = \{(3, 5, -3), (6, 10, -6), (7, -8, -6)\}$.

2.4 Change of Basis

Let V be a finite dimensional vector space over a field $\mathbb{F}$ and let $B = \{v_1, v_2, \ldots, v_n\}$ be a basis of V. Fix the order of elements in the basis B. If $v \in V$ then there exist unique scalars $\alpha_1, \alpha_2, \ldots, \alpha_n \in \mathbb{F}$ such that $v = \alpha_1 v_1 + \alpha_2 v_2 + \cdots + \alpha_n v_n$. Thus $v \in V$ determines a unique $n \times 1$ matrix $\begin{bmatrix} \alpha_1 \\ \alpha_2 \\ \vdots \\ \alpha_n \end{bmatrix}$. This $n \times 1$ matrix is known as the coordinate vector of v relative to the ordered basis B, denoted as $[v]_B$. Notice that basis B and the order of the elements in B play a very important role in determining the coordinate vector of any arbitrary element in V. Throughout this section we shall consider the ordered basis.

Definition 2.56 Let V be a vector space over a field $\mathbb{F}$ and let $B_1 = \{u_1, u_2, \ldots, u_n\}$ and $B_2 = \{v_1, v_2 \ldots, v_n\}$ be two ordered bases of V. Now if we consider B_2 as a basis of V, then each vector in V can be uniquely written as a linear combination of $v_1, v_2 \ldots, v_n$, i.e., $u_i = \sum_{j=1}^{n} \alpha_{ji} v_j$, for each i; $1 \le i \le n$, where $\alpha_{ji} \in \mathbb{F}$. Then the $n \times n$ matrix $P = (\alpha_{ji})$ is called the transition matrix of B_1 relative to the basis B_2. Similarly, if we consider B_1 as a basis of V, then we can write $v_j = \sum_{i=1}^{n} \beta_{ij} u_i$, for each j; $1 \le j \le n$, and the $n \times n$ matrix $Q = (\beta_{ij})$ is known as the transition matrix of B_2 relative to the basis B_1.

Remark 2.57 (i) Notice that in the above definition the coordinate vector of u_i relative to the basis B_2, i.e., $[u_i]_{B_2} = \begin{bmatrix} \alpha_{1i} \\ \alpha_{2i} \\ \vdots \\ \alpha_{ni} \end{bmatrix}$ is the ith column vector in the matrix P relative to the basis B_2.

(ii) In $\mathbb{F}^n$, the standard ordered basis is $\{e_1, e_2, \ldots, e_n\}$, where e_i is the n-tuple with ith component 1, and all the other components are zero. For example, in $\mathbb{R}^3$, $B_1 = \{(1, 0, 0), (0, 1, 0), (0, 0, 1)\}$ is the standard ordered basis, but the ordered basis $B_2 = \{(0, 1, 0), (1, 0, 0), (0, 0, 1)\}$ is not the standard ordered basis of $\mathbb{R}^3$.

Example 2.58 Let $B_1 = \{e_1, e_2, e_3\}$ and $B_2 = \{(1, 1, 0), (1, -1, 0), (0, 1, 1)\}$ be two ordered bases of vector space $\mathbb{R}^3$. Then

$$(1, 1, 0) = 1e_1 + 1e_2 + 0e_3$$
$$(1, -1, 0) = 1e_1 + (-1)e_2 + 0e_3$$
$$(0, 1, 1) = 0e_1 + 1e_2 + 1e_3.$$

Thus the transition matrix of B_2 relative to the basis B_1 is $P = \begin{bmatrix} 1 & 1 & 0 \\ 1 & -1 & 1 \\ 0 & 0 & 1 \end{bmatrix}$.

Similarly

$$e_1 = \tfrac{1}{2}(1, 1, 0) + \tfrac{1}{2}(1, -1, 0) + 0(0, 1, 1)$$
$$e_2 = \tfrac{1}{2}(1, 1, 0) + \tfrac{-1}{2}(1, -1, 0) + 0(0, 1, 1)$$
$$e_3 = \tfrac{-1}{2}(1, 1, 0) + \tfrac{1}{2}(1, -1, 0) + 1(0, 1, 1).$$

Hence the transition matrix of B_1 relative to the basis B_2 is $Q = \begin{bmatrix} \tfrac{1}{2} & \tfrac{1}{2} & \tfrac{-1}{2} \\ \tfrac{1}{2} & \tfrac{-1}{2} & \tfrac{1}{2} \\ 0 & 0 & 1 \end{bmatrix}$.

It is easy to see that $PQ = QP = I$ and hence P and Q are inverses of each other.

Theorem 2.59 *Let $B_1 = \{u_1, u_2, \ldots, u_n\}$ and $B_2 = \{v_1, v_2, \ldots, v_n\}$ be two ordered bases of a vector space V over a field $\mathbb{F}$. If P is the transition matrix of B_1 relative* to the basis B_2, then P is nonsingular and P^{-1} is the transition matrix of B_2 relative to the basis B_1.

Proof Let $P = (\alpha_{ij})$ and $Q = (\beta_{ij})$ be the transition matrices of B_1 relative to the basis B_2 and of B_2 relative to the basis B_1, respectively. Hence we find that $u_j = \sum_{i=1}^{n} \alpha_{ij} v_i$, for each j, $1 \le j \le n$, where $\alpha_{ij} \in \mathbb{F}$ and $v_i = \sum_{k=1}^{n} \beta_{ki} u_k$, for each i, $1 \le i \le n$. This shows that

$$
\begin{aligned}
u_j &= \sum_{i=1}^{n} \alpha_{ij} \left(\sum_{k=1}^{n} \beta_{ki} u_k \right) \\
&= \sum_{k=1}^{n} \left(\sum_{i=1}^{n} \beta_{ki} \alpha_{ij} \right) u_k \\
&= \sum_{k=1}^{n} \delta_{kj} u_k,
\end{aligned}
$$

where $\delta_{kj} = \sum_{i=1}^{n} \beta_{ki} \alpha_{ij}$ is the (k, j)-th entry of the product QP. But since B_1 is linearly independent, we find that $\delta_{kj} = 1$ for $k = j$ and $\delta_{kj} = 0$ for $k \ne j$. This yields that $QP = I$. In a similar manner, it can be shown that $PQ = I$. Hence P is nonsingular and $Q = P^{-1}$.

Theorem 2.60 Let $B = \{u_1, u_2, \ldots, u_n\}$ be an ordered basis of a nonzero finite dimensional vector space V over a field $\mathbb{F}$ and $P = (\alpha_{ij})$ be any invertible $n \times n$ matrix over $\mathbb{F}$. For each j, $1 \le j \le n$, if $v_j = \sum_{i=1}^{n} \alpha_{ij} u_i$, then $B' = \{v_1, v_2, \ldots, v_n\}$ is a basis of V.

Proof Since $dim V = n$, it is enough to prove that B' spans V. Let $W = L(B')$ and $P^{-1} = (\beta_{ij})$. Then for each t, $1 \le t \le n$ we find that

$$
\begin{aligned}
\sum_{j=1}^{n} \beta_{jt} v_j &= \sum_{j=1}^{n} \beta_{jt} \left(\sum_{i=1}^{n} \alpha_{ij} u_i \right) \\
&= \sum_{i=1}^{n} \left(\sum_{j=1}^{n} \alpha_{ij} \beta_{jt} \right) u_i \\
&= \sum_{i=1}^{n} \delta_{it} u_i, \quad \text{where } \delta_{it} = 1 \in \mathbb{F} \text{ for } i = t \text{ and } \delta_{it} = 0 \in \mathbb{F} \text{ for } i \ne t \\
&= u_t.
\end{aligned}
$$

This yields that $u_t \in W$ and hence $B \subseteq W$. Since W is a subspace of V, $V \subseteq W \subseteq V$ and V is spanned by B'.

Theorem 2.61 Let V be a vector space of dimension n over a field $\mathbb{F}$. If $P = (\alpha_{ij})$ is the transition matrix of the ordered basis $\{f_1, f_2, \ldots, f_n\}$ relative to the ordered basis $\{e_1, e_2, \ldots, e_n\}$, and $Q = (\beta_{ij})$ is the transition matrix of the ordered basis $\{g_1, g_2, \ldots, g_n\}$ relative to the ordered basis $\{f_1, f_2, \ldots, f_n\}$. Then the transition matrix of $\{g_1, g_2, \ldots, g_n\}$ relative to the basis $\{e_1, e_2, \ldots, e_n\}$ is PQ.

Proof We find that $f_k = \sum_{j=1}^{n} \alpha_{jk} e_j$, for each k, $1 \leq k \leq n$ and $g_i = \sum_{k=1}^{n} \beta_{ki} f_k$, for each i, $1 \leq i \leq n$. This yields that

$$
\begin{aligned}
g_i &= \sum_{k=1}^{n} \beta_{ki} \left(\sum_{j=1}^{n} \alpha_{jk} e_j \right) \\
&= \sum_{j=1}^{n} \left(\sum_{k=1}^{n} \alpha_{jk} \beta_{ki} \right) e_j \\
&= \sum_{j=1}^{n} \gamma_{ji} e_j,
\end{aligned}
$$

where $\gamma_{ji} = \sum_{k=1}^{n} \alpha_{jk} \beta_{ki}$. Hence the transition matrix of $\{g_1, g_2, \ldots, g_n\}$ relative to the ordered basis $\{e_1, e_2, \ldots, e_n\}$ is $(\gamma_{ij}) = (\alpha_{ij})(\beta_{ij}) = PQ$. This completes the proof.

Theorem 2.62 *Let $B_1 = \{u_1, u_2, \ldots, u_n\}$ and $B_2 = \{v_1, v_2 \ldots, v_n\}$ be two ordered bases of a vector space V over a field $\mathbb{F}$. Then for any $v \in V$, $[v]_{B_1} = Q[v]_{B_2}$ where Q is the transition matrix of B_2 relative to the basis B_1.*

Proof Let $Q = (\beta_{ij})$ be the transition matrix of B_2 relative to the basis B_1. Then $v_j = \sum_{k=1}^{n} \beta_{kj} u_k$, for each j, $1 \leq j \leq n$. Suppose that $[v]_{B_1} = \begin{bmatrix} a_1 \\ a_2 \\ \vdots \\ a_n \end{bmatrix}$ and $[v]_{B_2} = \begin{bmatrix} b_1 \\ b_2 \\ \vdots \\ b_n \end{bmatrix}$. Since $v \in V$, $v = \sum_{j=1}^{n} b_j v_j = \sum_{j=1}^{n} b_j \left(\sum_{k=1}^{n} \beta_{kj} u_k \right)$. This implies that

$v = \sum_{k=1}^{n} \left(\sum_{j=1}^{n} \beta_{kj} b_j \right) u_k$. Also since $[v]_{B_1} = \begin{bmatrix} a_1 \\ a_2 \\ \vdots \\ a_n \end{bmatrix}$, we find that $v = \sum_{k=1}^{n} a_k u_k$.

Now comparing the coefficients of u_k in the latter two relations for v, we arrive at $a_k = \sum_{j=1}^{n} \beta_{kj} b_j$. This shows that

$$
[v]_{B_1} = \begin{bmatrix} a_1 \\ a_2 \\ \vdots \\ a_n \end{bmatrix} = \begin{bmatrix} \sum_{j=1}^{n} \beta_{1j} b_j \\ \sum_{j=1}^{n} \beta_{2j} b_j \\ \vdots \\ \sum_{j=1}^{n} \beta_{nj} b_j \end{bmatrix} = \begin{bmatrix} \beta_{11} & \beta_{12} & \cdots & \beta_{1n} \\ \beta_{21} & \beta_{22} & \cdots & \beta_{2n} \\ \vdots & \vdots & \vdots & \vdots \\ \beta_{n1} & \beta_{n2} & \cdots & \beta_{nn} \end{bmatrix} \begin{bmatrix} b_1 \\ b_2 \\ \vdots \\ b_n \end{bmatrix}
$$

$$= Q[v]_{B_2}.$$

This completes the proof of the result.

Theorem 2.63 *Let* $B = \{u_1, u_2, \ldots, u_n\}$ *be an ordered basis of a vector space* V *over a field* $\mathbb{F}$. *Then the map* $\xi : V \to M_{n \times 1}(\mathbb{F})$ *such that* $\xi(v) = [v]_B$, $v \in V$ *satisfies the following:*

(i) $\xi(\alpha v_1 + \beta v_2) = \alpha \xi(v_1) + \beta \xi(v_2)$, *for any* $\alpha, \beta \in \mathbb{F}$ *and* $v_1, v_2 \in V$
(ii) ξ *is one-to-one and onto.*

Proof Let $[v_1]_B = \begin{bmatrix} a_1 \\ a_2 \\ \vdots \\ a_n \end{bmatrix}$ and $[v_2]_B = \begin{bmatrix} b_1 \\ b_2 \\ \vdots \\ b_n \end{bmatrix}$. Since B is an ordered basis of V,

we find that $\alpha v_1 = \sum_{i=1}^{n} \alpha a_i u_i$, $\beta v_2 = \sum_{i=1}^{n} \beta b_i u_i$. This yields that $\alpha v_1 + \beta v_2 = \sum_{i=1}^{n} (\alpha a_i + \beta b_i) u_i$. Hence we find that

$$[\alpha v_1 + \beta v_2]_B = \begin{bmatrix} \alpha a_1 + \beta b_1 \\ \alpha a_2 + \beta b_2 \\ \vdots \\ \alpha a_n + \beta b_n \end{bmatrix} = \begin{bmatrix} \alpha a_1 \\ \alpha a_2 \\ \vdots \\ \alpha a_n \end{bmatrix} + \begin{bmatrix} \beta b_1 \\ \beta b_2 \\ \vdots \\ \beta b_n \end{bmatrix} = \alpha \begin{bmatrix} a_1 \\ a_2 \\ \vdots \\ a_n \end{bmatrix} + \beta \begin{bmatrix} b_1 \\ b_2 \\ \vdots \\ b_n \end{bmatrix}.$$

This shows that $\xi(\alpha v_1 + \beta v_2) = \alpha \xi(v_1) + \beta \xi(v_2)$, for any $\alpha, \beta \in \mathbb{F}$ and $v_1, v_2 \in V$.

(ii) For any $v_1, v_2 \in V$, $\xi(v_1) = \xi(v_2)$ implies that $[v_1]_B = [v_2]_B$, which yields that $a_i = b_i$ for each i, $1 \leq i \leq n$. Hence $v_1 = v_2$ and the map ξ is one-to-one. Also

for any $c = \begin{bmatrix} c_1 \\ c_2 \\ \vdots \\ c_n \end{bmatrix} \in M_{n \times 1}(\mathbb{F})$ there exists $v = \sum_{i=1}^{n} c_i u_i \in V$ such that $\xi(v) = c$

and hence ξ is onto.

Example 2.64 Let $V = \mathbb{R}_2[x]$ be the vector space of all polynomials with real coefficients of degree less than or equal to 2. Consider the following three ordered bases of V, $B_1 = \{1, 1 + x, 1 + x + x^2\}$, $B_2 = \{x, 1, 1 + x^2\}$ and $B_3 = \{x, 1 + x^2, 1\}$.

(1) For $p(x) = 1 - 2x + 2x^2$, find $[p(x)]_{B_1}, [p(x)]_{B_2}, [p(x)]_{B_3}$.
(2) Find the matrix M_1 of B_3 relative to B_2.
(3) Find the matrix M_2 of B_2 relative to B_1.
(4) Find the matrix M of B_3 relative to B_1 and verify that $M = M_2 M_1$.

(1) Express $p(x)$ as a linear combination of ordered bases B_1, B_2 and B_3, given as below:
$$1 - 2x + 2x^2 = 3(1) + (-4)(1 + x) + 2(1 + x + x^2)$$
$$1 - 2x + 2x^2 = (-2)x + (-1)1 + 2(1 + x^2)$$

$1 - 2x + 2x^2 = (-2)x + 2(1 + x^2) + (-1)1$

Thus we get $[p(x)]_{B_1} = \begin{bmatrix} 3 \\ -4 \\ 2 \end{bmatrix}$, $[p(x)]_{B_2} = \begin{bmatrix} -2 \\ -1 \\ 2 \end{bmatrix}$, $[p(x)]_{B_3} = \begin{bmatrix} -2 \\ 2 \\ -1 \end{bmatrix}$.

(2) As we have $x = 1x + 01 + 0(1 + x^2)$,
$1 + x^2 = 0x + 01 + 1(1 + x^2)$
$1 = 0x + 1(1) + 0(1 + x^2)$.
This shows that matrix of B_3 relative to the ordered basis B_2 will be given by

$$M_1 = \begin{bmatrix} 1 & 0 & 0 \\ 0 & 0 & 1 \\ 0 & 1 & 0 \end{bmatrix}.$$

(3) Similarly, $x = (-1)(1) + 1(1 + x) + 0(1 + x + x^2)$,
$1 = 1(1) + 0(1 + x) + 0(1 + x + x^2)$,
$1 + x^2 = 1(1) + (-1)(1 + x) + 1(1 + x + x^2)$.
Thus the matrix of B_2 relative to the ordered basis B_1 will be given by

$$M_2 = \begin{bmatrix} -1 & 1 & 1 \\ 1 & 0 & -1 \\ 0 & 0 & 1 \end{bmatrix}.$$

(4) As above, we can write $x = -1(1) + 1(1 + x) + 0(1 + x + x^2)$,
$1 + x^2 = 1(1) + (-1)(1 + x) + 1(1 + x + x^2)$,
$1 = 1(1) + 0(1 + x) + 0(1 + x + x^2)$. Hence the matrix of B_3 relative to the ordered basis B_1 will be given by

$$M = \begin{bmatrix} -1 & 1 & 1 \\ 1 & -1 & 0 \\ 0 & 1 & 0 \end{bmatrix}.$$

It can be easily verified that $M = M_2 M_1$.

Exercises

1. In the vector space $\mathbb{R}^3$, find the transition matrix of the ordered basis $\{(1, cosx, sinx), (1, 0, 0), (1, -sinx, cosx)\}$ relative to the standard ordered basis $\{(1, 0, 0), (0, 1, 0), (0, 0, 1)\}$ of $\mathbb{R}^3$.

2. In the vector space $\mathbb{R}^3$, find the transition matrix of the ordered basis $\{(2, 1, 0), (0, 2, 1), (0, 1, 2)\}$ relative to the standard ordered basis $\{(1, 0, 0), (0, 1, 0), (0, 0, 1)\}$ of $\mathbb{R}^3$.

3. Let $B_1 = \{(1, 0), (0, 1)\}$ and $B_2 = \{(2, 3), (3, 2)\}$ be two ordered bases of $\mathbb{R}^2$. Then find the transition matrix P of B_2 relative to the basis B_1 and the transition matrix Q of B_1 relative to the basis B_2 and show that $PQ = QP = I_2$.

4. In the vector space $\mathbb{R}^3$, let $B_1 = \{(1, 0, 0), (0, 1, 0), (0, 0, 1)\}$ and $B_2 = \{(1, 1, 0), (1, -1, 0), (0, 1, 1)\}$ be two ordered bases. Then find the transition matrix P of B_2 relative to the basis B_1 and transition matrix Q of B_1 relative to the basis B_2 and show that $PQ = QP = I_3$.

5. Suppose the X and Y axes in the plane $\mathbb{R}^2$ are rotated counterclockwise 45° so that the new X and Y axes are along the line $y = x$ and the line $y = -x$, respectively.

   (a) Find the change of basis matrix.
   (b) Find the coordinates of the point $P(5, 6)$ under the given rotation.

6. Let $B = \{u_1, u_2, \ldots, u_n\}$ be an ordered basis of a nonzero finite dimensional vector space V over a field $\mathbb{F}$. For each $j : 1 \leq j \leq n$, define $v_j = \sum_{i=1}^{n} \alpha_{ij} u_j$, where $\alpha_{ij} \in \mathbb{F}$. If the ordered set $B' = \{v_1, v_2, \ldots, v_n\}$ is a basis of V, then prove that $P = (\alpha_{ij})$ is an invertible matrix over $\mathbb{F}$.

7. Let W be the subspace of $\mathbb{C}^3$ over $\mathbb{C}$ spanned by $\alpha_1 = (1, 0, i)$, $\alpha_2 = (i, 0, 1)$, where $\mathbb{C}$ is the field of complex numbers. Prove the following:

   (a) The set $B = \{\alpha_1, \alpha_2\}$ is a basis of W.
   (b) The set $B' = \{\beta_1, \beta_2\}$, where $\beta_1 = (1 + i, 0, 1 + i)$, $\beta_2 = (1 - i, 0, i - 1)$ is also a basis of W.
   (c) Find the matrix of the ordered basis $B' = \{\beta_1, \beta_2\}$ relative to the ordered basis $B = \{\alpha_1, \alpha_2\}$.

8. Find the total number of ordered bases of the vector space $V = \mathbb{F}^n$, where $\mathbb{F}$ is a finite field containing p elements.

9. Let $\{\alpha_1, \alpha_2, \ldots, \alpha_n\}$ be a basis of an n-dimensional vector space V. Show that $\{\lambda_1 \alpha_1, \lambda_2 \alpha_2, \ldots, \lambda_n \alpha_n\}$ is also a basis of V for any nonzero scalars $\lambda_1, \lambda_2, \ldots, \lambda_n$. If the coordinate of a vector v under the basis $\{\alpha_1, \alpha_2, \ldots, \alpha_n\}$ is $x = \{x_1, x_2, \ldots, x_n\}$, then find the coordinate of v under $\{\lambda_1 \alpha_1, \lambda_2 \alpha_2, \ldots, \lambda_n \alpha_n\}$. What are the coordinates of $w = \alpha_1 + \alpha_2 + \cdots + \alpha_n$ with respect to the bases $\{\alpha_1, \alpha_2, \ldots, \alpha_n\}$ and $\{\lambda_1 \alpha_1, \lambda_2 \alpha_2, \ldots, \lambda_n \alpha_n\}$?

10. Let $W = \left\{ \begin{bmatrix} a & b \\ b & c \end{bmatrix} \mid a, b, c \in \mathbb{R} \right\}$. Show that W is a subspace of $M_{2 \times 2}(\mathbb{R})$ over $\mathbb{R}$ and $\left\{ \begin{bmatrix} 1 & 0 \\ 0 & 0 \end{bmatrix}, \begin{bmatrix} 0 & 1 \\ 1 & 0 \end{bmatrix}, \begin{bmatrix} 0 & 0 \\ 0 & 1 \end{bmatrix} \right\}$ forms a basis of W. Find the coordinate of the matrix $\begin{bmatrix} 1 & -2 \\ -2 & 3 \end{bmatrix}$ under this basis.

11. Consider the vector space $\mathbb{R}^n$ over $\mathbb{R}$ with usual operations. Consider the bases $B_1 = \{e_1, e_2, \ldots, e_n\}$ where

$$e_1 = (1, 0, \ldots, 0), e_2 = (0, 1, \ldots, 0), \ldots, e_n = (0, 0, \ldots, 1)$$

and $B_2 = \{f_1, f_2, \ldots, f_n\}$ where

$$f_1 = (1, 0, \ldots, 0), \, f_2 = (1, 1, \ldots, 0), \ldots, f_n = (1, 1, \ldots, 1).$$

(a) Find the transition matrix of B_2 relative to the basis B_1.
(b) Find the transition matrix of B_1 relative to the basis B_2.

(c) If $v \in \mathbb{R}^n$ such that $[v]_{B_1} = \begin{bmatrix} 1 \\ 2 \\ \vdots \\ n \end{bmatrix}$, then find $[v]_{B_2}$.

Chapter 3
Linear Transformations

A map between any two algebraic structures (say groups, rings, fields, modules or algebra) of same kind is said to be an isomorphism if it is one-to-one, onto and homomorphism; roughly speaking, it preserves the operations in the underlying algebraic structures. If any two vector spaces over the same field are given, then one can study the relationship between two vector spaces. In this chapter, we will define the notion of a linear transformation between two vector spaces U and V which are defined over the same field and discuss the basic properties of linear transformations. Throughout, the vector spaces are considered over a field $\mathbb{F}$ unless otherwise stated.

Definition 3.1 Let U and V be vector spaces over the same field $\mathbb{F}$. A map $T : U \to V$ is said to be a linear transformation (a vector space homomorphism or a linear map) if it satisfies the following:

(1) $T(u_1 + u_2) = T(u_1) + T(u_2)$
(2) $T(\alpha u_1) = \alpha T(u_1)$

for all $u_1, u_2 \in U$ and $\alpha \in \mathbb{F}$.

Remark 3.2 (i) If $T : U \to V$ is a linear transformation, then it is obvious to see that $T(0) = 0$, where 0 on the left side denotes the zero vector in U while the 0 on the right side represents zero vector of V. Moreover, $T(-u) = -T(u)$, holds for all $u \in U$.

(ii) For any $u_1, u_2 \in U$, $u_1 - u_2 = u_1 + (-u_2)$ and hence if $T : U \to V$ is a linear transformation, then $T(u_1 - u_2) = T(u_1 + (-u_2)) = T(u_1) + T(-u_2) = T(u_1) - T(u_2)$.

(iii) Note that the vector spaces U and V are defined over the same field $\mathbb{F}$ to make the condition (2) in the above definition meaningful. A linear transformation $T : U \to V$ is said to be *epimorphism* if T is onto. If T is one-to-one it is said to be *monomorphism*. A linear transformation $T : U \to V$ which is both one-to-one and onto is called an *isomorphism*; if this being the case then the vector spaces U and V are said to be isomorphic and we write $V \cong W$.

(iv) In particular, a linear transformation $T : U \longrightarrow U$ is called a *linear operator* on U.

(v) It can be easily seen that a map $T : U \to V$ is a linear transformation if and only if $T(\alpha u_1 + \beta u_2) = \alpha T(u_1) + \beta T(u_2)$ for any $\alpha, \beta \in \mathbb{F}, u_1, u_2 \in U$.

(vi) If $T : U \to V$ is a linear transformation, then using the above remark and induction on n, it can be seen that $T(\sum_{i=1}^{n}(\alpha_i u_i)) = \sum_{i=1}^{n} \alpha_i T(u_i)$ for any $u_i \in U, \alpha_i \in \mathbb{F}, i = 1, 2, \ldots, n$.

Example 3.3 (1) Consider the vector space $V = \mathbb{R}[x]$ of polynomials over the field $\mathbb{R}$. Define a map $T : V \to V$ such that $T(f(x)) = f'(x)$, the usual derivative of polynomial $f(x)$. It can be easily seen that $T(f(x) + g(x)) = T(f(x)) + T(g(x))$ and $T(\alpha f(x)) = \alpha T(f(x))$, for any $f(x), g(x) \in V$ and $\alpha \in \mathbb{R}$. Hence T is a linear mapping which is onto but not one-to-one.

(2) Let $T : \mathbb{R}^2 \to \mathbb{R}^2$ such that $T(a, b) = (a + b, b)$ for all $a, b \in \mathbb{R}$. It can be easily seen that T is a linear transformation which is also an isomorphism.

(3) Let $T : \mathbb{R}^3 \to \mathbb{R}^3$ such that $T(a, b, c) = (0, b, c)$ for any $(a, b, c) \in \mathbb{R}^3$. One can easily verify that T is a linear transformation which is neither one-to-one nor onto.

(4) Consider $V = \mathscr{C}[a, b]$, the vector space of all continuous real-valued functions on the closed interval $[a, b]$. Define map $T : V \to \mathbb{R}$ such that $T(f(x)) = \int_a^b f(x)dx$. For any $\alpha, \beta \in \mathbb{R}$ and $f(x), g(x) \in \mathscr{C}[a, b]$, $T(\alpha f(x) + \beta g(x)) = \int_a^b (\alpha f(x) + \beta g(x))dx = \alpha \int_a^b f(x)dx + \beta \int_a^b f(x)dx = \alpha T(f(x)) + \beta T(g(x))$. This shows that T is a linear transformation which is onto but not one-to-one.

(5) Let $\mathbb{R}_2[x]$ denote the vector space of all polynomials of degree less than or equal to two, over the field $\mathbb{R}$. Then there exits a natural map $T : \mathbb{R}_2[x] \to \mathbb{R}^3$ defined by $T(\alpha_0 + \alpha_1 x + \alpha_2 x^2) = (\alpha_0, \alpha_1, \alpha_2)$, where $\alpha_0, \alpha_1, \alpha_2 \in \mathbb{R}$. This is a linear transformation which is one-to-one and onto.

(6) The map $T : \mathbb{R}^2 \to \mathbb{R}^2$ defined by $T(a, b) = (a + 1, b)$ (or, $T(a, b) = (|a|, |b|)$) is not a linear transformation.

(7) The map $T : \mathbb{R}^4 \to \mathbb{R}^2$ defined by $T(a_1, a_2, a_3, a_4) = (a_1 + a_2, a_3 + a_4)$ is a linear transformation. This is onto but not one-to-one.

(8) The map $T : M_{m \times n}(\mathbb{R}) \to M_{n \times m}(\mathbb{R})$ defined by $T(A) = A^t$, the transpose of the matrix A, is a linear transformation. It can be easily shown that this is an isomorphism.

(9) Let $T : \mathbb{R}^n \to \mathbb{R}^{n+1}$ such that $T(a_1, a_2, \ldots, a_n) = (a_1, a_2, \ldots, a_n, 0)$. Then T is a linear transformation called natural inclusion. It is an injective linear transformation but not surjective.

(10) Let $\mathbb{C}$ be the vector space over the field of real numbers $\mathbb{R}$. Let $T : \mathbb{C} \to \mathbb{C}$ be a map defined by $T(z) = \bar{z}$, where $\bar{z}$ is the conjugate of complex num-

ber z. Obviously T is a linear transformation. It can be verified that T is an isomorphism.

(11) Let $V = M_n(\mathbb{F})$ denote the vector space of all $n \times n$ matrices with entries from $\mathbb{F}$. Define a map $T : V \to \mathbb{F}$ such that $T(A) = tr(A)$, where $tr(A)$ denotes the trace of the matrix A. It can be easily proved that T is a linear transformation which is onto but not one-to-one in general. However, this homomorphism T will be an isomorphism if and only if $n = 1$.

(12) Consider the vector space $V = \mathbb{F}[x]$ of all polynomials in x with coefficients from the field $\mathbb{F}$, where characteristic of $\mathbb{F}$ is 0. Define a map $\int : V \to V$ such that $\int(f(x)) = \alpha_0 x + \alpha_1 \frac{x^2}{2} + \cdots + \alpha_n \frac{x^{n+1}}{n+1}$ if $f(x) = \alpha_0 + \alpha_1 x + \cdots + \alpha_n x^n$. It can be easily seen that $\int(f(x) + g(x)) = \int(f(x)) + \int(g(x))$ and $\int(\alpha f(x)) = \alpha \int(f(x))$, for any $f(x), g(x) \in V$ and $\alpha \in \mathbb{F}$. Hence $\int$ is a linear mapping which is one-to-one but not onto. This linear transformation $\int$ is called integration transformation.

(13) Let $\alpha_{ij} \in \mathbb{F}$ for each i, j such that $1 \leq i \leq m, 1 \leq j \leq n$. Define a map $T : \mathbb{F}^m \to \mathbb{F}^n$ such that

$$T(a_1, a_2, \ldots, a_m) = \left(\sum_{i=1}^{m} \alpha_{i1} a_i, \sum_{i=1}^{m} \alpha_{i2} a_i, \ldots, \sum_{i=1}^{m} \alpha_{in} a_i \right).$$

It can be easily verified that T is a linear transformation.

(14) Let $A = (\alpha_{ij})_{n \times m}$ be an $n \times m$ matrix over $\mathbb{F}$. Define a map $T : \mathbb{F}^m \to \mathbb{F}^n$ such that if $X = (x_1, x_2, \ldots, x_m) \in \mathbb{F}^m$, then $T(X) = AX = (\alpha_{ij})_{n \times m}(x_j)_{m \times 1} = (\beta_i)_{n \times 1}$, where $\beta_i = \sum_{j=1}^{m} \alpha_{ij} x_j$. Here X is being treated as a column matrix. Thus $T(X) = AX$ is an $n \times 1$ matrix which may be considered as an n-tuple belonging to $\mathbb{F}^n$. It can be proved that T is a linear transformation.

Some Standard Linear Transformations

(1) Let $T : U \to V$ be the mapping which assigns the zero vector of V to every vector $u \in U$, i.e., $T(u) = 0$ for all $u \in U$. Then it can be verified that T is a linear transformation, which is known as the *zero linear transformation*, usually denoted by $\mathbf{0}$.

(2) The identity mapping $I : U \to U$ such that $I(u) = u$ for all $u \in U$ is a linear transformation and is known as the *identity linear transformation*, denoted as I_U. It is an isomorphism also.

(3) Let V be any vector space and W be a subspace of V. The inclusion mapping $i : W \to V$ defined as $i(w) = w$ for all $w \in W$ is a linear transformation. This is known as *inclusion linear transformation*, which is injective also. It is an isomorphism if and only if $W = V$.

(4) Let V be a vector space and W a subspace of V. Let $T : V \to V/W$ be the map defined by $T(v) = v + W$ for every $v \in V$. It is easy to see that T is linear transformation which is known as the *quotient linear transformation*. It is a

surjective linear transformation which is not an injective linear transformation in general, but it is an isomorphism if and only if $W = \{0\}$.

(5) The map $T_i : \mathbb{F}^n \to \mathbb{F}$ given by $T_i(\alpha_1, \alpha_2, \ldots, \alpha_i, \ldots, \alpha_n) = \alpha_i$ is a linear transformation for every i; $1 \leq i \leq n$. This is known as ith *projection* of $\mathbb{F}^n$ to $\mathbb{F}$, which is surjective linear transformation but not injective linear transformation in general. This is an isomorphism if and only if $n = 1$.

Remark 3.4 (i) If $T : U \to V$ is a linear transformation, it can be easily seen that T may not map a linearly independent subset of U into a linearly independent subset of V. For example, consider the linear transformation $T : \mathbb{R}^2 \to \mathbb{R}^2$ such that $T(a_1, a_2) = (0, a_2)$. The set $\{(1, 0), (0, 1)\}$ is linearly independent in $\mathbb{R}^2$ while $\{T(1, 0), T(0, 1)\}$ is linearly dependent.

(ii) If $T : U \to V$ is a linear transformation, then T maps a linearly dependent subset of U into a linearly dependent subset of V. In fact, if $\{u_1, u_2, \ldots, u_n\}$ is linearly dependent in U, then there exist scalars $\alpha_1, \alpha_2, \ldots, \alpha_n$ (not all zero) such that $\alpha_1 u_1 + \alpha_2 u_2 + \cdots + \alpha_n u_n = 0$. This yields that

$$\alpha_1 T(u_1) + \alpha_2 T(u_2) + \cdots + \alpha_n T(u_n) = T(\alpha_1 u_1 + \alpha_2 u_2 + \cdots + \alpha_n u_n) = 0,$$

which shows that $\{T(u_1), T(u_2), \ldots, T(u_n)\}$ is linearly dependent in V.

(iii) In view of Definition 2.16, notice that there should be no confusion in internal direct sum and external direct sum. In fact, it can be easily seen that if V is an internal direct sum of $V_1, V_2, \ldots, V_n$, then V is isomorphic to the external direct sum of $V_1, V_2, \ldots, V_n$. Each $v \in V$ can be uniquely written as $v = v_1 + v_2 + \cdots + v_n$ where $v_i \in V_i$. Define a map $f : V \to V_1 \oplus V_2 \oplus \cdots \oplus V_n$ such that $f(v) = (v_1, v_2, \ldots, v_n)$. The map f is well defined because of the uniqueness of the representation of v. One can easily verify that f is one-to-one, onto and homomorphism.

The following example shows that there exists no nonzero linear transformation $T : \mathbb{R}^2 \longrightarrow \mathbb{R}^2$, which maps the straight line $ax + by + c = 0$, where $c \neq 0$ to $(0, 0) \in \mathbb{R}^2$.

Example 3.5 It is clear that out of a and b at least one must be nonzero, let it be a, i.e., $a \neq 0$. The straight line can also be represented as $L = \{(\frac{-bt-c}{a}, t) \mid t \in \mathbb{R}\} \subseteq \mathbb{R}^2$. But $T(l) = (0, 0)$ for all $l \in L$. This implies that $T(\frac{-bt-c}{a}, t) = (0, 0)$ for all $t \in \mathbb{R}$. In particular $T(\frac{-c}{a}, 0) = (0, 0)$, i.e., $\frac{(-c)}{a} T(1, 0) = (0, 0)$. Thus we arrive at $T(1, 0) = (0, 0)$. $T(\frac{-bt-c}{a}, t) = (0, 0)$ for all $t \in \mathbb{R}$ also gives us $T(\frac{-bt}{a}, t) + T(\frac{-c}{a}, 0) = (0, 0)$, i.e., $T(\frac{-bt}{a}, t) = (0, 0)$. In particular, putting $t = 1$, we get $T(\frac{-b}{a}, 1) = (0, 0)$, i.e., $\frac{(-b)}{a} T(1, 0) + T(0, 1) = (0, 0)$. This implies that $T(0, 1) = 0$. Finally we have $T(x, y) = T(x, 0) + T(0, y) = xT(1, 0) + yT(0, 1) = (0, 0)$ for all $(x, y) \in \mathbb{R}^2$, i.e., $T = 0$.

Exercises

1. Let V be the vector space of all continuous functions $f : \mathbb{R} \longrightarrow \mathbb{R}$ over the field of reals and define a mapping $\phi : V \longrightarrow V$ by $[\phi(f)](x) = \int_0^x f(t)dt$. Prove that ϕ is a linear transformation.

2. Let V be the vector space over $\mathbb{R}$ of polynomials with real coefficients. Define ψ :
   $V \longrightarrow V$ by $\psi\left(\sum\limits_{i=0}^k a_i x^i\right) = \sum\limits_{i=1}^k i a_i x^{i-1}$. Prove that ψ is a linear transformation.

3. Let $V_n = \{p(x) \in F[x] \mid degp(x) < n\}$, where n is any positive integer. Define $T : V_n \longrightarrow V_n$ by $T(p(x)) = p(x + 1)$. Show that T is an automorphism of V_n.

4. Let $U = \mathbb{F}^{n+1}$ and $V = \{p(x) \in F[x] \mid degp(x) \leq n\}$, where n is any positive integer, be the vector spaces over $\mathbb{F}$. Define $T : U \longrightarrow V$ by $T(\alpha_0, \alpha_1, \ldots, \alpha_n) = \alpha_0 + \alpha_1 x + \cdots + \alpha_n x^n$. Then prove that T is an isomorphism from U to V.

5. Let $V = \mathbb{R}^2$ be the 2-dimensional Euclidean space. Show that rotation through an angle θ^0 is a linear transformation on V.

6. Let V be the vector space of all twice differentiable functions in $[0, 1]$. Show that the mappings $T_1 : V \longrightarrow V$ and $T_2 : V \longrightarrow V$ defined by $T_1(f) = \frac{df}{dx}$ and $T_2(f) = xf$ are linear transformations.

7. $T : V \longrightarrow V$ be a linear transformation which is not onto. Show that there exists some $v \in V$, $v \neq 0$, such that $T(v) = 0$.

8. Show that the following mappings are linear transformations:

   (a) $T : \mathbb{R}^2 \longrightarrow \mathbb{R}^2$ defined by $T(x, y) = (2x - 3y, y)$.
   (b) $T : \mathbb{R}^3 \longrightarrow \mathbb{R}^2$ defined by $T(x, y, z) = (x + 2y + z, 3x - 4y - 2z)$.
   (c) $T : \mathbb{R}^4 \longrightarrow \mathbb{R}^3$ defined by $T(x, y, z, t) = (x - y + 2z + t, x - y - 2z + 3t, x + 3y + z - 2t)$.

9. Show that the following mappings are not linear mappings:

   (a) $T : \mathbb{R}^2 \longrightarrow \mathbb{R}^2$ defined by $T(x, y) = (xy, y^2)$.
   (b) $T : \mathbb{R}^2 \longrightarrow \mathbb{R}^3$ defined by $T(x, y) = (x + 3, 3y, 2x - y)$.
   (c) $T : \mathbb{R}^3 \longrightarrow \mathbb{R}^3$ defined by $T(x, y, z) = (|x + y|, 2x - 3y + z, x + y + 2)$.

10. Let V be the vector space of n-square real matrices. Let M be an arbitrary but fixed matrix in V. Let $T_1, T_2 : V \longrightarrow V$ be defined by $T_1(A) = AM + MA$, $T_2(A) = AM - MA$, where A is any matrix in V. Show that T_1 and T_2 both are linear transformations on V.

11. Prove that a mapping $T : U \longrightarrow V$ is a linear transformation if and only if $T(x + \alpha y) = T(x) + \alpha T(y)$ for all $x, y \in U$ and $\alpha \in \mathbb{F}$.

12. Prove that any linear functional $f : \mathbb{R}^n \to \mathbb{R}$ is a continuous function.

13. Let $f : \mathbb{R}^n \to \mathbb{R}$ be a continuous function which is also additive. Prove that f is a linear functional on $\mathbb{R}^n$.

3.1 Kernel and Range of a Linear Transformation

Definition 3.6 Let $T : U \rightarrow V$ be a linear transformation. The kernel or null space of T, denoted as $N(T)$ or $KerT$, is the set of all those elements of U which are mapped to the zero of V, that is,

$$N(T) = \{u \in U \mid T(u) = 0\}.$$

The range of T or the image of T or the rank space of T, denoted as $R(T)$, or $T(U)$ or Im (T) is the set of images in V, that is

$$R(T) = \{T(u) \mid u \in U\}.$$

Remark 3.7 (i) If $T : U \rightarrow V$ is a linear transformation, then $T(0) = 0$ and hence $0 \in N(T)$ and $N(T) \neq \emptyset$. Now if $u_1, u_2 \in N(T)$ and $\alpha, \beta \in \mathbb{F}$ then $T(\alpha u_1 + \beta u_2) = \alpha T(u_1) + \beta T(u_2) = \alpha 0 + \beta 0 = 0$ and hence $\alpha u_1 + \beta u_2 \in N(T)$. This shows that $N(T)$ is a subspace of U. This is also an easy exercise that $R(T)$ is a subspace of V.

(ii) If $T : U \rightarrow V$ is a linear transformation and the set $\{u_1, u_2, \ldots, u_n\}$ spans U, then the set $\{T(u_1), T(u_2), \ldots, T(u_n)\}$ spans $R(T)$. Let $v \in R(T)$. Then there exists $u \in U$ such that $T(u) = v$. In fact, if $\{u_1, u_2, \ldots, u_n\}$ spans U, then for any $u \in U$ there exist scalars $\alpha_1, \alpha_2, \ldots, \alpha_n \in \mathbb{F}$ such that $u = \alpha_1 u_1 + \alpha_2 u_2 + \cdots + \alpha_n u_n$. Hence, we find that $v = T(u) = T(\alpha_1 u_1 + \alpha_2 u_2 + \cdots + \alpha_n u_n) = \alpha_1 T(u_1) + \alpha_2 T(u_2) + \cdots + \alpha_n T(u_n)$. Therefore $R(T)$ is spanned by the set $\{T(u_1), T(u_2), \ldots, T(u_n)\}$.

(iii) If $T : U \rightarrow V$ is a linear transformation, then T is one-to-one if and only if $N(T) = \{0\}$. In fact, if T is one-to-one and $u \in N(T)$, then $T(u) = 0 = T(0)$ implies that $u = 0$ and hence $N(T) = \{0\}$. Conversely, assume that $N(T) = \{0\}$ and for some $u_1, u_2 \in U$, $T(u_1) = T(u_2)$. Then $T(u_1 - u_2) = T(u_1) - T(u_2) = 0$ and hence $u_1 - u_2 \in N(T) = \{0\}$ and therefore $u_1 = u_2$. This shows that T is one-to-one.

Definition 3.8 Let $T : U \rightarrow V$ be a linear transformation. If $N(T)$ and $R(T)$ are finite dimensional, then dimension of $N(T)$ is called nullity of T, denoted as $n(T)$ while the dimension of $R(T)$ (or the range of T) is called the rank of T denoted as $r(T)$.

Theorem 3.9 (Rank-Nullity Theorem or Sylvester Law) *Let U, V be vector spaces over a field $\mathbb{F}$ and $T : U \rightarrow V$ be a linear transformation. If U is finite dimensional then $n(T) + r(T) = dimU$.*

Proof Let $dimU = n$ and $dimN(T) = k$. If $\{u_1, u_2, \ldots, u_k\}$ is a basis of $N(T)$, then being a linearly independent subset of U, it can be extended to a basis of U, say $\{u_1, u_2, \ldots, u_k, u_{k+1}, \ldots, u_n\}$. Now our claim is that the set

$$B = \{T(u_{k+1}), T(u_{k+2}), \ldots, T(u_n)\}$$

is a basis of $R(T)$. First we show that B spans $R(T)$. Since $\{u_1, u_2, \ldots, u_k, u_{k+1}, \ldots, u_n\}$ spans U and $T(u_i) = 0$ for each i, $1 \leq i \leq k$, by Remark 3.7(ii)

$$R(T) = span\{T(u_1), T(u_2), \ldots, T(u_k), T(u_{k+1}), \ldots, T(u_n)\}$$
$$= \{T(u_{k+1}), \ldots, T(u_n)\}$$
$$= span\, B.$$

In order to show that B is linearly independent, suppose there exist scalars $\alpha_{k+1}, \alpha_{k+2}, \ldots, \alpha_n \in \mathbb{F}$ such that $\sum\limits_{i=k+1}^{n} \alpha_i T(u_i) = 0$. This implies that $T(\sum\limits_{i=k+1}^{n} \alpha_i u_i) = 0$, and hence $\sum\limits_{i=k+1}^{n} \alpha_i u_i \in N(T)$. But since $\{u_1, u_2, \ldots, u_k\}$ spans $N(T)$, there exist $\beta_1, \beta_2, \ldots, \beta_k \in \mathbb{F}$ such that $\sum\limits_{i=k+1}^{n} \alpha_i u_i = \sum\limits_{i=1}^{k} \beta_i u_i$, which yields that $\sum\limits_{i=k+1}^{n} \alpha_i u_i + \sum\limits_{i=1}^{k} (-\beta_i) u_i = 0$. But since $\{u_1, u_2, \ldots, u_k, u_{k+1}, \ldots, u_n\}$ is a basis of U, we find that each of $\alpha_i = 0$ and hence the set B is linearly independent and forms a basis of $R(T)$. This also shows that elements in B are distinct and $r(T) = n - k$, that is $dim\, U = r(T) + n(T)$.

Theorem 3.10 *Let U, V be vector spaces over a field $\mathbb{F}$ and $T : U \to V$ be a linear transformation. If U' is a finite dimensional subspace of U such that $U' \cap N(T) = \{0\}$, then the following hold:*

(i) *If $\{u_1, u_2, \ldots, u_n\}$ is a basis of U', then $\{T(u_1), T(u_2), \ldots, T(u_n)\}$ is a basis of $T(U')$,*

(ii) $dim\, U' = dim\, T(U')$,

(iii) *If T is one-to-one and U is finite dimensional, then $r(T) = dim\, U$.*

Proof (i) By Remark 3.7 (ii), if $\{u_1, u_2, \ldots, u_n\}$ spans U', then $\{T(u_1), T(u_2), \ldots, T(u_n)\}$ spans $T(U')$. In order to show that $\{T(u_1), T(u_2), \ldots, T(u_n)\}$ is a basis of $T(U')$, suppose there exist scalars $\alpha_1, \alpha_2, \ldots, \alpha_n \in \mathbb{F}$ such that $\sum\limits_{i=1}^{n} \alpha_i T(u_i) = 0$. This implies that $T(\sum\limits_{i=1}^{n} \alpha_i u_i) = 0$. But since $U' \cap N(T) = \{0\}$, we arrive at $\sum\limits_{i=1}^{n} \alpha_i u_i \in U' \cap N(T) = \{0\}$, and hence $\sum\limits_{i=1}^{n} \alpha_i u_i = 0$. Given that the set $\{u_1, u_2, \ldots, u_n\}$ being a basis of U' is linearly independent, we find that $\alpha_i = 0$ for each $i = 1, 2, \ldots, n$. This shows that $\{T(u_1), T(u_2), \ldots, T(u_n)\}$ is a basis of $T(U')$ if $\{u_1, u_2, \ldots, u_n\}$ is a basis of U'.

(ii) In view of the above we find that $dim\, T(U') = dim\, U'$.

(iii) If T is one-to-one, then $N(T) = \{0\}$. But if $N(T) = \{0\}$, then $U \cap N(T) = \{0\}$, and by (ii) we find that $dim\, U = dim\, T(U)$. This implies that $r(T) = dim\, U$.

Theorem 3.11 *Let U and V be any two finite dimensional vector spaces over the same field $\mathbb{F}$. If $f : U \to V$ is an isomorphism and $B = \{u_1, u_2, \ldots, u_n\}$ is a basis of U, then $B' = \{f(u_1), f(u_2), \ldots, f(u_n)\}$ is a basis of V.*

Proof We shall show that B' is linearly independent. Suppose there exist scalars $\alpha_1, \alpha_2, \ldots, \alpha_n$ such that $\sum_{i=1}^{n} \alpha_i f(u_i) = 0$. This implies that $\sum_{i=1}^{n} f(\alpha_i u_i) = 0$, i.e., $f(\sum_{i=1}^{n} (\alpha_i u_i)) = 0$. But since f is one-to-one map, we arrive at $\sum_{i=1}^{n} \alpha_i u_i = 0$. Since B is a basis of U, we find that $\alpha_i = 0$ for each $i = 1, 2, \ldots, n$. This shows that B' is linearly independent.

Now we show that B' spans V, let $v \in V$. Since f is onto, for $v \in V$ there exists $u \in U$ such that $f(u) = v$. Every vector $u \in U$ can be written as $u = \sum_{i=1}^{n} \beta_i u_i$, where $\beta_i \in \mathbb{F}$. This shows that $f(u) = \sum_{i=1}^{n} f(\beta_i u_i) = \sum_{i=1}^{n} \beta_i f(u_i)$, i.e., $v = \sum_{i=1}^{n} \beta_i f(u_i)$. Therefore, B' spans V, and hence a basis of V.

Remark 3.12 The above theorem holds even if U and V are infinite dimensional vector spaces over the same field $\mathbb{F}$. Accordingly, theorem can be stated as: Let U and V be any two infinite dimensional vector spaces over the same field $\mathbb{F}$. If $f : U \to V$ is an isomorphism and $B = \{u_1, u_2, \ldots, u_n, \ldots\}$ is a basis of U, then $B' = \{f(u_1), f(u_2), \ldots, f(u_n), \ldots\}$ is a basis of V. Proof of this fact follows the same pattern as above.

Theorem 3.13 *Let U, V be vector spaces over a field $\mathbb{F}$ and $\{u_1, u_2, \ldots, u_n\}$ be a basis of U. Let $\{v_1, v_2, \ldots, v_n\}$ be any set of vectors (not necessarily distinct) in V. Then there exists a unique linear transformation $T : U \to V$ such that $T(u_i) = v_i$, for each $i, 1 \leq i \leq n$.*

Proof Let u be an arbitrary element in U. Then since $\{u_1, u_2, \ldots, u_n\}$ be a basis of U, there exist $\alpha_1, \alpha_2, \ldots, \alpha_n \in \mathbb{F}$ such that $\sum_{i=1}^{n} \alpha_i u_i = u$. Define a map $T : U \to V$ such that $T(u) = \sum_{i=1}^{n} \alpha_i v_i$.

First we show that T is linear. Suppose that $x, y \in U$ and $\alpha, \beta \in \mathbb{F}$. Then $x = \sum_{i=1}^{n} \alpha_i u_i$ and $y = \sum_{i=1}^{n} \beta_i u_i$ for some $\alpha_i, \beta_i \in \mathbb{F}$ and

$$\alpha x + \beta y = \sum_{i=1}^{n} (\alpha \alpha_i) u_i + \sum_{i=1}^{n} (\beta \beta_i) u_i = \sum_{i=1}^{n} (\alpha \alpha_i + \beta \beta_i) u_i.$$

This implies that

$$T(\alpha x + \beta y) = \sum_{i=1}^{n}(\alpha \alpha_i + \beta \beta_i)v_i = \alpha \left(\sum_{i=1}^{n} \alpha_i v_i \right) + \beta \left(\sum_{i=1}^{n} \beta_i v_i \right) = \alpha T(x) + \beta T(y).$$

Clearly $T(u_i) = v_i$, for each i, $1 \le i \le n$. To prove the uniqueness, suppose there exists a linear transformation $T' : U \to V$ such that $T'(u_i) = v_i$, for each i, $1 \le i \le n$. Then, any $x \in U$ can be written as $x = \sum_{i=1}^{n} \alpha_i u_i$ and hence

$$T'(x) = \sum_{i=1}^{n} \alpha_i T'(u_i) = \sum_{i=1}^{n} \alpha_i v_i = T(x)$$

and therefore $T = T'$.

The following corollary shows that if any two linear transformations agree on the basis of a vector space, then they will be the same.

Corollary 3.14 *Let U, V be vector spaces over a field $\mathbb{F}$ and $\{u_1, u_2, \ldots, u_n\}$ be a basis of U. If $T, T' : U \to V$ are linear transformations and $T(u_i) = T'(u_i)$ for each i, $1 \le i \le n$, then $T = T'$.*

Remark 3.15 (i) Theorem 3.13 can be restated as: Let U, V be vector spaces over a field $\mathbb{F}$ and $\{u_1, u_2, \ldots, u_n\}$ be a basis of U. A map $f : \{u_1, u_2, \ldots, u_n\} \longrightarrow V$ can be uniquely extended to a linear map $T : U \longrightarrow V$, such that $T(u_i) = f(u_i)$ for each i, $1 \le i \le n$.

(ii) Any map f from a basis of U to V will determine a unique linear map $T : U \longrightarrow V$, which is called extension of f by linearity.

(iii) Thus maps from different bases of U to V or different maps from the same basis of U to V will give different linear maps from $U \longrightarrow V$. Thus Theorem 3.13 gives us a method for determining linear maps from a finite dimensional vector space U to a vector space V.

Theorem 3.16 *Two finite dimensional vector spaces U and V over a field $\mathbb{F}$ are isomorphic if and only if they are of the same dimension.*

Proof Let $U \cong V$. This implies that there exists a bijective linear map $T : U \longrightarrow V$. Suppose that $dim U = n$. We have to prove that $dim V = n$. Let $\{u_1, u_2, \ldots, u_n\}$ be a basis of U. We claim that the set $B = \{T(u_1), T(u_2), \ldots, T(u_n)\}$ is a basis of V. First we prove that B spans V. For this let $v \in V$. Since T is onto, there exits $u \in U$, such that $v = T(u)$. As $u \in U$, this shows that there exist scalars $\alpha_1, \alpha_2, \ldots, \alpha_n \in \mathbb{F}$ such that $u = \alpha_1 u_1 + \alpha_2 u_2 + \cdots + \alpha_n u_n$. As a result, we get $v = T(\sum_{i=1}^{n} \alpha_i u_i) = \sum_{i=1}^{n} \alpha_i T(u_i)$ and hence B spans V. Next to prove that B is a linearly independent set, let $\sum_{i=1}^{n} \beta_i T(u_i) = 0$ for some scalars $\beta_1, \beta_2, \ldots, \beta_n \in \mathbb{F}$. This gives us $T(\sum_{i=1}^{n} \beta_i u_i) =$

0, i.e., $\sum_{i=1}^{n} \beta_i u_i \in Ker T$. Since T is injective, we conclude that $\sum_{i=1}^{n} \beta_i u_i = 0$. This implies that $\beta_i = 0$ for all i, $1 \leq i \leq n$. It proves that B is a linearly independent set and thus B will also contain n elements. Thus B is a basis of V and dim $V = n$.

Conversely, suppose that $dim U = dim V = n$. Then we have to show that $U \cong V$. Let $\{u_1, u_2, \ldots, u_n\}$ and $\{v_1, v_2, \ldots, v_n\}$ be bases of U and V, respectively. Define a map $f : \{u_1, u_2, \ldots, u_n\} \longrightarrow V$ such that $f(u_i) = v_i$ for each i, $1 \leq i \leq n$. By Remark 3.7 (i), this map f can be uniquely extended to a linear map $T : U \longrightarrow V$ such that $T(u_i) = f(u_i) = v_i$ for each i, $1 \leq i \leq n$. We show that the linear map T is bijective. Let $x, y \in U$. There exist scalars $\alpha_1, \alpha_2, \ldots, \alpha_n, \beta_1, \beta_2, \ldots, \beta_n \in \mathbb{F}$ such that $x = \sum_{i=1}^{n} \alpha_i u_i$ and $y = \sum_{i=1}^{n} \beta_i u_i$. Then $T(x) = T(y)$ implies that $T(\sum_{i=1}^{n} \alpha_i u_i) = T(\sum_{i=1}^{n} \beta_i u_i)$. Now we obtain that $\sum_{i=1}^{n} \alpha_i T(u_i) = \sum_{i=1}^{n} \beta_i T(u_i)$, i.e., $\sum_{i=1}^{n} \alpha_i v_i = \sum_{i=1}^{n} \beta_i v_i$. This shows that $\sum_{i=1}^{n} (\alpha_i - \beta_i) v_i = 0$. Since $\{v_1, v_2, \ldots, v_n\}$ is a basis of V, we conclude that $\alpha_i = \beta_i$ for all i, $1 \leq i \leq n$. This implies that $\sum_{i=1}^{n} \alpha_i u_i = \sum_{i=1}^{n} \beta_i u_i$, i.e., $x = y$. These arguments prove that T is one-to-one. To prove the ontoness of T, let $v \in V$, thus there exist scalars $\gamma_1, \gamma_2, \ldots, \gamma_n \in \mathbb{F}$ such that $v = \sum_{i=1}^{n} \gamma_i v_i$. This shows that $v = \sum_{i=1}^{n} \gamma_i T(u_i)$. Since T is linear, we have $v = T(\sum_{i=1}^{n} \gamma_i u_i) = T(u)$, where $u = \sum_{i=1}^{n} \gamma_i u_i \in U$. Thus there exists $u \in U$ such that $T(u) = v$, so T is onto and therefore T is an isomorphism, i.e., $U \cong V$.

Exercises

1. Find out Range, Kernel, Rank and Nullity of all the linear transformations given in the Problems 1–6, Problem 8 and 10 of the preceding section.
2. Let $T : V_1 \longrightarrow V_2$ be a linear map between finite dimensional vector spaces. Prove that T is an isomorphism if and only if $n(T) = 0$ and $r(T) = dim V_2$.
3. If $T : U \longrightarrow V$ is a linear map, where U is finite dimensional, prove that

    (a) $n(T) \leq dim U$,
    (b) $r(T) \leq \min(dim U, dim V)$.

4. Let Z be subspace of a finite-dimensional vector space U, and V a finite-dimensional vector space. Then prove that Z will be the kernel of a linear map $T : U \longrightarrow V$ if and only if $dim Z \geq dim U - dim V$.
5. Let $T : \mathbb{R}^4 \longrightarrow \mathbb{R}^3$ be a linear map defined by $T(e_1) = (1, 1, 1)$, $T(e_2) = (1, -1, 1)$, $T(e_3) = (1, 0, 0)$, $T(e_4) = (1, 0, 1)$. Then verify Rank-Nullity Theorem, where $\{e_1, e_2, e_3, e_4\}$ is the standard basis of $\mathbb{R}^4$.
6. Let T be a nonzero linear transformation from $\mathbb{R}^5$ to $\mathbb{R}^2$ such that T is not onto. Find $r(T)$ and $n(T)$.

7. Let $T : P_3(\mathbb{R}) \longrightarrow P_2(\mathbb{R})$ be a map defined by $T(\alpha_0 + \alpha_1 x + \alpha_2 x^2 + \alpha_3 x^3) = \alpha_1 + 2\alpha_2 x + 3\alpha_3 x^2$, where $P_3(\mathbb{R})$ is the vector space of all real polynomials of degree less than or equal to 3 over field of real numbers.

   (a) Prove that T is a linear transformation.
   (b) Find N(T) and $R(T)$.
   (c) Verify that $r(T) + n(T) = 4$.

8. Let $P_2(x, y)$ be the vector space of all real polynomials of degree less than or equal to 2 in x and y over the field of real numbers. Let $L : P_2(x, y) \longrightarrow P_2(x, y)$ be defined by $L(P(x, y)) = \frac{\partial}{\partial x}(P(x, y)) + \frac{\partial}{\partial y}(P(x, y))$, $P(x, y) \in P_2(x, y)$. Find a basis for kernel of L.

9. Let $P_n[x]$ be the space of all real polynomials of degree at most n over the field of real numbers. Consider the derivation operator $\frac{d}{dx}$. Find the dimension of Kernel and Image of $\frac{d}{dx}$.

10. Consider $P_n[x]$ over $\mathbb{R}$. Define $T(P(x)) = xP'(x) - P(x)$, $P(x) \in P_n[x]$.

    (a) Show that T is a linear transformation on $P_n[x]$.
    (b) Find $N(T)$ and $R(T)$.

11. Let V be a finite dimensional vector space and T be a linear transformation on V. Show that there exists an integer $k \geq 0$ such that $V = R(T^k) \oplus N(T^k)$.

12. Let T be a linear transformation on a finite dimensional vector space V. Show that $dim(Im T^2) = dim(Im T)$ if and only if $V = Im T \oplus Ker T$. Specifically if $T^2 = T$ then $V = Im T \oplus Ker T$. Is the converse true?

13. Let $T : \mathbb{R}^5 \to \mathbb{R}^2$ be a nonzero linear transformation such that T is not onto. Find $r(T)$ and $n(T)$.

14. Let $V_1, V_2, \ldots, V_n$; $n \geq 2$ be any finite dimensional vector spaces over a field $\mathbb{F}$. Let $T_i : V_i \to V_{i+1}$; $(1 \leq i \leq n-1)$ be linear transformations such that

    (a) $Ker T_1 = \{0\}$,
    (b) $Ker T_{i+1} = R(T_i)$, for $1 \leq i \leq n-2$,
    (c) $R(T_{n-1}) = V_n$,

    then show that $\sum_{i=1}^{n}(-1)^i dim V_i = 0$.

3.2   Basic Isomorphism Theorems

In this section we prove some isomorphism theorems, which have vast applications in linear algebra.

Theorem 3.17 (Fundamental Theorem of Vector Space Homomorphisms or First Isomorphism Theorem) *Let $T : U \longrightarrow V$ be a linear transformation. Then the quotient space of U with regard to its subspace K, where $K = Ker T$ is isomorphic to the homomorphic image of U under T, i.e., $\frac{U}{K} \cong T(U)$.*

Proof Define a map $f : \frac{U}{K} \longrightarrow T(U)$ such that $f(u + K) = T(u)$. The map f is well defined because if we take $u_1 + K = u_2 + K$, then it implies that $u_1 - u_2 \in K$, i.e., $T(u_1 - u_2) = 0$ and hence $T(u_1) = T(u_2)$, i.e., $f(u_1 + K) = f(u_2 + K)$. Also, we have $f(\alpha(u_1 + K) + \beta(u_2 + K)) = f((\alpha u_1 + \beta u_2) + K) = T(\alpha u_1 + \beta u_2) = \alpha T(u_1) + \beta T(u_2) = \alpha f(u_1 + K) + \beta f(u_2 + K)$ for every $u_1 + K, u_2 + K \in \frac{U}{K}, \alpha, \beta \in \mathbb{F}$. This shows that f is a linear transformation. Let $f(u_1 + K) = f(u_2 + K)$. It shows that $T(u_1) = T(u_2)$, i.e., $T(u_1 - u_2) = 0$. As a result, $(u_1 - u_2) \in K = Ker T$, i.e., $u_1 + K = u_2 + K$ and therefore f is one-to-one. Obviously f is onto. Hence f is an isomorphism and we have shown that $\frac{U}{K} \cong T(U)$.

Corollary 3.18 *Let $T : U \longrightarrow V$ be an onto linear transformation. Then the quotient space of U with regard to its subspace K, where $K = Ker T$ is isomorphic to V, i.e., $\frac{U}{K} \cong V$.*

Theorem 3.19 (Second Isomorphism Theorem) *Let V_1 and V_2 be subspaces of the vector space V. Then $\frac{V_1 + V_2}{V_2} \cong \frac{V_1}{V_1 \cap V_2}$.*

Proof Clearly V_2 is a subspace of $V_1 + V_2$ and $V_1 \cap V_2$ is a subspace of V_1, Therefore $\frac{V_1 + V_2}{V_2}$ and $\frac{V_1}{V_1 \cap V_2}$ are quotient spaces. Define a map $f : \frac{V_1 + V_2}{V_2} \longrightarrow \frac{V_1}{V_1 \cap V_2}$ such that $f((v_1 + v_2) + V_2) = v_1 + (V_1 \cap V_2)$ for every $(v_1 + v_2) + V_2 \in \frac{V_1 + V_2}{V_2}$. The map f is well defined because if we take $(v_1 + v_2) + V_2 = (v_1' + v_2') + V_2$, then it implies that $(v_1 - v_1') + (v_2 - v_2') \in V_2$, i.e., $(v_1 - v_1') \in V_2$. As a result, we obtain that $(v_1 - v_1') \in (V_1 \cap V_2)$ and thus $v_1 + (V_1 \cap V_2) = v_1' + (V_1 \cap V_2)$, i.e., $f((v_1 + v_2) + V_2) = f((v_1' + v_2') + V_2)$. Now for every $\alpha, \beta \in \mathbb{F}$, we have $f(\alpha((v_1 + v_2) + V_2) + \beta((v_1' + v_2') + V_2) = f((\alpha v_1 + \beta v_1') + (\alpha v_2 + \beta v_2') + V_2) = (\alpha v_1 + \beta v_1') + (V_1 \cap V_2) = (\alpha v_1 + (V_1 \cap V_2)) + (\beta v_1' + (V_1 \cap V_2)) = \alpha(v_1 + (V_1 \cap V_2)) + \beta(v_1' + (V_1 \cap V_2)) = \alpha f((v_1 + v_2) + V_2) + \beta f((v_1' + v_2') + V_2)$. The previous arguments show that f is a linear transformation. To prove that f is one-to-one, let $f((v_1 + v_2) + V_2) = f((v_1' + v_2') + V_2)$. This implies that $v_1 + (V_1 \cap V_2) = v_1' + (V_1 \cap V_2)$, i.e., $(v_1 - v_1') \in (V_1 \cap V_2)$. This shows that $(v_1 - v_1') + V_2 = (-v_2 + v_2') + V_2$, i.e., $(v_1 + v_2) - (v_1' + v_2') \in V_2$. This implies that $(v_1 + v_2) + V_2 = (v_1' + v_2') + V_2$. For ontoness of f, let $v_1'' + (V_1 \cap V_2) \in \frac{V_1}{V_1 \cap V_2}$. Clearly $f((v_1'' + 0) + V_2) = v_1'' + (V_1 \cap V_2)$, and hence f is onto. This shows that f is an isomorphism, i.e., $\frac{V_1 + V_2}{V_2} \cong \frac{V_1}{V_1 \cap V_2}$.

Theorem 3.20 (Third Isomorphism Theorem) *Let V_1 and V_2 be subspaces of the vector space V such that $V_2 \subseteq V_1$. Then $\frac{V/V_2}{V_1/V_2} \cong \frac{V}{V_1}$.*

Proof It is easy to observe that $V/V_1, V/V_2, V_1/V_2$ are quotient spaces and V_1/V_2 is a subspace of V/V_2. As a result, $\frac{V/V_2}{V_1/V_2}$ is a quotient space. Define a map $f : \frac{V/V_2}{V_1/V_2} \longrightarrow \frac{V}{V_1}$ such that $f((v + V_2) + V_1/V_2) = v + V_1$ for every $(v + V_2) + V_1/V_2 \in \frac{V/V_2}{V_1/V_2}$. We prove that f is well defined. For this, let $(v + V_2) + V_1/V_2 = (v' + V_2) + V_1/V_2$. It implies that $(v + V_2) - (v' + V_2) \in V_1/V_2$, i.e., $(v - v') + V_2 \in V_1/V_2$. Now we conclude that $(v - v') \in V_1$, i.e., $v + V_1 = v' + V_1$, showing that $f((v + V_2) + V_1/V_2) = f((v' + V_2) + V_1/V_2)$ and hence f is well defined. Also we have

$f(\alpha((v + V_2) + V_1/V_2)) + \beta((v' + V_2) + V_1/V_2)) = f((\alpha(v + V_2) + V_1/V_2) + (\beta(v' + V_2) + V_1/V_2)) = f(((\alpha v + \beta v') + V_2) + V_1/V_2) = (\alpha v + \beta v') + V_1 = (\alpha v + V_1) + (\beta v' + V_1) = \alpha(v + V_1) + \beta(v' + V_1) = \alpha f((v + V_2) + V_1/V_2) + \beta f((v' + V_2) + V_1/V_2)$ for every $\alpha, \beta \in F$, $v, v' \in V$. The previous arguments show that f is a linear transformation. To prove that f is one-to-one, let $f((v + V_2) + V_1/V_2) = f((v' + V_2) + V_1/V_2)$. This implies that $v + V_1 = v' + V_1$, i.e., $(v - v') \in V_1$. Now we have $(v - v') + V_2 \in V_1/V_2$, i.e., $(v + V_2) - (v' + V_2) \in V_1/V_2$. This shows that $(v + V_2) + V_1/V_2 = (v' + V_2) + V_1/V_2$, since $v + V_2, v' + V_2$ are the members of the vector space V/V_2 and V_1/V_2 is a subspace of V/V_2. Thus f is one-to-one. To show the ontoness of f, let $v + V_1 \in V/V_1$. Obviously $f((v + V_2) + V_1/V_2) = v + V_1$, which shows that f is onto. Finally we conclude that f is an isomorphism and $\frac{V/V_2}{V_1/V_2} \cong \frac{V}{V_1}$.

Example 3.21 If U and W are vector spaces. Let $V = U \times W = \{(u, w) | u \in U, w \in W\}$. Show that V forms a vector space with regard to component-wise operations. Using first isomorphism theorem, prove that $\frac{V}{\{0\} \times W} \cong U$ and $\frac{V}{U \times \{0\}} \cong W$. Further if both the vector spaces U and W are finite dimensional, then also prove that V is finite dimensional and $dim V = dim U + dim W$.

Here $V = U \times W = \{(u, w) | u \in U, w \in W\}$. It can be easily verified that V forms a vector space with regard to component-wise operations. Define a map $T : V \longrightarrow U$ such that $T(u, w) = u$ for every $(u, w) \in V$. It is easy to observe that $T(\alpha(u_1, w_1) + \beta(u_2, w_2)) = T(\alpha u_1 + \beta u_2, \alpha w_1 + \beta w_2) = \alpha u_1 + \beta u_2 = \alpha T(u_1, w_1) + \beta T(u_2, w_2)$. This shows that T is a linear transformation, which is onto obviously. Here we also have Kernel $T = \{(0, w) | w \in W\} = \{0\} \times W$. Now using first isomorphism theorem, we get $\frac{V}{\{0\} \times W} \cong U$. Similarly, we can also show that $\frac{V}{U \times \{0\}} \cong W$. If both U and W are finite dimensional vector spaces, then let $dim U = m$ and $dim V = n$. Next suppose that $B_1 = \{u_1, u_2, \ldots, u_m\}$ and $B_2 = \{v_1, v_2, \ldots, v_n\}$ be bases of U and W, respectively. Then it can be easily verified that the set $B_1 \times \{0\} \bigcup \{0\} \times B_2$ is a basis of the vector space V. This proves that V is finite dimensional. Since we have $\frac{V}{\{0\} \times W} \cong U$, and U is finite dimensional vector space, then $dim \frac{V}{\{0\} \times W} = dim U$. Now using the fact that V is a finite dimensional vector, we arrive at $dim V - dim(\{0\} \times W) = dim U$. But $(\{0\} \times W) \cong W$, we also get $dim(\{0\} \times W) = dim W$. Finally, we conclude that $dim V - dim W = dim U$, i.e., $dim V = dim U + dim W$.

Exercises

1. Using first isomorphism theorem, prove the second and third isomorphism Theorems.
2. Let $V_n = \{p(x) \in \mathbb{Q}[x] \mid deg\, p(x) < n\}$ and $V_{n-1} = \{p(x) \in \mathbb{Q}(x) \mid deg\, p(x) < (n-1)\}$, where $n > 1$. Define $T : V_n \longrightarrow V_{n-1}$ by $T(p(x)) = \frac{d}{dx} p(x)$. Show that T is a linear transformation and using first isomorphism theorem also prove that $V_n/\mathbb{Q} \cong V_{n-1}$.
3. Let $V = V_1 \bigoplus V_2$ be the direct sum of its subspaces V_1 and V_2. Show that the mappings $p_1 : V \longrightarrow V_1$ and $p_2 : V \longrightarrow V_2$ defined by $p_1(v) = v_1, p_2(v) =$

v_2, where $v = v_1 + v_2, v_1 \in V_1, v_2 \in V_2$, are linear transformations from V to V_1 and V_2, respectively. With the help of first isomorphism theorem prove that $V/V_1 \cong V_2$ and $V/V_2 \cong V_1$

4. Using first isomorphism theorem, prove the Rank-Nullity Theorem.

5. If V_1 and V_2 are finite dimensional subspaces of a vector space V. Using second isomorphism theorem prove that $V_1 + V_2$ is finite dimensional and dim $(V_1 + V_2)$ = dim V_1+ dim V_2-dim $(V_1 \cap V_2)$.

6. If $V_1, V_2, \ldots, V_n$ are vector spaces over the field $\mathbb{F}$, where n is any integer greater than 1. Let $V = V_1 \times V_2 \times \cdots \times V_n = \{(v_1, v_2, \ldots, v_n) | v_1 \in V_1, v_2 \in V_2, \ldots, v_n \in V_n\}$. Show that V forms a vector space over $\mathbb{F}$ with regard to component-wise operations. Using first isomorphism theorem, prove that $\frac{V}{V_1 \times V_2 \times \cdots \times V_{i-1} \times \{0\} \times V_{i+1} \times V_n} \cong V_i$. Further if all the vector spaces $V_1, V_2, \ldots, V_n$ are finite dimensional, then also prove that V is finite dimensional and $dim V = dim V_1 + dim V_2 + \cdots + dim V_n$.

7. Prove that $M_2(\mathbb{R}) \cong \mathbb{R}^4$. Give two different isomorphisms of $M_2(\mathbb{R})$ onto $\mathbb{R}^4$.

3.3 Algebra of Linear Transformations

In the set of linear transformations one can combine any two linear transformations in various ways in order to obtain a new linear transformation. The study of this set is important because it forms a natural vector space structure. It is more important in case if we consider the set of linear transformations from a vector space into itself, because in that case, it is also possible to define the composition of two linear mappings.

Let $T_1, T_2 : U \to V$ be linear transformations from a vector space U to a vector space V over the same field $\mathbb{F}$. The sum $T_1 + T_2$ and the scalar product kT_1, where $k \in \mathbb{F}$, are defined to be the following mappings from U into V.

$$(T_1 + T_2)(u) = T_1(u) + T_2(u) \text{ and } (kT_1)(u) = k\, T_1(u).$$

It can be easily seen that $T_1 + T_2$ and kT_1 are also linear. In fact, for any $u_1, u_2 \in U$ and $\alpha, \beta \in \mathbb{F}$,

$$\begin{aligned}
(T_1 + T_2)(\alpha u_1 + \beta u_2) &= T_1(\alpha u_1 + \beta u_2) + T_2(\alpha u_1 + \beta u_2) \\
&= \alpha T_1(u_1) + \beta T_1(u_2) + \alpha T_2(u_1) + \beta T_2(u_2) \\
&= \alpha \{T_1(u_1) + T_2(u_1)\} + \beta \{T_1(u_2) + T_2(u_2)\} \\
&= \alpha (T_1 + T_2)(u_1) + \beta (T_1 + T_2)(u_2).
\end{aligned}$$

Similarly, it can be seen that for any $k \in \mathbb{F}$,

$$\begin{aligned}
kT_1(\alpha u_1 + \beta u_2) &= k\{T_1(\alpha u_1 + \beta u_2)\} \\
&= k\{\alpha T_1(u_1) + \beta T_1(u_2)\} \\
&= (k\alpha)T_1(u_1) + (k\beta)T_1(u_2) \\
&= (\alpha k)T_1(u_1) + (\beta k)T_1(u_2) \\
&= \alpha(kT_1(u_1)) + \beta(kT_1(u_2)) \\
&= \alpha(kT_1)(u_1) + \beta(kT_1)(u_2).
\end{aligned}$$

This shows that $T_1 + T_2$ and kT_1 are also linear.

Example 3.22 Let $f, g : \mathbb{R}^3 \to \mathbb{R}^2$ such that $f(a, b, c) = (2a, b - c)$ and $g(a, b, c) = (a, b + c)$ for all $a, b, c \in \mathbb{R}$. Then it can be easily seen that f, g are linear transformations. It is also easy to see that

$$\begin{aligned}
(f + g)(a, b, c) &= f(a, b, c) + g(a, b, c) \\
&= (2a, b - c) + (a, b + c) \\
&= (3a, 2b)
\end{aligned}$$

This is again a linear transformation.

Remark 3.23 For any two vector spaces U and V over the same field $\mathbb{F}$, the set $Hom(U, V)$ will denote the set of all linear transformations from U to V. There should be no confusion concerning the underlying field of scalars. In fact, under the above addition and scalar multiplication the set $Hom(U, V)$ forms a vector space.

Theorem 3.24 *Let U and V be vector spaces over the same field $\mathbb{F}$. Then Hom (U, V), the set of all linear transformations from U into V is a vector space over $\mathbb{F}$ under the operation of addition and scalar multiplication as follows:*

For any $T_1, T_2 \in Hom(U, V)$ and $k \in \mathbb{F}$

$$(T_1 + T_2)(u) = T_1(u) + T_2(u), (kT_1)(u) = kT_1(u) \text{ for any } u \in U.$$

Proof We have already seen that for any $T_1, T_2 \in Hom(U, V)$, $T_1 + T_2 \in Hom$ (U, V). Define a map $\bar{0} : U \to V$ such that $\bar{0}(u) = 0$, the zero vector in V. Then it is easy to see that $\bar{0}$ is a linear transformation called the zero linear transformation satisfying $T_1 + \bar{0} = \bar{0} + T_1 = T_1$. Also for any $T_1 \in Hom(U, V)$, define $(-T_1)(u) = -(T_1(u))$. It can also be seen that $(-T_1) \in Hom(U, V)$ and $T_1 + (-T_1) = (-T_1) + T_1 = \bar{0}$. Commutativity and associativity of addition in $Hom(U, V)$ is obvious and hence $Hom(U, V)$ forms an abelian group under the addition. Further, for any $T_1 \in Hom(U, V)$ and $k \in \mathbb{F}$, we have already seen that $kT_1 \in Hom(U, V)$. It is an easy exercise to see that for any $T_1, T_2 \in Hom(U, V)$ and $\alpha, \beta \in \mathbb{F}$

$$\alpha(T_1 + T_2) = \alpha T_1 + \alpha T_2, (\alpha + \beta)T_1 = \alpha T_1 + \beta T_1, (\alpha\beta)T_1 = \alpha(\beta T_1), 1T_1 = T_1.$$

This shows that $Hom(U, V)$ is a vector space over the field $\mathbb{F}$.

Theorem 3.25 *If U and V are vector spaces over a field $\mathbb{F}$ of dimension m and n, respectively, then the dimension of the vector space $Hom(U, V)$ over $\mathbb{F}$ is mn.*

Proof Let $B = \{u_1, u_2, \ldots, u_m\}$ and $B' = \{v_1, v_2, \ldots, v_n\}$ be bases of U and V, respectively. Define mn mappings, $f_{ij} : B \longrightarrow V$ for each $i = 1, 2, \ldots, m$ and $j = 1, 2, \ldots, n$ such that $f_{ij}(u_k) = 0$ for all $k \neq i$ and $f_{ij}(u_k) = v_j$ for all $k = i$, where $k = 1, 2, \ldots, m$. Now by Theorem 3.13, one can find mn linear transformations $T_{ij} : U \to V$, such that $T_{ij}|_B = f_{ij}$ where $i = 1, 2, \ldots, m$ and $j = 1, 2, \ldots, n$. Our result follows if we can show that the set $\{T_{ij} \mid i = 1, 2, \ldots, m, j = 1, 2, \ldots, n\}$ is a basis of $Hom(U, V)$. Let T be an arbitrary member of $Hom(U, V)$. Since for each $u_i \in U, T(u_i) \in V$ and $\{v_1, v_2, \ldots, v_n\}$ is a basis of V, corresponding to each $T(u_i)$ we can find n scalars $\alpha_{ij} \in \mathbb{F}, j = 1, 2, \ldots, n$ such that $T(u_i) = \sum_{j=1}^{n} \alpha_{ij} v_j$. Now it is clear that for every i and j, $T_{ij}(u_k) = f_{ij}(u_k) = v_j$ for all $k = i$ and $T_{ij}(u_k) = f_{ij}(u_k) = 0$ for all $k \neq i$. Next we claim that $T = \sum_{i=1}^{m} \sum_{j=1}^{n} \alpha_{ij} T_{ij}$. Since

$$
\begin{aligned}
\Big(\sum_{i=1}^{m} \sum_{j=1}^{n} \alpha_{ij} T_{ij} \Big)(u_k) &= \Big(\sum_{i=1}^{m} \Big(\sum_{j=1}^{n} \alpha_{ij} T_{ij} \Big)(u_k) \Big) \\
&= \sum_{i=1}^{m} \Big(\sum_{j=1}^{n} \alpha_{ij} T_{ij}(u_k) \Big) \\
&= \sum_{j=1}^{n} \alpha_{kj} T_{kj}(u_k) \\
&= \sum_{j=1}^{n} \alpha_{kj} v_j \\
&= T(u_k),
\end{aligned}
$$

$\sum_{i=1}^{m} \sum_{j=1}^{n} \alpha_{ij} T_{ij} \in Hom(U, V)$ and both $T, \Big(\sum_{i=1}^{m} \sum_{j=1}^{n} \alpha_{ij} T_{ij} \Big)$ agree on all basis elements of U. This shows that $T = \sum_{i=1}^{m} \sum_{j=1}^{n} \alpha_{ij} T_{ij}$, i.e., $T \in Hom(U, V)$ is a linear combination of T_{ij}s. Now in order to show that the set $\{T_{ij} \mid i = 1, 2, \ldots, m, j = 1, 2, \ldots, n\}$ is linearly independent, suppose that there exist scalars $\beta_{ij} \in \mathbb{F}$ such that $\sum_{i=1}^{m} \sum_{j=1}^{n} \beta_{ij} T_{ij} = \bar{0}$, where $\bar{0}$ stands for the zero linear transformation from U to V. This implies that $\Big(\sum_{i=1}^{m} \sum_{j=1}^{n} \beta_{ij} T_{ij} \Big)(u_k) = \bar{0}(u_k)$ for all $k = 1, 2, \ldots, m$. This yields that $\sum_{j=1}^{n} \Big(\sum_{i=1}^{m} \beta_{ij} T_{ij}(u_k) \Big) = 0$. But since $T_{ij}(u_k) = v_j$ for all $k = i$ and $T_{ij}(u_k) = 0$ for all $k \neq i$, the latter expression reduces to $\sum_{j=1}^{n} \beta_{kj} v_j = 0$ for all $k = 1, 2, \ldots, m$. Since $\{v_1, v_2, \ldots, v_n\}$ is a basis of V, we find that $\beta_{kj} = 0$. This yields that the set $\{T_{ij} \mid i = 1, 2, \ldots, m, j = 1, 2, \ldots, n\}$ is linearly independent and thus forms a basis of $Hom(U, V)$. Finally, we get $dim Hom(U, V) = mn = dim U \, dim V$.

Remark 3.26 Let U be a vector space over a field $\mathbb{F}$ such that $dim U = m$, then $dim Hom(U, U) = m^2$. Since every field $\mathbb{F}$ is a vector space over $\mathbb{F}$ of dimension one, $dim Hom(U, \mathbb{F}) = m$.

Example 3.27 Let $U = \mathbb{R}^3$ and $V = \mathbb{R}^2$ and let $\{e_1, e_2, e_3\}$ and $\{f_1, f_2\}$ be standard bases of U and V, respectively. This yields that $dim Hom(U, V) = 6$. In fact, we find the set $\{T_{ij} \mid i = 1, 2, 3, j = 1, 2\}$ as a basis of $Hom(U, V)$. Any $u = (a, b, c) \in \mathbb{R}^3$ can be written as $u = ae_1 + be_2 + ce_3$. Now $T_{11}(e_1) = f_1, T_{11}(e_2) = (0, 0), T_{11}(e_3) = (0, 0)$. This yields that $T_{11}(u) = aT_{11}(e_1) + bT_{11}(e_2) + cT_{11}(e_3) = af_1 = (a, 0)$.   Similarly $T_{12}(u) = af_2 = (0, a), T_{21}(u) = bf_1 = (b, 0), T_{22}(u) = bf_2 = (0, b), T_{31}(u) = cf_1 = (c, 0), T_{32}(u) = cf_2 = (0, c)$.

Lemma 3.28 *Let U, V, W, be vector spaces over a field $\mathbb{F}$. Then the following hold:*

(i) *For any linear transformations $T_1 : U \to V$ and $T_2 : V \to W$ the composite map $T_2 T_1 : U \to W$ is again a linear transformation.*

(ii) *For any linear transformations $T_1 : U \to V, T_2 : V \to W$ and $\alpha \in \mathbb{F}$, $\alpha(T_2 T_1) = (\alpha T_2)T_1 = T_2(\alpha T_1)$.*

(iii) *For any linear transformations $T_2, T_3 : U \to V$ and $T_1 : V \to W, T_1(T_2 + T_3) = T_1 T_2 + T_1 T_3$.*

(iv) *For any linear transformations $T_1, T_2 : V \to W$ and $T_3 : U \to V$ $(T_1 + T_2)T_3 = T_1 T_3 + T_2 T_3$.*

Proof (i) For any $u_1, u_2 \in U$ and $\alpha, \beta \in \mathbb{F}$

$$
\begin{aligned}
T_2 T_1(\alpha u_1 + \beta u_2) &= T_2(T_1(\alpha u_1 + \beta u_2)) \\
&= T_2(\alpha T_1(u_1) + \beta T_1(u_2)) \\
&= \alpha T_2(T_1(u_1)) + \beta T_2(T_1(u_2)) \\
&= \alpha(T_2 T_1)(u_1) + \beta(T_2 T_1)(u_2)
\end{aligned}
$$

This completes the proof.

   (ii) Since $T_2 T_1 : U \to W$, for any $u \in U$,

$$
(\alpha(T_2 T_1))(u) = \alpha((T_2 T_1)(u)).
$$

$$
((\alpha T_2))(T_1)(u) = (\alpha T_2)(T_1(u)) = \alpha(T_2(T_1(u))) = \alpha((T_2 T_1)(u))
$$

$$
(T_2(\alpha T_1))(u) = T_2((\alpha T_1)(u)) = T_2(\alpha(T_1(u))) = \alpha((T_2 T_1)(u)).
$$

The above three expressions ensure the validity of (ii).

   (iii) Since $T_1(T_2 + T_3), T_1 T_2 + T_1 T_3 : U \to W$ are linear transformations, for any $u \in U$, $(T_1(T_2 + T_3))(u) = T_1(T_2(u) + T_3(u)) = T_1(T_2(u)) + T_1(T_3(u)) = (T_1 T_2)(u) + (T_1 T_3)(u) = (T_1 T_2 + T_1 T_3)(u)$. This yields the required result.

   (iv) Proof is similar as (iii).

Remark 3.29 (i) The set of all linear transformations of V into itself, i.e., Hom (V, V) is usually denoted by $\mathscr{A}(V)$.

(ii) A linear operator T on the vector space U is called an *idempotent linear operator* on U if $T^2 = T$.

(iii) A linear operator T on the vector space U is called a *nilpotent linear operator* on U if $T^k = 0$ for some integer $k \geq 1$.

Theorem 3.30 *If V is a vector space over a field $\mathbb{F}$, then $\mathscr{A}(V)$ is an algebra with identity over the field $\mathbb{F}$, which is not commutative in general.*

Proof (*i*) By Theorem 3.24, $\mathscr{A}(V)$ is a vector space over $\mathbb{F}$ under the addition and scalar multiplication operations on linear mappings.

(*ii*) $\mathscr{A}(V)$ is a ring. In fact, $\mathscr{A}(V)$ is an abelian group under the operation addition of linear transformations. Also by Lemma 3.28 (i), $\mathscr{A}(V)$ is closed with respect to the operation of product (composition). Since the composition of functions is associative in general, the product in $\mathscr{A}(V)$ is associative. The distributivity of product on the operation of addition of linear transformations follows by Lemma 3.28(iii) and (iv). Hence $\mathscr{A}(V)$ forms ring under the operations of addition and composition of linear transformations.

(*iii*) By Lemma 3.28 (ii), for any $\alpha \in \mathbb{F}$ and $T_1, T_2 \in \mathscr{A}(V)$

$$\alpha(T_1 T_2) = (\alpha T_1)T_2 = T_1(\alpha T_2).$$

This shows that $\mathscr{A}(V)$ is an algebra over the field $\mathbb{F}$.

(*iv*) Since $I_V T = T I_V = T$ for all $T \in \mathscr{A}(V)$, where I_V is the identity linear map on V. As the composition of functions is not commutative in general, the product in $\mathscr{A}(V)$ is not commutative. These facts prove that $\mathscr{A}(V)$ is an algebra with identity I_V over the field $\mathbb{F}$, which is not commutative in general.

Exercises

1. Show that the set $\{T_1, T_2, T_3\}$ in the corresponding vector space is linearly independent, where

   (a) $T_1, T_2, T_3 \in \mathscr{A}(\mathbb{R}^2)$ defined by $T_1(x, y) = (x, 2y)$, $T_2(x, y) = (y, x + y)$, $T_3(x, y) = (0, x)$,

   (b) $T_1, T_2, T_3 \in Hom(\mathbb{R}^3, \mathbb{R})$ defined by $T_1(x, y, z) = x + y + z$, $T_2(x, y, z) = y + z$, $T_3(x, y, z) = x - z$.

2. Find the condition under which $dim Hom(V, U) = dim V$?

3. Suppose $V = U \oplus W$, where U and W are subspaces of V. Let T_1 and T_2 be linear operators on V defined by $T_1(v) = u$, $T_2(v) = w$, where $v = u + w$, $u \in U, w \in W$. Show that

   (a) $T_1^2 = T_1$ and $T_2^2 = T_2$, i.e., T_1 and T_2 are projections.

   (b) $T_1 + T_2 = I$, the identity mapping.

   (c) $T_1 T_2 = 0$ and $T_2 T_1 = 0$.

4. Let T_1 and T_2 be linear operators on V satisfying parts (a), (b), (c) of the above Problem 3. Prove that $V = T_1(V) \oplus T_2(V)$.

5. Give an example to show that the set of all nonzero elements of $\mathscr{A}(V)$ is not a group under composition of linear operators.

6. Determine the group of units of the ring $\mathscr{A}(V)$.

7. Let $T : V \longrightarrow W$ be an isomorphism, where V and W are vector spaces over the field $\mathbb{F}$. Prove that the mapping $f : \mathscr{A}(V) \longrightarrow \mathscr{A}(W)$ defined by $f(S) = TST^{-1}$ is an isomorphism.

8. If $T \in \mathscr{A}(V)$, prove that the set of all linear transformations S on V such that $TS = 0$ is a subspace and a right ideal of $\mathscr{A}(V)$.

9. If $dim\mathscr{A}(V) > 1$, then prove that $\mathscr{A}(V)$ is not commutative.

10. Let T_1, T_2 be two linear maps on V such that $T_1 T_2 = T_2 T_1$. Then prove that

    (a) $(T_1 + T_2)^2 = T_1^2 + 2(T_1 T_2) + T_2^2$;
    (b) $(T_1 + T_2)^n = {}^nC_0 T_1^n + {}^nC_1 T_1^{(n-1)} T_2 + \cdots + {}^nC_n T_2^n$.

11. Let V_1 be a subspace of a vector space V. Then prove that the set of all linear transformations from V to V that vanish on V_1 is a subspace of $\mathscr{A}(V)$.

12. Let T be an idempotent linear operator on any vector space U over $\mathbb{F}$. Then prove that $U = R(T) \oplus N(T)$ and for any $v \in R(T)$, $T(v) = v$.

13. Let V be an n-dimensional vector space, $n \geq 1$, and $T : V \to V$ be a linear transformation such that $T^n = 0$ but for some $v \in V$, $T^{n-1}(v) \neq 0$. Show that the set $\{v, T(v), T^2(v), \ldots, T^{n-1}(v)\}$ forms a basis of V.

14. Let $T : V \to V$ be a nilpotent linear operator. Show that $I + T$ is invertible. (Hint: Any linear transformation is said to be nilpotent if $T^k = 0$ for some integer $k \geq 1$. If $T^k = 0$, then $I - T + T^2 - T^3 + \cdots + (-1)^{k-1} T^{k-1}$ is the inverse of $(I + T)$.

15. Let $B = \{v_1, v_2, \ldots, v_n\}$ be a basis of V. Suppose that for each $1 \leq i, j \leq n$ define

$$f_{i,j}(v_k) = \begin{cases} v_k, & for\ k \neq i \\ v_i + v_j, & for\ k = i. \end{cases}$$

Prove that $f_{i,j}{}^s$ are differentiable and form a basis of $\mathscr{A}(V)$.

3.4 Nonsingular, Singular and Invertible Linear Transformations

Definition 3.31 A linear transformation $T : V \longrightarrow U$ is called nonsingular if $T(v) = 0$, where $v \in V$ implies that $v = 0$, i.e., if $KerT = \{0\}$. If, on the other hand, there exists a nonzero vector $v \in V$ such that $T(v) = 0$, i.e., $KerT \neq \{0\}$, then T is called singular.

Remark 3.32 (i) From the Remark 3.7(iii), T is nonsingular if and only if T is injective or one-to-one.

(ii) If V is a finite dimensional vector space then $T : V \longrightarrow U$ is nonsingular if and only if $r(T) = dim V$ follows by Theorem 3.9.

Theorem 3.33 *A linear transformation $T : V \longrightarrow U$ is nonsingular if and only if the image of any linearly independent set is linearly independent.*

Proof Let $T : V \longrightarrow U$ be nonsingular. Let S be any linearly independent subset of V. Now we have two cases.

Case I: Let S be finite, i.e., $S = \{v_1, v_2, \ldots, v_n\}$. We have to prove that $\{T(v_1), T(v_2), \ldots, T(v_n)\}$ is linearly independent. For this, let $\alpha_1 T(v_1) + \alpha_2 T(v_2) + \cdots + \alpha_n T(v_n) = 0$, where $\alpha_1, \alpha_2, \ldots, \alpha_n \in \mathbb{F}$. This implies that $T(\sum\limits_{i=1}^{n} \alpha_i v_i) = 0$, i.e., $\sum\limits_{i=1}^{n} \alpha_i v_i \in KerT$. As T is nonsingular, we have $KerT = \{0\}$. Now we conclude that $\sum\limits_{i=1}^{n} \alpha_i v_i = 0$, but as S is linearly independent, we get $\alpha_i = 0$ for all i, $1 \le i \le n$. Thus the image of S is linearly independent.

Case II: Let $S = \{w_1, w_2, \ldots, w_m, \ldots\}$ be infinite. We have to show that the image set $\{T(w_1), T(w_2), \ldots, T(w_m), \ldots\}$ is linearly independent. For this choose an arbitrary finite subset P of $\{T(w_1), T(w_2), \ldots, T(w_m), \ldots\}$, then clearly $P = \{T(w_{i_1}), T(w_{i_2}), \ldots, T(w_{i_m})\}$ for some $i_1, i_2, \ldots, i_m \in \mathbb{N}$. The set $\{w_{i_1}, w_{i_2}, \ldots, w_{i_m}\}$, being a subset of S, is a linearly independent set in V. Now using Case I, it follows that P is a linearly independent subset of U. As a result, the image set $\{T(w_1), T(w_2), \ldots, T(w_m), \ldots\}$ is linearly independent.

Conversely, suppose that the image of linearly independent set is linearly independent. We have to prove that T is nonsingular; equivalently we have to prove that $KerT = \{0\}$. Let $v \in KerT$. We claim that $v = 0$, for otherwise, if $v \ne \{0\}$, then $\{v\}$ will be a linearly independent set. By our hypothesis $\{T(v)\}$ will be linearly independent set, leading to a contradiction, because $T(v) = 0$.

Theorem 3.34 *Let U, V, W be finite dimensional vector spaces over the same field $\mathbb{F}$. Let $T_1 : U \longrightarrow V$ and $T_2 : V \longrightarrow W$ be linear transformations. Then*

(i) $r(T_2 T_1) \le Min \, [r(T_2), r(T_1)]$,
(ii) $n(T_2 T_1) \ge n(T_1)$.

Proof (i) $T_2 T_1 : U \longrightarrow W$ is a linear transformation. We have $T_1(U) \subseteq V$; this implies that $T_2(T_1(U)) \subseteq T_2(V)$, i.e., $T_2 T_1(U) \subseteq T_2(V)$. Since $T_2 T_1(U)$ and $T_2(V)$ are subspaces of W, $T_2 T_1(U)$ is a subspace of $T_2(V)$ and hence $dim T_2 T_1(U) \le dim T_2(V)$, i.e., $r(T_2 T_1) \le r(T_2)$. Again $T_1(U)$ is a subspace of V. Hence the restriction of T_2 on the subspace $T_1(U)$ will be a linear transformation, i.e., $T_2 |_{T_1(U)} : T_1(U) \longrightarrow W$. Hence by Theorem 3.9, $dim T_2 T_1(U) \le dim T_1(U)$, i.e., $r(T_2 T_1) \le r(T_1)$. Including the previous two inequalities we conclude that $r(T_2 T_1) \le Min \, [r(T_2), r(T_1)]$.

(*ii*) Let $u \in Ker T_1$. This implies that $T_1(u) = 0$, i.e., $T_2(T_1(u)) = T_2(0)$. Thus we get $T_2 T_1(u) = 0$, i.e., $u \in Ker T_2 T_1$. Therefore, $Ker T_1 \subseteq Ker T_2 T_1$. Since both $Ker T_1$ and $Ker T_2 T_1$ are subspaces of U, $Ker T_1$ is a subspace of $Ker T_2 T_1$. Now we get $dim Ker T_1 \leq dim Ker T_2 T_1$, i.e., $n(T_1) \leq n(T_2 T_1)$.

Theorem 3.35 *Let V be a finite dimensional vector space and $T, S : V \longrightarrow V$ be linear transformations, where S is nonsingular. Then $r(ST) = r(TS) = r(T)$.*

Proof Clearly, $ST : V \longrightarrow V$ is a linear transformation. $Im(ST) = (ST)(V) = S(T(V))$. Therefore, $r(ST) = dim S(T(V))$. Since $T(V)$ is a subspace of V, the restriction of S on the subspace $T(V)$ is also a nonsingular linear transformation, i.e., $S|_{T(V)}: T(V) \longrightarrow V$. Hence $r(S|_{T(V)}) = dim T(V)$ follows from Remark 3.32(ii). Now this implies that $dim S(T(V)) = dim T(V)$, and hence $r(ST) = r(T)$. Again since V is finite dimensional and $S : V \longrightarrow V$ is injective, S is surjective also. Hence $S(V) = V$, i.e., $T(S(V) = T(V)$. This relation shows that $(TS)(V) = T(V)$, i.e., $Im(TS) = Im(T)$. As a result, $dim Im(TS) = dim Im T$, i.e., $r(TS) = r(T)$. Finally we have proved that $r(ST) = r(TS) = r(T)$.

Definition 3.36 A linear transformation $T : U \longrightarrow V$ is called invertible if T is bijective as a map. In other words, if T is an isomorphism, then T is said to be invertible.

Remark 3.37 (*i*) If a linear transformation $T : U \longrightarrow V$ is invertible, then T is a bijective map and hence the inverse of the map T, which is usually denoted by T^{-1}, exists and which is a map from V to U. Now we prove that $T^{-1}(\alpha v_1 + \beta v_2) = \alpha T^{-1}(v_1) + \beta T^{-1}(v_2)$ for all $\alpha, \beta \in \mathbb{F}$, $v_1, v_2 \in V$. Let us suppose that $T^{-1}(\alpha v_1 + \beta v_2) = u \in U$ and $T^{-1}(v_1) = u_1 \in U$, $T^{-1}(v_2) = u_2 \in U$. This implies that $T(u) = \alpha v_1 + \beta v_2$ and $T(u_1) = v_1$, $T(u_2) = v_2$. Now we have $T(u) = \alpha T(u_1) + \beta T(u_2)$. But as T is a linear transformation, we have $T(u) = T(\alpha u_1 + \beta u_2)$. Since T is one-to-one, we get $u = \alpha u_1 + \beta u_2$, i.e., $T^{-1}(\alpha v_1 + \alpha v_2) = \alpha T^{-1}(v_1) + \beta T^{-1}(v_2)$. Thus T^{-1} is also linear transformation, which is known as the inverse of linear transformation T. We know that if T is a bijective map, then T^{-1} is also a bijective map. This shows that if T is invertible, i.e., T is an isomorphism, then its inverse T^{-1} is also an isomorphism and hence invertible also.

(*ii*) By the properties of invertible maps, we can say that the linear transformation $T : U \longrightarrow V$ is invertible if and only if there exists the linear transformation $T^{-1} : V \longrightarrow U$ such that $T T^{-1} = I_V$ and $T^{-1} T = I_U$, where I_U and I_V are identity linear transformations on vector spaces U and V, respectively. In particular, if $T : U \longrightarrow U$ is a linear operator on U. Then T is invertible if and only if there exists the linear operator T^{-1} on U, i.e., $T^{-1} : U \longrightarrow U$ such that $T T^{-1} = I_U = T^{-1} T$.

(*iii*) Let U be a finite dimensional vector space and $T : U \longrightarrow U$ be an invertible linear operator on U. This shows that T is bijective, i.e., $Ker T = \{0\}$ or T is nonsingular. Hence $n(T) = 0$. Using Rank-Nullity theorem, we have $r(T) = dim U$. This implies that T is surjective. It is easy to observe that if U is a finite dimensional vector space, then the linear map $T : U \longrightarrow U$ is invertible if and only if T is nonsingular, i.e., injective or alternatively T is surjective. Hence nonsingular and invertible linear transformations on a vector space U are synonyms when U is finite dimensional.

(*iv*) If U is not finite dimensional, then a linear transformation on U may be nonsingular, i.e., injective and yet not invertible as illustrated by the following example: Consider the map $T : F[x] \longrightarrow F[x]$, given by $T(\alpha_0 + \alpha_1 x + \alpha_2 x^2 + \cdots + \alpha_n x^n) = \alpha_0 x + \alpha_1 x^2 + \alpha_2 x^3 + \cdots + \alpha_n x^{n+1}$. We know that $F[x]$ is an infinite dimensional vector space and it can be easily verified that T is linear. Now we prove that T is nonsingular or injective. Let $\alpha_0 + \alpha_1 x + \alpha_2 x^2 + \cdots + \alpha_n x^n \in \text{Ker } T$. Now we get $T(\alpha_0 + \alpha_1 x + \alpha_2 x^2 + \cdots + \alpha_n x^n) = 0$ (the zero polynomial over F). This implies that $\alpha_0 x + \alpha_1 x^2 + \alpha_2 x^3 + \cdots + \alpha_n x^{n+1} = 0$. But $\{x, x^2, x^3, \ldots, x^{n+1}\}$ is a linearly independent set in the vector space $F[x]$, thus we get $\alpha_0 = \alpha_1 = \cdots = \alpha_n = 0$. Hence Ker $T = \{0\}$. This shows that T is injective. But T is not surjective because the polynomials of zero degree, i.e., scalars are not the image of any polynomial under T. Hence T is not invertible.

Theorem 3.38 *Let U and V be finite dimensional vector spaces over the same field $\mathbb{F}$ such that dim $U=$ dim V. If $T : U \longrightarrow V$ is a linear transformation, then the following statements are equivalent:*

(*i*) *T is invertible.*
(*ii*) *T is nonsingular.*
(*iii*) *The range of T is V.*
(*iv*) *If $\{u_1, u_2, \ldots, u_n\}$ is any basis of U, then $\{T(u_1), T(u_2), \ldots, T(u_n)\}$ is a basis of V.*
(*v*) *There is a basis $\{u_1, u_2, \ldots, u_n\}$ for U such that $\{T(u_1), T(u_2), \ldots, T(u_n)\}$ is a basis of V.*

Proof (*i*) $\Longrightarrow$ (*ii*) Since T is invertible, T is injective. Thus $Ker T = \{0\}$. This shows that T is nonsingular.

(*ii*) $\Longrightarrow$ (*iii*) T is nonsingular, it implies that $Ker T = \{0\}$, i.e., $n(T) = 0$. Using Rank-Nullity theorem, we get $r(T) = dim U$. But it is given that $dim U = dim V$, thus we get $r(T) = dim V$, i.e., $dim T(U) = dim V$. Since $T(U)$ is a subspace of V, we conclude that $T(U) = V$, i.e., the range of T is V.

(*iii*) $\Longrightarrow$ (*iv*) It is given that $T(U) = V$. This shows that $dim T(U) = dim V$, i.e., $r(T) = dim U$. Using again Rank-Nullity Theorem, we get $n(T) = 0$, i.e., T is nonsingular. Let us suppose that $\{u_1, u_2, \ldots, u_n\}$ is any basis of U, i.e., $\{u_1, u_2, \ldots, u_n\}$ is a linearly independent subset of U. By Theorem 3.33, it follows that $\{T(u_1), T(u_2), \ldots, T(u_n)\}$ is a linearly independent set of V. Now we show that the set $\{T(u_1), T(u_2), \ldots, T(u_n)\}$ spans V. For this let $v \in V$. Thus there exists $u \in U$ such that $T(u) = v$. Since $\{u_1, u_2, \ldots, u_n\}$ is a basis of U, there exist scalars $\alpha_1, \alpha_2, \ldots, \alpha_n \in \mathbb{F}$ such that $\alpha_1 u_1 + \alpha_2 u_2 + \cdots + \alpha_n u_n = u$. Now we can write $v = T(\alpha_1 u_1 + \alpha_2 u_2 + \cdots + \alpha_n u_n) = \alpha_1 T(u_1) + \alpha_2 T(u_2) + \cdots + \alpha_n T(u_n)$. These facts prove that the set $\{T(u_1), T(u_2), \ldots, T(u_n)\}$ is a basis of V.

(*iv*) $\Longrightarrow$ (*v*) Since U is a finite dimensional vector space, it has a finite basis. Let $\{u_1, u_2, \ldots, u_n\}$ be a basis of U. Now by hypothesis $\{T(u_1), T(u_2), \ldots, T(u_n)\}$ is a basis of V.

$(v) \implies (i)$ Suppose that there is a basis $\{u_1, u_2, \ldots, u_n\}$ for U such that $\{T(u_1), T(u_2), \ldots, T(u_n)\}$ is a basis of V. It is clear that Im (T) is generated by $\{T(u_1), T(u_2), \ldots, T(u_n)\}$ and also V is generated by $\{T(u_1), T(u_2), \ldots, T(u_n)\}$. Now we conclude that $T(U) = V$, i.e., T is onto. This implies that $r(T) = \dim V$, i.e., $r(T) = \dim U$. Using Rank-Nullity theorem, we conclude that $n(T) = 0$, i.e., T is one-to-one. This proves the fact that T is invertible.

Theorem 3.39 *Let U, V and W be vector spaces over the same field $\mathbb{F}$ and $T_1 : U \longrightarrow V$, $T_2 : V \longrightarrow W$ be invertible linear transformations. Then the following hold:*

(i) *$T_2 T_1$ is invertible and $(T_2 T_1)^{-1} = T_1^{-1} T_2^{-1}$.*
(ii) *If $\alpha(\neq 0) \in \mathbb{F}$, then αT_1 is invertible and $(\alpha T_1)^{-1} = \alpha^{-1} T_1^{-1}$.*

Proof (i) We know that the composition of two linear transformations is a linear transformation, $T_2 T_1 : U \longrightarrow W$ is a linear transformation. Both the maps T_1 and T_2 are bijective, and hence $T_2 T_1$ is also a bijective map. As a result, $T_2 T_1$ is invertible. Since T_1 and T_2 are invertible, it implies that there exist linear transformations $T_1^{-1} : V \longrightarrow U$, $T_2^{-1} : W \longrightarrow V$. Thus we have a linear transformation $T_1^{-1} T_2^{-1} : W \longrightarrow U$. Now we claim that $(T_2 T_1)^{-1} = T_1^{-1} T_2^{-1}$. To prove the claim, we have to show that $(T_2 T_1) T_1^{-1} T_2^{-1} = I_W$, $T_1^{-1} T_2^{-1} (T_2 T_1) = I_U$, where I_U and I_W are identity linear transformations on vector spaces U and W, respectively. Consider $(T_2 T_1) T_1^{-1} T_2^{-1} = T_2 (T_1 T_1^{-1}) T_2^{-1} = T_2 I_V T_2^{-1} = I_W$. Similarly, it can be shown that $T_1^{-1} T_2^{-1} (T_2 T_1) = I_U$.

(ii) We know that $\alpha T_1 : U \longrightarrow V$ is a linear transformation. We prove that αT_1 is one-to-one and onto. Let $(\alpha T_1)(u_1) = (\alpha T_1)(u_2)$, where $u_1, u_2 \in U$. This implies that $\alpha(T_1(u_1)) = \alpha(T_1(u_2))$. Since $\alpha(\neq 0) \in \mathbb{F}$, multiplying both sides of the previous relation by α^{-1}, we obtain that $T_1(u_1) = T_1(u_2)$, which implies that $u_1 = u_2$ because T_1 is one-to-one. In this way the above arguments prove that αT_1 is one-to-one. For ontoness of αT_1, let $v \in V$. Due to ontoness of T_1, there exists $u \in U$, such that $T_1(u) = v$. It can be easily verified that $(\alpha T_1)(\alpha^{-1} u) = v$. This shows that αT_1 is onto. Thus αT_1 is invertible. Since T_1 is invertible, it implies that there exists a linear transformation $T_1^{-1} : V \longrightarrow U$. Thus we have a linear transformation $\alpha^{-1} T_1^{-1} : V \longrightarrow U$. Now we claim that $(\alpha T_1)^{-1} = \alpha^{-1} T_1^{-1}$. To prove the claim, we have to show that $(\alpha T_1)(\alpha^{-1} T_1^{-1}) = I_V$, $(\alpha^{-1} T_1^{-1})(\alpha T_1) = I_U$, where I_U and I_V are identity linear transformations on vector spaces U and V, respectively. Consider $(\alpha T_1) \alpha^{-1} T_1^{-1} = (\alpha \alpha^{-1})(T_1 T_1^{-1}) = I_V$. Similarly, it can be shown that $\alpha^{-1} T_1^{-1} (\alpha T_1) = I_U$.

Theorem 3.40 *If T, T_1, T_2 are linear operators on a vector space U such that $T T_1 = T_2 T = I_U$, where I_U is the identity linear operator on U, then T is invertible and $T^{-1} = T_1 = T_2$.*

Proof We prove that T is a bijective map. For one-to-oneness of T, let us suppose that $T(u_1) = T(u_2)$. This implies that $T_2(T(u_1)) = T_2(T(u_2))$, i.e., $(T_2 T)(u_1) = (T_2 T)(u_2)$. Using hypothesis we get $I_U(u_1) = I_U(u_2)$, i.e., $u_1 = u_2$. Hence T is injective. For ontoness of T, let $u \in U$, clearly $T_1(u) \in U$ and obviously we have $T(T_1(u)) = I_U(u) = u$. Hence T is onto. Thus we conclude that T is invertible. Now let the inverse of T be T^{-1}, i.e., $TT^{-1} = T^{-1}T = I_U$. But it is given that $TT_1 = I_U$. This implies that $T^{-1}(TT_1) = T^{-1}I_U$, i.e., $T^{-1} = T_1$. Similarly, it can be proved that $T^{-1} = T_2$.

Theorem 3.41 *If U is a finite dimensional vector space, then a linear operator T on U is invertible if and only if there exists a linear operator T_1 on U such that (i) $TT_1 = I_U$ or alternatively (ii) $T_1T = I_U$.*

Proof Suppose that T is invertible, i.e., T^{-1} exists. Obviously, $TT^{-1} = T^{-1}T = I_U$. This implies that if we take $T_1 = T^{-1}$, then we get our required result.

Conversely, suppose that (i) holds. We show that T is onto. For this, let $y \in U$ so that $T_1(y) \in U$. Suppose that $T_1(y) = x \in U$. Consider $T(x) = T(T_1(y)) = (TT_1)(y) = I_U(y) = y$. This shows that T is onto. Since T is defined on a finite dimensional vector space, T is one-to-one also. Thus T is invertible.

If (ii) holds, then we have $T_1T = I_U$. We show that T is one-to-one. For this let $T(x) = T(y)$. This implies that $T_1(T(x)) = T_1(T(y))$, i.e., $(T_1T)(x) = (T_1T)(y)$. This shows that $I_U(x) = I_U(y)$, i.e., $x = y$. Thus T is injective. But U is finite dimensional therefore T will be onto also. As a result, T becomes invertible.

Exercises

1. Determine all nonsingular linear transformations $T : \mathbb{R}^4 \longrightarrow \mathbb{R}^3$.

2. Let $\mathbb{C}^2$ be a vector space over $\mathbb{C}$, the field of complex numbers. Let the linear transformation $T : \mathbb{C}^2 \longrightarrow \mathbb{C}^2$ be defined by $T(z_1, z_2) = (\alpha z_1 + \beta z_2, \gamma z_3 + \delta z_4)$, where $z_1, z_2 \in \mathbb{C}$ and $\alpha, \beta, \gamma, \delta$ are fixed scalars. Prove that T is nonsingular if and only if $\alpha\delta - \beta\gamma \neq 0$.

3. If α is a nonzero scalar, then prove that the linear transformation $T : V \longrightarrow U$ is singular if and only if αT is singular.

4. Suppose that V is a finite dimensional vector space. Let T be a linear transformation on V such that $r(T^2) = r(T)$. Show that Ker $T \bigcap$ Im $T = \{0\}$.

5. Let $S : U \longrightarrow V$ and $T : V \longrightarrow U$ are linear maps. Show that if S and T are nonsingular, then TS is also nonsingular. Give an example such that TS is nonsingular but T is singular.

6. Let $S, T : U \longrightarrow V$ be linear transformations, where V is a finite dimensional vector space. Then prove that

   (a) $|r(S) - r(T)| \leq r(S + T) \leq r(S) + r(T)$;
   (b) $r(\alpha T) = r(T)$ for each nonzero $\alpha \in \mathbb{F}$.

7. Draw the image of unit square in $\mathbb{R}^2$ under the linear transformation $T : \mathbb{R}^2 \longrightarrow \mathbb{R}^2$ defined by $T(x, y) = (x - y, 2x - 3y)$.

8. Let $T \in A(V)$ be such that $T^2 - T + I = 0$, where I is the identity linear operator on V. Then show that T is invertible, also find the following:

(a) T^{-1}

(b) $2T - T^{-1}$.

9. Let S and T be linear transformations on a finite dimensional vector space. Show that if $ST = I$, then $TS = I$. Is this true for infinite dimensional vector spaces?

10. Let $A, B, C, D \in M_{n \times n}(\mathbb{C})$. Define T on $M_{n \times n}(\mathbb{C})$ by $T(X) = AXB + CX + XD$, $X \in M_{n \times n}(\mathbb{C})$. Show that T is a linear transformation on $M_{n \times n}(\mathbb{C})$ and that when $C = D = 0$, T has an inverse if and only if A and B are invertible.

3.5 Matrix of a Linear Transformation

Throughout this section all the vector spaces will be finite dimensional. Consider an ordered basis $B = \{u_1, u_2, \ldots, u_m\}$ of a vector space V over a field $\mathbb{F}$. For any vector $u \in V$, there exist unique scalars $\alpha_1, \alpha_2, \ldots, \alpha_m$ such that $u = \alpha_1 u_1 + \alpha_2 u_2 + \cdots + \alpha_m u_m$. The coordinate vector of u relative to the ordered basis B, i.e., $[u]_B$ is a column vector denoted by $[u]_B = [\alpha_1, \alpha_2, \ldots, \alpha_m]^T$. Now, let V be an n-dimensional vector space with an ordered basis $B' = \{v_1, v_2, \ldots, v_n\}$ and $T : U \to V$ be a linear transformation. Then, for any $u_i \in U, T(u_i) \in V$. Therefore, for any $1 \le i \le m$ if $T(u_i) = \sum_{j=1}^{n} \alpha_{ji} v_j$, $\alpha_{ij} \in \mathbb{F}$, then it can be expressed as

$$T(u_1) = \alpha_{11} v_1 + \alpha_{21} v_2 + \cdots + \alpha_{n1} v_n$$
$$T(u_2) = \alpha_{12} v_1 + \alpha_{22} v_2 + \cdots + \alpha_{n2} v_n$$

$$\vdots$$

$$T(u_m) = \alpha_{1m} v_1 + \alpha_{2m} v_2 + \cdots + \alpha_{nm} v_n.$$

The transpose matrix of the coefficients matrix of the above equations of order $n \times m$ is called the matrix of T relative to the ordered bases B and B' of the vector spaces U and V, respectively, and is denoted as $m(T)_{(B,B')}$ or $[T]_B^{B'}$. The ith column of this matrix is the coefficients of $T(u_i)$, when it is expressed as a linear combination of v_j, $j = 1, 2, 3, \ldots, n$.

Example 3.42 Let $T : \mathbb{R}^3 \to \mathbb{R}^2$ be a linear transformation such that $T(a, b, c) = (a + b, b - c)$ for all $a, b, c \in \mathbb{R}$. We shall find the matrix of T relative to the ordered standard basis of $\mathbb{R}^3$ and $\mathbb{R}^2$, respectively. Now we can write

$$T(1, 0, 0) = 1(1, 0) + 0(0, 1)$$
$$T(0, 1, 0) = 1(1, 0) + 1(0, 1)$$
$$T(0, 0, 1) = 0(1, 0) + (-1)(0, 1).$$

Thus the matrix of T relative to the ordered standard basis of $\mathbb{R}^3$ and $\mathbb{R}^2$ is

$$[T] = \begin{bmatrix} 1 & 1 & 0 \\ 0 & 1 & -1 \end{bmatrix}.$$

Observe that $T(a, b, c) = (a + b, b - c) = \begin{bmatrix} 1 & 1 & 0 \\ 0 & 1 & -1 \end{bmatrix} \begin{bmatrix} a \\ b \\ c \end{bmatrix}$. Hence if we take

the ordered standard bases of both $\mathbb{R}^3$ and $\mathbb{R}^2$, then it is clear that the matrix of T relative to the ordered standard basis is the matrix in the above equation. Now we shall find the matrix of T relative to another basis say $\{(1, 1, 0), (1, 0, 1), (0, 1, 1)\}$ and $\{(1, 2), (2, 1)\}$ of $\mathbb{R}^3$ and $\mathbb{R}^2$, respectively. In fact,

$$T(1, 1, 0) = 0(1, 2) + 1(2, 1)$$
$$T(1, 0, 1) = -1(1, 2) + 1(2, 1)$$
$$T(0, 1, 1) = \tfrac{-1}{3}(1, 2) + \tfrac{2}{3}(2, 1).$$

Thus $[T] = \begin{bmatrix} 0 & -1 & \frac{-1}{3} \\ 1 & 1 & \frac{2}{3} \end{bmatrix}.$

Remark 3.43 (i) It is straightforward to see that the matrix of a linear transformation changes according to the choice of a basis of the underlying vector space. In fact, if we change the order of element in a given basis even then we get a different matrix of the linear transformation. Since the basis of a vector space is not unique, the order of element and the basis both play important role in finding the matrix of a linear transformation.

(ii) The matrix of the zero linear transformation is always the zero matrix for any choice of basis. However, the matrix of the identity linear transformation is the identity matrix with respect to the same basis, otherwise it may be different from the identity matrix.

Theorem 3.44 *Let $A = (\delta_{ji})$ be an $n \times m$ matrix. Fix bases $B = \{u_1, u_2, \ldots, u_m\}$ and $B' = \{v_1, v_2, \ldots, v_n\}$ of U and V, respectively. Let $u \in U$ such that $u = \sum\limits_{i=1}^{m} \alpha_i u_i$. Define a map $T : U \to V$ such that $T(u) = \sum\limits_{j=1}^{n} \beta_j v_j$, where each of β_j is defined as $\beta_j = \sum\limits_{k=1}^{m} \delta_{jk} \alpha_k$. Then $T : U \to V$ is a linear transformation such that $[T]_B^{B'} = A$.*

Proof First, we show that T is well defined. For this let $u = u'$, where $u = \sum\limits_{k=1}^{m} \alpha_k u_k$ and $u' = \sum\limits_{k=1}^{m} \alpha'_k u_k$. Thus we have $T(u) = \sum\limits_{j=1}^{n} \beta_j v_j$, where each of β_j is defined as $\beta_j = \sum\limits_{k=1}^{m} \delta_{jk} \alpha_k$ and $T(u') = \sum\limits_{j=1}^{n} \beta'_j v_j$, where each of β'_j is defined as $\beta'_j = \sum\limits_{k=1}^{m} \delta_{jk} \alpha'_k$. As B is a basis of U and $u = u'$, we conclude that $\alpha_k = \alpha'_k$ for each $k = 1, 2, 3, \ldots, n$. As a result, $\beta_j = \beta'_j$ for each $j = 1, 2, 3, \ldots, n$. Thus $T(u) = T(u')$ stands proved.

Let $\alpha, \beta \in \mathbb{F}$. Then $\alpha u + \beta u' = \sum_{k=1}^{m} (\alpha \alpha_k + \beta \alpha_k') u_k$ and let $T(\alpha u + \beta u') = \sum_{j=1}^{n} \gamma_j v_j$, where each of γ_j is defined as $\gamma_j = \sum_{k=1}^{m} \delta_{jk}(\alpha \alpha_k + \beta \alpha_k')$.

$$
\begin{aligned}
T(\alpha u + \beta u') &= \sum_{j=1}^{n} \gamma_j v_j \\
&= \sum_{j=1}^{n} \left(\sum_{k=1}^{m} \delta_{jk}(\alpha \alpha_k + \beta \alpha_k') \right) v_j \\
&= \alpha \sum_{j=1}^{n} \left(\sum_{k=1}^{m} \delta_{jk}\alpha_k \right) v_j + \beta \sum_{j=1}^{n} \left(\sum_{k=1}^{m} \delta_{jk}\alpha_k' \right) v_j \\
&= \alpha T(u) + \beta T(u').
\end{aligned}
$$

Thus T is a linear transformation.

Here we also have

$$
\begin{aligned}
T(u_1) &= \delta_{11}v_1 + \delta_{21}v_2 + \cdots + \delta_{n1}v_n \\
T(u_2) &= \delta_{12}v_1 + \delta_{22}v_2 + \cdots + \delta_{n2}v_n
\end{aligned}
$$

$$\vdots$$

$$
T(u_m) = \delta_{1m}v_1 + \delta_{2m}v_2 + \cdots + \delta_{nm}v_n.
$$

Obviously $[T]_B^{B'} = (\delta_{ji}) = A$. This completes the proof of the theorem.

Theorem 3.45 *Let $T_1, T_2 : U \to V$ be linear transformations, where U and V are vector spaces over $\mathbb{F}$ of dimensions m and n, respectively, with ordered bases B and B'. Then the following hold:*

(i) If $T_1 = T_2$, then $[T_1]_B^{B'} = [T_2]_B^{B'}$.
(ii) $[T_1 + T_2]_B^{B'} = [T_1]_B^{B'} + [T_2]_B^{B'}$.
(iii) $[\alpha T_1]_B^{B'} = \alpha [T_1]_B^{B'}$.

Proof (i) Let $B = \{x_1, x_2, \ldots, x_m\}$, $B' = \{y_1, y_2, \ldots, y_n\}$ and $[T_1]_B^{B'} = (\alpha_{ji})$, $[T_2]_B^{B'} = (\beta_{ji})$. We have

$$
\begin{aligned}
T_1 = T_2 &\Leftrightarrow T_1(x_i) = T_2(x_i), \quad \text{for } i = 1, 2, 3, \ldots, m \\
&\Leftrightarrow \sum_{j=1}^{n} \alpha_{ji} y_j = \sum_{j=1}^{n} \beta_{ji} y_j \\
&\Leftrightarrow \sum_{j=1}^{n} (\alpha_{ji} - \beta_{ji}) y_j = 0 \\
&\Leftrightarrow \alpha_{ji} = \beta_{ji} \text{ for each } i \text{ and each } j \\
&\Leftrightarrow (\alpha_{ji}) = (\beta_{ji}) \\
&\Leftrightarrow [T_1]_B^{B'} = [T_2]_B^{B'}.
\end{aligned}
$$

(ii) Since $T_1 + T_2 : U \to V$ is also a linear transformation and also,

$$(T_1 + T_2)(x_i) = T_1(x_i) + T_2(x_i), \quad \text{for} \quad i = 1, 2, 3, \ldots, m$$
$$= \sum_{j=1}^{n} \alpha_{ji} y_j + \sum_{j=1}^{n} \beta_{ji} y_j$$
$$= \sum_{j=1}^{n} (\alpha_{ji} + \beta_{ji}) y_j.$$

Therefore $[T_1 + T_2]_B^{B'} = [(\alpha_{ji} + \beta_{ji})] = (\alpha_{ji}) + (\beta_{ji}) = [T_1]_B^{B'} + [T_2]_B^{B'}$.

(iii) As $\alpha T_1 : U \to V$ is also a linear transformation and also,

$$(\alpha T_1)(x_i) = \alpha T_1(x_i), \quad \text{for} \quad i = 1, 2, 3, \ldots, m$$
$$= \alpha \left(\sum_{j=1}^{n} \alpha_{ji} y_j \right)$$
$$= \sum_{j=1}^{n} (\alpha \alpha_{ji}) y_j.$$

Thus $[\alpha T_1]_B^{B'} = (\alpha \alpha_{ji}) = \alpha(\alpha_{ji}) = \alpha [T_1]_B^{B'}$.

Theorem 3.46 *Let $T_1 : U \to V$ and $T_2 : V \to W$ be linear transformations, where U, V and W are vector spaces over $\mathbb{F}$ of dimensions m, n and p, respectively, with ordered bases B, B' and B''. Then $[T_2 T_1]_B^{B''} = [T_2]_{B'}^{B''} [T_1]_B^{B'}$.*

Proof Let $B = \{x_1, x_2, \ldots, x_m\}$, $B' = \{y_1, y_2, \ldots, y_n\}$, $B'' = \{z_1, z_2, \ldots, z_p\}$ and $[T_1]_B^{B'} = (\alpha_{ji})$, $[T_2]_{B'}^{B''} = (\beta_{kj})$. Clearly, $T_2 T_1 : U \to W$ is a linear transformation. Also, we have

$$(T_2 T_1)(x_i) = T_2(T_1(x_i)), \quad \text{for} \quad i = 1, 2, 3, \ldots, m$$
$$= T_2 \left(\sum_{j=1}^{n} \alpha_{ji} y_j \right)$$
$$= \sum_{j=1}^{n} \alpha_{ji} T_2(y_j)$$
$$= \sum_{j=1}^{n} \alpha_{ji} \left(\sum_{k=1}^{p} \beta_{kj} z_k \right)$$
$$= \sum_{k=1}^{p} \left(\sum_{j=1}^{n} \beta_{kj} \alpha_{ji} \right) z_k$$
$$= \sum_{k=1}^{p} \gamma_{ki} z_k, \quad \text{where} \quad \gamma_{ki} = \sum_{j=1}^{n} \beta_{kj} \alpha_{ji}.$$

Thus $[T_2 T_1]_B^{B''} = (\gamma_{ki}) = (\beta_{kj})(\alpha_{ji}) = [T_2]_{B'}^{B''} [T_1]_B^{B'}$.

Remark 3.47 If in the above theorem, we take $U = V = W$ and $B = B' = B''$, then it is obvious to observe the following fact: $[T_2 T_1]_B^B = [T_2]_B^B [T_1]_B^B$.

Theorem 3.48 *If U and V are vector spaces over $\mathbb{F}$ of dimensions n and m, respectively, then $Hom(U, V) \cong M_{m \times n}(\mathbb{F})$, where $M_{m \times n}(\mathbb{F})$ denotes the vector space of all $m \times n$ matrices with entries from $\mathbb{F}$ and hence dimension of the vector space $M_{m \times n}(\mathbb{F}) = mn$.*

Proof Let B and B' be ordered bases of U and V, respectively. Define a map $f : Hom(U, V) \to M_{m \times n}(\mathbb{F})$ such that $f(T) = [T]_B^{B'}$. In the light of Theorem 3.45 f is well defined and one-to-one. We also have $f(\alpha T_1 + \beta T_2) = [\alpha T_1 + \beta T_2]_B^{B'} = [\alpha T_1]_B^{B'} + [\beta T_2]_B^{B'} = \alpha [T_1]_B^{B'} + \beta [T_2]_B^{B'} = \alpha f(T_1) + \beta f(T_2)$ for every $\alpha, \beta \in \mathbb{F}$ and $T_1, T_2 \in Hom(U, V)$. Thus f is a linear transformation. Using Theorem 3.44, it is easy to observe that f is onto. Now we conclude that f is an isomorphism and hence $Hom(U, V) \cong M_{m \times n}(\mathbb{F})$. Therefore, $dim M_{m \times n}(\mathbb{F}) = dim Hom(U, V) = dim U \; dim V = mn$.

Theorem 3.49 *Let V be an n-dimensional vector space over $\mathbb{F}$. Then the algebras $\mathscr{A}(V)$ and $M_{n \times n}(\mathbb{F})$ are isomorphic.*

Proof Let B be an ordered basis of V. Define a map $f : \mathscr{A}(V) \to M_{n \times n}(\mathbb{F})$ such that $f(T) = [T]_B^B$. Using Theorems 3.44, 3.45 and Remark 3.47, one can easily verify the following. For any $T_1, T_2 \in L(V, V), \alpha, \beta \in \mathbb{F}$: (*i*) f is well defined, (*ii*) $f(\alpha T_1 + \beta T_2) = \alpha f(T_1) + \beta f(T_2)$, (*iii*) $f(T_1 T_2) = f(T_1) f(T_2)$, (*iv*) f is a bijective map. Thus f is an algebra isomorphism and hence the algebras $\mathscr{A}(V)$ and $M_{n \times n}(\mathbb{F})$ are isomorphic.

3.6 Effect of Change of Bases on a Matrix Representation of a Linear Transformation

Theorem 3.50 *Let $T : U \longrightarrow V$ be a linear transformation, where U and V are vector spaces over $\mathbb{F}$ of dimensions n and m, respectively. Let B and B' be two ordered bases of U and B_1, B_1' be two ordered bases of V. Then $[T]_B^{B_1}$ and $[T]_{B'}^{B_1'}$ are equivalent matrices and $[T]_{B'}^{B_1'} = P[T]_B^{B_1} S^{-1}$, where P is the transition matrix of B_1 relative to B_1' and S is the transition matrix of B relative to B'.*

Proof Let I_U and I_V denote the identity linear operators on the vector spaces U and V, respectively. It is clear that $T = I_V T I_U$. Using Theorem 3.46, we have $[T]_{B'}^{B_1'} = [I_V T I_U]_{B'}^{B_1'} = [I_V]_{B_1}^{B_1'} [T]_B^{B_1} [I_U]_{B'}^B$. Now let us put $P = [I_V]_{B_1}^{B_1'}$ and $Q = [I_U]_{B'}^B$. Thus we can write $[T]_{B'}^{B_1'} = P[T]_B^{B_1} Q$, where clearly P and Q are matrices of orders $m \times m$ and $n \times n$, respectively. Now we claim that P and Q are invertible matrices. Since $I_V = I_V I_V$, we find that $[I_V]_{B_1}^{B_1} = [I_V I_V]_{B_1}^{B_1} = [I_V]_{B_1'}^{B_1} [I_V]_{B_1}^{B_1'}$. Similarly, we also have $[I_V]_{B_1'}^{B_1'} = [I_V I_V]_{B_1'}^{B_1'} = [I_V]_{B_1}^{B_1'} [I_V]_{B_1'}^{B_1}$. But we know that $[I_V]_{B_1}^{B_1} = [I_V]_{B_1'}^{B_1'} = I_{n \times n}$, the identity matrix of order $n \times n$. Hence we conclude that $[I_V]_{B_1'}^{B_1} [I_V]_{B_1}^{B_1'} =$

$[I_V]_{B_1}^{B_1'}[I_V]_{B_1'}^{B_1} = I_{n \times n}$, i.e., $[I_V]_{B_1'}^{B_1} P = P[I_V]_{B_1'}^{B_1} = I_{n \times n}$. This shows that P is an invertible matrix. In the similar lines one can show that Q is also an invertible matrix. This proves that $[T]_B^{B_1}$ and $[T]_{B'}^{B_1'}$ are equivalent matrices.

As we have proved that $[T]_{B'}^{B_1'} = P[T]_B^{B_1} Q$, where $P = [I_V]_{B_1}^{B_1'}$, i.e., the transition matrix of B_1 relative to B_1' and $Q = [I_U]_{B'}^{B}$. If we put $S = Q^{-1}$, then $[T]_{B'}^{B_1'} = P[T]_B^{B_1} S^{-1}$, where $S = Q^{-1} = [I_U]_B^{B'}$, i.e., the transition matrix of B relative to B'. This completes the proof.

Corollary 3.51 *Let T be a linear operator on an n-dimensional vector space U. If B and B' are ordered bases of U, then $[T]_B^B$ and $[T]_{B'}^{B'}$ are similar matrices and also $[T]_{B'}^{B'} = P[T]_B^B P^{-1}$, where P is the transition matrix of B relative to B'.*

Proof Let I_U be the identity linear operator on the vector space U. If in particular, in the above theorem, we put $V = U$, $B = B_1$ and $B' = B_1'$, then we get $[T]_{B'}^{B'} = P[T]_B^B Q$, where $P = [I_U]_B^{B'}$ and $Q = [I_U]_{B'}^B$ are invertible matrices of order $n \times n$, i.e., $[T]_B^B$ and $[T]_{B'}^{B'}$ are equivalent matrices. Now we claim that $Q = P^{-1}$. Since $I_U = I_U I_U$, we find that $[I_U]_B^B = [I_U I_U]_B^B = [I_U]_{B'}^B[I_U]_B^{B'}$. In the similar way, $[I_U]_{B'}^{B'} = [I_U I_U]_{B'}^{B'} = [I_U]_B^{B'}[I_U]_{B'}^B$. But $[I_U]_B^B = [I_U]_{B'}^{B'} = I_{n \times n}$, the identity matrix of order $n \times n$. Hence we conclude that $[I_U]_{B'}^B[I_U]_B^{B'} = [I_U]_B^{B'}[I_U]_{B'}^B = I_{n \times n}$, i.e., $PQ = QP = I_{n \times n}$. This implies that $Q = P^{-1}$. Finally we have shown that $[T]_{B'}^{B'} = P[T]_B^B P^{-1}$. This implies that $[T]_B^B$ and $[T]_{B'}^{B'}$ are similar matrices.

As we have obtained that $[T]_{B'}^{B'} = P[T]_B^B P^{-1}$, where $P = [I_U]_B^{B'}$, which is obviously the transition matrix of B relative to B'.

Let $T : U \longrightarrow V$ be a linear transformation, where U and V are finite dimensional vector spaces over $\mathbb{F}$. Theorem 3.50 and Corollary 3.51 show that matrices of T relative to two different sets of bases are equivalent or similar. The next two results prove their converse.

Theorem 3.52 *Let $T : U \longrightarrow V$ be a linear transformation, where U and V are vector spaces over $\mathbb{F}$ of dimensions n and m, respectively. Let B and B_1 be ordered bases of U and V, respectively, and let $C = [T]_B^{B_1}$. If C is equivalent to D, then there exist ordered bases B', B_1' of U and V, respectively, such that $D = [T]_{B'}^{B_1'}$.*

Proof As C is equivalent to D, there exists nonsingular $m \times m$ matrix P and nonsingular $n \times n$ matrix Q such that $D = PCQ$. Let $S = Q^{-1}$. Then clearly $D = PCS^{-1}$. By Theorem 3.50, there exists an ordered basis B_1' of V such that the transition matrix of B_1' relative to B_1 is P^{-1}, i.e., $[I_V]_{B_1'}^{B_1} = P^{-1}$ and there exists an ordered basis B' of U such that Q is the transition matrix of B' relative to B, i.e., $[I_U]_{B'}^B = Q$. As a result, P is the transition matrix of B_1 relative to B_1' and $Q^{-1} = S$ is the transition matrix of B relative to B'. But by Theorem 3.50, $[T]_{B'}^{B_1'} = P[T]_B^{B_1} S^{-1} = PCQ = D$. This completes the proof.

Corollary 3.53 *Let T be a linear operator on an n-dimensional vector space U. For some ordered basis B of U, let $A = [T]_B^B$. Let A be similar to D. Then there exists an ordered basis B' of U such that $[T]_{B'}^{B'} = D$.*

Proof Since A is similar to D, there exists an invertible matrix P such that $D = PAP^{-1}$. By Theorem 3.50, there exists an ordered basis B' of U such that the transition matrix of B' relative to B is P^{-1}, i.e., $[I_U]_{B'}^B = P^{-1}$. This implies that $[I_U]_B^{B'} = P$, the transition matrix of B relative to B'. But Corollary 3.51, provides us $[T]_{B'}^{B'} = P[T]_B^B P^{-1} = PAP^{-1} = D$.

Theorem 3.54 *Let A be any $m \times n$ matrix over $\mathbb{F}$, B and B' be the standard bases of $\mathbb{F}^{n \times 1}$ and $\mathbb{F}^{m \times 1}$, respectively. Then, for $T : \mathbb{F}^{n \times 1} \longrightarrow \mathbb{F}^{m \times 1}$, given by $T(X) = AX$, $[T]_B^{B'} = A$. For $m = n$, $[T]_B^B = A$.*

Proof Let $B = \{e_1, e_2, \ldots, e_n\}$, $B' = \{e_1', e_2', \ldots, e_m'\}$, where $e_j = (\alpha_{t1})_{n \times 1}$, $j = 1, 2, \ldots, n$ and $\alpha_{t1} = 1$ if $t = j$ for otherwise it equals 0, $e_i' = (\alpha_{l1}')_{m \times 1}$, $i = 1, 2, \ldots, m$ and $\alpha_{\ell 1}' = 1$ if $\ell = i$ for otherwise it equals 0. Then, for any $m \times n$

$$\text{matrix } A = (a_{ij}), \text{ over } \mathbb{F}, \ T(e_j) = Ae_j = A \begin{bmatrix} 0 \\ 1 \\ \vdots \\ 0 \end{bmatrix} = \begin{bmatrix} a_{1j} \\ a_{2j} \\ \vdots \\ a_{mj} \end{bmatrix} = \sum_{i=1}^{m} a_{ij} e_i', \text{ where}$$

$j = 1, 2, \ldots, n$. Hence $[T]_B^{B'} = (a_{ij}) = A$. Obviously for $m = n$, $B' = B$, so $A = (a_{ij}) = [T]_B^B$.

Theorem 3.55 *Let $T : U \longrightarrow V$ be a linear transformation, where U and V are vector spaces over $\mathbb{F}$ of dimensions n and m, respectively. Let B and B' be ordered bases of U and V, respectively, and let $A = [T]_B^{B'}$. Then*

(i) *for any $u \in U$, $[T(u)]_{B'}^{B'} = A[u]_B^B$,*
(ii) *rank (T) = rank (A).*

Proof (i) Let $B = \{x_1, x_2, \ldots, x_n\}$, $B' = \{y_1, y_2, \ldots, y_m\}$ and $A = [T]_B^{B'} = (a_{ij})$. By definition $T(x_j) = \sum_{i=1}^{m} a_{ij} y_i$, $j = 1, 2, \ldots, n$. Consider $v \in V$. Let

$$[u]_B = \begin{bmatrix} \alpha_1 \\ \alpha_2 \\ \vdots \\ \alpha_n \end{bmatrix}.$$

Then $u = \sum_{j=1}^{n} \alpha_j x_j$. This implies that $T(u) = \sum_{j=1}^{n} \alpha_j T(x_j) = \sum_{j=1}^{n} \alpha_j [\sum_{i=1}^{m} a_{ij} y_i]$, i.e., $T(u) = \sum_{i=1}^{m} [\sum_{j=1}^{n} a_{ij}] y_i$. Thus, in $[T(u)]_{B'}^{B'}$, the $(i, 1)$th entry $= \sum_{j=1}^{n} a_{ij}$. It can be easily seen that the $(i, 1)$th entry of $A[u]_B^B = \sum_{j=1}^{n} a_{ij}$. This implies that $(i, 1)$th entry of $[T(u)]_{B'}^{B'} = (i, 1)$th entry of $A[u]_B^B$ for each $i = 1, 2, \ldots, m$. Thus $[T(u)]_{B'}^{B'} = A[u]_B^B$. This completes the proof of (i).

(*ii*) Consider the following homogeneous system of linear equations, given in

matrix form, i.e., $AX = O$, where $X = \begin{bmatrix} x_1 \\ x_2 \\ \vdots \\ x_n \end{bmatrix}$. Let W denote the solution space of

this system.

Next suppose that $Dim\,W = n - r$, where $r = rank(A)$, because the above system contains $(n - r)$ linearly independent solutions. Now define a map $f : U \longrightarrow \mathbb{F}^{n \times 1}$ such that $f(u) = [u]_B^B$, it can be easily verified that f is an isomorphism. Let $u \in Ker\,T \Longleftrightarrow T(u) = 0 \Longleftrightarrow [T(u)]_{B'}^{B'} = 0 \Longleftrightarrow A[u]_B^B = 0 \Longleftrightarrow [u]_B^B \in W$. But as $f(u) = [u]_B^B$. We conclude that $u \in Ker\,T \Longleftrightarrow f(u) \in W$. Therefore $N(T) = f^{-1}(W)$, i.e., $n(T) = dim[f^{-1}(W)]$. Since $f^{-1} : \mathbb{F}^{n \times 1} \longrightarrow U$ is also an isomorphism, therefore its restriction on subspace W, i.e., $\frac{f^{-1}}{W} : W \longrightarrow f^{-1}(W)$ is also an isomorphism. Thus by Rank-Nullity theorem $dim\,W = dim f^{-1}(W)$. Now we conclude that $n(T) = dim\,W$, i.e., $dim(U) - r(T) = n - r = dim(U) - r(A)$. This implies that $r(T) = r(A)$, i.e., $rank(T) = rank(A)$

Corollary 3.56 *Let T be a linear operator on the vector space U of dimension n and B be an ordered basis of U. Then T is an automorphism of U if and only if $A = [T]_B^B$ is nonsingular.*

Proof We know that T is an automorphism if and only if $Rank\,T = dim\,U = n$. Now using the above theorem, we also have T is an automorphism if and only if $Rank\,T = Rank\,A$. Now we conclude that $Rank\,A = n$. But we know that A is nonsingular if and only if $Rank\,A = n$. Thus it follows that T is an automorphism of U if and only if $A = [T]_B^B$ is nonsingular.

Example 3.57 Let T be the linear operator on $\mathbb{C}^2$ defined by $T(x_1, x_2) = (x_1, 0)$. Let B be the standard ordered basis for $\mathbb{C}^2$ and let $B' = \{\alpha_1, \alpha_2\}$ be the ordered basis defined by $\alpha_1 = (1, i), \alpha_2 = (-i, 2)$.

(1) The matrix of T relative to the pair of bases B, B'.
      We have $B = \{(1, 0), (0, 1)\}$. Since

$$T(1, 0) = (1, 0) = 2(1, i) + -i(-i, 2), \quad T(0, 1) = (0, 0) = 0(1, i) + 0(-i, 2),$$

we find that $[T]_B^{B'} = \begin{bmatrix} 2 & 0 \\ -i & 0 \end{bmatrix}$.

(2) The matrix of T relative to the pair of bases B', B.
      We have

$$T(1, i) = (1, 0) = 1(1, 0) + 0(0, 1), \quad T(-i, 2) = (-i, 0) = -i(1, 0) + 0(0, 1).$$

Thus we get $[T]_{B'}^B = \begin{bmatrix} 1 & -i \\ 0 & 0 \end{bmatrix}$.

(3) The matrix of T relative to the ordered basis B'.
      Since

$$T(1, i) = (1, 0) \quad = \quad 2(1, i) + (-i)(-i, 2),$$
$$T(-i, 2) = (-i, 0) = (-2i)(1, i) + (-1)(-i, 2)$$

we find that $[T]_{B'}^{B'} = \begin{bmatrix} 2 & -2i \\ -i & -1 \end{bmatrix}$.

(4) The matrix of T relative to the ordered basis $\{\alpha_2, \alpha_1\}$.
Since
$$T(-i, 2) = (-i, 0) = (-1)(-i, 2) + (-2i)(1, i),$$
$$T(1, i) = (1, 0) \quad = \quad (-i)(-i, 2) + 2(1, i)$$

the matrix of T relative to ordered basis $\{\alpha_2, \alpha_1\}$ is $[T] = \begin{bmatrix} -1 & -i \\ -2i & 2 \end{bmatrix}$.

Exercises

1. Let $V = \mathbb{R}^3$ and suppose that $\begin{bmatrix} 1 & 0 & 3 \\ -1 & -4 & 3 \\ 6 & 2 & 1 \end{bmatrix}$ is the matrix of $T \in \mathscr{A}(V)$ relative to the basis $\{e_1, e_2, e_3\}$, where $e_1 = (1, 0, 0)$, $e_2 = (0, 1, 0)$, $e_3 = (0, 0, 1)$. Find the matrix of T relative to the basis $\{e_1', e_2', e_3'\}$, where $e_1' = (1, 1, 0)$, $e_2' = (1, 0, 1)$ and $e_3' = (0, 1, 1)$.

2. Let V be the vector space of all polynomials of degree less than or equal to 3 over $\mathbb{R}$. In $\mathscr{A}(V)$, define T by $T(\alpha_0 + \alpha_1 x + \alpha_2 x^2 + \alpha_3 x^3) = \alpha_0 + \alpha_1 (x + 1) + \alpha_2 (x + 1)^2 + \alpha_3 (x + 1)^3$. Compute the matrix of T relative to bases

   (i) $\{1, x, x^2, x^3\}$;
   (ii) $\{1, 1 + x, 1 + x^2, 1 + x^3\}$.

   If the matrix in part (i) is A and that in part (ii) is B, then prove that A and B are similar matrices.

3. Let $T : \mathbb{R}^2 \longrightarrow \mathbb{R}^2$ be the linear operator such that $T(x, y) = (x - y, 2x + y)$. Let $B = \{(1, 0), (0, 1)\}$ and $B' = \{(1, 2), (2, 1)\}$ be ordered bases of $\mathbb{R}^2$.

   (a) Find $[T]_B^B, [T]_{B'}^{B'}, [T]_B^{B'}, [T]_{B'}^B$.
   (b) Find the transition matrix P of B relative to B' and verify that $[T]_{B'}^{B'} = P[T]_B^B P^{-1}$.
   (c) Find the formula for T^{-1}, and find $[T^{-1}]_{B'}^{B'}$. Also verify that $[T^{-1}]_{B'}^{B'} = ([T]_{B'}^{B'})^{-1}$.
   (d) Find $[T^{-1}]_{B'}^B$ and also verify $[T^{-1}]_{B'}^B = ([T]_B^{B'})^{-1}$.

4. Let $P_2(\mathbb{R})$ be the vector space of all polynomials of degree less than or equal to 2 over $\mathbb{R}$ and let $T : \mathbb{R}^2 \longrightarrow P_2(\mathbb{R})$ be given by $T(\alpha, \beta) = \beta x + \alpha x^2$. Consider the following bases $B = \{(1, -2), (-3, 0)\}$ of $\mathbb{R}^2$ and $B' = \{1, x, x^2\}$ of $P_2(\mathbb{R})$.

   (a) Find $[T]_B^{B'}$.
   (b) Verify that $[T(3, -4)]_{B'}^{B'} = [T]_B^{B'}[(3, -4)]_B^B$.
   (c) Verify that $Rank\, T = Rank[T]_B^{B'}$.

5. Let $T : \mathbb{R}^3 \longrightarrow \mathbb{R}^3$ be the linear operator whose matrix relative to the standard
   ordered basis $B = \{e_1=(1, 0, 0), e_2 = (0, 1, 0), e_3 = (0, 0, 1)\}$ is $\begin{bmatrix} 1 & 3 & 1 \\ -1 & -3 & 1 \\ 0 & 2 & 4 \end{bmatrix}$.

   (a) Find $T(e_1), T(e_2), T(e_3)$ and determine T. Prove that T is invertible and
       determine T^{-1}.
   (b) Find the matrix of each of the following relative to the standard ordered
       basis $T^2, T^2 + T, T^2 + I, (-2T)^3 + 6T^2 - I$.
   (c) Find $[T]_{B'}^{B'}$, where $B' = \{e_1' = (1, 1, 1), e_2' = (1, 1, 0), e_3' = (1, 0, 0)\}$.
   (d) Find a basis of $Ker T$ and a basis of $Range T$.
   (e) Show that both matrices $[T]_B^B$ and $[T]_{B'}^{B'}$ have the same rank.
   (f) Prove that $[T]_B^B$ and $[T]_{B'}^{B'}$ are similar matrices.

6. Let $P_3(\mathbb{R})$ be the vector space of all polynomials of degree less than or equal to
   3 over $\mathbb{R}$ and let $D : P_3(\mathbb{R}) \longrightarrow P_3(\mathbb{R})$ be the differentiation operator. Let B be
   the standard basis $\{1, x, x^2, x^3\}$ and $B' = \{1, 1 + x, (1 + x)^2, (1 + x)^3\}$.

   (a) Find $[D]_B^B, [D]_{B'}^{B'}, [D]_B^{B'}, [D]_{B'}^B$.
   (b) For $A = [D]_{B'}^{B'}$, verify that $A^4 = 0$, but $A^3 \neq 0$.
   (c) For any $\alpha \neq 0$ in $\mathbb{R}$, prove that $\alpha I + D$ is invertible.

7. Let V be the 2-dimensional vector space of solutions of the differential equation
   $y'' - 3y' + 2y = 0$ over $\mathbb{C}$ and let $B = \{y_1 = e^x, y_2 = e^{2x}\}$ be a basis of V and
   $D : V \longrightarrow V$ be the differentiation operator. Find $[D]_B^B$.
8. Let V be the vector space of all 2×2 matrices with real entries and let $T :$
   $V \longrightarrow \mathbb{R}$ be a map defined by $T(A) =$ the trace of A.

   (a) Show that the set $B = \left\{ \begin{bmatrix} 1 & 0 \\ 0 & 0 \end{bmatrix}, \begin{bmatrix} 0 & 1 \\ 0 & 0 \end{bmatrix}, \begin{bmatrix} 0 & 0 \\ 1 & 0 \end{bmatrix}, \begin{bmatrix} 0 & 0 \\ 0 & 1 \end{bmatrix} \right\}$ is a basis for V.
   (b) Prove that T is a linear transformation and determine $[T]_B^{B'}$, where $B' = \{5\}$.
       Determine the dimension and a basis for the Kernel of T.

9. Let T be a linear operator on $\mathbb{R}^3$ defined by $T(x, y, z) = (2y + z, x - 4z, 3x - 6z)$.

   (a) Find $[T]_B^B$, where $B = \{(1, 1, 0), (1, 0, 1), (0, 1, 1)\}$.
   (b) Verify that $[T]_B^B [v]_B = [T(v)]_B$ for any $v \in \mathbb{R}^3$.

10. For each of the following linear operators T on $\mathbb{R}^2$, find the matrix, that is
    represented by T relative to the standard basis $B = \{(1, 0), (0, 1)\}$ of $\mathbb{R}^2$.

    (a) T is defined by $T(1, 0) = (2, 3), T(0, 1) = (3, -4)$.
    (b) T is the rotation in $\mathbb{R}^2$ counterclockwise by 90^0.
    (c) T is the reflection in $\mathbb{R}^2$ about the line $y = x$.

11. The set $B = \{e^{3t}, te^{3t}, t^2e^{3t}\}$ is a basis of a vector space V of functions $f :$
    $\mathbb{R} \longrightarrow \mathbb{R}$. Let D be the differential operator on V, i.e., $D(f) = \frac{df}{dt}$. Find $[D]_B^B$.

12. Let A be the matrix representation of a linear operator T on a vector space V over the field $\mathbb{F}$. Prove that, for any polynomial $f(x)$ over $\mathbb{F}$, we have that $f(A)$ is the matrix representation of $f(T)$.

13. Let $T : \mathbb{R}^n \longrightarrow \mathbb{R}^m$ be the linear mapping defined by $T(x_1, x_2, \ldots, x_n) = (\alpha_{11}x_1 + \cdots + \alpha_{1n}x_n, \alpha_{21}x_1 + \cdots + \alpha_{2n}x_n, \ldots, \alpha_{m1}x_1 + \cdots + \alpha_{mn}x_n)$, where $\alpha_{ij} \in \mathbb{R}, i = 1, 2, \ldots, m, j = 1, 2, \ldots, n$. Show that the rows of the matrix $[T]$ representing T relative to the usual bases of $\mathbb{R}^n$ and $\mathbb{R}^m$ are the coefficients of x_i in the components of $T(x_1, x_2, \ldots, x_n)$.

14. Let $T : U \longrightarrow V$ be a linear transformation, where U and V are vector spaces of dimensions m and n, respectively. Prove that there exist bases of U and V such that matrix representation of T relative to these bases has the form $\begin{bmatrix} I_r & 0 \\ 0 & 0 \end{bmatrix}$, where I_r is the r-square identity matrix.

15. Let V be the vector space of all 2×2 matrices with real entries. Consider the following matrix M and usual basis B of V; $M = \begin{bmatrix} a & b \\ c & d \end{bmatrix}$ and $B = \left\{ \begin{bmatrix} 1 & 0 \\ 0 & 0 \end{bmatrix}, \begin{bmatrix} 0 & 1 \\ 0 & 0 \end{bmatrix}, \begin{bmatrix} 0 & 0 \\ 1 & 0 \end{bmatrix}, \begin{bmatrix} 0 & 0 \\ 0 & 1 \end{bmatrix} \right\}$. Find $[T]_B^B$, represented by each of the following linear operators T on V.

(a) $T(A) = MA$,
(b) $T(A) = AM$,
(c) $T(A) = AM - MA$.

16. Suppose $T : V \longrightarrow V$ is linear. A subspace W of V is called invariant under T if $T(W) \subseteq W$. Suppose W is invariant under T and $dim W = r$. Show that T has a block triangular representation $M = \begin{bmatrix} A & B \\ 0 & C \end{bmatrix}$, where A is $r \times r$ submatrix.

17. Let $U = V + W$, where V and W are subspaces of U, which are invariant under a linear operator $T : U \longrightarrow U$. Also suppose that $dim V = r$ and $dim W = s$. Show that T has a block diagonal representation $M = \begin{bmatrix} A & 0 \\ 0 & B \end{bmatrix}$, where A and B are $r \times r$ and $s \times s$ submatrices.

18. Consider $\mathbb{C}$ as a vector space over $\mathbb{R}$. For $a \in \mathbb{C}$, let $T_a : \mathbb{C} \longrightarrow \mathbb{C}$ be a linear operator given by $T_a(x) = xa$ for all $x \in \mathbb{C}$. Find $[T_a]_B^B$, where $B = \{1, i\}$ and so get an isomorphic representation of the complex numbers as 2×2 matrices over real field.

19. Consider Q as a vector space over $\mathbb{R}$, where Q is the division ring of real quaternions. For $a \in Q$, let $T_a : Q \longrightarrow Q$ be a linear operator defined as $T_a(x) = xa$ for all $x \in Q$. Find $[T_a]_B^B$, where $B = \{1, i, j, k\}$ and so get an isomorphic representation of Q as 4×4 matrices over real field.

20. Suppose that x- and y-axes in the plane $\mathbb{R}^2$ are rotated counterclockwise 30^0 to yield new X- and Y-axes for the plane. Consider this rotation as a map $T : \mathbb{R}^2 \longrightarrow \mathbb{R}^2$ and prove that T is a linear transformation, also find the following:

(a) The change of basis matrix or transition matrix for the new coordinate system.

(b) $T(2, 3), T(-6, 8), T(-2, 6), T(3, 5), T(a, b)$.

21. Let T be a linear operator on $\mathbb{R}^3$ defined by $T(x, y, z) = (2x - y + z, x - y + 2z, 2x - 3y + z)$.

   (a) Find $[T]_B^B$ and $[T]_{B'}^{B'}$, where $B = \{(1, 1, 1), (1, 1, 0), (1, 0, 0)\}$ and $B' = \{(1, 1, 0), (1, 2, 3), (1, 3, 5)\}$.

   (b) Verify that Determinant $[T]_B^B$ = Determinant $[T]_{B'}^{B'}$.

22. Let $T : \mathbb{R}^3 \longrightarrow \mathbb{R}^2$ be a linear transformation given by $T(x, y, z) = (2x - y + z, -3x + 2y - z)$. Let $B = \{(1, 0, 0), (0, 1, 0), (0, 0, 1)\}$, $B' = \{(1, 1, 0), (0, 1, 2), (0, 1, 1)\}$ and $B_1 = \{(1, 1), (1, 2)\}$, $B_1' = \{(1, 0), (0, 2)\}$ be ordered bases of $\mathbb{R}^3$ and $\mathbb{R}^2$, respectively. Then

   (a) Find $[T]_B^{B_1}$ and $[T]_{B'}^{B_1'}$.

   (b) Verify that $[T]_B^{B_1}$ and $[T]_{B'}^{B_1'}$ are equivalent matrices.

   (c) Verify that Rank $[T]_B^{B_1}$ = Rank $[T]_{B'}^{B_1'}$.

   (d) Find out nonsingular matrices P and Q such that $[T]_{B'}^{B_1'} = P[T]_B^{B_1} Q$.

   (e) Verify that for any $v \in \mathbb{R}^3$, $[T(v)]_{B_1} = [T]_B^{B_1}[v]_B$ and $[T(v)]_{B_1'} = [T]_{B'}^{B_1'}[v]_{B'}$.

23. Let T be a linear transformation on an n-dimensional vector space V. If $T^{n-1}(v) \neq 0$ but $T^n(v) = 0$, for some $v \in V$, then $v, T(v), \ldots, T^{n-1}(v)$ are linearly independent, and thus form a basis of V. Find the matrix representation of T under this basis.

24. Let $C_\infty(\mathbb{R})$ be the vector space of real valued functions on $\mathbb{R}$ having derivative of all orders. Consider the differential operator $D(y) = -y'' + ay' + by$, $y \in C_\infty(\mathbb{R})$, where a and b are real constants. Show that $y = e^{\lambda x}$ lies in the $Ker D$ if and only if λ is a root of the quadratic equation $t^2 + at + b = 0$.

25. Let T be a linear transformation on $\mathbb{R}^2$ associated with the matrix $\begin{bmatrix} 2 & 1 \\ 0 & 2 \end{bmatrix}$ under the basis $\{\alpha_1 = (1, 0), \alpha_2 = (0, 1)\}$. Let W_1 be the subspace of $\mathbb{R}^2$ spanned by α_1. Show that W_1 is invariant under T and that there does not exist a subspace W_2 invariant under T such that $\mathbb{R}^2 = W_1 \oplus W_2$.

26. Let S and T be linear transformations on $\mathbb{R}^2$. Given that the matrix representation of S under the basis $\{\alpha_1 = (1, 2), \alpha_2 = (2, 1)\}$ is $\begin{bmatrix} 1 & 2 \\ 2 & 3 \end{bmatrix}$, and the matrix representation of T under the basis $\{\beta_1 = (1, 1), \beta_2 = (1, 2)\}$ is $\begin{bmatrix} 3 & 3 \\ 2 & 4 \end{bmatrix}$. Let $u = (3, 3) \in \mathbb{R}^2$. Find

   (a) The matrix of $S + T$ under the basis $\{\beta_1, \beta_2\}$.

   (b) The matrix of ST under the basis $\{\alpha_1, \alpha_2\}$.

   (c) The coordinate vector of $S(u)$ under the basis $\{\alpha_1, \alpha_2\}$.

   (d) The coordinate vector of $T(u)$ under the basis $\{\beta_1, \beta_2\}$.

Chapter 4
Dual Spaces

Duality is a very important tool in mathematics. In this chapter, we explore some instances of duality. Let V be a vector space over a field $\mathbb{F}$. Since every field is a vector space over itself, one can consider the set of all linear transformations $Hom(V, \mathbb{F})$. The set $Hom(V, \mathbb{F})$ forms a vector space known as dual space to V or conjugate space of V and is denoted as $\widehat{V} = Hom(V, \mathbb{F})$. It is also denoted by V^*. In the present chapter, we shall study the properties of this vector space. Throughout, we assume that the underlying vector spaces are finite dimensional unless otherwise mentioned.

4.1 Linear Functionals and the Dual Space

Definition 4.1 Let V be a vector space over a field $\mathbb{F}$. A linear transformation $T : V \to \mathbb{F}$ is called a linear functional and the set of all linear functionals on V denoted as $\widehat{V} = Hom(V, \mathbb{F})$ forms a vector space over the field $\mathbb{F}$.

Example 4.2 (1) Let V be a vector space over a field F. Define a map $f : V \longrightarrow \mathbb{F}$ given by $f(x) = 0$ for all $x \in V$. It is obvious to observe that f is a linear functional on V, which is known as the zero linear functional on V.

(2) Consider the vector space $\mathbb{F}^n$ over the field $\mathbb{F}$. Define the map $f : \mathbb{F}^n \longrightarrow \mathbb{F}$ by $f(x_1, x_2, \ldots, x_n) = a_1 x_1 + a_2 x_2 + \cdots + a_n x_n$, where $x_1, x_2, \ldots, x_n \in \mathbb{F}$ and $a_1, a_2, \ldots, a_n$ are n fixed scalars. It can be easily verified that f is a linear functional on $\mathbb{F}^n$.

(3) Consider the vector space $\mathbb{F}^n$ over the field $\mathbb{F}$. Let $\pi_i : \mathbb{F}^n \longrightarrow \mathbb{F}$ be the ith projection, i.e., $\pi_i(x_1, x_2, \ldots, x_n) = x_i$. It can be easily shown that π_i is a linear functional on $\mathbb{F}^n$.

© The Author(s), under exclusive license to Springer Nature Singapore Pte Ltd. 2022   111
M. Ashraf et al., *Advanced Linear Algebra with Applications*,
https://doi.org/10.1007/978-981-16-2167-3_4

(4) Let $\mathbb{R}[x]$ be the vector space of all polynomials in x over the real field $\mathbb{R}$. Define the map $I : \mathbb{R}[x] \longrightarrow \mathbb{R}$ by $I(f(x)) = \int_0^1 f(x)dx$. It can be easily proved that I is a linear functional on $\mathbb{R}[x]$.

(5) Let $M_{n \times n}(\mathbb{F})$ be the vector space of all $n \times n$ matrices over the field $\mathbb{F}$. Consider the map $f : M_{n \times n}(\mathbb{F}) \longrightarrow \mathbb{F}$, defined as $f(A) = trace(A) = a_{11} + a_{22} + \cdots + a_{nn}$, where $A = (a_{ij})_{n \times n}$. It can be easily seen that f is a linear functional on $M_{n \times n}(\mathbb{F})$.

(6) Let $\mathbb{F}[x]$ be the vector space of all polynomials in x over the field $\mathbb{F}$. Define the map $L : \mathbb{F}[x] \longrightarrow \mathbb{F}$ by $L(f(x)) =$ the value of $f(x)$ at some fixed element $t \in \mathbb{F}$, i.e., $f(t)$. It can be easily seen that L is a linear functional on $\mathbb{F}[x]$.

(7) Let $\mathscr{C}[a, b]$ be the vector space of all real-valued continuous functions over the field $\mathbb{R}$ of real numbers. Define the map $L : \mathscr{C}[a, b] \longrightarrow \mathbb{R}$ by $L(f(x)) = \int_a^b f(x)dx$. It is obvious that L is a linear functional on $\mathscr{C}[a, b]$.

Definition 4.3 Let V be a vector space over $\mathbb{F}$ and let $B = \{v_1, v_2, \ldots, v_n\}$ be a basis of V. Then a subset $\widehat{B} = \{\widehat{v_1}, \widehat{v_2}, \ldots, \widehat{v_n}\}$ of $\widehat{V}$ is said to be the dual basis of $\widehat{V}$ with respect to the basis B if $\widehat{B}$ is a basis of $\widehat{V}$ and $\widehat{v_i}(v_i) = 1$, $\widehat{v_i}(v_j) = 0$ for $i \neq j$.

Remark 4.4 (i) Since $\mathbb{F}$ is a vector space of dimension one over itself, $dim Hom (V, \mathbb{F}) = dim V$.

(ii) It is clear that the dual basis with respect to any basis of a vector space V is always unique while the change of basis of V will always produce a different dual basis.

Theorem 4.5 *Let V be a vector space over a field $\mathbb{F}$. Then $\widehat{V}$ has a dual basis.*

Proof Let $B = \{v_1, v_2, \ldots, v_n\}$ be a basis of V. If $v \in V$, then there exist unique $\alpha_1, \alpha_2, \ldots, \alpha_n \in \mathbb{F}$ such that $v = \alpha_1 v_1 + \alpha_2 v_2 + \cdots + \alpha_n v_n$. Now, for each i; $1 \leq i \leq n$, define a mapping $\widehat{v_i} : V \to \mathbb{F}$ such that $\widehat{v_i}(v) = \alpha_i$. Clearly, $\widehat{v_i}$ for each i; $1 \leq i \leq n$, is well-defined. It can be easily seen that $\widehat{v_i}$, for each i; $1 \leq i \leq n$, is linear, i.e., $\widehat{v_i} \in \widehat{V}$ for each i; $1 \leq i \leq n$. Indeed, if $a, b \in V$, then there exist unique scalars $\beta_1, \beta_2, \ldots, \beta_n$ and $\delta_1, \delta_2, \ldots, \delta_n$ such that $a = \sum\limits_{i=1}^{n} \beta_i v_i$ and $b = \sum\limits_{i=1}^{n} \delta_i v_i$. This shows that for each i; $1 \leq i \leq n$,

$$\widehat{v_i}(a + b) = \widehat{v_i}\left(\sum_{i=1}^{n}(\beta_i + \delta_i)v_i \right) = \beta_i + \delta_i = \widehat{v_i}(a) + \widehat{v_i}(b)$$

and for any $\alpha \in \mathbb{F}$, $\widehat{v_i}(\alpha a) = \widehat{v_i}\left(\sum_{i=1}^{n} \alpha \beta_i v_i \right) = \alpha \beta_i = \alpha \widehat{v_i}(a)$. Hence, $\widehat{v_i} \in \widehat{V}$ for each i; $1 \leq i \leq n$ and it can be seen that for each i; $1 \leq i \leq n$, $\widehat{v_i}(v_i) = 1$, $\widehat{v_i}(v_j){=}0$ for $i \neq j$; $j = 1, 2, \ldots, n$. Since $dim V {=} dim \widehat{V}$, in order to show that $\{\widehat{v_1}, \widehat{v_2}, \ldots, \widehat{v_n}\}$ is a basis of $\widehat{V}$, it remains only to show that it is linearly independent. For any $\gamma_i \in \mathbb{F}$, where $i = 1, 2, \ldots, n$ whenever $\sum_{i=1}^{n} \gamma_i \widehat{v_i} = 0$, then $\sum_{i=1}^{n} \gamma_i \widehat{v_i}(v_j) = 0$, for each $j = 1, 2, \ldots, n$. This implies that $\gamma_i = 0$ for each $i = 1, 2, \ldots, n$. Hence we conclude that $\{\widehat{v_1}, \widehat{v_2}, \ldots, \widehat{v_n}\}$ is a basis of $\widehat{V}$.

Theorem 4.6 *Let $B = \{v_1, v_2, \ldots, v_n\}$ be an ordered basis of V and $\widehat{B} = \{\widehat{v_1}, \widehat{v_2}, \ldots, \widehat{v_n}\}$ be the ordered dual basis of $\widehat{V}$ dual to the basis B. Then*

(i) *any $v \in V$ can be uniquely written as $v = \widehat{v_1}(v)v_1 + \widehat{v_2}(v)v_2 + \cdots + \widehat{v_n}(v)v_n$,*

(ii) *any $\widehat{v} \in \widehat{V}$ can be uniquely written as $\widehat{v} = \widehat{v}(v_1)\widehat{v_1} + \widehat{v}(v_2)\widehat{v_2} + \cdots + \widehat{v}(v_n)\widehat{v_n}$,*

(iii) *the mapping $\mu : \widehat{V} \to \mathbb{F}^n$ such that $\mu(\widehat{v}) = (\widehat{v}(v_1), \widehat{v}(v_2), \ldots, \widehat{v}(v_n))$ is a linear transformation which is both one-to-one and onto.*

Proof (i) Since B is a basis of V, $v \in V$ can be uniquely written as $v = \alpha_1 v_1 + \alpha_2 v_2 + \cdots + \alpha_n v_n$. Now for any $1 \leq i \leq n$, $\widehat{v_i}(v) = \alpha_1 \widehat{v_i}(v_1) + \alpha_2 \widehat{v_i}(v_2) + \cdots + \alpha_n \widehat{v_i}(v_n) = \alpha_i$. This shows that $v = \widehat{v_1}(v)v_1 + \widehat{v_2}(v)v_2 + \cdots + \widehat{v_n}(v)v_n$.

(ii) Since $\widehat{v} \in \widehat{V}$ and $\widehat{B}$ is a basis of $\widehat{V}$, there exist unique scalars $\alpha_1, \alpha_2, \ldots, \alpha_n$ such that $\widehat{v} = \alpha_1 \widehat{v_1} + \alpha_2 \widehat{v_2} + \cdots + \alpha_n \widehat{v_n}$. Now for any $v_j \in V$, $1 \leq j \leq n$;

$$\widehat{v}(v_j) = \alpha_1 \widehat{v_1}(v_j) + \alpha_2 \widehat{v_2}(v_j) + \cdots + \alpha_n \widehat{v_n}(v_j) = \alpha_j; \ 1 \leq j \leq n.$$

This yields that $\widehat{v} = \widehat{v}(v_1)\widehat{v_1} + \widehat{v}(v_2)\widehat{v_2} + \cdots + \widehat{v}(v_n)\widehat{v_n}$.

(iii) It can be easily seen that μ is a linear transformation. Indeed, for $\phi, \xi \in \widehat{V}$ and $\alpha, \beta \in \mathbb{F}$,

$$\begin{aligned}
\alpha\phi + \beta\xi &= (\alpha\phi + \beta\xi)(v_1)\widehat{v_1} + (\alpha\phi + \beta\xi)(v_2)\widehat{v_2} + \cdots + (\alpha\phi + \beta\xi)(v_n)\widehat{v_n} \\
&= \big(\alpha\phi(v_1) + \beta\xi(v_1)\big)\widehat{v_1} + \big(\alpha\phi(v_2) + \beta\xi(v_2)\big)\widehat{v_2} + \cdots + \\
&\quad \big(\alpha\phi(v_n) + \beta\xi(v_n)\big)\widehat{v_n}.
\end{aligned}$$

Therefore,

$$\begin{aligned}
\mu(\alpha\phi + \beta\xi) &= (\alpha\phi(v_1) + \beta\xi(v_1), \alpha\phi(v_2) + \beta\xi(v_2), \ldots, \alpha\phi(v_n) + \beta\xi(v_n)) \\
&= (\alpha\phi(v_1), \alpha\phi(v_2), \ldots, \alpha\phi(v_n)) + (\beta\xi(v_1), \beta\xi(v_2), \ldots, \beta\xi(v_n)) \\
&= \alpha(\phi(v_1), \phi(v_2), \ldots, \phi(v_n)) + \beta(\xi(v_1), \xi(v_2), \ldots, \xi(v_n)) \\
&= \alpha\mu(\phi) + \beta\mu(\xi).
\end{aligned}$$

This shows that μ is linear. Given that B is a basis of V and let $f \in \widehat{V}$. Then $f = 0$ if and only if $f(v_i) = 0$ for every i with $1 \leq i \leq n$. Therefore

$$\begin{aligned}
\mu(f) = 0 &\Leftrightarrow (f(v_1), f(v_2), \ldots, f(v_n)) = 0 \\
&\Leftrightarrow f(v_i) = 0 \text{ for each } 1 \leq i \leq n \\
&\Leftrightarrow f = 0.
\end{aligned}$$

Hence, $Ker\,\mu = \{0\}$ and therefore μ is one-to-one. Also since $dim\,\widehat{V} = dim\,\mathbb{F}^n$, μ is onto. This completes the proof.

Theorem 4.7 *Let V be a vector space over $\mathbb{F}$. Then the following hold:*

(i) *For any nonzero vector $v \in V$, there exists a linear functional $\widehat{v} \in \widehat{V}$ such that $\widehat{v}(v) \neq 0$.*

(ii) *A vector $v \in V$ is zero if and only if $\widehat{v}(v) = 0$ for all $\widehat{v} \in \widehat{V}$.*

(iii) *If $v_1, v_2 \in V$ such that $\widehat{v}(v_1) = \widehat{v}(v_2)$ for all $\widehat{v} \in \widehat{V}$, then $\{v_1, v_2\}$ is linearly dependent.*

(iv) *Let $\widehat{v} \in \widehat{V}$. If $\widehat{v}(x) \neq 0$, then $V = \langle x \rangle \oplus Ker\widehat{v}$.*

(v) *Two nonzero linear functionals $\widehat{v_1}, \widehat{v_2} \in \widehat{V}$ have the same kernel if and only if there is a nonzero scalar λ such that $\widehat{v_1} = \lambda \widehat{v_2}$.*

Proof (i) Since $v \neq 0$ in V, one can find a basis $B = \{v = v_1, v_2, \ldots, v_n\}$ of V. Hence, there exists a dual basis $\{\widehat{v} = \widehat{v_1}, \widehat{v_2}, \ldots, \widehat{v_n}\}$ of $\widehat{V}$ relative to the basis B of V such that $\widehat{v}(v) = \widehat{v_1}(v_1) = 1 \neq 0$.

(ii) In view of (i), it is obvious.

(iii) If $\{v_1, v_2\}$ is linearly independent, then $\{v_1, v_2\}$ can be extended to a basis of V say $B = \{v_1, v_2, v_3, \ldots, v_n\}$. Let $\{\widehat{v_1}, \widehat{v_2}, \ldots, \widehat{v_n}\}$ be the dual basis of $\widehat{V}$ relative to B. Then $\widehat{v_2}(v_1) = 0$, but it does not imply that $\widehat{v_2}(v_2) = 0(\widehat{v_2}(v_2) = 1)$. Thus $\{v_1, v_2\}$ is linearly dependent.

(iv) If $v \in \langle x \rangle \cap Ker\widehat{v}$, then $\widehat{v}(v) = 0$ and $v = \alpha x$ for some nonzero α, whence $\widehat{v}(v) = \alpha\widehat{v}(x) = 0$ implies that $\widehat{v}(x) = 0$, a contradiction. Hence $\langle x \rangle \cap Ker\widehat{v} = \{0\}$. Now for any $v \in V$,

$$v = \frac{\widehat{v}(v)}{\widehat{v}(x)}x + \left(v - \frac{\widehat{v}(v)}{\widehat{v}(x)}x\right) \in \langle x \rangle + Ker\widehat{v}$$

which yields that $V = \langle x \rangle + Ker\widehat{v}$ and hence $V = \langle x \rangle \oplus Ker\widehat{v}$.

(v) If $\widehat{v_1} = \lambda\widehat{v_2}$ for $\lambda \neq 0$, then $Ker\widehat{v_1} = Ker\widehat{v_2}$. Conversely, if $K = Ker\widehat{v_1} = Ker\widehat{v_2}$, then by (iv), for any $x \notin K$, $V = \langle x \rangle \oplus K$ and hence, $\widehat{v_1}|_K = \lambda\widehat{v_2}|_K$ for any scalar λ. Further, if $\lambda = \frac{\widehat{v_1}(x)}{\widehat{v_2}(x)}$, it follows that $\lambda\widehat{v_2}(x) = \widehat{v_1}(x)$ for all $x \notin K$. Therefore $\widehat{v_1} = \lambda\widehat{v_2}$.

Example 4.8 $\mathbb{R}^3 = \{(x, y, z) \mid x, y, z \in \mathbb{R}\}$ is a vector space over $\mathbb{R}$. If $B = \{(-1, 1, 1), (1, -1, 1), (1, 1, -1)\}$ is a basis of $\mathbb{R}^3$, then we shall find the dual basis of B. Any $(x, y, z) \in \mathbb{R}^3$ can be written as

$$(x, y, z) = \frac{y+z}{2}(-1, 1, 1) + \frac{x+z}{2}(1, -1, 1) + \frac{x+y}{2}(1, 1, -1).$$

Define

$$\phi_1(x, y, z) = \frac{y+z}{2}, \phi_2(x, y, z) = \frac{x+z}{2}, \phi_3(x, y, z) = \frac{x+y}{2}.$$

Obviously, $\phi_1, \phi_2, \phi_3 \in \widehat{\mathbb{R}^3}$, and $\phi_1(-1, 1, 1) = 1$, $\phi_1(1, -1, 1) = 0$, $\phi_1(1, 1, -1) = 0$, $\phi_2(-1, 1, 1) = 0$, $\phi_2(1, -1, 1) = 1$, $\phi_2(1, 1, -1) = 0$, $\phi_3(-1, 1, 1) = 0$, $\phi_3(1, -1, 1) = 0$, $\phi_3(1, 1, -1) = 1$. Hence, $\{\phi_1, \phi_2, \phi_3\}$ is the dual basis of B.

Remark 4.9 Define a function $f : \mathbb{F}^n \to \mathbb{F}$ such that $f(x_1, x_2, \ldots, x_n) = \sum_{i=1}^{n} a_i x_i$, then it is a linear functional defined on the vector space $\mathbb{F}^n$, which is determined uniquely by $(a_1, a_2, \ldots, a_n) \in \mathbb{F}^n$ relative to the standard ordered basis $B = \{e_1, e_2, \ldots, e_n\}$; $f(e_i) = a_i$, $1 \leq i \leq n$. Every linear functional on $\mathbb{F}^n$ is of this form for some scalars $a_1, a_2, \ldots, a_n$, because if B is the standard basis of $\mathbb{F}^n$, $f \in \widehat{\mathbb{F}^n}$ and $f(e_i) = a_i$, then for any $(x_1, x_2, \ldots, x_n) \in \mathbb{F}^n$, $(x_1, x_2, \ldots, x_n) = \sum_{i=1}^{n} x_i e_i$ yields that

$$f(x_1, x_2, \ldots, x_n) = \sum_{i=1}^{n} x_i f(e_i) = \sum_{i=1}^{n} a_i x_i.$$

Example 4.10 If $B = \{(1, 1, 0), (1, 0, 1), (0, 1, 1)\}$ is a basis of $\mathbb{R}^3$, then in order to find the dual basis of B, define $\phi_1, \phi_2, \phi_3 : \mathbb{R}^3 \to \mathbb{R}$ such that

$$\phi_1(x, y, z) = a_1 x + b_1 y + c_1 z,$$

$$\phi_2(x, y, z) = a_2 x + b_2 y + c_2 z,$$

$$\phi_3(x, y, z) = a_3 x + b_3 y + c_3 z.$$

In order to find the dual basis we, need

$$\phi_1(1, 1, 0) = 1, \phi_1(1, 0, 1) = 0, \phi_1(0, 1, 1) = 0,$$

$$\phi_2(1, 1, 0) = 0, \phi_2(1, 0, 1) = 1, \phi_2(0, 1, 1) = 0,$$

$$\phi_3(1, 1, 0) = 0, \phi_3(1, 0, 1) = 0, \phi_3(0, 1, 1) = 1.$$

Now solving the equations $\phi_1(1, 1, 0) = a_1 + b_1 = 1$, $\phi_1(1, 0, 1) = a_1 + c_1 = 0$, $\phi_1(0, 1, 1) = b_1 + c_1 = 0$, we find that $a_1 = \frac{1}{2}, b_1 = \frac{1}{2}, c_1 = \frac{-1}{2}$, which shows that $\phi_1(x, y, z) = \frac{x+y-z}{2}$. Similarly by solving the equations

$$\phi_2(1, 1, 0) = a_2 + b_2 = 0, \phi_2(1, 0, 1) = a_2 + c_2 = 1, \phi_2(0, 1, 1) = b_2 + c_2 = 0$$

and

$$\phi_3(1, 1, 0) = a_3 + b_3 = 0, \phi_3(1, 0, 1) = a_3 + c_3 = 0, \phi_3(0, 1, 1) = b_3 + c_3 = 1,$$

we find that $a_2 = \frac{1}{2}, b_2 = \frac{-1}{2}, c_2 = \frac{1}{2}$ and $a_3 = \frac{-1}{2}, b_3 = \frac{1}{2}, c_3 = \frac{1}{2}$, respectively. This yields that $\phi_2(x, y, z) = \frac{x-y+z}{2}, \phi_3(x, y, z) = \frac{-x+y+z}{2}$. Hence, $\{\phi_1, \phi_2, \phi_3\}$ is a basis of $\mathbb{R}^3$ dual to the basis B.

Exercises

1. If V is finite dimensional and $v_1, v_2 \in V$, where $v_1 \neq v_2$, then prove that there exists $f \in \widehat{V}$ such that $f(v_1) \neq f(v_2)$.
2. If $u, v \in V$ such that $f(u) = 0$ implies $f(v) = 0$, for all $f \in \widehat{V}$ then prove that $v = \lambda u$ for some $\lambda \in \mathbb{F}$.
3. For $f, g \in \widehat{V}$ if $f(v) = 0$ implies that $g(v) = 0$ for all $v \in V$, then show that $\{f, g\}$ is linearly dependent.
4. Let W be a proper subspace of a finite dimensional vector space V and let $v \in V \setminus W$. Show that there is a linear functional $f \in \widehat{V}$ for which $f(v) = 1$ and $f(w) = 0$ for all $w \in W$.
5. Find out the dual basis of each of the following basis of $\mathbb{R}^3$:

   (a) $\{(1, 1, 1), (1, 1, 0), (1, 0, 0)\}$,
   (b) $\{(1, 2, 3), (2, 3, 1), (3, 2, 1)\}$ and
   (c) $\{(1, -2, 7), (-3, 4, 5), (6, -1, 3)\}$.

6. Let $\mathbb{R}_2[x]$ be the vector space of all real polynomials of degree less than or equal to two. Find the dual basis of $\widehat{\mathbb{R}_2[x]}$ induced by the following bases of $\mathbb{R}_2[x]$:

   (a) $\{1, x, x^2\}$,
   (b) $\{1 + x, 1 - x, 1 + x^2\}$ and
   (c) $\{x, 2 + x, 1 - x - x^2\}$.

7. Let V be the vector space of all polynomials over $\mathbb{R}$ of degree less than or equal to 2. Let ϕ_1, ϕ_2, ϕ_3 be the linear functionals on V defined by $\phi_1(f(t)) = \int_0^1 f(t)dt$, $\phi_2(f(t)) = f'(1)$, $\phi_3(f(t)) = f(0)$. Here $f'(t)$ denotes the derivative of $f(t)$. Find a basis $\{f_1(t), f_2(t), f_3(t)\}$ of V such that its dual is $\{\phi_1, \phi_2, \phi_3\}$.
8. Let V be the vector space of all polynomials over $\mathbb{F}$ of degree less than or equal to 2. Let $a, b, c \in \mathbb{F}$ be distinct scalars. Let ϕ_a, ϕ_b, ϕ_c be the linear functionals on V defined by $\phi_a(f(t)) = f(a), \phi_b(f(t)) = f(b), \phi_c(f(t)) = f(c)$. Show that $\{\phi_a, \phi_b, \phi_c\}$ is a basis of $\widehat{V}$ and also find its dual basis.
9. Let $\{e_1, e_2, \ldots, e_n\}$ be the usual basis of $\mathbb{F}^n$. Show that its dual basis is $\{\pi_1, \pi_2, \ldots, \pi_n\}$, where π_i is the ith projection mapping: $\pi_i(a_1, a_2, \ldots, a_n) = a_i$.
10. Let W be a subspace of V. For any linear functional ϕ on W, show that there is a linear functional f on V such that $f(w) = \phi(w)$ for any $w \in W$; that is, ϕ is the restriction of f to W.
11. Let V be a vector space over $\mathbb{R}$. Let $\phi_1, \phi_2 \in \widehat{V}$ and suppose $f : V \longrightarrow \mathbb{R}$, defined by $f(v) = \phi_1(v)\phi_2(v)$, also belongs to $\widehat{V}$. Show that either $\phi_1 = 0$ or $\phi_2 = 0$.
12. Let V be the vector space of all polynomials over $\mathbb{R}$ of degree less than or equal to 3. Find the dual basis of a basis $B = \{1, 1 + x, (1 + x)^2, (1 + x)^3\}$ of V.

13. Consider $\mathbb{R}^2$ as a vector space over $\mathbb{R}$. Find the formulae for linear functionals f and g on $\mathbb{R}^2$ such that for a fixed θ, $f(cos\theta, sin\theta) = 1$, $f(-sin\theta, cos\theta) = 2$, $g(cos\theta, sin\theta) = 2$ and $g(-sin\theta, cos\theta) = 1$.

   (a) Prove that $B' = \{f, g\}$ is a basis of $\widehat{\mathbb{R}^2}$.
   (b) Find an ordered basis $B = \{u, v\}$ of $\mathbb{R}^2$ such that B' is the dual of B.

14. Let n and m be any two positive integers.

   (a) For any m linear functionals $f_1, f_2, \ldots, f_m$ on $\mathbb{F}^n$, prove that $\sigma : \mathbb{F}^n \longrightarrow \mathbb{F}^m$ given by $\sigma(u) = (f_1(u), f_2(u), \ldots, f_m(u))$ is a linear transformation.
   (b) Given any linear transformation $T : \mathbb{F}^n \longrightarrow \mathbb{F}^m$. Prove that there exist uniquely determined linear functionals $g_1, g_2, \ldots, g_m$ depending upon T such that $T(u) = (g_1(u), g_2(u), \ldots, g_m(u))$.

15. Let V be the vector space of all polynomials over $\mathbb{R}$ of degree less than or equal to 2. Define ϕ_1, ϕ_2, ϕ_3 in $\widehat{V}$ such that $\phi_1(f(t)) = \int_0^1 f(t)dt$, $\phi_2(f(t)) = \int_0^2 f(t)dt$ and $\phi_3(f(t)) = \int_0^{-1} f(t)dt$. Show that $B' = \{\phi_1, \phi_2, \phi_3\}$ is a basis of $\widehat{V}$. Find a basis B of V, of which B' is dual.

16. Let $\mathbb{F}$ be a field of characteristic zero and let V be a finite dimensional vector space over $\mathbb{F}$. If $v_1, v_2, \ldots, v_m$ are finitely many vectors in V, each different from the zero vector, prove that there is a linear functional f on V such that $f(v_i) \neq 0$, $i = 1, 2, \ldots, m$.

17. In $\mathbb{R}^3$, let $v_1 = (1, 0, 1)$; $v_2 = (0, 1, -2)$; $v_3 = (-1, -1, 0)$.

   (a) If f is a linear functional on $\mathbb{R}^3$ such that $f(v_1) = 1$, $f(v_2) = -1$, $f(v_3) = 3$ and if $u = (x, y, z)$, then find $f(u)$.
   (b) Describe explicitly a linear functional f on $\mathbb{R}^3$ such that $f(v_1) = f(v_2) = 0$ but $f(v_3) \neq 0$.
   (c) If f is any linear functional such that $f(v_1) = f(v_2) = 0$ but $f(v_3) \neq 0$ and if $u = (2, 3, -1)$, then show that $f(u) \neq 0$.

18. Let $B = \{v_1, v_2, v_3\}$ be a basis of $\mathbb{C}^3$ defined by $v_1 = (1, 0, -1)$, $v_2 = (1, 1, 1)$, $v_3 = (2, 2, 0)$. Find the dual basis of B.

4.2 Second Dual Space

For any vector space V, one can consider its dual space $\widehat{V}$ which contains all the linear functionals on V. Since $\widehat{V}$ is also a vector space, consider $\widehat{\widehat{V}}$ the dual of $\widehat{V}$, which contains all the linear functionals on $\widehat{V}$. If V is finite dimensional then $\widehat{V}$ and $\widehat{\widehat{V}}$ are also finite dimensional and $dim V = dim \widehat{V} = dim \widehat{\widehat{V}}$.

Theorem 4.11 (Principal of duality) *If V is a vector space over $\mathbb{F}$, then there exists a canonical isomorphism from V onto $\widehat{\widehat{V}}$.*

Proof We shall show that each $v \in V$ determines a specific element $\bar{v} \in \widehat{\widehat{V}}$. Define $\bar{v} : \widehat{V} \to \mathbb{F}$ such that $\bar{v}(f) = f(v)$ for all $f \in \widehat{V}$. We show that this map is linear. For scalars $\alpha, \beta \in \mathbb{F}$, $f, g \in \widehat{V}$, we have

$$\bar{v}(\alpha f + \beta g) = (\alpha f + \beta g)(v) = \alpha f(v) + \beta g(v) = \alpha \bar{v}(f) + \beta \bar{v}(g).$$

Thus, $\bar{v}$ is a linear functional on $\widehat{V}$ and hence $\bar{v} \in \widehat{\widehat{V}}$. Now define the canonical map $\sigma : V \to \widehat{\widehat{V}}$ such that $\sigma(v) = \bar{v}$. For any $\alpha, \beta \in \mathbb{F}$ and $v_1, v_2 \in V$, $\sigma(\alpha v_1 + \beta v_2) = \overline{\alpha v_1 + \beta v_2}$. It can be easily seen that for any $f \in \widehat{V}$, $\overline{\alpha v_1 + \beta v_2}(f) = f(\alpha v_1 + \beta v_2) = \alpha f(v_1) + \beta f(v_2) = \alpha \overline{v_1}(f) + \beta \overline{v_2}(f) = (\alpha \overline{v_1} + \beta \overline{v_2})(f)$.
This shows that $\overline{\alpha v_1 + \beta v_2} = \alpha \overline{v_1} + \beta \overline{v_2}$. Hence, using this relation in the preceding definition of σ, it can be seen that $\sigma(\alpha v_1 + \beta v_2) = \alpha \sigma(v_1) + \beta \sigma(v_2)$ and σ is a linear transformation from V to $\widehat{\widehat{V}}$. Now if $v \in Ker\sigma$, then $\sigma(v) = \bar{v} = 0$ or $\bar{v}(f) = 0$ for all $f \in \widehat{V}$. This shows that $f(v) = 0$ for all $f \in \widehat{V}$. By Theorem 4.7(i), $v = 0$ and consequently $Ker\sigma = \{0\}$ and σ is a monomorphism. This yields that $dim V = dim\sigma(V)$. However, $dim V = dim\widehat{V} = dim\widehat{\widehat{V}}$ so that $\sigma(V)$ is a subspace of $\widehat{\widehat{V}}$ such that $dim\sigma(V) = dim\widehat{\widehat{V}}$. This can happen only if $\sigma(V) = \widehat{\widehat{V}}$ and hence σ is onto. This completes the proof of our result.

Remark 4.12 (i) It is to be noted that in the above theorem $\sigma : V \longrightarrow \widehat{\widehat{V}}$ does not depend upon any particular choice of basis of V, that is why it is called canonical isomorphism.

(ii) If V is any arbitrary vector space (need not be finite dimensional), even then $\sigma : V \longrightarrow \widehat{\widehat{V}}$ will exist which is an injective homomorphism, but need not be onto.

Theorem 4.13 *Let V be a vector space over a field $\mathbb{F}$ and let $\widehat{B} = \{f_1, f_2, \ldots, f_n\}$ be a basis of $\widehat{V}$. Then there exists a basis $B = \{v_1, v_2, \ldots, v_n\}$ of V such that $\widehat{B}$ is a dual basis of B.*

Proof Since $\widehat{B} = \{f_1, f_2, \ldots, f_n\}$ is a basis of $\widehat{V}$, by Theorem 4.5, there exists a basis $\widehat{\widehat{B}} = \{g_1, g_2, \ldots, g_n\}$ of $\widehat{\widehat{V}}$ dual to $\widehat{B}$. Hence $g_i(f_j) = 0$ for $i \neq j$ and $g_i(f_j) = 1$ for $i = j$, $1 \leq i, j \leq n$. Now let $\sigma : V \to \widehat{\widehat{V}}$ be a canonical isomorphism. Then for any $g_j \in \widehat{\widehat{V}}$, $1 \leq j \leq n$, there exists $v_j \in V$ such that $\sigma(v_j) = g_j$. Then $\{v_1, v_2, \ldots, v_n\}$ is a basis of V. Now for $1 \leq i, j \leq n$,

$$f_i(v_j) = \big(\sigma(v_j)\big)(f_i) = g_j(f_i) = \delta_{ij},$$

where δ_{ij} is the Kronecker delta, i.e., $\delta_{ij} = \begin{cases} 1 & if \ i = j \\ 0 & if \ i \neq j \end{cases}$. This yields that $\widehat{B}$ is the dual basis of B.

Example 4.14 Consider the vector space $\mathbb{R}_2[x]$, i.e., the vector space of all polynomials of degree less than or equal to two over $\mathbb{R}$ and let $\psi_1, \psi_2, \psi_3 : \mathbb{R}_2[x] \to \mathbb{R}$

such that $\psi_1(f(x)) = \int_0^1 f(x)dx$, $\psi_2(f(x)) = f'(x)$, $\psi_3(f(x)) = f'(0)$. Now we find the basis of $\mathbb{R}_2[x]$ dual to $\{\psi_1, \psi_2, \psi_3\}$.

Let $\{a_1 + b_1x + c_1x^2, a_2 + b_2x + c_2x^2, a_3 + b_3x + c_3x^2\}$ be the required basis of $\mathbb{R}_2[x]$, then

$$\psi_1(a_1 + b_1x + c_1x^2) = a_1 + \frac{b_1}{2} + \frac{c_1}{3} = 1,$$

$$\psi_1(a_2 + b_2x + c_2x^2) = a_2 + \frac{b_2}{2} + \frac{c_2}{3} = 0,$$

$$\psi_1(a_3 + b_3x + c_3x^2) = a_3 + \frac{b_3}{2} + \frac{c_3}{3} = 0,$$

$$\psi_2(a_1 + b_1x + c_1x^2) = b_1 + 2c_1 = 0,$$

$$\psi_2(a_2 + b_2x + c_2x^2) = b_2 + 2c_2 = 1,$$

$$\psi_2(a_3 + b_3x + c_3x^2) = b_3 + 2c_3 = 0,$$

$$\psi_3(a_1 + b_1x + c_1x^2) = b_1 = 0,$$

$$\psi_3(a_2 + b_2x + c_2x^2) = b_2 = 0,$$

$$\psi_3(a_3 + b_3x + c_3x^2) = b_3 = 1.$$

This yields that $b_1 = 0, c_1 = 0, a_1 = 1, b_2 = 0, c_2 = \frac{1}{2}, a_2 = \frac{-1}{6}, b_3 = 1, c_3 = \frac{-1}{2}, a_3 = \frac{-b_3}{2} - \frac{c_3}{2} = -\frac{1}{2} + \frac{1}{6} = -\frac{1}{3}$. Hence, $\{1, -\frac{1}{6} + \frac{x^2}{2}, -\frac{1}{3} + x - \frac{1}{2}x^2\}$ is the required basis of $\mathbb{R}_2[x]$.

4.3 Annihilators

Definition 4.15 Let V be a vector space over a field $\mathbb{F}$ and S be a subset of V. Then annihilator S° of S in $\widehat{V}$ is defined as the collection of all $f \in \widehat{V}$ such that $f(s) = 0$ for all $s \in S$, i.e., $S^\circ = \{f \in \widehat{V} \mid f(s) = 0 \text{ for all } s \in S\}$.

Remark 4.16 (i) For a subset S of $\widehat{V}$, the annihilator S° of S is defined as the set of all $v \in V$ such that $f(v) = 0$ for all $f \in S$.

(ii) In a finite dimensional vector space V if $0 \neq v \in V$, then we have seen that there exists $f \in \widehat{V}$ such that $f(v) \neq 0$. This shows that V° contains no nonzero functional, i.e., $V^\circ = \{0\}$. It is also clear that if a subset S of V contains the zero vector alone, then $S^\circ = \widehat{V}$.

(iii) For any subset S of a vector space V, $(S^\circ)^\circ$ can be viewed as a subspace of V under the identification of V and $\widehat{\widehat{V}}$, i.e., $(S^\circ)^\circ = \{v \in V \mid f(v) = 0 \text{ for every } f \in S^\circ\}$.

Lemma 4.17 *Let V be a vector space over a field $\mathbb{F}$. Then*

(i) *For any subsets S_1, S_2 of V if $S_1 \subseteq S_2$, then $S_2^\circ \subseteq S_1^\circ$.*
(ii) *For any subset S of V, S° is a subspace of $\widehat{V}$ and $S \subseteq (S^\circ)^\circ$.*
(iii) *For any subset S of V, $S^\circ = (L(S))^\circ$.*

Proof (i) Suppose that $S_1 \subseteq S_2$ and $f \in S_2^\circ$. Then for any $v \in S_1$, $f(v) = 0$ and consequently, $f \in S_1^\circ$ which completes the required proof.

(ii) Since for any $v \in S$, $0(v) = 0$, we find that $0 \in S^\circ$ and therefore $S^\circ \neq \phi$. Let $f, g \in S^\circ$ and $\alpha, \beta \in \mathbb{F}$. Then for every $v \in S$

$$(\alpha f + \beta g)v = \alpha f(v) + \beta g(v) = \alpha 0 + \beta 0 = 0,$$

which shows that S° is a subspace of $\widehat{V}$. Now let $v \in S$. Then for every linear functional $f \in S^\circ$, $\widehat{v}(f) = f(v) = 0$. Hence $\widehat{v} \in (S^\circ)^\circ$ and under the identification of V and $\widehat{\widehat{V}}$, $v \in (S^\circ)^\circ$.

(iii) Since $S \subseteq L(S)$, we find that $(L(S))^\circ \subseteq S^\circ$. Conversely, suppose that $f \in S^\circ$, i.e., $f(s) = 0$ for all $s \in S$. Now for any $\alpha_1, \alpha_2, \ldots, \alpha_n \in \mathbb{F}$, $\alpha_1 v_1 + \alpha_2 v_2 + \cdots + \alpha_n v_n \in L(S)$ for all $v_1, v_2, \ldots, v_n \in S$. Then $f(\alpha_1 v_1 + \alpha_2 v_2 + \cdots + \alpha_n v_n) = \alpha_1 f(v_1) + \alpha_2 f(v_2) + \cdots + \alpha_n f(v_n) = 0$ and hence $f \in (L(S))^\circ$. This completes the proof.

Theorem 4.18 *Let V be a vector space over a field $\mathbb{F}$ and W a subspace of V. Then*

(i) *$dimW + dimW^\circ = dimV$,*
(ii) *$\widehat{W} \cong \frac{\widehat{V}}{W^\circ}$ and*
(iii) *$(W^\circ)^\circ = W$.*

Proof (i) Let $\{w_1, w_2, \ldots, w_m\}$ be a basis of W. Then it can be extended to a basis of V say $B = \{w_1, w_2, \ldots, w_m, w_{m+1}, \ldots, w_n\}$. Now let $\widehat{B} = \{\widehat{w}_1, \widehat{w}_2, \ldots, \widehat{w}_m, \widehat{w}_{m+1}, \ldots, \widehat{w}_n\}$ be a basis of $\widehat{V}$ dual to B. Then for $m + 1 \leq k \leq n$ and $1 \leq j \leq m$, $\widehat{w}_k(w_j) = 0$. This shows that $\widehat{w}_k(w) = 0$ for all $w \in W$ and $m + 1 \leq k \leq n$, and hence $\{\widehat{w}_{m+1}, \ldots, \widehat{w}_n\} \subseteq W^\circ$. Now we show that $\{\widehat{w}_{m+1}, \ldots, \widehat{w}_n\}$ is a basis of W°. This is a linearly independent subset with $n - m$ elements. Suppose that $\widehat{w}$ is an arbitrary member of W°. Since $\widehat{w} \in \widehat{V}$, and $\widehat{B}$ is a basis of $\widehat{V}$, we find that $\widehat{w} = \sum_{i=1}^{n} \widehat{w}(w_i)\widehat{w}_i$. But since $w_j \in W$ for all $1 \leq j \leq m$, we arrive at $\widehat{w} = \sum_{i=m+1}^{n} \widehat{w}(w_i)\widehat{w}_i$. This yields that $\{\widehat{w}_{m+1}, \ldots, \widehat{w}_n\}$ spans W°. Hence $dimW^\circ = n - m = n - dimW$, i.e., $dimW + dimW^\circ = dimV$.

(ii) Given that W is a subspace of V. Suppose that $f \in \widehat{V}$ and $f|_W$, the restriction of f to W. Then it is straightforward to see that $f|_W \in \widehat{W}$. Now define a map $\psi : \widehat{V} \to \widehat{W}$ such that $\psi(f) = f|_W$. It is clear that for any $\alpha, \beta \in \mathbb{F}$ and $f, g \in \widehat{V}$,

$$\psi(\alpha f + \beta g) = (\alpha f + \beta g)|_W = \alpha f|_W + \beta g|_W = \alpha \psi(f) + \beta \psi(g).$$

This shows that ψ is a vector space homomorphism. Now if $f \in Ker\psi$, then the restriction of f to W must be zero, i.e., $f(w) = 0$ for all $w \in W$ or $f \in W^\circ$. Conversely, if $f \in W^\circ$, i.e., $f(w) = 0$ for all $w \in W$, then $f|_W = 0$ and $f \in Ker\psi$. Hence $Ker\psi = W^\circ$. Now we show that ψ is onto. Then we show that any given $h \in \widehat{W}$ is the restriction of some $f \in \widehat{V}$. Let $\{w_1, w_2, \ldots, w_m\}$ be a basis of W. Then it can be extended to a basis of V say $\{w_1, w_2, \ldots, w_m, u_1, u_2, \ldots, u_r\}$, where $m + r = \dim V$. Hence, we can write $V = W \oplus U$, where U is a subspace of V spanned by $\{u_1, u_2, \ldots, u_r\}$. Now for any $h \in \widehat{W}$, define $\xi \in \widehat{V}$ such that for any $v \in V, v = w + u$ and $\xi(v) = h(w)$, where $w \in W, u \in U$. Let $v' = v''$ and suppose that $v' = w' + u'$, $v'' = w'' + u''$ where $w', w'' \in W$ and $u', u'' \in U$. This implies that $w' = w''$ and $u' = u''$. As h is a map, we get $h(w') = h(w'')$. This implies that $\xi(v') = \xi(v'')$ and hence ξ is well defined. It can be easily seen that ξ is a linear functional whose restriction on W is h, i.e., $\psi(\xi) = \xi|_W = h$. Hence, ψ is onto and by fundamental theorem of vector space homomorphism $\widehat{W} \cong \dfrac{\widehat{V}}{Ker\psi}$, i.e., $\widehat{W} \cong \dfrac{\widehat{V}}{W^\circ}$.

(iii) Suppose that $\dim W = m$ and $\dim V = n$. Then by the above (i), we find that $\dim W^\circ = n - m$. Therefore,

$$\dim(W^\circ)^\circ = \dim\widehat{V} - \dim W^\circ = \dim V - \dim W^\circ = n - (n - m) = m = \dim W.$$

Since by Lemma 4.17 $W \subseteq (W^\circ)^\circ$, we find that $W = (W^\circ)^\circ$.

Proposition 4.19 *If W_1, W_2 are subspaces of a finite dimensional vector space V, then*

(i) $(W_1 + W_2)^\circ = W_1^\circ \cap W_2^\circ$;
(ii) $(W_1 \cap W_2)^\circ = W_1^\circ + W_2^\circ$.

Proof (i) Since $W_1 \subseteq W_1 + W_2$ and $W_2 \subseteq W_1 + W_2$, by Lemma 4.17, $(W_1 + W_2)^\circ \subseteq W_1^\circ \cap W_2^\circ$. Now, on the other hand, suppose that $\varphi \in (W_1^\circ \cap W_2^\circ)$. Then φ annihilates both W_1 and W_2. If $v \in W_1 + W_2$, then $v = w_1 + w_2$, where $w_1 \in W_1$ and $w_2 \in W_2$. Now $\varphi(v) = \varphi(w_1) + \varphi(w_2) = 0$. This shows that φ annihilates $W_1 + W_2$, i.e., $\varphi \in (W_1 + W_2)^\circ$. Therefore, $(W_1^\circ \cap W_2^\circ) \subseteq (W_1 + W_2)^\circ$ and hence $(W_1 + W_2)^\circ = W_1^\circ \cap W_2^\circ$.

(ii) Replacing W_1 by W_1° and W_2 by W_2° in (i) and using Theorem 4.18(iii), we get $(W_1^\circ + W_2^\circ)^\circ = (W_1^\circ)^\circ \cap (W_2^\circ)^\circ = W_1 \cap W_2$ and hence $((W_1^\circ + W_2^\circ)^\circ)^\circ = (W_1 \cap W_2)^\circ$. This implies that $(W_1 \cap W_2)^\circ = W_1^\circ + W_2^\circ$.

Remark 4.20 Observe that no dimension argument is employed in the proof (i), hence the above result (i) holds for vector spaces of finite or infinite dimensions.

4.4 Hyperspaces or Hyperplanes

Definition 4.21 Let V be a vector space over $\mathbb{F}$. Then a maximal proper subspace of V is called a hyperspace or hyperplane of V.

Example 4.22 (1) Consider $\mathbb{R}^n$, $n \geq 2$, as a vector space over $\mathbb{R}$. Then the subspaces $W_1 = \{(\alpha_1, \alpha_2, \ldots, \alpha_{n-1}, 0) \mid \alpha_i \in \mathbb{R}, i = 1, 2, \ldots, n - 1\}$ and $W_2 = \{(0, \beta_1, \ldots, \beta_{n-1}) \mid \beta_i \in \mathbb{R}, i = 1, 2, \ldots, n - 1\}$ are hyperspaces of $\mathbb{R}^n$.

(2) Consider $P_n(x)$, $n \geq 1$, as the vector space of all real polynomials of degree at most of degree n over the field $\mathbb{R}$. Then the subspaces $W_1 = \{\alpha_1 x + \alpha_2 x^2 + \cdots + \alpha_n x^n \mid \alpha_i \in \mathbb{R}, i = 1, 2, \ldots, n\}$ and $W_2 = \{\beta_1 + \beta_2 x + \cdots + \beta_{n-1} x^{n-1} \mid \beta_i \in \mathbb{R}, i = 1, 2, \ldots, n - 1\}$ are hyperspaces of $P_n(x)$.

Theorem 4.23 *Let V be an n-dimensional vector space over a field $\mathbb{F}$, where $n \geq 2$. Then a subspace W of V will be a hyperspace of V if and only if $\dim W = n - 1$.*

Proof Suppose that the subspace W is a hyperspace of V. As W is a proper subspace of V, hence $0 < \dim W < \dim V = n$. We claim that $\dim W = n - 1$. Suppose on contrary, $0 < \dim W = m < n - 1$. Since W is a proper subspace of V, there exists an element $v \in V \setminus W$. Consider W_1, the subspace of V generated by v, i.e., $W_1 = [v]$. It is obvious that $W \subset W + W_1 \subseteq V$. Since $W + W_1$ is a subspace of V and $W \cap W_1 = \{0\}$, we have $\dim W + W_1 = \dim W + \dim W_1 - \dim W \cap W_1 = \dim W + \dim W_1 = m + 1 < n$. But W is a maximal subspace of V also; this forces us to conclude that $W + W_1 = V$ and hence $\dim W + W_1 = \dim V = n$, which leads to a contradiction. Thus we conclude that $\dim W = n - 1$.

Conversely suppose that $\dim W = n - 1$. We have to show that W is a hyperspace of V. By our hypothesis, it is clear that $0 < \dim W < n$. This shows that W is a proper subspace of V. Next we prove that W is also a maximal subspace of V. For this, let W_2 be a subspace of V such that $W \subseteq W_2 \subseteq V$. Then we have to prove that either $W_2 = W$ or $W_2 = V$. If $W_2 = W$, then there is nothing to do. If $W_2 \neq W$, then there exists $v' \in W_2 \setminus W$. Consider W_3, the subspace of V generated by v', i.e., $W_3 = [v']$ and arguing in the similar way as above one can show that $\dim W + W_3 = n - 1 + 1 = n$. But as $W + W_3$ is a subspace of V, we conclude that $W + W_3 = V$. We also have $W + W_3 \subset W_2$. Hence, we conclude that $V \subset W_2$. Finally we get $V = W_2$. Thus W is a hyperspace of V.

Theorem 4.24 *If f is a nonzero linear functional on a vector space V, then the null space of f is a hyperspace of V. Conversely, every hyperspace of V is the null space of a (not unique) nonzero linear functional on V.*

Proof Let f be a nonzero linear functional on the vector space V and W the null space of f. We have to show that W is a hyperspace of V. It is obvious that $W \neq V$. Also $W \neq \{0\}$ as $\dim V > 1$ if V is finite dimensional. This shows that W is a proper subspace of V. To prove that W is also a maximal subspace of V, let W_1 be a subspace of V such that $W \subseteq W_1 \subseteq V$. Then we have to prove that either $W_1 = W$ or $W_1 = V$. If $W_1 = W$, then there is nothing to do. If $W_1 \neq W$, then there exists

$v \in W_1 \setminus W$. Now consider the subspace $W + [v] = \{w + \lambda v \mid w \in W, \lambda \in \mathbb{F}\}$. It is clear that $0 \neq f(v) \in \mathbb{F}$. Let $x \in V$. We can write $x = (x - f(x)f(v)^{-1}v) + f(x)f(v)^{-1}v$. We also have $f(x - f(x)f(v)^{-1}v) = f(x) - f(x)f(v)^{-1}f(v) = 0$. Thus $(x - f(x)f(v)^{-1}v) \in$ null space of $f = W$. This ensures that $x \in W + [v]$, i.e., $V \subseteq W + [v]$. But it is given that $W + [v] \subseteq W_1$. Hence, we conclude that $V \subseteq W_1$, i.e., $W_1 = V$. Thus W is a hyperspace of V.

Conversely, suppose that W is a hyperspace of V. Then we have to construct a nonzero linear functional g on V such that null space of $g = W$, as $\{0\} \neq W \neq V$. Thus there exists $v' \in V \setminus W$. Since $W + [v']$ is a subspace of V and $W \subset W + [v']$, therefore using the maximality of W, we conclude that $V = W + [v']$. Let $z \in V$. Then $z = w + \alpha v'$ for some $w \in W, \alpha \in \mathbb{F}$. We claim that for any $z \in V$, corresponding $w \in W$ and $\alpha \in \mathbb{F}$ are unique. To prove this claim, let us suppose that $z = w_1 + \alpha_1 v'$ and $z = w_2 + \alpha_2 v'$, where $w_1, w_2 \in W$ and $\alpha_1, \alpha_2 \in \mathbb{F}$. This implies that $w_1 + \alpha_1 v' = w_2 + \alpha_2 v'$, i.e., $w_1 - w_2 = (\alpha_2 - \alpha_1)v'$. From previous relations, we conclude that $\alpha_1 = \alpha_2$, for otherwise we have $v' = (\alpha_2 - \alpha_1)^{-1}(w_1 - w_2) \in W$, leading to a contradiction. It is now obvious that $w_1 = w_2$. Now define a map $g : V \longrightarrow \mathbb{F}$ such that $g(z) = \alpha$, which is obviously well defined. Now we show that g is a linear functional on V and null space of $g = W$. Let $z_1 = w_1 + \alpha_1 v'$ and $z_2 = w_2 + \alpha_2 v'$, where $w_1, w_2 \in W$ and $\alpha_1, \alpha_2 \in \mathbb{F}$. Then $\lambda z_1 + \mu z_2 = (\lambda w_1 + \mu w_2) + (\lambda \alpha_1 + \mu \alpha_2)v'$, for any $\lambda, \mu \in \mathbb{F}$. Hence $g(\lambda z_1 + \mu z_2) = \lambda \alpha_1 + \mu \alpha_2 = \lambda g(z_1) + \lambda g(z_2)$. This proves that g is a linear functional on V. It is obvious to observe that null space of g is W and $g \neq 0$.

Lemma 4.25 *If f and g are linear functionals on a vector space V, then g is a scalar multiple of f if and only if the null space of g contains the null space of f, i.e., if and only if $f(v) = 0$ implies $g(v) = 0$, where $v \in V$.*

Proof We divide the proof in two cases.
Case I: If at least one of f and g is the zero linear functional, then lemma holds trivially.
Case II: $f \neq 0$ and $g \neq 0$. Suppose that $g = \alpha f$, for some $0 \neq \alpha \in \mathbb{F}$. Let $f(v) = 0$ for some $v \in V$, as we are given $g(v) = \alpha f(v)$. This shows that $g(v) = \alpha 0 = 0$. Thus we obtain $g(v) = 0$. Conversely, suppose that the null space of g contains the null space of f. Since $f \neq 0$, there exists $v' \in V$ such that $f(v') \neq 0$. Also assume that the null space of f is N and therefore $v' \in V \setminus N$. By the previous theorem N will be a maximal subspace of V. Let $\beta = g(v')(f(v'))^{-1}$. Now define the linear functional h on V, given by $h = g - \beta f$. Next we show that h is the zero linear functional on V. Consider the subspace W of V, which is spanned by $N \cup \{v'\}$, i.e., $[N \cup \{v'\}]$. Since N is a maximal subspace of V and $v' \in V - N$, therefore we conclude that $V = [N \cup \{v'\}] = N + [v'] = \{x + \lambda v' \mid x \in N, \lambda \in \mathbb{F}\}$. Thus $h(x + \lambda v') = (g - \beta f)(x + \lambda v') = g(x + \lambda v') - \beta f(x + \lambda v')$. Now using the fact that the null space of g contains the null space of f, we have $h(x + \lambda v') = \lambda g(v') - \beta \lambda f(v') = \lambda g(v') - \lambda (f(v'))^{-1} f(v')g(v') = 0$ for all $x \in N$ and for all $\lambda \in \mathbb{F}$. This implies that h equals the zero linear functional on V. Finally, we conclude that $g - \beta f = 0$, i.e., $g = \beta f$. Hence, we have shown that g is a scalar multiple of f.

Theorem 4.26 *Let* $f, f_1, f_2, \ldots, f_r$ *be linear functionals on a vector space* V *with respective null spaces* $N, N_1, N_2, \ldots, N_r$. *Then* f *is a linear combination of* $f_1, f_2, \ldots, f_r$ *if and only if* $N_1 \cap N_2 \cap \cdots \cap N_r \subseteq N$.

Proof Let $f = \alpha_1 f_1 + \alpha_2 f_2 + \cdots + \alpha_r f_r$, for some $\alpha_1, \alpha_2, \ldots, \alpha_r \in \mathbb{F}$. For any $v \in N_1 \cap N_2 \cap \cdots \cap N_r$, we have $f(v) = (\alpha_1 f_1 + \alpha_2 f_2 + \cdots + \alpha_r f_r)(v) = \alpha_1 f_1(v) + \alpha_2 f_2(v) + \cdots + \alpha_r f_r(v) = 0$. This implies that $v \in N$. Thus we proved $N_1 \cap N_2 \cap \cdots \cap N_r \subseteq N$.

We shall prove the converse by induction on the number r. If $r = 1$, then the result holds by the previous lemma. Suppose the result holds for $r = k - 1$, and let $f_1, f_2, \ldots, f_k$ be linear functionals with null spaces $N_1, N_2, \ldots, N_k$ such that $N_1 \cap N_2 \cap \cdots \cap N_k \subseteq N$. Let $f', f_1', f_2', \ldots, f_{k-1}'$ be the restrictions of $f, f_1, f_2, \ldots, f_{k-1}$ to the subspace N_k. Then $f', f_1', f_2', \ldots, f_{k-1}'$ are linear functionals on the vector space N_k. Furthermore, if $v \in N_k$ and v is an element of the null space of f_i', $i = 1, 2, \ldots, k - 1$, i.e., $f_i'(v) = 0, i = 1, 2, \ldots, k - 1$, then $v \in N_1 \cap N_2 \cap \cdots \cap N_k$ because the null space of $f_i' \subseteq N_i, i = 1, 2, \ldots, k - 1$ and thus by our hypothesis $v \in N$, i.e., $f(v) = 0$. We also get $f'(v) = f(v) = 0$. Now by the induction hypothesis, there exist scalars $\beta_i, i = 1, 2, \ldots, k - 1$ such that $f' = \beta_1 f_1' + \beta_2 f_2' + \cdots + \beta_{k-1} f_{k-1}'$. Now define a map $h : V \longrightarrow \mathbb{F}$ such that $h(x) = (f - \sum_{i=1}^{k-1} \beta_i f_i)(x)$ for all $x \in V$. It is easy to verify that h is a linear functional on V. Let u be an element of the null space of f_k. Thus $h(u) = (f - \sum_{i=1}^{k-1} \beta_i f_i)(u) = f(u) - \sum_{i=1}^{k-1} \beta_i f_i(u)$. Now using the facts that $u \in N_k$, $f(u) = f'(u)$, $f_1(u) = f_1'(u), \ldots, f_{k-1}(u) = f_{k-1}'(u)$, we have $h(u) = f'(u) - \sum_{i=1}^{k-1} \beta_i f_i'(u) = 0$. This proves that u is an element of the null space of h also. By the preceding lemma, h is a scalar multiple of f_k. If $h = \beta_k f_k$ for some $\beta_k \in \mathbb{F}$, then $f - \sum_{i=1}^{k-1} \beta_i f_i = \beta_k f_k$, i.e., $f = \sum_{i=1}^{k} \beta_i f_i$.

Example 4.27 Let n be a positive integer and $\mathbb{F}$ a field. Suppose that W is the set of all vectors $(x_1, x_2, \ldots, x_n) \in \mathbb{F}^n$ such that $x_1 + x_2 + \cdots + x_n = 0$. Then it can be seen that W° consists of all linear functionals f of the form $f(x_1, x_2, \ldots, x_n) = c \sum_{i=1}^{n} x_i$.

Case I : Suppose that $char(\mathbb{F}) \neq 2$. If $f \in W^\circ$, then clearly $f \in \widehat{\mathbb{F}^n}$. But we know that $f(x_1, x_2, \ldots, x_n) = \sum_{i=1}^{n} c_i x_i$, for some fixed $c_1, c_2, \ldots, c_n \in \mathbb{F}$ precisely. It is obvious that $(1, -1, 0, 0, \ldots, 0, 0) \in W$. Hence $f(1, -1, 0, 0, \ldots, 0, 0) = 0$. It implies that $c_1 = c_2$. Similarly $(0, 1, -1, 0, \ldots, 0, 0) \in W$. As a result $f(0, 1, -1, 0, \ldots, 0, 0) = 0$, which implies that $c_2 = c_3$. Arguing in the same way, we conclude that $c_1 = c_2 = \cdots = c_n$. Let us say each of the previous c_i to be $c \in \mathbb{F}$. Thus, we have proved that $f(x_1, x_2, \ldots, x_n) = c \sum_{i=1}^{n} x_i$.

Case II : Suppose that $char(\mathbb{F}) = 2$. Now in this case $x = -x$ for all $x \in \mathbb{F}$. It is easy to observe that the proof given in Case I holds in Case II also.

Exercises

1. If $\mathbb{F}$ is the field of real numbers, find W° and its dimension, where

   (a) W is the subspace of $\mathbb{R}^3$ spanned by $(1, 2, -1)$ and $(1, 1, 0)$.
   (b) W is the subspace of $\mathbb{R}^4$ spanned by $(0, 0, 1, 1)$, $(-1, -2, 3, 1)$ and $(1, 0, 0, 3)$.

2. If $V = W_1 \oplus W_2$, then prove that $\widehat{V} = W_1^\circ \oplus W_2^\circ$, and hence deduce that $\frac{\widehat{V}}{W_1^\circ} \cong W_2^\circ$.

3. Let W_1, W_2 be subspaces of a vector space V. Then show that $\widehat{W_1 \oplus W_2} \cong \widehat{W_1} \oplus \widehat{W_2}$.

4. Let W be a subspace of V. Prove that $\frac{V}{W} \cong W^\circ$.

5. Let $V = \mathbb{R}^4$ and W be the subspace spanned by $v_1 = (2, -2, 3, 2)$, $v_2 = (3, -3, 1, 1)$.

   (a) Find $dim\ W$ and $dim\ W^\circ$, by explicitly giving their respective basis. Verify that $4 = dim\ W + dim\ W^\circ$.
   (b) Let $f \in \widehat{V}$ be given as $f(x_1, x_2, x_3, x_4) = x_1 + x_2 + x_3 - x_4$. Is $f \in W^\circ$?

6. Let V be the vector space of all polynomials over $\mathbb{R}$ of degree less than or equal to 3 and W be the subspace of V, consisting of those polynomials $p(x) \in V$ such that $p(1) = 0$, $p(-1) = 0$. Find $dim\ W$ and $dim\ W^\circ$.

7. Let W_1 and W_2 be the row space and the column space of $A = \begin{bmatrix} 1 & 2 & 2 & 1 \\ 1 & 2 & 2 & 2 \\ 2 & 4 & 3 & 3 \\ 0 & 0 & 1 & -1 \end{bmatrix}$, respectively.

   (a) Find the general formula for those $f \in \widehat{\mathbb{R}^4}$ that belong to W_1°.
   (b) Find the general formula for those $f \in \widehat{\mathbb{R}^4}$ that belong to W_2°.

8. For any $A, B \in M_n(\mathbb{F})$, the vector space of $n \times n$ matrices over a field $\mathbb{F}$, prove that

   (a) $tr(AB) = tr(BA)$,
   (b) if A and B are similar, then $tr(A) = tr(B)$,
   (c) there exist no two matrices A and B in $M_2(\mathbb{R})$, such that $AB - BA = I_2$, where I_2 is the identity matrix of order 2.

9. Let $V = M_2(\mathbb{R})$ be the vector space of all 2×2 matrices with real entries and W is the subspace of V consisting of those $A \in V$ such that $AB = BA$, where $B = \begin{bmatrix} 1 & 3 \\ 2 & 6 \end{bmatrix}$. Find $dim\ W$ and $dim\ W^\circ$. Does there exist a nonzero $f \in \widehat{V}$ such that $f(I_2) = 0$, $f\left(\begin{bmatrix} 0 & 0 \\ 0 & 1 \end{bmatrix}\right) = 0$ and $f \in W^\circ$.

10. Find a basis of the annihilator W° of the subspace W of $\mathbb{R}^4$ spanned by $(1, 2, -3, 4)$ and $(0, 1, 4, -1)$.

11. Let W be the subspace of $\mathbb{R}^5$, which is spanned by the vectors $v_1 = e_1 + 2e_2 + e_3$, $v_2 = e_2 + 3e_3 + 3e_4 + e_5$, $v_3 = e_1 + 4e_2 + 6e_3 + 4e_4 + e_5$, where $\{e_1, e_2, e_3, e_4, e_5\}$ is the standard basis of $\mathbb{R}^5$. Find a basis for W°.

12. Let $V = M_2(\mathbb{R})$ be the vector space of all 2×2 matrices with real entries and let $B = \begin{bmatrix} 2 & -2 \\ -1 & 1 \end{bmatrix}$. Let W be the subspace of V consisting of those $A \in V$ such that $AB = 0_{2\times2}$. Let f be a linear functional on V which is in the annihilator of W. Suppose that $f(I_2) = 0$ and $f(C) = 3$, where I_2 is the identity matrix of order 2×2 and $C = \begin{bmatrix} 0 & 0 \\ 0 & 1 \end{bmatrix}$. Find $f(B)$.

13. Let S be a set, $\mathbb{F}$ a field and $V(S; \mathbb{F})$ the vector space of all functions from S into $\mathbb{F}$, where operations are defined as follows: $(f + g)(x) = f(x) + g(x)$; $(\alpha f)(x) = \alpha f(x)$. Let W be any n-dimensional subspace of $V(S; \mathbb{F})$. Show that there exist points $x_1, x_2, \ldots, x_n$ in S and functions $f_1, f_2, \ldots, f_n$ in W such that $f_i(x_j) = \delta_{ij}$.

14 If W is a subspace of a finite dimensional vector space V and if $\{f_1, f_2, \ldots, f_r\}$ is any basis for W°, then prove that $W = \cap_{i=1}^r N_i$, where N_i is the null space of f_i, $i = 1, 2, \ldots, r$.

4.5 Dual (or Transpose) of Linear Transformation

For a given vector space V over a field $\mathbb{F}$, one can always find its dual space $\widehat{V}$. For a given linear transformation $T : V \rightarrow W$, is it possible to find a linear map $\widehat{T} : \widehat{W} \rightarrow \widehat{V}$? In the present section, we shall discuss properties of such linear maps. In fact, if $T : V \rightarrow W$ is a linear transformation, then for any $f \in \widehat{W}$, the composition $f \circ T : V \rightarrow \mathbb{F}$ given by $f \circ T(v) = f(T(v))$ for all $v \in V$ defines a linear transformation. In fact, for any $u, v \in V$ and $\alpha, \beta \in \mathbb{F}$, $f \circ T(\alpha u + \beta v) = \alpha f \circ T(u) + \beta f \circ T(v)$ and hence $f \circ T \in \widehat{V}$.

Definition 4.28 If V, W are vector spaces over $\mathbb{F}$ and $T : V \rightarrow W$ is a linear transformation, and $f \in \widehat{W}$, then a map $\widehat{T} : \widehat{W} \rightarrow \widehat{V}$ given by $\widehat{T}(f) = f \circ T$ is called dual or transpose of T; it is also denoted by T^t.

For any $f, g \in \widehat{W}, \alpha, \beta \in \mathbb{F}$, and $v \in V$

$$(\widehat{T}(\alpha f + \beta g))(v) = ((\alpha f + \beta g) \circ T)(v)$$
$$= (\alpha f + \beta g)(T(v))$$
$$= \alpha f(T(v)) + \beta g(T(v))$$
$$= \alpha(f \circ T)(v) + \beta(g \circ T)(v)$$
$$= (\alpha(f \circ T) + \beta(g \circ T))(v)$$
$$= (\alpha\widehat{T}(f) + \beta\widehat{T}(g))(v).$$

This shows that $\widehat{T}(\alpha f + \beta g) = \alpha\widehat{T}(f) + \beta\widehat{T}(g)$, and hence $\widehat{T} : \widehat{W} \to \widehat{V}$ is a linear transformation. This implies that if $T \in Hom(V, W)$, then $\widehat{T} \in Hom(\widehat{W}, \widehat{V})$.

Example 4.29 Let f be the linear functional on $\mathbb{R}^3$ defined by $f(x, y, z) = x - 2y + 2z$. Evaluate $\widehat{T}(f)$, in each case if $T : \mathbb{R}^2 \longrightarrow \mathbb{R}^3$ is a linear transformation defined as

(1) $T(x, y) = (3x - y, x + 2y, x - y)$,
(2) $T(x, y) = (-2x + y, 3y, x + 6y)$ and
(3) $T(x, y) = (2x - y, x - 2y, -x)$.

Solution:

(1) $\widehat{T}(f)(x, y) = (f \circ T)(x, y) = f(T(x, y)) = f(3x - y, x + 2y, x - y)$
 $= 3x - y - 2x - 4y + 2x - 2y = 3x - 7y$.
(2) $\widehat{T}(f)(x, y) = (f \circ T)(x, y) = f(T(x, y)) = f(-2x + y, 3y, x + 6y)$
 $= -2x + y - 6y + 2x + 12y = 7y$.
(3) $\widehat{T}(f)(x, y) = (f \circ T)(x, y) = f(T(x, y)) = f(2x - y, x - 2y, -x)$
 $= 2x - y - 2x + 4y - 2x = -2x + 3y$.

Lemma 4.30 *If U, V are vector spaces over $\mathbb{F}$ and $T_1, T_2 \in Hom(U, V); \alpha \in \mathbb{F}$, then*

 (i) $\widehat{0} = 0$, *where 0 stands for the zero linear transformation.*
 (ii) $\widehat{I_U} = I_{\widehat{U}}$, *where I_U and $I_{\widehat{U}}$ stand for the identity linear operators on the vector spaces U and $\widehat{U}$, respectively.*
 (iii) $\widehat{(T_1 + T_2)} = \widehat{T_1} + \widehat{T_2}$.
 (iv) $\widehat{(\alpha T_1)} = \alpha\widehat{T_1}$.

Proof *(i)* Since $0 : U \longrightarrow V$, hence $\widehat{0} : \widehat{V} \longrightarrow \widehat{U}$. If $f \in \widehat{V}$, and $u \in U$, then $(\widehat{0}(f))(u) = (f \circ 0)(u) = f(0(u)) = f(0) = 0 = 0(u)$. This shows that $\widehat{0}(f) = 0$, i.e., $\widehat{0} = 0$.
(ii) We know that $I_U : U \longrightarrow U$, thus $\widehat{I_U} : \widehat{U} \longrightarrow \widehat{U}$. For any $f \in \widehat{U}$ and $u \in U$, we have $(\widehat{I_U}(f))(u) = (f \circ I_U)(u) = f(u)$. This implies that $(\widehat{I_U}(f)) = f$, i.e., $(\widehat{I_U}(f)) = I_{\widehat{U}}(f)$. Finally, we have obtained $\widehat{I_U} = I_{\widehat{U}}$.

(iii) If $T_1 + T_2 : U \longrightarrow V$, then $\widehat{T_1 + T_2} : \widehat{V} \longrightarrow \widehat{U}$. If $f \in \widehat{V}$, and $u \in U$, then $((\widehat{T_1 + T_2})(f))(u) = (f \circ (T_1 + T_2))(u) = f((T_1 + T_2)(u)) = f(T_1(u) + T_2(u)) = (f \circ T_1 + f \circ T_2)(u)$. This implies that $(\widehat{T_1 + T_2})(f) = (f \circ T_1 + f \circ T_2)$, i.e.,

$(\widehat{T_1 + T_2})(f) = \widehat{T_1}(f) + \widehat{T_2}(f)$. Finally, we get $(\widehat{T_1 + T_2})(f) = (\widehat{T_1} + \widehat{T_2})(f)$, i.e., $(\widehat{T_1 + T_2}) = \widehat{T_1} + \widehat{T_2}$.

(iv) Since $\alpha T_1 : U \longrightarrow V$, hence $\widehat{\alpha T_1} : \widehat{V} \longrightarrow \widehat{U}$. If $f \in \widehat{V}$, and $u \in U$, then $((\widehat{\alpha T_1})(f))(u) = (f \circ (\alpha T_1))(u) = f(\alpha T_1)(u) = f(\alpha (T_1(u))) = \alpha(f(T_1(u)))$. This gives us $((\widehat{\alpha T_1})(f))(u) = (\alpha(f \circ T_1))(u)$, i.e., $\widehat{\alpha T_1}(f) = (\alpha(f \circ T_1))$. Thus, we conclude that $(\widehat{\alpha T_1})(f) = (\alpha(\widehat{T_1}(f)))$, i.e., $(\widehat{\alpha T_1})(f) = (\alpha \widehat{T_1})(f)$. This shows that $(\widehat{\alpha T_1}) = \alpha \widehat{T_1}$.

Theorem 4.31 *Let U, V, W be vector spaces over $\mathbb{F}$. If $T : U \longrightarrow V$ and $S : V \longrightarrow W$ are linear transformations, then $(\widehat{S \circ T}) = \widehat{T} \circ \widehat{S}$.*

Proof Since $S \circ T : U \longrightarrow W$, $\widehat{S \circ T} : \widehat{W} \longrightarrow \widehat{U}$. We also have $\widehat{T} : \widehat{V} \longrightarrow \widehat{U}$ and $\widehat{S} : \widehat{W} \longrightarrow \widehat{V}$, as a result we get $\widehat{T} \circ \widehat{S} : \widehat{W} \longrightarrow \widehat{U}$. Thus, both $\widehat{S \circ T}$ and $\widehat{T} \circ \widehat{S}$ are linear transformations from $\widehat{W}$ to $\widehat{U}$. To prove the equality of these two maps, let $\widehat{w} \in \widehat{W}$. Then $(\widehat{S \circ T})(\widehat{w}) = \widehat{w} \circ (S \circ T) = (\widehat{w} \circ S) \circ T = (\widehat{S}(\widehat{w})) \circ T$. If we assume $\widehat{S}(\widehat{w}) = \widehat{v} \in \widehat{V}$, then we arrive at $(\widehat{S \circ T})(\widehat{w}) = \widehat{v} \circ T = \widehat{T}(\widehat{v}) = \widehat{T}(\widehat{S}(\widehat{w})) = (\widehat{T} \circ \widehat{S})(\widehat{w})$. Finally, we conclude that $(\widehat{S \circ T}) = \widehat{T} \circ \widehat{S}$.

Lemma 4.32 *If U, V are vector spaces over $\mathbb{F}$ and $T : U \longrightarrow V$ is an invertible linear transformation, then $\widehat{T}$ is also an invertible linear transformation and $(\widehat{T})^{-1} = \widehat{(T^{-1})}$.*

Proof Since $T : U \longrightarrow V$ is an invertible linear transformation, there exists the invertible linear transformation $T^{-1} : V \longrightarrow U$ such that $T^{-1}T = I_U$, $TT^{-1} = I_V$, where I_U and I_V are the identity linear operators on the vector spaces U and V, respectively. We also have $\widehat{T} : \widehat{V} \longrightarrow \widehat{U}$ and $\widehat{(T^{-1})} : \widehat{U} \longrightarrow \widehat{V}$. Using Theorem 4.31 and Lemma 4.30, we have $\widehat{T}\widehat{(T^{-1})} = I_{\widehat{U}}$; $\widehat{(T^{-1})}\widehat{T} = I_{\widehat{V}}$. Previous relations show that $\widehat{T}$ is invertible and $(\widehat{T})^{-1} = \widehat{(T^{-1})}$.

Theorem 4.33 *Let U and V be vector spaces over a field $\mathbb{F}$ and let T be a linear transformation from U to V. Then null space or kernel of $\widehat{T}$ is the annihilator of range of T. Further if U and V are finite dimensional, then*

(i) $r(\widehat{T}) = r(T)$.
(ii) *the range of $\widehat{T}$ is the annihilator of the null space of T.*

Proof We first prove that $N(\widehat{T}) = (T(U))°$. Since $\widehat{T} : \widehat{V} \longrightarrow \widehat{U}$, we find that

$$f \in N(\widehat{T}) \Leftrightarrow \widehat{T}(f) = 0, \text{ the zero linear functional on U}$$
$$\Leftrightarrow f \circ T = 0$$
$$\Leftrightarrow (f \circ T)(u) = 0 \in \mathbb{F}, \text{ for all } u \in U$$
$$\Leftrightarrow f(T(u)) = 0, \text{ for all } T(u) \in T(U)$$
$$\Leftrightarrow f \in (T(U))°.$$

Hence $N(\widehat{T}) = (T(U))°$.

(*i*) Now let $dim\ U = n$, $dim\ V = m$ and $r(T) = dim\ (T(U)) = p$. Since $T(U)$ is a subspace of V, using Theorem 4.18, we find that $dim\ (T(U))° = dim\ V - dim\ (T(U)) = m - p$. But since $N(\widehat{T}) = (T(U))°$, we arrive at $dim\ N(\widehat{T}) = dim\ (T(U))°$, i.e., $n(\widehat{T}) = m - p$. Now, using the relation $r(\widehat{T}) + n(\widehat{T}) = dim\ \widehat{V}$, we obtain $r(\widehat{T}) = dim\widehat{V} - n(\widehat{T}) = m - (m - p) = p = r(T)$.

(*ii*) We have to show that $\widehat{T}(\widehat{V}) = (N(T))°$. For this, let $f \in \widehat{T}(\widehat{V})$. This implies that $f = \widehat{T}(g)$, for some $g \in \widehat{V}$, i.e., $f = g \circ T$. If $u \in N(T)$, then $f(u) = (\widehat{T}(g))(u) = (g \circ T)(u) = g(T(u)) = g(0) = 0$. This shows that $f \in (N(T))°$ and hence $\widehat{T}(\widehat{V}) \subseteq (N(T))°$. But since $dim\ \widehat{T}(\widehat{V}) = r(\widehat{T}) = r(T) = p$ and $N(T)$ is a subspace of U, by Theorem 4.18, we find that $dim\ (N(T))° = dim\ U - dim\ N(T) = n - (n - p) = p$. Finally, it follows that $\widehat{T}(\widehat{V}) = (N(T))°$.

Theorem 4.34 *Let U and V be finite dimensional vector spaces over a field* $\mathbb{F}$. *Let* B_1 *be an ordered basis for U with dual basis* $\widehat{B_1}$, *and let* B_2 *be an ordered basis for V with dual basis* $\widehat{B_2}$. *Let* $T : U \longrightarrow V$ *be a linear transformation, let* $[T]^{B_2}_{B_1}$ *be the matrix of T relative to* B_1, B_2 *and let* $[\widehat{T}]^{\widehat{B_2}}_{\widehat{B_1}}$ *be the matrix of* $\widehat{T}$ *relative to* $\widehat{B_1}, \widehat{B_2}$. *Then* $[\widehat{T}]^{\widehat{B_2}}_{\widehat{B_1}} =$ *the transpose of* $[T]^{B_2}_{B_1}$.

Proof Let $B_1 = \{u_1, u_2, \ldots, u_n\}$, $B_2 = \{v_1, v_2, \ldots, v_m\}$ and $\widehat{B_1} = \{f_1, f_2, \ldots, f_n\}$; $\widehat{B_2} = \{g_1, g_2, \ldots, g_m\}$. Suppose that $[T]^{B_2}_{B_1} = (\alpha_{ij})_{m \times n}$ and $[\widehat{T}]^{\widehat{B_2}}_{\widehat{B_1}} = (\beta_{ij})_{m \times n}$. By definition, we have $T(u_j) = \sum_{i=1}^{m} \alpha_{ij} v_i$ and $\widehat{T}(g_j) = \sum_{i=1}^{m} \beta_{ij} f_i$, $j = 1, 2, \ldots, n$. On the other hand, $(\widehat{T}(g_s))(u_r) = g_s(T(u_r)) = g_s\left(\sum_{k=1}^{m} \alpha_{kr} v_k\right) = \sum_{k=1}^{m} \alpha_{kr} g_s(v_k) = \sum_{k=1}^{m} \alpha_{kr} \delta_{sk}$ $= \alpha_{sr}$ where $r = 1, 2, \ldots, n$; $s = 1, 2, \ldots, m$. For any linear functional f on U, we have $f = \sum_{s=1}^{m} f(u_r)(f_s)$. If we apply this formula to the functional $f = \widehat{T}(g_s)$ and use the fact that $(\widehat{T}(g_s))(u_r) = \alpha_{sr}$, we have $\beta_{rs} = \alpha_{sr}$, because $\widehat{T}(g_s) = \sum_{r=1}^{m} \beta_{rs} f_s$ also, where $r = 1, 2, \ldots, n$; $s = 1, 2, \ldots, m$. This implies that $(\beta_{ij})_{m \times n} =$ the transpose of $(\alpha_{ij})_{m \times n}$, i.e., $[\widehat{T}]^{\widehat{B_2}}_{\widehat{B_1}} =$ the transpose of $[T]^{B_2}_{B_1}$.

Exercises

1. Let f be the linear functional on $\mathbb{R}^3$ defined by $f(x, y, z) = 2x - 3y + z$. For each of the following linear operators on $\mathbb{R}^3$, find $(\widehat{T}(f))(x, y, z)$:

   (a) $T(x, y, z) = (x - 3y + z, -2x + y + z, 2x - z)$,
   (b) $T(x, y, z) = (-2x + y + z, x - 2y, x + z)$ and
   (c) $T(x, y, z) = (-2x + y, x - 2z, y + z)$.

2. Let V be a finite dimensional vector space. Then prove that the linear transformation $T : V \longrightarrow V$ is nonsingular if and only if its transpose $\widehat{T} : \widehat{V} \longrightarrow \widehat{V}$ is nonsingular.

3. Let V be the vector space of all polynomial functions over the field of real numbers. Let a and b be fixed real numbers and let f be the linear functional on V defined by $f(p(x)) = \int_a^b p(x)dx$, where $p(x) \in V$. If D is the differentiation operator on V, then find $\widehat{D}(f)$.

4. Let V be the vector space of all $n \times n$ matrices over a field $\mathbb{F}$ and let B be a fixed $n \times n$ matrix. If T is the linear operator on V defined by $T(A) = AB - BA$, and if f is the trace function, then find $\widehat{T}(f)$.

5. Let V be a finite dimensional vector space over the field $\mathbb{F}$ and let T be a linear operator on V. Let α be a scalar and suppose there is a nonzero vector $v \in V$ such that $T(v) = \alpha v$. Prove that there is a nonzero linear functional f on V such that $\widehat{T}(f) = \alpha f$.

6. Let n be a positive integer and let V be the vector space of all polynomial functions over the field of real numbers which have degree atmost n, i.e., functions of the form $f(x) = \alpha_o + \alpha_1 x + \cdots + \alpha_n x^n$. Let D be the differential operator on V. Find a basis for the null space of the transpose operator $\widehat{D}$.

7. Let V be a finite dimensional vector space over a field $\mathbb{F}$. Show that $T \longrightarrow \widehat{T}$ is an isomorphism from $\mathscr{A}(V)$ to $\mathscr{A}(\widehat{V})$.

Chapter 5
Inner Product Spaces

In the previous chapters, we have considered vector space V over an arbitrary field $\mathbb{F}$. In the present chapter, we shall restrict ourselves over the field of reals $\mathbb{R}$ or the complex field $\mathbb{C}$. One can see that the concept of "length" and "orthogonality" did not appear in the investigation of vector space over arbitrary field. In this chapter, we place an additional structure on a vector space V to obtain an inner product space. If V is a vector space over $\mathbb{R}$ then V is called real vector space. On the other hand, if V is a vector space over $\mathbb{C}$ then V is called complex vector space.

5.1 Inner Products

Definition 5.1 A vector space V over $\mathbb{F}$ is said to be an inner product space if there exists a function $\langle , \rangle : V \times V \to \mathbb{F}$ satisfying the following axioms:

(1) $\langle u, v \rangle = \overline{\langle v, u \rangle}$ for all $u, v \in V$.
(2) $\langle u, u \rangle \geq 0$ and $\langle u, u \rangle = 0 \Leftrightarrow u = 0$ for all $u \in V$.
(3) $\langle \alpha u + \beta v, w \rangle = \alpha \langle u, w \rangle + \beta \langle v, w \rangle$ for all $u, v, w \in V$ and $\alpha, \beta \in \mathbb{F}$.

Remark 5.2 (i) The function $\langle , \rangle$ satisfying the axioms (1), (2) and (3) is called inner product on V.
(ii) If $\mathbb{F} = \mathbb{R}$, then the complex conjugate $\overline{\langle v, u \rangle} = \langle v, u \rangle$, and hence the axiom (1) can be written as $\langle u, v \rangle = \langle v, u \rangle$.
(iii) $\langle u, v \rangle$ is generally denoted as (u, v), $u.v$ or $\langle u|v \rangle$. Throughout we shall denote it by $\langle u, v \rangle$.
(iv) If $\mathbb{F} = \mathbb{C}$, the field of complex numbers, then axiom (1) implies that $\langle u, u \rangle$ is real and hence the axiom (2) makes sense. For any $\alpha, \beta \in \mathbb{C}$ and $u, v, w \in V$, applying (1) and (2) we see that

$$\begin{aligned}
\langle u, \alpha v + \beta w \rangle &= \overline{\langle \alpha v + \beta w, u \rangle} \\
&= \overline{\alpha \langle v, u \rangle + \beta \langle w, u \rangle} \\
&= \overline{\alpha \langle v, u \rangle} + \overline{\beta \langle w, u \rangle} \\
&= \overline{\alpha}\,\overline{\langle v, u \rangle} + \overline{\beta}\,\overline{\langle w, u \rangle} \\
&= \overline{\alpha} \langle u, v \rangle + \overline{\beta} \langle u, w \rangle.
\end{aligned}$$

Example 5.3 (1) In the vector space $V = \mathbb{F}^2$, for any $u = (\alpha_1, \alpha_2), v = (\beta_1, \beta_2) \in V$ define

$$\langle u, v \rangle = 2\alpha_1 \overline{\beta_1} + \alpha_1 \overline{\beta_2} + \alpha_2 \overline{\beta_1} + \alpha_2 \overline{\beta_2}.$$

It can be easily seen that

$$\begin{aligned}
\overline{\langle u, v \rangle} &= \overline{2\alpha_1 \overline{\beta_1} + \alpha_1 \overline{\beta_2} + \alpha_2 \overline{\beta_1} + \alpha_2 \overline{\beta_2}} \\
&= 2\overline{\alpha_1}\beta_1 + \overline{\alpha_1}\beta_2 + \overline{\alpha_2}\beta_1 + \overline{\alpha_2}\beta_2 \\
&= 2\beta_1 \overline{\alpha_1} + \beta_1 \overline{\alpha_2} + \beta_2 \overline{\alpha_1} + \beta_2 \overline{\alpha_2} \\
&= \langle v, u \rangle,
\end{aligned}$$

$$\begin{aligned}
\langle u, u \rangle &= 2\alpha_1 \overline{\alpha_1} + \alpha_1 \overline{\alpha_2} + \alpha_2 \overline{\alpha_1} + \alpha_2 \overline{\alpha_2} \\
&= \alpha_1 \overline{\alpha_1} + \alpha_1 (\overline{\alpha_1} + \overline{\alpha_2}) + \alpha_2 (\overline{\alpha_1} + \overline{\alpha_2}) \\
&= |\alpha_1|^2 + (\alpha_1 + \alpha_2)(\overline{\alpha_1} + \overline{\alpha_2}) \\
&= |\alpha_1|^2 + |\alpha_1 + \alpha_2|^2 \geq 0.
\end{aligned}$$

This shows that $\langle u, u \rangle = 0$ if and only if $\alpha_1 = \alpha_2 = 0$, i.e., $u = 0$.
For any $u = (\alpha_1, \alpha_2), v = (\beta_1, \beta_2), w = (\gamma_1, \gamma_2) \in V$ and $\delta, \lambda \in \mathbb{F}$, it is also straightforward to see that

$$\langle \delta u + \lambda v, w \rangle = \delta \langle u, w \rangle + \lambda \langle v, w \rangle.$$

This shows that the above product defines an inner product on V.

(2) In the vector space $V = \mathbb{F}^n$, for any $u = (\alpha_1, \alpha_2, \ldots, \alpha_n), v = (\beta_1, \beta_2, \ldots, \beta_n) \in V$ define

$$\langle u, v \rangle = \alpha_1 \overline{\beta_1} + \alpha_2 \overline{\beta_2} + \cdots + \alpha_n \overline{\beta_n}.$$

It can be easily seen that the above product defines an inner product on V. The above inner product is called standard inner product in $\mathbb{R}^n$ and $\mathbb{C}^n$, and the resulting inner product space is called Euclidean and Unitary space, respectively.

(3) Let $V = \mathscr{C}[a, b]$, the vector space of all continuous complex valued functions defined on the closed interval $[a, b]$. For any $f(t), g(t) \in V$ define

$$\langle f(t), g(t) \rangle = \int_a^b f(t)\overline{g(t)}dt.$$

This defines an inner product on V. For any $f(t), g(t) \in V$, it can be seen that

$$\overline{\int_a^b g(t)\overline{f(t)}dt} = \int_a^b \overline{g(t)}f(t)dt$$
$$= \int_a^b f(t)\overline{g(t)}dt.$$

This shows that $\langle f(t), g(t)\rangle = \overline{\langle g(t), f(t)\rangle}$. Also

$$\int_a^b f(t)\overline{f(t)}dt = \int_a^b |f(t)|^2 dt \geq 0$$

and equality holds if and only if $f(t) = 0$.
For any $\alpha, \beta \in \mathbb{C}$, $f(t), g(t), h(t) \in V$

$$\langle \alpha f(t) + \beta g(t), h(t)\rangle = \int_a^b \{\alpha f(t) + \beta g(t)\}\overline{h(t)}dt$$
$$= \alpha \int_a^b f(t)\overline{h(t)}dt + \beta \int_a^b g(t)\overline{h(t)}dt$$
$$= \alpha \langle f(t), h(t)\rangle + \beta \langle g(t), h(t)\rangle.$$

(4) For any diagonal matrix $A = \begin{bmatrix} \lambda_1 & 0 & 0 \\ 0 & \lambda_2 & 0 \\ 0 & 0 & \lambda_3 \end{bmatrix}$ with all reals $\lambda_i \geq 0, 1 \leq i \leq 3$. For any $u, v \in \mathbb{R}^3$, let $u = (u_1, u_2, u_3), v = (v_1, v_2, v_3)$

$$\langle u, v\rangle = \begin{bmatrix} u_1 & u_2 & u_3 \end{bmatrix} \begin{bmatrix} \lambda_1 & 0 & 0 \\ 0 & \lambda_2 & 0 \\ 0 & 0 & \lambda_3 \end{bmatrix} \begin{bmatrix} v_1 \\ v_2 \\ v_3 \end{bmatrix}$$
$$= \lambda_1 u_1 v_1 + \lambda_2 u_2 v_2 + \lambda_3 u_3 v_3.$$

Then the above product defines an inner product on $\mathbb{R}^3$. Hence there are infinitely many inner product that can be defined on $\mathbb{R}^3$. Readers are advised to show that the above product uAv^t cannot be an inner product if A has a negative diagonal entry $\lambda_i < 0$.

(5) Consider the vector space $V = M_2(\mathbb{R})$. For any $A = \begin{bmatrix} a_{11} & a_{12} \\ a_{21} & a_{22} \end{bmatrix}$ and $B = \begin{bmatrix} b_{11} & b_{12} \\ b_{21} & b_{22} \end{bmatrix}$ in V, define

$$\langle A, B\rangle = a_{11}b_{11} + a_{12}b_{12} + a_{21}b_{21} + a_{22}b_{22}.$$

Then it can be easily seen that the above product defines an inner product on V. In fact $\langle A, A\rangle = a_{11}^2 + a_{12}^2 + a_{21}^2 + a_{22}^2 \geq 0$ and $\langle A, A\rangle = 0$ if and only if $a_{11} = 0, a_{12} = 0, a_{21} = 0, a_{22} = 0$, i.e., $A = \begin{bmatrix} 0 & 0 \\ 0 & 0 \end{bmatrix}$. Moreover, since all the entries of the matrix are real, $\langle A, B\rangle = \langle B, A\rangle$. Further for any $\alpha, \beta \in \mathbb{R}$ and $A, B, C \in V$,

$$\langle \alpha A + \beta B, C \rangle = \left\langle \begin{bmatrix} \alpha a_{11} + \beta b_{11} & \alpha a_{12} + \beta b_{12} \\ \alpha a_{21} + \beta b_{21} & \alpha a_{22} + \beta b_{22} \end{bmatrix}, \begin{bmatrix} c_{11} & c_{12} \\ c_{21} & c_{22} \end{bmatrix} \right\rangle$$
$$= (\alpha a_{11} + \beta b_{11})c_{11} + (\alpha a_{12} + \beta b_{12})c_{12}$$
$$+ (\alpha a_{21} + \beta b_{21})c_{21} + (\alpha a_{22} + \beta b_{22})c_{22}$$
$$= \alpha(a_{11}c_{11} + a_{12}c_{12} + a_{21}c_{21} + a_{22}c_{22}) +$$
$$\beta(b_{11}c_{11} + b_{12}c_{12} + b_{21}c_{21} + b_{22}c_{22})$$
$$= \alpha \langle A, C \rangle + \beta \langle B, C \rangle.$$

This shows that the above product $\langle , \rangle$ is an inner product on V.

(6) Let $A, B \in M_2(\mathbb{R})$. Define $\langle A, B \rangle = tr(A^t B)$.

   (i) If $A = \begin{pmatrix} a_{11} & a_{12} \\ a_{21} & a_{22} \end{pmatrix}$, then $\langle A, A \rangle = tr(A^t A) = a_{11}^2 + a_{12}^2 + a_{21}^2 + a_{22}^2 \geq 0$,
   and $\langle A, A \rangle = 0 \Leftrightarrow a_{11} = 0, a_{12} = 0, a_{21} = 0, a_{22} = 0$ or $A = 0$.

   (ii) $\langle A, B \rangle = tr(A^t B) = tr(A^t B)^t = tr(B^t A) = \langle B, A \rangle$.

   (iii) For any scalars α, β and $A, B, C \in M_{2\times 2}(\mathbb{R})$

$$\langle \alpha A + \beta B, C \rangle = tr((\alpha A + \beta B)^t C)$$
$$= tr((\alpha A^t + \beta B^t)C)$$
$$= tr(\alpha A^t C) + tr(\beta B^t C)$$
$$= \alpha \, tr(A^t C) + \beta \, tr((B^t C)$$
$$= \alpha \langle A, C \rangle + \beta \langle B, C \rangle.$$

This shows that $\langle , \rangle$ is an inner product on $M_2(\mathbb{R})$. Moreover, if A and B are matrices of order $n \times n$, even then $\langle A, B \rangle = tr(A^t B)$ defines an inner product on $M_n(\mathbb{R})$.

Lemma 5.4 *Let V be an inner product space and let $u, v, w, x \in V$ and $\alpha, \beta, \gamma, \delta \in \mathbb{F}$. Then the following hold:*

   (i) $\langle 0, v \rangle = 0 = \langle v, 0 \rangle$, *for all $v \in V$.*
   (ii) $\langle \alpha u + \beta v, \gamma w + \delta x \rangle = \alpha \overline{\gamma} \langle u, w \rangle + \alpha \overline{\delta} \langle u, x \rangle + \beta \overline{\gamma} \langle v, w \rangle + \beta \overline{\delta} \langle v, x \rangle.$
   (iii) *If for any fixed $u \in V$, $\langle u, v \rangle = 0$ for all $v \in V$, then $u = 0$.*
   (iv) *For any $v_1, v_2 \in V$ if $\langle u, v_1 \rangle = \langle u, v_2 \rangle$ for all $u \in V$, then $v_1 = v_2$.*

Proof (i) Obvious.

   (ii) For any $u, v, w, x \in V$ and $\alpha, \beta, \gamma, \delta \in \mathbb{F}$

$$\langle \alpha u + \beta v, \gamma w + \delta x \rangle = \alpha \langle u, \gamma w + \delta x \rangle + \beta \langle v, \gamma w + \delta x \rangle$$
$$= \alpha \overline{\gamma} \langle u, w \rangle + \alpha \overline{\delta} \langle u, x \rangle + \beta \overline{\gamma} \langle v, w \rangle + \beta \overline{\delta} \langle v, x \rangle.$$

   (iii) Suppose that $\langle u, v \rangle = 0$ for all $v \in V$. Then in particular $\langle u, u \rangle = 0$ and hence $u = 0$.

   (iv) Since $\overline{\langle v_1, u \rangle} = \langle u, v_1 \rangle = \langle u, v_2 \rangle = \overline{\langle v_2, u \rangle}$, we find that $\langle v_1, u \rangle = \langle v_2, u \rangle$ for all $u \in V$. This implies that $\langle v_1 - v_2, u \rangle = \langle v_1, u \rangle - \langle v_2, u \rangle = 0$ for all $u \in V$, and hence by (iii), we get the required result.

Exercises

1. Let $\mathbb{R}^2$ be the vector space over the real field. Find all 4-tuples of real numbers (a, b, c, d) such that for $u = (\alpha_1, \alpha_2)$, $v = (\beta_1, \beta_2) \in \mathbb{R}^2$, $\langle u, v \rangle = a\alpha_1\beta_1 + b\alpha_2\beta_2 + c\alpha_1\beta_2 + d\alpha_2\beta_1$ defines an inner product on $\mathbb{R}^2$.

2. Let V be the vector space of all real functions $y = f(x)$ over the field of reals, satisfying $\frac{d^2y}{dx^2} + 4y = 0$. In V define $\langle u, v \rangle = \int_\pi^0 uv\,dx$. Prove that this defines an inner product on V.

3. Let V be the vector space of all real functions $y = f(x)$ over the field of reals, satisfying $\frac{d^3y}{dx^3} - 12\frac{d^2y}{dx^2} + 44\frac{dy}{dx} - 48y = 0$. In V if $\langle u, v \rangle = \int_{-\infty}^0 uv\,dx$, then prove that this defines an inner product on V.

4. Let $V = \{(a_1, a_2, a_3, \ldots), \ a_i \in \mathbb{R} \mid \sum_{i=1}^{\infty} a_i^2 \text{ is convergent }\}$. Then V is a vector space over $\mathbb{R}$ with addition and scalar multiplication defined component-wise. Prove that the map $\langle, \rangle : V \times V \longrightarrow \mathbb{R}$, given by $\langle (a_1, a_2, \ldots), (b_1, b_2, \ldots) \rangle = a_1b_1 + a_2b_2 + \cdots$ is well defined and also prove that it is an inner product on V.

5. Let V be a finite dimensional vector space over $\mathbb{F}$, and $\{e_1, e_2, \ldots, e_n\}$ be a basis of V. If $u, v \in V$, then $u = a_1e_1 + a_2e_2 + \cdots + a_ne_n$, $v = b_1e_1 + b_2e_2 + \cdots + b_ne_n$, where a_i, b_i are uniquely determined scalars. Define $\langle u, v \rangle = a_1\overline{b_1} + a_2\overline{b_2} + \cdots + a_n\overline{b_n}$. Prove that this map is an inner product on V.

6. Let $\mathbb{R}^2$ be the vector space over the real field. For $u = (\alpha_1, \alpha_2)$, $v = (\beta_1, \beta_2) \in \mathbb{R}^2$, prove that $\langle u, v \rangle = \frac{(\alpha_1 - \alpha_2)(\beta_1 - \beta_2)}{4} + \frac{(\alpha_1 + \alpha_2)(\beta_1 + \beta_2)}{4}$ defines an inner product on $\mathbb{R}^2$.

7. Let V be a n-dimensional vector space over the field of complex numbers $\mathbb{C}$. Let B be a basis of V. Define, for arbitrary $u, v \in V$, $\langle u, v \rangle = [u]_B[v]_B$, where the inner product of the coordinate vectors on the right hand side is the natural inner product of the vector space $\mathbb{C}^n$ over $\mathbb{C}$. Prove that $\langle, \rangle$ is an inner product on V.

8. Let V be the vector space of $m \times n$ matrices over $\mathbb{R}$. Prove that $\langle A, B \rangle = tr(B^t A)$ defines an inner product in V.

9. Suppose $f(u, v)$ and $g(u, v)$ are inner products on a vector space V over $\mathbb{R}$. Prove that

   (a) The sum $f + g$ is an inner product on V, where $(f + g)(u, v) = f(u, v) + g(u, v)$.

   (b) The scalar product kf, for $k > 0$, is an inner product on V, where $(kf)(u, v) = kf(u, v)$.

10. Find the values of k so that the following is an inner product on $\mathbb{R}^2$, where $u = (x_1, x_2)$ and $v = (y_1, y_2)$; $\langle u, v \rangle = x_1y_1 - 3x_1y_2 - 3x_2y_1 + kx_2y_2$.

11. Show that the formula $\langle \sum_j a_j x^j, \sum_k b_k x^k \rangle = \sum_{j,k} \frac{a_j b_k}{j+k+1}$ defines an inner product on the space $\mathbb{R}[x]$ of polynomials over the real field $\mathbb{R}$.

12. Let $\langle, \rangle$ be the standard inner product on $\mathbb{R}^2$ and let T be the linear operator $T(x_1, x_2) = (-x_2, x_1)$. Now T is "rotation through 90^0" anti-clockwise and has

the property that $\langle u, T(u) \rangle = 0$ for all $u \in \mathbb{R}^2$. Find all the inner products $\langle, \rangle$ on $\mathbb{R}^2$ such that $\langle u, T(u) \rangle = 0$ for all u.

13. Consider any $u, v \in \mathbb{R}^2$. Prove that

$$\begin{vmatrix} \langle u, u \rangle & \langle u, v \rangle \\ \langle u, v \rangle & \langle v, v \rangle \end{vmatrix} = 0$$

if and only if u and v are linearly independent.

5.2   The Length of a Vector

Definition 5.5 Let V be an inner product space. If $v \in V$, then the length of v (or norm of v) denoted as $\|v\|$ and is defined as $\|v\| = \sqrt{\langle v, v \rangle}$.

Lemma 5.6 *If V is an inner product space, then for any $u, v \in V, \alpha \in \mathbb{F}$*

(i) $\|u\| \geq 0, \|u\| = 0 \Leftrightarrow u = 0$,
(ii) $\|\alpha u\| = |\alpha| \|u\|$,
(iii) $\|u + v\|^2 = \|u\|^2 + \|v\|^2 + 2Re\langle u, v \rangle$,
(iv) $\|u + v\|^2 + \|u - v\|^2 = 2(\|u\|^2 + \|v\|^2)$,
(v) *for any $u, v \in V$*

$$\langle u, v \rangle = \begin{cases} \frac{1}{4}\|u + v\|^2 - \frac{1}{4}\|u - v\|^2, & \text{if } \mathbb{F} = \mathbb{R} \\ \frac{1}{4}\|u + v\|^2 - \frac{1}{4}\|u - v\|^2 + \frac{i}{4}\|u + iv\|^2 - \frac{i}{4}\|u - iv\|^2, & \text{if } \mathbb{F} = \mathbb{C}. \end{cases}$$

Proof (i) Clear from the definition of the inner product space.

(ii) For any $\alpha \in \mathbb{F}, u \in V$

$$\|\alpha u\|^2 = \langle \alpha u, \alpha u \rangle = \alpha \overline{\alpha} \langle u, u \rangle = \alpha \overline{\alpha} \|u\|^2 = |\alpha|^2 \|u\|^2 = (|\alpha| \|u\|)^2.$$

This implies that $\|\alpha u\| = |\alpha| \|u\|$.

(iii) For any $u, v \in V$

$$\begin{aligned} \|u + v\|^2 &= \langle u + v, u + v \rangle \\ &= \langle u, u \rangle + \langle u, v \rangle + \langle v, u \rangle + \langle v, v \rangle \\ &= \|u\|^2 + \|v\|^2 + \langle u, v \rangle + \overline{\langle u, v \rangle} \\ &= \|u\|^2 + \|v\|^2 + 2Re\langle u, v \rangle. \end{aligned}$$

(iv) In view of (iii), we find that $\|u - v\|^2 = \|u\|^2 + \|v\|^2 - 2Re\langle u, v \rangle$. This yields that $\|u + v\|^2 + \|u - v\|^2 = 2(\|u\|^2 + \|v\|^2)$. This equality is also known as parallelogram equality.

(v) In view of (iii) we have $\|u + v\|^2 - \|u - v\|^2 = 4Re\langle u, v \rangle$. If $\mathbb{F} = \mathbb{R}$, then $Re\langle u, v \rangle = \langle u, v \rangle$ and hence $\langle u, v \rangle = \frac{1}{4}\|u + v\|^2 - \frac{1}{4}\|u - v\|^2$. On the other hand if $\mathbb{F} = \mathbb{C}$, then replacing v by iv we arrive at $4Re\langle u, iv \rangle = \|u + iv\|^2 - \|u - iv\|^2$. But since $\langle u, iv \rangle = -i\langle u, v \rangle$ and $Re\langle u, iv \rangle = Im\langle u, v \rangle$, the above relation yields that $4Im\langle u, v \rangle = \|u + iv\|^2 - \|u - iv\|^2$. But

$$
\begin{aligned}
\langle u, v \rangle &= Re\langle u, v \rangle + iIm\langle u, v \rangle \\
&= \tfrac{1}{4}\|u + v\|^2 - \tfrac{1}{4}\|u - v\|^2 + \tfrac{i}{4}\|u + iv\|^2 - \tfrac{i}{4}\|u - iv\|^2.
\end{aligned}
$$

The above identities (iii), (iv) and (v) are known as Polarization identities.

Definition 5.7 Let V be a vector space over $\mathbb{F}$. A function $\|.\| : V \to \mathbb{F}$ is said to be a norm on V if it satisfies the following axioms:

(1) $\|v\| \geq 0$ and $\|v\| = 0 \Leftrightarrow v = 0$ for all $v \in V$,
(2) $\|\alpha v\| = |\alpha|\|v\|$, for any $\alpha \in \mathbb{F}$ and $v \in V$,
(3) $\|u + v\| \leq \|u\| + \|v\|$ for all $u, v \in V$.

A vector space equipped with a norm is said to be a normed space.

Theorem 5.8 (Cauchy-Schwartz inequality) *If V is an inner product space and $u, v \in V$, then $|\langle u, v \rangle| \leq \|u\|\|v\|$.*

Proof If $u = 0$ or $v = 0$, in both the cases we find that $|\langle u, v \rangle| \leq \|u\|\|v\|$. Therefore, we assume that neither $u = 0$ nor $v = 0$. Then in this case $\|u\| \neq 0$. If $\|u\| = 0$, then $\langle u, u \rangle = 0$ and hence $u = 0$, a contradiction. Now for any scalar λ

$$
\langle u - \lambda v, u - \lambda v \rangle = \langle u, u \rangle - \lambda\langle v, u \rangle - \bar{\lambda}\{\langle u, v \rangle - \lambda\langle u, v \rangle\langle v, v \rangle\} \geq 0.
$$

Now if we choose $\lambda = \frac{\langle u, v \rangle}{\langle v, v \rangle}$, then we find that $\|u\|^2 - \frac{\langle u, v \rangle}{\langle v, v \rangle}\langle v, u \rangle \geq 0$. This implies that $\|u\|^2 - \frac{\langle u, v \rangle\langle u, v \rangle}{\|v\|^2} \geq 0$, i.e., $|\langle u, v \rangle|^2 \leq \|u\|^2\|v\|^2$. This yields that $|\langle u, v \rangle| \leq \|u\|\|v\|$.

Remark 5.9 (i) If we consider the Example 5.3(2), the above theorem gives that for any $u = (\alpha_1, \alpha_2, \ldots, \alpha_n), v = (\beta_1, \beta_2, \ldots, \beta_n) \in V$

$$
\left| \sum_{i=1}^{n} \alpha_i \overline{\beta_i} \right| \leq \sqrt{\sum_{i=1}^{n} |\alpha_i|^2} \sqrt{\sum_{i=1}^{n} |\beta_i|^2}.
$$

(ii) In case of the inner product given in Example 5.3(3) we find that for any $f(t), g(t) \in \mathscr{C}[a, b]$

$$\left| \int_a^b f(t)\overline{g(t)}dt \right| \leq \sqrt{\int_a^b |f(t)|^2 dt} \sqrt{\int_a^b |g(t)|^2 dt}.$$

(iii) From Cauchy-Schwartz inequality, we have $-1 \leq \frac{\langle u,v \rangle}{\|u\|\|v\|} \leq 1$ for any two nonzero vector u and v. This ensures that there exists a unique $\theta \in [0, \pi]$ such that $cos\theta = \frac{\langle u,v \rangle}{\|u\|\|v\|}$ or $\langle u, v \rangle = \|u\|\|v\|cos\theta$. This angle θ is called the angle between u and v.

Lemma 5.10 *Let V be an inner product space. Then for any $u, v \in V$, $|\langle u, v \rangle| = \|u\|\|v\|$ if and only if u, v are linearly dependent.*

Proof If any one of u, v is zero, then the result follows. Hence, we assume that neither u nor v is zero. If $|\langle u, v \rangle| = \|u\|\|v\|$, then for scalar $\lambda = \frac{\langle u,v \rangle}{\langle v,v \rangle}$

$$\begin{aligned}
\langle u - \lambda v, u - \lambda v \rangle &= \langle u, u \rangle - \overline{\lambda}\langle u, v \rangle - \lambda\langle v, u \rangle + \lambda\overline{\lambda}\langle v, v \rangle \\
&= \|u\|^2 - \frac{\langle u,v \rangle \langle v,u \rangle}{\langle v,v \rangle} - \overline{\lambda}\{\langle u, v \rangle - \frac{\langle u,v \rangle}{\langle v,v \rangle}\langle v, v \rangle\} \\
&= \|u\|^2 - \frac{|\langle u,v \rangle|^2}{\|v\|^2} \\
&= 0.
\end{aligned}$$

This yields that $u - \lambda v = 0$, i.e., u, v are linearly dependent.
Conversely, assume that $\{u, v\}$ is linearly dependent. Then $u = \alpha v$ for some scalar α, and hence

$$\|u\|\|v\| = |\alpha|\|v\|^2 = |\alpha|\langle v, v \rangle = |\langle \alpha v, v \rangle| = |\langle u, v \rangle|.$$

Theorem 5.11 *Every inner product space is normed space together with the norm $\|u\| = \sqrt{\langle u, u \rangle}$.*

Proof Suppose that V is an inner product space.
(i) Since $\|u\| = \sqrt{\langle u, u \rangle}$, we find that $\|u\|^2 = \langle u, u \rangle \geq 0$. This implies that $\|u\| \geq 0$. Furthermore,

$$\begin{aligned}
\|u\| = 0 &\Leftrightarrow \sqrt{\langle u, u \rangle} = 0 \\
&\Leftrightarrow u = 0.
\end{aligned}$$

Hence $\|u\| = 0 \Leftrightarrow u = 0$.

(ii) For any $\alpha \in \mathbb{F}$ and $u \in V$

$$\|\alpha u\|^2 = \langle \alpha u, \alpha u \rangle = \alpha\overline{\alpha}\langle u, u \rangle = (|\alpha|\|u\|)^2.$$

This implies that $\|\alpha u\| = |\alpha|\|u\|$.
(iii) For any $u, v \in V$

$$\begin{aligned}
\|u + v\|^2 &= \langle u + v, u + v \rangle \\
&= \langle u, u \rangle + \langle u, v \rangle + \langle v, u \rangle + \langle v, v \rangle \\
&= \langle u, u \rangle + \langle u, v \rangle + \overline{\langle u, v \rangle} + \langle v, v \rangle \\
&= \|u\|^2 + 2Re(\langle u, v \rangle) + \|v\|^2 \\
&\leq \|u\|^2 + 2|\langle u, v \rangle| + \|v\|^2 \quad (\text{since } \alpha \in \mathbb{F} \text{ and } Rel\,\alpha \leq |\alpha|) \\
&\leq \|u\|^2 + 2\|u\|\|v\| + \|v\|^2 \quad (\text{by Cauchy-Schwarz inequality}) \\
&\leq (\|u\| + \|v\|)^2.
\end{aligned}$$

This shows that $\|u + v\| \leq \|u\| + \|v\|$ for all $u, v \in V$ and hence V is a normed space.

Definition 5.12 If S is any set, a function $d : S \times S \to \mathbb{R}$ is said to be a metric on S, if for any $a, b, c \in S$ it satisfies the following:

(1) $d(a, b) \geq 0, d(a, b) = 0 \Leftrightarrow a = b,$
(2) $d(a, b) = d(b, a),$
(3) $d(a, c) \leq d(a, b) + d(b, c).$

The set S equipped with a metric d is said to be a metric space, generally denoted as (S, d).

Theorem 5.13 *Let V be an inner product space over $\mathbb{R}$. Then the function $d : V \times V \to \mathbb{R}$ such that $d(u, v) = \|u - v\|$, where $u, v \in V$ is a metric on V.*

Proof Obviously, $d(u, v) = \|u - v\| \geq 0$, and $d(u, v) = 0$ if and only if $u = v$. It can also be seen that $d(u, v) = \|u - v\| = \| - (v - u)\| = \|v - u\| = d(v, u)$ for all $u, v \in V$. For any $u, v, w \in V$

$$\begin{aligned}
d(u, v) &= \|u - v\| \\
&= \|(u - w) + (w - v)\| \\
&\leq \|u - w\| + \|w - v\|.
\end{aligned}$$

This shows that $d(u, v) \leq d(u, w) + d(w, v)$ for all $u, v, w \in V$ and hence d is a metric on V.

Exercises

1. Consider $f(x) = 4x^3 - 6x + 5$ and $g(x) = -2x^2 + 7$ in the polynomial space $P(x)$ with inner product $\langle f, g \rangle = \int_0^1 f(x)g(x)dx$. Find $\|f\|$ and $\|g\|$.
2. Let V be a real inner product space. Show that

    (a) $\|u\| = \|v\|$ if and only if $\langle u + v, u - v \rangle = 0,$
    (b) $\|u + v\|^2 = \|u\|^2 + \|v\|^2$ if and only if $\langle u, v \rangle = 0$. Show by counter examples that the above statements are not true for $\mathbb{C}^2$.

3. Let $\mathbb{R}^n$ and $\mathbb{C}^n$ be the vector spaces over $\mathbb{R}$ and $\mathbb{C}$, respectively. Prove that

    (a) $\|(a_1, a_2, \ldots, a_n)\|_\infty = \max(|a_i|),$
    (b) $\|(a_1, a_2, \ldots, a_n)\|_1 = |a_1| + |a_2| + \cdots + |a_n|,$

(c) $\|(a_1, a_2, \ldots, a_n)\|_2 = \sqrt{|a_1|^2 + |a_2|^2 + \cdots + |a_n|^2}$

are norms on $\mathbb{R}^n$ and $\mathbb{C}^n$. These are known as infinity-norm, one-norm and two-norm, respectively.

4. Solve the above problem 3., for $u = (1 + i, -2i, 1 - 6i)$ and $v = (1 - i, 2 + 3i, -3i)$ in $\mathbb{C}^3$.

5. Consider vectors $u = (1, -2, -4, 3, -6)$ and $v = (3, -2, 1, -4, -1)$ in $\mathbb{R}^5$. Find

   (a) $\|u\|_\infty$ and $\|v\|_\infty$,
   (b) $\|u\|_1$ and $\|v\|_1$,
   (c) $\|u\|_2$ and $\|v\|_2$,
   (d) $d_\infty(u, v), d_1(u, v), d_2(u, v)$,

   where the norms $\|.\|_\infty$, $\|.\|_1$ and $\|.\|_2$ are the infinity-norm, one-norm and two-norm, respectively, on $\mathbb{R}^5$ and d_∞, d_1, d_2 are metric functions induced by these norms, respectively.

6. Let $\mathscr{C}[a, b]$ be the vector space of real continuous functions on $[a, b]$ over $\mathbb{R}$. Prove that (i) $\|f\|_1 = \int_a^b |f(t)| dt$, (ii) $\|f\|_2 = \int_a^b [f(t)]^2 dt$, (iii) $\|f\|_\infty = max(|f(t)|)$ are norms on $\mathscr{C}[a, b]$. Further consider the functions $f(t) = 2t^2 - 6t$ and $g(t) = t^3 + 6t^2$ in $\mathscr{C}[1, 3]$ and hence find

   (a) $d_1(f, g)$,
   (b) $d_2(f, g)$,
   (c) $d_\infty(f, g)$,

   where d_1, d_2, d_∞ are metric functions induced by the above norms, respectively.

7. Find out norms and metrics induced by the inner products defined in the problems $1 - 8$ and 10 of the preceding section.

8. Prove the Apollonius identity

$$\|w - u\|^2 + \|w - v\|^2 = \frac{1}{2}\|u - v\|^2 + 2\|w - \frac{1}{2}(u + v)\|^2.$$

9. Let $u = (r_1, r_2, \ldots, r_n)$ and $v = (s_1, s_2, \ldots, s_n)$ be in $\mathbb{R}^n$. The Cauchy-Schwartz inequality states that
   $|r_1 s_1 + r_2 s_2 + \cdots + r_n s_n|^2 \le (r_1^2 + r_2^2 + \cdots + r_n^2)(s_1^2 + s_2^2 + \cdots +^2 s_n^2)$.
   Prove that

$$(|r_1 s_1| + |r_2 s_2| + \cdots + |r_n s_n|)^2 \le (r_1^2 + r_2^2 + \cdots + r_n^2)(s_1^2 + s_2^2 + \cdots + s_n^2).$$

5.3 Orthogonality and Orthonormality

Definition 5.14 Let V be an inner product space over $\mathbb{F}$. Let $u, v \in V$. Then u is said to be orthogonal to v if $\langle u, v \rangle = 0$, whenever $u \ne v$.

Remark 5.15 (i) A vector v is said to be orthogonal to a subset S of an inner product space V if v is orthogonal to each vector in S. Also any two subspaces are called orthogonal if every vector in one subspace is orthogonal to each vector in other.

(ii) A subset S of an inner product space V is said to be an orthogonal set if any two distinct vectors in S are orthogonal.

(iii) The zero vector is the only vector which is orthogonal to itself. Moreover, since for any $v \in V$, $\langle 0, v \rangle = \langle 0v, v \rangle = 0\langle v, v \rangle = 0$, we find that $0 \in V$ is orthogonal to every $v \in V$.

(iv) The relation of orthogonality is in fact symmetric, i.e., if u is orthogonal to v, then $\langle u, v \rangle = 0$. This implies that $\overline{\langle v, u \rangle} = \overline{0}$, which shows that $\langle v, u \rangle = 0$, and hence v is orthogonal to u.

(v) If u is orthogonal to v, then it can be easily seen that every scalar multiple of u is orthogonal to v. In fact, for any scalar α, $\langle \alpha u, v \rangle = \alpha \langle u, v \rangle = \alpha 0 = 0$.

Theorem 5.16 *Let $S = \{u_1, u_2, \ldots, u_n\}$ be a pairwise orthogonal set of nonzero vectors in an inner product space V. Then the following hold:*

(i) S is linearly independent.

(ii) If $v \in V$ is in the linear span of S, then $v = \sum_{k=1}^{n} \frac{\langle v, u_k \rangle}{\|u_k\|^2} u_k$.

(iii) $\|\sum_{i=1}^{n} u_i\|^2 = \sum_{i=1}^{n} \|u_i\|^2$ (Pythagoras theorem).

Proof (i) Suppose there exist scalars $\alpha_1, \alpha_2, \ldots, \alpha_n$ such that $\sum_{i=1}^{n} \alpha_i u_i = 0$. Thus for each $1 \le k \le n$

$$0 = \langle 0, u_k \rangle$$
$$= \langle \sum_{i=1}^{n} \alpha_i u_i, u_k \rangle$$
$$= \sum_{i=1}^{n} \alpha_i \langle u_i, u_k \rangle$$
$$= \alpha_k \langle u_k, u_k \rangle$$
$$= \alpha_k \|u_k\|^2.$$

But since $\|u_k\|^2 \ne 0$ we find that $\alpha_k = 0$ for each $1 \le k \le n$, and hence S is linearly independent.

(ii) Since $v \in L(S)$, there exist scalars $\alpha_1, \alpha_2, \ldots, \alpha_n$ such that $v = \sum_{i=1}^{n} \alpha_i u_i$. Thus for each $1 \le k \le n$, $\langle v, u_k \rangle = \langle \sum_{i=1}^{n} \alpha_i u_i, u_k \rangle = \sum_{i=1}^{n} \alpha_i \langle u_i, u_k \rangle$. But since S is orthogonal, we find that $\langle v, u_k \rangle = \alpha_k \langle u_k, u_k \rangle$. Now since $u_k \ne 0$, $\|u_k\|^2 \ne 0$, the latter relation yields that $\alpha_k = \frac{\langle v, u_k \rangle}{\|u_k\|^2}$, $1 \le k \le n$, and hence $v = \sum_{k=1}^{n} \frac{\langle v, u_k \rangle}{\|u_k\|^2} u_k$.

(iii) For any $u_i \in S$, $1 \le i \le n$

$$\| \sum_{i=1}^{n} u_i \|^2 = \left\langle \sum_{i=1}^{n} u_i, \sum_{i=1}^{n} u_i \right\rangle$$
$$= \sum_{1 \le i,j \le n} \langle u_i, u_j \rangle$$
$$= \sum_{i=1}^{n} \langle u_i, u_i \rangle$$
$$= \sum_{i=1}^{n} \| u_i \|^2.$$

Remark 5.17 Any two vectors u, v in an Euclidean space are orthogonal if and only if $\| u + v \|^2 = \| u \|^2 + \| v \|^2$. This result is not true in unitary space. In fact, for any $u = (u_1, u_2), v = (v_1, v_2) \in \mathbb{C}^2$, $\langle u, v \rangle = u_1 \overline{v_1} + u_2 \overline{v_2}$ defines an inner product on $\mathbb{C}^2$. If we consider $u = (0, i), v = (0, 1) \in \mathbb{C}^2$, then $\langle u, v \rangle = 0 \overline{0} + i \overline{1} = i \neq 0$, and hence u and v are not orthogonal. But

$$\begin{aligned} \| u + v \|^2 &= \| (0, i) + (0, 1) \|^2 \\ &= \| (0, 1 + i) \|^2 \\ &= \langle (0, 1 + i), (0, 1 + i) \rangle \\ &= 0 \overline{0} + (1 + i) \overline{(1 + i)} \\ &= (1 + i)(1 - i) \\ &= 2, \end{aligned}$$

$\| u \|^2 = \langle (0, i), (0, i) \rangle = 1$ and $\| v \|^2 = \langle (0, 1), (0, 1) \rangle = 1$ yield that $\| u + v \|^2 = \| u \|^2 + \| v \|^2$.

Theorem 5.18 *Let $\{ v_1, v_2, \ldots, v_n \}$ be any subset of linearly independent vectors in an inner product V. Define $u_1, u_2, \ldots, u_n \in V$ inductively as $u_1 = v_1$ and*

$$u_k = v_k - \sum_{j=1}^{k-1} \frac{\langle v_k, u_j \rangle}{\| u_j \|^2} u_j; \ for \ 2 \le k \le n.$$

Then $\{ u_1, u_2, \ldots, u_n \}$ is pairwise orthogonal subset in V, and $L(\{ v_1, v_2, \ldots, v_n \}) = L(\{ u_1, u_2, \ldots, u_n \})$.

Proof We apply induction on n. For $n = 1$, $u_1 = v_1$ and the result holds trivially. Now assume that $n > 1$ and the result holds for all $m < n$ linearly independent vectors in V. By applying the induction on the sets $\{ v_1, v_2, \ldots, v_{n-1} \}$, we find that $\{ u_1, u_2, \ldots, u_{n-1} \}$ is linearly independent and $L(\{ v_1, v_2, \ldots, v_{n-1} \}) = L(\{ u_1, u_2, \ldots, u_{n-1} \})$. Since

$$u_n = v_n - \sum_{j=1}^{n-1} \frac{\langle v_n, u_j \rangle}{\| u_j \|^2} u_j,$$

if $u_n = 0$, we find that $v_n = \sum_{j=1}^{n-1} \frac{\langle v_n, u_j \rangle}{\|u_j\|^2} u_j$ which is spanned by $\{v_1, v_2, \ldots, v_{n-1}\}$. This shows that $\{v_1, v_2, \ldots, v_n\}$ is linearly dependent, a contradiction to our assumption. Hence $u_n \neq 0$. Further, for $1 \leq i \leq n-1$

$$\langle u_n, u_i \rangle = \langle v_n, u_i \rangle - \sum_{j=1}^{n-1} \frac{\langle v_n, u_j \rangle}{\|u_j\|^2} \langle u_j, u_i \rangle.$$

But by the induction hypothesis $\langle u_j, u_i \rangle = 0$, for $i \neq j; 1 \leq j \leq n-1$. Hence the above yields that $\langle u_n, u_i \rangle = \langle v_n, u_i \rangle - \frac{\langle v_n, u_i \rangle}{\|u_i\|^2} \langle u_i, u_i \rangle = 0$. This shows that the set $\{u_1, u_2, \ldots, u_n\}$ is pairwise orthogonal and by Theorem 5.16, it is linearly independent. Hence the subspace W spanned by $\{u_1, u_2, \ldots, u_n\}$ has dimension n, and by the above relation for u_n ensures that $v_n \in W$. Consequently, n linearly independent vectors $v_1, v_2, \ldots, v_n \in W$ and hence $W = L(\{v_1, v_2, \ldots, v_n\})$. This completes the proof of our theorem.

Lemma 5.19 *If u and v are any two vectors in a real inner product space, then $u + v$ and $u - v$ are orthogonal if and only if $\|u\| = \|v\|$.*

Proof Suppose that $u + v$ and $u - v$ are orthogonal. Then we find that $\langle u + v, u - v \rangle = 0$. This implies that $\|u\|^2 = \|v\|^2$, and hence $\|u\| = \|v\|$. Conversely, assume that $\|u\| = \|v\|$. Then

$$\begin{aligned}
\langle u + v, u - v \rangle &= \langle u, u \rangle - \langle u, v \rangle + \langle v, u \rangle - \langle v, v \rangle \\
&= \langle u, u \rangle - \langle u, v \rangle + \overline{\langle u, v \rangle} - \langle v, v \rangle \\
&= \langle u, u \rangle - \langle u, v \rangle + \langle u, v \rangle - \langle v, v \rangle \\
&= \|u\|^2 - \|v\|^2 \\
&= 0,
\end{aligned}$$

and hence $u + v$ and $u - v$ are orthogonal.

Remark 5.20 If vectors u and v represent the adjacent sides of a parallelogram, then $u + v$ and $u - v$ represent the diagonals of the parallelogram. The above lemma shows that the diagonals are perpendicular if and only if the lengths of sides are same. In other words, the above lemma ensures that a parallelogram is a rhombus if and only if the diagonals are perpendicular.

Consider $\mathbb{R}^2$, geometrically as the cartesian plane and algebraically as a vector space over $\mathbb{R}$. Let $u, v \in \mathbb{R}^2$ such that $\|u = 1$. Next suppose that the vectors v and u represent the position vectors of the points P and Q, respectively, in the plane. Thus $\overrightarrow{OP} = v$, $\overrightarrow{OQ} = u$, where O is the origin. If θ is the angle between the vector $\overrightarrow{OP}$, and $\overrightarrow{OQ}$, then it is obvious that the orthogonal projection of vector $\overrightarrow{OP}$ along the vector $\overrightarrow{OQ}$ will be given by the vector $\overrightarrow{OR} = (|\overrightarrow{OP}| cos\theta)u$, where R is the foot of the perpendicular drawn from P on the vector OQ. Thus $\overrightarrow{OR} = (v.u)u$. Since the scalar product $v.u$ in $\mathbb{R}^2$ coincides with an inner product in $\mathbb{R}^2$, $\overrightarrow{OR} = \langle u, v \rangle u$.

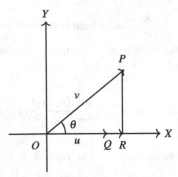

The above observation suggests the following:

Definition 5.21 Let V be an inner product space and let $v \in V$, and $u \in V$ such that $\|u\| = 1$. Then the projection $P_u(v)$ of v along u is defined as $P_u(v) = \langle u, v \rangle u$. If $u \in V$ is any nonzero vector, then

$$P_u(v) = \left\langle \frac{u}{\|u\|}, v \right\rangle \frac{u}{\|u\|} = \frac{\langle u, v \rangle}{\langle u, u \rangle} u.$$

Remark 5.22 We observe that for any u, v in a real inner product space

$$
\begin{aligned}
\langle v - P_u(v), u \rangle &= \langle v, u \rangle - \langle \tfrac{\langle u, v \rangle}{\langle u, u \rangle} u, u \rangle \\
&= \langle v, u \rangle - \langle u, v \rangle \\
&= \langle v, u \rangle - \overline{\langle v, u \rangle} \\
&= \langle v, u \rangle - \langle v, u \rangle \ \text{since} \langle v, u \rangle \in \mathbb{R} \\
&= 0.
\end{aligned}
$$

This shows that $v - P_u(v)$ is orthogonal to u.

Lemma 5.23 *Let V be an inner product space defined over $\mathbb{R}$. For a unit vector u and any vector $v \in V$, let $P_u(v) = \langle u, v \rangle u$. Then $d(P_u(v), v) \le d(\alpha u, v)$ for any $\alpha \in \mathbb{R}$.*

Proof By the above remark, it is clear that $v - P_u(v)$ is orthogonal to u. Therefore, for any $\alpha \in \mathbb{R}$, $v - P_u(v)$ is orthogonal to αu or $P_u(v)$, and hence $v - P_u(v)$ is orthogonal to $P_u(v) - \alpha u$. Thus by Theorem 5.16(iii)

$$
\begin{aligned}
\|v - \alpha u\|^2 &= \|v - P_u(v) + P_u(v) - \alpha u\|^2 \\
&= \|v - P_u(v)\|^2 + \|P_u(v) - \alpha u\|^2.
\end{aligned}
$$

This yields that $d(v, \alpha u)^2 \ge d(v, P_u(v))^2$, and hence we find that $d(P_u(v), v) \le d(\alpha u, v)$ for any $\alpha \in \mathbb{R}$.

Definition 5.24 Let S be a set of vectors in an inner product space V. Then S is said to be orthonormal set if

(1) each vector in S has unit norm, i.e., $\|u\| = 1$,
(2) for any $u, v \in S$, $u \neq v$, $\langle u, v \rangle = 0$.

In addition, if S is a basis of V, then S is called an *orthonormal basis* of V.

Remark 5.25 (i) If S is a finite orthonormal subset of an inner product space V, then it is clear that S is linearly independent.

(ii) A set consisting of mutually orthogonal unit vectors is an orthonormal set.

Theorem 5.26 *Let $\{v_1, v_2, \ldots, v_n\}$ be an orthonormal subset of an inner product space V. If $v = \sum_{i=1}^{n} \alpha_i v_i$, then $\alpha_i = \langle v, v_i \rangle$ and $\|v\|^2 = \sum_{i=1}^{n} |v_i|^2 = \sum_{i=1}^{n} |\langle v, v_i \rangle|^2$.*

Proof Since the set $\{v_1, v_2, \ldots, v_n\}$ is orthonormal,

$$\langle v, v_i \rangle = \left\langle \sum_{j=1}^{n} \alpha_j v_j, v_i \right\rangle = \sum_{j=1}^{n} \alpha_j \langle v_j, v_i \rangle = \alpha_i.$$

Hence $\alpha_i \bar{\alpha}_i = |\alpha_i|^2 = |\langle v, v_i \rangle|^2$. Therefore

$$\begin{aligned}
\|v\|^2 &= \langle v, v \rangle \\
&= \left\langle \sum_{i=1}^{n} \alpha_i v_i, \sum_{j=1}^{n} \alpha_j v_j \right\rangle \\
&= \sum_{j=1}^{n} \sum_{i=1}^{n} \alpha_i \bar{\alpha}_j \langle v_i, v_j \rangle \\
&= \sum_{i=1}^{n} \alpha_i \bar{\alpha}_i \\
&= \sum_{i=1}^{n} |\langle v, v_i \rangle|^2.
\end{aligned}$$

Definition 5.27 Let S be any subset of an inner product space V. The orthogonal complement of S denoted as $S^{\perp}$ consists of those vectors in V which are orthogonal to every vector of S, i.e.,

$$S^{\perp} = \{v \in V \mid \langle v, w \rangle = 0 \text{ for all } w \in S\}.$$

Remark 5.28 (i) For any given vector $u \in V$, $u^{\perp}$ consists of all vectors in V which are orthogonal to u, i.e., $u^{\perp} = \{v \in V \mid \langle v, u \rangle = 0\}$. Also for any subset S of V, $(S^{\perp})^{\perp} = \{v \in V \mid \langle v, w \rangle = 0, \text{ for all } w \in S^{\perp}\}$ will be denoted by $S^{\perp\perp}$.

(ii) Since for any $0 \neq v \in V$, $\langle v, v \rangle \neq 0$, we find that $v \notin V^{\perp}$, and hence $V^{\perp} = \{0\}$. Obviously $\{0\}^{\perp} = V$.

(iii) Since zero vector in V is orthogonal to every vector in V, clearly $0 \in S^{\perp}$ and hence $S^{\perp} \neq \emptyset$. For any $u, v \in S^{\perp}$, $\alpha, \beta \in \mathbb{F}$ and any vector $w \in S$

$$\langle \alpha u + \beta v, w \rangle = \alpha \langle u, w \rangle + \beta \langle v, w \rangle = \alpha 0 + \beta 0 = 0$$

and hence $S^\perp$ is a subspace of V. This shows that if S is a nonempty subset of an inner product space V, then $S^\perp$ is a subspace of V. Moreover, if $v \in S \cap S^\perp$, then $\langle v, v \rangle = 0$ yields that $v = 0$. Hence $S \cap S^\perp = \{0\}$.

(iv) Suppose that S_1, S_2 are any two subsets of V such that $S_1 \subseteq S_2$. Then it can be easily seen that $S_2^\perp \subseteq S_1^\perp$. In fact, if $v \in S_2^\perp$, then we find that $\langle v, w \rangle = 0$ for all $w \in S_2$, which yields that $\langle v, w \rangle = 0$ for all $w \in S_1$, i.e., $v \in S_1^\perp$.

Theorem 5.29 *Let V be a finite dimensional inner product space and W a subspace of V. Then $V = W \oplus W^\perp$, and $W = W^{\perp\perp}$.*

Proof If $W = \{0\}$, then obviously $W^\perp = V$. Hence $V = \{0\} \oplus V = W \oplus W^\perp$. Henceforth assume that $W \neq \{0\}$. Since W is a subspace of an inner product space V, W is also an inner product space which has an orthonormal basis say $B = \{w_1, w_2, \ldots, w_n\}$. Let $v \in V$, and $w = \sum_{i=1}^{n} \langle v, w_i \rangle w_i \in W$. Then it can be seen that $\langle v - w, w_j \rangle = 0$ for each $1 \leq j \leq n$. But since B spans W, we find that $v - w \in W^\perp$. Now $v = w + (v - w) \in W + W^\perp$, and hence $V = W + W^\perp$. By the above Remark 5.28(iii) $W \cap W^\perp = \{0\}$ and consequently $V = W \oplus W^\perp$.

Now if $w \in W$, then for any $v \in W^\perp$, $\langle w, v \rangle = 0$ implies that $\overline{\langle w, v \rangle} = \overline{0} = 0$ i.e., $\langle v, w \rangle = 0$. This implies that $w \in W^{\perp\perp}$ and hence $W \subseteq W^{\perp\perp}$. Since $V = W^\perp \oplus W^{\perp\perp}$ and if $dim V = n$, $dim W = m$, we find that $dim W^\perp = dim V - dim W = n - m$. This yields that $dim W^{\perp\perp} = dim V - dim W^\perp = n - (n - m) = m$, and hence $dim W = dim W^{\perp\perp}$ and therefore $W = W^{\perp\perp}$. ∎

Corollary 5.30 *If V is a finite dimensional inner product space and W_1, W_2 are subspaces of V, then*

  (i) *if $W_1 \subseteq W_2$, then $W_2^\perp \subseteq W_1^\perp$,*
  (ii) *$(W_1 + W_2)^\perp = W_1^\perp \cap W_2^\perp$,*
  (iii) *$(W_1 \cap W_2)^\perp = W_1^\perp + W_2^\perp$,*
  (iv) *if $V = W_1 \oplus W_2$, then $V = W_1^\perp \oplus W_2^\perp$.*

Proof *((i))* Obvious in view of the above Remark 5.28(iv).

(ii) Since W_1 and W_2 both are subsets of the subspace $W_1 + W_2$, we find that $(W_1 + W_2)^\perp \subseteq W_1^\perp$ and $(W_1 + W_2)^\perp \subseteq W_2^\perp$. This yields that $(W_1 + W_2)^\perp \subseteq W_1^\perp \cap W_2^\perp$. Now conversely, assume that $u \in W_1^\perp \cap W_2^\perp$. Then $u \in W_1^\perp$ and $u \in W_2^\perp$. Therefore u is orthogonal to every vector in W_1 and also to every vector in W_2. Let w be any vector in $W_1 + W_2$, then $w = w_1 + w_2$, where $w_1 \in W_1, w_2 \in W_2$. It can be easily seen that $\langle u, w \rangle = \langle u, w_1 + w_2 \rangle = \langle u, w_1 \rangle + \langle u, w_2 \rangle = 0$. Therefore, $u \in (W_1 + W_2)^\perp$ and hence $W_1^\perp \cap W_2^\perp \subseteq (W_1 + W_2)^\perp$. Combining this with the above fact we find that $(W_1 + W_2)^\perp = W_1^\perp \cap W_2^\perp$.

(iii) Since $W_1^\perp$ and $W_2^\perp$ are also subspaces of V, taking $W_1^\perp$ in place of W_1 and $W_2^\perp$ in place of W_2 in (ii) we find that $(W_1^\perp + W_2^\perp)^\perp = W_1^{\perp\perp} \cap W_2^{\perp\perp} = W_1 \cap W_2$. This implies that $(W_1^\perp + W_2^\perp)^{\perp\perp} = (W_1 \cap W_2)^\perp$ and hence $(W_1 \cap W_2)^\perp = W_1^\perp + W_2^\perp$.

(iv) Since $V = W_1 \oplus W_2$, we find that $\{0\} = V^\perp = (W_1 + W_2)^\perp = W_1^\perp \cap W_2^\perp$. But $W_1 \cap W_2 = 0$ implies that $V = \{0\}^\perp = (W_1 \cap W_2)^\perp = W_1^\perp + W_2^\perp$. This yields the required result.

Suppose that V is an inner product space over $\mathbb{F}$. If $u \in V$, then the map that sends v to $\langle v, u \rangle$ is a linear functional on V. The following result shows that every linear functional on V is of this form.

Definition 5.31 Let V be an inner product space and let $W_1, W_2, \ldots, W_n$ be subspaces of V. Then V is the orthogonal direct sum of $W_1, W_2, \ldots, W_n$, written as $V = W_1 \odot W_2 \odot \cdots \odot W_n$ if

(1) $V = W_1 \oplus W_2 \oplus \cdots \oplus W_n$,
(2) $W_i \perp W_j$ for $i \neq j$; $1 \leq i, j \leq n$.

Theorem 5.32 *Let V be an inner product space. Then the following are equivalent:*

  (i) $V = U \odot W$,
  (ii) $V = U \oplus W$ and $W = U^\perp$,
  (iii) $V = U \oplus W$ and $W \subseteq U^\perp$.

Proof *(i)* $\implies$ *(ii)* If *(i)* holds, then $V = U \oplus W$ and $U \perp W$, which implies that $W \subseteq U^\perp$. But if $v \in U^\perp$, then $v = u + w$, for $u \in U, w \in W$. Hence

$$0 = \langle u, v \rangle = \langle u, u \rangle + \langle u, w \rangle = \langle u, u \rangle.$$

This yields that $u = 0$ and $v \in W$, which implies that $U^\perp \subseteq W$. Hence, $U^\perp = W$.

(ii) $\implies$ *(iii)* Obvious.

(iii) $\implies$ *(i)* If *(iii)* holds then $W \subseteq U^\perp$, which implies that $U \perp W$ and *(i)* holds.

Theorem 5.33 (Riesz Representation Theorem) *Let V be a finite dimensional inner product space over $\mathbb{F}$ and $\varphi : V \to \mathbb{F}$ be a linear functional on V. Then there exists a unique vector $u \in V$ such that $\varphi(v) = \langle v, u \rangle$ for every $v \in V$.*

Proof First we show there exists a vector $u \in V$ such that $\varphi(v) = \langle v, u \rangle$ for every $v \in V$. Let $\{v_1, v_2, \ldots, v_n\}$ be an orthonormal basis of V. Then by Theorem 5.16*(ii)*, $v = \sum_{i=1}^{n} \langle v, v_i \rangle v_i$, and hence

$$
\begin{aligned}
\varphi(v) &= \varphi\big(\langle v, v_1 \rangle v_1\big) + \varphi(\langle v, v_2 \rangle v_2) + \cdots + \varphi(\langle v, v_n \rangle v_n) \\
&= \langle v, v_1 \rangle \varphi(v_1) + \langle v, v_2 \rangle \varphi(v_2) + \cdots + \langle v, v_n \rangle \varphi(v_n) \\
&= \langle v, \overline{\varphi(v_1)} v_1 \rangle + \langle v, \overline{\varphi(v_2)} v_2 \rangle + \cdots + \langle v, \overline{\varphi(v_n)} v_n \rangle \\
&= \langle v, \overline{\varphi(v_1)} v_1 + \overline{\varphi(v_2)} v_2 + \cdots + \overline{\varphi(v_n)} v_n \rangle
\end{aligned}
$$

for every $v \in V$. Now setting $u = \overline{\varphi(v_1)} v_1 + \overline{\varphi(v_2)} v_2 + \cdots + \overline{\varphi(v_n)} v_n$, we arrive at $\varphi(v) = \langle v, u \rangle$ for every $v \in V$. Now in order to show that there exists a unique vector $u \in V$ with the desired behavior suppose that there exist $u_1, u_2 \in V$ such that $\varphi(v) = \langle v, u_1 \rangle = \langle v, u_2 \rangle$ for every $v \in V$. Then we find that

$$\langle v, u_1 \rangle - \langle v, u_2 \rangle = \langle v, u_1 - u_2 \rangle = 0, \quad \text{for every } v \in V.$$

Now in particular, for $v = u_1 - u_2$ we find that $u_1 - u_2 = 0$ or $u_1 = u_2$ which proves the uniqueness of u.

Definition 5.34 Let V and W be any two inner product spaces over the same field $\mathbb{F}$. Then V is said to be isomorphic to W if there exists a vector isomorphism $\psi : V \to W$ such that for any $v_1, v_2 \in V$, $\langle v_1, v_2 \rangle = \langle \psi(v_1), \psi(v_2) \rangle$. Such an isomorphism is called inner product space isomorphism.

Theorem 5.35 *Any inner product space V over the field $\mathbb{F}$ of dimension n is isomorphic to $\mathbb{R}^n$ or $\mathbb{C}^n$ according as $\mathbb{F} = \mathbb{R}$ or $\mathbb{C}$.*

Proof Since V is finite dimensional, it has an orthonormal basis say $B = \{v_1, v_2, \ldots, v_n\}$. Define a map $\psi : \mathbb{F}^n \to V$ such that

$$\psi(\alpha_1, \alpha_2, \ldots, \alpha_n) = \sum_{i=1}^{n} \alpha_i v_i \text{ for every} (\alpha_1, \alpha_2, \ldots, \alpha_n) \in \mathbb{F}^n.$$

Clearly ψ is a vector space isomorphism. For any $\alpha = (\alpha_1, \alpha_2, \ldots, \alpha_n), \beta = (\beta_1, \beta_2, \ldots, \beta_n) \in \mathbb{F}^n$

$$\langle \alpha, \beta \rangle = \sum_{i=1}^{n} \alpha_i \overline{\beta_i}.$$

Also

$$\langle \psi(\alpha), \psi(\beta) \rangle = \left\langle \sum_{i=1}^{n} \alpha_i v_i, \sum_{j=1}^{n} \beta_j v_j \right\rangle$$

$$= \sum_{i=1}^{n} \sum_{j=1}^{n} \alpha_i \overline{\beta_j} \langle v_i, v_j \rangle$$

$$= \sum_{i=1}^{n} \alpha_i \overline{\beta_i}$$

$$= \langle \alpha, \beta \rangle$$

for all $v \in V$. This shows that V is isomorphic to $\mathbb{F}^n$.

Lemma 5.36 *If S is a subset of an inner product space V, then*

(i) $S^{\perp} = (L(S))^{\perp}$,
(ii) $L(S) \subseteq S^{\perp\perp}$,
(iii) $L(S) = S^{\perp\perp}$, if V is finite dimensional.

Proof (i) It is clear that $S \subseteq L(S)$, and hence $(L(S))^{\perp} \subseteq S^{\perp}$. Now suppose that $v \in S^{\perp}$. Then v is orthogonal to each vector on S. let $w \in L(S)$, then w is a linear combination of finite number of vectors in S, i.e., $w = \sum_{i=1}^{n} \alpha_i v_i$, where $v_i \in S$. Thus

$$
\begin{aligned}
\langle v, w \rangle &= \langle v, \sum_{i=1}^{n} \alpha_i v_i \rangle \\
&= \sum_{i=1}^{n} \alpha_i \langle v, v_i \rangle \\
&= 0 \text{ (since } v \text{ is orthogonal to each vector in } S).
\end{aligned}
$$

Therefore v is orthogonal to every vector in $L(S)$, i.e., $v \in (L(S))^{\perp}$. This shows that $S^{\perp} \subseteq (L(S))^{\perp}$ and hence $S^{\perp} = (L(S))^{\perp}$.

(ii) Let $v \in L(S)$. If w is any vector in $S^{\perp}$, then w is orthogonal to each vector in S. Thus w is orthogonal to v which is a linear combination of a finite number of vectors in S. This ensures that v is orthogonal to every vector w in $S^{\perp}$ and hence $v \in S^{\perp\perp}$. This implies that $L(S) \subseteq S^{\perp\perp}$.

(iii) From (i), we have $S^{\perp} = (L(S))^{\perp}$, which implies that $S^{\perp\perp} = (L(S))^{\perp\perp}$. But since $L(S)$ is a subspace of V, by Theorem 5.29, we find that $(L(S))^{\perp\perp} = L(S)$. This shows that $S^{\perp\perp} = L(S)$.

Theorem 5.37 (Bessel's inequality) *Let $S = \{v_1, v_2, \ldots, v_n\}$ be any nonempty orthonormal finite subset of an inner product space V. Then for any $v \in V$,*

$$
\sum_{i=1}^{n} |\langle v, v_i \rangle|^2 \leq \|v\|^2.
$$

Proof For any $v \in V$, observe that

$$
\begin{aligned}
\left\| v - \sum_{i=1}^{n} \langle v, v_i \rangle v_i \right\|^2 &= \left\langle v - \sum_{i=1}^{n} \langle v, v_i \rangle v_i, v - \sum_{j=1}^{n} \langle v, v_j \rangle v_j \right\rangle \\
&= \langle v, v \rangle - \sum_{i=1}^{n} \langle v, v_i \rangle \langle v_i, v \rangle \\
&\quad - \sum_{j=1}^{n} \overline{\langle v, v_j \rangle} \langle v, v_j \rangle + \sum_{j=1}^{n} \sum_{i=1}^{n} \langle v, v_i \rangle \overline{\langle v, v_j \rangle} \langle v_i, v_j \rangle.
\end{aligned}
$$

But since S is orthonormal set, i.e., $\langle v_i, v_j \rangle = 0$ if $i \neq j$ and $\langle v_i, v_j \rangle = 1$ if $i = j$, the above relation yields that

$$\|v\|^2 - \sum_{i=1}^{n} |\langle v, v_i \rangle|^2 - \sum_{i=1}^{n} |\langle v, v_i \rangle|^2 + \sum_{i=1}^{n} |\langle v, v_i \rangle|^2 \geq 0,$$

which implies that $\sum_{i=1}^{n} |\langle v, v_i \rangle|^2 \leq \|v\|^2$.

Theorem 5.38 *If $S = \{v_1, v_2, \ldots, v_n\}$ is an orthonormal subset of a finite dimensional inner product space V, then for any vector $v \in V$, $v - \sum_{i=1}^{n} \langle v, v_i \rangle v_i$ is orthogonal to each of $v_1, v_2, \ldots, v_n$. Moreover, if $u \notin L(S)$, then for $w = u - \sum_{i=1}^{n} \langle u, v_i \rangle v_i$, $\|w\| \neq 0$ and for $v_{n+1} = \frac{w}{\|w\|}$, the set $\{v_1, v_2, \ldots, v_n, v_{n+1}\}$ is an orthonormal subset of V.*

Proof For any $1 \leq j \leq n$ we observe that

$$\begin{aligned}
\langle v - \sum_{i=1}^{n} \langle v, v_i \rangle v_i, v_j \rangle &= \langle v, v_j \rangle - \sum_{i=1}^{n} \langle v, v_i \rangle \langle v_i, v_j \rangle \\
&= \langle v, v_j \rangle - \langle v, v_j \rangle \langle v_j, v_j \rangle \\
&= \langle v, v_j \rangle - \langle v, v_j \rangle \\
&= 0.
\end{aligned}$$

This shows that $v - \sum_{i=1}^{n} \langle v, v_i \rangle v_i$ is orthogonal to each $v_1, v_2, \ldots, v_n$. Now assume that $u \notin L(S)$ and $\|w\| \neq 0$, where $w = u - \sum_{i=1}^{n} \langle u, v_i \rangle v_i$, because if $\|w\| = 0$ then $w = 0$ and hence the above relation yields that $u = \sum_{i=1}^{n} \langle u, v_i \rangle v_i$ or u is in the linear combination of $v_1, v_2, \ldots, v_n$, i.e., $u \in L(S)$, a contradiction and hence $\|w\| \neq 0$. Further, we see that $\{v_1, v_2, \ldots, v_n, v_{n+1}\}$ is also an orthonormal subset of V. For each $1 \leq i \leq n$,

$$\begin{aligned}
\langle v_{n+1}, v_i \rangle &= \langle \tfrac{w}{\|w\|}, v_i \rangle \\
&= \tfrac{1}{\|w\|} \langle u - \sum_{j=1}^{n} \langle u, v_j \rangle v_j, v_i \rangle \\
&= \tfrac{1}{\|w\|} \{ \langle u, v_i \rangle - \sum_{j=1}^{n} \langle u, v_j \rangle \langle v_j, v_i \rangle \} \\
&= \tfrac{1}{\|w\|} \{ \langle u, v_i \rangle - \langle u, v_i \rangle \} \\
&= 0.
\end{aligned}$$

Finally we see that $\langle v_{n+1}, v_{n+1} \rangle = \langle \tfrac{w}{\|w\|}, \tfrac{w}{\|w\|} \rangle = \tfrac{1}{\|w\|^2} \langle w, w \rangle = 1$, which completes the proof.

If we consider the standard basis $B = \{e_1, e_2, \ldots, e_n\}$ of $\mathbb{R}^n$ (or $\mathbb{C}^n$), it can be easily observed that the basis B is an orthonormal basis, i.e., vectors in B are pairwise orthogonal and each vector in B has unit length. For $n \geq 2$ this vector space has infinitely many bases whose members are pairwise orthogonal, and has length 1. We shall now discuss a method for obtaining orthonormal basis for a finite dimensional inner product space.

Gram-Schmidt orthonormalization process

Consider any sequence of linearly independent vectors say $(v_1, v_2, \ldots, v_n, \ldots)$ in an inner product space V. If W_i is the subspace spanned by $v_1, v_2, \ldots, v_i$ and W be the subspace spanned by all the given vectors, then obviously W is the union of all W_i. By Theorem 5.18 if we define $u_1 = v_1$ and

$$u_k = v_k - \sum_{j=1}^{k-1} \frac{\langle v_k, u_j \rangle}{\|u_j\|^2} u_j; \text{ for } k \geq 2,$$

then $\{u_1, u_2, \ldots, u_i\}$ forms a basis of W_i. This yields that $(u_1, u_2, \ldots, u_n, \ldots)$ forms an orthogonal basis of W. Now for any $k \geq 1$, define $w_k = \frac{u_k}{\|u_k\|}$. Then $\|w_k\| = 1$, and $(w_1, w_2, \ldots, w_n, \ldots)$ forms an orthonormal basis of W. This process of constructing an orthonormal basis, starting from a basis of a countably generated subspace of V is called Gram-Schmidt orthonormalization process.

Example 5.39 (1) Apply Gram-Schmidt orthonormalization process to find an orthonormal basis of the subspace W of $\mathbb{C}^3$ generated by $\{(1, i, 0), (1, 2, 1 - i)\}$.

Suppose that $v_1 = (1, i, 0)$, $v_2 = (1, 2, 1 - i)$. Put $u_1 = v_1 = (1, i, 0)$,
$u_2 = v_2 - \frac{\langle v_2, v_1 \rangle}{\langle v_1, v_1 \rangle} v_1$
$= (1, 2, 1 - i) - \frac{1.1 + 2.(-i) + (1-i).0}{1.1 + i.(-i) + 0.0}(1, i, 0) = (\frac{1+2i}{2}, \frac{2-i}{2}, \frac{2-2i}{2})$.
Assume $w_1 = \frac{u_1}{\|u_1\|} = (\frac{1}{\sqrt{2}}, \frac{i}{\sqrt{2}}, 0)$,
$w_2 = \frac{u_2}{\|u_2\|} = \frac{(\frac{1+2i}{2}, \frac{2-i}{2}, \frac{2-2i}{2})}{\frac{\sqrt{18}}{2}} = (\frac{1+2i}{\sqrt{18}}, \frac{2-i}{\sqrt{18}}, \frac{2-2i}{\sqrt{18}})$.
Thus the required orthonormal basis of W is given by

$$\{w_1, w_2\} = \{(\frac{1}{\sqrt{2}}, \frac{i}{\sqrt{2}}, 0), (\frac{1+2i}{\sqrt{18}}, \frac{2-i}{\sqrt{18}}, \frac{2-2i}{\sqrt{18}})\}.$$

(2) Let $\mathbb{R}[x]$, the vector space of all real polynomials with inner product on $\mathbb{R}[x]$ given by

$$\langle p(x), q(x) \rangle = \int_{-1}^{1} p(x)q(x)dx.$$

Applying Gram-Schmidt orthonormalization process to the sequence $B = (1, x, x^2, x^3, \ldots)$ yields $v_1 = 1, v_2 = x, v_3 = x^2, v_4 = x^3$ and so on. Put $u_1 = v_1 = 1$,
$u_2 = v_2 - \frac{\langle v_2, v_1 \rangle}{\langle v_1, v_1 \rangle} v_1 = x - \frac{\int_{-1}^{1} (1.x)dx}{\int_{-1}^{1} (1.1)dx} 1 = x,$

$$u_3 = v_3 - \frac{\langle v_3, u_1 \rangle}{\langle u_1, u_1 \rangle} u_1 - \frac{\langle v_3, u_2 \rangle}{\langle u_2, u_2 \rangle} u_2 = x^2 - \frac{\int_{-1}^{1} x^2 dx}{\int_{-1}^{1} 1 dx} 1 - \frac{\int_{-1}^{1} x^3 dx}{\int_{-1}^{1} x^2 dx} x = x^2 - \frac{1}{3},$$

$$u_4 = v_4 - \frac{\langle v_4, u_1 \rangle}{\langle u_1, u_1 \rangle} u_1 - \frac{\langle v_4, u_2 \rangle}{\langle u_2, u_2 \rangle} u_2 - \frac{\langle v_4, u_3 \rangle}{\langle u_3, u_3 \rangle} u_3,$$

$$= x^3 - \frac{\int_{-1}^{1} x^3 dx}{\int_{-1}^{1} 1 dx} 1 - \frac{\int_{-1}^{1} x^4 dx}{\int_{-1}^{1} x dx} x - \frac{\int_{-1}^{1} x^3 (x^2 - \frac{1}{3}) dx}{\int_{-1}^{1} (x^2 - \frac{1}{3})^2 dx} (x^2 - \frac{1}{3}) = x^3 - \frac{3}{5} x,$$

and so on. Now let

$$w_1 = \frac{u_1}{\|u_1\|} = \frac{1}{\sqrt{\int_{-1}^{1} dx}} = \frac{1}{\sqrt{2}},$$

$$w_2 = \frac{u_2}{\|u_2\|} = \frac{x}{\sqrt{\int_{-1}^{1} x^2 dx}} = \frac{\sqrt{3}}{\sqrt{2}} x,$$

$$w_3 = \frac{u_3}{\|u_3\|} = \frac{x^2 - \frac{1}{3}}{\sqrt{\int_{-1}^{1} (x^2 - \frac{1}{3})^2 dx}} = \frac{\sqrt{10}}{4} (3x^2 - 1),$$

$$w_4 = \frac{u_4}{\|u_4\|} = \frac{x^3 - \frac{3}{5} x}{\sqrt{\int_{-1}^{1} (x^3 - \frac{3}{5} x)^2 dx}} = \frac{2\sqrt{2}}{5\sqrt{7}} (x^3 - \frac{3}{5} x)$$

and so on. Hence the required orthonormal sequence in $\mathbb{R}[x]$ is given by
$(\frac{1}{\sqrt{2}}, \frac{\sqrt{3}}{\sqrt{2}} x, \frac{\sqrt{10}}{4} (3x^2 - 1), \frac{2\sqrt{2}}{5\sqrt{7}} (x^3 - \frac{3}{5} x), \ldots)$.

Lemma 5.40 *Let $\{v_1, v_2, \ldots, v_n\}$ be the set of nonzero pairwise orthogonal vectors in an inner product space V. Then the vectors $u_1, u_2, \ldots, u_n$ with $u_1 = v_1$ and*
$$u_k = v_k - \sum_{j=1}^{k-1} \frac{\langle v_k, u_j \rangle}{\|u_j\|^2} u_j; \ \text{for } 2 \le k \le n \text{ are such that } v_i = u_i; \ 1 \le i \le n.$$

Proof Given that $u_1 = v_1$. The result can be proved easily by using induction. Assume that for some k; $1 \le k < n$, $u_i = v_i$, $1 \le i \le k$. Then for $1 \le j \le k$; $\langle v_{k+1}, u_j \rangle = \langle v_{k+1}, v_j \rangle = 0$ and hence

$$u_{k+1} = v_{k+1} - \sum_{j=1}^{k} \frac{\langle v_{k+1}, u_j \rangle}{\|u_j\|^2} u_j = v_{k+1},$$

and the result follows.

It is well-known that in a finite dimensional vector space every linearly independent subset can be extended to a basis of the vector space. The following theorem shows that the similar result follows for orthonormal subset of a finite dimensional inner product space:

Theorem 5.41 *Let V be a finite dimensional inner product space with $\dim V = n \ge 1$. Then every orthonormal subset of V can be extended to an orthonormal basis of V.*

Proof Suppose $S = \{v_1, v_2, \ldots, v_k\}$ is an orthonormal subset of V. Since S is orthonormal, by Theorem 5.16, it is linearly independent. Hence S can be extended to a basis of V say $\{v_1', v_2', \ldots, v_n'\}$ such that $v_i' = v_i$, $1 \le i \le k$. Now define $u_1 = v_1$ and $u_i = v_i' - \sum_{j=1}^{i-1} \frac{\langle v_i', u_j \rangle}{\|u_j\|^2} u_j$; for $2 \le i \le n$. Hence by Lemma 5.40, $u_i = v_i' = v_i$ for $1 \le i \le k$. Now let $w_j = \frac{u_j}{\|u_j\|}$ for $1 \le j \le k$. But since the set S is orthonormal, $\|v_j\| = 1$, $w_j = v_j$ for $1 \le j \le k$, and by using the Gram-Schmidt process we find that $B = \{w_1, w_2, \ldots, w_n\}$ is an orthonormal basis of V.

Exercises

1. Let $B = \{e_1, e_2, \ldots, e_n\}$ be an orthonormal basis of V. Prove that

   (a) For any $v \in V$, we have $v = \langle v, e_1 \rangle e_1 + \langle v, e_2 \rangle e_2 + \cdots + \langle v, e_n \rangle e_n$,
   (b) $\langle a_1 e_1 + a_2 e_2 + \cdots + a_n e_n, b_1 e_1 + b_2 e_2 + \cdots + b_n e_n \rangle = a_1 b_1 + a_2 b_2 + \cdots + a_n b_n$,
   (c) For any $u, v \in V$, we have $\langle u, v \rangle = \langle u, e_1 \rangle \langle v, e_1 \rangle + \cdots + \langle u, e_n \rangle \langle v, e_n \rangle$.

2. Let $B = \{e_1, e_2, \ldots, e_n\}$ be an orthogonal basis of V. Then prove that for any $v \in V$, $v = \frac{\langle v, e_1 \rangle}{\langle e_1, e_1 \rangle} e_1 + \frac{\langle v, e_2 \rangle}{\langle e_2, e_2 \rangle} e_2 + \cdots + \frac{\langle v, e_n \rangle}{\langle e_n, e_n \rangle} e_n$.

3. Consider the subspace U of $\mathbb{R}^4$ spanned by the vectors: $v_1 = (1, 1, 1, 1)$, $v_2 = (1, 1, 2, 4)$, $v_3 = (1, 2, -4, -3)$, using Gram-Schmidt algorithm, find

   (a) an orthogonal basis of U,
   (b) an orthonormal basis of U.

4. Let $\mathbb{R}_3[x]$ be the inner product of all polynomials of degree at most 3, under the inner product

$$\langle p(x), q(x) \rangle = \int_{-\infty}^{\infty} p(x) q(x) e^{-x^2} dx.$$

   Apply Gram-Schmidt process to the basis $\{1, x, x^2, x^3\}$ thereby computing the first four Hermite polynomials (at least upto the multiplicative constant).

5. Let $w \neq 0$. Suppose that v is any vector in V. Prove that $c = \frac{\langle v, w \rangle}{\langle w, w \rangle} = \frac{\langle v, w \rangle}{\|w\|^2}$ is the unique scalar such that $v' = v - cw$ is orthogonal to w.

6. Suppose $v = (1, -3, 5, -6)$. Find the projection of v onto W or, in other words, find $w \in W$ that minimizes $\|v - w\|$, where W is the subspace of $\mathbb{R}^4$ spanned by

   (a) $u_1 = (1, 2, 3, 1)$ and $u_2 = (1, -3, 4, -6)$,
   (b) $v_1 = (-2, 0, 4, 5)$ and $v_2 = (-3, -6, 1, 0)$.

7. Suppose $B = \{e_1, e_2, \ldots, e_r\}$ is an orthogonal basis for a subspace W of a finite dimensional inner product space V. Prove that B can be extended to an orthogonal basis for V, that is one may find vectors $e_{r+1}, e_{r+2}, \ldots, e_n$ such that $B = \{e_1, e_2, \ldots, e_n\}$ is an orthogonal basis of V.

8. Using Gram-Schmidt algorithm, find an orthonormal basis of the subspace W of $\mathbb{C}^3$ spanned by $v_1 = (1, i, i + 3)$, $v_2 = (1, -i + 2, i + 4)$.

9. Let U be a subspace of $\mathbb{R}^4$ spanned by the vectors: $v_1 = (1, 1, 1, 1)$, $v_2 = (1, -1, 2, 2)$, $v_3 = (1, 2, -3, -4)$. Apply Gram-Schmidt algorithm, to find

   (a) an orthogonal basis of U and an orthonormal basis of U,
   (b) the projection of $v = (-1, 2, 3, -4)$ onto U.

10. Show that an orthonormal subset $\{v_1, v_2, \ldots, v_m\}$ of an inner product space V is an orthonormal basis of V if and only if $\sum_{i=1}^{m} |\langle v, v_i \rangle|^2 = \|v\|^2$ for every $v \in V$.

11. Let the set $\{v_1, v_2, \ldots, v_n\}$ be linearly dependent. What happens when the Gram-Schmidt process of orthogonalization is applied to it?

12. Apply Gram-Schmidt algorithm to orthonormalize the set of linearly indepen-
    dent vectors $\{(1, 0, 1, 1), (-1, 0, -1, 1), (0, -1, 1, 1)\}$ of $\mathbb{R}^4$.
13. Find k so that $u = (-1, 2, -3, 4)$ and $v = (-3, 5, k, 6)$ in $\mathbb{R}^4$ are orthogonal.
14. Let W be the subspace of $\mathbb{R}^3$ spanned by $u = (2, -3, 5)$ and $v = (2, 4, -7)$.
    Find a basis of the orthogonal complement $W^\perp$ of W.
15. If W is a subspace of $\mathbb{C}^3$ spanned by $(1, i, 1 - i)$ and $(i, -1, 0)$, then under the
    standard inner product, find $W^\perp$.
16. Let $u = (-2, 3, 1, 5, 0)$ be a vector in $\mathbb{R}^5$. Find an orthogonal basis for $u^\perp$.
17. Let S consist of the following vectors in $\mathbb{R}^4$: $u_1 = (1, 1, 0, -1)$, $u_2 = (1, 2, 1, 3)$,
    $u_3 = (1, 1, -9, 2)$, $u_4 = (16, -13, 1, 3)$.

    (a) Show that S is orthogonal and a basis of $\mathbb{R}^4$.
    (b) Find the coordinates of an arbitrary vector $v = (a, b, c, d)$ in $\mathbb{R}^4$ relative to
        the basis S.

18. Suppose that S, S_1, S_2 are the subsets of an inner product space V. Prove the
    following:

    (a) $S \subseteq S^{\perp\perp}$,
    (b) if $S_1 \subseteq S_2$, then $S_1^{\perp\perp} \subseteq S_2^{\perp\perp}$.

19. Let V be a complex inner product space and let W be a subspace of V. Suppose
    that $v \in V$ is a vector for which $\langle v, w \rangle + \langle w, v \rangle \leq \langle w, w \rangle$ for all $w \in W$. Prove
    that $v \in W^\perp$.
20. Suppose $w_1, w_2, \ldots, w_n$ form an orthogonal set of nonzero vectors in V. Let
    $v \in V$. Define $v' = v - (c_1 w_1 + c_2 w_2 + \cdots + c_n w_n)$, where $c_i = \frac{\langle v, w_i \rangle}{\langle w_i, w_i \rangle}$. Then
    prove that v' is orthogonal to $w_1, w_2, \ldots, w_n$.
21. Suppose $w_1, w_2, \ldots, w_n$ form an orthogonal set of nonzero vectors in V. Let v be
    any vector of V and let c_i be the component of v along w_i. Then, prove that for any
    scalars $a_1, a_2, \ldots, a_n$, we have $\left\| v - \sum_{k=1}^{n} c_k w_k \right\| \leq \left\| v - \sum_{k=1}^{n} a_k w_k \right\|$, i.e., $\sum c_i w_i$
    is the closest approximation to v as a linear combination of $w_1, w_2, \ldots, w_n$.
22. Let V be the vector space of polynomials over $\mathbb{R}$ of degree ≤ 3 with inner product
    defined by $\langle f, g \rangle = \int_0^1 f(t)g(t)dt$. Find a basis of the subspace W orthogonal
    to $h(t) = -3 + 2t + 5t^2 + 6t^3$.
23. Let $M_2(\mathbb{R})$ be the vector space with inner product $\langle A, B \rangle = tr(B^t A)$. Find an
    orthogonal basis for the orthogonal complement of

    (a) diagonal matrices,
    (b) symmetric matrices,
    (c) skew symmetric matrices.

24. Suppose $\{u_1, u_2, \ldots, u_n\}$ is an orthogonal set of vectors. Show that
    $\{k_1 u_1, k_2 u_2, \ldots, k_n u_n\}$ is an orthogonal set for any scalars $k_1, k_2, \ldots, k_n$.
25. Let V be the real inner product space consisting of the real valued continuous
    function on $[-1, 1]$ with the inner product $\langle f, g \rangle = \int_{-1}^{+1} f(t)g(t)dt$. Let W be

the subspace of odd functions, i.e., functions satisfying $f(-t) = -f(t)$. Find orthogonal complement of W.

26. Suppose $V = W_1 \oplus W_2$ and that f_1 and f_2 are inner products on W_1 and W_2, respectively. Show that there is a unique inner product f on V such that

    (a) $W_2 = W_1^\perp$,
    (b) $f(u, v) = f_k(u, v)$ when $u, v \in W_k$, $k = 1, 2$.

5.4 Operators on Inner Product Spaces

Let U, V be inner product spaces over $\mathbb{F}$ and $T : V \to U$ be a linear transformation. Fix $u \in U$ and consider a linear functional φ on V such that $\varphi(v) = \langle T(v), u \rangle, v \in V$. This linear functional depends on T and u. By Riesz representation theorem there exists a unique vector in V such that this linear functional is given by taking the inner product with it. If $T^*(u)$ is the unique vector in V, then $\langle T(v), u \rangle = \langle v, T^*(u) \rangle$.

Definition 5.42 Let U and V be inner product spaces over the field $\mathbb{F}$ and $T : U \to V$ be a linear transformation. The adjoint of T, denoted as T^*, is the function $T^* : V \to U$ such that $\langle T(u), v \rangle = \langle u, T^*(v) \rangle$, for every $u \in U$ and $v \in V$.

Example 5.43 Let $T : \mathbb{R}^2 \to \mathbb{R}^3$ such that $T(a, b) = (a + b, a - b, b)$ be a linear transformation. Find the adjoint of T.

Proof Adjoint of T, i.e., T^* is a function from $\mathbb{R}^3$ to $\mathbb{R}^2$. Now let $(x, y, z) \in \mathbb{R}^3$ be an arbitrary point in $\mathbb{R}^3$. Then for every $(a, b) \in \mathbb{R}^2$

$$
\begin{aligned}
\langle (a, b), T^*(x, y, z) \rangle &= \langle T(a, b), (x, y, z) \rangle \\
&= \langle (a + b, a - b, b), (x, y, z) \rangle \\
&= ax + bx + ay - by + bz \\
&= \langle (a, b), (x + y, x - y + z) \rangle.
\end{aligned}
$$

This shows that $T^*(x, y, z) = (x + y, x - y + z)$, which is the required adjoint of T.

Remark 5.44 The above example shows that if $T : \mathbb{R}^2 \to \mathbb{R}^3$ is a linear transformation, then $T^* : \mathbb{R}^3 \to \mathbb{R}^2$ is also a linear transformation. The following result shows that this is indeed true in general.

Theorem 5.45 *If U and V are inner product spaces over the same field $\mathbb{F}$ and $T : U \to V$ is a linear transformation, then T^* is a linear transformation from V to U.*

Proof Given that $T : U \to V$ is a linear transformation. For let $v_1, v_2 \in V$ and $\alpha, \beta \in \mathbb{F}$. If $u \in U$, then

$$\langle u, T^*(\alpha v_1 + \beta v_2)\rangle = \langle T(u), \alpha v_1 + \beta v_2\rangle$$
$$= \langle T(u), \alpha v_1\rangle + \langle T(u), \beta v_2\rangle$$
$$= \overline{\alpha}\langle T(u), v_1\rangle + \overline{\beta}\langle T(u), v_2\rangle$$
$$= \overline{\alpha}\langle u, T^*(v_1)\rangle + \overline{\beta}\langle u, T^*(v_2)\rangle$$
$$= \langle u, \alpha T^*(v_1)\rangle + \langle u, \beta T^*(v_2)\rangle$$
$$= \langle u, \alpha T^*(v_1) + \beta T^*(v_2)\rangle.$$

Hence the above yields that $T^*(\alpha v_1 + \beta v_2) = \alpha T^*(v_1) + \beta T^*(v_2)$ for all $v_1, v_2 \in V$ and $\alpha, \beta \in \mathbb{F}$.

Theorem 5.46 *Let U, V, W be inner product spaces over the same field $\mathbb{F}$. Then*

 (i) $(S + T)^* = S^* + T^*$ *for all* $S, T \in Hom(U, V)$,
 (ii) $(\alpha T)^* = \overline{\alpha} T^*$ *for all* $\alpha \in \mathbb{F}, T \in Hom(U, V)$,
(iii) $(ST)^* = T^* S^*$ *for all* $T \in Hom(U, V), S \in Hom(V, W)$,
(iv) $(T^*)^* = T$ *for all* $T \in Hom(U, V)$,
 (v) $I^* = I$ *(resp. $O^* = O$), where I (rep. O) is the identity (resp. zero) linear transformation on U*
(vi) *if $T \in Hom(U, V)$ and T is invertible, then $(T^*)^{-1} = (T^{-1})^*$.*

Proof (i) Let $S, T \in Hom(U, V)$. If $u \in U$ and $v \in V$, then

$$\langle u, (S + T)^*(v)\rangle = \langle (S + T)(u), v\rangle$$
$$= \langle S(u), v\rangle + \langle T(u), v\rangle$$
$$= \langle u, S^*(v)\rangle + \langle u, T^*(v)\rangle$$
$$= \langle u, S^*(v) + T^*(v)\rangle$$

for all $u \in U, v \in V$. This shows that $(S + T)^* = S^* + T^*$ for all $S, T \in Hom$ (U, V).

(ii) For any $\alpha \in \mathbb{F}$ and $u \in U, v \in V$, we find that

$$\langle u, (\alpha T)^*(v)\rangle = \langle (\alpha T)(u), v\rangle = \alpha\langle T(u), v\rangle = \alpha\langle u, T^*(v)\rangle = \langle u, \overline{\alpha} T^*(v)\rangle.$$

This shows that $(\alpha T)^* = \overline{\alpha} T^*$ for all $\alpha \in \mathbb{F}, T \in Hom(U, V)$.

(iii) Assume that $T \in Hom(U, V)$, $S \in Hom(V, W)$. Then for any $u \in U, w \in W$

$$\langle u, (ST)^*(w)\rangle = \langle (ST)(u), w\rangle = \langle S(T(u)), w\rangle = \langle T(u), S^*(w)\rangle = \langle u, T^*(S^*(w))\rangle.$$

This implies that $(ST)^* = T^* S^*$ for all $T \in Hom(U, V), S \in Hom(V, W)$.

(iv) Let $T \in Hom(U, V)$. For any $u \in U, v \in V$, we have $\langle T^*(u), v\rangle = \langle u, (T^*)^*(v)\rangle$. It is also easy to see that $\langle T^*(u), v\rangle = \overline{\langle v, T^*(u)\rangle} = \overline{\langle T(v), u\rangle} = \langle u, T(v)\rangle$. Combining this with the above relation, we find that $(T^*)^* = T$ for all $T \in Hom(U, V)$.

(v) For any $u, v \in U$ we have $\langle u, v \rangle = \langle I(u), v \rangle = \langle u, I^*(v) \rangle$. This shows that $I^* = I$. Similarly one can prove that $0^* = 0$.

(vi) If $T \in Hom(U, V)$ and is invertible, then $TT^{-1} = T^{-1}T = I$. This implies that $(TT^{-1})^* = I^*$, i.e., $(TT^{-1})^* = I$. This yields that $(T^{-1})^*T^* = I$. Similarly, it can be seen that $T^*(T^{-1})^* = I$ which shows that $(T^*)^{-1} = (T^{-1})^*$.

Remark 5.47 (i) One can find the relationship between $N(T)$, the null space and $R(T)$, the range of a linear map T and its adjoint T^*. If $T \in Hom(U, V)$, then for any $v \in V$, $v \in N(T^*)$ if and only if $T^*(v) = 0$, and $T^*(v) = 0$ if and only if $\langle u, T^*(v) \rangle = 0$ for all $u \in U$ or $\langle T(u), v \rangle = 0$ for all $u \in U$. This shows that $v \in N(T^*)$ if and only if $v \in (R(T))^\perp$ and hence $N(T^*) = (R(T))^\perp$. By taking the orthogonal complement both the sides one can find that $R(T) = N(T^*)^\perp$, and hence replacing T by T^* we find that $R(T^*) = (N(T))^\perp$.

(ii) If T is a linear transformation from U to V, then the following theorem shows that it is easy to find T^* if we can find the relation between $m(T)$ and $m(T^*)$.

Theorem 5.48 *Let U and V be nonzero finite dimensional inner product spaces over the same field $\mathbb{F}$, and let $B_1 = \{u_1, u_2, \ldots, u_m\}$ and $B_2 = \{v_1, v_2, \ldots, v_n\}$ be ordered orthonormal basis of U and V, respectively. If T is a linear transformation from U into V, and $m(T) = (\alpha_{ji})$ be the matrix of T of order $n \times m$ relative to the basis B_1 of U, then the matrix of T^* relative to the basis B_2 is the conjugate transpose of $m(T)$ of order $m \times n$.*

Proof Given that U and V are finite dimensional inner product spaces with orthonormal basis $B_1 = \{u_1, u_2, \ldots, u_m\}$ and $B_2 = \{v_1, v_2, \ldots, v_n\}$, respectively. Suppose that $m(T^*) = (\beta_{ij})$ represents the matrix of T^* relative to the basis B_2. Then for $1 \leq i \leq m, i \leq j \leq n$,

$$T(u_i) = \sum_{j=1}^{n} \alpha_{ji} v_j \text{ and } T^*(v_j) = \sum_{i=1}^{m} \beta_{ij} u_i.$$

Since B_1 and B_2 are orthonormal bases, we find that

$$\langle T(u_i), v_j \rangle = \left\langle \sum_{j=1}^{n} \alpha_{ji} v_j, v_j \right\rangle = \alpha_{ji}.$$

Similarly,

$$\langle T^*(v_j), u_i \rangle = \left\langle \sum_{i=1}^{m} \beta_{ij} u_i, u_i \right\rangle = \beta_{ij}.$$

But the latter relations show that

$$\beta_{ij} = \overline{\langle u_i, T^*(v_j) \rangle} = \overline{\langle T(u_i), v_j \rangle} = \overline{\alpha_{ji}},$$

and hence $m(T^*) = \overline{(m(T))}'$, i.e., $m(T^*)$ is the conjugate transpose of $m(T)$. This relation is very useful in determining T^* if T is known.

Example 5.49 Let $T : \mathbb{R}^2 \to \mathbb{R}^3$ be a linear transformation given by $T(a, b) = (a + b, a - b, b)$. Then the matrix of T relative to the standard bases is

$$m(T) = \begin{bmatrix} 1 & 1 \\ 1 & -1 \\ 0 & 1 \end{bmatrix}.$$

Thus the matrix of T^* relative to the standard bases is

$$m(T^*) = \begin{bmatrix} 1 & 1 & 0 \\ 1 & -1 & 1 \end{bmatrix}.$$

Hence, $T^*(a, b, c) = (a + b, a - b + c)$.

Example 5.50 Let $T : \mathbb{C}^3 \to \mathbb{C}^3$ be a linear transformation defined by $T(a, b, c) = (2a + (1 - i)b, (4 + i)a - 5ic, 3ia + (2 + i)b + 3c)$. Then the matrix of T with respect to the standard basis of $\mathbb{C}^3$ is given by

$$m(T) = \begin{bmatrix} 2 & 1 - i & 0 \\ 4 + i & 0 & -5i \\ 3i & 2 + i & 3 \end{bmatrix}.$$

Thus the matrix of T^* relative to the standard basis of $\mathbb{C}^3$ is

$$m(T^*) = \begin{bmatrix} 2 & 4 - i & -3i \\ 1 + i & 0 & 2 - i \\ 0 & 5i & 3 \end{bmatrix},$$

and $T^*(a, b, c) = (2a + (4 - i)b - 3ic, (1 + i)a + (2 - i)c, 5ib + 3c)$.

Now we consider linear operator T on a finite dimensional inner product space V, i.e., $T \in \mathcal{A}(V)$ instead of $T \in Hom(V, V)$.

Definition 5.51 An operator $T \in \mathcal{A}(V)$ is called self-adjoint if $T = T^*$, and if $T^* = -T$, then T is called skew-adjoint.

Remark 5.52 (i) $T \in \mathcal{A}(V)$ is self-adjoint if and only if $\langle T(u), v \rangle = \langle u, T(v) \rangle$, for all $u, v \in V$.

(ii) In an Euclidean space, the self-adjoint linear transformation is called symmetric and in case of unitary space it is called Hermitian.

(iii) If $T \in \mathcal{A}(V)$ is a self-adjoint operator on an inner product space V, then $(S^*TS)^* = S^*T^*(S^*)^* = S^*TS$, i.e., S^*TS is self-adjoint for all $S \in \mathcal{A}(V)$. On the other hand, if S is invertible and S^*TS is self-adjoint, then it can be seen that T is self-adjoint. In fact, if S^*TS is self-adjoint, then $(S^*TS)^* = S^*TS$. This implies that $S^*T^*S = S^*TS$. Since S invertible, S^* is also invertible and the latter relation shows that $T^*S = TS$, and hence $T^* = T$ i.e., T is self-adjoint.

Example 5.53 Let $T : \mathbb{C}^3 \to \mathbb{C}^3$ be a linear transformation such that $T(a, b, c) = (5a + (2 + 3i)b + (3 - 2i)c, (2 - 3i)a + 2b + (1 + i)c, (3 + 2i)a + (1 - i)b + 7c)$. Then the matrix of T with respect to the standard basis of $\mathbb{C}^3$ is given by

$$m(T) = \begin{bmatrix} 5 & 2 + 3i & 3 - 2i \\ 2 - 3i & 2 & 1 + i \\ 3 + 2i & 1 - i & 7 \end{bmatrix}.$$

Thus

$$m(T^*) = \begin{bmatrix} 5 & 2 + 3i & 3 - 2i \\ 2 - 3i & 2 & 1 + i \\ 3 + 2i & 1 - i & 7 \end{bmatrix}.$$

This shows that $T = T^*$ and T is a self-adjoint operator.

Example 5.54 Let $T : \mathbb{C}^2 \to \mathbb{C}^2$ such that $T(a, b) = (2ia + (4 + i)b, (-4 + i)a + 3ib)$, then $m(T) = \begin{bmatrix} 2i & 4 + i \\ -4 + i & 3i \end{bmatrix}$ and hence $m(T^*) = \begin{bmatrix} -2i & -4 - i \\ 4 - i & -3i \end{bmatrix} = -\begin{bmatrix} 2i & 4 + i \\ -4 + i & 3i \end{bmatrix} = -m(T)$. Thus $T = -T^*$ and T is skew-adjoint.

If $S, T \in \mathscr{A}(V)$ are self-adjoint operators on a finite dimensional inner product space V, then $S^* = S$ and $T^* = T$. Now using the results given in Theorem 5.46, one can easily prove the following:

Theorem 5.55 *Let $S, T \in \mathscr{A}(V)$ be self-adjoint operators on a finite dimensional inner product space V. Then*

(i) *$S + T$ is self-adjoint,*
(ii) *if T is invertible, then T^* is invertible and $(T^*)^{-1} = (T^{-1})^*$,*
(iii) *TS is self-adjoint if and only if $TS = ST$,*
(iv) *for any real number α, αT is self-adjoint.*

Definition 5.56 An operator $T \in \mathscr{A}(V)$ on an inner product space V is called normal if $TT^* = T^*T$.

Remark 5.57 If $T \in \mathscr{A}(V)$ is self-adjoint, then $T^* = T$ and hence every self-adjoint operator is normal, but there exist normal operators which are not self-adjoint.

Example 5.58 Let $T : \mathbb{R}^2 \to \mathbb{R}^2$ such that $T(a, b) = (2a - 3b, 3a + 2b)$. Then with respect to standard basis, $m(T) = \begin{bmatrix} 2 & -3 \\ 3 & 2 \end{bmatrix}$. Thus $m(T^*) = \begin{bmatrix} 2 & 3 \\ -3 & 2 \end{bmatrix}$, hence $T^*(a, b) = (2a + 3b, -3a + 2b)$. It can be easily verified that $TT^* = T^*T$, i.e., T is normal which is not self-adjoint.

Remark 5.59   (i) If $T \in \mathscr{A}(V)$ is self-adjoint, then $\langle T(v), v \rangle = \langle v, T^*(v) \rangle = \langle v, T(v) \rangle = \overline{\langle T(v), v \rangle}$, for all $v \in V$, i.e., $\langle T(v), v \rangle$ is real. Similarly it can also be seen that if $T \in \mathscr{A}(V)$ is skew-adjoint, then $\langle T(v), v \rangle$ is purely imaginary for all $v \in V$.

(ii) $T \in \mathscr{A}(V)$ is normal if and only if $(TT^* - T^*T)(v) = 0$ for all $v \in V$. In fact,

$$T \text{ is normal} \iff \langle TT^*(v), v \rangle - \langle T^*T(v), v \rangle = 0$$
$$\iff \langle T(v), T(v) \rangle - \langle T^*(v), T^*(v) \rangle = 0$$

for all $v \in V$. This shows that T is normal if and only if $\|T(v)\|^2 = \|T^*(v)\|^2$, i.e., $\|T(v)\| = \|T^*(v)\|$.

(iii) If $T \in \mathscr{A}(V)$ is normal, then it can be easily seen that $N(T) = N(T^*)$. In fact, if $v \in N(T)$, then $T(v) = 0$. This implies that $\langle T^*(v), T^*(v) \rangle = \langle v, TT^*(v) \rangle = \langle v, T^*T(v) \rangle = \langle v, 0 \rangle = 0$, and hence $T^*(v) = 0$, i.e., $v \in N(T^*)$. This yields that $N(T) \subseteq N(T^*)$. In a similar manner it can be seen that if $u \in N(T^*)$, then $\langle T(u), T(u) \rangle = 0$, which shows that $T(u) = 0$ and hence $u \in N(T)$, i.e., $N(T^*) \subseteq N(T)$. This gives that $N(T) = N(T^*)$.

(iv) If V is a complex inner product space and $T \in \mathscr{A}(V)$ such that $\langle T(v), v \rangle = 0$ for all $v \in V$, then $T = 0$. In fact, $\langle T(v), v \rangle = 0$ for all $v \in V$ yields that

$$\begin{aligned} 0 &= \tfrac{1}{4}\{\langle T(u+w), u+w \rangle - \langle T(u-w), u-w \rangle + i(\langle T(u+iw), u+iw \rangle \\ &\quad - \langle T(u-iw), u-iw \rangle)\} \\ &= \tfrac{1}{4}\{2\langle T(u), w \rangle + 2\langle T(w), u \rangle + i(-2i\langle T(u), w \rangle + 2i\langle T(w), u \rangle)\} \\ &= \langle T(u), w \rangle, \end{aligned}$$

that is, $\langle T(u), w \rangle = 0$, for all $u, w \in V$. Hence, in particular $\langle T(u), T(u) \rangle = 0$, i.e., $T(u) = 0$ for all $u \in V$.

(v) If V is a real inner product space, a nonzero operator T might satisfy $\langle T(v), v \rangle = 0$ for all $v \in V$. However, this cannot happen for a self-adjoint operator, i.e., if $T = T^*$ and $\langle T(v), v \rangle = 0$ for all $v \in V$, then $T = 0$. In fact, if T is self-adjoint, then for all $u, w \in V$, $\langle T(w), u \rangle = \langle w, T(u) \rangle = \langle T(u), w \rangle$. This yields that

$$\begin{aligned} 0 &= \tfrac{1}{4}\{\langle T(u+w), u+w \rangle - \langle T(u-w), u-w \rangle\} \\ &= \tfrac{1}{4}\{2\langle T(u), w \rangle + 2\langle T(w), u \rangle\} \\ &= \langle T(u), w \rangle, \end{aligned}$$

for all $u, w \in V$ and hence again $T = 0$.

(vi) If $T \in \mathscr{A}(V)$ is self-adjoint and $T \neq 0$, then it can be seen that T cannot be nilpotent. In fact, if T is nilpotent, then there exists a positive integer n such that $T^n = 0$, but $T^{n-1} \neq 0$. Thus there exists $u \in V$ such that $T^{n-1}(u) \neq 0$. Now $\langle T^{n-1}(u), v \rangle = \langle u, (T^{n-1})^*(v) \rangle$ for all $v \in V$. Since T is self-adjoint $\langle T^{n-1}(u), T^{n-1}(u) \rangle = \langle u, T^{2n-2}(u) \rangle$. But since $2n - 2 \geq n$ for $n > 1$, we find that $\langle T^{n-1}(u), T^{n-1}(u) \rangle = 0$ and hence $T^{n-1}(u) = 0$, a contradiction.

Assume that every inner product space is a normed space and every normed space is a metric space. In fact, if V is an inner product space then V is a normed space under the norm $\|v\| = \sqrt{\langle v, v \rangle}$, and it has also been seen in the beginning of this chapter that every normed space is a metric space with respect to the metric $d : V \times V \to \mathbb{R}$ defined by $d(u, v) = \|u - v\|$. In case of isomorphism, it is also known that the linear map preserves the operations of the vector space. Now we define a linear transformation on V which do not change the distance or do not change the length of a vector in an inner product space V.

Definition 5.60 Let V be an inner product space over a field $\mathbb{F}$, $d : V \times V \to \mathbb{R}$ be a metric on V and $T \in \mathscr{A}(V)$. Then T is said to be an isometry if for all $u, v \in V$, $d(u, v) = d(T(u), T(v))$.

Example 5.61 Consider the linear transformation $T : \mathbb{R}^2 \to \mathbb{R}^2$ such that $T(a, b) = (b, a)$, where $\mathbb{R}^2$ is an inner product space with standard inner product. Let $z = (x, y) \in \mathbb{R}^2$. Then $\|z\| = \sqrt{x^2 + y^2}$. Moreover, $T(z) = (y, x)$ yields that $\|T(z)\| = \sqrt{y^2 + x^2}$. Thus $\|T(z)\| = \|z\|$ and T preserves length of vector. In this case let $d : \mathbb{R}^2 \times \mathbb{R}^2 \to \mathbb{R}$ such that $d(z_1, z_2) = \|z_1 - z_2\|$ be a metric on $\mathbb{R}^2$. Then for $z_1 = (a_1, b_1), z_2 = (a_2, b_2) \in \mathbb{R}^2$

$$d(z_1, z_2) = \sqrt{(a_1 - a_2)^2 + (b_1 - b_2)^2}$$

$$d(T(z_1), T(z_2)) = \sqrt{(b_1 - b_2)^2 + (a_1 - a_2)^2}.$$

This shows that $d(z_1, z_2) = d(T(z_1), T(z_2))$, i.e., d preserves the distance. However, if we take $S : \mathbb{R}^2 \to \mathbb{R}^2$ such that $S(a, b) = (0, b)$, then it can be easily seen that $\|z\| \neq \|S(z)\|, z \in \mathbb{R}^2$.

Remark 5.62 ((i) $T \in \mathscr{A}(V)$ is an isometry if and only if $\|T(x)\| = \|x\|$ for all $x \in V$. If T is an isometry, then for any $x, y \in V$ $\|x - y\| = \|T(x) - T(y)\| = \|T(x - y)\|$, and hence in particular $\|x\| = \|T(x)\|$. Conversely, if $\|T(x)\| = \|x\|$ for all $x \in V$, then for any $x, y \in V$, $\|x - y\| = \|T(x - y)\| = \|T(x) - T(y)\|$, i.e., T is an isometry.

(ii)   The sum of two isometries need not be an isometry. For example, let $S, T : \mathbb{R}^2 \to \mathbb{R}^2$ such that $S(a, b) = (b, a)$, $T(a, b) = (-b, -a)$ for any $a, b \in \mathbb{R}$. Then S and T are isometries on $\mathbb{R}^2$, but $S + T = 0$, which is not an isometry.

(iii)   In a finite dimensional inner product space V, isometries are non-singular. In fact, if $T \in \mathscr{A}(V)$ is an isometry, then $a \in N(T)$ if and only if $T(a) = 0$, i.e., $\|a\| = 0$ or $a = 0$. Hence T is one-to-one and since T is onto, T is non-singular.

Theorem 5.63 *Let V be an inner product space over $\mathbb{F}$. Then $T \in \mathscr{A}(V)$ is an isometry if and only if $\langle T(u), T(v) \rangle = \langle u, v \rangle$, for all $u, v \in V$.*

Proof Let T be an isometry. We separate the proof in two cases:
Case I. Assume that V is an Euclidean space. Then $\langle u, v \rangle = \overline{\langle v, u \rangle} = \langle v, u \rangle$. Since

$\|u + v\|^2 = \|u\|^2 + \|v\|^2 + \langle u, v \rangle + \langle v, u \rangle$, we find that $\langle u, v \rangle = \frac{\|u+v\|^2 - \|u\|^2 - \|v\|^2}{2}$ for all $u, v \in V$. Now since $T \in \mathscr{A}(V)$, the latter relation yields that

$$\begin{aligned}
\langle T(u), T(v) \rangle &= \frac{\|T(u)+T(v)\|^2 - \|T(u)\|^2 - \|T(v)\|^2}{2} \\
&= \frac{\|T(u+v)\|^2 - \|T(u)\|^2 - \|T(v)\|^2}{2} \\
&= \frac{\|u+v\|^2 - \|u\|^2 - \|v\|^2}{2} \\
&= \langle u, v \rangle
\end{aligned}$$

for all $u, v \in V$.

Case II. Suppose that V is unitary space. In this case $\langle u, v \rangle = \overline{\langle v, u \rangle}$ implies that

$$\langle u, v \rangle + \overline{\langle u, v \rangle} = \|u + v\|^2 - \|u\|^2 - \|v\|^2.$$

It can be easily seen that

$$\begin{aligned}
\|u + iv\|^2 &= \langle u + iv, u + iv \rangle \\
&= \langle u, u \rangle - i\langle u, v \rangle + i\langle v, u \rangle - i^2\langle v, v \rangle \\
&= \|u\|^2 - i\langle u, v \rangle + i\overline{\langle u, v \rangle} + \|v\|^2.
\end{aligned}$$

The above relation yields that

$$\langle u, v \rangle - \overline{\langle u, v \rangle} = i\{\|u + iv\|^2 - \|u\|^2 - \|v\|^2\}.$$

The above two relations reduce to

$$2\langle u, v \rangle = \|u + v\|^2 + i\|u + iv\|^2 - (1 + i)(\|u\|^2 + \|v\|^2) \quad \text{for} \quad \text{all } u, v \in V.$$

Replacing u by $T(u)$, v by $T(v)$, and using the fact that T is an isometry with $T(iu) = iT(u)$, we find that

$$\begin{aligned}
\langle T(u), T(v) \rangle &= \tfrac{1}{2}\{\|T(u) + T(v)\|^2 + i\|T(u) + iT(v)\|^2 \\
&\quad -(1 + i)(\|T(u)\|^2 + \|T(v)\|^2)\} \\
&= \tfrac{1}{2}\{\|T(u + v)\|^2 + i\|T(u + iv)\|^2 - (1 + i)(\|T(u)\|^2 + \|T(v)\|^2)\} \\
&= \tfrac{1}{2}\{\|u + v\|^2 + i\|u + iv\|^2 - (1 + i)(\|u\|^2 + \|v\|^2)\} \\
&= \langle u, v \rangle.
\end{aligned}$$

Conversely if $\langle T(u), T(v) \rangle = \langle u, v \rangle$ for all $u, v \in V$, then in particular $\langle T(u), T(u) \rangle = \langle u, u \rangle$ for all $u \in V$, that is, $\|T(u)\|^2 = \|u\|^2$. Hence $\|T(u)\| = \|u\|$ and T is an isometry.

One can also find another kind of linear transformation on an inner product space V which preserves all the structure of V.

Definition 5.64 Let V be an inner product space over the field $\mathbb{F}$. Then $T \in \mathscr{A}(V)$ is said to be unitary if

$$\langle T(u), T(v) \rangle = \langle u, v \rangle \quad \text{for all} \quad u, v \in V.$$

Remark 5.65 If $T \in \mathscr{A}(V)$ satisfies $\langle T(u), T(u) \rangle = \langle u, u \rangle$ for all $u \in V$, then it can be seen that T is unitary. In fact, if $\langle T(u), T(u) \rangle = \langle u, u \rangle$, then by linearizing this relation on u we find that

$$\langle T(u), T(v) \rangle + \langle T(v), T(u) \rangle = \langle u, v \rangle + \langle v, u \rangle \quad \text{for all} \quad u, v \in V.$$

Since the above relation holds for all $v \in V$, replacing v by iv, where $i^2 = -1$ we arrive at $-i \langle T(u), T(v) \rangle + i \langle T(v), T(u) \rangle = -i \langle u, v \rangle + i \langle v, u \rangle$ for all $u, v \in V$. Now multiplying by i yields that

$$\langle T(u), T(v) \rangle - \langle T(v), T(u) \rangle = \langle u, v \rangle - \langle v, u \rangle \quad \text{for all} \quad u, v \in V.$$

Combining the above two relations we find that $\langle T(u), T(v) \rangle = \langle u, v \rangle$ for all $u, v \in V$, and therefore T is unitary.

Example 5.66 Let V be a finite dimensional inner product space and $\lambda_1, \lambda_2, \ldots, \lambda_n$ be scalars with absolute value 1. If $T \in \mathscr{A}(V)$ satisfies $T(v_j) = \lambda_j v_j$ for some orthonormal basis $\{v_1, v_2, \ldots, v_n\}$ of V, then it can be seen that T is an isometry. Suppose that $v \in V$. Then there exist scalars $\alpha_1, \alpha_2, \ldots, \alpha_n$ such that $v = \sum_{i=1}^{n} \alpha_i v_i$.

Since the given basis is orthonormal, $\langle v, v_i \rangle = \alpha_i$. Hence $v = \sum_{i=1}^{n} \langle v, v_i \rangle v_i$.

$$
\begin{aligned}
\|v\|^2 &= \left\langle \sum_{i=1}^{n} \alpha_i v_i, \sum_{i=1}^{n} \alpha_i v_i \right\rangle \\
&= \sum_{i=1}^{n} \alpha_i \overline{\alpha_i} \langle v_i, v_i \rangle \\
&= \sum_{i=1}^{n} |\alpha_i|^2 \\
&= \sum_{i=1}^{n} |\langle v, v_i \rangle|^2.
\end{aligned}
$$

Since $v = \sum_{i=1}^{n} \langle v, v_i \rangle v_i$, we find that $T(v) = \sum_{i=1}^{n} \langle v, v_i \rangle T(v_i)$. This yields that $T(v) = \sum_{i=1}^{n} \langle v, v_i \rangle \lambda_i v_i$. Now

$$
\begin{aligned}
\|T(v)\|^2 &= \left\langle \sum_{i=1}^{n} \langle v, v_i \rangle \lambda_i v_i, \sum_{i=1}^{n} \langle v, v_i \rangle \lambda_i v_i \right\rangle \\
&= \sum_{i=1}^{n} |\langle v, v_i \rangle|^2 |\lambda_i|^2 \\
&= \sum_{i=1}^{n} |\langle v, v_i \rangle|^2.
\end{aligned}
$$

The above two relations yield that $\|T(v)\|^2 = \|v\|^2$, i.e., $\|T(v)\| = \|v\|$, and T is isometry.

The following theorem characterizes an isometry in terms of orthonormal basis of an inner product space.

Theorem 5.67 *Let V be a finite dimensional inner product space over $\mathbb{F}$. $T \in \mathscr{A}(V)$ is an isometry if and only if T maps an orthonormal basis of V into an orthonormal basis of V.*

Proof Let $\{v_1, v_2, \ldots, v_n\}$ be an orthonormal basis of V. Hence $\langle v_i, v_j \rangle = 0$ or 1 according as $i \neq j$ or $i = j$. But since T is an isometry, $\langle T(v_i), T(v_j) \rangle = \langle v_i, v_j \rangle$, $1 \leq i, j \leq n$. This yields that $\langle T(v_i), T(v_j) \rangle = 0$ or 1 according as $i \neq j$ or $i = j$. This shows that $\{T(v_1), T(v_2), \ldots, T(v_n)\}$ is an orthonormal set and since orthonormal set is linearly independent set, $\{T(v_1), T(v_2), \ldots, T(v_n)\}$ is a basis of V. Conversely, assume that T maps orthonormal basis of V into an orthonormal basis of V. If $\{v_1, v_2, \ldots, v_n\}$ is an orthonormal basis of V, then for any $v \in V$, $v = \sum_{i=1}^{n} \alpha_i v_i$, and

$$\langle v, v \rangle = \left\langle \sum_{i=1}^{n} \alpha_i v_i, \sum_{j=1}^{n} \alpha_j v_j \right\rangle$$
$$= \sum_{i=1}^{n} \alpha_i \sum_{j=1}^{n} \overline{\alpha_j} \langle v_i, v_j \rangle$$
$$= \sum_{i=1}^{n} |\alpha_i|^2.$$

Similarly,

$$\langle T(v), T(v) \rangle = \left\langle \sum_{i=1}^{n} \alpha_i T(v_i), \sum_{j=1}^{n} \alpha_j T(v_j) \right\rangle$$
$$= \sum_{i=1}^{n} \alpha_i \sum_{j=1}^{n} \overline{\alpha_j} \langle T(v_i), T(v_j) \rangle$$
$$= \sum_{i=1}^{n} |\alpha_i|^2.$$

This shows that $\langle v, v \rangle = \langle T(v), T(v) \rangle$, for all $v \in V$, i.e., $\|T(v)\| = \|v\|$ or T is isometry.

Theorem 5.68 *Let V be a finite dimensional inner product space. Then $T \in \mathscr{A}(V)$ is isometry if there exists a unique isometry $S \in \mathscr{A}(V)$ such that $TS = ST = I$.*

Proof Let $v \in V$ and $T \in \mathscr{A}(V)$. Then define a map $\varphi_v : V \to \mathbb{F}$, such that $\varphi_v(u) = \langle T(u), v \rangle$. For any $u_1, u_2 \in V$, $\alpha, \beta \in \mathbb{F}$

$$\varphi_v(\alpha u_1 + \beta u_2) = \langle T(\alpha u_1 + \beta u_2), v \rangle$$
$$= \alpha \langle T(u_1), v \rangle + \beta \langle T(u_2), v \rangle$$
$$= \alpha \varphi_v(u_1) + \beta \varphi_v(u_2).$$

Thus φ_v is a linear functional on V and hence by Riesz representation theorem for each $u \in V$ there exists a unique $v' \in V$ such that $\varphi_v(u) = \langle u, v' \rangle$. Now define $S : V \to V$ such that $S(v) = v'$ for all $v \in V$. Hence $\langle T(u), v \rangle = \langle u, S(v) \rangle$ for all $u, v \in V$. Now

$$\begin{aligned} \langle T(u), \alpha v_1 + \beta v_2 \rangle &= \overline{\alpha} \langle T(u), v_1 \rangle + \overline{\beta} \langle T(u), v_2 \rangle \\ &= \overline{\alpha} \langle u, S(v_1) \rangle + \overline{\beta} \langle u, S(v_2) \rangle \\ &= \langle u, \alpha S(v_1) + \beta S(v_2) \rangle. \end{aligned}$$

But since $\langle T(u), \alpha v_1 + \beta v_2 \rangle = \langle u, S(\alpha v_1 + \beta v_2) \rangle$, the above relation yields that $S(\alpha v_1 + \beta v_2) = \alpha S(v_1) + \beta S(v_2)$ and $S \in \mathscr{A}(V)$. Now our claim is that S is unique. Suppose there exists $S' \in \mathscr{A}(V)$ such that $\langle T(u), v \rangle = \langle u, S'(v) \rangle$. Then $\langle u, (S - S')(v) \rangle = 0$ for all $v \in V$. This shows that $S - S' = 0$. It can also be seen that S is one-to-one. If $S(v) = 0$, then $\langle T(u), v \rangle = \langle u, S(v) \rangle = 0$, for all $u \in V$. Since T is an isometry, T^{-1} exists, and hence $\langle T(T^{-1})(v), v \rangle = 0$ which yields that $\langle v, v \rangle = 0$ or $v = 0$. Since V is finite dimensional, S is non-singular. As $\langle T(u), v \rangle = \langle u, S(v) \rangle$ for all $u, v \in V$, we find that $\langle T(u), T(v) \rangle = \langle u, S(T(v)) \rangle$ for all $u, v \in V$. But since T is an isometry, we have $\langle u, v \rangle = \langle u, ST(v) \rangle$. This reduces to $\langle u, (ST - I)(v) \rangle = 0$, for all $u, v \in V$. Hence, in particular $\langle (ST - I)(v), (ST - I)(v) \rangle = 0$, and $ST = I$. Similarly, it can also be seen that $TS = I$ and hence S is an isometry. In fact, for any isometry T there exists $S \in \mathscr{A}(V)$ such that $ST = TS = I$. Now for any $v \in V$, $\|TS(v)\| = \|T(S(v))\| = \|S(v)\|$, i.e., $\|v\| = \|I(v)\| = \|S(v)\|$ for all $v \in V$ or S is an isometry.

Theorem 5.69 *Let V be a finite dimensional inner product space and $T \in \mathscr{A}(V)$. Then there exists a unique linear operator T^* on V such that $\langle T(u), v \rangle = \langle u, T^*(v) \rangle$, for all $u, v \in V$.*

Proof Let $v \in V$. Define a map $\varphi_v : V \to \mathbb{F}$ such that $\varphi_v(u) = \langle T(u), v \rangle$. It is straightforward to show that φ_v is a linear functional on V. Thus by Riesz representation theorem there exists a unique $v' \in V$ depending on v such that $\varphi_v(u) = \langle u, v' \rangle$. Now let $T^*(v) = v'$ so that $\langle T(u), v \rangle = \langle u, T^*(v) \rangle$. For any $u, v_1, v_2 \in V, \alpha, \beta \in \mathbb{F}$

$$\begin{aligned} \langle u, T^*(\alpha v_1 + \beta v_2) \rangle &= \langle T(u), \alpha v_1 + \beta v_2 \rangle \\ &= \overline{\alpha} \langle T(u), v_1 \rangle + \overline{\beta} \langle T(u), v_2 \rangle \\ &= \overline{\alpha} \langle u, T^*(v_1) \rangle + \overline{\beta} \langle u, T^*(v_2) \rangle \\ &= \langle u, \alpha T^*(v_1) + \beta T^*(v_2) \rangle. \end{aligned}$$

This shows that $T^*(\alpha v_1 + \beta v_2) = \alpha T^*(v_1) + \beta T^*(v_2)$ and hence T^* is a linear operator on V. For uniqueness, let there exist another linear operator on V such that $\langle T(u), v \rangle = \langle u, T'(v) \rangle$. This yields that $\langle u, T^*(v) \rangle = \langle u, T'(v) \rangle$ or $\langle u, T^*(v) - T'(v) \rangle = 0$ and hence in particular we find that $T^*(u) - T'(u) = 0$ for all $u \in V$ i.e., $T^* = T'$.

Exercises

1. If $T : \mathbb{R}^3 \to \mathbb{R}^2$ is a linear transformation, then find the adjoint of T.

2. Let V, W be inner product spaces over $\mathbb{F}$. For a fix $u \in V$ and $x \in W$ let $T :$ $V \to W$ be linear transformation such that $T(v) = \langle v, u \rangle x$ for all $v \in V$. Find the adjoint T^*. (Hint: $\langle v, T^*(w) \rangle = \langle v, \langle w, x \rangle u \rangle$ for $w \in W, v \in V$).

3. Let $T : \mathbb{R}^3 \to \mathbb{R}^3$ be a linear transformation such that $T(a, b, c) = (a + b, b, a + b + c)$. Find T^*.

4. Let $T : \mathbb{C}^3 \to \mathbb{C}^3$ be a linear transformation such that $T(a, b, c) = (2a - ib, b - 5ic, a + (1 - i)b + 3c)$. Find T^*.

5. If $T \in \mathbb{C}^2$, then find T^* in each of the following cases:

   (a) $T(a, b) = (ia + (1 - i)b, (1 + i)a - ib)$,
   (b) $T(a, b) = (2a + ib, (i - 1)a + b)$,
   (c) $T(a, b) = ((i - 1)a + b, a - b)$.

6. If V is an inner product space over the $\mathbb{F}$, and $T \in \mathscr{A}(V)$, then show that the following conditions are equivalent:

   (a) $\langle T(u), T(v) \rangle = \langle u, v \rangle$,
   (b) $T^*T = I$,
   (c) T is an isometry, i.e., $\|T(u)\| = \|u\|$.

7. In the above problem if $T \in \mathscr{A}(V)$ is one-to-one and onto, then show that (a) and (c) are equivalent to $(b)'$ $T^*T = TT^* = I$.

8. If $\{v_1, v_2, \ldots, v_n\}$ is an orthonormal basis of an inner product space V, then for any $u, v \in V$ show that $\langle u, v \rangle = \sum_{i=1}^{n} \langle u, v_i \rangle \langle v_i, v \rangle$.

9. Show that the set of all isometries on an inner product space forms a group.

10. If $T \in \mathscr{A}(V)$, then show that $T = S + K$, where $S, K \in \mathscr{A}(V)$ such that S is self-adjoint and K is skew-adjoint and the above decomposition is unique.(Hint: Consider $S = \frac{T + T^*}{2}$ and $K = \frac{T - T^*}{2}$).

11. Let $T : \mathbb{C}^3 \to \mathbb{C}^3$ be a linear transformation such that $T(a, b, c) = (a + ib + (1 + i)c, ia - b, 2ia + 3b + (2i - 1)c)$. Find T^*.

Chapter 6
Canonical Forms of an Operator

In this chapter, we study the structure of linear operators. In all that follows, V will be a finite dimensional vector space and $T : V \rightarrow V$ a linear operator from V to itself. We recall that the kernel $N(T)$ and the image $R(T)$ of T are both subspaces of V and, in the light of the rank-nullity theorem, the following conditions are equivalent: (i) T is bijective, (ii) $N(T) = \{0\}$, (iii) $R(T) = V$. A linear operator which satisfies these equivalent conditions is said to be *invertible*, its inverse function is also an operator and its matrix with respect to an arbitrary basis of V is an invertible matrix. An operator that is not invertible is said to be a *singular operator* and its matrix with respect to an arbitrary basis of V is a singular matrix.

We already have discussed that if V is a vector space of dimension n over the field $\mathbb{F}$ and B is a basis for V, then we can associate a matrix $A \in M_n(\mathbb{F})$ to T. More precisely, the column coordinate vectors of A are the images of the elements of the basis B. Thus any linear operator on V is represented by an appropriate matrix, whose scalar entries depend precisely on the choice of the basis for V.

We remind the reader of the basic effect of the change of basis in V. If B and B' are two different bases for V and A and A' are the matrices of T relative to the bases B and B', respectively, then A and A' are similar to each other. In particular $A' = P^{-1}AP$, where P is the transition matrix. Hence A and A' represent the same operator T and A' is obtained from A by conjugation. In other words, any matrix which is similar to A represents the linear operator T.

The first question one might ask oneself is, how to determine a basis of V with respect to which the linear operator is represented by a particularly simple matrix. To answer this question, we first introduce the concepts of *eigenvalue, eigenvector* and *eigenspace* of a linear operator T. They will be the main tools for analyzing the structure of the matrix of T. Then we describe the most relevant classes of linear operators, in accordance with the study of their representing matrices. More precisely, we focus our attention on operators which are represented in terms of some

M. Ashraf et al., *Advanced Linear Algebra with Applications*,
https://doi.org/10.1007/978-981-16-2167-3_6

suitable bases for V, by triangular, diagonal or block diagonal matrices (usually called *canonical forms of a matrix*). Then, we conclude this chapter by studying the class of normal operators.

6.1 Eigenvalues and Eigenvectors

Definition 6.1 An *eigenvector* of T is a nonzero vector $v \in V$ such that $T(v) = \lambda v$, for some $\lambda \in \mathbb{F}$. Analogously, let $B = \{e_1, \ldots, e_n\}$ be a basis of V, A the matrix of T with respect to the basis B and v a nonzero vector of V. If X is the nonzero coordinates column vector of v in terms of B, then v is said to be an *eigenvector* of A if $AX = \lambda X$ for some $\lambda \in \mathbb{F}$. The scalar $\lambda \in \mathbb{F}$ is called the *eigenvalue* associated with the eigenvector v.

To determine whether $\lambda \in \mathbb{F}$ is an eigenvalue of the linear operator $T : V \to V$, we must determine whether there are any nonzero solutions to the matrix equation $AX = \lambda X$, where $A \in M_n(\mathbb{F})$ is the $n \times n$ matrix of T. If we denote $X = [x_1, x_2, \ldots \ldots, x_n]^T$ and $A = (a_{ij})$, the above matrix equation reduces to the linear system

$$a_{11}x_1 + a_{12}x_2 + \cdots + a_{1n}x_n = \lambda x_1$$
$$a_{21}x_1 + a_{22}x_2 + \cdots + a_{2n}x_n = \lambda x_2$$
$$\cdots \quad \cdots \quad \cdots \quad \cdots$$
$$a_{n1}x_1 + a_{n2}x_2 + \cdots + a_{nn}x_n = \lambda x_n,$$

that is, the homogeneous linear system $(A - \lambda I)X = 0$, where I denotes the $n \times n$ identity matrix. We know that the homogeneous linear system admits nontrivial solutions only if its rank is less than the number of indeterminates. In particular, to determine the nonzero solutions of $(A - \lambda I)X = 0$ we need that $|A - \lambda I| = 0$. The polynomial $p(\lambda) = |A - \lambda I|$ is said to be the *characteristic polynomial* of T (of A). Here

$$p(\lambda) = \begin{vmatrix} a_{11} - \lambda & a_{12} & \ldots & a_{1n} \\ & \ddots & & \\ & & \ddots & \\ a_{n1} & a_{n2} & \ldots & a_{nn} - \lambda \end{vmatrix}$$
$$= (-1)^n \lambda^n + \alpha_{n-1}\lambda^{n-1} + \cdots + \alpha_1 \lambda + \alpha_0,$$

where $\alpha_i \in \mathbb{F}$, for any $i = 0, \ldots, n - 1$. Of course, its degree is equal to n. Moreover, $p(\lambda)$ has exactly n roots in the algebraic closure of $\mathbb{F}$. Any of these roots which belongs to $\mathbb{F}$, is an eigenvalue of T (of A). Analogously, the equation $p(\lambda) = 0$ is said to be the *characteristic equation* of T (of A) and has n solutions in the algebraic closure of $\mathbb{F}$. Any of these solutions which belongs to $\mathbb{F}$, is an eigenvalue of T (of A). Without loss of generality, in all that follows we refer to $p(\lambda)$ as the characteristic polynomial of the matrix A.

Now we assume that $\lambda_0 \in \mathbb{F}$ is an eigenvalue of A. To find an eigenvector of A associated with λ_0, we have to solve the matrix equation $AX = \lambda_0 X$. Since $|A - \lambda_0 I| = 0$, the rank of the matrix $A - \lambda_0 I$ is equal to $r < n$. Therefore, the above homogeneous linear system admits nontrivial solutions. More precisely, we can describe the set of all solutions as follows:

$$V_0 = \{0 \neq X \in V \mid (A - \lambda_0 I)X = 0\}.$$

The vector subspace of V defined as $W_0 = V_0 \cup \{0\}$ is said to be the *eigenspace* of A associated with the eigenvalue λ_0. The dimension of W_0 is equal to $n - r$.

Example 6.2 Let $T : \mathbb{R}^3 \to \mathbb{R}^3$ be the linear operator with associated matrix

$$A = \begin{bmatrix} 1 & 2 & 2 \\ 1 & 3 & 1 \\ 2 & 2 & 1 \end{bmatrix}$$

with respect to the canonical basis of $\mathbb{R}^3$. The characteristic polynomial is the following determinant:

$$p(\lambda) = \begin{vmatrix} 1 - \lambda & 2 & 2 \\ 1 & 3 - \lambda & 1 \\ 2 & 2 & 1 - \lambda \end{vmatrix} = (\lambda - 1)(-\lambda^2 + 4\lambda + 5).$$

There are three distinct roots of $p(\lambda)$, that is, three distinct eigenvalues $\lambda_1 = -1$, $\lambda_2 = 5$, $\lambda_3 = 1$ of T. Let us find the corresponding eigenspaces.

For $\lambda_1 = -1$, we have

$$A + I = \begin{bmatrix} 2 & 2 & 2 \\ 1 & 4 & 1 \\ 2 & 2 & 2 \end{bmatrix}$$

which has rank equal to 2. Thus, the associated homogeneous linear system

$$\begin{aligned} 2x_1 + 2x_2 + 2x_3 &= 0 \\ x_1 + 4x_2 + x_3 &= 0 \\ 2x_1 + 2x_2 + 2x_3 &= 0 \end{aligned}$$

has solutions $X = (\alpha, 0, -\alpha)$, for any $\alpha \in \mathbb{R}$. Hence, the eigenspace corresponding to λ_1 has dimension 1 and is generated by the eigenvector $(1, 0, -1)$.

For $\lambda_2 = 5$, we have

$$A - 5I = \begin{bmatrix} -4 & 2 & 2 \\ 1 & -2 & 1 \\ 2 & 2 & -4 \end{bmatrix}$$

having rank equal to 2, so that the associated homogeneous linear system

$$-4x_1 + 2x_2 + 2x_3 = 0$$
$$x_1 - 2x_2 + x_3 \quad = 0$$
$$2x_1 + 2x_2 - 4x_3 \quad = 0$$

has solutions $X = (\alpha, \alpha, \alpha)$, for any $\alpha \in \mathbb{R}$. In this case, the eigenspace corresponding to λ_2 is generated by the eigenvector $(1, 1, 1)$.

Finally, for $\lambda_3 = 1$,

$$A - I = \begin{bmatrix} 0 & 2 & 2 \\ 1 & 2 & 1 \\ 2 & 2 & 0 \end{bmatrix}$$

has rank equal to 2 and the associated homogeneous linear system

$$2x_2 + 2x_3 \quad = 0$$
$$x_1 + 2x_2 + x_3 = 0$$
$$2x_1 + 2x_2 \quad = 0$$

has solutions $X = (\alpha, -\alpha, \alpha)$, for any $\alpha \in \mathbb{R}$. The eigenspace corresponding to λ_3 is generated by the eigenvector $(1, -1, 1)$.

Example 6.3 Let $T : \mathbb{R}^3 \to \mathbb{R}^3$ be the linear operator with associated matrix

$$A = \begin{bmatrix} 0 & -1 & 1 \\ 1 & 0 & 2 \\ 0 & 0 & 2 \end{bmatrix}$$

with respect to the canonical basis of $\mathbb{R}^3$. The characteristic polynomial is

$$p(\lambda) = \begin{vmatrix} -\lambda & -1 & 1 \\ 1 & -\lambda & 2 \\ 0 & 0 & 2-\lambda \end{vmatrix} = (2 - \lambda)(\lambda^2 + 1)$$

having three distinct roots $\lambda_1 = 2$, $\lambda_2 = i$, $\lambda_3 = -i$. In the light of our definition, we say that $\lambda_1 = 2$ is eigenvalue of T, but $\lambda_2 = i$, $\lambda_3 = -i$ are not, since $i, -i \notin \mathbb{R}$.

Example 6.4 Let now $T : \mathbb{C}^3 \to \mathbb{C}^3$ be the linear operator with associated matrix

$$A = \begin{bmatrix} 0 & -1 & 1 \\ 1 & 0 & 2 \\ 0 & 0 & 2 \end{bmatrix}$$

with respect to the canonical basis of $\mathbb{C}^3$. We repeat the same computations as above and find that $\lambda_1 = 2$, $\lambda_2 = i$ and $\lambda_3 = -i$ are the eigenvalues of T. For $\lambda_1 = 2$, we have

$$A - 2I = \begin{bmatrix} -2 & -1 & 1 \\ 1 & -2 & 2 \\ 0 & 0 & 0 \end{bmatrix}$$

which has rank equal to 2. Thus, the associated homogeneous linear system

$$-2x_1 - x_2 + x_3 = 0$$
$$x_1 - 2x_2 + 2x_3 = 0$$

has solutions $X = (0, \alpha, \alpha)$, for any $\alpha \in \mathbb{R}$. Hence, the eigenspace corresponding to λ_1 has dimension 1 and is generated by the eigenvector $(0, 1, 1)$.

For $\lambda_2 = i$, we have

$$A - iI = \begin{bmatrix} -i & -1 & 1 \\ 1 & -i & 2 \\ 0 & 0 & 2-i \end{bmatrix}$$

having rank equal to 2, so that the associated homogeneous linear system

$$-ix_1 - x_2 + x_3 = 0$$
$$x_1 - ix_2 + 2x_3 = 0$$
$$(2 - i)x_3 = 0$$

has solutions $X = (i\alpha, \alpha, 0)$, for any $\alpha \in \mathbb{R}$. In this case, the eigenspace corresponding to λ_2 is generated by the eigenvector $(i, 1, 0)$.

Finally, for $\lambda_3 = -i$,

$$A + iI = \begin{bmatrix} i & -1 & 1 \\ 1 & i & 2 \\ 0 & 0 & 2+i \end{bmatrix}$$

has rank equal to 2 and the associated homogeneous linear system has solutions $X = (-i\alpha, \alpha, 0)$, for any $\alpha \in \mathbb{R}$. The eigenspace corresponding to λ_3 is generated by the eigenvector $(-i, 1, 0)$.

Lemma 6.5 *Let $A \in M_n(\mathbb{F})$. If λ is an eigenvalue of A, then λ^m is an eigenvalue of A^m, for any $m \geq 2$. Moreover, any eigenvector of A associated with λ is an eigenvector of A^m associated with λ^m.*

Proof It follows directly from the definition of eigenvalue and associated eigenvector. In fact, multiplying on the left by A in the matrix equation $AX = \lambda X$, we get $A^2 X = \lambda A X = \lambda^2 X$, that is A^2 admits the eigenvalue λ^2, moreover X is an eigenvector associated with λ^2. In a similar way $A^3 X = \lambda^3 X$, and continuing this process one has that $A^m X = \lambda^m X$.

Lemma 6.6 *Let $A \in M_n(\mathbb{C})$. If λ_0 is an eigenvalue of A, then the conjugate element $\overline{\lambda_0}$ is an eigenvalue of the adjoint A^*.*

Proof It is known that, for any matrix $B \in M_n(\mathbb{C})$, $|B^*| = \overline{|B|}$. Thus

$$\overline{|A - \lambda I|} = |(A - \lambda I)^*| = |(A^* - \overline{\lambda} I|$$

so that $|(A^* - \overline{\lambda_0} I| = 0$ follows from $|A - \lambda_0 I| = 0$.

Lemma 6.7 *Let $A \in M_n(\mathbb{F})$. If λ_0 is an eigenvalue of A, then λ_0 is an eigenvalue of the transpose A^T.*

Proof It follows from

$$|A - \lambda I| = |(A - \lambda I)^T| = |A^T - \lambda I|.$$

Lemma 6.8 *Let $T : V \to V$ be a linear operator on a finite dimensional vector space V. Let $A \in M_n(\mathbb{F})$ be the matrix of T with respect to a basis of V, λ_0 an eigenvalue of A and $X \in V$ an eigenvector of T corresponding to λ_0. Then, for any $\alpha_0 \in \mathbb{F}$, the scalar $\lambda_0 + \alpha_0$ is an eigenvalue of the matrix $A + \alpha_0 I$. Moreover, X is eigenvector of $T + \alpha_0 I$ corresponding to the eigenvalue $\lambda_0 + \alpha_0$.*

Proof Let $X \in V$ be an eigenvector of A associated with λ_0. One can easily see that

$$(A + \alpha_0 I)X = AX + \alpha_0 X = \lambda_0 X + \alpha_0 X = (\lambda_0 + \alpha_0)X$$

as required.

Example 6.9 Let $T : \mathbb{C}^3 \to \mathbb{C}^3$ be the linear operator as in Example 6.4 and introduce the following operator $G : \mathbb{C}^3 \to \mathbb{C}^3$ defined as $G(X) = T(X) + 3X$, for any $X \in \mathbb{C}^3$. Thus, the matrix of G with respect to the canonical basis of $\mathbb{C}^3$ is

$$A = \begin{bmatrix} 3 & -1 & 1 \\ 1 & 3 & 2 \\ 0 & 0 & 5 \end{bmatrix}.$$

The characteristic polynomial is

$$p(\lambda) = \begin{vmatrix} 3 - \lambda & -1 & 1 \\ 1 & 3 - \lambda & 2 \\ 0 & 0 & 5 - \lambda \end{vmatrix} = (5 - \lambda)\big((3 - \lambda)^2 + 1\big)$$

having three distinct roots $\lambda_1 = 5$, $\lambda_2 = 3 + i$, $\lambda_3 = 3 - i$. Easy computations show that

(i)   The eigenspace corresponding to $\lambda_1 = 5$ has dimension 1 and is generated by the eigenvector $(0, 1, 1)$.

(ii)   The eigenspace corresponding to $\lambda_2 = 3 + i$ has dimension 1 and is generated by the eigenvector $(i, 1, 0)$.

(iii)   The eigenspace corresponding to $\lambda_3 = 3 - i$ has dimension 1 and is generated by the eigenvector $(-i, 1, 0)$.

Theorem 6.10 *Let $A \in M_n(\mathbb{F})$. Suppose λ_1, λ_2 are distinct eigenvalues of A and X_1 and X_2 are nonzero eigenvectors corresponding to λ_1 and λ_2, respectively. Then $\{X_1, X_2\}$ is linearly independent.*

Proof On the contrary we assume that there exists a nonzero $\alpha \in \mathbb{F}$, such that $X_1 = \alpha X_2$. By the hypothesis, we have that $AX_1 = \lambda_1 X_1$ and $AX_2 = \lambda_2 X_2$. The facts that $AX_1 = \lambda_1 X_1$ and $X_1 = \alpha X_2$ imply that $\alpha AX_2 = \lambda_1 \alpha X_2$, that is, $\alpha \lambda_2 X_2 = \lambda_1 \alpha X_2$. Hence $\alpha(\lambda_2 - \lambda_1)X_2 = 0$, which is a contradiction.

Remark 6.11 The above result implies easily that if $\{\lambda_1, \ldots, \lambda_t\}$ are all the distinct eigenvalues of A and $X_1, X_2, \ldots, X_t$ are nonzero eigenvectors corresponding to $\lambda_1, \lambda_2, \ldots, \lambda_t$, respectively, then $\{X_1, \ldots, X_t\}$ is linearly independent. Moreover, if $W_1, \ldots, W_t$ are the eigenspaces associated with $\lambda_1, \ldots, \lambda_t$, respectively, the subspace $W = W_1 + \cdots + W_t$ is a direct sum and we write $W = W_1 \oplus \cdots \oplus W_t$.

Example 6.12 Let $T : \mathbb{R}^3 \to \mathbb{R}^3$ be the linear operator having matrix

$$A = \begin{bmatrix} 2 & 0 & 4 \\ 0 & 4 & 1 \\ 1 & 0 & 2 \end{bmatrix}$$

with respect to the canonical basis of $\mathbb{R}^3$. The characteristic polynomial is

$$p(\lambda) = -\lambda(4 - \lambda)^2$$

having two distinct roots $\lambda_1 = 0$, $\lambda_2 = 4$. The eigenspace W_1 corresponding to $\lambda_1 = 0$ has dimension 1 and is generated by the eigenvector $(8, -1, 4)$. The eigenspace W_2 corresponding to $\lambda_2 = 4$ has dimension 1 and is generated by the eigenvector $(0, 1, 0)$. It is easy to see that $W_1 \cap W_2 = \{0\}$, so that $W_1 \oplus W_2$ is a direct sum.

Remark 6.13 Let $A \in M_n(\mathbb{F})$ be an upper-triangular matrix. Then, the eigenvalues of A consist precisely of the entries on the main diagonal of A.

In fact, if

$$A = \begin{bmatrix} a_{11} & a_{12} & \ldots & a_{1n} \\ & a_{22} & \ldots & a_{2n} \\ & & \ddots & \\ & & & a_{nn} \end{bmatrix}$$

then

$$A - \lambda I = \begin{bmatrix} a_{11} - \lambda & a_{12} & \ldots & a_{1n} \\ & a_{22} - \lambda & \ldots & a_{2n} \\ & & \ddots & \\ & & & a_{nn} - \lambda \end{bmatrix}$$

and $|A - \lambda I| = (a_{11} - \lambda)(a_{22} - \lambda) \cdots (a_{nn} - \lambda)$, whose roots are precisely

$$\lambda_1 = a_{11}, \lambda_2 = a_{22}, \ldots, \lambda_n = a_{nn}.$$

Theorem 6.14 *Let A, $B \in M_n(\mathbb{F})$ be similar matrices, that is, there exists an invertible matrix $P \in M_n(\mathbb{F})$ such that $B = P^{-1}AP$. Then A and B have the same eigenvalues. Moreover, if Y is an eigenvector of B associated with the eigenvalue λ, then $X = PY$ is an eigenvector of A associated with λ.*

Proof Let λ be an eigenvalue of A and X one of its associated eigenvectors. Since $A = PBP^{-1}$, it follows that $(PBP^{-1})X = \lambda X$, that is, $B(P^{-1}X) = \lambda(P^{-1}X)$, as required.

Example 6.15 Let $T : \mathbb{R}^2 \to \mathbb{R}^2$ be the linear operator having the following matrix, with respect to the canonical basis B of $\mathbb{R}^2$,

$$A = \begin{bmatrix} 1 & 1 \\ -2 & 4 \end{bmatrix}.$$

Let $B' = \{(1, 0), (1, 1)\}$ a basis for $\mathbb{R}^2$. The matrix of T with respect to B' is $P^{-1}AP = A'$, where

$$P = \begin{bmatrix} 1 & 1 \\ 0 & 1 \end{bmatrix}$$

is the transition matrix having vectors of B' in the columns. By computations, we have that

$$A' = P^{-1}AP = \begin{bmatrix} 1 & -1 \\ 0 & 1 \end{bmatrix}\begin{bmatrix} 1 & 1 \\ -2 & 4 \end{bmatrix}\begin{bmatrix} 1 & 1 \\ 0 & 1 \end{bmatrix} = \begin{bmatrix} 3 & 0 \\ -2 & 2 \end{bmatrix}.$$

The matrices A, A' have the same eigenvalues $\lambda_1 = 3$, $\lambda_2 = 2$. The eigenspace of A corresponding to $\lambda_1 = 3$ is generated by $X_1 = (1, 2)^t$; the eigenspace of A corresponding to $\lambda_2 = 2$ is generated by $X_2 = (1, 1)^t$.

Moreover, the eigenspace of A' corresponding to $\lambda_1 = 3$ is generated by the coordinates vector $Y_1 = (-1, 2)^t$ in terms of the basis B'; the eigenspace of A' corresponding to $\lambda_2 = 2$ is generated by the coordinates vector $Y_2 = (0, 1)^t$ in terms of the basis B'. It is easy to see that $X_1 = PY_1$ and $X_2 = PY_2$.

Now, we define algebraic and geometric multiplicity of an eigenvalue as follows:

Definition 6.16 Let $A \in M_n(\mathbb{F})$ and $p(\lambda)$ its characteristic polynomial. Assume that $\mathbb{F}$ contains the splitting field of $p(\lambda)$ and let $S = \{\lambda_1, \ldots, \lambda_t\}$ be the set of all distinct roots of $p(\lambda)$. Hence, we get the following decomposition

$$p(\lambda) = (\lambda_1 - \lambda)^{a_1}(\lambda_2 - \lambda)^{a_2} \cdots (\lambda_t - \lambda)^{a_t}$$

where $\sum_{i=1}^{t} a_i = n$. For any eigenvalue $\lambda_k \in S$, we say that

(*i*) a_k is its *algebraic multiplicity*, i.e., it is the number of times λ_k occurs as a root of $p(\lambda)$.

(*ii*) The dimension of the eigenspace associated with λ_k is the *geometric multiplicity* of λ_k.

The algebraic multiplicity and geometric multiplicity of an eigenvalue can differ. However, the geometric multiplicity can never exceed the algebraic multiplicity, as proved in the following:

Theorem 6.17 *Let V be a vector space of dimension n and $T : V \rightarrow V$ the linear transformation from V to itself, B a basis of V and $A \in M_n(\mathbb{F})$ the matrix of T with respect to the basis B. Assume that λ_0 is an eigenvalue of A having algebraic multiplicity equal to a and geometric multiplicity equal to g. Then $g \leq a$.*

Proof Let $p(\lambda)$ be the characteristic polynomial of A and V_0 the eigenspace of A associated with the eigenvalue λ_0. By the hypothesis $dim(V_0) = g$ and let $\{X_1, \ldots, X_g\}$ be a basis for V_0, where each X_i is an eigenvector corresponding to λ_0.

Extending $\{X_1, \ldots, X_g\}$ to a basis $B' = \{X_1, \ldots, X_g, Y_1, \ldots, Y_{n-g}\}$ of V, we can now determine the matrix A' of T with respect to the basis B'. As with any linear transformation, the columns of A' are the coordinate vectors of the images $T(X_1), \ldots, T(Y_{n-g})$ in terms of the basis B'. Since $T(X_i) = \lambda_0 X_i$, for any $i = 1, \ldots, g$, we have that

$$A' = \begin{bmatrix} D & A_1 \\ 0_{n-g,g} & A_2 \end{bmatrix}$$

where $0_{n-g,g} \in M_{n-g,g}(\mathbb{F})$, $A_1 \in M_{g,n-g}(\mathbb{F})$, $A_2 \in M_{n-g}(\mathbb{F})$ and $D \in M_g(\mathbb{F})$ is a diagonal block defined as:

$$D = \begin{bmatrix} \lambda_0 & & & \\ & \lambda_0 & & \\ & & \ddots & \\ & & & \lambda_0 \end{bmatrix}.$$

Since A and A' are similar matrices, they have the same eigenvalues. In particular, A and A' have the same characteristic polynomial. On the other hand, the characteristic polynomial of A' is $p(\lambda) = (\lambda - \lambda_0)^g q(\lambda)$, where $q(\lambda)$ is the characteristic polynomial of the matrix A_2. Therefore λ_0 occurs, as a root of $p(\lambda)$, at least g-times. As a consequence, the algebraic multiplicity of λ_0 is $a \geq g$.

Example 6.18 Let $T : \mathbb{R}^4 \rightarrow \mathbb{R}^4$ be the linear operator having matrix

$$A = \begin{bmatrix} 2 & 1 & 0 & 0 \\ -1 & 0 & 1 & 0 \\ 1 & 3 & 1 & 0 \\ 0 & 0 & 0 & -1 \end{bmatrix}$$

with respect to the canonical basis of $\mathbb{R}^4$. The characteristic polynomial is $p(\lambda) = (-1 - \lambda)^2(2 - \lambda)^2$, having two distinct roots $\lambda_1 = 2$ (with algebraic multiplicity equal to 2) and $\lambda_2 = -1$ (with algebraic multiplicity equal to 2).

For $\lambda_1 = 2$, we have

$$A - 2I = \begin{bmatrix} 0 & 1 & 0 & 0 \\ -1 & -2 & 1 & 0 \\ 1 & 3 & -1 & 0 \\ 0 & 0 & 0 & -3 \end{bmatrix}$$

having rank equal to 3, so that the associated homogeneous linear system

$$\begin{aligned} x_2 &= 0 \\ -x_1 - 2x_2 + x_3 &= 0 \\ x_1 + 3x_2 - x_3 &= 0 \\ -3x_4 &= 0 \end{aligned}$$

has solutions $X = (\alpha, 0, \alpha, 0)$, for any $\alpha \in \mathbb{R}$. The eigenspace corresponding to λ_1 is generated by the eigenvector $(1, 0, 1, 0)$ and the geometric multiplicity of λ_1 is equal to 1.

For $\lambda_2 = -1$, we have

$$A + I = \begin{bmatrix} 3 & 1 & 0 & 0 \\ -1 & 1 & 1 & 0 \\ 1 & 3 & 2 & 0 \\ 0 & 0 & 0 & 0 \end{bmatrix}$$

having rank equal to 2, so that the associated homogeneous linear system

$$\begin{aligned} 3x_1 + x_2 &= 0 \\ -x_1 + x_2 + x_3 &= 0 \\ x_1 + 3x_2 + 2x_3 &= 0 \end{aligned}$$

has solutions $X = (\alpha, -3\alpha, 4\alpha, \beta)$, for any $\alpha, \beta \in \mathbb{R}$. In this case, the eigenspace corresponding to λ_2 is generated by the eigenvectors $\{(1, -3, 4, 0), (0, 0, 0, 1)\}$ and the geometric multiplicity of λ_2 is equal to 2.

Exercises

1. Let $\mathbb{F}$ be a field and $A \in M_n(\mathbb{F})$ be such that its characteristic polynomial is

$$p(\lambda) = (-1)^n \lambda^n + \alpha_{n-1} \lambda^{n-1} + \cdots + \alpha_1 \lambda + \alpha_0, \qquad \alpha_i \in \mathbb{F},$$

where $i = 0, 1, \ldots, n - 1$. Prove that

(a) $\alpha_0 = |A|$.

(b) $\alpha_{n-1} = (-1)^{n-1} tr(A)$.

2. Let $T : V \to V$ be a nonsingular linear operator on a n-dimensional vector space V, A the matrix of T with respect to a basis for V. Let λ be an eigenvalue of T and $X \in V$ an eigenvector of T corresponding to λ. Prove that $\lambda \neq 0$ (trivial) and λ^{-1} is an eigenvalue of A^{-1} and X is an eigenvector of T^{-1} corresponding to λ^{-1}.

3. Let $T : V \to V$ be a linear operator on a n-dimensional vector space V and let $A, B \in M_n(\mathbb{F})$ two matrices of T with respect two different bases for V (i.e., A and B are similar). Let $\lambda \in \mathbb{F}$ be an eigenvalue of A and B. Prove that:

(a) The geometric multiplicity of λ as eigenvalue of A coincides with its geometric multiplicity as eigenvalue of B.

(b) The trace, the determinant and the rank of A are, respectively, equal to the trace, the determinant and the rank of B.

4. Let $T : \mathbb{R}^3 \to \mathbb{R}^3$ be the linear operator with associated matrix

$$A = \begin{bmatrix} 0 & 1 & 0 \\ 0 & 0 & 1 \\ 9 & 9 & 1 \end{bmatrix}.$$

Determine eigenvalues, eigenvectors and a basis for each eigenspace of T.

5. Redo Exercise 4, in the case the linear operator $T : \mathbb{R}^3 \to \mathbb{R}^3$ is represented by the following matrix

$$A = \begin{bmatrix} 0 & 0 & 0 \\ 1 & 0 & -1 \\ 3 & 1 & 2 \end{bmatrix}.$$

6. Redo Exercise 4, in the case the linear operator $T : \mathbb{C}^4 \to \mathbb{C}^4$ is represented by the following matrix

$$A = \begin{bmatrix} 1 & -2 & 0 & 0 \\ 2 & 1 & 0 & 0 \\ 0 & 0 & 2 & -2 \\ 0 & 0 & 3 & -2 \end{bmatrix}.$$

7. Let $T : \mathbb{R}^3 \to \mathbb{R}^3$ be the linear operator with associated matrix

$$A = \begin{bmatrix} -1 & 7 & 4 \\ 3 & -3 & 4 \\ 2 & 1 & 4 \end{bmatrix}.$$

Determine $det(e^A)$ and $trace(e^A)$.

8. Let A be a 3×3 matrix with real entries such that det $(A) = 6$ and the trace of A is 0. If det $(A + I) = 0$, where I denotes the 3×3 identity matrix, then determine eigen values of A.

9. Let J denotes a 101×101 matrix with all the entries equal to 1 and let I denotes the identity matrix of order 101. Then find det $(J - I)$ and trace $(J - I)$.

10. Let A be a 4×4 matrix with real entries such that $-1, 1, 2, -2$ are its eigen values. If $B = A^4 - 5A^2 + 5I$, then find:

   (a) det $(A + B)$
   (b) det (B)
   (c) trace $(A - B)$
   (d) trace $(A + B)$.

11. Show that all eigen values of a real skew-symmetric orthogonal matrix are of unit modulus.

12. Let α, β be two distinct eigenvalues of a square matrix A, and W_1, W_2 be the corresponding eigenspaces associated with α, β, respectively. Then show that $W_1 \cap W_2 = \{0\}$.

6.2 Triangularizable Operators

Definition 6.19 Let V be a vector space of dimension n. A linear operator $T : V \rightarrow V$ is said to be *triangularizable* if there exists a basis B of V such that the matrix A of T with respect to B is upper triangular, i.e.,

$$A = \begin{bmatrix} \lambda_1 & a_{12} & \cdots & a_{1n} \\ & \lambda_2 & \cdots & a_{2n} \\ & & \ddots & \\ & & & \lambda_n \end{bmatrix}$$

where $\lambda_1, \ldots, \lambda_n \in \mathbb{F}$ are precisely the eigenvalues of T (of A), see Remark 6.13. Analogously, we say that a matrix is triangularizable if it is similar to an upper triangular matrix.

Theorem 6.20 (Schur's Theorem) *Let V be an inner product space of dimension n over a field $\mathbb{F}$, where $\mathbb{F} = \mathbb{R}$ or $\mathbb{C}$ and $T : V \rightarrow V$ a linear transformation from V to itself. The characteristic polynomial $p(\lambda)$ of T splits over $\mathbb{F}$, that is,*

$$p(\lambda) = (\lambda_1 - \lambda)^{a_1} (\lambda_2 - \lambda)^{a_2} \cdots (\lambda_t - \lambda)^{a_t},$$

where $\sum_{i=1}^{t} a_i = n$ and any $\lambda_i \in \mathbb{F}$, if and only if T has an upper-triangular matrix with respect to some orthonormal basis of V.

Proof Let $A \in M_n(\mathbb{F})$ be the matrix of T. The fact that T has an upper-triangular matrix with respect to some orthonormal basis of V is equivalent to saying that A is unitarily similar to an upper triangular matrix A', namely $H^{-1}AH = A'$, where H is an unitary matrix.

Firstly, we assume that the characteristic polynomial $p(\lambda)$ of T splits over $\mathbb{F}$ and prove that T is unitary triangularizable. We prove the result by induction on the dimension of V. Of course, it is trivial in the case $dim(V) = 1$. Thus, we assume that the result holds for any inner product space of dimension less than n. Let $\lambda_0 \in \mathbb{F}$ be an eigenvalue of T and X_1 an eigenvector corresponding to λ_0 having unit norm, that is $\|X_1\| = 1$. Extending $\{X_1\}$ to an orthonormal basis $B = \{X_1, \ldots, X_n\}$ of V, we may compute the matrix A' of T with respect to B. If we denote by P the transition matrix, then it follows that

$$A' = P^{-1}AP,$$

where $P = [X_1, \ldots, X_n]$. On the other hand, since X_1 is eigenvector associated with λ_0, A' is a diagonal block defined as

$$A' = \begin{bmatrix} \lambda_0 & u^t \\ 0_{n-1,1} & A_1 \end{bmatrix}$$

for $0_{n-1,1} = [\ \underbrace{0, \ldots, 0}_{(n-1)-times}\ ]^t$, $u \in \mathbb{F}^{n-1}$ and $A_1 \in M_{n-1}(\mathbb{F})$. Since A and A' have the same characteristic polynomial, a fortiori any eigenvalue of A_1 is an eigenvalue of A. Therefore, the characteristic polynomial of A_1 splits over $\mathbb{F}$. By induction hypothesis, there exists an unitary matrix $Q \in M_{n-1}(\mathbb{F})$ such that $Q^{-1}A_1Q = A_2$ is an upper triangular matrix. Hence, if

$$R = \begin{bmatrix} 1 & 0_{1,n-1} \\ 0_{n-1,1} & Q \end{bmatrix}$$

for $0_{1,n-1} = [\ \underbrace{0, \ldots, 0}_{(n-1)-times}\ ]$, then R is invertible and

$$R^{-1}A'R = \begin{bmatrix} 1 & 0_{1,n-1} \\ 0_{n-1,1} & Q^{-1} \end{bmatrix} \begin{bmatrix} \lambda_0 & u^t \\ 0_{n-1,1} & A_1 \end{bmatrix} \begin{bmatrix} 1 & 0_{1,n-1} \\ 0_{n-1,1} & Q \end{bmatrix}$$

$$= \begin{bmatrix} \lambda_0 & u^t \\ 0_{n-1,1} & A_2 \end{bmatrix}$$

is upper triangular. Moreover,

$$R^{-1}A'R = R^{-1}P^{-1}APR = (PR)^{-1}A(PR)$$

where both P and R are unitary matrices. Thus, the product PR is an unitary matrix, and its columns are the coordinate vectors of an unitary basis with respect to which the linear operator T is represented by the upper triangular matrix

$$\begin{bmatrix} \lambda_0 & u^T \\ 0_{n-1,1} & A_2 \end{bmatrix}.$$

Conversely, assume that T has an upper-triangular matrix with respect to some orthonormal basis of V, that is, there exists an unitary matrix $U \in M_n(\mathbb{F})$ such that $U^{-1}AU = A'$ is an upper triangular matrix having entries from $\mathbb{F}$. By the fact that the eigenvalues of A' consist precisely of the entries on the main diagonal of A' (see Remark 6.13) and since A, A' have the same eigenvalues (they are similar to each other), we obtain the required conclusion.

Corollary 6.21 *Let $T : V \to V$ be a linear operator on a finite dimensional complex vector space V. There is an unitary basis B of V such that the matrix of T with respect to B is upper-triangular. This is equivalent to saying that every complex matrix is unitarily similar to an upper triangular matrix.*

Example 6.22 Let $T : \mathbb{R}^4 \to \mathbb{R}^4$ be the linear operator having matrix

$$A = \begin{bmatrix} 0 & 1 & 1 & 1 \\ -2 & 1 & 0 & 2 \\ -1 & 0 & 2 & 1 \\ -2 & 1 & 1 & 3 \end{bmatrix}$$

with respect to the canonical basis of $\mathbb{R}^4$. The characteristic polynomial is $p(\lambda) = (2 - \lambda)^2(1 - \lambda)^2$ having two distinct roots $\lambda_1 = 1$ and $\lambda_2 = 2$.

For $\lambda_1 = 1$, we have

$$A - I = \begin{bmatrix} -1 & 1 & 1 & 1 \\ -2 & 0 & 0 & 2 \\ -1 & 0 & 1 & 1 \\ -2 & 1 & 1 & 2 \end{bmatrix}$$

so that the associated homogeneous linear system has solutions: $X = (\alpha, 0, 0, \alpha)$, for any $\alpha \in \mathbb{R}$. The corresponding eigenspace is $N_1 = \langle (1, 0, 0, 1) \rangle = \langle (\frac{1}{\sqrt{2}}, 0, 0, \frac{1}{\sqrt{2}}) \rangle$. The orthogonal complement is

$$N_1^\perp = \langle (\frac{1}{\sqrt{2}}, 0, 0, -\frac{1}{\sqrt{2}}), (0, 1, 0, 0), (0, 0, 1, 0) \rangle$$

so that, an orthonormal basis for $\mathbb{R}^4$ is

$$B_1 = \{(\frac{1}{\sqrt{2}}, 0, 0, \frac{1}{\sqrt{2}}), (\frac{1}{\sqrt{2}}, 0, 0, -\frac{1}{\sqrt{2}}), (0, 1, 0, 0), (0, 0, 1, 0)\}.$$

The vectors in B_1 are the columns of the orthonormal transition matrix

$$Q_1 = \begin{bmatrix} \frac{1}{\sqrt{2}} & \frac{1}{\sqrt{2}} & 0 & 0 \\ 0 & 0 & 1 & 0 \\ 0 & 0 & 0 & 1 \\ \frac{1}{\sqrt{2}} & -\frac{1}{\sqrt{2}} & 0 & 0 \end{bmatrix}.$$

Then, the matrix of T with respect to B_1 is

$$A_1 = Q_1^T A Q_1 = \begin{bmatrix} 1 & -3 & 1 & 1 \\ 0 & 2 & 0 & 0 \\ 0 & -\frac{4}{\sqrt{2}} & 1 & 0 \\ 0 & -\frac{2}{\sqrt{2}} & 0 & 2 \end{bmatrix}.$$

Since the method is by induction, this leads to a recursive algorithm.

At this point, we delete the first row and column of A_1 and consider the following 3×3 submatrix:

$$C_1 = \begin{bmatrix} 2 & 0 & 0 \\ -\frac{4}{\sqrt{2}} & 1 & 0 \\ -\frac{2}{\sqrt{2}} & 0 & 2 \end{bmatrix}$$

having $\mu = 1$ as eigenvalue, with corresponding unit eigenvector $(0, 1, 0) \in \mathbb{R}^3$. We extend the set $\{(0, 1, 0)\}$, by adding the vectors $(1, 0, 0)$, $(0, 0, 1)$, to obtain an orthonormal basis $B_2 = \{(0, 1, 0), (1, 0, 0), (0, 0, 1)\}$ for $\mathbb{R}^3$. We use the elements of B_2 to construct the orthonormal matrix

$$Q_2 = \begin{bmatrix} 0 & 1 & 0 \\ 1 & 0 & 0 \\ 0 & 0 & 1 \end{bmatrix}.$$

Hence we compute

$$A_2 = Q_2^T C_1 Q_2 = \begin{bmatrix} 1 & -\frac{4}{\sqrt{2}} & 0 \\ 0 & 2 & 0 \\ 0 & -\frac{2}{\sqrt{2}} & 2 \end{bmatrix}.$$

Once again, we delete the first row and column of A_2 and consider the following 2×2 submatrix:

$$C_2 = \begin{bmatrix} 2 & 0 \\ -\frac{2}{\sqrt{2}} & 2 \end{bmatrix}$$

having eigenvalue $\nu = 2$ with corresponding unit eigenvector $(0, 1) \in \mathbb{R}^2$. Starting from this eigenvector we obtain the orthonormal basis $B_3 = \{(0, 1), (1, 0)\}$ for $\mathbb{R}^2$. Using the vectors in B_3, we have the orthonormal matrix

$$Q_3 = \begin{bmatrix} 0 & 1 \\ 1 & 0 \end{bmatrix}$$

and compute

$$A_3 = Q_3^T C_2 Q_3 = \begin{bmatrix} 2 & -\frac{2}{\sqrt{2}} \\ 0 & 2 \end{bmatrix}.$$

Actually the process is finished. We have now to construct the final upper-triangular matrix which is similar to A. In other words, we have to find the orthonormal matrix $U \in M_4(\mathbb{R})$ such that $U^T A U$ is an upper-triangular matrix. This matrix is the product of 3 orthonormal matrices $U_1, U_2, U_3 \in M_4(\mathbb{R})$. Each of them is corresponding to one step of the above process. More precisely:

(i)   The first matrix is $U_1 = Q_1$.
(ii) Let U_2 be the matrix

$$\begin{bmatrix} 1 & 0 \\ 0 & Q_2 \end{bmatrix} = \begin{bmatrix} 1 & 0 & 0 & 0 \\ 0 & 0 & 1 & 0 \\ 0 & 1 & 0 & 0 \\ 0 & 0 & 0 & 1 \end{bmatrix}.$$

(iii) Analogously, we construct U_3 as

$$\begin{bmatrix} I_2 & 0 \\ 0 & Q_3 \end{bmatrix} = \begin{bmatrix} 1 & 0 & 0 & 0 \\ 0 & 1 & 0 & 0 \\ 0 & 0 & 0 & 1 \\ 0 & 0 & 1 & 0 \end{bmatrix}.$$

Thus

$$U = U_1 U_2 U_3 = \begin{bmatrix} \frac{1}{\sqrt{2}} & 0 & 0 & \frac{1}{\sqrt{2}} \\ 0 & 1 & 0 & 0 \\ 0 & 0 & 1 & 0 \\ \frac{1}{\sqrt{2}} & 0 & 0 & -\frac{1}{\sqrt{2}} \end{bmatrix}$$

and

$$U^t A U = \begin{bmatrix} 1 & \frac{2}{\sqrt{2}} & \frac{2}{\sqrt{2}} & -3 \\ 0 & 1 & 0 & -\frac{4}{\sqrt{2}} \\ 0 & 0 & 2 & -\frac{2}{\sqrt{2}} \\ 0 & 0 & 0 & 2 \end{bmatrix}.$$

Exercises

1. Let $T : \mathbb{R}^4 \to \mathbb{R}^4$ be the linear operator having matrix

$$A = \begin{bmatrix} 1 & 2 & 0 & 1 \\ 2 & 1 & 1 & 1 \\ 0 & 0 & 1 & 2 \\ 0 & 0 & 2 & 1 \end{bmatrix}$$

with respect to the canonical basis of $\mathbb{R}^4$. Determine an orthonormal basis for $\mathbb{R}^4$ with respect to which the matrix of T is upper-triangular.

2. Repeat Exercise 1 for the linear operator $T : \mathbb{R}^4 \to \mathbb{R}^4$ having matrix

$$A = \begin{bmatrix} 0 & 0 & 1 & 0 \\ 0 & 1 & 0 & 0 \\ 2 & 2 & 1 & 0 \\ 2 & 0 & 0 & 0 \end{bmatrix}$$

with respect to the canonical basis of $\mathbb{R}^4$.

3. Repeat Exercise 1 for the linear operator $T : \mathbb{R}^4 \to \mathbb{R}^4$ having matrix

$$A = \begin{bmatrix} 2 & 2 & 0 & 0 \\ 1 & -2 & 1 & 0 \\ 0 & 0 & 2 & -6 \\ 1 & 0 & 1 & -2 \end{bmatrix}$$

with respect to the canonical basis of $\mathbb{R}^4$.

4. Repeat Exercise 1 for the linear operator $T : \mathbb{C}^4 \to \mathbb{C}^4$ having matrix

$$A = \begin{bmatrix} 2 & -3 & 0 & 0 \\ 2 & -2 & 0 & 0 \\ 0 & 0 & 2 & -3 \\ 0 & 0 & 3 & -2 \end{bmatrix}$$

with respect to the canonical basis of $\mathbb{C}^4$.

5. Repeat Exercise 1 for the linear operator $T : \mathbb{C}^3 \to \mathbb{C}^3$ having matrix

$$A = \begin{bmatrix} 0 & 1 & 1 \\ -1 & 0 & 2 \\ -1 & -2 & 0 \end{bmatrix}$$

with respect to the canonical basis of $\mathbb{C}^3$.

6.3   Diagonalizable Operators

Definition 6.23 Let V be a vector space of dimension n. A linear operator $T : V \to V$ is said to be *diagonalizable* if there exists a basis B of V such that the matrix A of T with respect to B is diagonal, i.e.,

$$A = \begin{bmatrix} \lambda_1 & & & \\ & \ddots & & \\ & & \ddots & \\ & & & \lambda_n \end{bmatrix}$$

where $\lambda_1, \ldots, \lambda_n \in \mathbb{F}$ are precisely the eigenvalues of T (of A)(see Remark 6.13). Usually, B is said to be a diagonalizing basis of T. Analogously, we say that a matrix is diagonalizable if it is similar to a diagonal matrix.

Assume that V has dimension n. One may note that, in light of the definition of the matrix associated with a linear operator, the fact that $B = \{e_1, \ldots, e_n\}$ is a diagonalizing basis of V coincides with the one that $T(e_1) = \lambda_1 e_1, \ldots, T(e_n) = \lambda_n e_n$, for suitable elements $\lambda_i \in \mathbb{F}$.

Nevertheless, not all matrices could be diagonalized, in the sense that, in some cases it is not possible to find a basis of V such that the matrix of T with respect to that basis is diagonal. Here, we give an answer to the question of which matrices are similar to diagonal matrices or, analogously, which linear operators are represented by diagonal matrices. The diagonalizable linear operators are characterized by the following:

Theorem 6.24 *Let V be a vector space of dimension n and $T : V \to V$ a linear transformation from V to itself. If the characteristic polynomial $p(\lambda)$ of T splits over $\mathbb{F}$, i.e., $p(\lambda)$ breaks over $\mathbb{F}$ in linear factors, then the following conditions are equivalent:*

- (i)   *T is diagonalizable.*
- (ii) *For any eigenvalue of T, its algebraic multiplicity matches with its geometric multiplicity.*
- (iii) *There exists a basis for V that consists entirely of eigenvectors of T.*

Proof Let $A \in M_n(\mathbb{F})$ be the matrix of T. The fact that T has a diagonal matrix with respect to some basis of V is equivalent to saying that A is similar to a diagonal matrix A', namely $P^{-1}AP = A'$, where the columns of P are the coordinate vectors of a basis with respect to which the linear operator T is represented by the diagonal matrix $A.'$ Thus, we actually show that the following conditions are equivalent:

- (i) A is diagonalizable.
- (ii) For any eigenvalue of A, its algebraic multiplicity matches with its geometric multiplicity.
- (iii) There exists a basis for V that consists entirely of eigenvectors of A.

Let $\lambda_1, \ldots, \lambda_m$ be the list of the distinct eigenvalues of A. Further suppose that $a_1, a_2, \ldots, a_m$ and $g_1, \ldots, g_m$, respectively, be the algebraic and geometric multiplicities of $\lambda_1, \ldots, \lambda_m$. We recall that $a_1 + a_2 + \cdots + a_m = n$.

$(i) \Rightarrow (ii)$ Since $A \in M_n(\mathbb{F})$ is a diagonalizable matrix, there exists an invertible matrix $P \in M_n(\mathbb{F})$ such that

$$A' = P^{-1}AP = \begin{bmatrix} \lambda_1 & & & \\ & \ddots & & \\ & & \ddots & \\ & & & \lambda_m \end{bmatrix},$$

where the diagonal entries λ_i are not necessarily distinct and any eigenvalue λ_i repeatedly occurs on the main diagonal as many times as it occurs as a root of the characteristic polynomial of A'. Let λ_k be any eigenvalue of A'. Since the rank of $A' - \lambda_k I$ is equal to $n - a_k$, the dimension of the eigenspace V_k associated with λ_k is $n - (n - a_k) = a_k$, as required.

$(ii) \Rightarrow (iii)$ We now assume that, for any eigenvalue λ_k of A, its algebraic multiplicity a_k matches with its geometric multiplicity g_k. Thus, the dimension of any eigenspace V_k associated with λ_k is equal to a_k. Hence we have

$$dim V_1 + dim V_2 + \cdots + dim V_m = n = dim V$$

that is

$$V_1 \oplus V_2 \oplus \cdots \oplus V_m = V.$$

Therefore, the union of the bases of any V_k is a basis for V that consists entirely of eigenvectors of A.

$(iii) \Rightarrow (i)$ We finally assume that $B = \{X_1, \ldots, X_n\}$ is a basis for V that consists entirely of eigenvectors of A. Even if the eigenvalues of A are not necessarily distinct, we can list them in the set $\{\lambda_1, \ldots, \lambda_n\}$, such that any eigenvalue λ_i repeatedly occurs as many times as it occurs as a root of the characteristic polynomial of A and

$$AX_1 = \lambda_1 X_1, \quad AX_2 = \lambda_2 X_2, \quad AX_3 = \lambda_3 X_3, \ldots, AX_n = \lambda_n X_n.$$

Here, we compute the coordinate vectors of the images $T(X_i)$ in terms of B :

$$T(X_1) = AX_1 = \lambda_1 X_1 = (\lambda_1, 0, 0, 0, \ldots, 0)_B$$

$$T(X_2) = AX_2 = \lambda_2 X_2 = (0, \lambda_2, 0, 0, \ldots, 0)_B$$

$$T(X_3) = AX_3 = \lambda_3 X_3 = (0, 0, \lambda_3, 0, \ldots, 0)_B$$

$$\cdots\cdots$$

$$T(X_n) = AX_n = \lambda_n X_n = (0, 0, 0, 0, \ldots, \lambda_n)_B.$$

If we denote by A' the matrix of T with respect to the basis B, then the column coordinate vectors of A' are precisely the images $T(X_i)$, that is,

$$A' = \begin{bmatrix} \lambda_1 & & & \\ & \ddots & & \\ & & \ddots & \\ & & & \lambda_n \end{bmatrix}.$$

Moreover, $A' = P^{-1}AP$, where $P = \begin{bmatrix} X_1 \ldots\ldots\ldots X_n \end{bmatrix}$ is the transition matrix having in any ith column the coordinates of the eigenvector X_i.

Example 6.25 Let $T : \mathbb{R}^3 \to \mathbb{R}^3$ be the linear operator having matrix

$$A = \begin{bmatrix} 2 & -3 & 3 \\ 0 & 1 & 1 \\ 0 & -2 & 4 \end{bmatrix}$$

with respect to the canonical basis of $\mathbb{R}^3$. The characteristic polynomial is $p(\lambda) = (2 - \lambda)^2(3 - \lambda)$, having two distinct roots $\lambda_1 = 2$ (with algebraic multiplicity equal to 2), $\lambda_2 = 3$ (with algebraic multiplicity equal to 1).

For $\lambda_1 = 2$, we have

$$A - 2I = \begin{bmatrix} 0 & -3 & 3 \\ 0 & -1 & 1 \\ 0 & -2 & 2 \end{bmatrix}$$

having rank equal to 1, so that the associated homogeneous linear system reduces to $x_2 - x_3 = 0$ and has solutions $X = (\alpha, \beta, \beta)$, for any $\alpha, \beta \in \mathbb{R}$. The eigenspace corresponding to λ_1 is generated by the eigenvectors $\{(1, 0, 0), (0, 1, 1)\}$ and the geometric multiplicity of λ_1 is equal to 2.

For $\lambda_2 = 3$, we have

$$A - 3I = \begin{bmatrix} -1 & -3 & 3 \\ 0 & -2 & 1 \\ 0 & -2 & 1 \end{bmatrix}$$

having rank equal to 2, so that the associated homogeneous linear system

$$-x_1 - 3x_2 + 3x_3 = 0$$
$$-2x_2 + x_3 = 0$$

has solutions $X = (3\alpha, \alpha, 2\alpha)$, for any $\alpha \in \mathbb{R}$. In this case, the eigenspace corresponding to λ_2 is generated by the eigenvector $(3, 1, 2)$ and the geometric multiplicity of λ_2 is equal to 1.

Now consider the basis $B = \{(1, 0, 0), (0, 1, 1), (3, 1, 2)\}$ for $\mathbb{R}^3$, consisting entirely of eigenvectors of T. With respect to this basis, T has a diagonal form. The matrix of T with respect to B can be computed as $A' = P^{-1}AP$, where

$$P = \begin{bmatrix} 1 & 0 & 3 \\ 0 & 1 & 1 \\ 0 & 1 & 2 \end{bmatrix}$$

is the transition matrix having vectors of B in the columns. Hence,

$$A' = \begin{bmatrix} 1 & 0 & 3 \\ 0 & 1 & 1 \\ 0 & 1 & 2 \end{bmatrix}^{-1} \begin{bmatrix} 2 & -3 & 3 \\ 0 & 1 & 1 \\ 0 & -2 & 4 \end{bmatrix} \begin{bmatrix} 1 & 0 & 3 \\ 0 & 1 & 1 \\ 0 & 1 & 2 \end{bmatrix}$$

$$= \begin{bmatrix} 1 & 3 & -3 \\ 0 & 2 & -1 \\ 0 & -1 & 1 \end{bmatrix} \begin{bmatrix} 2 & -3 & 3 \\ 0 & 1 & 1 \\ 0 & -2 & 4 \end{bmatrix} \begin{bmatrix} 1 & 0 & 3 \\ 0 & 1 & 1 \\ 0 & 1 & 2 \end{bmatrix}$$

$$= \begin{bmatrix} 2 & 0 & 0 \\ 0 & 2 & 0 \\ 0 & 0 & 3 \end{bmatrix}.$$

Example 6.26 Let $T : \mathbb{C}^4 \to \mathbb{C}^4$ be the linear operator having matrix

$$A = \begin{bmatrix} 1 & -1 & 1 & 0 \\ 1 & 1 & 1 & 1 \\ 0 & 0 & 1 & -1 \\ 0 & 0 & 1 & 1 \end{bmatrix}$$

with respect to the canonical basis of $\mathbb{C}^4$. The characteristic polynomial is $p(\lambda) = (\lambda - 1 - i)^2(\lambda - 1 + i)^2$, having two distinct roots $\lambda_1 = 1 + i$ (with algebraic multiplicity equal to 2), $\lambda_2 = 1 - i$ (with algebraic multiplicity equal to 2).

For $\lambda_1 = 1 + i$, we have

$$A - (1 + i)I = \begin{bmatrix} -i & -1 & 1 & 0 \\ 1 & -i & 1 & 1 \\ 0 & 0 & -i & -1 \\ 0 & 0 & 1 & -i \end{bmatrix}$$

having rank equal to 3, so that the associated homogeneous linear system

$$\begin{aligned}
-ix_1 - x_2 + x_3 &= 0 \\
x_1 - ix_2 + x_3 + x_4 &= 0 \\
-ix_3 - x_4 &= 0 \\
x_3 - ix_4 &= 0
\end{aligned}$$

has solutions $X = (\alpha, -i\alpha, 0, 0)$, for any $\alpha \in \mathbb{R}$. The eigenspace corresponding to λ_1 is generated by the eigenvector $(1, -i, 0, 0)$ and the geometric multiplicity of λ_1 is equal to 1. We may conclude that, in this case, T is not diagonalizable.

Example 6.27 Let $T : \mathbb{R}^5 \to \mathbb{R}^5$ be the linear operator having matrix

$$A = \begin{bmatrix}
1 & 2 & -1 & 1 & 0 \\
2 & 1 & 1 & -1 & 0 \\
0 & 0 & 2 & 1 & 0 \\
0 & 0 & 1 & 2 & 0 \\
0 & 0 & 0 & 0 & 3
\end{bmatrix}$$

with respect to the canonical basis of $\mathbb{R}^5$. The characteristic polynomial is $p(\lambda) = (3 - \lambda)^3(1 - \lambda)(-1 - \lambda)$, having three distinct roots $\lambda_1 = 3$ (with algebraic multiplicity equal to 3), $\lambda_2 = 1$ (with algebraic multiplicity equal to 1), $\lambda_3 = -1$ (with algebraic multiplicity equal to 1).

For $\lambda_1 = 3$, we have

$$A - 3I = \begin{bmatrix}
-2 & 2 & -1 & 1 & 0 \\
2 & -2 & 1 & -1 & 0 \\
0 & 0 & -1 & 1 & 0 \\
0 & 0 & 1 & -1 & 0 \\
0 & 0 & 0 & 0 & 0
\end{bmatrix}$$

having rank equal to 2, so that the associated homogeneous linear system

$$\begin{aligned}
2x_1 - 2x_2 + x_3 - x_4 &= 0 \\
x_3 - x_4 &= 0
\end{aligned}$$

has solutions $X = (\alpha, \alpha, \beta, \beta, \gamma)$, for any $\alpha, \beta, \gamma \in \mathbb{R}$.

The eigenspace corresponding to λ_1 is generated by the set of eigenvectors given by $\{(1, 1, 0, 0, 0), (0, 0, 1, 1, 0), (0, 0, 0, 0, 1)\}$ and the geometric multiplicity of λ_1 is equal to 3.

For $\lambda_2 = 1$, we have

$$A - I = \begin{bmatrix}
0 & 2 & -1 & 1 & 0 \\
2 & 0 & 1 & -1 & 0 \\
0 & 0 & 1 & 1 & 0 \\
0 & 0 & 1 & 1 & 0 \\
0 & 0 & 0 & 0 & 2
\end{bmatrix}$$

having rank equal to 4, so that the associated homogeneous linear system

$$
\begin{aligned}
2x_2 - x_3 + x_4 &= 0 \\
2x_1 + x_3 - x_4 &= 0 \\
x_3 + x_4 &= 0 \\
x_5 &= 0
\end{aligned}
$$

has solutions $X = (-\alpha, \alpha, \alpha, -\alpha, 0)$, for any $\alpha \in \mathbb{R}$. The eigenspace corresponding to λ_2 is generated by the eigenvector $(-1, 1, 1, -1, 0)$ and the geometric multiplicity of λ_2 is equal to 1. For $\lambda_3 = -1$, we have

$$
A + I = \begin{bmatrix}
2 & 2 & -1 & 1 & 0 \\
2 & 2 & 1 & -1 & 0 \\
0 & 0 & 3 & 1 & 0 \\
0 & 0 & 1 & 3 & 0 \\
0 & 0 & 0 & 0 & 4
\end{bmatrix}
$$

having rank equal to 4, so that the associated homogeneous linear system

$$
\begin{aligned}
2x_1 + 2x_2 - x_3 + x_4 &= 0 \\
2x_1 + +2x_2 + x_3 - x_4 &= 0 \\
3x_3 + x_4 &= 0 \\
x_3 + 3x_4 &= 0 \\
x_5 &= 0
\end{aligned}
$$

has solutions $X = (-\alpha, \alpha, 0, 0, 0)$, for any $\alpha \in \mathbb{R}$. The eigenspace corresponding to λ_3 is generated by the eigenvector $(-1, 1, 0, 0, 0)$ and the geometric multiplicity of λ_3 is equal to 1.

Therefore, there exists a basis B for $\mathbb{R}^5$ that consists entirely of eigenvectors of T, i.e.,

$$
B = \{(1, 1, 0, 0, 0), (0, 0, 1, 1, 0), (0, 0, 0, 0, 1), (-1, 1, 1, -1, 0), (-1, 1, 0, 0, 0)\}.
$$

The matrix A' of T with respect to B has the following diagonal form:

$$
A' = \begin{bmatrix}
3 & 0 & 0 & 0 & 0 \\
0 & 3 & 0 & 0 & 0 \\
0 & 0 & 3 & 0 & 0 \\
0 & 0 & 0 & 1 & 0 \\
0 & 0 & 0 & 0 & -1
\end{bmatrix}.
$$

Exercises

1. Let $A \in M_n(\mathbb{F})$ be a block diagonal matrix, having the form

$$A = \begin{bmatrix} A_1 & 0 \\ 0 & A_2 \end{bmatrix}.$$

   Prove that A is diagonalizable if and only if both A_1 and A_2 are diagonalizable.
2. Let $T : V \to V$ be a linear operator on a finite dimensional vector space V. Prove that if there exists $k \geq 1$ such that T^k is the identity map on V, then T is diagonalizable.
3. Let $T : \mathbb{R}^3 \to \mathbb{R}^3$ be the linear operator having matrix

$$A = \begin{bmatrix} 0 & 0 & 1 \\ 0 & 0 & -1 \\ 1 & 1 & 2 \end{bmatrix}$$

   in terms of the canonical basis for $\mathbb{R}^3$. Determine, if possible, the basis for $\mathbb{R}^3$ with respect to which the matrix of T is diagonal.
4. Redo Exercise 3 for the linear operator $T : \mathbb{R}^4 \to \mathbb{R}^4$ having matrix

$$A = \begin{bmatrix} 0 & -1 & 1 & 1 \\ -1 & 0 & 1 & 1 \\ 1 & 1 & -2 & 1 \\ 0 & 0 & 0 & 1 \end{bmatrix}.$$

5. For which values of constants $\alpha, \beta, \gamma \in \mathbb{R}$ are the matrices

$$A = \begin{bmatrix} 3 & \alpha & \beta \\ 0 & 2 & \gamma \\ 0 & 0 & 2 \end{bmatrix}, \quad B = \begin{bmatrix} 2 & \alpha & \beta \\ 0 & 2 & \gamma \\ 0 & 0 & 3 \end{bmatrix}, \quad C = \begin{bmatrix} 2 & 1 & \alpha \\ 0 & \beta & \gamma \\ 0 & 0 & 3 \end{bmatrix}$$

   diagonalizable?
6. Suppose that $A \in M_n(\mathbb{F})$ has two distinct eigenvalues λ and μ such that $dim(E_\lambda) = (n - 1)$, where E_λ is the eigenspace associated with λ. Prove that A is diagonalizable.
7. For each of the following linear operators T on a vector space V, test for diagonalizability and if T is diagonalizable find a basis B for V such that $[T]_B$ is a diagonal matrix.

   (a) $V = P_3(\mathbb{R})$ and T is defined by $T(f(x)) = f'(x) + f''(x)$.
   (b) $V = P_2(\mathbb{R})$ and T is defined by $T(ax^2 + bx + c) = cx^2 + bx + a$.
   (c) $V = \mathbb{R}^3$ and T is defined by

$$T \begin{pmatrix} a_1 \\ a_2 \\ a_3 \end{pmatrix} = \begin{pmatrix} a_2 \\ -a_1 \\ 2a_3 \end{pmatrix}.$$

(d) $V = P_2(\mathbb{R})$ and T is defined by $T(f(x)) = f(0) + f(1)(x + x^2)$.
(e) $V = \mathbb{C}^2$ and T is defined by $T(z, w) = (z + iw, iz + w)$.
(f) $V = M_2(\mathbb{R})$ and T is defined by $T(A) = A^t$.

6.4 Jordan Canonical Form of an Operator

We know that T is diagonalizable if and only if there exists a basis for V that consists entirely of eigenvectors of T. On the other hand, not all linear operators could be diagonalized. Here, we analyze what can be done in the case T is not diagonalizable. In this case and under suitable assumptions, we shall see that there exists a basis of V with respect to which the matrix of T assumes a fairly simple form, called the Jordan canonical form.

More precisely, the $p \times p$ Jordan block associated with the scalar $\alpha \in \mathbb{F}$ is defined as the following $p \times p$ matrix:

$$J_p(\alpha) = \begin{bmatrix} \alpha & 1 & & \\ & \ddots & \ddots & \\ & & \alpha & 1 \\ & & & \alpha \end{bmatrix}$$

that is, any entry a_{ij} of $J_p(\alpha)$ is equal to

$$a_{ij} = \begin{cases} \alpha & \text{if} & i = j \\ 1 & \text{if} & j = i + 1, 1 \le i \le p - 1 \\ 0 & \text{elsewhere} \end{cases}$$

Notice that the characteristic polynomial of the Jordan block $J_p(\alpha)$ is $(\alpha - \lambda)^p$. Therefore, $J_p(\alpha)$ has an unique eigenvalue, i.e., $\alpha \in \mathbb{F}$, having algebraic multiplicity p. Moreover, since

$$J_p(\alpha) - \alpha I_p = \begin{bmatrix} 0 & 1 & & \\ & \ddots & \ddots & \\ & & 0 & 1 \\ & & & 0 \end{bmatrix}$$

has rank $p - 1$, the geometric multiplicity of α is 1. This means that $J_p(\alpha)$ is not diagonalizable, unless the trivial case $p = 1$ and $J_1(\alpha) = [\alpha] \in M_1(\mathbb{F})$.
We say that $A \in M_n(\mathbb{F})$ has Jordan form if A is made up of diagonal Jordan blocks, that is

$$A = \begin{bmatrix} J_{n_1}(\alpha_1) & & & \\ & \ddots & & \\ & & \ddots & \\ & & & J_{n_r}(\alpha_r) \end{bmatrix},$$

where any $J_{n_i}(\alpha_i)$ is a $n_i \times n_i$ Jordan block associated with the scalar α_i such that $\sum_i n_i = n$. It is easy to see that $\alpha_1, \ldots, \alpha_r$ are the eigenvalues of A which are not necessarily distinct, and the characteristic polynomial of A is

$$p_A(\lambda) = (\lambda - \alpha_1)^{n_1}(\lambda - \alpha_2)^{n_2} \cdots (\lambda - \alpha_r)^{n_r}.$$

Example 6.28 The matrix

$$A_1 = \begin{bmatrix} 3 & 1 & 0 & 0 & 0 \\ 0 & 3 & 1 & 0 & 0 \\ 0 & 0 & 3 & 0 & 0 \\ 0 & 0 & 0 & 2 & 1 \\ 0 & 0 & 0 & 0 & 2 \end{bmatrix} = \begin{bmatrix} J_3(3) & 0 \\ 0 & J_2(2) \end{bmatrix}$$

has Jordan form, in fact it consists of 2 diagonal Jordan blocks associated with the eigenvalues 3 and 2, respectively.

Example 6.29 The matrix

$$A_2 = \begin{bmatrix} 2 & 1 & 0 & 0 & 0 & 0 \\ 0 & 2 & 1 & 0 & 0 & 0 \\ 0 & 0 & 2 & 0 & 0 & 0 \\ 0 & 0 & 0 & 4 & 1 & 0 \\ 0 & 0 & 0 & 0 & 4 & 0 \\ 0 & 0 & 0 & 0 & 0 & 3 \end{bmatrix} = \begin{bmatrix} J_3(2) & 0 & 0 \\ 0 & J_2(4) & 0 \\ 0 & 0 & J_1(3) \end{bmatrix}$$

has Jordan form, in fact, it consists of 3 diagonal Jordan blocks associated with the eigenvalues 2, 4 and 3, respectively.

Example 6.30 The matrix

$$A_3 = \begin{bmatrix} 3 & 0 & 0 & 0 \\ 0 & 2 & 0 & 0 \\ 0 & 0 & 4 & 0 \\ 0 & 0 & 0 & 5 \end{bmatrix} = \begin{bmatrix} J_1(3) & 0 & 0 & 0 \\ 0 & J_1(2) & 0 & 0 \\ 0 & 0 & J_1(4) & 0 \\ 0 & 0 & 0 & J_1(5) \end{bmatrix}$$

has Jordan form, in fact, it consists of 4 diagonal 1×1 Jordan blocks associated with the eigenvalues 3, 2, 4 and 5, respectively.

The operator T is said to be *Jordanizable* if there exists a basis of V with respect to which the matrix of T has a Jordan form. In other words, the matrix $A \in M_n(\mathbb{F})$ is said to be *Jordanizable* if it is similar to a matrix having Jordan form. Example 6.30 shows that the diagonalizable matrices (operators) represent a special case of the more general class of matrices which are similar to some Jordan canonical form.

Definition 6.31 Let W be a subspace of the vector space V and $T : V \rightarrow V$ be a linear operator. We say that W is *invariant* under T (*T-invariant*) if $T(W) \subset W$. In this case, if $T_{|W}$ denotes the restriction of T to the smaller domain W, then $T_{|W} : W \rightarrow W$ is a linear operator on W.

Remark 6.32 Any eigenspace of a linear operator $T : V \rightarrow V$ is a T-invariant subspace of V.

Remark 6.33 It is easy to see that, if $T : V \rightarrow V$ is a linear operator and W_1, W_2 are T-invariant subspaces of V, then both $W_1 + W_2$ and $W_1 \cap W_2$ are T-invariants.

Remark 6.34 Let $T : V \rightarrow V$ be a linear operator, $I : V \rightarrow V$ the identity map on V, W a T-invariant subspace of V and $\lambda \in \mathbb{F}$. Then W is $(T - \lambda I)$-invariant.

Lemma 6.35 *Let V be a vector space of dimension n, $T : V \rightarrow V$ a linear operator and U, W are T-invariant subspaces of V such that $V = U \oplus W$. If we denote by $p_T(\lambda)$, $p_{T_{|U}}(\lambda)$ and $p_{T_{|W}}(\lambda)$ the characteristic polynomials of T, $T_{|U}$ and $T_{|W}$, respectively, the following condition holds:*

$$p_T(\lambda) = p_{T_{|U}}(\lambda) p_{T_{|W}}(\lambda).$$

Proof Let $B_U = \{u_1, \ldots, u_k\}$ and $B_W = \{w_1, \ldots, w_{n-k}\}$ be bases of U and W, respectively. Thus $B_U \cup B_W = B$ is a basis of V. Since $T(U) \subset U$ and $T(W) \subset W$, the matrix A of T with respect to B has the following block diagonal form

$$A = \begin{bmatrix} A_U & 0_{k,n-k} \\ 0_{n-k,k} & A_W \end{bmatrix},$$

where $0_{k,n-k}$ is the zero matrix in $M_{k,n-k}(\mathbb{F})$, $0_{n-k,k}$ is the zero matrix in $M_{n-k,k}(\mathbb{F})$, A_U is the matrix of the restriction $T_{|U}$ with respect to the basis B_U, A_W is the matrix of the restriction $T_{|W}$ with respect to the basis B_W. Therefore,

$$p_T(\lambda) = |A - \lambda I_n| = |A_U - \lambda I_k||A_W - \lambda I_{n-k}| = p_{T_{|U}}(\lambda) p_{T_{|W}}(\lambda)$$

as required.

Definition 6.36 Let V be a vector space of dimension n. A linear operator $T : V \rightarrow V$ is called *nilpotent* if there exists a positive integer k such that T^k is zero, i.e., $T^k(v) = 0$, for any $v \in V$. The smallest such integer k is called the *index of nilpotency* of T. In other words, k is the index of nilpotency if $T^h \neq 0$ for any positive integer $h < k$. Analogously, a matrix $A \in M_n(\mathbb{F})$ is called *nilpotent* if there exists a positive integer k such that $A^k = 0$. The smallest such k is called the *index of nilpotency* of A, in the sense that $A^h \neq 0$ for any positive integer $h < k$. Of course, the linear operator T is nilpotent if and only if its matrix A is nilpotent.

Lemma 6.37 *Let V be a vector space of dimension n, $T : V \to V$ a linear operator and $A \in M_n(\mathbb{F})$ the matrix of T. If A is nilpotent, then its characteristic polynomial is $p(\lambda) = (-1)^n \lambda^n$.*

Proof Let $k \geq 1$ be the smallest integer such that $A^k = 0$. Suppose that $0 \neq \lambda \in \mathbb{F}$ is an eigenvalue of A. Then there exists an eigenvector $0 \neq v \in V$ such that $Av = \lambda v$. By Lemma 6.5, it follows that $A^k v = \lambda^k v \neq 0$, which contradicts the nilpotency of A. Therefore $\lambda = 0$ is the only eigenvalue of A so that its characteristic polynomial is precisely $p(\lambda) = (-1)^n \lambda^n$.

Definition 6.38 A *generalized eigenvector* corresponding to an eigenvalue λ of a linear operator $T : V \to V$ is a nonzero vector $v \in V$ such that $(T - \lambda I_V)^k v = 0$, for some integer $k \geq 1$. The *exponent* of the generalized eigenvector v is the smallest integer $h \geq 1$ such that $(T - \lambda I_V)^h v = 0$.

Let now $A \in M_n(\mathbb{F})$ be the matrix of $T : V \to V$, and λ an eigenvalue of T. Denote $(T - \lambda I)^0 = I, (T - \lambda I)^1 = T - \lambda I, (T - \lambda I)^2 = (T - \lambda I)(T - \lambda I), \ldots, (T - \lambda I)^h = (T - \lambda I)(T - \lambda I)^{h-1}$ and consider both kernel and image of any $(T - \lambda I)^h : V \to V$, respectively:

$$N_{h,\lambda} = N\big((T - \lambda I)^h\big) = \{v \in V \mid (T - \lambda I)^h(v) = 0\},$$

$$R_{h,\lambda} = R\big((T - \lambda I)^h\big) = \{(T - \lambda I)^h(v) \mid v \in V\}.$$

It is easy to see that:

$$(0) = N_{0,\lambda} \subset N_{1,\lambda} \subset N_{2,\lambda} \subset \ldots \ldots \subset N_{s,\lambda} \subset \ldots$$

is a chain of subspaces of V. Since V is finite dimensional, this chain cannot be infinite. In a similar way, the chain

$$V = R_{0,\lambda} \supset R_{1,\lambda} \supset R_{2,\lambda} \supset \ldots \ldots \supset R_{s,\lambda} \supset \ldots$$

cannot be infinite. Since both chains terminate after finitely many steps, we can find a minimum integer $m \geq 1$ such that

$$N_{m,\lambda} = N_{m+1,\lambda} = N_{m+2,\lambda} = \cdots \cdots = N_{t,\lambda} \qquad \text{for all } t \geq m$$

and

$$R_{m,\lambda} = R_{m+1,\lambda} = R_{m+2,\lambda} = \cdots \cdots = R_{t,\lambda} \qquad \text{for all } t \geq m.$$

Notice that if $v \in N_{m,\lambda}$ then $0 = (T - \lambda I)^m(v) = (T - \lambda I)^{m-1}(T - \lambda I)(v)$, which means that $(T - \lambda I)(v) \in N_{m-1,\lambda} \subset N_{m,\lambda}$. Since $v \in N_{m,\lambda}$, then also $T(v) \in$

$N_{m,\lambda}$. Moreover, if $w \in R_{m,\lambda}$ then $w = (T - \lambda I)^m(u)$, for some $u \in V$. Hence $(T - \lambda I)(w) = (T - \lambda I)^{m+1}(u) \in R_{m+1,\lambda} \subset R_{m,\lambda}$. As above, since $w \in R_{m,\lambda}$, we also have $T(w) \in R_{m,\lambda}$. Hence both $N_{m,\lambda}$ and $R_{m,\lambda}$ are T-invariant subspaces of V.

Remark 6.39 The linear operator $T : V \to V$ is nilpotent if and only if there exists $k \geq 1$ (which is precisely the index of nilpotency of T) such that $N_{k,0} = V$ and $R_{k,0} = \{0\}$.

Definition 6.40 Let $T : V \to V$ be a linear operator, λ an eigenvalue of T and $s \geq 1$ the smallest integer such that $N_{s,\lambda} = N_{t,\lambda}$, for any $t \geq s$. The subspace

$$N_{s,\lambda} = \{X \in V \mid \text{ there exists } i \leq s, (T - \lambda I)^i(X) = 0\}$$

is called *generalized eigenspace* of T corresponding to the eigenvalue λ. Any element of $N_{s,\lambda}$ is a generalized eigenvector corresponding to the eigenvalue λ. The integer s is called *index* of the eigenvalue λ. It is the greatest integer among the exponents of the generalized eigenvectors corresponding to the eigenvalue λ.

Lemma 6.41 *Let $T : V \to V$ be a linear operator of the finite dimensional vector space V, λ an eigenvalue of T, t the index of λ. Then $N_{t,\lambda}$ is invariant under the action of the linear operator $T - \lambda I$.*

Proof Let $X \in N_{t,\lambda}$ be a generalized eigenvector corresponding to λ, that is $0 = (T - \lambda I)^t X = (T - \lambda I)^{t-1}(T - \lambda I)X$. Hence $(T - \lambda I)X \in N_{t-1,\lambda} \subset N_{t,\lambda}$, which means that $N_{t,\lambda}$ is invariant under the action of the operator $T - \lambda I$.

Lemma 6.42 *Let $T : V \to V$ be a linear operator of the finite dimensional vector space V, $\lambda \neq \mu$ distinct eigenvalues of T, t the index of λ. Then $N_{t,\lambda}$ is invariant under the action of the linear operator $T - \mu I$.*

Proof Let $X \in N_{t,\lambda}$ be a generalized eigenvector corresponding to λ, that is $(T - \lambda I)^t X = 0$. Moreover, by $\mu \neq \lambda$, we have $(T - \mu I)X = T(X) - \mu X \neq 0$. Since, by the previous lemma, $(T - \lambda I)X \in N_{t,\lambda}$, there is $Y \in N_{t,\lambda}$ such that $T(X) = \lambda X + Y \in N_{t,\lambda}$, which implies that $(T - \mu I)X = T(X) - \mu X \in N_{t,\lambda}$, as required.

Lemma 6.43 *Let $A \in M_n(\mathbb{F})$ be the matrix of a linear operator $T : V \to V$, λ an eigenvalue of T, s be the minimum integer such that $N_{s,\lambda} = N_{t,\lambda}$ and $R_{s,\lambda} = R_{t,\lambda}$, for any $t \geq s$. Then $dim(N_{s,\lambda})$ is equal to the algebraic multiplicity of the eigenvalue λ.*

Proof We divide the proof into two steps. Firstly, we consider the case when T has the eigenvalue $\lambda = 0$. Let $G = T_{|N_{s,0}} : N_{s,0} \to N_{s,0}$ be the restriction of T to $N_{s,0}$, $H = T_{|R_{s,0}} : R_{s,0} \to R_{s,0}$ the restriction of T to $R_{s,0}$. Denote by $p_T(t)$, $p_G(t)$ and $p_H(t)$ the characteristic polynomials of T, G and H, respectively. Since $V = N_{s,0} \oplus R_{s,0}$ and by Lemma 6.35, we have

$$p_T(t) = p_G(t)p_H(t).$$

On the other hand, since G is nilpotent, $p_G(t) = (-1)^k t^k$, where $k = dim(N_{s,0})$. Moreover H is bijective, so that $t = 0$ cannot be eigenvalue of H, that is $t = 0$ is not a root of the polynomial $p_H(t)$, then the algebraic multiplicity of $\lambda = 0$ is precisely the dimension of its generalized eigenspace.

Let now $\lambda \neq 0$ be an eigenvalue of T and let $h \geq 1$ be its algebraic multiplicity. It is clear that 0 is an eigenvalue of the linear operator $T - \lambda I : V \to V$, having algebraic multiplicity equal to h. By the previous argument, the dimension of the generalized eigenspace of 0 (as eigenvalue of $T - \lambda I$) is precisely h. On the other hand, the generalized eigenspace of 0 as eigenvalue of $T - \lambda I$ coincides with the generalized eigenspace of λ as eigenvalue of T, and the proof is complete.

Theorem 6.44 *Let $T : V \to V$ be a linear operator of a finite dimensional vector space V. Then V is the direct sum of all generalized eigenspaces corresponding to the distinct eigenvalues of T.*

Proof Let $\lambda_1, \ldots, \lambda_r$ be the distinct eigenvalues of T, $t_1, \ldots, t_r$ the indices and $N_{t_1,\lambda_1}, \ldots, N_{t_r,\lambda_r}$ the generalized eigenspaces of $\lambda_1, \ldots, \lambda_r$, respectively. Firstly, we prove that $N_{t_1,\lambda_1} + \cdots + N_{t_r,\lambda_r}$ is a direct sum. To do this, we show that, if

$$X_1 + \cdots + X_r = 0, \qquad X_i \in N_{t_i,\lambda_i} \tag{6.1}$$

then $X_i = 0$ for any $i = 1, \ldots, r$.

Since $(T - \lambda_r I)^{t_r}(X_1 + \cdots + X_r) = 0$ and $(T - \lambda_r I)^{t_r} X_r = 0$, then $(T - \lambda_r I)^{t_r}$ $(X_1 + \cdots + X_{r-1}) = 0$, that is $Y_1 + \cdots + Y_{r-1} = 0$, where $Y_i = (T - \lambda_r I)^{t_r} X_i$, for $i = 1, \ldots, r - 1$. By the previous lemma, $Y_i \in N_{t_i,\lambda_i}$.

Analogously, $(T - \lambda_{r-1} I)^{t_{r-1}}(Y_1 + \cdots + Y_{r-1}) = 0$ implies $(T - \lambda_{r-1} I)^{t_{r-1}}(Y_1 + \cdots + Y_{r-2}) = 0$, that is $Z_1 + \cdots + Z_{r-2} = 0$, where $Z_i = (T - \lambda_{r-1} I)^{t_{r-1}} Y_i = (T - \lambda_{r-1} I)^{t_{r-1}}(T - \lambda_r I)^{t_r} X_i \in N_{t_i,\lambda_i}$, for any $i = 1, \ldots, r - 2$. Continuing this process, we finally get

$$(T - \lambda_{r-1} I)^{t_2}(T - \lambda_{r-1} I)^{t_3} \cdots (T - \lambda_{r-1} I)^{t_r} X_1 = 0. \tag{6.2}$$

On the other hand, $(T - \lambda_{r-1} I)^{t_k} \cdots (T - \lambda_{r-1} I)^{t_r} X_1 \in N_{t_1,\lambda_1} \setminus N_{t_s,\lambda_s}$, for any $k = 2, \ldots, r$ and $s \neq 1$. Therefore, relation (6.2) is true only if $X_1 = 0$. In this case, (6.1) reduces to

$$X_2 + \cdots + X_r = 0, \qquad X_i \in N_{t_i,\lambda_i}. \tag{6.3}$$

So, starting from (6.3) and repeating the same above process $r - 1$ times, one may prove that $X_i = 0$, for any $i = 1, \ldots, r$.

Let now $W = N_{t_1,\lambda_1} \oplus \cdots \oplus N_{t_r,\lambda_r}$. By Lemma 6.43, we know that $dim(N_{t_i,\lambda_i})$ is equal to the algebraic multiplicity of the eigenvalue λ_i, that is $dim_{\mathbb{F}}(W) = dim_{\mathbb{F}}(V)$ and $V = W$.

Lemma 6.45 *Let $A \in M_n(\mathbb{F})$ be the matrix of a linear operator $T : V \to V$, λ an eigenvalue of T, s be the minimum integer such that $N_{s,\lambda} = N_{t,\lambda}$ and $R_{s,\lambda} = R_{t,\lambda}$, for any $t \geq s$. Then:*

(i)   *The restriction of $T - \lambda I$ to $N_{s,\lambda}$ is nilpotent.*
(ii) *The restriction of $T - \lambda I$ to $R_{s,\lambda}$ is an isomorphism.*
(iii) $V = N_{s,\lambda} \oplus R_{s,\lambda}$.

Proof Let $G = (T - \lambda I)_{|N_{s,\lambda}} : N_{s,\lambda} \to N_{s,\lambda}$ be the restriction of $T - \lambda I$ to $N_{s,\lambda}$. For any $X \in N_{s,\lambda}$, we have $G^s(X) = (T - \lambda I)^s(X) = 0$. Thus G is nilpotent.

Now let $H = (T - \lambda I)_{|R_{s,\lambda}} : R_{s,\lambda} \to R_{s,\lambda}$ be the restriction of $T - \lambda I$ to $R_{s,\lambda}$. Let $X \in N(H) = \{v \in R_{s,\lambda} | H(v) = 0\}$ be the kernel of H. Since $X \in R_{s,\lambda}$, there exists $Y \in V$ such that $X = (T - \lambda I)^s(Y)$. As a consequence, $0 = H(X) = H((T - \lambda I)^s(Y)) = (T - \lambda I)^{s+1}(Y)$, that is $Y \in N_{s+1,\lambda}$. On the other hand $N_{s,\lambda} = N_{s+1,\lambda}$, so that $Y \in N_{s,\lambda}$, that is $0 = (T - \lambda I)^s(Y) = X$. It is proved that $N(H) = \{0\}$, so that the image of H is equal to $R_{s,\lambda}$ and the endomorphism H is bijective.

We finally show that $V = N_{s,\lambda} \oplus R_{s,\lambda}$. Firstly, we prove that $N_{s,\lambda} \cap R_{s,\lambda} = \{0\}$. On the contrary we suppose that there exists $0 \neq X \in N_{s,\lambda} \cap R_{s,\lambda}$. Since $X \in N_{s,\lambda}$, we have $(T - \lambda I)^s(X) = 0$. On the other hand, since $X \in R_{s,\lambda}$, there is some $0 \neq Y \in V$ such that $X = (T - \lambda I)^s(Y)$. Therefore $(T - \lambda I)^{2s}(Y) = 0$, that is $Y \in N_{2s,\lambda} = N_{s,\lambda}$. This implies the contradiction $0 = (T - \lambda I)^s(Y) = X$.

To complete the proof, we recall that $(T - \lambda I)^s : V \to V$ is a linear operator of V, so that $dim(N_{s,\lambda}) + dim(R_{s,\lambda}) = dim(V)$ and we are done.

Remark 6.46 Suppose that $\lambda \neq 0$ is an eigenvalue of a linear operator $T : V \to V$, A the matrix of T with respect to a basis B of V, and replace the operator T by $T - \lambda I$. The matrix of $T - \lambda I$ with respect to the basis B is $A - \lambda I$. Moreover, if one of A or $A - \lambda I$ is in Jordan form, so is the other. More precisely, A contains the Jordan block

$$J_p(\lambda) = \begin{bmatrix} \lambda & 1 & & \\ & \ddots & \ddots & \\ & & \lambda & 1 \\ & & & \lambda \end{bmatrix}$$

if and only if $A - \lambda I$ contains the Jordan block

$$J_p(0) = \begin{bmatrix} 0 & 1 & & \\ & \ddots & \ddots & \\ & & 0 & 1 \\ & & & 0 \end{bmatrix}.$$

Thus, in order to analyze the Jordan form of T, we replace T by $T - \lambda I$ and reduce to the case of singular operators (that is, operators having at least one eigenvalue $\lambda = 0$).

To simplify the notation, we will write N_s instead of $N_{s,0}$ and R_s instead of $R_{s,0}$.

Definition 6.47 Let $A \in M_n(\mathbb{F})$ be the matrix of a singular linear operator $T : V \to V$, s be the minimum integer such that $N_s = N_t$ and $R_s = R_t$, for any $t \geq s$. Let

$\{e_1, \ldots, e_t\}$ be a basis for N_s. Since $V = N_s \oplus R_s$, there exist vectors $e_{t+1}, \ldots, e_n \in R_s$ such that $B = \{e_1, \ldots, e_n\}$ is a basis for V. With respect to B, the matrix A of T has the following form:

$$A = \begin{bmatrix} A_1 & 0_{t,n-t} \\ 0_{n-t,t} & A_2 \end{bmatrix} \tag{6.4}$$

where:

(i) A_1 is a nilpotent matrix of index s, corresponding to the restriction of T to N_s;
(ii) A_2 is an invertible matrix, corresponding to the restriction of T to R_s.

More precisely:

(iii) For $i = 1, \ldots, t$, the coordinate column vectors of the image $T(e_i)$ define the submatrix $\begin{bmatrix} A_1 \\ 0_{n-t,t} \end{bmatrix}$.

(iv) For $i = t + 1, \ldots, n$, the coordinate column vectors of the image $T(e_i)$ define the submatrix $\begin{bmatrix} 0_{t,n-t} \\ A_2 \end{bmatrix}$.

Thus, according to the decomposition $V = N_s \oplus R_s$ we have the decomposition (6.4). It is called *Fitting decomposition of A* (usually $V = N_s \oplus R_s$ is also called Fitting decomposition of T).

Now, we can definitively describe the characteristic polynomial of a nilpotent operator:

Theorem 6.48 *Let V be a vector space of dimension n, $T : V \to V$ a linear operator and $A \in M_n(\mathbb{F})$ the matrix of T. A is nilpotent if and only if its characteristic polynomial is $p(\lambda) = (-1)^n \lambda^n$.*

Proof In case A is nilpotent, then $p(\lambda) = (-1)^n \lambda^n$ is proved in Lemma 6.37. Hence, we assume $p(\lambda) = (-1)^n \lambda^n$, that is 0 is the only eigenvalue of A. By contradiction, suppose that A is not nilpotent, that is T is not nilpotent. Then there is some vector $v \in V$ such that $T^s(v) \neq 0$, in particular $R_s \neq \{0\}$ (see Remark 6.39). Thus, by Fitting decomposition rule, $V = N_s \oplus R_s$ and

$$A = \begin{bmatrix} A_1 & 0_{t,n-t} \\ 0_{n-t,t} & A_2 \end{bmatrix},$$

where A_2 is an invertible matrix. On the other hand, since we have assumed that 0 is the only eigenvalue of A, a fortiori it is the only eigenvalue of the nonsingular matrix A_2, which is a contradiction.

We are now able to construct a basis of V with respect to which the matrix of T has Jordan form. It will be an inductive process and consists of several steps. We start with the following:

Theorem 6.49 *Let V be a vector space of dimension n, $T : V \to V$ a singular linear operator. Let $X_r \in V$ be such that $X_r \in N_r \setminus N_{r-1}$, in other words let X_r be a generalized eigenvector of T and r be the exponent of X_r. If we consider the following chain of vectors $X_{r-i} = T(X_{r-i+1}) = T^i(X_r)$, for $i = 1, \ldots, r - 1$, then the following properties hold:*

   (*i*) *$X_{r-i} \in N_{r-i} \setminus N_{r-i-1}$, for any $i = 1, \ldots, r - 1$.*
  (*ii*) *$\{X_1, X_2, \ldots, X_r\}$ is a linearly independent set.*
 (*iii*) *If $W = Span\{X_1, X_2, \ldots, X_r\}$ is the subspace of V generated by $\{X_1, X_2, \ldots, X_r\}$, then $T(W) \subseteq W$.*

Proof Consider the vector $X_{r-1} = T(X_r)$. Hence $T^{r-1}(X_{r-1}) = T^{r-1}T(X_r) = T^r(X_r) = 0$, because of $X_r \in N_r$. Then $X_{r-1} \in N_{r-1}$. Now, suppose that $X_{r-1} \in N_{r-2}$. This should imply that $0 = T^{r-2}(X_{r-1}) = T^{r-2}T(X_r) = T^{r-1}(X_r)$, which contradicts the assumption that $X_r \notin N_{r-1}$. Thus we have proved that $X_{r-1} \in N_{r-1} \setminus N_{r-2}$. The inductive construction of the chain of vectors $X_{r-i} = T(X_{r-i+1}) = T^i(X_r)$, indicates that we may repeat the previous argument in order to obtain the required conclusion $X_{r-i} \in N_{r-i} \setminus N_{r-i-1}$, for any $i = 1, \ldots, r - 1$.

Now assume

$$\alpha_1 X_1 + \alpha_2 X_2 + \cdots + \alpha_r X_r = 0 \qquad (6.5)$$

for some elements $\alpha_1, \ldots, \alpha_r \in \mathbb{F}$. By applying T^{r-1} to (6.5), we have

$$0 = \alpha_1 T^{r-1}(X_1) + \alpha_2 T^{r-1}(X_2) + \cdots + \alpha_r T^{r-1}(X_r). \qquad (6.6)$$

Since $X_j \in N_j$, for any $j = 1, \ldots, r - 1$, and $X_r \notin N_{r-1}$, by (6.6) it follows $0 = \alpha_r T^{r-1}(X_r)$, implying $\alpha_r = 0$. Thus, we may rewrite (6.5) as follows:

$$\alpha_1 X_1 + \alpha_2 X_2 + \cdots + \alpha_{r-1} X_{r-1} = 0. \qquad (6.7)$$

Hence, application of T^{r-2} to (6.7), and since $X_{r-1} \notin N_{r-2}$, implies $\alpha_{r-1} = 0$. By repeating this process we may prove that $\alpha_i = 0$, for any $i = 1, \ldots, r$, that is $\{X_1, X_2, \ldots, X_r\}$ is a linearly independent set.

Finally, we prove that $W = Span\{X_1, X_2, \ldots, X_r\}$ is T-invariant. At first we notice that $T(X_1) = TT^{r-1}(X_r) = T^r(X_r) = 0 \in W$. We now focus our attention on the image $T(X_j)$, when $1 < j \le r$. In this case $T(X_j) = TT^{r-j}(X_r) = T^{r+1-j}(X_r) = X_{j-1} \in W$. Hence, the image of any linear combination $\alpha_1 X_1 + \alpha_2 X_2 + \cdots + \alpha_r X_r$ is an element of W, as desired.

Remark 6.50 Let $X_r \in V$ be such that $X_r \in N_r \setminus N_{r-1}$. The chain of vectors $X_{r-i} = T(X_{r-i+1}) = T^i(X_r)$, for $i = 1, \ldots, r - 1$, as in Theorem 6.49, can be defined as follows:

  (*i*) $X_{k-1} = T(X_k)$, for $k = 2, \ldots, r$;
 (*ii*) $0 = T(X_1)$.

Remark 6.51 Let $T : V \to V$ be a nilpotent operator (that is, 0 is the only eigenvalue of T), $X_1, \ldots, X_k$ generalized eigenvectors of T such that r_i is the exponent of X_i, for any $i = 1, \ldots, k$. Starting from any X_i we now construct the chain of vectors as in Theorem 6.49:

(i) $X_i = X_{i,r_i}$;
(ii) $X_{i,r_i-j} = T^j(X_{i,r_i})$, for $j = 1, \ldots, r_i - 1$.

The set $B_i = \{X_{i,1}, X_{i,2}, \ldots, X_{i,r_i}\}$ is linearly independent and the subspace $W_i = Span(B_i)$ is T-invariant. Moreover, in light of Remark 6.50:

(iii) $X_{i,k-1} = T(X_{i,k})$, for $k = 2, \ldots, r_i$;
(iv) $T(X_{i,1}) = 0$, that is, $X_{i,1}$ is an eigenvector of T.

If $V = \oplus_{i=1}^{k} W_i$, where $W_i's$ are subspaces of V having bases $B_i's$ respectively, then the set $B = \bigcup_{i=1}^{k} B_i$ is a basis for V. Write

$$B = \{X_{1,1}, \ldots, X_{1,r_1}; X_{2,1}, \ldots, X_{2,r_2}; \ldots, \ldots; X_{k,1}, \ldots, X_{k,r_k}\}$$

and let A be the matrix of T with respect to B. The column coordinate vectors of A are the images of the elements of the basis B. By computing these images, we have that

(v) $T(X_{i,j}) = X_{i,j-1}$ in case $X_{i,j}$ is not the first vector in the chain B_i;
(vi) $T(X_{i,j}) = 0$ in case $X_{i,j}$ is the first vector in the chain B_i.

Therefore, the coordinate vector of $T(X_{i,j})$ with respect to B is either
$(0, 0, \ldots, 0, 1, 0, \ldots, 0)$, where 1 is the $\left(\sum_{h=1}^{i-1} r_h + (j - 1)\right)$-th entry, or
$(0, 0, \ldots, 0, 0)$, respectively. Therefore, the matrix A has the following block diagonal form

$$\begin{bmatrix} J_1(0) & & & \\ & \ddots & & \\ & & \ddots & \\ & & & J_k(0) \end{bmatrix}$$

where

$$J_i(0) = \begin{bmatrix} 0 & 1 & & \\ & \ddots & \ddots & \\ & & 0 & 1 \\ & & & 0 \end{bmatrix}$$

is a $r_i \times r_i$ Jordan block. In this case $\{X_1, \ldots, X_k\}$ is called a *set of Jordan generators* and B is called a *Jordan basis* for V.

Theorem 6.52 *Let $T : V \to V$ be a linear operator of a finite dimensional vector space V. If T is nilpotent, then there exists a basis B for V such that the matrix of T with respect to B has Jordan form.*

Proof Let $dim_\mathbb{F} V = n$. We prove the result by induction on the dimension of V. It is trivial, in case $dim_\mathbb{F} V = 1$.

Let N and R be the kernel and the image of T, respectively. Since T is nilpotent, $N \neq \{0\}$, $dim_\mathbb{F} R < dim_\mathbb{F} V$, and every nonzero vector of V is a generalized eigenvector with eigenvalue 0. By our induction assumption, the result is true for $T_{|R}$, the restriction of T to R. So there exists a set of Jordan generators $\{X_1, \ldots, X_k\}$ for $T_{|R}$. Let t_i be the exponent of any X_i and let U_i be the subspace of R spanned by $\{X_{i,1}, \ldots, X_{i,t_i}\}$, where

(i) $X_{i,t_i} = X_i$

(ii) $X_{i,t_i-j} = T_{|R}^j(X_{i,t_i})$, for $j = 1, \ldots, t_i - 1$.

By Theorem 6.49, $\{X_{i,1}, \ldots, X_{i,t_i}\}$ is a linearly independent set, then it is a basis for U_i. Moreover U_i is $T_{|R}$-invariant and $R = U_1 \oplus \cdots \oplus U_k$ by the induction assumption.

Let now $Y_i \in V$ be such that $T(Y_i) = X_i$. Hence the exponent of Y_i is equal to $t_i + 1$. Denote by V_i the subspace of V spanned by $\{Y_{i,1}, \ldots, Y_{i,t_i+1}\}$, where

(i) $Y_{i,t_i+1} = Y_i$

(ii) $Y_{i,t_i+1-j} = T^j(Y_{i,t_i+1})$, for $j = 1, \ldots, t_i$

and $W = V_1 + \cdots + V_k$. Since each V_i is T-invariant, W is T-invariant. Moreover $T(V_i) = U_i$, and hence $T(W) = R$.

If $X \in V_i \cap N$, then $T(X) = 0$ and there exist $\alpha_1, \ldots, \alpha_{t_i+1} \in \mathbb{F}$ such that

$$X = \alpha_1 Y_{i,1} + \cdots + \alpha_{t_i+1} Y_{i,t_i+1}. \tag{6.8}$$

By applying T to relation (6.8), one has

$$0 = T(X) = \alpha_1 T(Y_{i,1}) + \cdots + \alpha_{t_i+1} T(Y_{i,t_i+1}). \tag{6.9}$$

Notice that
$$T(Y_{i,1}) = T\left(T^{t_i}(Y_{i,t_i+1})\right) = T^{t_i+1}(Y_{i,t_i+1}) = 0.$$

Moreover
$$Y_{im} = T^{t_i+1-m}(Y_{i,t_i+1}) = T^{t_i-m}(X_{i,t_i})$$

implies
$$T(Y_{im}) = T^{t_i+1-m}(X_{i,t_i}) \neq 0, \quad \forall 1 < m \leq t_i + 1.$$

Therefore, (6.9) reduces to

$$0 = \sum_{m=2}^{t_i+1} \alpha_m T^{t_i+1-m}(X_{i,t_i}) = \sum_{m=2}^{t_i+1} (\alpha_m X_{i,m-1}). \tag{6.10}$$

Since $\{X_{i,1}, \ldots, X_{i,t_i}\}$ is a linearly independent set, relation (6.10) implies $\alpha_m = 0$, for any $m = 2, \ldots, t_i + 1$. Thus, by (6.8), it follows

$$X = \alpha_1 Y_{i,1} = \alpha_1 T^{t_i}(Y_{i,t_i+1}) = \alpha_1 T^{t_i}(Y_i) = \alpha_1 T^{t_i-1}(X_i)$$

that is, X is a nonzero element of U_i. In other words, we have showed that $V_i \cap N \subseteq U_i$.

Let now $v_1 \in V_1, \ldots, v_k \in V_k$ be such that $v_1 + \cdots + v_k = 0$. Hence

$$0 = T(v_1 + \cdots + v_k) = T(v_1) + \cdots + T(v_k). \tag{6.11}$$

Moreover, since $T(V_i) = U_i$, we also have $T(v_i) \in U_i$, for all $i = 1, \ldots, k$. Therefore, by the fact that $U_1, \ldots, U_k$ are independent and by (6.11), it follows $T(v_i) = 0$, thus $v_i \in V_i \cap N \subseteq U_i$, for any $i = 1, \ldots, k$. Once again since $U_1, \ldots, U_k$ are independent and by the assumption that $v_1 + \cdots + v_k = 0$, we conclude that $v_i = 0$, for any $i = 1, \ldots, k$. In other words we have proved that $W = V_1 \oplus \cdots \oplus V_k$, so that $\{Y_1, \ldots, Y_k\}$ is a set of Jordan generators for $T_{|W}$.

Finally, let $v_1, v_2 \in V$ be such that $T(v_1) = v_2$. Since $T(W) = R$, there exists an element $u \in W$ such that $T(u) = v_2 = T(v_1)$, that is $T(u - v_1) = 0$ and $u - v_1 \in N$. Hence, there is $x \in N$ such that $v_1 = u + x$. Repeating this process for any element of V, we have that $V = W + N$. Thus, we can extend a basis of W to a basis of V by adding elements of N. In this sense, let $x_1, \ldots, x_m \in N$ be such that

$$V = W \oplus Span\{x_1, \ldots, x_m\} = V_1 \oplus \cdots \oplus V_k \oplus Span\{x_1, \ldots, x_m\}.$$

Notice that $Span\{x_1, \ldots, x_m\}$ is trivially T-invariant, moreover the restriction of T to $Span\{x_1, \ldots, x_m\}$ is the zero operator and its matrix is the zero matrix.

Therefore, since $\{Y_1, \ldots, Y_k\}$ is a set of Jordan generators for $T_{|W}$, $\{Y_1, \ldots, Y_k, x_1, \ldots, x_m\}$ is a set of Jordan generators for T.

Theorem 6.53 *Let $T : V \to V$ be a linear operator of a finite dimensional vector space V. If the characteristic polynomial $p(\lambda)$ of T splits over $\mathbb{F}$, then there exists a basis B for V such that the matrix of T with respect to B has Jordan form.*

Proof We prove the theorem by induction on the dimension of V. The result is trivial if $dim_\mathbb{F} V = 1$. Assume $dim_\mathbb{F} V = n \geq 2$ and the theorem holds for any T-invariant proper subspace of V.

We firstly consider the case $p(\lambda) = (\lambda - \lambda_1)^n$, that is there exists a unique eigenvalue λ_1 of T, having algebraic multiplicity equal to n. Let A be the matrix of T with respect to a basis B of V. Now replace T by $T - \lambda_1 I$, so that $A - \lambda_1 I$ is the matrix of $T - \lambda_1 I$ with respect to the same basis B. Moreover, if $A - \lambda_1 I$ is similar to a Jordan diagonal blocks matrix, so is A. However, $T - \lambda_1 I$ is a nilpotent operator and we can conclude by Theorem 6.52.

Hence, we may assume that there exist at least two distinct eigenvalues λ_1, λ_2 of T. The linear operator $T - \lambda_1 I$ is singular, having eigenvalue zero, and $A - \lambda_1 I$ is its matrix. As above, if $A - \lambda_1 I$ admits a Jordan canonical form, so does A.

Let $s \geq 1$ be the index of λ_1. That is, s is the minimum integer such that $N_{s,0} = N_{t,0}$ and $R_{s,0} = R_{t,0}$, for any $t \geq s$. We write N_s instead of $N_{s,0}$ and R_s instead of $R_{s,0}$.

Notice that $R_s \neq 0$, since $(T - \lambda_1 I)X \neq 0$ for any eigenvector X corresponding to λ_2 and and $N_s \neq 0$, since $T - \lambda_1 I$ is singular. Thus, we can consider the Fitting decomposition of $(T - \lambda_1 I)$, i.e., $V = N_s \oplus R_s$. Both the restrictions $(T - \lambda_1 I)_{|N_s}$ and $(T - \lambda_1 I)_{|R_s}$ are linear operators over proper vector subspace of V. Hence, by our induction assumption, there exist $B_1 = \{X_1, \dots, X_k\}$ set of Jordan generators for $(T - \lambda_1 I)_{|N_s}$ and $B_2 = \{X_{k+1}, \dots, X_n\}$ set of Jordan generators for $(T - \lambda_1 I)_{|R_s}$. So the matrix A_1 of $(T - \lambda_1 I)_{|N_s}$ with respect to the basis B_1 has Jordan form, as well as the matrix A_2 of $(T - \lambda_1 I)_{|R_s}$ in terms of the basis B_2. As in Lemma 6.35, the matrix of $T - \lambda_1 I$ with respect to the basis $B_1 \cup B_2$ of V has the following block diagonal form

$$\begin{bmatrix} A_1 & 0_{k,n-k} \\ 0_{n-k,k} & A_2 \end{bmatrix}$$

and the proof is complete.

Lemma 6.54 *Let $T : V \to V$ be a singular linear operator of the finite dimensional vector space V, and assume that there exists a basis B for V such that the matrix A of T with respect to B has Jordan form. If $J_k(0)$ is a $k \times k$ Jordan block of A corresponding to the eigenvalue zero, then*

$$rank\left(J_k(0)^d\right) = \begin{cases} k - d, & 0 \leq d < k \\ 0, & d \geq k \end{cases}$$

Proof Consider the linear operator $H : \mathbb{F}^k \to \mathbb{F}^k$ whose matrix is

$$J_k(0) = \begin{bmatrix} 0 & 1 & & & \\ & \ddots & \ddots & & \\ & & \ddots & \ddots & \\ & & & 0 & 1 \\ & & & & 0 \end{bmatrix}.$$

The fact that $J_k(0)^k =$ the zero matrix, is trivial, so we can consider the case $d < k$. We denote $R, R^2, \dots, R^d$ the images of the operators $H, H^2, \dots, H^d$ respectively. Let $X_1, \dots, X_k$ be generalized eigenvectors of T corresponding to the eigenvalue zero, such that $J_k(0)$ is the matrix of H with respect to the basis $C = \{X_1, \dots, X_k\}$ for $\mathbb{F}^k$. Since the columns of $J_k(0)$ are the coordinate vectors of $H(X_1), \dots, H(X_k)$ with respect to the basis C, then $H(X_1) = 0$ and $H(X_i) = X_{i-1}$, for any $i = 2, \dots, k$. Since $\{H(X_1), \dots, H(X_k)\}$ is a generator set for R, then $\{H(X_2), \dots, H(X_k)\} = \{X_1, \dots, X_{k-1}\}$ is a basis for R. Here we also notice that $H^2(X_1) = 0$, $H^2(X_2) = H(H(X_2)) = H(X_1) = 0$ and $H^2(X_i) = H(X_{i-1}) = X_{i-2}$, for any $i = 3, \dots, k$. As above, this means that $\{H(X_3), \dots, H(X_k)\} = \{X_2, \dots, X_{k-1}\}$ is a basis for R^2.

Repeating this process, we can see that $H^d(X_1) = 0$, $H^d(X_i) = 0$ for any $i \leq d$ and $H^d(X_i) = X_{i-d}$, for any $d < i \leq k$, so that $\{H(X_{d+1}), \dots, H(X_k)\} = \{X_d, \dots, X_{k-1}\}$ is a basis for R^d, that is $dim_{\mathbb{F}}(R^d) = k - d$. Moreover $J_k(0)^d$ is the matrix of any H^d, in other words the rank of $J_k(0)^d$ is equal to the dimension of R^d, and we are done.

Corollary 6.55 *Let $T : V \to V$ be a singular linear operator of the finite dimensional vector space V and assume that there exists a basis B for V such that the matrix A of T with respect to B has Jordan form. If $J_k(0)$ is a $k \times k$ Jordan block of A corresponding to the eigenvalue zero, then*

$$rank\big(J_k(0)^{d+1}\big) - 2rank\big(J_k(0)^d\big) + rank\big(J_k(0)^{d-1}\big) = \begin{cases} 1, & d = k \\ 0, & 1 \le d \ne k \end{cases}$$

Proof It follows by easy computations, as a consequence of Lemma 6.54.

Lemma 6.56 *Let $A, B \in M_n(\mathbb{F})$ be similar and λ an eigenvalue of A and B. Then $rank\big((A - \lambda I)^k\big) = rank\big((B - \lambda I)^k\big)$ for any $k = 1, \dots, n$.*

Proof Let $C \in M_n(\mathbb{F})$ be the nonsingular matrix such that $B = C^{-1}AC$. For any $k \ge 1$,

$$C^{-1}(A - \lambda I)^k C = \big(C^{-1}(A - \lambda I)C\big)^k = (B - \lambda I)^k$$

that is, $(A - \lambda I)^k$ and $(B - \lambda I)^k$ are similar. Thus, $rank(A - \lambda I)^k = rank(B - \lambda I)^k$.

Corollary 6.57 *$A, B \in M_n(\mathbb{F})$ are similar if and only if $(A - \lambda I)$ and $(B - \lambda I)$ are similar.*

Theorem 6.58 *Let $T : V \to V$ be a linear operator of a finite dimensional vector space V over the field $\mathbb{F}$, $A \in M_n(\mathbb{F})$ be the matrix of T with respect to a basis for V. Assume that there exists a basis B for V such that the matrix A' of T with respect to B has Jordan form. If $J_k(\lambda)$ is a $k \times k$ Jordan block of A' corresponding to the eigenvalue λ, then the number of times $J_k(\lambda)$ occurs as a diagonal block in A' is equal to*

$$rank\big((A - \lambda I)^{k+1}\big) - 2rank\big((A - \lambda I)^k\big) + rank\big((A - \lambda I)^{k-1}\big).$$

Proof Let $B = \{X_1, \dots, X_n\}$ be the Jordan basis for V, with respect to which the matrix A' of T has Jordan form. Hence $A' = C^{-1}AC$, where C is the transition matrix. By Lemma 6.56, $rank\big((A - \lambda I)^k\big) = rank\big((A' - \lambda I)^k\big)$, for any $k = 1, \dots, n$. Hence, in order to prove our formula, we replace A by A'. Let

$$A' = \begin{bmatrix} A_1 & & & \\ & \ddots & & \\ & & \ddots & \\ & & & A_t \end{bmatrix}$$

be the Jordan block diagonal form, where any $n_i \times n_i$ block A_i is associated with an eigenvalue λ_i of T. Thus

$$A' - \lambda I = \begin{bmatrix} A_1 - \lambda I_{n_1} & & & \\ & \ddots & & \\ & & \ddots & \\ & & & A_t - \lambda I_{n_t} \end{bmatrix}$$

and

$$(A' - \lambda I)^k = \begin{bmatrix} (A_1 - \lambda I_{n_1})^k & & & \\ & \ddots & & \\ & & \ddots & \\ & & & (A_t - \lambda I_{n_t})^k \end{bmatrix}$$

so that

$$rank\big((A' - \lambda I)^k\big) = \sum_{i=1}^{t} rank\big((A_i - \lambda I_{n_i})^k\big). \tag{6.12}$$

To simplify the notation we denote $r_k(A) = rank\big((A - \lambda I)^k\big)$ and $r_k(A_i) = rank\big((A_i - \lambda I_{n_i})^k\big)$. By (6.12) it follows that

$$r_{k+1}(A) - 2r_k(A) + r_{k-1}(A) = \sum_{i=1}^{t} \Big(r_{k+1}(A_i) - 2r_k(A_i) + r_{k-1}(A_i) \Big). \tag{6.13}$$

If $\lambda_i \neq \lambda$, then λ is not the eigenvalue corresponding to A_i, then $A_i - \lambda I_{n_i}$ is an invertible $n_i \times n_i$ matrix, having rank equal to n_i. Hence any power of $A_i - \lambda I_{n_i}$ has rank equal to n_i. In this case

$$\sum_{i=1}^{t} \Big(r_{k+1}(A_i) - 2r_k(A_i) + r_{k-1}(A_i) \Big) = 0.$$

On the other hand, for $\lambda_i = \lambda$, the matrix $A - \lambda I_{n_i}$ is a $n_i \times n_i$ Jordan block having zero on the main diagonal (its form is $J_{n_i}(0)$). In this case, by Corollary 6.55,

$$r_{k+1}(A_i) - 2r_k(A_i) + r_{k-1}(A_i) = \begin{cases} 1, \ k = n_i \\ 0, \ k \neq n_i \end{cases}$$

Hence, in light of relation (6.13), we obtain that $r_{k+1}(A) - 2r_k(A) + r_{k-1}(A)$ represents precisely the number of Jordan blocks A_i having dimension k and associated to the eigenvalue λ.

Theorem 6.59 *Let $T : V \to V$ be a linear operator of a finite dimensional vector space V, $A \in M_n(\mathbb{F})$ be the matrix of T with respect to a basis for V. Assume that there exists a basis B for V such that the matrix A' of T with respect to B has Jordan form. If λ is an eigenvalue of T having index k, then there exists at least one $k \times k$*

Jordan block in A' corresponding to the eigenvalue λ. Moreover, every Jordan block in A' associated with λ has order less than or equal to k.

Proof Let $N_h = Ker(A - \lambda I)^h$, for any $h \geq 1$. In particular N_k is the generalized eigenspace of λ. For any $i < k$, we know that $N_i \subset N_k$, and hence $dim_{\mathbb{F}}(N_i) < dim_{\mathbb{F}}(N_k)$ and $rank((A - \lambda I)^i) > rank((A - \lambda I)^k)$. Moreover, for any $t \geq k$, $N_k = N_t$, that is $rank((A - \lambda I)^t) > rank((A - \lambda I)^k)$. Hence, if $J_k(\lambda)$ is a $k \times k$ Jordan block corresponding to the eigenvalue λ, we can compute the number m of times $J_k(\lambda)$ occurs as a block in A' by using the result in Theorem 6.58, that is

$$m = rank((A - \lambda I)^{k+1}) - 2rank((A - \lambda I)^k) + rank((A - \lambda I)^{k-1}).$$

In light of previous comments, it is easy to see that $m \geq 1$. Now let $J_h(\lambda)$ an $h \times h$ Jordan block corresponding to λ, with $h \neq k$. If we suppose $h > k$, then $h - 1 \geq k$, thus $N_{h-1} = N_k$ and a fortiori $N_{h+1} = N_h = N_k$. Thus

$$rank((A - \lambda I)^{h+1}) = rank((A - \lambda I)^h) = rank((A - \lambda I)^{h-1})$$

and an application of Theorem 6.58 shows that the number of times $J_h(\lambda)$ occurs as a block in A' is

$$rank((A - \lambda I)^{h+1}) - 2rank((A - \lambda I)^h) + rank((A - \lambda I)^{h-1}) = 0,$$

which leads to a contradiction.

Theorem 6.60 *Let $T : V \to V$ be a linear operator of the finite dimensional vector space V over the field $\mathbb{F}$, $A \in M_n(\mathbb{F})$ be the matrix of T with respect to a basis for V. Assume that there exists a basis B for V such that the matrix A' of T with respect to B has Jordan form. If λ is an eigenvalue of T having index k, then the total number of Jordan blocks in A' corresponding to the eigenvalue λ is equal to the geometric multiplicity of λ.*

Proof As above, let $N_h = Ker(A - \lambda I)^h$, for any $h \geq 1$, and N_k be the generalized eigenspace of λ (that is k is the index of λ). To obtain the total number t of blocks corresponding to λ, one can apply the formula in Theorem 6.58. Moreover, the greatest order for a block corresponding to λ is equal to k, therefore

$$t = \sum_{i=1}^{k} \{rank((A - \lambda I)^{i+1}) - 2rank((A - \lambda I)^i) + rank((A - \lambda I)^{i-1})\}.$$

By computations, it follows that

$$t = rank((A - \lambda I)^0) - rank(A - \lambda I) + rank((A - \lambda I)^{k+1}) - rank((A - \lambda I)^k).$$

Since $N_{k+1} = N_k$, we have that

$$rank\big((A - \lambda I)^{k+1}\big) = rank\big((A - \lambda I)^k\big)$$

so that

$$t = rank\big((A - \lambda I)^0\big) - rank\big(A - \lambda I\big) = n - rank\big(A - \lambda I\big)$$

which is precisely the geometric multiplicity of λ.

Example 6.61 Let $T : \mathbb{R}^5 \to \mathbb{R}^5$ be the linear operator having matrix

$$A = \begin{bmatrix} 3 & 1 & 2 & 1 & 2 \\ 0 & 3 & 0 & 2 & -2 \\ 0 & 0 & 3 & -1 & 1 \\ 0 & 0 & 0 & 2 & 1 \\ 0 & 0 & 0 & 0 & 2 \end{bmatrix}$$

with respect to the canonical basis of $\mathbb{R}^5$. The characteristic polynomial is $p(\lambda) = (3 - \lambda)^3(2 - \lambda)^2$, having two distinct roots $\lambda_1 = 3$ (with algebraic multiplicity equal to 3), $\lambda_2 = 2$ (with algebraic multiplicity equal to 2).

For $\lambda_1 = 3$, we have

$$A - 3I = \begin{bmatrix} 0 & 1 & 2 & 1 & 2 \\ 0 & 0 & 0 & 2 & -2 \\ 0 & 0 & 0 & -1 & 1 \\ 0 & 0 & 0 & -1 & 1 \\ 0 & 0 & 0 & 0 & -1 \end{bmatrix}$$

having rank $r_1 = 3$, so that the associated homogeneous linear system has solutions $X = (\alpha, -2\beta, \beta, 0, 0)$, for any $\alpha, \beta \in \mathbb{R}$. The eigenspace corresponding to λ_1 is $N_{1,\lambda_1} = \langle (1, 0, 0, 0, 0), (0, -2, 1, 0, 0) \rangle$ and the geometric multiplicity of λ_1 is equal to 2. Therefore, there are 2 Jordan blocks corresponding to the eigenvalue $\lambda_1 = 3$. More precisely, one block has dimension 2 and one block has dimension 1. Now

$$(A - 3I)^2 = \begin{bmatrix} 0 & 0 & 0 & -1 & -1 \\ 0 & 0 & 0 & -2 & 4 \\ 0 & 0 & 0 & 1 & -2 \\ 0 & 0 & 0 & 1 & -2 \\ 0 & 0 & 0 & 0 & 1 \end{bmatrix}$$

has rank $r_2 = 2$, that is the associated homogeneous linear system has solutions $(\alpha, \beta, \gamma, 0, 0)$, for any $\alpha, \beta, \gamma \in \mathbb{R}$. Thus N_{2,λ_1} has dimension 3 and $N_{2,\lambda_1} = \langle (1, 0, 0, 0, 0), (0, 1, 0, 0, 0), (0, 0, 1, 0, 0) \rangle$. Notice that the dimension of N_{2,λ_1} coincides with the algebraic multiplicity of λ_1, so that N_{2,λ_1} is the generalized eigenspace of λ_1. In fact, it is easy to see that $N_{k,\lambda_1} = N_{2,\lambda_1}$, for any $k \geq 3$. To obtain a set of Jordan generators corresponding to λ_1, we start from $X_2 \in N_{2,\lambda_1} \setminus N_{1,\lambda_1}$. We choose $X_2 = (0, 1, 0, 0, 0)$. Then compute $X_1 = (A - 3I)X_2 = (1, 0, 0, 0, 0)$. Hence $B_1 = \{X_1, X_2\}$ is a set of Jordan generators associated with the block of dimension 2. The set of Jordan generators associated with the block of dimension 1 must be $B_2 = \{X_3\}$, where $X_3 \in N_{1,\lambda_1} \setminus \langle X_1, X_2 \rangle$, that is $X_3 = (0, -2, 1, 0, 0)$,

For $\lambda_2 = 2$, we have

$$A - 2I = \begin{bmatrix} 1 & 1 & 2 & 1 & 2 \\ 0 & 1 & 0 & 2 & -2 \\ 0 & 0 & 1 & -1 & 1 \\ 0 & 0 & 0 & 0 & 1 \\ 0 & 0 & 0 & 0 & 0 \end{bmatrix}$$

having rank equal to 4, so that the associated homogeneous linear system has solutions $(-\alpha, -2\alpha, \alpha, \alpha, 0)$, for any $\alpha \in \mathbb{R}$. The eigenspace corresponding to λ_2 is $N_{1,\lambda_2} = \langle (-1, -2, 1, 1, 0) \rangle$ and the geometric multiplicity of λ_2 is equal to 1. Therefore, there is 1 Jordan block corresponding to the eigenvalue $\lambda_2 = 2$, having dimension 2. Since

$$(A - 2I)^2 = \begin{bmatrix} 1 & 2 & 4 & 1 & 3 \\ 0 & 1 & 0 & 2 & 0 \\ 0 & 0 & 1 & -1 & 0 \\ 0 & 0 & 0 & 0 & 0 \\ 0 & 0 & 0 & 0 & 0 \end{bmatrix}$$

has rank $r_2 = 3$, then the associated homogeneous linear system has solutions $(-\alpha - 3\beta, -2\alpha, \alpha, \alpha, \beta)$, for any $\alpha, \beta \in \mathbb{R}$. Thus, the generalized eigenspace N_{2,λ_2} has dimension 2 and $N_{2,\lambda_2} = \langle (-1, -2, 1, 1, 0), (-3, 0, 0, 0, 1) \rangle$. To obtain a set of Jordan generators corresponding to λ_2, we start from $X_5 \in N_{2,\lambda_2} \setminus N_{1,\lambda_2}$. Of course $X_5 = (-3, 0, 0, 0, 1)$. Then compute $X_4 = (A - 2I)X_5 = (-1, -2, 1, 1, 0)$. Hence $B_3 = \{X_4, X_5\}$ is a set of Jordan generators associated with the block of dimension 2, corresponding to λ_2.

Finally, we can write a Jordan basis for T :

$$\begin{aligned} B &= B_1 \cup B_2 \cup B_3 \\ &= \{X_1, X_2, X_3, X_4, X_5\} \\ &= \{(1, 0, 0, 0, 0), (0, 1, 0, 0, 0), (0, -2, 1, 0, 0), (-1, -2, 1, 1, 0), (-3, 0, 0, 0, 1)\}. \end{aligned}$$

Let A' be the matrix of T with respect to B. The column coordinate vectors of A' are the images of the elements of B, more precisely:

$$\begin{aligned} T(X_1) &= AX_1 \\ &= (3, 0, 0, 0, 0), \text{ with coordinates in terms of } B \text{ as } (3, 0, 0, 0, 0); \end{aligned}$$

$$T(X_2) = AX_2$$
$$= (1, 3, 0, 0, 0), \text{ with coordinates in terms of } B \text{ as } (1, 3, 0, 0, 0);$$

$$T(X_3) = AX_3$$
$$= (0, -6, 3, 0, 0), \text{ with coordinates in terms of } B \text{ as } (0, 0, 3, 0, 0);$$

$$T(X_4) = AX_4$$
$$= (-2, -4, 2, 2, 0), \text{ with coordinates in terms of } B \text{ as } (0, 0, 0, 2, 0);$$

$$T(X_5) = AX_5$$
$$= (-7, -2, 1, 1, 2), \text{ with coordinates in terms of } B \text{ as } (0, 0, 0, 1, 2).$$

Hence, if P is the following transition matrix

$$P = \begin{bmatrix} 1 & 0 & 0 & -1 & -3 \\ 0 & 1 & -2 & -2 & 0 \\ 0 & 0 & 1 & 1 & 0 \\ 0 & 0 & 0 & 1 & 0 \\ 0 & 0 & 0 & 0 & 1 \end{bmatrix}$$

then

$$A' = P^{-1}AP = \begin{bmatrix} 3 & 1 & 0 & 0 & 0 \\ 0 & 3 & 0 & 0 & 0 \\ 0 & 0 & 3 & 0 & 0 \\ 0 & 0 & 0 & 2 & 1 \\ 0 & 0 & 0 & 0 & 2 \end{bmatrix}$$

is the canonical Jordan form of A.

Example 6.62 Let $T : \mathbb{R}^5 \to \mathbb{R}^5$ be the linear operator having matrix

$$A = \begin{bmatrix} 0 & 0 & 0 & -1 & 1 \\ -1 & 1 & 1 & -1 & 1 \\ 0 & 0 & 1 & 0 & 0 \\ 1 & 0 & 0 & 2 & 0 \\ 0 & 0 & 0 & 0 & 1 \end{bmatrix}$$

with respect to the canonical basis of $\mathbb{R}^5$. The characteristic polynomial is $p(\lambda) = (1 - \lambda)^5$, having one root $\lambda = 1$ (with algebraic multiplicity equal to 5).
We have

$$A - I = \begin{bmatrix} -1 & 0 & 0 & -1 & 1 \\ -1 & 0 & 1 & -1 & 1 \\ 0 & 0 & 0 & 0 & 0 \\ 1 & 0 & 0 & 1 & 0 \\ 0 & 0 & 0 & 0 & 0 \end{bmatrix}$$

$$(A - I)^2 = \begin{bmatrix} 0 & 0 & 0 & 0 & -1 \\ 0 & 0 & 0 & 0 & -1 \\ 0 & 0 & 0 & 0 & 0 \\ 0 & 0 & 0 & 0 & 1 \\ 0 & 0 & 0 & 0 & 0 \end{bmatrix}$$

and $(A - I)^3 = 0$. In particular, this means that there exists a Jordan block corresponding to λ of dimension 3 and any other Jordan block of λ has dimension less than 3.

Therefore, $A - I$ has rank $r_1 = 3$, so that the associated homogeneous linear system has solutions $X = (\alpha, \beta, 0, -\alpha, 0)$, for any $\alpha, \beta \in \mathbb{R}$. So $N_{1,\lambda} = \langle (1, 0, 0, -1, 0), (0, 1, 0, 0, 0) \rangle$, the geometric multiplicity of λ is equal to 2 and there are 2 Jordan blocks corresponding to the eigenvalue λ. Since $(A - I)^2$ has rank $r_2 = 1$, the associated homogeneous linear system has solutions $X = (\alpha, \beta, \gamma, \delta, 0)$, for any $\alpha, \beta, \gamma, \delta \in \mathbb{R}$. So

$$N_{2,\lambda} = \langle (1, 0, 0, 0, 0), (0, 1, 0, 0, 0), (0, 0, 1, 0, 0), (0, 0, 0, 1, 0) \rangle.$$

Moreover, $(A - I)^3 = 0$ implies $N_{3,\lambda} = \mathbb{R}^5$. We also remark that, the number of blocks corresponding to λ, having dimension 1 is equal to $n - 2r_1 + r_2 = 0$. Analogously, the number of blocks corresponding to λ, having dimension 2 is equal to $r_1 - 2r_2 + r_3 = 1$, as expected. In other words, we have one block of dimension 2 and one block of dimension 3.

To obtain a set of Jordan generators $\{X_1, X_2, X_3\}$ corresponding to λ and associated with the block of dimension 3, we start from $X_3 \in N_{3,\lambda} \setminus N_{2,\lambda}$. We choose $X_3 = (0, 0, 0, 0, 1)$. Then compute $X_2 = (A - I)X_3 = (1, 1, 0, 0, 0)$, $X_1 = (A - I)X_2 = (-1, -1, 0, 1, 0)$. Hence $B_1 = \{X_1, X_2, X_3\}$ is a set of Jordan generators associated with the block of dimension 3. The set of Jordan generators associated with the block of dimension 2 must be $B_2 = \{X_4, X_5\}$, where $X_5 \in N_{2,\lambda} \setminus N_{1,\lambda}$ and $X_5 \notin \langle X_1, X_2, X_3 \rangle$. We choose $X_5 = (0, 0, 1, 0, 0)$, so $X_4 = (A - I)X_5 = (0, 1, 0, 0, 0)$.

Hence, we can write a Jordan basis for T :

$B = B_1 \cup B_2$
$= \{X_1, X_2, X_3, X_4, X_5\}$
$= \{(-1, -1, 0, 1, 0), (1, 1, 0, 0, 0), (0, 0, 0, 0, 1), (0, 1, 0, 0, 0), (0, 0, 1, 0, 0)\}.$

Let A' be the matrix of T with respect to B. If P is the following transition matrix

$$P = \begin{bmatrix} -1 & 1 & 0 & 0 & 0 \\ -1 & 1 & 0 & 1 & 0 \\ 0 & 0 & 0 & 0 & 1 \\ 1 & 0 & 0 & 0 & 0 \\ 0 & 0 & 1 & 0 & 0 \end{bmatrix}$$

then

$$A' = P^{-1}AP = \begin{bmatrix} 1 & 1 & 0 & 0 & 0 \\ 0 & 1 & 1 & 0 & 0 \\ 0 & 0 & 1 & 0 & 0 \\ 0 & 0 & 0 & 1 & 1 \\ 0 & 0 & 0 & 0 & 2 \end{bmatrix}$$

is the canonical Jordan form of A.

As in the previous example, one can also compute A' by using the fact that the column coordinate vectors of A' are the images of the element of B (the reader can easily verify it).

Exercises

1. For each of the following linear operators T on the vector space V, determine whether the given subspace W is a T-invariant subspace of V :

    (a) $V = P_3(\mathbb{R})$, $T(f(x)) = f'(x)$, and $W = P_2(\mathbb{R})$.
    (b) $V = P(\mathbb{R})$, $T(f(x)) = xf(x)$, and $W = P_2(\mathbb{R})$.
    (c) $V = \mathbb{R}^3$, $T(a, b, c) = T(a + b + c, a + b + c, a + b + c)$ and $W = \{(t, t, t) : t \in \mathbb{R}\}$.
    (d) $V = C([0, 1])$, $T(f(t)) = [\int_0^1 f(x)dx]t$, and $W = \{f \in V : f(t) = at + b \text{ for some } a \text{ and } b\}$.
    (e) $V = M_2(\mathbb{R})$, $T(A) = \begin{pmatrix} 0 & 1 \\ 1 & 0 \end{pmatrix} A$, and $W = \{A \in V : A^t = A\}$.

2. Let $T : \mathbb{R}^4 \to \mathbb{R}^4$ be the linear operator having matrix

$$A = \begin{bmatrix} 2 & 0 & 0 & 0 \\ 1 & 3 & 1 & 0 \\ 0 & -1 & 1 & 1 \\ 0 & 1 & 2 & 1 \end{bmatrix}$$

with respect to the canonical basis of $\mathbb{R}^4$. Determine the (diagonal or Jordan) canonical form A' of A and the basis for $\mathbb{R}^4$ with respect to which the matrix of T is A'.

3. Repeat Exercise 2, for the linear operator $T : \mathbb{R}^4 \to \mathbb{R}^4$ having matrix

$$A = \begin{bmatrix} 1 & 0 & 0 & 0 \\ 1 & -2 & 0 & -4 \\ 0 & 0 & -1 & 0 \\ 0 & 1 & 0 & 3 \end{bmatrix}$$

with respect to the canonical basis of $\mathbb{R}^4$.

4. Repeat Exercise 2, for the linear operator $T : \mathbb{R}^5 \to \mathbb{R}^5$ having matrix

$$A = \begin{bmatrix} 1 & 4 & 1 & 1 & 0 \\ 1 & 1 & 2 & 1 & 0 \\ 0 & 0 & 1 & 4 & 0 \\ 0 & 0 & 1 & 1 & 0 \\ 0 & 1 & 0 & 1 & 3 \end{bmatrix}$$

with respect to the canonical basis of $\mathbb{R}^5$.

5. Repeat Exercise 2, for the linear operator $T : \mathbb{R}^6 \to \mathbb{R}^6$ having matrix

$$A = \begin{bmatrix} 0 & 1 & 1 & 0 & 0 & 0 \\ 0 & 1 & 0 & 0 & 1 & -1 \\ -1 & 1 & 2 & 0 & 0 & 0 \\ 0 & 0 & 0 & 1 & 0 & 0 \\ 1 & 0 & -1 & 1 & 1 & 1 \\ 0 & 0 & 0 & 1 & 0 & 1 \end{bmatrix}$$

with respect to the canonical basis of $\mathbb{R}^6$.

6. Let $T : \mathbb{C}^3 \to \mathbb{C}^3$ be the linear operator having matrix

$$A = \begin{bmatrix} i & 1 & 0 \\ 1 & i & i \\ 0 & 1 & i \end{bmatrix}$$

with respect to the canonical basis of $\mathbb{C}^3$. Determine the (diagonal or Jordan) canonical form A' of A and the basis for $\mathbb{C}^3$ with respect to which the matrix of T is A'.

7. Let $T : V \to V$ be a linear operator on a finite dimensional complex vector space V. Prove that T is diagonalizable if and only if every generalized eigenvector of T is an eigenvector of T.

8. Let $\mathbb{F}$ be a field and $A, B \in M_n(\mathbb{F})$ be such that their characteristic polynomials split over $\mathbb{F}$. Prove that A and B are similar if and only if they have the same Jordan canonical form.

6.5   Cayley-Hamilton Theorem and Minimal Polynomial

Let $\mathbb{F}$ be a field, $p(X) \in \mathbb{F}[X]$ a polynomial of degree $d \geq 1$, $A \in M_n(\mathbb{F})$ a $n \times n$ matrix. If we write

$$p(X) = a_0 + a_1 X + a_2 X^2 + \cdots + a_d X^d, \quad a_0, a_1, \ldots, a_d \in \mathbb{F},$$

we can substitute the matrix A for the indeterminate X and obtain a matrix $p(A) \in M_n(\mathbb{F})$, more precisely

$$p(A) = a_0 I + a_1 A + a_2 A^2 + \cdots + a_d A^d \in M_n(\mathbb{F}).$$

In particular, we say that A is a root for the polynomial $p(X)$, in case $p(A)$ is the zero matrix in $M_n(\mathbb{F})$ and write $p(A) = 0$. Starting from these comments, we may now state the following:

Theorem 6.63 (Cayley-Hamilton Thoerem) *Let* $T : V \to V$ *be a linear operator of the finite dimensional vector space* V, $A \in M_n(\mathbb{F})$ *be the matrix of* T *with respect to a basis for* V. *If* $p(\lambda) \in \mathbb{F}[\lambda]$ *is the characteristic polynomial of* T, *then* $p(A) = 0$.

Proof Let $\lambda_1, \ldots, \lambda_k$ be the distinct eigenvalues of T, t_i the index of λ_i, for $i = 1, \ldots, k$. Let $V = N_1 \oplus \cdots \oplus N_k$, where $N_i = Ker\big((A - \lambda_i I)^{t_i}\big) = N_{t_i, \lambda_i}$ is the generalized eigenspace corresponding to the eigenvalue λ_i. For any $X \in V$, there exist $X_1 \in N_1, X_2 \in N_2, \ldots, X_k \in N_k$ such that

$$X = X_1 + \cdots + X_k. \tag{6.14}$$

Recall that, if a_i is the algebraic multiplicity of λ_i, then $a_i \geq t_i$, so $(A - \lambda_i I)^{a_i} X_i = 0$, for any $X_i \in N_i$. Therefore, multiplying on the left the relation (6.14) by the matrix $p(A)$, we have

$$p(A)X = p(A)X_1 + \cdots + p(A)X_k. \tag{6.15}$$

On the other hand, $p(\lambda) = (-1)^n (\lambda - \lambda_1)^{a_1} \cdots (\lambda - \lambda_k)^{a_k}$, that is

$$p(A) = (-1)^n (A - \lambda_1 I)^{a_1} \cdots (A - \lambda_k I)^{a_k}. \tag{6.16}$$

Since $(A - \lambda_i I)$ and $(A - \lambda_j I)$ commute for any $\lambda_i \neq \lambda_j$, and starting from (6.16), it is easy to see that

$$p(A)X_i = (-1)^n (A - \lambda_1 I)^{a_1} \cdots (A - \lambda_k I)^{a_k} X_i = 0, \quad \forall X_i \in N_i, \forall i = 1, \ldots, k. \tag{6.17}$$

Hence combining (6.15) and (6.17) we get $p(A)X = 0$, for any $X \in V$. Thus $p(A)$ is the matrix which represents the zero operator over V, that is, $p(A)$ is the zero matrix in $M_n(\mathbb{F})$.

In the following example, we show how Cayley-Hamilton Theorem can be used in order to solve some problems related to the computation of the inverse of a matrix, the power of a matrix and their eigenvalues.

Example 6.64 Let $T : \mathbb{R}^2 \to \mathbb{R}^2$ be the linear operator having matrix

$$A = \begin{bmatrix} 2 & 3 \\ 3 & 2 \end{bmatrix}$$

with respect to the canonical basis of $\mathbb{R}^2$. The characteristic polynomial of T is $p(\lambda) = \lambda^2 - 4\lambda - 5$. The eigenvalues of T are $-1, 5$ and, by Cayley-Hamilton Theorem, $A^2 - 4A - 5I = 0$, where I is the identity matrix of order 2.

We firstly use the Cayley-Hamilton Theorem to compute the power A^6. Starting from $A^2 = 4A + 5I$, we get

$$A^3 = 4A^2 + 5A = 4(4A + 5I) + 5A = 21A + 20I.$$

Then
$A^6 = (A^3)^2 = (21A + 20I)^2 = 441A^2 + 840A + 400I = 441(4A + 5I) + 840A + 400I = 2604A + 2605I$. This implies that

$$A^6 = \begin{bmatrix} 7813 & 7812 \\ 7812 & 7813 \end{bmatrix}.$$

We now obtain the inverse of A, starting again from $A^2 - 4A = 5I$. Multiplying by A^{-1} we get $A - 4I = 5A^{-1}$, so that

$$A^{-1} = \frac{1}{5}(A - 4I) = \frac{1}{5}\begin{bmatrix} -2 & 3 \\ 3 & -2 \end{bmatrix}.$$

Finally, we compute the eigenvalues for the matrix $B = A^4 - 3A^3 + 2A^2 - A + I$.

If we divide $x^4 - 3x^3 + 2x^2 - x + 1$ by the characteristic polynomial $x^2 - 4x - 5$, we get

$$x^4 - 3x^3 + 2x^2 - x + 1 = (x^2 - 4x - 5)(x^2 + x + 11) + (48x + 56).$$

In particular, since $A^2 - 4A - 5I = 0$, one has

$$B = A^4 - 3A^3 + 2A^2 - A + I = (A^2 - 4A - 5I)(A^2 + A + 11I) + (48A + 56I) = 48A + 56I$$

that is

$$B = \begin{bmatrix} 152 & 144 \\ 144 & 152 \end{bmatrix}$$

having eigenvalues $8, 296$.

Definition 6.65 Let A be the matrix of a linear operator $T : V \to V$, $p(\lambda) \in \mathbb{F}[\lambda]$ the characteristic polynomial of T. A polynomial $m(\lambda) \in \mathbb{F}[\lambda]$ is said to be the minimal polynomial of T if it is the monic polynomial of smallest degree such that $m(A)$ is the zero matrix.

Theorem 6.66 *Let A be the matrix of a linear operator $T : V \to V$, $f(\lambda) \in \mathbb{F}[\lambda]$ a polynomial. Then $f(A) = 0$ if and only if the minimal polynomial $m(\lambda)$ of T divides $f(\lambda)$.*

Proof Let $f(A) = 0$. By the division algorithm for polynomials, there exist polynomials $q(\lambda), r(\lambda) \in \mathbb{F}[\lambda]$ such that $f(\lambda) = q(\lambda)m(\lambda) + r(\lambda)$, where $r(\lambda) = 0$ or $deg\, r(\lambda) < deg\, m(\lambda)$. Thus $0 = f(A) = q(A)m(A) + r(A) = r(A)$. Since $m(\lambda)$ is the minimal polynomial of T and $deg\, r(\lambda) < deg\, m(\lambda)$, then the polynomial $r(\lambda)$ must be identically zero. Thus $m(\lambda)$ divides $f(\lambda)$.

Assume now that $f(\lambda) = m(\lambda)q(\lambda)$, for some $q(\lambda) \in \mathbb{F}[\lambda]$. Then $f(A) = m(A)q(A) = 0$, as required.

Corollary 6.67 *The minimal polynomial of a linear operator T divides the characteristic polynomial of T.*

Theorem 6.68 *Let A be the matrix of a linear operator $T : V \to V$, $m(\lambda) \in \mathbb{F}[\lambda]$ the minimal polynomial of T. $\lambda_0 \in \mathbb{F}$ is a root of $m(\lambda)$ if and only if λ_0 is an eigenvalue of T.*

Proof Let $p(\lambda) \in \mathbb{F}[\lambda]$ be the characteristic polynomial of T. The result is trivial in case $m(\lambda) = p(\lambda)$. Thus we can consider $deg(m) < deg(p)$ and $p(\lambda) = m(\lambda)q(\lambda)$ for some $q(\lambda) \in \mathbb{F}[\lambda]$.

Assume first that λ_0 is a root of $m(\lambda)$. In this case it is easy to see that $p(\lambda_0) = m(\lambda_0)q(\lambda_0) = 0$, that is λ_0 is a root of the characteristic polynomial, i.e., λ_0 is an eigenvalue of T.

Suppose, now $p(\lambda_0) = 0$ and suppose by contradiction that $(\lambda - \lambda_0)$ does not divide $m(\lambda)$. Hence, there exists a polynomial $q(\lambda) \in \mathbb{F}[\lambda]$ such that $m(\lambda) = (\lambda - \lambda_0)q(\lambda) + r$, where $deg(q) = deg(m) - 1$ and $0 \neq r \in \mathbb{F}$. Hence $0 = m(A) = (A - \lambda_0 I)q(A) + rI$, that is $-rI = (A - \lambda_0 I)q(A)$. By computing the determinants of the matrices in the last identity, we get $|-rI| = |A - \lambda_0 I||q(A)|$, which means $(-r)^n = |A - \lambda_0 I||q(A)|$. Since λ_0 is an eigenvalue of T, $|A - \lambda_0 I| = p(\lambda_0) = 0$ and $(-r)^n = 0$, which is a contradiction.

Theorem 6.69 *The minimal polynomial of a linear operator $T : V \to V$ has the following form*

$$m(\lambda) = (\lambda - \lambda_1)^{t_1} \cdots (\lambda - \lambda_k)^{t_k},$$

where $\lambda_1, \ldots, \lambda_k$ are the distinct eigenvalues of T and t_i is the index of λ_i, for any $i = 1, \ldots, k$.

Proof Let $N_1, \ldots, N_k$ be the generalized eigenspace of $\lambda_1, \ldots, \lambda_k$ respectively. Then $V = N_1 \oplus \cdots \oplus N_k$ and, for any $X \in V$ there are $X_1 \in N_1, \ldots, X_k \in N_k$ such that $X = X_1 + \cdots + X_k$. Let A be the matrix of the linear operator T, with regard to a basis B of V.

Since $(A - \lambda_i I)$ and $(A - \lambda_j I)$ commute for any $\lambda_i \neq \lambda_j$, we see that

$$(A - \lambda_1 I)^{t_1} \cdots (A - \lambda_k I)^{t_k} X_i = 0, \quad \text{for all } X_i \in N_i, \forall i = 1, \ldots, k$$

that is

$$(A - \lambda_1 I)^{t_1} \cdots (A - \lambda_k I)^{t_k} X = 0, \quad \text{for all } X \in V.$$

Therefore, the polynomial $m(\lambda)$ represents the zero operator on V, so that $m(A) = 0$. Moreover, the polynomial $m(\lambda)$ divides the characteristic polynomial $p(\lambda)$, since any $t_i \leq a_i$, for any $i = 1, \ldots, k$, where a_i is the algebraic multiplicity of λ_i.

To conclude the proof, it is sufficient to show that $m(\lambda)$ is the monic polynomial of smallest degree such that $m(A)$ is the zero matrix.

Denote

$$f(\lambda) = (\lambda - \lambda_1)^{s_1} \cdots (\lambda - \lambda_k)^{s_k},$$

the minimal polynomial of T and suppose by contradiction that $f(\lambda) \neq m(\lambda)$. Since $m(A) = 0$, by Theorem 6.66, $f(\lambda)$ divides $m(\lambda)$. In other words, there is at least one $s_j \in \{s_1, \ldots, s_k\}$ such that $s_j < t_j$. Since t_j is the index of λ_j, then there exists $X \in N_j$ such that $0 \neq (A - \lambda_j I)^{s_j} X = Y \in N_j$. On the other hand, for any $Z \in N_j$ and $l \neq j$, we know that $Z \notin N_l$ and, by Lemma 6.42, $0 \neq (A - \lambda_l I)^{s_l} Z \in N_j$. Moreover, since $(A - \lambda_i I)$ and $(A - \lambda_j I)$ commute for any $\lambda_i \neq \lambda_j$, we can write the factorization of $f(A)$ as follows:

$$f(A) = (A - \lambda_1 I)^{s_1} \cdots (A - \lambda_{j-1} I)^{s_{j-1}} (A - \lambda_{j+1} I)^{s_{j+1}} \cdots (A - \lambda_j I)^{s_k} (A - \lambda_j I)^{s_j}.$$

It is obvious that $f(A) = 0$. Therefore, $f(A)(X) = 0$ and $f(A)(Y) = 0$ for all $X, Y \in N_j$, $j = 1, 2, \cdots, k$. On the other hand, we arrive at

$$
\begin{aligned}
0 &= f(A)(X) \\
&= (A - \lambda_1 I)^{s_1} \cdots (A - \lambda_{j-1} I)^{s_{j-1}} (A - \lambda_{j+1} I)^{s_{j+1}} \cdots (A - \lambda_j I)^{s_k} (A - \lambda_j I)^{s_j} X \\
&= (A - \lambda_1 I)^{s_1} \cdots (A - \lambda_{j-1} I)^{s_{j-1}} (A - \lambda_{j+1} I)^{s_{j+1}} \cdots (A - \lambda_j I)^{s_k} Y \\
&\neq 0
\end{aligned}
$$

which leads to a contradiction. Thus, we conclude that $s_i = t_i$ for all $i = 1, \ldots, k$ that is, $m(\lambda)$ is precisely the minimal polynomial of T.

Example 6.70 Let $T : \mathbb{R}^5 \to \mathbb{R}^5$ be the linear operator as in the Example 6.61. The characteristic polynomial of T is $p(\lambda) = (3 - \lambda)^3 (2 - \lambda)^2$ and the Jordan canonical form of T is represented by the matrix

$$
A' = \begin{bmatrix}
3 & 1 & 0 & 0 & 0 \\
0 & 3 & 0 & 0 & 0 \\
0 & 0 & 3 & 0 & 0 \\
0 & 0 & 0 & 2 & 1 \\
0 & 0 & 0 & 0 & 2
\end{bmatrix}
$$

with respect to the basis

$$B = \{(1, 0, 0, 0, 0), (0, 1, 0, 0, 0), (0, -2, 1, 0, 0), (-1, -2, 1, 1, 0), (-3, 0, 0, 0, 1)\}$$

for $\mathbb{R}^5$. In this case, the minimal polynomial of T is $m(\lambda) = (3 - \lambda)^2(2 - \lambda)^2$.

Example 6.71 Let $T : \mathbb{R}^5 \to \mathbb{R}^5$ be the linear operator as in Example 6.62. The characteristic polynomial of T is $p(\lambda) = (1 - \lambda)^5$ and the Jordan canonical form of T is represented by the matrix

$$A' = \begin{bmatrix} 1 & 1 & 0 & 0 & 0 \\ 0 & 1 & 1 & 0 & 0 \\ 0 & 0 & 1 & 0 & 0 \\ 0 & 0 & 0 & 1 & 1 \\ 0 & 0 & 0 & 0 & 2 \end{bmatrix}$$

with respect to the basis

$$B = \{(-1, -1, 0, 1, 0), (1, 1, 0, 0, 0), (0, 0, 0, 0, 1), (0, 1, 0, 0, 0), (0, 0, 1, 0, 0)\}$$

for $\mathbb{R}^5$. The minimal polynomial of T is $m(\lambda) = (1 - \lambda)^3$.

Exercises

1. Let $T : V \to V$ be a linear operator on a vector space V of dimension 3 over the field $\mathbb{F}$, such that the characteristic polynomial of T splits over $\mathbb{F}$. Describe all possibilities for the Jordan canonical form and for the minimal polynomial of T.

2. Repeat Exercise 1, in case $T : V \to V$ is a linear operator on a vector space V of dimension 4 over the field $\mathbb{F}$.

3. Let $T : \mathbb{R}^3 \to \mathbb{R}^3$ be the linear operator having matrix

$$A = \begin{bmatrix} 2 & 1 & 2 \\ 1 & 2 & 2 \\ 0 & 0 & 1 \end{bmatrix}$$

with respect to the canonical basis for $\mathbb{R}^3$. Applying the Cayley-Hamilton Theorem, find the eigenvalues and the eigenvectors of the matrix $B = A^5 - 5A^4 + 8A^3 - 8A^2 + 8A - 7I$.

4. Let $T : \mathbb{R}^2 \to \mathbb{R}^2$ be the linear operator having matrix

$$A = \begin{bmatrix} 0 & 1 \\ 2 & 1 \end{bmatrix}$$

with respect to the canonical basis for $\mathbb{R}^2$. Compute A^{10} using the Cayley-Hamilton Theorem.

5. Let $T : \mathbb{R}^3 \to \mathbb{R}^3$ be the linear operator having matrix

$$A = \begin{bmatrix} 3 & 1 & 1 \\ 1 & 3 & 1 \\ 0 & 0 & 3 \end{bmatrix}$$

with respect to the canonical basis for $\mathbb{R}^3$. Find the inverse matrix of A using the Cayley-Hamilton Theorem.

6. Let $T : \mathbb{R}^3 \to \mathbb{R}^3$ be the linear operator having matrix

$$A = \begin{bmatrix} 3 & 1 & 1 \\ 0 & -1 & 1 \\ 0 & 0 & 2 \end{bmatrix}$$

with respect to the canonical basis for $\mathbb{R}^3$. Applying the Cayley-Hamilton Theorem,

   (a) Find the eigenvalues and the eigenvectors of the matrix $B = A^5 - 2A^4 - 7A^3 + 8A^2 + 11A - 2I$.
   (b) Find the inverse matrix of A.
   (c) Compute the matrix A^8.

7. Let $T : \mathbb{R}^9 \to \mathbb{R}^9$ be the linear operator having characteristic polynomial equal to $p(\lambda) = (3 - \lambda)^4(2 - \lambda)^2(-1 - \lambda)^3$ and minimal polynomial equal to $m(\lambda) = (3 - \lambda)^2(2 - \lambda)(-1 - \lambda)^2$. Describe all possibilities for the Jordan canonical form of T.

6.6   Normal Operators on Inner Product Spaces

In this final Section, V is an inner product space over $\mathbb{F}$, $T : V \to V$ a linear operator. Here, we always assume that $T : V \to V$ is normal ($TT^* = T^*T$) and $\mathbb{F}$ denote either $\mathbb{R}$ or $\mathbb{C}$. We remind the reader of some operators which belong to the more general class of normal operators:

   (i)   *self-adjoint* operator: $T = T^*$ and $\langle T(u), v \rangle = \langle u, T(v) \rangle$, for any $u, v \in V$;
   (ii) *skew-adjoint* operator: $T = -T^*$ and $\langle T(u), v \rangle = -\langle u, T(v) \rangle$, for any $u, v \in V$;
   (iii) *unitary (orthogonal in case $\mathbb{F} = \mathbb{R}$)* operator: $T^* = T^{-1}$ and $\langle T(u), T(v) \rangle = \langle u, v \rangle$, for any $u, v \in V$.

Let B be a basis of V with respect to which $A \in M_n(\mathbb{F})$ is the matrix of T and A^* is the matrix of T^*. Thus, for any $X \in V$, $TT^*(X) = T^*T(X)$ and $AA^*X = A^*AX$, that is $(AA^* - A^*A)X = 0$. Hence $AA^* = A^*A$, which means that the matrix A of a normal operator is also normal, in the sense that it commutes with its conjugate transpose A^*.

Remark 6.72 Symmetric, Hermitian, skew-symmetric, skew-Hermitian, orthogonal and unitary operators are normal operators. As a consequence,

(*i*)   the matrix of a symmetric (Hermitian) operator is symmetric (Hermitian);
(*ii*) the matrix of a skew-symmetric (skew-Hermitian) operator is skew-symmetric (skew-Hermitian);
(*iii*) the matrix of an orthogonal (unitary) operator is orthogonal (unitary).

The above remark is being justified as below.
   Let $T : V \to V$ the operator and A, A^* be the matrices of T and T^* respectively.

(*i*)   If $T = T^*$ then $T(X) = T^*(X)$, for any $X \in V$. Thus $AX = A^*X$ and $(A - A^*)X = 0$ for any $X \in V$, implying that $A = A^*$.
(*ii*) By using the same above argument, one can prove that $T = -T^*$ implies $A = -A^*$.
(*iii*) If $T^* = T^{-1}$ then $TT^*(X) = X$, that is $AA^*X = X$, for any $X \in V$. Thus AA^* is the identity in $M_n(\mathbb{F})$ so that $A^* = A^{-1}$.

Here, we develop a detailed description of the class of normal operators (matrices). We start with the following:

Lemma 6.73 *Let* $T : V \to V$ *be normal and let* $X \in V$ *be an eigenvector of* T *corresponding to the eigenvalue* λ. *Then* X *is an eigenvector of* T^* *corresponding to the eigenvalue* $\overline{\lambda}$.

Proof Since T is normal, so is $T - \lambda I$. We are also given $T(X) = \lambda X$, i.e., $(T - \lambda I)X = 0$. This implies that

$$0 = \|(T - \lambda I)X\| = \|(T - \lambda I)^*X\| = \|(T^* - \overline{\lambda}I)X\|$$

that is $(T^* - \overline{\lambda}I)X = 0$, as desired.

Lemma 6.74 *Let* $T : V \to V$ *be normal. Then, eigenvectors of* T *corresponding to distinct eigenvalues are orthogonal.*

Proof Let λ, μ be distinct eigenvalues of T and $X, Y \in V$ be such that $T(X) = \lambda X$, $T(Y) = \mu Y$. Then
$$\langle T(X), Y \rangle = \langle \lambda X, Y \rangle = \lambda \langle X, Y \rangle$$

and also
$$\langle T(X), Y \rangle = \langle X, T^*(Y) \rangle = \langle X, \overline{\mu}Y \rangle = \mu \langle X, Y \rangle.$$

   Hence $\lambda \langle X, Y \rangle = \mu \langle X, Y \rangle$, so that $\langle X, Y \rangle = 0$ follows from $\lambda \neq \mu$.

Lemma 6.75 *Let* $T : V \to V$ *be a self-adjoint operator on a inner product space* V *over the field* $\mathbb{F}$. *Every eigenvalue of* T *is real.*

Proof Let λ be an eigenvalue of T and $X \in V$ be an eigenvector of T corresponding to λ, i.e., $T(X) = \lambda X$. Then

$$\lambda \|X\|^2 = \lambda \langle X, X \rangle = \langle \lambda X, X \rangle = \langle T(X), X \rangle =$$
$$\langle X, T(X) \rangle = \langle X, \lambda X \rangle = \overline{\lambda} \langle X, X \rangle = \overline{\lambda} \|X\|^2 .$$

Hence, $\lambda = \overline{\lambda} \in \mathbb{R}$.

Lemma 6.76 *Let* $T : V \to V$ *be a skew-adjoint operator on a inner product space* V *over the field* $\mathbb{F}$. *Every nonzero eigenvalue of* T *is purely imaginary.*

Proof Assume there exists a nonzero eigenvalue λ of T and $X \in V$ be an eigen vector corresponding to the eigen value λ. Thus $T(X) = \lambda X$. In this case

$$\lambda \|X\|^2 = \lambda \langle X, X \rangle = \langle \lambda X, X \rangle = \langle T(X), X \rangle =$$
$$\langle -X, T(X) \rangle = \langle -X, \lambda X \rangle = -\overline{\lambda} \langle X, X \rangle = -\overline{\lambda} \|X\|^2 .$$

Hence $\lambda = -\overline{\lambda}$, that is the real part of λ is zero.

Remark 6.77 Immediate consequences are that:

(*i*) If T is a symmetric operator on a real inner product space, or T is Hermitian on a complex inner product space, then every eigenvalue of T is real.
(*ii*) If T is a skew-symmetric operator on a real inner product space, or T is skew-Hermitian on a complex inner product space, then every eigenvalue of T is purely imaginary.

Moreover, we have the following:

Remark 6.78 If T is unitary (orthogonal), then every eigenvalue has modulus (absolute value) equal to 1. In fact, if $\lambda \in \mathbb{C}$ is an eigenvalue of T and $X \in V$ is an eigenvector of T corresponding to λ, such that $\|X\| = 1$, then

$$1 = \langle X, X \rangle = \langle T(X), T(X) \rangle = \langle \lambda X, \lambda X \rangle = \lambda \overline{\lambda} \langle X, X \rangle = \lambda \overline{\lambda}$$

as desired.

Theorem 6.79 (Spectral Theorem) *Let* $T : V \to V$ *be an operator on the inner product space* V *over a field* $\mathbb{F}$, *A the matrix of* T. *Then A is orthogonally similar to a diagonal matrix if and only if T is self-adjoint. In other words, there exists an orthonormal basis of V consisting of eigenvectors of T if and only if T is self-adjoint.*

Proof Suppose that V has an orthonormal basis B consisting of eigenvectors of T. Then the matrix A' of T with respect to B is diagonal. Since $A' = (A')^*$, then A' is self-adjoint as well as T.

We now prove the other direction of the theorem and assume that T is self-adjoint. By Lemma 6.75, every eigenvalue of T is real, in particular the characteristic polynomial of T splits over $\mathbb{F}$. By Theorem 6.20, there exists an orthonormal basis B of V with respect to which the matrix A' of T is upper-triangular, that is there exists an orthonormal matrix $P \in M_n(\mathbb{F})$ such that $P^*AP = A'$. Since $A^* = A$, $(A')^* = P^*AP = A'$. Therefore, A' is simultaneously self-adjoint and upper-triangular. In this case, it is easy to see that A' must be a diagonal matrix.

Remark 6.80 The Spectral Theorem can be stated from two different points of view:

(i) If V is a real inner product space, then the matrix A of T is orthogonally similar to a diagonal matrix if and only if A is symmetric.

(ii) If V is a complex inner product space, then the matrix A of T is unitarily similar to a diagonal matrix if and only if A is Hermitian.

Example 6.81 Let $T : \mathbb{R}^4 \to \mathbb{R}^4$ be the symmetric linear operator having matrix

$$A = \begin{bmatrix} 2 & 1 & 1 & 0 \\ 1 & 2 & 1 & 0 \\ 1 & 1 & 2 & 0 \\ 0 & 0 & 0 & 2 \end{bmatrix}$$

with respect to the canonical basis of $\mathbb{R}^4$. The characteristic polynomial is $p(\lambda) = (1 - \lambda)^2(2 - \lambda)(4 - \lambda)$.

For $\lambda_1 = 1$, we have

$$A - I = \begin{bmatrix} 1 & 1 & 1 & 0 \\ 1 & 1 & 1 & 0 \\ 1 & 1 & 1 & 0 \\ 0 & 0 & 0 & 1 \end{bmatrix}$$

so that the associated homogeneous linear system has solutions: $X = (-\alpha - \beta, \alpha, \beta, 0)$, for any $\alpha, \beta \in \mathbb{R}$. So the corresponding eigenspace is $N_{1,\lambda_1} = \langle (-1, 1, 0, 0), (-1, 0, 1, 0) \rangle$. Starting from vectors $(-1, 1, 0, 0)$, $(-1, 0, 1, 0)$, we can easily construct the following orthonormal basis for N_{1,λ_1}:

$$N_{1,\lambda_1} = \left\langle \left(-\frac{1}{\sqrt{2}}, \frac{1}{\sqrt{2}}, 0, 0 \right), \left(\frac{1}{\sqrt{6}}, \frac{1}{\sqrt{6}}, -\frac{2}{\sqrt{6}}, 0 \right) \right\rangle.$$

For $\lambda_2 = 2$, we have

$$A - 2I = \begin{bmatrix} 0 & 1 & 1 & 0 \\ 1 & 0 & 1 & 0 \\ 1 & 1 & 0 & 0 \\ 0 & 0 & 0 & 0 \end{bmatrix}$$

so that the associated homogeneous linear system has solutions: $X = (0, 0, 0, \alpha)$, for any $\alpha \in \mathbb{R}$. So the corresponding eigenspace is $N_{1,\lambda_2} = \langle (0, 0, 0, 1) \rangle$. Finally, for $\lambda_3 = 4$, we have

$$A - 4I = \begin{bmatrix} -2 & 1 & 1 & 0 \\ 1 & -2 & 1 & 0 \\ 1 & 1 & -2 & 0 \\ 0 & 0 & 0 & -2 \end{bmatrix}$$

so that the associated homogeneous linear system has solutions: $X = (\alpha, \alpha, \alpha, 0)$, for any $\alpha \in \mathbb{R}$. The corresponding eigenspace is
$N_{1,\lambda_3} = \langle (1, 1, 1, 0) \rangle = \langle (\frac{1}{\sqrt{3}}, \frac{1}{\sqrt{3}}, \frac{1}{\sqrt{3}}, 0) \rangle$.

Therefore, with respect to the orthonormal basis

$$\left\{ \left(-\frac{1}{\sqrt{2}}, \frac{1}{\sqrt{2}}, 0, 0 \right), \left(\frac{1}{\sqrt{6}}, \frac{1}{\sqrt{6}}, -\frac{2}{\sqrt{6}}, 0 \right), (0, 0, 0, 1), \left(\frac{1}{\sqrt{3}}, \frac{1}{\sqrt{3}}, \frac{1}{\sqrt{3}}, 0 \right) \right\}$$

the matrix of T is

$$A' = \begin{bmatrix} 1 & 0 & 0 & 0 \\ 0 & 1 & 0 & 0 \\ 0 & 0 & 2 & 0 \\ 0 & 0 & 0 & 4 \end{bmatrix}.$$

Example 6.82 Let $T : \mathbb{C}^4 \to \mathbb{C}^4$ be the Hermitian linear operator having matrix

$$A = \begin{bmatrix} 1 & -i & 0 & 0 \\ i & 0 & -1 & 0 \\ 0 & -1 & 1 & 0 \\ 0 & 0 & 0 & 2 \end{bmatrix}$$

with respect to the canonical basis of $\mathbb{C}^4$. The characteristic polynomial is
$p(\lambda) \doteq (2 - \lambda)^2 (1 - \lambda)(-1 - \lambda)$.

For $\lambda_1 = 1$, we have

$$A - I = \begin{bmatrix} 0 & -i & 0 & 0 \\ i & -1 & -1 & 0 \\ 0 & -1 & 0 & 0 \\ 0 & 0 & 0 & 1 \end{bmatrix}$$

so that the associated homogeneous linear system has solutions: $X = (\alpha, 0, i\alpha, 0)$, for any $\alpha \in \mathbb{R}$. The corresponding eigenspace is $N_{1,\lambda_1} = \langle (1, 0, i, 0) \rangle = \langle (\frac{1}{\sqrt{2}}, 0, \frac{i}{\sqrt{2}}, 0) \rangle$. For $\lambda_2 = -1$, we have

$$A + I = \begin{bmatrix} 2 & -i & 0 & 0 \\ i & 1 & -1 & 0 \\ 0 & -1 & 2 & 0 \\ 0 & 0 & 0 & 3 \end{bmatrix}$$

so that the associated homogeneous linear system has solutions:
$X = (i\alpha, 2\alpha, \alpha, 0)$, for any $\alpha \in \mathbb{R}$. The corresponding eigenspace is
$N_{1,\lambda_2} = \langle (i, 2, 1, 0) \rangle = \langle (\frac{i}{\sqrt{6}}, \frac{2}{\sqrt{6}}, \frac{1}{\sqrt{6}}, 0) \rangle$.

Finally, for $\lambda_3 = 2$ we have

$$A - 2I = \begin{bmatrix} -1 & -i & 0 & 0 \\ i & -2 & -1 & 0 \\ 0 & -1 & -1 & 0 \\ 0 & 0 & 0 & 0 \end{bmatrix}$$

so that the associated homogeneous linear system has solutions:
$X = (-i\alpha, \alpha, -\alpha, \beta)$, for any $\alpha, \beta \in \mathbb{R}$. The corresponding eigenspace is
$N_{1,\lambda_3} = \langle (-i, 1, -1, 0), (0, 0, 0, 1) \rangle = \langle (-\frac{i}{\sqrt{3}}, \frac{1}{\sqrt{3}}, -\frac{1}{\sqrt{3}}, 0), (0, 0, 0, 1) \rangle$.

Therefore, with respect to the unitary basis

$$\left\{ \left(-\frac{i}{\sqrt{3}}, \frac{1}{\sqrt{3}}, -\frac{1}{\sqrt{3}}, 0 \right), (0, 0, 0, 1), \left(\frac{1}{\sqrt{2}}, 0, \frac{i}{\sqrt{2}}, 0 \right), \left(\frac{i}{\sqrt{6}}, \frac{2}{\sqrt{6}}, \frac{1}{\sqrt{6}}, 0 \right) \right\}$$

the matrix of T is

$$A' = \begin{bmatrix} 2 & 0 & 0 & 0 \\ 0 & 2 & 0 & 0 \\ 0 & 0 & 1 & 0 \\ 0 & 0 & 0 & -1 \end{bmatrix}.$$

Theorem 6.83 *Let $T : V \to V$ be an operator on a inner product space V over a field $\mathbb{F}$, A the matrix of T and assume that the characteristic polynomial $p(\lambda)$ of T splits over $\mathbb{F}$. Then T is normal (A is normal) if and only if T is diagonalizable (resp. A is diagonalizable). More precisely, if $\mathbb{F} = \mathbb{R}$ then A is orthogonally similar to a diagonal matrix; if $\mathbb{F} = \mathbb{C}$ then A is unitarily similar to a diagonal matrix.*

Proof If V has an orthonormal basis B consisting of eigenvectors of T, then the matrix A' of T with respect to B is diagonal and so it is normal, as well as T.

Now assume that T is normal. Since the characteristic polynomial of T splits over $\mathbb{F}$, again by Theorem 6.20, there exists an orthonormal basis B of V with respect to which the matrix A' of T is upper-triangular.

In particular, if $\mathbb{F} = \mathbb{R}$, then there is an orthogonal matrix $C \in M_n(\mathbb{R})$ such that $A' = C^T A C$ is upper triangular; if $\mathbb{F} = \mathbb{C}$, then there is an unitary matrix $P \in M_n(\mathbb{C})$ such that $A' = P^* A P$ is upper triangular.

Thus A is either orthogonally or unitarily similar to the upper-triangular A'. Moreover A' is normal, as A is. We prove that A' is diagonal by induction on the dimension of V.

If $dim_{\mathbb{F}} V = 1$, it is trivial. Now assume $dim_{\mathbb{F}} V = n \geq 2$ and suppose that the desired result holds for any vector space of smaller dimension. Let $B = \{e_1, \ldots, e_n\}$ be the basis of V with respect to which A' is the matrix of T. Since A' is upper-triangular, we can write its block diagonal form as follows:

$$A' = \begin{bmatrix} \alpha & u^T \\ 0_{n-1,1} & E \end{bmatrix}$$

for $0_{n-1,1} = [\ \underbrace{0,\dots,0}_{(n-1)-times}\ ]^T$, $u \in \mathbb{F}^{n-1}$ and $E \in M_{n-1}(\mathbb{F})$. Moreover $(A')^* A' = A'(A')^*$, so that, for any $X \in V$, we have

$$\|T(X)\|^2 = (A'X)^*(A'X) = X^*(A')^* A'X = X^* A'(A')^* X = ((A')^* X)^*(A')^* X = \|T^*(X)\|^2. \tag{6.18}$$

We notice that

$$T(e_1) = \begin{bmatrix} \alpha & u^T \\ 0_{n-1,1} & E \end{bmatrix} \begin{bmatrix} 1 \\ 0_{n-1,1} \end{bmatrix} = \begin{bmatrix} \alpha \\ 0_{n-1,1} \end{bmatrix}$$

and

$$T^*(e_1) = \begin{bmatrix} \overline{\alpha} & 0_{1,n-1} \\ \overline{u} & E^* \end{bmatrix} \begin{bmatrix} 1 \\ 0_{n-1,1} \end{bmatrix} = \begin{bmatrix} \overline{\alpha} \\ u \end{bmatrix}$$

that is

$$\|T(e_1)\|^2 = (A'X)^*(A'X) = \begin{bmatrix} \overline{\alpha} & 0_{1,n-1} \end{bmatrix} \begin{bmatrix} \alpha \\ 0_{n-1,1} \end{bmatrix} = \overline{\alpha}\alpha$$

and

$$\|T^*(e_1)\|^2 = ((A')^* X)^*((A')^* X) = \begin{bmatrix} \alpha & u^* \end{bmatrix} \begin{bmatrix} \overline{\alpha} \\ u \end{bmatrix} = \overline{\alpha}\alpha + u^* u.$$

By relation (6.18), $\|T(e_1)\|^2 = \|T^*(e_1)\|^2$ so that $0 = u^* u = \|u\| = 0$, that is $u = 0$. Hence

$$A' = \begin{bmatrix} \alpha & 0_{1,n-1} \\ 0_{n-1,1} & E \end{bmatrix}$$

and

$$A'(A')^* = \begin{bmatrix} \alpha & 0_{1,n-1} \\ 0_{n-1,1} & E \end{bmatrix} \begin{bmatrix} \overline{\alpha} & 0_{1,n-1} \\ 0_{n-1,1} & E^* \end{bmatrix} = \begin{bmatrix} \alpha\overline{\alpha} & 0_{1,n-1} \\ 0_{n-1,1} & EE^* \end{bmatrix}$$

$$(A')^* A' = \begin{bmatrix} \overline{\alpha} & 0_{1,n-1} \\ 0_{n-1,1} & E^* \end{bmatrix} \begin{bmatrix} \alpha & 0_{1,n-1} \\ 0_{n-1,1} & E \end{bmatrix} = \begin{bmatrix} \alpha\overline{\alpha} & 0_{1,n-1} \\ 0_{n-1,1} & E^* E \end{bmatrix}.$$

Since $A'(A')^* = (A')^* A'$, we get $EE^* = E^* E$, i.e., E is a $(n-1) \times (n-1)$ normal matrix. Moreover, any eigenvalue of E is an eigenvalue of A', and hence the characteristic polynomial of E splits over $\mathbb{F}$. By the induction hypothesis, E is orthogonally (resp. unitary) similar to a diagonal matrix, that is there exists an orthogonal (resp. unitary) matrix $Q \in M_{n-1}(\mathbb{F})$ such that $Q^{-1} E Q = D$ is a diagonal matrix. Hence, for

$$U = \begin{bmatrix} 1 & 0 \\ 0 & Q \end{bmatrix}$$

we have

$$U^{-1} A' U = \begin{bmatrix} \alpha & 0 \\ 0 & D \end{bmatrix},$$

where U is either orthogonal or unitary with respect to the fact that Q is orthogonal or unitary.

Corollary 6.84 (Complex Spectral Theorem) *Suppose that V is a complex inner product space, $T : V \to V$ a linear operator and A the matrix of T. Then A is unitarily similar to a diagonal matrix if and only if T is normal.*

Corollary 6.85 *Suppose that V is a complex inner product space, $T : V \to V$ and A the matrix of T.*

  (*i*) *If T is skew-Hermitian, then A is unitarily similar to a diagonal matrix.*
 (*ii*) *If T is unitary, then A is unitarily similar to a diagonal matrix.*

We cannot expect the same conclusion in the case V is a real inner product space:

Example 6.86 Let $T : \mathbb{R}^3 \to \mathbb{R}^3$ be the normal (skew-symmetric) linear operator having matrix

$$A = \begin{bmatrix} 0 & -2 & 2 \\ 2 & 0 & -1 \\ -2 & 1 & 0 \end{bmatrix}$$

with respect to the canonical basis of $\mathbb{R}^3$. The characteristic polynomial is $p(\lambda) = -\lambda^3 - 9\lambda$, having roots $0, 3i, -3i$.

For $\lambda_1 = 0$, the homogeneous linear system associated with A has solutions: $X = (\alpha, 2\alpha, 2\alpha)$, for any $\alpha \in \mathbb{R}$. The corresponding eigenspace is $N_{1,\lambda_1} = \langle (1, 2, 2) \rangle$. For $\lambda_2 = 3i$, we have

$$A - (3i)I = \begin{bmatrix} -3i & -2 & 2 \\ 2 & -3i & -1 \\ -2 & 1 & -3i \end{bmatrix}$$

so that the associated homogeneous linear system has solutions: $X = ((-6i - 2)\alpha, (3i - 4)\alpha, 5\alpha)$, for any $\alpha \in \mathbb{R}$. The corresponding eigenspace is $N_{1,\lambda_2} = \langle (-6i - 2, 3i - 4, 5) \rangle$.

Finally, for $\lambda_3 = -3i$, we have

$$A + (3i)I = \begin{bmatrix} 3i & -2 & 2 \\ 2 & 3i & -1 \\ -2 & 1 & 3i \end{bmatrix}$$

so that the associated homogeneous linear system has solutions:
$X = ((6i - 2)\alpha, (-3i - 4)\alpha, 5\alpha)$, for any $\alpha \in \mathbb{R}$. The corresponding eigenspace is
$N_{1,\lambda_3} = \langle (6i - 2, -3i - 4, 5) \rangle$.

There is no basis for $\mathbb{R}^3$ consisting of real eigenvectors of T, and hence A is not similar to a diagonal matrix.

It is clear from the above discussion that one has to ask oneself what could be the canonical form of real normal operators, in the case it is not diagonal. To answer this question, we need to fix some results on invariant subspaces of V.

Lemma 6.87 *Let $T : V \to V$ be a normal operator on the inner product space V over the field $\mathbb{F}$, U an T-invariant subspace of V. Then,*

(*i*) *$U^{\perp}$ is T-invariant.*
(*ii*) *U is T^*-invariant.*
(*iii*) *$(T_{|U})^* = T^*_{|U}$.*
(*iv*) *$T_{|U}$ is normal.*
(*v*) *$T_{|U^{\perp}}$ is normal.*

Proof

(*i*) Let $V = U \oplus U^{\perp}$ and $B_1 = \{e_1, \ldots, e_k\}$, $B_2 = \{c_1, \ldots, c_{n-k}\}$ be bases of U and $U^{\perp}$, respectively. Since $T(e_i) \in U$, for any $i = 1, \ldots, k$, then the matrix A of T with respect to the basis $B_1 \cup B_2$ for V is

$$A = \begin{bmatrix} A_1 & E \\ 0_{n-k,k} & A_2 \end{bmatrix},$$

where $0_{n-k,k}$ is the zero matrix in $M_{n-k,k}(\mathbb{F})$, $A_1 \in M_k(\mathbb{F})$, $A_2 \in M_{n-k}(\mathbb{F})$, $E \in M_{k,n-k}(\mathbb{F})$. Here, we denote $\alpha_{ij} \in \mathbb{F}$ and $\eta_{ij} \in \mathbb{F}$ the elements of A_1 and E, respectively. For $i = 1, \ldots k$, $T(e_i)$ is the coordinate vector of the entries in the ith column of A.

The matrix A^* of T^* with respect to the same basis $B_1 \cup B_2$ is

$$A^* = \begin{bmatrix} A_1^* & 0_{k,n-k} \\ E^* & A_2^* \end{bmatrix},$$

where $0_{k,n-k}$ is the zero matrix in $M_{k,n-k}(\mathbb{F})$. For $i = 1, \ldots, k$, $T^*(e_i)$ is the coordinate vector of the entries in the i-th column of $(A_1)^*$ and E^*, that is the conjugates of the entries in the i-th row of A_1 and E.
Since T is normal, $\|T(e_i)\|^2 = \|T^*(e_i)\|^2$ for any i, thus

$$\sum_{i=1}^{k} \|T(e_i)\|^2 = \sum_{i=1}^{k} \|T^*(e_i)\|^2$$

that is

$$\sum_{i=1}^{k}\sum_{j=1}^{k}\alpha_{ji}\overline{\alpha_{ji}} = \sum_{i=1}^{k}\sum_{j=1}^{k}\alpha_{ij}\overline{\alpha_{ij}} + \sum_{i=1}^{k}\sum_{j=k+1}^{n}\eta_{ij}\overline{\eta_{ij}}.$$

This implies that $\sum_{i=1}^{k}\sum_{j=k+1}^{n}\eta_{ij}\overline{\eta_{ij}} = 0$ that is each η_{ij} must be zero. Therefore $E = 0$ and

$$A = \begin{bmatrix} A_1 & 0_{k,n-k} \\ 0_{n-k,k} & A_2 \end{bmatrix}.$$

It is easy to see that $AX \in U^{\perp}$, for any $X \in U^{\perp}$, that is $U^{\perp}$ is T-invariant.

(ii) It follows from the first result, since the matrix of T^* is

$$A^* = \begin{bmatrix} A_1^* & 0_{k,n-k} \\ 0_{n-k,k} & A_2^* \end{bmatrix}.$$

(iii) Denote $G = T_{|U}$ and let $X, Y \in U$. Then

$$\langle X, G^*(Y) \rangle = \langle G(X), Y \rangle = \langle T(X), Y \rangle = \langle X, T^*(Y) \rangle$$

that is $G^*(Y) - T^*(Y) \in U \cap U^{\perp} = \{0\}$, for any $Y \in U$. Thus $G^*(Y) = T^*(Y)$, where $T^*(Y) \in U$ by (ii). Hence $G^* = T_{|U}^*$ as desired.

(iv) Since $TT^* = T^*T$, in particular $T_{|U}T_{|U}^* = T_{|U}^*T_{|U}$ and, by (iii), $T_{|U}(T_{|U})^* = (T_{|U})^*T_{|U}$.

(v) It is a consequence of the previous result, since $U^{\perp}$ is T-invariant.

Lemma 6.88 *Let $T : V \to V$ be a normal operator on a real inner product space V. Then V has an invariant subspace U of dimension 1 or 2.*

Proof Suppose that T has a real eigenvalue λ. Let $X \in V$ be an eigenvector corresponding to λ and $U = Span\{X\}$. Then $T(X) = \lambda X \in U$ and U is trivially T-invariant.

Assume now that any eigenvalue of T is complex (not real). Of course the characteristic polynomial $p(\lambda)$ of T has coefficients in the real field. Hence, if $\lambda_0 \in \mathbb{C}$ is a root of $p(\lambda)$ then its complex conjugate $\overline{\lambda_0}$ is also a root of $p(\lambda)$. Thus $p(\lambda)$ has $n = 2m$ roots, say $\lambda_k = \alpha_k + i\beta_k$ and $\overline{\lambda_k} = \alpha_k - i\beta_k$, for $\alpha_k, \beta_k \in \mathbb{R}$ and $k = 1, \ldots m$. The decomposition of $p(\lambda)$ is

$$p(\lambda) = \prod_{k=1}^{m}(\lambda - (\alpha_k + i\beta_k))(\lambda - (\alpha_k - i\beta_k)) = \prod_{k=1}^{m}(\lambda^2 + a_k\lambda + b_k),$$

where $\alpha_k = -\frac{a_k}{2}$ and $\beta_k = \pm\frac{\sqrt{4b_k - a_k^2}}{2}$. If A is the matrix of T, then by Cayley-Hamilton Theorem,

$$0 = p(A) = \prod_{k=1}^{m}(A^2 + a_kA + b_kI).$$

In particular, there is at least one $j \in \{1, \ldots, m\}$ such that the matrix $A^2 + a_j A + b_j I$ is not invertible, that is the operator $T^2 + a_j T + b_j I$ is not injective on V. Thus there exists a nonzero vector $X \in V$ such that $(T^2 + a_j T + b_j I)X = 0$. Consider now the 2-dimensional subspace W of V spanned by the set $\{X, T(X)\}$. For any element $Y \in W$, there exist $\mu_1, \mu_2 \in \mathbb{R}$ such that $Y = \mu_1 X + \mu_2 T(X)$. Hence

$$T(Y) = \mu_1 T(X) + \mu_2 T^2(X) = \mu_1 T(X) + \mu_2(-a_j T(X) - b_j X) \in W$$

that is W is T-invariant, as required.

We are now ready to prove the following:

Theorem 6.89 *Suppose that V is a real inner product space, $T : V \to V$ a linear operator, A the matrix of T with respect to a basis for V. T is normal if and only if A is orthogonally similar to a block diagonal matrix*

$$\begin{bmatrix} A_1 & & & \\ & \ddots & & \\ & & \ddots & \\ & & & A_k \end{bmatrix} \tag{6.19}$$

where each A_i is either a real number or a block of dimension 2, having the form

$$\begin{bmatrix} \alpha & -\beta \\ \beta & \alpha \end{bmatrix} \tag{6.20}$$

where $\alpha, \beta \in \mathbb{R}$ and $\beta > 0$.

Proof Firstly, we suppose there exists an orthogonal matrix $U \in M_n(\mathbb{F})$ such that

$$A' = U^{-1}AU = \begin{bmatrix} A_1 & & & \\ & \ddots & & \\ & & \ddots & \\ & & & A_k \end{bmatrix},$$

where each A_i has the form described in the statement of the theorem. In this case, we can see that each A_i commutes with its transpose, that is each A_i is normal. Hence the fact that A' commutes with its transpose follows from easy computations. Thus A' is normal, as well as A.

Assume now that A is a normal matrix. In case $n = 1$ it is trivial.

We now prove the result for $n = 2$. Write

$$A = \begin{bmatrix} a & b \\ c & d \end{bmatrix}$$

the matrix of T with respect to some orthonormal basis $\{X_1, X_2\}$ for V. Thus

$$\|T(X_1)\|^2 = (AX_1)^T(AX_1) = \left(\begin{bmatrix} a & b \\ c & d \end{bmatrix}\begin{bmatrix} 1 \\ 0 \end{bmatrix}\right)^T \left(\begin{bmatrix} a & b \\ c & d \end{bmatrix}\begin{bmatrix} 1 \\ 0 \end{bmatrix}\right) = a^2 + c^2$$

and

$$\|T^*(X_1)\|^2 = (A^T X_1)^T(A^T X_1)$$
$$= \left(\begin{bmatrix} a & c \\ b & d \end{bmatrix}\begin{bmatrix} 1 \\ 0 \end{bmatrix}\right)^T \left(\begin{bmatrix} a & c \\ b & d \end{bmatrix}\begin{bmatrix} 1 \\ 0 \end{bmatrix}\right)$$
$$= a^2 + b^2.$$

Since A is normal, $\|T(X_1)\|^2 = \|T^*(X_1)\|^2$, that is $b^2 = c^2$.

If $b = c$, then A is a symmetric matrix. By the Spectral Theorem, A is orthogonally similar to a diagonal matrix and we are done.

Let now $c = -b$, then

$$A = \begin{bmatrix} a & b \\ -b & d \end{bmatrix}.$$

Of course, we can assume $b \neq 0$, if not there is nothing to prove. Since $A^T A = AA^T$, by computing the $(1, 2)$-entries of both $A^T A$ and AA^T, we have $-ab + bd = ab - bd$, that is $ab = bd$. Hence $a = d$ and

$$A = \begin{bmatrix} a & b \\ -b & a \end{bmatrix}.$$

If $b < 0$ the result is proved. If $b > 0$, we compute the matrix A'' of T with respect to the orthonormal basis $\{X_1, -X_2\}$. It is

$$A'' = \begin{bmatrix} a & -b \\ b & a \end{bmatrix}$$

as required.

Hence we can suppose in what follows $n \geq 3$ and prove the result by induction on n. Assume that the theorem holds for any normal matrix having order less than n.

Let U be a T-invariant subspace of V having dimension 1 or 2. If $dim_{\mathbb{R}} U = 1$, then any vector in U with norm 1 is an orthonormal basis of U and the matrix A_1 of $T_{|U}$ has order 1. If $dim_{\mathbb{R}} U = 2$, then $T_{|U}$ is a normal operator on U (see Lemma 6.87) and the matrix A_1 of $T_{|U}$ has the form (6.20) with respect to an orthonormal basis C_1 of U. Since $T_{|U^\perp}$ is also a normal operator on $U^\perp$ (see again Lemma 6.87), by induction hypothesis there exists an orthonormal basis C_2 of $U^\perp$ with respect to which the matrix A_2 of $T_{|U^\perp}$ has the desired form.

Therefore, considering the basis $C_1 \cup C_2$ for V, the matrix A of T with respect to $C_1 \cup C_2$ is

$$\begin{bmatrix} A_1 & 0 \\ 0 & A_2 \end{bmatrix}$$

having the form (6.19).

Remark 6.90 Let

be the block diagonal form of the matrix of a normal operator T as in Theorem 6.89. Then each 1×1 block A_i is precisely a real eigenvalue of T and any 2×2 block having form (6.20) is corresponding to the pair of complex conjugate eigenvalues $\alpha + i\beta$, $\alpha - i\beta$ of T.

Now we give a method to construct canonical form of a real normal operator:

Suppose that V is a n-dimensional real inner product space, $T : V \to V$ an operator, A the matrix of T with respect to a basis for V. Here, we describe the way to obtain an orthonormal basis of V with respect to which the matrix of T is similar to a block diagonal matrix of the form (6.19). To do this, we firstly prove the following facts:

(a) Let $\lambda = a + ib$ and $\overline{\lambda} = a - ib$ $(b \neq 0)$ be a pair of complex (not real) eigenvalues of T, $X \in \mathbb{C}^n$ a complex eigenvector corresponding to λ. Then $\overline{X}$ is an eigenvector of T corresponding to $\overline{\lambda}$.

Proof From $AX = \lambda X$ it follows

$$\overline{AX} = \overline{\lambda X} = \overline{\lambda}\,\overline{X}.$$

On the other hand, since A is real,

$$\overline{AX} = \overline{A}\,\overline{X} = A\overline{X}.$$

Thus, $A\overline{X} = \overline{\lambda}\,\overline{X}$ as desired.

(b) Let $\lambda = a + ib$ be a complex (not real) eigenvalue of T, $X \in \mathbb{C}^n$ a complex eigenvector corresponding to λ. Then X and $\overline{X}$ are orthogonal.

Proof It follows from the fact that $\lambda \neq \overline{\lambda}$ (since λ is not real). Hence, X and $\overline{X}$ are corresponding to the distinct eigenvalues λ and $\overline{\lambda}$, respectively.

(c) Let $X \in \mathbb{C}^n$ be an eigenvector of T. Since X is a complex vector, it is defined as a combination of two real vectors $x \in \mathbb{R}^n$ the real part, and $y \in \mathbb{R}^n$ the imaginary part of X, that is $X = x + iy$. Then x, y are orthogonal vectors and $\|x\|^2 = \|y\|^2$.

Proof Since $X, \overline{X}$ are orthogonal, we have $0 = \overline{X}^* X = X^T X = (x + iy)^T (x + iy) = x^T x + 2ix^T y - y^T y = \|x\|^2 + 2ix^T y - \|y\|^2$ which implies both $\|x\|^2 = \|y\|^2$ and $x^T y = 0$ as required.

(d) Let $Z, Y \in \mathbb{C}^n$ be such that both $Z^* Y = 0$ and $\overline{Z}^* Y = 0$. If $Z = a + ib$ and $Y = c + id$, for $a, b, c, d \in V$, then $a^T c = a^T d = b^T c = b^T d = 0$, that is the real and imaginary parts of Z are orthogonal to both the real and imaginary parts of Y.

Proof By $Z^* Y = 0$ we get

$$0 = (a + ib)^*(c + id) = (a - ib)^T (c + id) = a^T c + b^T d + i(a^T d - b^T c)$$

that is

$$a^T c + b^T d = 0, \qquad a^T d - b^T c = 0. \tag{6.21}$$

Analogously, by $\overline{Z}^* Y = 0$, it follows

$$0 = (a - ib)^*(c + id) = (a + ib)^T (c + id) = a^T c - b^T d + i(a^T d + b^T c)$$

that is

$$a^T c - b^T d = 0, \qquad a^T d + b^T c = 0. \tag{6.22}$$

Comparing (6.21) with (6.22), we have the required conclusion.

(e) Let $\lambda = a + ib$ be a complex, not real, eigenvalue of T, $X = x + iy$ a complex eigenvector of T corresponding to λ, where $x, y \in V$. Then $T(x) = ax - by$ and $T(y) = bx + ay$.

Proof It is sufficient to compute the image of X :

$$T(X) = AX = \lambda X = (a + ib)(x + iy) = (ax - by) + i(bx + ay).$$

On the other hand, $AX = A(x + iy) = Ax + iAy$, hence $Ax = ax - by$ and $Ay = bx + ay$.

(f) If $X = x + iy$ is a complex eigenvector of T having length equal to 1, then both $\sqrt{2}x$ and $\sqrt{2}y$ have length equal to 1.

Proof It follows from:

$$1 = \|X\| = \|x\|^2 + \|y\|^2 = 2\|x\|^2 = 2\|y\|^2.$$

We are now ready to construct the required orthonormal basis for V. Let $\alpha_1, \ldots, \alpha_r$ be the real eigenvalues of T, $w_1, \ldots, w_r$ real eigenvectors of T corresponding to $\alpha_1, \ldots, \alpha_r$, respectively, and let $W = \langle w_1, \ldots, w_r \rangle$. By standard computations, we obtain an orthonormal basis for W. Let $\{z_1, \ldots, z_r\}$ be such a basis. Let

$\lambda_1, \overline{\lambda_1}, \ldots, \lambda_k, \overline{\lambda_k}$ be the complex, not real, eigenvalues of T and $X_1, \overline{X_1}, \ldots, X_k, \overline{X_k}$ complex eigenvectors corresponding to $\lambda_1, \overline{\lambda_1}, \ldots, \lambda_k, \overline{\lambda_k}$ respectively. For any $j = 1, \ldots, k$, we choose $\lambda_j = a_j - ib_j$, for $a_j, b_j \in \mathbb{R}$ and $b_j > 0$, and write $X_j = x_j + iy_j$, for $x_j, y_j \in V$.

Consider the following set of real vectors:

$$B = \{\sqrt{2}x_1, \sqrt{2}y_1, \ldots, \sqrt{2}x_k, \sqrt{2}y_k, z_1, \ldots, z_r\}.$$

In light of our previous comments, we see that B is a set of n orthonormal real vectors of V, that is B is an orthonormal basis for V. Let A' be the matrix of T with respect to B. Then, the column coordinate vectors of A' are the images of the elements of B :

$T(\sqrt{2}x_1) = A(\sqrt{2}x_1) = \sqrt{2}(a_1x_1 + b_1y_1)$, having coordinates in terms of B :

$$[a_1, b_1, \underbrace{0 \ldots, 0}_{(n-2)-times}],$$

$T(\sqrt{2}y_1) = A(\sqrt{2}y_1) = \sqrt{2}(-b_1x_1 + a_1y_1)$, having coordinates in terms of B :

$$[-b_1, a_1, \underbrace{0 \ldots, 0}_{(n-2)-times}],$$

$T(\sqrt{2}x_2) = A(\sqrt{2}x_2) = \sqrt{2}(a_2x_2 + b_2y_2)$, having coordinates in terms of B :

$$[0, 0, a_2, b_2, \underbrace{0 \ldots, 0}_{(n-4)-times}],$$

$T(\sqrt{2}y_2) = A(\sqrt{2}y_2) = \sqrt{2}(-b_2x_2 + a_2y_2)$, having coordinates in terms of B :

$$[0, 0, -b_2, a_2, \underbrace{0 \ldots, 0}_{(n-4)-times}].$$

More generally, for any $j = 1, \ldots, k : T(\sqrt{2}x_j) = A(\sqrt{2}x_j) = \sqrt{2}(a_jx_j + b_jy_j)$, having coordinates in terms of B :

$$[\underbrace{0 \ldots, 0}_{(2j-2)-times}, a_j, b_j, \underbrace{0 \ldots, 0}_{(n-2j)-times}],$$

$T(\sqrt{2}y_j) = A(\sqrt{2}y_j) = \sqrt{2}(-b_jx_j + a_jy_j)$, having coordinates in terms of B :

$$[\underbrace{0 \ldots, 0}_{(2j-2)-times}, -b_j, a_j, \underbrace{0 \ldots, 0}_{(n-2j)-times}].$$

Moreover, for any $h = 1, \ldots, r$
$T(z_h) = \alpha_h z_h$, having coordinates in terms of B:

$$[\underbrace{0 \ldots, 0}_{(2k+h-1)-times}, \alpha_h, \underbrace{0 \ldots, 0}_{(n-2k-h)-times}].$$

Therefore, A' has precisely the form

$$\begin{bmatrix} A_1 & & & & & \\ & \ddots & & & & \\ & & A_k & & & \\ & & & \alpha_1 & & \\ & & & & \ddots & \\ & & & & & \alpha_r \end{bmatrix}$$

where each A_i has the form

$$\begin{bmatrix} a & -b \\ b & a \end{bmatrix}$$

where $\alpha, a, b \in \mathbb{R}$ and $b > 0$.

Two special cases of real normal operators are described in the following:

Corollary 6.91 *Let $A \in M_n(\mathbb{R})$ be an orthonormal matrix. Then, A is orthogonally similar to a block diagonal matrix having the form*

$$\begin{bmatrix} \lambda_1 & & & & & \\ & \ddots & & & & \\ & & \lambda_k & & & \\ & & & A_1 & & \\ & & & & \ddots & \\ & & & & & A_t \end{bmatrix},$$

where $\lambda_h = \pm 1$, for any $h = 1, \ldots, k$ and

$$A_j = \begin{bmatrix} cos(\theta_j) & -sin(\theta_j) \\ sin(\theta_j) & cos(\theta_j) \end{bmatrix}$$

for suitable $\theta_j \in [0, 2\pi)$, and for $j = 1, \ldots, t$.

Proof Since A is orthogonal, it is normal. The real eigenvalues of an orthonormal matrix are precisely ± 1. The complex, not real, eigenvalues are of the form $e^{i\theta} = cos(\theta) + isin(\theta)$, with $sin(\theta) \neq 0$. Hence the result follows from Theorem 6.89 and Remark 6.90.

Corollary 6.92 *Let $A \in M_n(\mathbb{R})$ be a skew-symmetric matrix. Then A is orthogonally similar to a block diagonal matrix having the form*

$$\begin{bmatrix} A_1 & & & & & \\ & \ddots & & & & \\ & & A_k & & & \\ & & & 0 & & \\ & & & & \ddots & \\ & & & & & 0 \end{bmatrix},$$

where

$$A_j = \begin{bmatrix} 0 & -\alpha_j \\ \alpha_j & 0 \end{bmatrix}$$

for suitable $\alpha_j \in \mathbb{R}$, and for $j = 1, \ldots, k$.

Proof Since the eigenvalues of a skew-symmetric matrix are precisely imaginary numbers, every eigenvalue of A is either zero or $\pm i\alpha$, for some $\alpha \in \mathbb{R}$. Once again the result follows from Theorem 6.89 and Remark 6.90.

Example 6.93 Let $T : \mathbb{R}^3 \to \mathbb{R}^3$ be the normal (skew-symmetric) linear operator as in Example 6.86. The matrix of T with respect to the canonical basis of $\mathbb{R}^3$ is

$$A = \begin{bmatrix} 0 & -2 & 2 \\ 2 & 0 & -1 \\ -2 & 1 & 0 \end{bmatrix}.$$

The eigenvalues are $3i$, $-3i$, 0. For $\lambda = 3i$, the corresponding eigenspace is $W_1 = \langle (-6i - 2, 3i - 4, 5) \rangle$. So we obtain an eigenvector $X_1 \in \mathbb{C}^3$ having length equal to 1, such that $W_1 = \langle X_1 \rangle$:

$$X_1 = \left(\frac{-6i - 2}{\sqrt{90}}, \frac{3i - 4}{\sqrt{90}}, \frac{5}{\sqrt{90}} \right) = \left(\frac{-2}{\sqrt{90}}, \frac{-4}{\sqrt{90}}, \frac{5}{\sqrt{90}} \right) + i \left(\frac{-6}{\sqrt{90}}, \frac{3}{\sqrt{90}}, 0 \right).$$

Analogously, for $\lambda = -3i$, the corresponding eigenspace is $W_2 = \langle (6i - 2, -3i - 4, 5) \rangle$ and we obtain $X_2 \in \mathbb{C}^3$ having length equal to 1, such that $W_2 = \langle X_2 \rangle$:

$$X_2 = \left(\frac{6i - 2}{\sqrt{90}}, \frac{-3i - 4}{\sqrt{90}}, \frac{5}{\sqrt{90}} \right) = \left(\frac{-2}{\sqrt{90}}, \frac{-4}{\sqrt{90}}, \frac{5}{\sqrt{90}} \right) + i \left(\frac{6}{\sqrt{90}}, \frac{-3}{\sqrt{90}}, 0 \right).$$

Finally, for $\lambda = 0$, the eigenspace is generated by the following eigenvector of length 1 :

$$X_3 = \left(\frac{1}{3}, \frac{2}{3}, \frac{2}{3} \right).$$

Notice that $\overline{X_1} = X_2$. In particular, we choose the vector X_2 (corresponding to the eigenvalue whose imaginary part is negative) and write $X_2 = x_2 + iy_2$, where

$$x_2 = \left(\frac{-2}{\sqrt{90}}, \frac{-4}{\sqrt{90}}, \frac{5}{\sqrt{90}}\right), \quad y_2 = \left(\frac{6}{\sqrt{90}}, \frac{-3}{\sqrt{90}}, 0\right).$$

Adding $\sqrt{2}x_2$ and $\sqrt{2}y_2$ to X_3, we construct the following orthonormal basis for $\mathbb{R}^3$:

$$B = \left\{\left(\frac{-2}{\sqrt{45}}, \frac{-4}{\sqrt{45}}, \frac{5}{\sqrt{45}}\right), \left(\frac{6}{\sqrt{45}}, \frac{-3}{\sqrt{45}}, 0\right), \left(\frac{1}{3}, \frac{2}{3}, \frac{2}{3}\right)\right\}.$$

Let A' be the matrix of T with respect to B. Then the column coordinate vectors of A' are the images of the elements of B. By computations, we get

$$T(\sqrt{2}x_2) = 3\sqrt{2}y_2$$
$$T(\sqrt{2}y_2) = -3\sqrt{2}x_2$$
$$T(X_3) = 0.$$

Therefore, A' has precisely the form

$$\begin{bmatrix} 0 & -3 & 0 \\ 3 & 0 & 0 \\ 0 & 0 & 0 \end{bmatrix}$$

as expected.

Exercises

1. Let $T : \mathbb{R}^4 \to \mathbb{R}^4$ be the symmetric operator having matrix

$$A = \begin{bmatrix} 1 & 1 & 0 & 0 \\ 1 & 1 & 0 & 0 \\ 0 & 0 & 2 & 3 \\ 0 & 0 & 3 & 2 \end{bmatrix}$$

   with respect to the canonical basis of $\mathbb{R}^4$. Determine an orthonormal basis for $\mathbb{R}^4$ with respect to which the matrix of T has diagonal form.

2. Repeat Exercise 1, for the symmetric linear operator $T : \mathbb{R}^4 \to \mathbb{R}^4$ having matrix

$$A = \begin{bmatrix} 1 & 1 & 1 & 1 \\ 1 & 1 & 1 & 1 \\ 1 & 1 & 1 & 1 \\ 1 & 1 & 1 & 1 \end{bmatrix}$$

with respect to the canonical basis of $\mathbb{R}^4$.

3. Let $T : \mathbb{C}^4 \to \mathbb{C}^4$ be the Hermitian operator having matrix

$$A = \begin{bmatrix} 1 & i & 0 & 0 \\ -i & 1 & 0 & 0 \\ 0 & 0 & 1 & i \\ 0 & 0 & -i & 1 \end{bmatrix}$$

with respect to the canonical basis of $\mathbb{C}^4$. Determine an unitary basis for $\mathbb{C}^4$ with respect to which the matrix of T has diagonal form.

4. Let $T : \mathbb{R}^4 \to \mathbb{R}^4$ be the normal (skew-symmetric) operator having matrix

$$A = \begin{bmatrix} 0 & -2 & 0 & 2 \\ 2 & 0 & 0 & 2 \\ 0 & 0 & 0 & 0 \\ -2 & -2 & 0 & 0 \end{bmatrix}$$

with respect to the canonical basis of $\mathbb{R}^4$. Determine an orthogonal basis for $\mathbb{R}^4$ with respect to which the matrix of T has block diagonal form.

5. Let $T : V \to V$ be a normal operator on the n-dimensional vector space V over the field $\mathbb{F}$, A the matrix of T with respect to a basis for V. Let λ be an eigenvalue of T and $X \in V$ an eigenvector of T corresponding to λ. Prove that $\lambda\bar{\lambda}$ is eigenvalue of the operator $T^*T = TT^*$ and X is eigenvector of $T^*T = TT^*$ corresponding to $\lambda\bar{\lambda}$.

6. For each linear operator T on an inner product space V, determine whether T is normal, self-adjoint or neither. If possible, produce an orthonormal basis of eigenvectors of T for V and list the corresponding eigenvalues.

   (a) $V = \mathbb{R}^2$ and T is defined by $T(a, b) = (2a - 2b, -2a + 5b)$.
   (b) $V = \mathbb{R}^3$ and T is defined by $T(a, b, c) = (-a + b, 5b, 4a - 2b + 5c)$.
   (c) $V = \mathbb{C}^2$ and T is defined by $T(a, b) = (2a + ib, a + 2b)$.
   (d) $V = P_2(\mathbb{R})$ and T is defined by $T(f) = f'$, where $\langle f, g \rangle = \int_0^1 f(t)g(t)dt$.
   (e) $V = M_2(\mathbb{R})$ and T is defined by $T(A) = A^t$.
   (f) $V = M_2(\mathbb{R})$ and T is defined by $T \begin{pmatrix} a & b \\ c & d \end{pmatrix} = \begin{pmatrix} c & d \\ a & b \end{pmatrix}$.

7. Let V be a complex inner product space and let T be a linear operator on V. Define $T_1 = \frac{1}{2}(T + T^*)$ and $T_2 = \frac{1}{2i}(T - T^*)$.

   (a) Prove that T_1 and T_2 are self-adjoint and that $T = T_1 + iT_2$.
   (b) Suppose also that $T = U_1 + iU_2$, where U_1 and U_2 are self-adjoint. Prove that $U_1 = T_1$ and $U_2 = T_2$.
   (c) Prove that T is normal if and only if $T_1T_2 = T_2T_1$.

8. Prove that if T is a unitary operator on a finite dimensional inner product space V, then T has a unitary square root; that is there exists a unitary operator U such that $T = U^2$.

9. Let A be an $n \times n$ real symmetric or complex normal matrix. Prove that $tr(A) = \sum_{i=1}^{n} \lambda_i$ and $tr(A^*A) = \sum_{i=1}^{n} |\lambda_i|^2$, where $\lambda_i's$ are the (not necessarily distinct) eigenvalues of A.

Chapter 7
Bilinear and Quadratic Forms

This chapter is devoted to the study of the properties of bilinear and quadratic forms, defined on a vector space V over a field $\mathbb{F}$. The main goal will be the construction of appropriate methods aimed at obtaining the canonical expression of the functions, in terms of suitable bases for V. To do this, we will introduce the concept of orthogonality with respect to a bilinear form and mostly make use of the orthogonalization Gram-Schmidt process. Unless otherwise stated, here any vector space V is a finite dimensional vector space over $\mathbb{F}$.

7.1 Bilinear Forms and Their Matrices

Let $\mathbb{F}$ be a field, V, W vector spaces over $\mathbb{F}$ and $V \times W$ the cartesian product of V and W (as sets). A function $f : V \times W \to \mathbb{F}$ is called *bilinear* if it is linear in each variable separately, that is,

$$f(\alpha_1 v_1 + \alpha_2 v_2, w) = \alpha_1 f(v_1, w) + \alpha_2 f(v_2, w)$$

$$f(v, \beta_1 w_1 + \beta_2 w_2) = \beta_1 f(v, w_1) + \beta_2 f(v, w_2)$$

for any $\alpha_1, \alpha_2, \beta_1, \beta_2 \in \mathbb{F}$, $v_1, v_2 \in V$ and $w_1, w_2 \in W$. A bilinear function $f : V \times W \to \mathbb{F}$ is usually called a *bilinear form* on $V \times W$.

Example 7.1 The inner product $f : \mathbb{R}^n \times \mathbb{R}^n \longrightarrow \mathbb{R}$ is a bilinear form on $\mathbb{R}^n \times \mathbb{R}^n$.

Example 7.2 Let $B = \{b_1, b_2\}$ be a basis for $\mathbb{R}^2$ and $C = \{c_1, c_2, c_3\}$ a basis for $\mathbb{R}^3$. Let $f : \mathbb{R}^2 \times \mathbb{R}^3 \longrightarrow \mathbb{R}$ be the function defined by

$$f\big((x_1, x_2), (y_1, y_2, y_3)\big) = x_1(y_1 + y_2) + x_2(y_1 - y_3),$$

© The Author(s), under exclusive license to Springer Nature Singapore Pte Ltd. 2022
M. Ashraf et al., *Advanced Linear Algebra with Applications*,
https://doi.org/10.1007/978-981-16-2167-3_7

where (x_1, x_2) and (y_1, y_2, y_3) are the coordinate vectors of any $v \in \mathbb{R}^2$ and $w \in \mathbb{R}^3$ in terms of B and C, respectively. Then f is a bilinear form on $\mathbb{R}^2 \times \mathbb{R}^3$. In fact, for any $\alpha, \beta \in \mathbb{R}$, $v_1 = (x_1, x_2)$, $v_2 = (x_1', x_2') \in \mathbb{R}^2$ and $w_1 = (y_1, y_2, y_3)$, $w_2 = (y_1', y_2', y_3') \in \mathbb{R}^3$, it is easy to see that

$$f\big(\alpha(x_1, x_2) + \beta(x_1', x_2'), (y_1, y_2, y_3)\big) = \alpha f\big((x_1, x_2), (y_1, y_2, y_3)\big)$$
$$+ \beta f\big((x_1', x_2'), (y_1, y_2, y_3)\big)$$

and

$$f\big((x_1, x_2), \alpha(y_1, y_2, y_3) + \beta(y_1', y_2', y_3')\big) = \alpha f\big((x_1, x_2), (y_1, y_2, y_3)\big)$$
$$+ \beta f\big((x_1, x_2), (y_1', y_2', y_3')\big).$$

Let $f : V \times W \to \mathbb{F}$ be a bilinear form on $V \times W$, where V and W are finite dimensional vector spaces over $\mathbb{F}$. Let $B = \{b_1, \ldots, b_n\}$ and $C = \{c_1, \ldots, c_m\}$ be ordered bases for V and W, respectively. Let $[v]_B$ and $[w]_C$ be the coordinate vectors of $v \in V$ in terms of B and $w \in W$ in terms of C, respectively. Say $[v]_B = [x_1, \ldots, x_n]^t$ and $[w]_C = [y_1, \ldots, y_m]^t$, that is, $v = \sum_{i=1}^{n} x_i b_i$ and $w = \sum_{j=1}^{m} y_j c_j$. Since f is compatible with linear combinations in each variable, we have

$$f(v, w) = f\left(\sum_{i=1}^{n} x_i b_i, \sum_{j=1}^{m} y_j c_j\right) = \sum_{i,j} x_i y_j f(b_i, c_j).$$

If we consider the coefficients matrix $A = (a_{ij})$, where $a_{ij} = f(b_i, c_j)$, for any $1 \leq i, j \leq n$ then it is easy to see that

$$f(v, w) = [v]_B^t A [w]_C.$$

The $n \times m$ matrix $A = (a_{ij})$ is said to be the *matrix of the bilinear form* with respect to the ordered bases B and C.

Conversely, let $A \in M_{nm}(\mathbb{F})$ and let, as above, $[v]_B$, $[w]_C$ be the coordinate vectors of $v \in V$, in terms of B, and $w \in W$, in terms of C, respectively. If we define a function $f : V \times W \to \mathbb{F}$ such that $f(v, w) = [v]_B^t A [w]_C$, then by computations it follows that $f(v, w) = \sum_{i,j} x_i y_j a_{ij}$. Of course, this map is compatible with linear combinations in each variable, that is, f is a bilinear map on the set $V \times W$.

Moreover, if we assume that there exist two $n \times m$ matrices A and A' of f, in terms of the same ordered bases B for V and C for W, then, for any $v \in V$ and $w \in W$, it follows that $[v]_B^t A [w]_C = f(v, w) = [v]_B^t A' [w]_C$, that is, $[v]_B^t (A - A')[w]_C = 0$. Denote by (α_{ij}) the coefficient entries of the matrix $A - A'$. But since $v \in V$ and $w \in W$ are arbitrary, for all $i = 1, \ldots, n$ and $j = 1, \ldots, m$, we get

$$[\underbrace{0, \ldots, 0}_{(i-1)-times}, 1, \underbrace{0, \ldots, 0}_{(n-i)-times}]_B (A - A') [\underbrace{0, \ldots, 0}_{(j-1)-times}, 1, \underbrace{0, \ldots, 0}_{(m-j)-times}]^t_C = 0 \Longrightarrow \alpha_{ij} = 0.$$

This means that $A - A'$ is the zero matrix, that is $A = A'$.

Thus, we have proved that there exists an unique $n \times m$ matrix having entries in $\mathbb{F}$, which represents the bilinear form with respect to the given ordered bases B and C. So, it is clear that the matrix of a bilinear form on the set $V \times W$ changes according to the choice of the bases of the underlying vector spaces.

Example 7.3 Let $f : \mathbb{R}^2 \times \mathbb{R}^3 \to \mathbb{R}$ be a bilinear form on $\mathbb{R}^2 \times \mathbb{R}^3$ defined by

$$f((x_1, x_2), (y_1, y_2, y_3)) = x_1(y_1 + y_2) + x_2(y_1 - y_3)$$

with respect to the canonical bases in $\mathbb{R}^2$ and $\mathbb{R}^3$. The matrix $A = (a_{ij})$ of f is obtained by the following computations:

$$a_{11} = f((1, 0), (1, 0, 0)) = 1, \quad a_{12} = f((1, 0), (0, 1, 0)) = 1,$$

$$a_{13} = f((1, 0), (0, 0, 1)) = 0; \quad a_{21} = f((0, 1), (1, 0, 0)) = 1,$$

$$a_{22} = f((0, 1), (0, 1, 0)) = 0, \quad a_{23} = f((0, 1), (0, 0, 1)) = -1.$$

Thus

$$A = \begin{bmatrix} 1 & 1 & 0 \\ 1 & 0 & -1 \end{bmatrix}.$$

Now consider two different ordered bases for $\mathbb{R}^2$ and $\mathbb{R}^3$, precisely:

$$D = \{d_1, d_2\} = \{(1, 1), (2, 1)\} \quad \text{for} \quad \mathbb{R}^2$$

$$\text{and} \quad E = \{e_1, e_2, e_3\} = \{(1, 1, 0), (0, 0, 1), (2, 0, 1)\} \quad \text{for} \quad \mathbb{R}^3.$$

Since

$$f(d_1, e_1) = 3, \quad f(d_1, e_2) = -1, \quad f(d_1, e_3) = 3;$$

$$f(d_2, e_1) = 5, \quad f(d_2, e_2) = -1, \quad f(d_2, e_3) = 5$$

the matrix A' of f with respect to the bases D and E is the following:

$$A' = \begin{bmatrix} 3 & -1 & 3 \\ 5 & -1 & 5 \end{bmatrix}.$$

and hence f can be represented by

$$f((x_1, x_2), (y_1, y_2, y_3)) = \begin{bmatrix} x_1 & x_2 \end{bmatrix} \begin{bmatrix} 3 & -1 & 3 \\ 5 & -1 & 5 \end{bmatrix} \begin{bmatrix} y_1 \\ y_2 \\ y_3 \end{bmatrix}$$
$$= x_1(3y_1 - y_2 + 3y_3) + x_2(5y_1 - y_2 + 5y_3)$$

with respect to the bases D and E.

Example 7.4 Let $f : \mathbb{R}^3 \times \mathbb{R}^2 \to \mathbb{R}$ be a bilinear form on $\mathbb{R}^3 \times \mathbb{R}^2$, having the matrix

$$\begin{bmatrix} 1 & 1 \\ 1 & 0 \\ 2 & 1 \end{bmatrix}$$

in terms of the bases $B = \{b_1, b_2, b_3\} = \{(1, 1, 0), (0, 0, 1), (0, 1, 1)\}$ for $\mathbb{R}^3$ and $C = \{c_1, c_2\} = \{(1, 1), (0, 1)\}$ for $\mathbb{R}^2$. Then f can be expressed as follows

$$f((x_1, x_2, x_3), (y_1, y_2)) = \begin{bmatrix} x_1 & x_2 & x_3 \end{bmatrix} \begin{bmatrix} 1 & 1 \\ 1 & 0 \\ 2 & 1 \end{bmatrix} \begin{bmatrix} y_1 \\ y_2 \end{bmatrix}$$
$$= x_1(y_1 + y_2) + x_2 y_1 + x_3(2y_1 + y_2).$$

We now obtain the matrix of f in terms of different ordered bases for V and W. Let $D = \{d_1, d_2, d_3\} = \{(0, 1, 0), (0, 1, 1), (2, 0, 1)\}$ be a basis for $\mathbb{R}^3$ and $E = \{e_1, e_2\} = \{(1, 1), (2, 1)\}$ be a basis for $\mathbb{R}^2$. In order to obtain the matrix relative to the bases D and E, one has to compute the coefficients $f(d_i, e_j)$. The first step is to determine the coordinate vectors of d_1, d_2, d_3 in terms of B, and e_1, e_2 in terms of C : $d_1 = [0, 1, 0]^t = [0, -1, 1]^t_B$, $d_2 = [0, 1, 1]^t = [0, 0, 1]^t_B$, $d_3 = [2, 0, 1]^t = [2, 3, -2]^t_B$

$$e_1 = [1, 1]^t = [1, 0]^t_C, \quad e_2 = [2, 1]^t = [2, -1]^t_C.$$

Thus

$$f(d_1, e_1) = \begin{bmatrix} 0 & -1 & 1 \end{bmatrix} \begin{bmatrix} 1 & 1 \\ 1 & 0 \\ 2 & 1 \end{bmatrix} \begin{bmatrix} 1 \\ 0 \end{bmatrix} = 1,$$

$$f(d_1, e_2) = \begin{bmatrix} 0 & -1 & 1 \end{bmatrix} \begin{bmatrix} 1 & 1 \\ 1 & 0 \\ 2 & 1 \end{bmatrix} \begin{bmatrix} 2 \\ -1 \end{bmatrix} = 1,$$

$$f(d_2, e_1) = 2, \quad f(d_2, e_2) = 3, \quad f(d_3, e_1) = 1, \text{ and } f(d_3, e_2) = 2.$$

Therefore, with respect to the bases D and E, f can be expressed as

$$f\big((x_1, x_2, x_3), (y_1, y_2)\big) = \begin{bmatrix} x_1 & x_2 & x_3 \end{bmatrix} \begin{bmatrix} 1 & 1 \\ 2 & 3 \\ 1 & 2 \end{bmatrix} \begin{bmatrix} y_1 \\ y_2 \end{bmatrix}$$
$$= x_1(y_1 + y_2) + x_2(2y_1 + 3y_2) + x_3(y_1 + 2y_2).$$

7.2 The Effect of the Change of Bases

We now consider two different ordered bases $B = \{b_1, \ldots, b_n\}$ and $B' = \{b'_1, \ldots, b'_n\}$ for V, as well as two different ordered bases $C = \{c_1, \ldots, c_m\}$ and $C' = \{c'_1, \ldots, c'_m\}$ for W. In view of the above, the bilinear form $f : V \times W \to \mathbb{F}$ can be represented by different matrices, in connection with the choice of a basis for V and W. For instance, let A be the matrix of f with respect to the ordered bases B for V and C for W, and A' the matrix of f with respect to the ordered bases B' for V and C' for W. Now let us describe the relationship between the matrices A and A'.

Let $P \in M_n(\mathbb{F})$ be the transition matrix of B' relative to B, whose i-th column is the coordinates vector $[b'_i]_B$, and $Q \in M_m(\mathbb{F})$ be the transition matrix of C' relative to C, whose i-th column is the coordinates vector $[c'_i]_C$. We recall that, for any vectors $v \in V$ and $w \in W$, the following hold:

$$[v]_B = P[v]_{B'} \quad \text{and} \quad [w]_C = Q[w]_{C'}.$$

Thus
$$f(v, w) = [v]_B^t A [w]_C$$
$$= \big(P[v]_{B'}\big)^t A \big(Q[w]_{C'}\big)$$
$$= [v]_{B'}^t \big(P^t A Q\big)[w]_{C'}.$$

On the other hand, $f(v, w) = [v]_{B'}^t A'[w]_{C'}$ and, by the uniqueness of A' in terms of the ordered bases B' and C', we get $A' = P^t A Q$.

Example 7.5 Let $f : \mathbb{R}^3 \times \mathbb{R}^2 \to \mathbb{R}$ be the bilinear form as in Example 7.4, having the matrix

$$A = \begin{bmatrix} 1 & 1 \\ 1 & 0 \\ 2 & 1 \end{bmatrix}$$

in terms of the ordered bases $B = \{b_1, b_2, b_3\} = \{(1, 1, 0), (0, 0, 1), (0, 1, 1)\}$ for $\mathbb{R}^3$ and $C = \{c_1, c_2\} = \{(1, 1), (0, 1)\}$ for $\mathbb{R}^2$.

We introduce $D = \{d_1, d_2, d_3\} = \{(0, 1, 0), (0, 1, 1), (2, 0, 1)\}$ a basis for $\mathbb{R}^3$ and $E = \{e_1, e_2\} = \{(1, 1), (2, 1)\}$ a basis for $\mathbb{R}^2$.

The transition matrices P, Q of D relative to B and that of E relative to C, respectively, are

$$P = \begin{bmatrix} 0 & 0 & 2 \\ -1 & 0 & 3 \\ 1 & 1 & -2 \end{bmatrix}, \quad Q = \begin{bmatrix} 1 & 2 \\ 0 & -1 \end{bmatrix}$$

so that, the matrix A' of f, in terms of the ordered bases D for $\mathbb{R}^3$ and E for $\mathbb{R}^2$, is

$$A' = P^t A Q$$

$$= \begin{bmatrix} 0 & -1 & 1 \\ 0 & 0 & 1 \\ 2 & 3 & -2 \end{bmatrix} \begin{bmatrix} 1 & 1 \\ 1 & 0 \\ 2 & 1 \end{bmatrix} \begin{bmatrix} 1 & 2 \\ 0 & -1 \end{bmatrix}$$

$$= \begin{bmatrix} 1 & 1 \\ 2 & 3 \\ 1 & 2 \end{bmatrix}.$$

Let us now investigate the special case when $V = W$, in other words, we consider the bilinear form $f : V \times V \to \mathbb{F}$. Under this assumption, we may always consider only one ordered basis $B = \{e_1, \ldots, e_n\}$ for V, so that the coefficients matrix $A = (a_{ij})$ of f is obtained by the computations $a_{ij} = f(e_i, e_j)$, for any $i, j = 1, \ldots, n$.

Therefore, if B and B' are two different bases for V, then f can be represented by two different matrices: A, the matrix of f with respect to the ordered basis B, and A', the matrix of f with respect to the ordered basis B'. In light of the above argument, $A' = P^t A P$, where P is the transition matrix of B' relative to B. Notice that A, A', P are $n \times n$ square matrices, and, in particular, P is an invertible matrix. At this point, we would like to recall the following:

Definition 7.6 Let A, A' be two $n \times n$ matrices having coefficients in some field $\mathbb{F}$. A, A' are called *congruent matrices* if there exists an invertible $n \times n$ matrix P with coefficients in $\mathbb{F}$, such that $A' = P^t A P$. The relationship $A' = P^t A P$ is usually called *congruence*. It is easy to see that the congruence between matrices is an equivalence relation.

More precisely, we have:

Theorem 7.7 *Let $f : V \times V \to \mathbb{F}$ be a bilinear form on $V \times V$ (equivalently we say that f is a bilinear form on V). Two matrices A, A' represent f, in terms of two different ordered bases for V, if and only if they are congruent.*

Proof We already have proved one direction: if both A and A' represent the same bilinear form f with respect to the ordered bases B and B', respectively, then there exists a nonsingular matrix P (which is precisely the transition matrix of B' relative to B) such that $A' = P^t A P$.

In order to prove the other direction, we assume that A is the matrix of f in terms of the ordered basis $B = \{b_1, \ldots, b_n\}$. Here suppose that $A' = P^t A P$ for some nonsingular $n \times n$ matrix P. Of course, the i-th column vector $[u_{1i}, \ldots, u_{ni}]^t$ in P can be viewed as a coordinate vector with respect to the ordered basis B. Let

$$u_i = \sum_{j=1}^{n} u_{ji} b_j$$ be the vector of V, having coordinates $[u_{1i}, \ldots, u_{ni}]^t$ in terms of B.

Hence, for any $i = 1, \ldots, n$, we obtain a sequence of n linearly independent vectors

$u_1, \ldots, u_n$. Hence, the set $B' = \{u_1, \ldots, u_n\}$ is an ordered basis for V and P is the transition matrix of B' relative to B. Therefore, A' is precisely the matrix of f with respect to the ordered basis B'.

7.3 Symmetric, Skew-Symmetric and Alternating Bilinear Forms

We focus our attention on three different types of bilinear forms:

Definition 7.8 Let $f : V \times V \to \mathbb{F}$ be a bilinear form on V.

(1) f is called *symmetric* if $f(u, v) = f(v, u)$ for all $u, v \in V$.
(2) f is called *skew-symmetric* if $f(u, v) = -f(v, u)$ for all $u, v \in V$.
(3) f is called *alternating* if $f(u, u) = 0$ for all $u \in V$.

Nevertheless, here we prove that any skew-symmetric form is precisely a symmetric or an alternating form, according to the fact that the characteristic of $\mathbb{F}$ is 2 or not, respectively.

Theorem 7.9 *Let $f : V \times V \to \mathbb{F}$ be a bilinear form on V.*

  (i) *If f is alternating form, then it is skew-symmetric.*
 (ii) *If $char(\mathbb{F}) \neq 2$, then f is skew-symmetric if and only if it is alternating.*
(iii) *If $char(\mathbb{F}) = 2$, then f is skew-symmetric if and only if it is symmetric.*

Proof (i) We firstly assume that f is alternating. Thus, by expanding the relation $f(u + v, u + v) = 0$ for any $u, v \in V$, we get

$$0 = f(u, u) + f(u, v) + f(v, u) + f(v, v) = f(u, v) + f(v, u)$$

so that $f(u, v) = -f(v, u)$ for any $u, v \in V$, i.e., f is skew-symmetric.

(ii) Consider now the case $char(\mathbb{F}) \neq 2$ and suppose f is skew-symmetric. Therefore, $f(u, u) = -f(u, u)$ for any $u \in V$, which implies $2f(u, u) = 0$, for any $u \in V$, that is, f is alternating.

(iii) Finally, if $char(\mathbb{F}) = 2$, then f is symmetric if and only if $f(u, v) = f(v, u) = -f(v, u)$ for any $u, v \in V$ (since $1 = -1$). This last relation holds if and only if f is skew-symmetric, as desired.

Theorem 7.10 *Let $f : V \times V \to \mathbb{F}$ be a bilinear form on V. Then*

(i) *f is symmetric if and only if the matrix A of f is symmetric ($A^t = A$), whatever the choice of ordered basis for V with respect to which A is related.*

(ii) *f is alternating if and only if the matrix A of f is skew-symmetric ($A^t = -A$)
    and the diagonal entries of A are zero, whatever the choice of ordered basis for
    V with respect to which A is related.*

Proof We remark that the same result holds for bilinear skew-symmetric forms, i.e.,
f is skew-symmetric if and only if the matrix A of f is skew-symmetric. Nevertheless, in light of the previous theorem, it is known that a skew-symmetric form is
either symmetric or alternating. Therefore, it is sufficient to prove the result in these
last two cases.

Let $B = \{e_1, \ldots, e_n\}$ be any ordered basis for V and $A = (a_{ij})$ be the matrix of
f in terms of B.

(i) Firstly, we assume that f is symmetric. Thus, $a_{ij} = f(e_i, e_j) = f(e_j, e_i) = a_{ji}$
for all $i \neq j$, i.e., A is a symmetric matrix.

Suppose now $A^t = A$. Hence, for any $u, v, \in V$, $f(u, v) = u^t A v$ and $f(v, u) =
v^t A u$. Here, we have identified u by $[u]_B$ and v by $[v]_B$, respectively. On the other
hand, since $u^t A v \in \mathbb{F}$ is a scalar element, we have

$$f(u, v) = u^t A v = (u^t A v)^t = v^t A^t u = v^t A u = f(v, u),$$

as required.

(ii) Let now f be alternating. Thus, $a_{ii} = f(e_i, e_i) = 0$ for any $i = 1, \ldots, n$. Moreover, for any $i \neq j$, $f(e_i + e_j, e_i + e_j) = 0$ implies $f(e_i, e_j) + f(e_j, e_i) = 0$, i.e.,
$a_{ij} = f(e_i, e_j) = -f(e_j, e_i) = -a_{ji}$ and hence A is skew-symmetric.

Conversely, let $A^t = -A$ be such that $a_{ii} = 0$ for any $i = 1, \ldots, n$. Hence, for
any $u \in V$, $f(u, u) = u^t A u$. As above, since $u^t A u \in \mathbb{F}$ is a scalar element,

$$f(u, u) = u^t A u = (u^t A u)^t = u^t A^t u = -u^t A u = -f(u, u).$$

Hence, in case of $char(\mathbb{F}) \neq 2$, it follows $f(u, u) = 0$ for any $u \in V$ and we are
done.

Finally, let $char(\mathbb{F}) = 2$ and $u = \sum_{i=1}^{n} \alpha_i e_i$ be any vector of V. Since $f(e_i, e_i) =
a_{ii} = 0$ for any i, and $f(e_i, e_j) = a_{ij} = -a_{ji} = -f(e_j, e_i) = f(e_j, e_i)$ for any $i \neq
j$, it follows that

$$f(u, u) = f\left(\sum_{i=1}^{n} \alpha_i e_i, \sum_{i=1}^{n} \alpha_i e_i\right) = 2 \sum_{i \neq j} \alpha_i \alpha_j f(e_i, e_j) = 0,$$

as required.

Definition 7.11 Let $f : V \times V \longrightarrow \mathbb{F}$ be either symmetric, skew-symmetric or
alternating on V. We will refer to (V, f) as a *metric vector space*.

7.4 Orthogonality and Reflexive Forms

A bilinear form is a generalization of the inner product, so the condition $f(u, v) = 0$ for some $u, v \in V$ can be viewed as a generalization of the orthogonality. More precisely, we give the following:

Definition 7.12 Let (V, f) be a metric vector space and $u, v \in V$. Then u is said to be f-*orthogonal* to v if $f(u, v) = 0$. In this case, we write $u \perp v$.

In general, the relation $\perp$ of orthogonality might not be symmetric, in the sense that we can have $u \perp v$ ($f(u, v) = 0$) but $v \not\perp u$ ($f(v, u) \neq 0$).

Example 7.13 Let $f : \mathbb{R}^2 \times \mathbb{R}^2 \to \mathbb{R}$ be a bilinear form defined as

$$f((x_1, x_2), (y_1, y_2)) = \begin{bmatrix} x_1 & x_2 \end{bmatrix} \begin{bmatrix} 3 & 2 \\ -1 & 1 \end{bmatrix} \begin{bmatrix} y_1 \\ y_2 \end{bmatrix}$$

$$= 3x_1 y_1 - x_2 y_1 + 2x_1 y_2 + x_2 y_2.$$

For $u = [1, 1]^t$ and $v = [3, -2]^t$, we have $f(u, v) = 0$ but $f(v, u) \neq 0$.

Remark 7.14 Of course, if f is either symmetric or alternating, then the orthogonality relation $\perp$ is symmetric. In fact, if $f(u, v) = f(v, u)$ (or $f(u, v) = -f(v, u)$) for any $u, v \in V$, then $u \perp v$ if and only if $v \perp u$.

When the orthogonality relation is symmetric, that is, $f(u, v) = 0$ if and only if $f(v, u) = 0$ for all $u, v \in V$, we say that f is *reflexive*.

Theorem 7.15 *Let* $f : V \times V \to \mathbb{F}$ *be a bilinear form on the vector space* V. *Then* f *is reflexive if and only if* f *is either symmetric or alternating.*

Proof In light of Remark 7.14, we now assume that f is reflexive and prove that it is either symmetric or alternating. Let $x, y, z \in V$. Of course, $f(x, y) f(x, z) = f(x, z) f(x, y)$. Thus, we have

$$0 = f(x, y) f(x, z) - f(x, z) f(x, y) = f\big(x, f(x, y)z - f(x, z)y\big). \quad (7.1)$$

Since f is reflexive, (7.1) implies

$$0 = f\big(f(x, y)z - f(x, z)y, x\big) = f(x, y) f(z, x) - f(x, z) f(y, x) \text{ for all } x, y, z \in V. \quad (7.2)$$

In particular, for $x = z$ in (7.2), one has

$$f(x, y) f(x, x) - f(x, x) f(y, x) = 0 \quad \text{for all } x, y \in V. \quad (7.3)$$

In other words, for any $x, y \in V$

$$\text{either} \quad f(x, x) = f(y, y) = 0 \quad \text{or} \quad f(y, x) = f(x, y). \quad (7.4)$$

Here, we suppose that f is neither symmetric nor alternating and show that a contradiction follows. In light of our last assumption, there exist $u, v, w \in V$ such that $f(v, w) \neq f(w, v)$ and $f(u, u) \neq 0$. By (7.4) and $f(v, w) \neq f(w, v)$, it follows that $f(w, w) = f(v, v) = 0$. Moreover, by (7.3) and $f(u, u) \neq 0$, we also have both $f(u, v) = f(v, u)$ and $f(u, w) = f(w, u)$.

On the other hand, for $x = v$, $y = w$ and $z = u$ in (7.2), it follows

$$0 = f(v, w)f(u, v) - f(v, u)f(w, v) = f(u, v)\big(f(v, w) - f(w, v)\big) \quad (7.5)$$

and analogously, for $x = w$, $y = v$ and $z = u$ in (7.2),

$$0 = f(w, v)f(u, w) - f(w, u)f(v, w) = f(u, w)\big(f(w, v) - f(v, w)\big). \quad (7.6)$$

Therefore, since $f(v, w) \neq f(w, v)$, relations (7.5) and (7.6) say that $f(u, v) = f(v, u) = 0$ and $f(u, w) = f(w, u) = 0$. Hence

$$f(u + v, w) = f(v, w) \quad \text{and} \quad f(w, u + v) = f(w, v),$$

that is, $f(u + v, w) \neq f(w, u + v)$. By using (7.4), we get $f(u + v, u + v) = 0$. This gives the contradiction

$$0 = f(u + v, u + v) = f(u, u) + f(u, v) + f(v, u) + f(v, v) = f(u, u).$$

If f is reflexive on V and W is a subspace of V, we set

$$W^\perp = \{v \in V \mid f(v, w) = 0 \text{ for all } w \in W\} = \{v \in V \mid f(w, v) = 0 \text{ for all } w \in W\}$$

and call $W^\perp$ the f-*orthogonal space* of W. One may notice that this definition is equivalent to the one of orthogonal complement in an inner product space. In this sense, we prefer to use the term f-*orthogonal space*, and not orthogonal complement for the set $W^\perp$, in order to distinguish the case of metric spaces and the other one of inner product spaces.

In fact, if $W^\perp$ is the orthogonal complement of the subspace W of an inner product space V, then $W \oplus W^\perp = V$. On the other hand, if $W^\perp$ is simply the f-orthogonal space of W in the metric space V (i.e., V is equipped by a bilinear form that is not an inner product), then it may happen that $W + W^\perp \neq V$.

Example 7.16 Let $f : \mathbb{R}^3 \times \mathbb{R}^3 \to \mathbb{R}$ be a (skew-symmetric) bilinear form defined as

$$f\big((x_1, x_2, x_3), (y_1, y_2, y_3)\big) = \begin{bmatrix} x_1 & x_2 & x_3 \end{bmatrix} \begin{bmatrix} 0 & 1 & 0 \\ -1 & 0 & 1 \\ 0 & -1 & 0 \end{bmatrix} \begin{bmatrix} y_1 \\ y_2 \\ y_3 \end{bmatrix}$$

$$= x_1 y_2 - x_2 y_1 + x_2 y_3 - x_3 y_2.$$

If $W = \langle (1, -1, 0) \rangle$, then $W^\perp = \langle (1, -1, 0), (1, 0, 1) \rangle$ and $W + W^\perp = W^\perp \neq \mathbb{R}^3$.

7.5   The Restriction of a Bilinear Form

If (V, f) is a metric vector space and W is a subspace of V, then we may introduce the restriction $f_{|W}$ of f to W, so we get the metric space $(W, f_{|W})$. It is clear that if f is either symmetric, skew-symmetric or alternating on V then so is the restriction $f_{|W}$ on W.

Example 7.17 Let $f : \mathbb{R}^3 \times \mathbb{R}^3 \to \mathbb{R}$ be a (symmetric) bilinear form defined as

$$f((x_1, x_2, x_3), (y_1, y_2, y_3)) = \begin{bmatrix} x_1 & x_2 & x_3 \end{bmatrix} \begin{bmatrix} 0 & 1 & 1 \\ 1 & 0 & 1 \\ 1 & 1 & 0 \end{bmatrix} \begin{bmatrix} y_1 \\ y_2 \\ y_3 \end{bmatrix}$$

$$= x_1 y_2 + x_2 y_1 + x_1 y_3 + x_3 y_1 + x_2 y_3 + x_3 y_2.$$

If $W = \langle (1, 1, 0), (0, 1, 1) \rangle$, then any pair of vectors $u, v \in W$ can be written as $u = \alpha_1(1, 1, 0) + \alpha_2(0, 1, 1)$, $v = \beta_1(1, 1, 0) + \beta_2(0, 1, 1)$, $\alpha_1, \alpha_2, \beta_1, \beta_2 \in \mathbb{R}$. Hence

$$\begin{aligned} f(u, v) &= \alpha_1 \beta_1 f((1, 1, 0), (1, 1, 0)) + \alpha_1 \beta_2 f((1, 1, 0), (0, 1, 1)) \\ &\quad + \alpha_2 \beta_1 f((0, 1, 1), (1, 1, 0)) + \alpha_2 \beta_2 f((0, 1, 1), (0, 1, 1)) \\ &= 2\alpha_1 \beta_1 + 3\alpha_1 \beta_2 + 3\alpha_2 \beta_1 + 2\alpha_2 \beta_2 \\ &= \begin{bmatrix} \alpha_1 & \alpha_2 \end{bmatrix} \begin{bmatrix} 2 & 3 \\ 3 & 2 \end{bmatrix} \begin{bmatrix} \beta_1 \\ \beta_2 \end{bmatrix}. \end{aligned}$$

Thus, the matrix $\begin{bmatrix} 2 & 3 \\ 3 & 2 \end{bmatrix}$ represents the restriction of f to W.

For instance, let $u \in W$ and the coordinate vector of u for V be $[2, 1, -1]^t$. Thus $u = 2(1, 1, 0) + (-1)(0, 1, 1)$. Similarly, if $v \in W$ and the coordinate vector of v for V be $[1, 4, 3]^t$ then $v = 1(1, 1, 0) + 3(0, 1, 1)$. Thus, the coordinate vector of u with respect to the basis $B = \{(1, 1, 0), (0, 1, 1)\}$ for W is $[2, -1]^t$. Analogously, the coordinate vector of v with respect to B is $[1, 3]^t$.

As vectors of V, we get

$$f(u, v) = \begin{bmatrix} 2 & 1 & -1 \end{bmatrix} \begin{bmatrix} 0 & 1 & 1 \\ 1 & 0 & 1 \\ 1 & 1 & 0 \end{bmatrix} \begin{bmatrix} 1 \\ 4 \\ 3 \end{bmatrix} = 13.$$

As vectors of W,

$$f(u, v) = \begin{bmatrix} 2 & -1 \end{bmatrix} \begin{bmatrix} 2 & 3 \\ 3 & 2 \end{bmatrix} \begin{bmatrix} 1 \\ 3 \end{bmatrix} = 13.$$

7.6 Non-degenerate Bilinear Forms

Given a vector space V and a reflexive bilinear form f on V, we define the *radical* of f as the subspace: $Rad(f) = V^\perp = \{u \in V \mid f(u, v) = 0 \text{ for all } v \in V\} = \{u \in V \mid f(v, u) = 0 \text{ for all } v \in V\}$. The bilinear form f is said to be *non-degenerate* if $V^\perp = \{0\}$. This is equivalent to say that $f(v, u) = 0$, for any $v \in V$, implies $u = 0$. In case $V^\perp \neq \{0\}$, we refer to f as a *degenerate* form.

Similarly, we may define the *radical* of the restriction $f_{|W}$, where W is a subspace of V,

$$Rad(f_{|W}) = W \cap W^\perp = \{u \in W \mid f(u, w) = 0 \text{ for all } w \in W\}$$

and we say that $f_{|W}$ is non-degenerate on W if $W \cap W^\perp = \{0\}$.

Lemma 7.18 *The reflexive bilinear form $f : V \times V \to \mathbb{F}$ is non-degenerate if and only if the matrix of f is invertible.*

Proof Let $A \in M_n(\mathbb{F})$ be the matrix of f with respect to the ordered basis $B = \{e_1, \dots, e_n\}$. It is clear that if $u \in V^\perp$, then $f(u, e_i) = 0$ for any $i = 1, \dots, n$. On the other hand, if $f(u, e_i) = 0$ for any vector $e_i \in B$, then $f(u, \sum_i \alpha_i e_i) = \sum_i \alpha_i f(u, e_i) = 0$ for any scalar elements $\alpha_1, \dots, \alpha_n$. Thus $f(u, v) = 0$ for any $v \in V$, that is, $u \in V^\perp$. In other words, we have proved that $v \in V^\perp$ if and only if $f(v, e_i) = 0$ for any vector e_i of the basis for V.

Now, let $X \in V^\perp$ and $[x_1, \dots, x_n]_B^t$ be the coordinate vector of X with respect to B. Hence

$$
\begin{aligned}
X \in V^\perp \quad &\Longleftrightarrow X^t A e_i = 0 \quad \text{for all } i = 1, \dots, n \\
&\Longleftrightarrow e_i^t A X = 0 \quad \text{for all } i = 1, \dots, n \\
&\Longleftrightarrow [\underbrace{0, \dots, 0}_{(i-1)-times}, 1, \underbrace{0, \dots, 0}_{(n-i)-times}]_B A [x_1, \dots, x_n]_B^t = 0 \quad (7.7)
\end{aligned}
$$

for all $i = 1, \dots, n$.

If (a_{ij}) are the coefficient entries of the matrix A, then the relation (7.7) means that

$$
\begin{aligned}
a_{11}x_1 + a_{12}x_2 + \cdots + a_{1n}x_n &= 0 \\
a_{21}x_1 + a_{22}x_2 + \cdots + a_{2n}x_n &= 0 \\
\cdots \quad \cdots \quad \cdots \quad \cdots & \\
a_{n1}x_1 + a_{n2}x_2 + \cdots + a_{nn}x_n &= 0
\end{aligned}
$$

in other words, $X \in V^\perp$ if and only if its coordinate vector $[x_1, \dots, x_n]_B^t$, in terms of B, is a solution of the homogeneous linear system associated with the matrix A.

Thus $V^\perp = \{0\}$ if and only if the homogeneous linear system associated with the matrix A has only the trivial solution. This happens if and only if the rank of A is equal to n, that is, A is invertible, as required.

Remark 7.19 A reflexive non-degenerate bilinear form on a vector space V might restrict to a degenerate bilinear form on a subspace W of V. For example, let $f : \mathbb{R}^3 \times \mathbb{R}^3 \to \mathbb{R}$ be the (symmetric) bilinear form having matrix

$$A = \begin{bmatrix} 1 & 1 & 1 \\ 1 & 0 & -1 \\ 1 & -1 & 0 \end{bmatrix}.$$

It is trivially a non-degenerate form, since A is invertible. Nevertheless, if $W = \langle (1, 1, 0), (0, 0, 1) \rangle$, then the matrix of $f_{|W}$ is

$$\begin{bmatrix} 3 & 0 \\ 0 & 0 \end{bmatrix}$$

so that $f_{|W}$ is a degenerate reflexive form on W, in fact, by easy computations we find that $Rad(f_{|W}) = \{(0, 0, \alpha) | \alpha \in \mathbb{R}\} = \langle (0, 0, 1) \rangle$.

Theorem 7.20 *Let $f : V \times V \to \mathbb{F}$ be a reflexive form, W a subspace of V. Then $f_{|W}$ is non-degenerate on W if and only if $V = W \oplus W^\perp$.*

Proof Of course, we may assume that W is a proper subspace of V, if not there is nothing to prove.

Trivially, the directness of the sum $V = W \oplus W^\perp$ implies $Rad(f_{|W}) = W \cap W^\perp = \{0\}$ and $f_{|W}$ is non-degenerate on W.

Conversely, assume that $f_{|W}$ is non-degenerate on W, that is, no nonzero element of W lies in $W^\perp$ and $W \cap W^\perp = Rad(f_{|W}) = \{0\}$. This means that W and $W^\perp$ are in direct sum. Let $dim(V) = n$, $dim(W) = k < n$ and $B = \{e_1, \ldots, e_k\}$ be a basis for W. Extending B to a basis B' for V, we find $e_{k+1}, \ldots, e_n$ vectors of V such that $B' = \{e_1, \ldots, e_n\}$.

Now, let A be the matrix of f in terms of B', $X \in W^\perp$ and $[x_1, \ldots, x_n]^t_{B'}$ be the coordinate vector of X with respect to B'. Hence:

$$
\begin{aligned}
X \in W^\perp \quad &\Longleftrightarrow X^t A e_i = 0 \quad \text{for all } i = 1, \ldots, k \\
&\Longleftrightarrow e_i^t A X = 0 \quad \text{for all } i = 1, \ldots, k \\
&\Longleftrightarrow [\underbrace{0, \ldots, 0}_{(i-1)-times}, 1, \underbrace{0, \ldots, 0}_{(n-i)-times}]_{B'} A [x_1, \ldots, x_n]^t_{B'} = 0 \quad (7.8)
\end{aligned}
$$

for all $i = 1, \ldots, k$.

If (a_{ij}) are the coefficient entries of the matrix A, then the relation (7.8) means that

$$
\begin{aligned}
a_{11}x_1 + a_{12}x_2 + \cdots + a_{1n}x_n &= 0 \\
a_{21}x_1 + a_{22}x_2 + \cdots + a_{2n}x_n &= 0 \\
\cdots \quad \cdots \quad \cdots \quad \cdots & \\
a_{k1}x_1 + a_{k2}x_2 + \cdots + a_{kn}x_n &= 0
\end{aligned}
$$

Therefore, $X \in W^{\perp}$ if and only if its coordinate vector $[x_1, \ldots, x_n]^t_{B'}$ is a solution of the homogeneous linear system associated with the submatrix A' of A consisting in the first top k rows of A. Since the rank of A' is $\leq k$, the null space of A' has dimension $\geq n - k$, that is, $dim(W^{\perp}) \geq n - k$. On the other hand, $dim(V) = n \geq dim(W \oplus W^{\perp}) = dim(W) + dim(W^{\perp}) = k + dim(W^{\perp}) \geq k + n - k = n$, implying that $dim(W \oplus W^{\perp}) = n$ and $W \oplus W^{\perp}$ is precisely equal to V.

Remark 7.21 Using the above argument one may prove that if $f : V \times V \to \mathbb{F}$ is a reflexive non-degenerate form and W is any subspace of V, then $dim(W) + dim(W^{\perp}) = dim(V)$.

Moreover, it is clear that the following holds:

Corollary 7.22 Let $f : V \times V \to \mathbb{F}$ be a reflexive non-degenerate form and W a subspace of V. Then $f_{|W}$ is non-degenerate on W if and only if $f_{|W^{\perp}}$ is non-degenerate on $W^{\perp}$.

Example 7.23 Let $f : \mathbb{R}^3 \times \mathbb{R}^3 \to \mathbb{R}$ be a (symmetric) bilinear form defined as

$$f((x_1, x_2, x_3), (y_1, y_2, y_3)) = \begin{bmatrix} x_1 & x_2 & x_3 \end{bmatrix} \begin{bmatrix} 0 & \frac{1}{2} & \frac{1}{2} \\ \frac{1}{2} & 0 & \frac{1}{2} \\ \frac{1}{2} & \frac{1}{2} & 0 \end{bmatrix} \begin{bmatrix} y_1 \\ y_2 \\ y_3 \end{bmatrix}$$

$$= \tfrac{1}{2}x_1 y_2 + \tfrac{1}{2}x_2 y_1 + \tfrac{1}{2}x_1 y_3 + \tfrac{1}{2}x_3 y_1 + \tfrac{1}{2}x_2 y_3 + \tfrac{1}{2}x_3 y_2.$$

Since the matrix of f is invertible, f is non-degenerate. Consider now the following decomposition of $\mathbb{R}^3$

$$\mathbb{R}^3 = \langle (1, 1, 0) \rangle \oplus \left\langle \left(-\frac{1}{2}, \frac{1}{2}, 0 \right), (-1, -1, 1) \right\rangle.$$

If we denote $W = \langle (1, 1, 0) \rangle$, it is easy to see that $\langle (-\frac{1}{2}, \frac{1}{2}, 0), (-1, -1, 1) \rangle$ is the f-orthogonal space of W. So we may write $W^{\perp} = \langle (-\frac{1}{2}, \frac{1}{2}, 0), (-1, -1, 1) \rangle$. Thus $\mathbb{R}^3 = W \oplus W^{\perp}$.

Moreover, any pair of vectors $u, v \in W$, can be written as

$$u = \alpha_1 (1, 1, 0), \quad v = \beta_1 (1, 1, 0), \quad \alpha_1, \beta_1 \in \mathbb{R}.$$

Hence, by the definition of f, $f(u, v) = \alpha_1 \beta_1$. Thus, the 1×1 matrix $[1]$ represents the restriction $f_{|W}$ of f to W. This restriction is clearly non-degenerate.

Similarly, any pair of vectors $u', v' \in W^{\perp}$ can be written as

$$u' = (-\frac{\alpha_1}{2} - \alpha_2, \frac{\alpha_1}{2} - \alpha_2, \alpha_2), \quad v' = (-\frac{\beta_1}{2} - \beta_2, \frac{\beta_1}{2} - \beta_2, \beta_2)$$

where (α_1, α_2) and (β_1, β_2) are the coordinates of u' and v', respectively (in terms of the above fixed basis for $W^\perp$). By computations, we have that

$$f(u', v') = -\frac{1}{4}\alpha_1\beta_1 - \alpha_2\beta_2.$$

Therefore, the matrix

$$\begin{bmatrix} -\frac{1}{4} & 0 \\ 0 & -1 \end{bmatrix}$$

represents the restriction $f_{|W^\perp}$ of f to $W^\perp$ and it is also non-degenerate.

Definition 7.24 Let $f : V \times V \to \mathbb{F}$ be a reflexive form and $B = \{e_1, \ldots, e_n\}$ a basis for V. We say that B is an f-*orthogonal basis* if $f(e_i, e_j) = 0$ for all $i \neq j$.

Definition 7.25 Let $f : V \times V \to \mathbb{F}$ be a reflexive form and $B = \{e_1, \ldots, e_n\}$ an f-orthogonal basis for V. We say that B is an f-*orthonormal basis* if $f(e_k, e_k) = 1$ for all $k = 1, \ldots, n$.

Example 7.26 Consider the symmetric form f defined in Example 7.23 and the basis

$$B = \left\{ (1, 1, 0), \left(-\frac{1}{2}, \frac{1}{2}, 0\right), (-1, -1, 1) \right\}$$

for $\mathbb{R}^3$. In light of Definition 7.24, B is an f-orthogonal basis for $\mathbb{R}^3$.

Definition 7.27 Let $f : V \times V \to \mathbb{F}$ be a bilinear form. A nonzero vector $v \in V$ is called f-*isotropic* if $f(v, v) = 0$; otherwise v is called f-*nonisotropic*. The vector space V is called f-*isotropic* if it contains at least one f-isotropic vector; otherwise V is called f-*nonisotropic*. V is said to be *totally* f-*isotropic* (or also f-*symplectic*) if every vector of V is isotropic.

Example 7.28 Let $f : \mathbb{R}^3 \times \mathbb{R}^3 \to \mathbb{R}$ be a (symmetric) bilinear form defined by the matrix

$$A = \begin{bmatrix} 1 & 2 & 1 \\ 2 & 2 & 0 \\ 1 & 0 & 0 \end{bmatrix}$$

in terms of the canonical basis for $\mathbb{R}^3$. Notice that f is non-degenerate. The vector $X = [1, 0, -\frac{1}{2}]^t$ is f-isotropic, in fact $f(X, X) = 0$. Hence $\mathbb{R}^3$ is f-isotropic. The vector $Y = [1, 2, 1]^t$ is f-nonisotropic, in fact $f(Y, Y) = 19$.

Example 7.29 Let $f : \mathbb{R}^3 \times \mathbb{R}^3 \to \mathbb{R}$ be a (symmetric) bilinear form defined by the matrix

$$A = \begin{bmatrix} 1 & 1 & 0 \\ 1 & 2 & 1 \\ 0 & 1 & 3 \end{bmatrix}$$

in terms of the canonical basis for $\mathbb{R}^3$. If we denote by $X = [x_1, x_2, x_3]^t$, any nonzero vector of $\mathbb{R}^3$ in terms of the canonical basis for $\mathbb{R}^3$, then

$$f(X, X) = x_1^2 + 2x_1x_2 + 2x_2^2 + 2x_2x_3 + 3x_3^2$$
$$= (x_1 + x_2)^2 + (x_2 + x_3)^2 + 2x_3^2 > 0.$$

Therefore, there is no nonzero vector in $\mathbb{R}^3$ which is f-isotropic. Hence, in the sense of Definition 7.27, $\mathbb{R}^3$ is a f-nonisotropic real vector space.

Example 7.30 Let $f : \mathbb{R}^3 \times \mathbb{R}^3 \to \mathbb{R}$ be a (skew-symmetric) bilinear form defined by the matrix

$$A = \begin{bmatrix} 0 & 1 & 2 \\ -1 & 0 & 1 \\ -2 & -1 & 0 \end{bmatrix}$$

in terms of the canonical basis for $\mathbb{R}^3$. If we denote by $X = [x_1, x_2, x_3]^t$, any vector of $\mathbb{R}^3$ in terms of the canonical basis for $\mathbb{R}^3$, then it is clear that $f(X, X) = 0$, i.e., $\mathbb{R}^3$ is a totally f-isotropic real vector space, in the sense of Definition 7.27.

7.7 Diagonalization of Symmetric Forms

Let V be equipped with the symmetric bilinear form f. Notice that any inner product on a real vector space (according to the definition in Chap. 5) is a symmetric bilinear form. The converse is not generally true. For instance, let $f : \mathbb{R}^2 \times \mathbb{R}^2 \to \mathbb{R}$ be a bilinear form having matrix

$$A = \begin{bmatrix} -4 & 2 \\ 2 & -2 \end{bmatrix}$$

in terms of the canonical basis for $\mathbb{R}^2$. One can verify that f is symmetric. On the other hand, for $u = [1, 1]^t \in \mathbb{R}^2$, we see that $f(u, u) = u^t Au = -2$. Thus, f is not an inner product (in the sense of our definition in Chap. 5).

In general, as stated in the previous section, if the bilinear form f is reflexive, one may refer to (V, f) as a metric space. In particular, if f is symmetric, (V, f) is called *symmetric space*. Nevertheless, any symmetric form having the additional property $f(v, v) > 0$ for all $0 \neq v \in V$, is an inner product. Of course, in this case, we may also refer to (V, f) as an inner product space.

Lemma 7.31 *Let* $\mathbb{F}$ *be a field of characteristic different from* 2 *and* (V, f) *a symmetric space. If* $f \neq 0$ *(that is* f *is not identically zero on* $V \times V$*), then there is* $w_0 \in V$ *such that* $f(w_0, w_0) \neq 0$.

Proof Assume that $f(v, v) = 0$ for all $v \in V$. Therefore, for any $x_0, y_0 \in V$, it follows that $f(x_0, x_0) = 0$, $f(y_0, y_0) = 0$ and $f(x_0 + y_0, x_0 + y_0) = 0$. Therefore, since the form is symmetric, we get the contradiction

$$0 = f(x_0 + y_0, x_0 + y_0) = f(x_0, y_0) + f(y_0, x_0) = 2f(x_0, y_0).$$

Theorem 7.32 *Let $\mathbb{F}$ be a field of characteristic different from 2 and (V, f) a symmetric space. Then there is an f-orthogonal basis for V.*

Proof Firstly, we remark that, in case either V is 1-dimensional over $\mathbb{F}$ or $f = 0$, any basis is f-orthogonal. Thus, we assume that $dim(V) = n \geq 2$ and there exist $x_0, y_0 \in V$ such that $f(x_0, y_0) \neq 0$. By Lemma 7.31, there is $w_0 \in V$ such that $f(w_0, w_0) \neq 0$.

Now, we prove our result by induction on the dimension $n \geq 2$, so we assume that the theorem is true for any symmetric space of smaller dimension.

Since $f(w_0, w_0) \neq 0$, the restriction of f to the subspace $W = \langle w_0 \rangle$ of V is nondegenerate. Hence, by Theorem 7.20, $V = W \oplus W^\perp$. Moreover, $W^\perp$ is a symmetric space having dimension $n - 1$ over $\mathbb{F}$. So, by the induction assumption, there is an f-orthogonal basis $\{e_1, \ldots, e_{n-1}\}$ for $W^\perp$ and then $\{e_1, \ldots, e_{n-1}, w_0\}$ is an f-orthogonal basis for V. $\blacksquare$

Remark 7.33 It is clear that if $A = (a_{ij})$ is the matrix representing a bilinear form in terms of an f-orthogonal basis, then A is a diagonal matrix, because of $a_{ij} = f(e_i, e_j) = 0$ for all $i \neq j$. So we may give another version of Theorem 7.32, from the matrix theory point of view: *Any symmetric matrix, over a field of characteristic different from 2, is congruent to a diagonal matrix.*

7.8 The Orthogonalization Process for Nonisotropic Symmetric Spaces

Here, we would like to describe the classical method for finding an f-orthogonal basis for a symmetric space (V, f), under the assumption that $f(v, v) \neq 0$ for any nonzero $v \in V$.

Actually, the present method does not differ from the one we have already outlined in the case of inner product spaces and is usually called the *Gram-Schmidt process*.

Let $B = \{b_1, \ldots, b_n\}$ be a basis of the symmetric space V and A_0 the matrix of f with respect to B. Using the Gram-Schmidt process, we may compute the basis $E = \{e_1, \ldots, e_n\}$ for V as follows:

$$e_1 = b_1$$

$$e_k = b_k - \sum_{i=1}^{k-1} \frac{f(b_k, e_i)}{f(e_i, e_i)} e_i \qquad \text{for all } k = 2, \ldots, n;$$

that is,

$$e_1 = b_1$$

$$e_2 = b_2 - \frac{f(b_2,e_1)}{f(e_1,e_1)}e_1$$

$$e_3 = b_3 - \frac{f(b_3,e_1)}{f(e_1,e_1)}e_1 - \frac{f(b_3,e_2)}{f(e_2,e_2)}e_2 \qquad\qquad (7.9)$$

$$\cdots\cdots\cdots$$

$$e_n = b_n - \frac{f(b_n,e_1)}{f(e_1,e_1)}e_1 - \frac{f(b_n,e_2)}{f(e_2,e_2)}e_2 - \cdots - \frac{f(b_n,e_{n-1})}{f(e_{n-1},e_{n-1})}e_{n-1}.$$

By easy computations, one can see that $e_k \perp e_i$, for any $k \neq i$. Moreover, since f is nonisotropic, $f(e_k,e_k) \neq 0$ for any $k = 1, \ldots, n$. Hence, E is an f-orthogonal basis for V. Let U be the transition matrix of E relative to B such that $U^t A_0 U$ is diagonal. Then, the column vectors of U are the coordinates of the elements of E in terms of B. In particular, U has the following form:

$$U = \begin{bmatrix} 1 & \alpha_{12} & \alpha_{13} & \cdots & & \alpha_{1n} \\ 0 & 1 & \alpha_{23} & \cdots & & \alpha_{2n} \\ 0 & 0 & \ddots & & & \vdots \\ \vdots & & & \ddots & & \vdots \\ \vdots & & & & \ddots & \alpha_{n-1,n} \\ 0 & 0 & & & 0 & 1 \end{bmatrix}, \quad \alpha_{ij} \in \mathbb{F}. \qquad (7.10)$$

We recall that such a matrix is usually called *upper unitriangular*, in the sense that it is an upper triangular matrix having all diagonal coefficients equal to 1.

The matrix A of f in terms of E is diagonal, namely

$$A = U^t A_0 U = \begin{bmatrix} a_{11} & & & \\ & a_{22} & & \\ & & \ddots & \\ & & & a_{nn} \end{bmatrix} \qquad (7.11)$$

where $a_{ii} = f(e_i,e_i) \neq 0$.

At this point, we have to split our argument into two different cases.

Firstly, we assume that the field $\mathbb{F}$ is algebraically closed. Then we know that there exist $\alpha_1, \ldots, \alpha_n \in \mathbb{F}$ such that $\alpha_i^2 = a_{ii}$, for any $i = 1, \ldots, n$. Hence, by using the vectors from the basis B, we compute a different basis $\tilde{B} = \{\tilde{e}_1, \ldots, \tilde{e}_n\}$ such that $\tilde{e}_i = \alpha_i^{-1} e_i$. It is clear that B' is an f-orthogonal basis for V. Moreover, we have

$$f(\tilde{e}_k, \tilde{e}_k) = f\left(\alpha_k^{-1} e_k, \alpha_k^{-1} e_k\right) = \alpha_k^{-2} f(e_k, e_k) = a_{kk}^{-1} a_{kk} = 1 \text{ for all } k = 1, \ldots, n.$$

Hence, the matrix $\tilde{A}$ of f in terms of the basis $\tilde{B}$ has the following diagonal form:

$$\tilde{A} = \begin{bmatrix} 1 & & & \\ & 1 & & \\ & & \ddots & \\ & & & 1 \end{bmatrix}.$$

If $\tilde{P}$ is the transition matrix having in any i-th column the coordinates of the vector $\tilde{e}_i$, then $\tilde{P}^t A \tilde{P} = \tilde{A}$.

Of course, if $\mathbb{F}$ is not algebraically closed, one cannot expect the same final result. In particular, here we describe the case $\mathbb{F} = \mathbb{R}$.

Starting from B, we obtain a basis $C = \{c_1, \ldots, c_n\}$ for V, as follows:

$$c_k = \frac{1}{\sqrt{|f(e_k, e_k)|}} e_k = \frac{1}{\sqrt{|a_{kk}|}} e_k \quad \text{for all } k = 1, \ldots, n$$

where $f(c_k, c_i) = 0$ for all $k \neq i$,

$$f(c_k, c_k) = f\left(\frac{1}{\sqrt{|a_{kk}|}} e_k, \frac{1}{\sqrt{|a_{kk}|}} e_k \right) = \frac{1}{|a_{kk}|} f(e_k, e_k) = \pm 1 \quad \text{for all } k = 1, \ldots, n.$$

Hence, the matrix A' of f in terms of the basis C has the following diagonal form:

$$A' = \begin{bmatrix} \pm 1 & & & \\ & \pm 1 & & \\ & & \ddots & \\ & & & \pm 1 \end{bmatrix},$$

where any diagonal (i, i)-entry is equal to $+1$ or -1 according to the fact that a_{ii} is positive or negative, respectively. Finally, by reordering the vectors in C, we get an f-orthogonal basis $D = \{w_1, \ldots, w_n\}$ for V, with respect to which the matrix A'' of f is

$$A'' = \begin{bmatrix} 1 & & & & & \\ & \ddots & & & & \\ & & 1 & & & \\ & & & -1 & & \\ & & & & \ddots & \\ & & & & & -1 \end{bmatrix}.$$

Moreover, if P is the transition matrix having in any i-th column the coordinates of the vector w_i, then $P^t A P = A''$.

The Gram-Schmidt process can be essentially viewed as a recursive algorithm, divided into several steps, which should allow us to obtain a sequence of bases for V. In any single step of the process, the basis of the sequence consists of vectors obtained as linear combinations of vectors from the basis given at the previous step. More precisely, at the i-th step, we get a basis $B_i = \{e_1^{(i)}, \ldots, e_n^{(i)}\}$ such that

$\{e_1^{(i)}, \ldots, e_i^{(i)}, e_k^{(i)}\}$ is an f-orthogonal set for any $k \geq i + 1$. We stop the process whenever the basis consists of all f-orthogonal vectors.

Denote by $B_0 = \{e_1^{(0)}, \ldots, e_n^{(0)}\}$ the starting basis for V and $A_0 = (a_{ij})$ the coefficients matrix of f with respect to B_0. Let us describe in detail any single step:

(i) **Step** 1 :

Set $B_1 = \{e_1^{(1)}, \ldots, e_n^{(1)}\}$, where

$$e_1^{(1)} = e_1^{(0)}$$

and

$$e_k^{(1)} = e_k^{(0)} - \frac{f(e_k^{(0)}, e_1^{(1)})}{f(e_1^{(1)}, e_1^{(1)})} e_1^{(1)}, \quad k = 2, \ldots, n.$$

Then $e_k^{(1)} \perp e_1^{(1)}$ for any $k \neq 1$, and $V = \langle e_1^{(1)} \rangle \oplus \langle e_1^{(1)} \rangle^{\perp}$, where $\langle e_1^{(1)} \rangle^{\perp} = \langle e_2^{(1)}, \ldots, e_n^{(1)} \rangle$. The matrix A_1 of f in terms of B_1 has any $(k, 1)$-entry and $(1, k)$-entry equal to zero for $k \neq 1$. Moreover, if P_1 is the transition matrix having in any i-th column the coordinates of the vector $e_i^{(1)}$ in terms of B_0, then $P_1^t A P_1 = A_1$.

(ii) **Step** 2 :

Set $B_2 = \{e_1^{(2)}, \ldots, e_n^{(2)}\}$, where

$$e_1^{(2)} = e_1^{(1)}$$

and

$$e_2^{(2)} = e_2^{(1)}$$

so that, by the previous step,

$$e_1^{(2)} \perp e_2^{(2)}$$

and

$$e_k^{(2)} = e_k^{(1)} - \frac{f(e_k^{(1)}, e_2^{(2)})}{f(e_2^{(2)}, e_2^{(2)})} e_2^{(2)}, \quad k = 3, \ldots, n.$$

Then $e_k^{(2)} \perp e_1^{(2)}$ and $e_k^{(2)} \perp e_2^{(2)}$, for any $k \neq 1, 2$ and $V = \langle e_1^{(2)}, e_2^{(2)} \rangle \oplus \langle e_1^{(2)}, e_2^{(2)} \rangle^{\perp}$, where $\langle e_1^{(2)}, e_2^{(2)} \rangle^{\perp} = \langle e_3^{(2)}, \ldots, e_n^{(2)} \rangle$. The matrix A_2 of f in terms of B_2 has any (k, j)-entry and (j, k)-entry equal to zero, for $j = 1, 2$ and $k \neq j$. Moreover, if P_2 is the transition matrix having in any i-th column the coordinates of the vector $e_i^{(2)}$ in terms of B_1, then $P_2^t A_1 P_2 = A_2$.

(iii) **Step** m : **for** $m \geq 2$.

We set $B_m = \{e_1^{(m)}, \ldots, e_n^{(m)}\}$, where

$$e_j^{(m)} = e_j^{(m-1)}, \quad j = 1, \ldots, m$$

so that, thanks to the previous steps,

$$e_j^{(m)} \perp e_i^{(m)} \quad \text{for all } i \neq j \quad \text{and} \quad i, j = 1, \ldots, m$$

and

$$e_k^{(m)} = e_k^{(m-1)} - \frac{f(e_k^{(m-1)}, e_m^{(m)})}{f(e_m^{(m)}, e_m^{(m)})} e_m^{(m)}, \quad k = m+1, \ldots, n.$$

Then $e_k^{(m)} \perp e_j^{(m)}$, for any $j = 1, \ldots, m$ and $k = m+1, \ldots, n$, and $V = \langle e_1^{(m)}, \ldots, e_m^{(m)} \rangle \oplus \langle e_1^{(m)}, \ldots, e_m^{(m)} \rangle^{\perp}$, where $\langle e_1^{(m)}, \ldots, e_m^{(m)} \rangle^{\perp} = \langle e_{m+1}^{(m)}, \ldots, e_n^{(m)} \rangle$. Thus, the matrix A_m of f in terms of B_m has any (k, j)-entry and (j, k)-entry equal to zero, for $j = 1, \ldots, m$ and $k \neq j$. Once again, if P_m is the transition matrix having in any i-th column the coordinates of the vector $e_i^{(m)}$ in terms of $B_{m-1} = \{e_1^{(m-1)}, \ldots, e_n^{(m-1)}\}$, then $P_m^t A_{m-1} P_m = A_m$.

In any step of the process, we may define the transition matrix P_i, associated with the transition from the basis B_i to B_{i-1} : the column vectors of P_i are the elements of B_i in terms of B_{i-1}.

Assume for example that the process consists of n steps: at the end, we obtain the matrix $P = P_1 P_2 \ldots P_n$, which is the transition matrix of B_n relative to B_0, such that $P^t A_0 P$ is diagonal. The column vectors of P are the elements of the f-orthogonal basis for V.

Example 7.34 Let $f : \mathbb{R}^3 \times \mathbb{R}^3 \to \mathbb{R}$ be the symmetric bilinear form defined by the matrix

$$A = \begin{bmatrix} -2 & 1 & 0 \\ 1 & -2 & 1 \\ 0 & 1 & -3 \end{bmatrix}$$

in terms of the canonical basis for $\mathbb{R}^3$. If we denote by $X = [x_1, x_2, x_3]^t$ any nonzero vector of $\mathbb{R}^3$ in terms of the canonical basis $B = \{e_1, e_2, e_3\}$ for $\mathbb{R}^3$, then

$$f(X, X) = -2x_1^2 + 2x_1 x_2 - 2x_2^2 + 2x_2 x_3 - 3x_3^2$$
$$= -\left\{ x_1^2 + (x_1 - x_2)^2 + (x_2 - x_3)^2 + 2x_3^2 \right\} < 0 \quad \text{for all } X \in \mathbb{R}^3.$$

Therefore, there is no nonzero vector in $\mathbb{R}^3$ which is f-isotropic. Hence, in the sense of Definition 7.27, $\mathbb{R}^3$ is a f-nonisotropic real vector space.

Starting from $B = \{e_1, e_2, e_3\}$, we construct the new basis $B' = \{e_1', e_2', e_3'\}$ for $\mathbb{R}^3$, defined as follows:

$$e_1' = e_1, \quad e_2' = e_2 - \frac{f(e_1, e_2)}{f(e_1, e_1)} e_1, \quad e_3' = e_3 - \frac{f(e_1, e_3)}{f(e_1, e_1)} e_1,$$

that is,

$$e_1' = e_1, \quad e_2' = e_2 + \frac{1}{2}e_1, \quad e_3' = e_3.$$

Therefore, the transition matrix is

$$C_1 = \begin{bmatrix} 1 & \frac{1}{2} & 0 \\ 0 & 1 & 0 \\ 0 & 0 & 1 \end{bmatrix}$$

and the matrix of f in terms of the basis B' is

$$A' = C_1^t A C_1 = \begin{bmatrix} -2 & 0 & 0 \\ 0 & \frac{-3}{2} & 1 \\ 0 & 1 & -3 \end{bmatrix}.$$

In the second step, we define the following new basis $B'' = \{e_1'', e_2'', e_3''\}$ for $\mathbb{R}^3$:

$$e_1'' = e_1', \quad e_2'' = e_2', \quad e_3'' = e_3' - \frac{f(e_2', e_3')}{f(e_2', e_2')}e_2',$$

that is,

$$e_1'' = e_1', \quad e_2'' = e_2', \quad e_3'' = e_3' + \frac{2}{3}e_2'.$$

The transition matrix is now

$$C_2 = \begin{bmatrix} 1 & 0 & 0 \\ 0 & 1 & \frac{2}{3} \\ 0 & 0 & 1 \end{bmatrix}$$

and the matrix of f in terms of the basis B'' is

$$A'' = C_2^t A' C_2 = \begin{bmatrix} -2 & 0 & 0 \\ 0 & -\frac{3}{2} & 0 \\ 0 & 0 & -\frac{7}{3} \end{bmatrix}.$$

In order to determine the coordinates of vectors e_1'', e_2'', e_3'', we make the composition of both changes of basis. So we obtain the transition matrix C of the final basis B'' relative to the starting canonical basis:

$$C = C_1 C_2 = \begin{bmatrix} 1 & \frac{1}{2} & \frac{1}{3} \\ 0 & 1 & \frac{2}{3} \\ 0 & 0 & 1 \end{bmatrix}.$$

Hence, the coordinates of e_1'', e_2'', e_3'' are precisely the columns of C, that is

$$B'' = \left\{(1, 0, 0), (\frac{1}{2}, 1, 0), (\frac{1}{3}, \frac{2}{3}, 1)\right\}.$$

A'' is the matrix of f in terms of B''. Finally, we construct the basis for $\mathbb{R}^3$ with respect to which the matrix of f has all nonzero entries equal to ± 1. To do this, we replace each e_i'' by $\dfrac{e_i''}{\sqrt{|f(e_i'', e_i'')|}}$. In particular, we have

$$\sqrt{|f(e_1'', e_1'')|} = \sqrt{2}, \quad \sqrt{|f(e_2'', e_2'')|} = \frac{\sqrt{3}}{\sqrt{2}}, \quad \sqrt{|f(e_3'', e_3'')|} = \frac{\sqrt{7}}{\sqrt{3}}$$

and obtain the basis

$$\tilde{B} = \left\{(\frac{1}{\sqrt{2}}, 0, 0), (\frac{1}{\sqrt{6}}, \frac{2}{\sqrt{6}}, 0), (\frac{1}{\sqrt{21}}, \frac{2}{\sqrt{21}}, \frac{3}{\sqrt{21}})\right\}.$$

The transition matrix of $\tilde{B}$ relative to the starting basis B is

$$\tilde{C} = \begin{bmatrix} \frac{1}{\sqrt{2}} & \frac{1}{\sqrt{6}} & \frac{1}{\sqrt{21}} \\ 0 & \frac{2}{\sqrt{6}} & \frac{2}{\sqrt{21}} \\ 0 & 0 & \frac{3}{\sqrt{21}} \end{bmatrix}$$

and the matrix of f in terms of $\tilde{B}$ is

$$\tilde{A} = \tilde{C}^t A \tilde{C} = \begin{bmatrix} -1 & 0 & 0 \\ 0 & -1 & 0 \\ 0 & 0 & -1 \end{bmatrix}.$$

Here, we would like to repeat the previous example, in this case, the bilinear form is defined on a complex vector space. More precisely:

Example 7.35 Let $f : \mathbb{C}^3 \times \mathbb{C}^3 \to \mathbb{C}$ be the symmetric bilinear form defined by the matrix

$$A = \begin{bmatrix} -2 & 1 & 0 \\ 1 & -2 & 1 \\ 0 & 1 & -3 \end{bmatrix}$$

in terms of the canonical basis B for $\mathbb{C}^3$.

By using the above argument, we find a basis

$$B'' = \left\{(1, 0, 0), (\frac{1}{2}, 1, 0), (\frac{1}{3}, \frac{2}{3}, 1)\right\}$$

in terms of which the matrix of f is the following diagonal one:

$$A'' = \begin{bmatrix} -2 & 0 & 0 \\ 0 & -\frac{3}{2} & 0 \\ 0 & 0 & -\frac{7}{3} \end{bmatrix}.$$

A'' is the matrix of f in terms of B''. Since the vector space is defined over a algebraically closed field, here we may construct a basis for $\mathbb{C}^3$ with respect to which the matrix of f has all nonzero entries equal to 1. To do this, we replace each e_i'' by $\frac{e_i''}{\sqrt{f(e_i'',e_i'')}}$. Thus, we have

$$\sqrt{f(e_1'', e_1'')} = i\sqrt{2}, \quad \sqrt{f(e_2'', e_2'')} = i\frac{\sqrt{3}}{\sqrt{2}}, \quad \sqrt{f(e_3'', e_3'')} = i\frac{\sqrt{7}}{\sqrt{3}}$$

and obtain the basis

$$\tilde{B} = \left\{ (\frac{1}{i\sqrt{2}}, 0, 0), (\frac{1}{i\sqrt{6}}, \frac{2}{i\sqrt{6}}, 0), (\frac{1}{i\sqrt{21}}, \frac{2}{i\sqrt{21}}, \frac{3}{i\sqrt{21}}) \right\}.$$

The transition matrix of $\tilde{B}$ relative to the starting basis B is

$$\tilde{C} = \begin{bmatrix} \frac{1}{i\sqrt{2}} & \frac{1}{i\sqrt{6}} & \frac{1}{i\sqrt{21}} \\ 0 & \frac{2}{i\sqrt{6}} & \frac{2}{i\sqrt{21}} \\ 0 & 0 & \frac{3}{i\sqrt{21}} \end{bmatrix}$$

and the matrix of f in terms of $\tilde{B}$ is

$$\tilde{A} = \tilde{C}^t A \tilde{C} = \begin{bmatrix} 1 & 0 & 0 \\ 0 & 1 & 0 \\ 0 & 0 & 1 \end{bmatrix}.$$

7.9   The Orthogonalization Process for Isotropic Symmetric Spaces

It is clear that the application of Gram-Schmidt process is strictly connected to the following needed condition: in order to compute the i-th step, the vector $e_i^{(i-1)} \in B_{i-1}$ must be nonisotropic. In other words, the matrix A_{i-1} of f, in terms of the basis B_{i-1}, must have a nonzero diagonal (i, i)-entry.

Hence, if we assume that (V, f) is isotropic, this condition could be not necessarily verified. Here, we would like to describe a simple method for taking forward the process, usually called *Lagrange orthogonalization process*.

So, after m steps, we suppose that $B_m = \{e_1^{(m)}, \ldots, e_n^{(m)}\}$ is the basis for V such that $\{e_1^{(m)}, \ldots, e_m^{(m)}, e_k^{(m)}\}$ is an f-orthogonal set, for any $k \geq m+1$, but $f(e_{m+1}^{(m)}, e_{m+1}^{(m)}) = 0$. Nevertheless, if there exists $j \geq m+2$ such that $f(e_j^{(m)}, e_j^{(m)}) \neq 0$, we may obtain a new basis B_m' where $e_{m+1}^{(m)}$ switches places with $e_j^{(m)}$. The outcome achieved enables us to apply the Gram-Schmidt method starting from B_m'.

Therefore, we have to consider the hardest case when $f(e_j^{(m)}, e_j^{(m)}) = 0$, for any $j \geq m+1$. Firstly, we notice that if $f(e_j^{(m)}, e_k^{(m)}) = 0$, for any $j \geq m+1$ and $k \geq j+1$, then B_m is already an f-orthogonal basis, and we are done. Thus, we suppose there are $j \geq m+1$ and $k \geq j+1$ such that $f(e_j^{(m)}, e_k^{(m)}) = \alpha \neq 0$. Easy computations show that

$$f(e_j^{(m)} + e_k^{(m)}, e_j^{(m)} + e_k^{(m)}) = 2\alpha \neq 0.$$

Once again, we may obtain a new basis B_m'' for V, by replacing $e_j^{(m)}$ with $e_j^{(m)} + e_k^{(m)}$ in the basis B_m. As above, we can apply the Gram-Schmidt algorithm, by using the vectors from B_m''. By repeating this argument, we will finally find an f-orthogonal basis for V.

Example 7.36 Let $f : \mathbb{R}^3 \times \mathbb{R}^3 \to \mathbb{R}$ be the symmetric bilinear form defined by the matrix

$$A = \begin{bmatrix} 0 & \frac{1}{2} & \frac{1}{2} \\ \frac{1}{2} & 0 & \frac{1}{2} \\ \frac{1}{2} & \frac{1}{2} & 0 \end{bmatrix}$$

in terms of the canonical basis $B = \{e_1, e_2, e_3\}$ for $\mathbb{R}^3$. Notice that, in this case, $\mathbb{R}^3$ is f-isotropic. We firstly proceed to construct a basis $B' = \{e_1', e_2', e_3'\}$ for $\mathbb{R}^3$ in terms of which the matrix of f has some nonzero element on the main diagonal. Since $f(e_1, e_1) = 0$ and $f(e_1, e_2) \neq 0$, we may define

$$e_1' = e_1 + \alpha e_2, \quad e_2' = e_2, \quad e_3' = e_3,$$

where $\alpha \in \mathbb{R}$ and $f(e_1', e_1') = \alpha$. It is easy to see that for $\alpha = 1$ we get the required condition $f(e_1', e_1') \neq 0$. Thus, the transition matrix is

$$C_1 = \begin{bmatrix} 1 & 0 & 0 \\ 1 & 1 & 0 \\ 0 & 0 & 1 \end{bmatrix}$$

so that the matrix of f in terms of the new basis is

$$A' = C_1^t A C_1 = \begin{bmatrix} 1 & \frac{1}{2} & 1 \\ \frac{1}{2} & 0 & \frac{1}{2} \\ 1 & \frac{1}{2} & 0 \end{bmatrix}.$$

The next change of basis will be the following:

$$e_1'' = e_1', \quad e_2'' = e_2' - \frac{f(e_1', e_2')}{f(e_1', e_1')} e_1', \quad e_3'' = e_3' - \frac{f(e_1', e_3')}{f(e_1', e_1')} e_1',$$

that is,

$$e_1'' = e_1', \quad e_2'' = e_2' - \frac{1}{2} e_1', \quad e_3'' = e_3' - e_1'$$

and the corresponding transition matrix is

$$C_2 = \begin{bmatrix} 1 & -\frac{1}{2} & -1 \\ 0 & 1 & 0 \\ 0 & 0 & 1 \end{bmatrix}.$$

Hence, the new expression of the matrix of f is

$$A'' = C_2^t A' C_2 = \begin{bmatrix} 1 & 0 & 0 \\ 0 & -\frac{1}{4} & 0 \\ 0 & 0 & -1 \end{bmatrix}.$$

The basis $B''' = \{e_1''', e_2''', e_3'''\}$ in terms of which the matrix of f is precisely A'', is obtained by the computation

$$C = C_1 \cdot C_2 = \begin{bmatrix} 1 & -\frac{1}{2} & -1 \\ 1 & \frac{1}{2} & -1 \\ 0 & 0 & 1 \end{bmatrix},$$

where C is the transition matrix from B''' to the starting basis B. Looking at the columns of C, we have

$$B''' = \left\{ (1, 1, 0), \left(-\frac{1}{2}, \frac{1}{2}, 0\right), (-1, -1, 1) \right\}.$$

Moreover, $C^t A C = A''$.

Finally, we construct the basis for $\mathbb{R}^3$, with respect to which the matrix of f has all nonzero entries equal to ± 1. To do this, we replace each e_i''' by $\frac{e_i'''}{\sqrt{|f(e_i''', e_i''')|}}$. In particular, we have

$$\sqrt{|f(e_1''', e_1''')|} = 1, \quad \sqrt{|f(e_2''', e_2''')|} = \frac{1}{2}, \quad \sqrt{|f(e_3''', e_3''')|} = 1$$

and obtain the basis

$$\tilde{B} = \left\{ (1, 1, 0), (-1, 1, 0), (-1, -1, 1) \right\}.$$

The transition matrix from $\tilde{B}$ to the starting basis B is

$$\tilde{C} = \begin{bmatrix} 1 & -1 & -1 \\ 1 & 1 & -1 \\ 0 & 0 & 1 \end{bmatrix}$$

and the matrix of f in terms of $\tilde{B}$ is

$$\tilde{A} = \tilde{C}^t A \tilde{C} = \begin{bmatrix} 1 & 0 & 0 \\ 0 & -1 & 0 \\ 0 & 0 & -1 \end{bmatrix}.$$

Putting Theorem 7.32 and the above orthogonalization process together, we are now able to state the following:

Theorem 7.37 *Let $\mathbb{F}$ be an algebraically closed field of characteristic different from 2 and (V, f) a symmetric vector space. Then there is an f-orthogonal basis B for V such that the matrix of f in terms of B has the following form:*

$$\begin{bmatrix} 1 & & & & & \\ & \ddots & & & & \\ & & 1 & & & \\ & & & 0 & & \\ & & & & \ddots & \\ & & & & & 0 \end{bmatrix}.$$

(Notice that we cannot say that B is orthonormal, because there is the chance that some diagonal element is zero, that is, $f(e_i, e_i) = 0$ for some $e_i \in B$).

Theorem 7.38 *Let (V, f) be a real symmetric vector space. Then there is an f-orthogonal basis B for V such that the matrix of f in terms of B has the following form:*

7.10 Quadratic Forms Associated with Bilinear Forms

Let $\mathbb{F}$ be a field, V a finite dimensional vector space over $\mathbb{F}$ and $B = \{e_1, \ldots, e_n\}$ a
basis for V. As usual, we denote by $X = [x_1, \ldots, x_n]^t$ the coordinate vector of any
$v \in V$ in terms of B.

A *quadratic form* on V is any map $q : V \to \mathbb{F}$ such that $q(X) = \sum\limits_{i,j=1}^{n} \alpha_{ij} x_i x_j$,
that is, $q(X)$ is a homogeneous polynomial of degree 2, involving the indeterminates
$\{x_1, \ldots, x_n\}$. This definition is strictly connected with the coordinates $x_1, \ldots, x_n$ of
vectors of V, with respect to the given basis.

Here, we give an equivalent definition of a quadratic form, which is coordinate-
free:

Definition 7.39 Let $q : V \to \mathbb{F}$. Define a map $f : V \times V \to \mathbb{F}$ such that

$$f(u, v) = \begin{cases} \frac{1}{2}\left\{q(u + v) - q(u) - q(v)\right\} & \text{when } char(\mathbb{F}) \neq 2 \\ q(u + v) - q(u) - q(v) & \text{when } char(\mathbb{F}) = 2. \end{cases} \qquad (7.12)$$

The map q is called a *quadratic form* on V if both the following conditions hold:

(1) the map (7.12) is bilinear;
(2) for any $\alpha \in \mathbb{F}$ and $v \in V$, $q(\alpha v) = \alpha^2 q(v)$.

It is easy to see that f is symmetric. The map f is called *the symmetric bilinear form
associated with q.*

Remark 7.40 For the rest of this section, we always assume that $char(\mathbb{F}) \neq 2$.

Proposition 7.41 *The map $q : V \to \mathbb{F}$ is a quadratic form on V if and only if there
exists a bilinear form $\varphi : V \times V \to \mathbb{F}$ such that, for any $v \in V$, $q(v) = \varphi(v, v)$.*

Proof If q is a quadratic form on V, then φ is precisely the bilinear symmetric map
f of Definition 7.39. Next, we show that $q(v) = f(v, v)$. Putting $\alpha = 0$ and $\alpha =$

-1 in $q(\alpha v) = \alpha^2 q(v)$, respectively, we get $q(0) = 0$ and $q(-v) = q(v)$. Taking $u = -v$ in $f(u, v)$, we obtain $f(-v, v) = \frac{1}{2}(q(-v + v) - q(-v) - q(v))$. As f is bilinear, we get $-f(v, v) = \frac{1}{2}(q(0) - q(v) - q(v))$. After simplification, we get $q(v) = f(v, v)$.

Conversely, assume there is a bilinear form $\varphi : V \times V \to \mathbb{F}$ such that, for any $v \in V$, $q(v) = \varphi(v, v)$. Then, for any $\alpha \in \mathbb{F}$ and $v \in V$, we have

$$q(\alpha v) = \varphi(\alpha v, \alpha v) = \alpha^2 \varphi(v, v) = \alpha^2 q(v). \tag{7.13}$$

Moreover, for any $u, v \in V$,

$$\frac{1}{2}\left\{q(u + v) - q(u) - q(v)\right\} = \frac{1}{2}\left\{\varphi(u + v, u + v) - \varphi(u, u) - \varphi(v, v)\right\}$$

$$= \frac{1}{2}\left\{\varphi(u, u) + \varphi(u, v) + \varphi(v, u) + \varphi(v, v)\right.$$

$$\left. -\varphi(u, u) - \varphi(v, v)\right\}$$

$$= \frac{1}{2}\left\{\varphi(u, v) + \varphi(v, u)\right\}. \tag{7.14}$$

Let now $f : V \times V \to \mathbb{F}$ be defined as $f(u, v) = \frac{1}{2}\left\{q(u + v) - q(u) - q(v)\right\}$, for any $u, v \in V$. By relation (7.14) it is clear that $f(u, v) = f(v, u)$, that is, f is symmetric, moreover, since φ is bilinear, so is also f. Thus, in light of Definition 7.39, we conclude that q is a quadratic form on V, having f as its associated symmetric bilinear form.

Remark 7.42 The previous result outlines the fact that the quadratic form q can be expressed both in terms of its associated symmetric bilinear form f, and in terms of any bilinear form φ having the property $\varphi(v, v) = q(v)$ for all $v \in V$. Nevertheless, φ is not required to be necessarily symmetric. On the other hand, in case we assume φ symmetric, by (7.14) it follows that $\varphi = f$.

In other words, there is an unique symmetric bilinear form associated with q.

Definition 7.43 Let $q : V \to \mathbb{F}$ be a quadratic form on V, $B = \{e_1, \ldots, e_n\}$ a basis for V and φ be any bilinear form associated with q, that is, $\varphi(v, v) = q(v)$ for any $v \in V$. If C is the matrix of φ in terms of the basis B, then $q(v) = \varphi(v, v) = v^t C v$. We say that C is a *matrix associated with q* in terms of the basis B.

In light of the previous remark, there is only one symmetric matrix associated with a quadratic form in terms of a fixed basis.

Now the question that arises is how we can compute the symmetric form associated with a given quadratic form. To provide an answer to this question, we prove the following:

Theorem 7.44 *Let $q : V \to \mathbb{F}$ be a quadratic form on V and $B = \{e_1, \ldots, e_n\}$ be an ordered basis for V and A the matrix of q in terms of B. Then the (unique) symmetric bilinear form associated with q is represented by the matrix $\frac{1}{2}(C + C^t)$ with respect to the basis B, where C is the matrix of any bilinear form associated with q.*

Proof Let φ be any bilinear form associated with q, that is, $\varphi(v, v) = q(v)$, for any $v \in V$. If C is the matrix of φ in terms of the basis B, then $q(v) = \varphi(v, v) = v^t C v$.

Now we define the bilinear form $\tilde{\varphi}$ having matrix C^t, that is, $\tilde{\varphi}(v, w) = v^t C^t w$ for any $v, w \in V$. So we may introduce the quadratic form associated with $\tilde{\varphi}$ as follows: $\tilde{q}(v) = v^t C^t v$ for any $v \in V$. Since both $v^t C^t v$ and $v^t C v$ are scalar elements of $\mathbb{F}$, it is clear that each of them coincides with its transpose. On the other hand, the transpose of $v^t C^t v$ is precisely $v^t C v$ (and viceversa). Hence $v^t C^t v = v^t C v$, i.e., $q(v) = \tilde{q}(v)$, for any $v \in V$ and

$$2q(v) = q(v) + \tilde{q}(v) = v^t(C + C^t)v \implies q(v) = \frac{1}{2}\{v^t(C + C^t)v\} = v^t\{\frac{1}{2}(C + C^t)\}v,$$

where $C + C^t$ is a symmetric matrix which represents q with respect to the basis B.

Moreover, if we assume there is another symmetric matrix D associated with q, then we have

$$q(v) = v^t\{\frac{1}{2}(C + C^t)\}v \quad \text{and} \quad q(v) = v^t D v \quad \text{for all} \quad v \in V,$$

which means $v^t\left(\frac{1}{2}(C + C^t) - D\right)v = 0$, for any $v \in V$. For clearness, we write $\frac{1}{2}(C + C^t) - D = E = (\alpha_{ij})$. Since D and $\frac{1}{2}(C + C^t)$ are symmetric matrices, then so E is. Thus, for any $e_i \in B$, one has $e_i^t E e_i = 0$, implying $\alpha_{ii} = 0$. Moreover, for any $i \neq j$ and $e_i, e_j \in B$, we also get

$$0 = (e_i + e_j)^t E(e_i + e_j) = e_i^t E e_j + e_j^t E e_i = \alpha_{ij} + \alpha_{ji} = 2\alpha_{ij}.$$

Therefore, $E = 0$ and $\frac{1}{2}(C + C^t) = D$. Thus, the uniqueness of the symmetric form associated with q is proved.

Example 7.45 Let $q : \mathbb{R}^2 \to \mathbb{R}$ defined by $q\big((x_1, x_2)\big) = x_1^2 + 5x_1 x_2 + 2x_2^2$ in terms of the canonical basis for $\mathbb{R}^2$. By easy computations, we can see that each of the following bilinear forms can be associated with q :

$$\vartheta : \mathbb{R}^2 \times \mathbb{R}^2 \to \mathbb{R}, \quad \vartheta\big((x_1, x_2), (y_1, y_2)\big) = x_1 y_1 + 2x_1 y_2 + 3x_2 y_1 + 2x_2 y_2;$$

$$\eta : \mathbb{R}^2 \times \mathbb{R}^2 \to \mathbb{R}, \quad \eta\big((x_1, x_2), (y_1, y_2)\big) = x_1 y_1 + x_1 y_2 + 4x_2 y_1 + 2x_2 y_2;$$

$$\psi : \mathbb{R}^2 \times \mathbb{R}^2 \to \mathbb{R}, \quad \psi\big((x_1, x_2), (y_1, y_2)\big) = x_1 y_1 + 5x_1 y_2 + 2x_2 y_2.$$

None of the above bilinear maps is symmetric. In order to obtain the symmetric form associated with q, we may choose arbitrarily one of the above bilinear form and compute its matrix in terms of the canonical basis for $\mathbb{R}^2$. For instance, the matrix of ϑ is

$$A = \begin{bmatrix} 1 & 2 \\ 3 & 2 \end{bmatrix}, \quad \text{so that} \quad A^t = \begin{bmatrix} 1 & 3 \\ 2 & 2 \end{bmatrix}$$

and the only symmetric bilinear map φ associated with q has matrix

$$\frac{1}{2}(A + A^t) = \begin{bmatrix} 1 & \frac{5}{2} \\ \frac{5}{2} & 2 \end{bmatrix},$$

that is,

$$\varphi\big((x_1, x_2), (y_1, y_2)\big) = x_1 y_1 + \frac{5}{2} x_1 y_2 + \frac{5}{2} x_2 y_1 + 2 x_2 y_2.$$

Of course, we get the same result starting from the matrix of η or ψ.

7.11 The Matrix of a Quadratic Form and the Change of Basis

Let $q : V \to \mathbb{F}$ be a quadratic form, $f : V \times V \to \mathbb{F}$ a bilinear form associated with q, $B = \{e_1, \ldots, e_n\}$ an ordered basis for V, and $A = (a_{ij})$ the matrix of f in terms of B. Thus, for any vectors $X, Y \in V$, we know that $f(X, Y) = X^t A Y$. In particular,

$$q(X) = f(X, X) = X^t A X,$$

where X is the column coordinate vector with respect to B. We say that A is the matrix of q in terms of the basis B.

Moreover, if B' is another basis for V, different from B, it is known that f can be represented also by the matrix $A' = P^t A P$ in terms of B', where P is the transition matrix of B' relative to B.

As above, we say that A' is the matrix of q with respect to the basis B' and

$$q(X') = X'^t A' X',$$

where X' is the column coordinate vector in terms of the basis B'.

In summary: the quadratic form q can be represented by any matrix, which represents a bilinear form associated with q. In particular, two matrices A, A' represent q, in terms of two different bases for V, if and only if they are congruent.

As pointed out in the Remark 7.42, we may associate different bilinear forms to the same quadratic form q. Hence, in general, once a basis for V has been established, we may represent q by different matrices, each of which is relative to a bilinear form associated with q. Only one of these forms is symmetric.

Hence, there is a one-to-one correspondence between symmetric bilinear forms on V and quadratic forms on V and knowing the quadratic form is equivalent to knowing the corresponding bilinear form.

In particular, the matrix of a quadratic form coincides with the matrix of the corresponding symmetric bilinear form and it can be obtained as follows: let B be a basis for V, $X = [x_1, \ldots, x_n]^t$ the generic column coordinate vector with respect to B and $q : V \to \mathbb{F}$ a quadratic form defined as $q(X) = \sum\limits_{i=1}^{n} \alpha_i x_i^2 + \sum\limits_{i<j} \alpha_{ij} x_i x_j$. Then the matrix of q is

$$
A = \begin{bmatrix}
\alpha_1 & \frac{\alpha_{12}}{2} & \cdots & \frac{\alpha_{1n}}{2} \\
\frac{\alpha_{12}}{2} & \alpha_2 & \cdots & \frac{\alpha_{2n}}{2} \\
\vdots & \vdots & \ddots & \vdots \\
\frac{\alpha_{1n}}{2} & \frac{\alpha_{2n}}{2} & \cdots & \alpha_n
\end{bmatrix}.
$$

7.12 Diagonalization of a Quadratic Form

In light of the equivalence between quadratic and symmetric bilinear forms, let us rephrase Theorems 7.37 and 7.38 as follows:

Theorem 7.46 *Let $\mathbb{F}$ be an algebraically closed field of characteristic different from 2 and $q : V \to \mathbb{F}$ a quadratic form on V. Then there is an f-orthogonal basis $B = \{e_1, \ldots, e_n\}$ for V such that the matrix of q in terms of B has the following form:*

$$
A = \begin{bmatrix}
1 & & & & & \\
& \ddots & & & & \\
& & 1 & & & \\
& & & 0 & & \\
& & & & \ddots & \\
& & & & & 0
\end{bmatrix}.
$$

In other words, if r is the rank of the matrix of q, then it follows that, for any $X \in V$, having coordinate vector $[x_1, \ldots, x_n]^t$ in terms of B,

$$q(X) = X^t A X$$

$$= [x_1, \ldots, x_n] \begin{bmatrix} 1 & & & & & & \\ & \ddots & & & & & \\ & & 1 & & & & \\ & & & 0 & & & \\ & & & & \ddots & & \\ & & & & & 0 & \end{bmatrix} \begin{bmatrix} x_1 \\ \vdots \\ x_n \end{bmatrix}$$

$$= x_1^2 + \cdots + x_r^2.$$

Theorem 7.47 *Let $q : V \to \mathbb{R}$ be a real quadratic form on V. Then there is an f-orthogonal basis $B = \{e_1, \ldots, e_n\}$ for V such that the matrix of q in terms of B has the following form:*

$$A = \begin{bmatrix} 1 & & & & & & & \\ & \ddots & & & & & & \\ & & 1 & & & & & \\ & & & -1 & & & & \\ & & & & \ddots & & & \\ & & & & & -1 & & \\ & & & & & & 0 & \\ & & & & & & & \ddots \\ & & & & & & & & 0 \end{bmatrix}$$

In other words, if r is the rank of the matrix of q, there exists an integer $0 \le p \le r$ such that, for any $X \in V$, having coordinate vector $[x_1, \ldots, x_n]^t$ in terms of B :

$$q(X) = X^t A X$$

$$= [x_1, \ldots, x_n] \begin{bmatrix} 1 & & & & & & & \\ & \ddots & & & & & & \\ & & 1 & & & & & \\ & & & -1 & & & & \\ & & & & \ddots & & & \\ & & & & & -1 & & \\ & & & & & & 0 & \\ & & & & & & & \ddots \\ & & & & & & & & 0 \end{bmatrix} \begin{bmatrix} x_1 \\ \vdots \\ x_n \end{bmatrix}$$

$$= x_1^2 + \cdots + x_p^2 - x_{p+1}^2 - \cdots - x_r^2.$$

7.13 Definiteness of a Real Quadratic Form

Here, we shall still process some properties of congruent matrices. To do this, we need the following lemmas:

Lemma 7.48 *Let A, P be two $n \times n$ matrices having coefficients in a field $\mathbb{F}$. If P is invertible, then A and AP have the same rank.*

Proof Each of A, P, AP represents a linear operator on the vector space $\mathbb{F}^n$. We will discuss the range of any such operator. Without loss of generality, we will refer to these ranges as $R(A)$, $R(P)$, $R(AP)$.

Let $X \in \mathbb{F}^n$ be such that $X \in R(AP)$. Hence, there is $Y \in \mathbb{F}^n$ such that $X = (AP)Y = A(PY)$, implying that $X \in R(A)$. Conversely, let now $X \in \mathbb{F}^n$ be such that $X \in R(A)$, that is, $X = AZ$ for some $Z \in \mathbb{F}^n$. Since P is invertible, the linear operator represented by P is an isomorphism. Therefore, there is precisely one vector $Y \in \mathbb{F}^n$ such that $Z = PY$, so $X = AZ = APY$ and $X \in R(AP)$.

Hence, the image of A is equal to the image of AP. Both of them are generated by the linearly $\mathbb{F}$-independent columns in A and AP, respectively. Since the number of linearly independent columns in a matrix is precisely its rank, we get the required conclusion.

Lemma 7.49 *Let A, P be two $n \times n$ matrices having coefficients in a field $\mathbb{F}$. If P is invertible, then A and PA have the same rank.*

Proof As above, we consider the linear operators represented by A, P, PA. Now, we will discuss also the kernel of any such operator and refer to these kernels as $N(A)$, $N(P)$, $N(PA)$.

Let $X \in \mathbb{F}^n$ be such that $X \in N(A)$, that is, $AX = 0$ and a fortiori $PAX = 0$, which implies $X \in N(PA)$. Conversely, let $X \in \mathbb{F}^n$ be such that $X \in N(PA)$, that is $(PA)X = 0$. Since P is invertible, we may left multiply the previous relation by P^{-1}, having $AX = 0$, that is, $X \in N(A)$. Therefore $N(A) = N(PA)$. On the other hand $n = dim(N(A)) + dim(R(A))$ and also $n = dim(N(PA)) + dim(R(PA))$, implying that $dim(R(A)) = dim(R(PA))$. As in the previous lemma, we conclude that the rank of A is equal to the rank of PA.

Lemma 7.50 *Let A, A' be two $n \times n$ matrices having coefficients in a field $\mathbb{F}$. If A, A' are congruent, then they have the same rank.*

Proof By our hypothesis, there exists an invertible $n \times n$ matrix P with coefficients in $\mathbb{F}$, such that $A = P^t A' P$. Since P^t is invertible, and by Lemma 7.49, the matrices A and $A'P$ have the same rank. Moreover, using Lemma 7.48, the rank of $A'P$ is equal to the rank of A', as desired.

Definition 7.51 Let $q : V \to \mathbb{R}$ be a real quadratic form, associated with the symmetric bilinear form f. The ordered pair $(p, r - p)$ such that, for any $X = (x_1, \ldots, x_n) \in \mathbb{R}^n$, $q(X) = x_1^2 + \cdots + x_p^2 - x_{p+1}^2 - \cdots - x_r^2$ with respect to an appropriate f-orthogonal basis for V, is called the *signature* of q.

In a similar way, we give the following:

Definition 7.52 Let A be a real $n \times n$ symmetric matrix, Q a real $n \times n$ invertible matrix such that $Q^t A Q = D$ is a diagonal matrix of the form

$$
D = \begin{bmatrix}
1 & & & & & & & & \\
 & \ddots & & & & & & & \\
 & & 1 & & & & & & \\
 & & & -1 & & & & & \\
 & & & & \ddots & & & & \\
 & & & & & -1 & & & \\
 & & & & & & 0 & & \\
 & & & & & & & \ddots & \\
 & & & & & & & & 0
\end{bmatrix}
$$

having p ones and $r - p$ negative ones on the main diagonal, where r is the rank of A (recall that A and D have the same rank, since they are congruent). The ordered pair $(p, r - p)$ is called the *signature* of A.

Remark 7.53 Let D be a real $n \times n$ diagonal matrix having the form

$$
D = \begin{bmatrix}
\alpha_1^2 & & & & & & & & \\
 & \ddots & & & & & & & \\
 & & \alpha_p^2 & & & & & & \\
 & & & -\alpha_{p+1}^2 & & & & & \\
 & & & & \ddots & & & & \\
 & & & & & -\alpha_r^2 & & & \\
 & & & & & & 0 & & \\
 & & & & & & & \ddots & \\
 & & & & & & & & 0
\end{bmatrix},
$$

where $\alpha_i \in \mathbb{R}$, for any $i = 1, \ldots, r$. The signature of D is precisely $(p, r - p)$.

In fact, if $q : V \to \mathbb{R}$ is the real quadratic form represented by D in terms of a basis B, then, by Theorem 7.47, there is an orthogonal basis $B' = \{e_1, \ldots, e_n\}$ for V such that the matrix of q in terms of B' has the following form:

$$
D' = \begin{bmatrix}
1 & & & & & & & & \\
& \ddots & & & & & & & \\
& & 1 & & & & & & \\
& & & -1 & & & & & \\
& & & & \ddots & & & & \\
& & & & & -1 & & & \\
& & & & & & 0 & & \\
& & & & & & & \ddots & \\
& & & & & & & & 0
\end{bmatrix}
$$

having p positive ones and $r - p$ negative ones on the main diagonal. In other words, there exists an invertible matrix P, that is, the transition matrix of B' relative to B, such that $P^t D P = D'$.

Definition 7.54 Let $q : V \to \mathbb{R}$ be a real quadratic form, associated with the symmetric bilinear form f. The quadratic form q is called:

(1) *Positive definite* if $q(X) > 0$ for all nonzero vectors $X \in V$; in this case, the signature is $(n, 0)$.
(2) *Negative definite* if $q(X) < 0$ for all nonzero vectors $X \in V$; in this case, the signature is $(0, n)$.
(3) *Indefinite* if it is neither positive-definite nor negative-definite, in the sense that q takes on V both positive and negative values; in this case, the signature is $(p, r - p)$ for $p \neq r$.
(4) *Positive semi-definite* if $q(X) \geq 0$ for all $X \in V$, but there is some nonzero vector $X_0 \in V$ so that $q(X_0) = 0$; in this case, the signature is $(r, 0)$.
(5) *Negative semi-definite* if $q(X) \leq 0$ for all $X \in V$, but there is some nonzero vector $X_0 \in V$ so that $q(X_0) = 0$; in this case, the signature is $(0, r)$.

Analogously, we may introduce the following:

Definition 7.55 A symmetric matrix $A \in M_n(\mathbb{R})$ is called:

(1) *Positive definite* if $X^t A X > 0$ for all nonzero vectors $X \in \mathbb{R}^n$; in this case, the signature of A is $(n, 0)$.
(2) *Negative definite* if $X^t A X < 0$ for all nonzero vectors $X \in \mathbb{R}^n$; in this case, the signature is $(0, n)$.
(3) *Indefinite* if it is neither positive-definite nor negative-definite, in the sense that $X^t A X$ takes both positive and negative values, depending on the choice of $X \in \mathbb{R}^n$; in this case, the signature is $(p, r - p)$, for $p \neq r$.
(4) *Positive semi-definite* if $X^t A X \geq 0$ for all $X \in \mathbb{R}^n$, but there is some nonzero vector $X_0 \in \mathbb{R}^n$ so that $X_0^t A X_0 = 0$; in this case, the signature is $(r, 0)$.
(5) *Negative semi-definite* if $X^t A X \leq 0$ for all $X \in \mathbb{R}^n$, but there is some nonzero vector $X_0 \in \mathbb{R}^n$ so that $X_0^t A X_0 = 0$; in this case, the signature is $(0, r)$.

Remark 7.56 Let (V, f) be a finite dimensional metric space over the field $\mathbb{R}$. If f is a symmetric bilinear form on V, having the additional property that $f(v, v) \geq 0$, for any $v \in V$ and $f(v, v) = 0$ if and only if $v = 0$, then we say that (V, f) is an inner product vector space. This is equivalent to say that the quadratic form q associated with f is positive definite.

Lemma 7.57 *Let D, E be two real $n \times n$ diagonal matrices. If D, E are congruent, then they have the same rank and signature.*

Proof Since D and E are congruent, they represent the same real quadratic form $q : V \to \mathbb{R}$, in terms of two different bases for V. The fact that D and E have the same rank is proved in Lemma 7.50, namely, $rank(D) = rank(E) = r \leq n$. Thus, there exists a basis $B = \{e_1, \ldots, e_n\}$ for V such that the matrix of q in terms of B is

$$D = \begin{bmatrix} \alpha_1^2 & & & & & & & \\ & \ddots & & & & & & \\ & & \alpha_p^2 & & & & & \\ & & & -\alpha_{p+1}^2 & & & & \\ & & & & \ddots & & & \\ & & & & & -\alpha_r^2 & & \\ & & & & & & 0 & \\ & & & & & & & \ddots \\ & & & & & & & & 0 \end{bmatrix},$$

where $\alpha_i \in \mathbb{R}$, for any $i = 1, \ldots, r$. Simultaneously, there exists a basis $B' = \{e_1', \ldots, e_n'\}$ for V such that the matrix of q in terms of B' is

$$E = \begin{bmatrix} \beta_1^2 & & & & & & & \\ & \ddots & & & & & & \\ & & \beta_s^2 & & & & & \\ & & & -\beta_{s+1}^2 & & & & \\ & & & & \ddots & & & \\ & & & & & -\beta_r^2 & & \\ & & & & & & 0 & \\ & & & & & & & \ddots \\ & & & & & & & & 0 \end{bmatrix},$$

where $\beta_i \in \mathbb{R}$, for any $i = 1, \ldots, r$. Hence, the pair $(p, r - p)$ is the signature of D and the pair $(s, r - s)$ is the signature of E (see Remark 7.53). Our aim is to prove that $p = s$. By our assumptions, we have

$$q(e_i) > 0 \quad \text{for all} \quad i = 1, \dots, p;$$
$$q(e_i) < 0 \quad \text{for all} \quad i = p+1, \dots, r;$$
$$q(e_i) = 0 \quad \text{for all} \quad i = r+1, \dots, n;$$
$$q(e'_i) > 0 \quad \text{for all} \quad i = 1, \dots, s;$$
$$q(e'_i) < 0 \quad \text{for all} \quad i = s+1, \dots, r;$$
$$q(e'_i) = 0 \quad \text{for all} \quad i = r+1, \dots, n.$$

Consider now the set $S = \{e_1 \dots, e_p, e'_{s+1}, \dots, e'_n\}$ and suppose there exist $\lambda_1, \dots, \lambda_p,$ $\lambda'_1, \dots, \lambda'_{n-s} \in \mathbb{R}$ such that

$$\lambda_1 e_1 + \cdots + \lambda_p e_p + \lambda'_1 e'_{s+1} + \cdots + \lambda'_{n-s} e'_n = 0. \tag{7.15}$$

If we denote $u = \lambda_1 e_1 + \cdots + \lambda_p e_p = -(\lambda'_1 e'_{s+1} + \cdots + \lambda'_{n-s} e'_n) \in V$, then we have both

$$q(u) = q\left(\lambda_1 e_1 + \cdots + \lambda_p e_p\right) = \sum_{i=1}^{p} \lambda_i^2 q(e_i) = \sum_{i=1}^{p} \lambda_i^2 \alpha_i^2 \geq 0$$

and

$$q(u) = q\left(-\lambda'_1 e'_{s+1} - \cdots - \lambda'_{n-s} e'_n\right) = \sum_{i=1}^{n-s} \lambda_i^2 q(e'_{s+i}) = -\sum_{i=1}^{r-s} \lambda_i^2 \beta_{s+i}^2 \leq 0.$$

This implies that $q(u) = 0$. Therefore,

$$\sum_{i=1}^{p} \lambda_i^2 \alpha_i^2 = \sum_{i=1}^{r-s} \lambda_i^2 \beta_{s+i}^2 = 0.$$

That is,

$$\lambda_i = 0 \quad \text{for all} \quad i = 1, \dots, p \quad \text{and} \quad \lambda'_j = 0 \quad \text{for all} \quad j = 1, \dots, r-s.$$

Thus, the relation (7.15) reduces to

$$\lambda'_1 e'_{s+1} + \cdots + \lambda'_{n-s} e'_n = 0$$

implying

$$\lambda'_j = 0 \quad \text{for all} \quad j = 1, \dots, n-s.$$

Hence, S is a linearly independent set, the vector subspaces $\langle e_1, \dots, e_p \rangle$ and $\langle e'_{s+1}, \dots, e'_n \rangle$ of V are direct summands, so that the dimension of the vector subspace spanned by S is equal to $p + n - s \leq n$, i.e., $p \leq s$.

Symmetrically, let $S' = \{e'_1 \dots, e'_s, e_{p+1}, \dots, e_n\}$. By the above arguments, one may prove that S' is a linearly independent set, the vector subspaces $\langle e'_1, \dots, e'_s \rangle$ and $\langle e_{p+1}, \dots, e_n \rangle$ of V are direct summands, so that the dimension of the vector subspace spanned by S' is equal to $s + n - p \le n$, i.e., $s \le p$. In summary, $p = s$, as required.

Theorem 7.58 *Let A, A' be two real $n \times n$ symmetric matrices. A, A' are congruent if and only if they have the same signature.*

Proof We firstly recall that A, A' are congruent if and only if they represent the same real quadratic form $q : V \to \mathbb{R}$, in terms of two different bases for V (see Theorem 7.7). Moreover, Remark 7.33 assures that any symmetric matrix, over a field of characteristic different from 2, is congruent to a diagonal matrix.

In light of these comments, there exist invertible $n \times n$ real matrices M, P, Q and diagonal $n \times n$ real matrices D, D' such that the following relations hold simultaneously:

$$A = M^t A' M, \qquad A' = P^t D' P, \qquad A = Q^t D Q.$$

Of course, by Remark 7.53, we may assume that D and D' represent the signature of A and A', respectively. Therefore, by

$$Q^t D Q = A = M^t A' M = M^t (P^t D' P) M,$$

it follows

$$D = (Q^t)^{-1} M^t P^t D' P M Q^{-1} = (P M Q^{-1})^t D' (P M Q^{-1}).$$

Hence, the diagonal matrices D, D' are congruent and, by Lemma 7.57, they have the same signature.

We now prove the other direction of this theorem and suppose that A, A' have the same signature. Thus, there exists a diagonal $n \times n$ real matrix D (representing the common signature of A and A') and two invertible $n \times n$ real matrices P, Q such that $D = P^t A P = Q^t A' Q$. Hence

$$A' = (Q^t)^{-1} P^t A P Q^{-1} = (P Q^{-1})^t A (P Q^{-1}),$$

i.e., A and A' are congruent.

Theorem 7.59 *Let A be a real $n \times n$ symmetric matrix having signature $(p, r - p)$. Then*

(i) *The number of positive eigenvalues of A is equal to p.*
(ii) *The number of negative eigenvalues of A is equal to $r - p$.*

Proof Since the signature of A is $(p, r - p)$, A is congruent to a diagonal matrix D having the form

$$D = \begin{bmatrix} 1 & & & & & & & \\ & \ddots & & & & & & \\ & & 1 & & & & & \\ & & & -1 & & & & \\ & & & & \ddots & & & \\ & & & & & -1 & & \\ & & & & & & 0 & \\ & & & & & & & \ddots \\ & & & & & & & & 0 \end{bmatrix}$$

with p positive ones and $r - p$ negative ones on the main diagonal. More precisely, there exists an invertible $n \times n$ real matrix Q such that $D = Q^t A Q$.

On the other hand, since A is symmetric, it is orthogonally similar to a diagonal matrix D' having the eigenvalues $\{\lambda_1, \ldots, \lambda_n\}$ of A on the main diagonal, that is, there exists an orthogonal real $n \times n$ matrix P such that

$$P^t A P = D' = \begin{bmatrix} \lambda_1 & & & \\ & \ddots & & \\ & & \ddots & \\ & & & \lambda_n \end{bmatrix}.$$

By using the eigenvalues $\{\lambda_1, \ldots, \lambda_n\}$ of A, we now construct the following diagonal matrix

$$C = \begin{bmatrix} \mu_1 & & & \\ & \ddots & & \\ & & \ddots & \\ & & & \mu_n \end{bmatrix},$$

where

$$\mu_i = \begin{cases} \frac{1}{\sqrt{\lambda_i}} & \text{if } \lambda_i > 0 \\ \frac{1}{\sqrt{-\lambda_i}} & \text{if } \lambda_i < 0 \\ 1 & \text{if } \lambda_i = 0. \end{cases}$$

Easy computations show that

$$C D' C = C D' C^t = \begin{bmatrix} \eta_1 & & & \\ & \ddots & & \\ & & \ddots & \\ & & & \eta_n \end{bmatrix} = D'',$$

where

$$\eta_i = \mu_i^2 \lambda_i = \begin{cases} 1 & \text{if } \lambda_i > 0 \\ -1 & \text{if } \lambda_i < 0 \\ 0 & \text{if } \lambda_i = 0 \end{cases}.$$

Hence, if we assume that A has rank equal to $t \leq n$, s positive eigenvalues and $t - s$ negative eigenvalues, the signature of D'' is precisely $(s, t - s)$. On the other hand, since P is orthogonal and C is diagonal invertible, we also have $A = PD'P^t$ and $D' = C^{-1}D''(C^t)^{-1}$, so that $D = Q^tAQ = Q^tPD'P^tQ = Q^tPC^{-1}D''(C^t)^{-1}P^tQ = ((C^t)^{-1}P^tQ)^t D''((C^t)^{-1}P^tQ)$. Therefore, D, D'' are congruent, having the same rank and signature (see Lemma 7.57), that is, $r = t$, $p = s$ and $r - p = t - s$, as required.

Corollary 7.60 *A real $n \times n$ symmetric matrix A is positive definite if and only if any eigenvalue of A is positive. Analogously, let $q : V \to \mathbb{R}$ be a real quadratic form and A be the matrix of q, in terms of a fixed basis for V. Then q is positive definite if and only if any eigenvalue of A is positive.*

We conclude this chapter by studying the relationship between the principal submatrices of a matrix associated with a quadratic form q and the definiteness of q. Here, we recall the following well-known objects:

Definition 7.61 Let A be a $n \times n$ matrix over $\mathbb{F}$, namely

$$A = \begin{bmatrix} a_{11} & a_{12} & \cdots & a_{1n} \\ \vdots & \vdots & \vdots & \vdots \\ \vdots & \vdots & \vdots & \vdots \\ a_{n1} & a_{n2} & \cdots & a_{nn} \end{bmatrix}.$$

The $k \times k$ *principal submatrix* of A is the square submatrix of A that is obtained by taking the elements common to the first k rows and columns of A. In other words, a matrix of this form arises from A by omission of the last $n - k$ rows and columns. A *principal minor* of A is the determinant of any principal submatrix of A. In other words:

$$[a_{11}], \begin{bmatrix} a_{11} & a_{12} \\ a_{21} & a_{22} \end{bmatrix}, \begin{bmatrix} a_{11} & a_{12} & a_{13} \\ a_{21} & a_{22} & a_{23} \\ a_{31} & a_{32} & a_{33} \end{bmatrix}, \ldots, \begin{bmatrix} a_{11} & a_{12} & \cdots & a_{1k} \\ \vdots & \vdots & \vdots & \vdots \\ \vdots & \vdots & \vdots & \vdots \\ a_{k1} & a_{k2} & \cdots & a_{kk} \end{bmatrix}$$

for any $k \leq n$; are the principal submatrices of A, and

$$a_{11}, \quad \begin{vmatrix} a_{11} & a_{12} \\ a_{21} & a_{22} \end{vmatrix}, \quad \begin{vmatrix} a_{11} & a_{12} & a_{13} \\ a_{21} & a_{22} & a_{23} \\ a_{31} & a_{32} & a_{33} \end{vmatrix}, \ldots, \quad \begin{vmatrix} a_{11} & a_{12} & \cdots & a_{1k} \\ \vdots & \vdots & \vdots & \vdots \\ \vdots & \vdots & \vdots & \vdots \\ a_{k1} & a_{k2} & \cdots & a_{kk} \end{vmatrix}$$

for any $k \leq n$; are its principal minors.

Theorem 7.62 *Let A be the symmetric $n \times n$ matrix of the quadratic form $q : V \to \mathbb{R}$. Then q is positive definite if and only if any principal minor of A is positive.*

Proof Firstly, we assume that any principal minor of A is positive and show, by induction on the dimension n, that q is positive definite. Remind that this is equivalent to say that any eigenvalue of A is positive.

Of course, in case $n = 1$, the conclusion is trivial, because $A = (a_{11})$, where $a_{11} > 0$ and $q(X) = a_{11}x_1^2 > 0$, for any $X \in V$, which has its coordinate vector $[X]$, where $[X] = [x_1]^t$.

Thus, we assume that the result holds for any quadratic form $q' : W \to \mathbb{R}$, where W is a real space of dimension $m < n$. In other words, we suppose that if any principal minor of the symmetric real matrix associated with q' is positive, then q' is positive definite.

Since a_{11} is the principal minor of order 1, then $a_{11} > 0$. By using this element, we may apply the orthogonalization process for symmetric spaces. We compute the following coefficients:

$$\beta_{12} = \frac{a_{12}}{a_{11}} \quad \beta_{13} = \frac{a_{13}}{a_{11}} \cdots \quad \cdots \quad \beta_{1i} = \frac{a_{1i}}{a_{11}} \quad \text{for all} \ i \geq 2$$

and obtain the matrix

$$U = \begin{bmatrix} 1 & -\beta_{12} & -\beta_{13} & \cdots & & -\beta_{1n} \\ 0 & 1 & 0 & \cdots & & 0 \\ 0 & 0 & \ddots & & & \vdots \\ \vdots & & & \ddots & & \vdots \\ \vdots & & & & \ddots & 0 \\ 0 & 0 & & & 0 & 1 \end{bmatrix}$$

such that

$$U^t A U = \begin{bmatrix} a_{11} & 0_{1,n-1} \\ 0_{n-1,1} & B \end{bmatrix} \tag{7.16}$$

where

$$0_{1,n-1} = \underbrace{(0, \ldots, 0)}_{(n-1)-times}, \quad 0_{n-1,1} = 0_{1,n-1}^t \quad \text{and} \quad B \in M_{n-1}(\mathbb{R}) \quad \text{is symmetric.}$$

Notice that, since U is upper unitriangular, U^t is lower unitriangular, in the sense that it is a lower triangular matrix having all diagonal coefficients equal to 1 :

$$U^t = \begin{bmatrix} 1 & 0 & 0 & \cdots & & 0 \\ -\beta_{12} & 1 & 0 & \cdots & & 0 \\ -\beta_{13} & 0 & \ddots & & & \vdots \\ \vdots & & & \ddots & & \vdots \\ \vdots & & & & \ddots & 0 \\ -\beta_{1n} & 0 & & & 0 & 1 \end{bmatrix}.$$

This matrix represents a finite number of elementary row operations. If we denote by $R_i = [a_{i1}, \ldots, a_{in}]$ the row coordinate vector consisting of the elements from the i-th row of A, we may write

$$A = \begin{bmatrix} R_1 \\ R_2 \\ \vdots \\ \vdots \\ R_n \end{bmatrix}.$$

Analogously, we denote by R_i' the row coordinate vector consisting of the elements from the i-th row of the product $U^t A$. By easy computations, we see that

$$U^t A = \begin{bmatrix} R_1 \\ R_2 - \beta_{12} R_1 \\ R_3 - \beta_{13} R_1 \\ \vdots \\ R_n - \beta_{1n} R_1 \end{bmatrix}.$$

In other words, $R_1' = R_1$ and, for any $k \geq 2$, $R_k' = R_k - \beta_{1k} R_1$. Therefore, any $k \times k$ principal submatrix of $U^t A$ is obtained by the corresponding $k \times k$ principal submatrix of A after using a finite number of elementary row operations. In particular, the operation is the addition of a multiple of a row to another. It is well known that this type of basic operation has no effect on the determinant. Hence any $k \times k$ principal minor of $U^t A$ is equal to the corresponding $k \times k$ principal minor of A.

By using the same argument, one can see that the matrix $U^t(AU)$ has the same principal minors of AU. On the other hand, since any matrix has the same principal minors of its transpose matrix, AU and $U^t A^t$ have the same principal minors. In summary, $U^t AU$, AU, $U^t A^t$, A^t and A have the same principal minors.

Looking at (7.16), it is clear that any principal minor of $U^t AU$ is the product of $a_{11} > 0$ with a principal minor of B. Therefore, any principal minor of B should be positive. On the other hand, B is a symmetric matrix of dimension $n - 1$, therefore, by induction hypothesis, the eigenvalues $\{\lambda_1, \ldots, \lambda_{n-1}\}$ of B are positive. Thus, the eigenvalues $\{a_{11}, \lambda_1, \ldots, \lambda_{n-1}\}$ of $U^t A U$ are positive, so that q is positive definite.

Suppose now that q is positive definite. Let A_k be any principal submatrix of A of dimension k and $Y = (\alpha_1, \ldots, \alpha_k) \in \mathbb{R}^k$. Hence, the coordinate vector of Y is $[Y] = [\alpha_1, \ldots, \alpha_k]^t$. Further, let $X \in V$ such that its coordinate vector is $[X] = [\alpha_1, \ldots, \alpha_k, \underbrace{0, \ldots, 0}_{(n-k)-times}]^t \in \mathbb{R}^n$. Since q is positive definite, $X^t A X > 0$. That is,

$$
X^t A \overset{\bullet}{X} = [\alpha_1, \ldots, \alpha_k, 0, \ldots, 0] \begin{bmatrix} A_k & B_{k,n-k} \\ C_{n-k,k} & E_{n-k,n-k} \end{bmatrix} \begin{bmatrix} \alpha_1 \\ \vdots \\ \alpha_k \\ 0 \\ \vdots \\ 0 \end{bmatrix}
$$

$$
= [\alpha_1, \ldots, \alpha_k] A_k \begin{bmatrix} \alpha_1 \\ \vdots \\ \alpha_k \end{bmatrix} > 0,
$$

where $B_{k,n-k} \in M_{k,n-k}(\mathbb{R})$, $C_{n-k,k} \in M_{n-k,k}(\mathbb{R})$ and $E_{n-k,n-k} \in M_{n-k,n-k}(\mathbb{R})$.

By the arbitrariness of $Y = (\alpha_1, \ldots, \alpha_k)$, it follows that $Y^t A_k Y > 0$ for any vector $Y \in \mathbb{R}^k$, that is, $A_k \in M_k(\mathbb{R})$ is the matrix of a positive definite quadratic form on $\mathbb{R}^k$. Therefore, any eigenvalue of A_k is positive, and hence the determinant of A_k is positive because the determinant of A_k is equal to the product of all eigenvalues of A_k. Thus, any principal minor of A is positive.

Exercises

1. Let $f : \mathbb{R}^4 \times \mathbb{R}^4 \to \mathbb{R}$ be a symmetric form associated with the matrix

$$
\begin{bmatrix} 0 & 1 & 1 & 1 \\ 1 & 0 & 1 & 1 \\ 1 & 1 & 0 & 1 \\ 1 & 1 & 1 & 0 \end{bmatrix}
$$

in terms of the canonical basis for $\mathbb{R}^4$. Determine an orthogonal basis for $\mathbb{R}^4$ in terms of which the matrix of f is diagonal.

2. Let $f : \mathbb{R}^4 \times \mathbb{R}^4 \to \mathbb{R}$ be the symmetric form defined as

$$f((x_1, x_2, x_3, x_4), (y_1, y_2, y_3, y_4)) = 2x_1y_1 - 4x_1y_3 + x_1y_4 - 2x_2y_2 - 2x_2y_3$$
$$-4x_2y_4 - 4x_3y_1 - 2x_3y_2 + x_4y_1 - 4x_4y_2$$
$$+x_4y_4.$$

in terms of the canonical basis for $\mathbb{R}^4$. Determine an orthogonal basis for $\mathbb{R}^4$ in terms of which the matrix of f is diagonal.

3. Let $f : \mathbb{C}^3 \times \mathbb{C}^3 \to \mathbb{C}$ be a symmetric form associated with the matrix

$$\begin{bmatrix} 1 & 1 & 2 \\ 1 & 2 & 3 \\ 2 & 3 & 3 \end{bmatrix}$$

in terms of the canonical basis for $\mathbb{C}^3$. Determine an orthogonal basis for $\mathbb{C}^3$ in terms of which the matrix of f is the 3×3 identity matrix.

4. Let $f : \mathbb{R}^n \times \mathbb{R}^n \to \mathbb{R}$ be a non-degenerate bilinear form. Prove that there exist nonzero f-isotropic vectors if and only if f is neither positive nor negative definite.

5. Let A be the symmetric $n \times n$ matrix of the quadratic form $q : V \to \mathbb{R}$. Let A_k be the $k \times k$ *principal submatrix* of A, obtained by taking the elements common to the first k rows and columns of A. Prove that q is negative definite if and only if, for any $k = 1, \ldots, n$, $(-1)^k det(A_k) > 0$.

6. Which of the following functions f on $\mathbb{C}^2 \times \mathbb{C}^2$ are bilinear forms? For $\alpha = (x_1, x_2)$, $\beta = (y_1, y_2)$ in $\mathbb{C}^2$;
   (a) $f(\alpha, \beta) = 1$;
   (b) $f(\alpha, \beta) = x_1 + x_2 + y_1 + y_2$;
   (c) $f(\alpha, \beta) = x_1y_1 + x_2y_2$;
   (d) $f(\alpha, \beta) = x_1\overline{y_1} + x_2\overline{y_2}$;
   (e) $f(\alpha, \beta) = x_1y_1 + x_2y_2 + 2x_2y_1 + 3x_2\overline{y_2}$;
   (f) $f(\alpha, \beta) = x_1{}^2 + y_1{}^2 + x_1y_1$.

7. For each of the following symmetric bilinear forms over $\mathbb{R}$, find the associated quadratic form:
   (a) $f(X, Y) = x_1y_1 + x_1y_2 + x_2y_1$ on $\mathbb{R}^2$, where $X = (x_1, x_2)$, $Y = (y_1, y_2)$;
   (b) $f(X, Y) = -2x_1y_2 + x_2y_2 - x_3y_3 + 4x_1y_2 + 4x_2y_1$ on $\mathbb{R}^3$, where $X = (x_1, x_2, x_3)$, $Y = (y_1, y_2, y_3)$;
   (c) $f(X, Y) = x_1y_2 + x_2y_1$ on $\mathbb{R}^2$, where $X = (x_1, x_2)$, $Y = (y_1, y_2)$.

8. For each of the following quadratic forms, find the corresponding symmetric bilinear form f :
   (a) $q(X) = x_1^2 + x_2^2 - x_3^2$ on $\mathbb{R}^3$, where $X = (x_1, x_2, x_3)$;
   (b) $q(X) = ax_1x_2 + bx_1x_3 + cx_3^2$ on $\mathbb{R}^3$, where $X = (x_1, x_2, x_3)$; $a, b, c \in \mathbb{R}$;
   (c) $q(X) = x_2^2$ on $\mathbb{R}^4$, where $X = (x_1, x_2, x_3, x_4)$;
   (d) $q(X) = 2x_1x_2 - x_2^2$ on $\mathbb{R}^2$, where $X = (x_1, x_2)$;
   (e) $q(X) = x_1x_2 + x_2x_3 + x_3x_4 + x_4x_1$ on $\mathbb{R}^4$, where $X = (x_1, x_2, x_3, x_4)$.

9. Let V be the real vector space of all 2×2 (complex) Hermitian matrices, i.e., 2×2 complex matrices $A = (a_{ij})$ which satisfy $a_{ij} = \overline{a_{ji}}$ for all $i, j = 1, 2$.

(*a*) Show that the equation $q(A) = \det A$ defines a quadratic form q on V.

(*b*) Let W be the subspace of V of matrices of trace 0. Show that the bilinear form f determined by q is negative definite on the subspace W.

10. Prove that matrix $A = \begin{pmatrix} 0 & b \\ b & 0 \end{pmatrix}$ over a field $\mathbb{F}$ with char $(\mathbb{F}) \neq 2$, is congruent to $A' = \begin{pmatrix} 2b & 0 \\ 0 & -\frac{b}{2} \end{pmatrix}$. Further, find a nonsingular matrix P over $\mathbb{F}$ such that $A' = P^t A P$.

Chapter 8
Sesquilinear and Hermitian Forms

In the previous chapter, we have studied and characterized the maps $V \times V \to \mathbb{F}$, having the property of being linear in each of their arguments, where V is a vector space over the field $\mathbb{F}$. However, not all forms of interest are bilinear. For example, on complex vector spaces, one may introduce functions that provide an extra complex conjugation. In this regard, we now consider a somewhat larger class of functions than the bilinear one.

8.1 Sesquilinear Forms and Their Matrices

Let $\mathbb{C}$ be the complex field, V, W complex vector spaces and $V \times W$ the cartesian product of V and W (as sets). A function $f : V \times W \to \mathbb{C}$ is called *sesquilinear* if it is conjugate-linear (complex semi-linear) in the first variable and linear in the second one, that is

$$f(\alpha_1 v_1 + \alpha_2 v_2, w) = \overline{\alpha_1} f(v_1, w) + \overline{\alpha_2} f(v_2, w) \tag{8.1}$$

and

$$f(v, \beta_1 w_1 + \beta_2 w_2) = \beta_1 f(v, w_1) + \beta_2 f(v, w_2) \tag{8.2}$$

for any $\alpha_1, \alpha_2, \beta_1, \beta_2 \in \mathbb{C}$, $v_1, v_2 \in V$ and $w_1, w_2 \in W$. A sesquilinear function $f : V \times W \to \mathbb{C}$ is usually called a *sesquilinear form* on $V \times W$.

Example 8.1 The inner product $f : \mathbb{C}^n \times \mathbb{C}^n \longrightarrow \mathbb{C}$ defined as

$$f\left((x_1, \ldots, x_n), (y_1, \ldots, y_n)\right) = \sum_{i=1}^{n} \overline{x_i} y_i$$

is a sesquilinear form on $\mathbb{C}^n \times \mathbb{C}^n$.

Example 8.2 Let A be a complex $n \times n$ matrix. The function $f : \mathbb{C}^n \times \mathbb{C}^n \longrightarrow \mathbb{C}$ defined as $f(X, Y) = X^*AY$ is a sesquilinear form on $\mathbb{C}^n \times \mathbb{C}^n$, where $X, Y \in \mathbb{C}^n$ and X^* denotes the transpose conjugate of the vector X, i.e., $X^* = (\overline{X})^t$. Here the vectors X and Y have been identified by column matrices with complex entries.

Let $f : V \times W \to \mathbb{C}$ be a sesquilinear form on $V \times W$, where V and W are finite dimensional complex vector spaces. Let $B = \{b_1, \ldots, b_n\}$ and $C = \{c_1, \ldots, c_m\}$ be ordered bases for V and W, respectively. Let $[v]_B$ and $[w]_C$ be the coordinate vectors of $v \in V$ in terms of B and $w \in W$ in terms of C, respectively. Say $[v]_B = [x_1, \ldots, x_n]^t$ and $[w]_C = [y_1, \ldots, y_m]^t$, i.e., $v = \sum_{i=1}^{n} x_i b_i$ and $w = \sum_{j=1}^{m} y_j c_j$. We have

$$f(v, w) = f(\sum_{i=1}^{n} x_i b_i, \sum_{j=1}^{m} y_j c_j) = \sum_{i,j} \overline{x_i} y_j f(b_i, c_j).$$

If we consider the coefficients matrix $A = (a_{ij})$, where $a_{ij} = f(b_i, c_j)$, for any $i = 1, \ldots, n$ and $j = 1, \ldots, m$, then it is easy to see that

$$f(v, w) = [v]_B^* A [w]_C.$$

The $n \times m$ matrix $A = (a_{ij})$ is said to be the *matrix of the sesquilinear form* f with respect to the ordered bases B and C.

   Conversely, let $A \in M_{nm}(\mathbb{C})$ and let, as above, $[v]_B$, $[w]_C$ be the coordinate vectors of $v \in V$, in terms of B, and $w \in W$, in terms of C, respectively. If we define the function $f : V \times W \to \mathbb{F}$ such that $f(v, w) = [v]_B^* A [w]_C$, then by computations it follows that $f(v, w) = \sum_{i,j} \overline{x_i} y_j a_{ij}$, i.e., f is a sesquilinear map on the set $V \times W$.

   Moreover, using an analogous argument to that of the matrix of a bilinear form, one can see that there is an unique $n \times m$ complex matrix, which represents the sesquilinear form with respect to the given ordered bases B and C.

Example 8.3 Let $f : \mathbb{C}^3 \times \mathbb{C}^4 \to \mathbb{C}$ be a sesquilinear form on $\mathbb{C}^3 \times \mathbb{C}^4$. Let $B = \{b_1, b_2, b_3\}$ and $C = \{c_1, c_2, c_3, c_4\}$ be bases for $\mathbb{C}^3$ and $\mathbb{C}^4$, respectively. Let $[X]_B$ and $[Y]_C$ be the coordinate vectors of $v \in V$ in terms of B and $w \in W$ in terms of C, respectively. Say $[X]_B = [x_1, x_2, x_3]^t$ and $[Y]_C = [y_1, y_2, y_3, y_4]^t$, i.e., $v = \sum_{i=1}^{3} x_i b_i$ and $w = \sum_{j=1}^{4} y_j c_j$. Let

$$\begin{aligned}
f(v, w) &= f(\sum_{i=1}^{3} x_i b_i, \sum_{j=1}^{4} y_j c_j) \\
&= i\overline{x_1} y_1 + \overline{x_1} y_2 + (2 + i)\overline{x_1} y_3 + (1 + i)\overline{x_1} y_4 - i\overline{x_2} y_1 + \overline{x_2} y_2 + \overline{x_2} y_3 \\
&\quad + (1 - i)\overline{x_2} y_4 + 3i\overline{x_3} y_1 + 2\overline{x_3} y_2 + 2\overline{x_3} y_3 + 3\overline{x_3} y_4.
\end{aligned}$$

This implies that $f(b_1, c_1) = i$, $f(b_1, c_2) = 1, \ldots, f(b_3, c_4) = 3$. In light of the above comments, the matrix of f in terms of bases B and C for $\mathbb{C}^3$ and $\mathbb{C}^4$ respectively, is the following:

$$\begin{bmatrix} i & 1 & 2+i & 1+i \\ -i & 1 & 1 & 1-i \\ 3i & 2 & 2 & 3 \end{bmatrix}.$$

As in the case of bilinear forms, we introduce the concept of the quadratic form associated with a sesquilinear form f. More precisely:

Definition 8.4 Let $f : V \times V \to \mathbb{C}$ be a sesquilinear form. The corresponding quadratic function $Q : V \to \mathbb{C}$ is defined as $Q(v) = f(v, v)$, for any $v \in V$.

By relations (8.1) and (8.2), it follows that the corresponding quadratic form $Q(v) = f(v, v)$ satisfies the following:

$$Q(u + v) + Q(u - v) = 2Q(u) + 2Q(v) \tag{8.3}$$

and

$$Q(\lambda u) = |\lambda|^2 Q(u) \tag{8.4}$$

for any $\lambda \in \mathbb{C}$, $u \in V$.

Moreover, relation (8.3) yields

$$Q(u + v) = Q(u) + Q(v) + f(u, v) + f(v, u) \tag{8.5}$$

for any $u, v \in V$.

Unlike the situation, we have previously described in the case of bilinear forms, here we may prove the following:

Proposition 8.5 Let $f : V \times V \to \mathbb{C}$ be a sesquilinear form and $Q : V \to \mathbb{C}$ the quadratic form defined as $Q(v) = f(v, v)$, for any $v \in V$. Then f is the unique sesquilinear form corresponding to Q.

Proof In (8.5) we replace v by iv, so it follows that

$$Q(u + iv) = Q(u) + Q(iv) + f(u, iv) + f(iv, u) = Q(u) + Q(v) + if(u, v) - if(v, u). \tag{8.6}$$

Multiplying (8.6) by i we get

$$iQ(u + iv) = iQ(u) + iQ(v) - f(u, v) + f(v, u). \tag{8.7}$$

Finally, a comparison of (8.5) and (8.7) yields

$$iQ(u + iv) - Q(u + v) = (i - 1)Q(u) + (i - 1)Q(v) - 2f(u, v)$$

that is

$$2f(u,v) = Q(u+v) - iQ(u+iv) + (i-1)Q(u) + (i-1)Q(v)$$

so that

$$f(u,v) = \frac{1}{2}\Big\{Q(u+v) - Q(u) - Q(v)\Big\} - \frac{i}{2}\Big\{Q(u+iv) - Q(u) - Q(v)\Big\}.$$
(8.8)

Therefore, f is uniquely determined by the function Q.

Remark 8.6 To underline the difference between bilinear and sesquilinear cases, we recall that if Q is a quadratic form associated with a bilinear function f, then f is uniquely determined only if it is symmetric.

In order to extend the concept of symmetric forms to complex vector spaces, we introduce the following:

Definition 8.7 Let $\mathbb{C}$ be the complex field and V a complex vector space. A sesquilinear form $f : V \times V \to \mathbb{C}$ is called *Hermitian* if

$$f(v,w) = \overline{f(w,v)}$$

for any $v, w \in V$.

Example 8.8 Let $f : \mathbb{C}^3 \times \mathbb{C}^3 \to \mathbb{C}$ be a sesquilinear form on $\mathbb{C}^3 \times \mathbb{C}^3$. Let $B = \{b_1, b_2, b_3\}$ and $C = \{c_1, c_2, c_3\}$ be ordered bases for $\mathbb{C}^3$. Next suppose that $[X]_B$ and $[Y]_C$ be the coordinate vectors of $v \in V$ and $w \in V$ in terms of B and C, respectively. Say $[X]_B = [x_1, x_2, x_3]^t$ and $[Y]_C = [y_1, y_2, y_3]^t$, i.e., $v = \sum_{i=1}^{3} x_i b_i$ and $w = \sum_{j=1}^{3} y_j c_j$. Let

$$\begin{aligned}
f(v,w) &= f(\sum_{i=1}^{3} x_i b_i, \sum_{j=1}^{3} y_j c_j) \\
&= \overline{x_1} y_1 + i\overline{x_1} y_2 + (2+i)\overline{x_1} y_3 - i\overline{x_2} y_1 + \overline{x_2} y_2 + (1+i)\overline{x_2} y_3 \\
&\quad + (2-i)\overline{x_3} y_1 + (1-i)\overline{x_3} y_2 + 2\overline{x_3} y_3.
\end{aligned}$$

Hence, $f(b_1, c_1) = 1$, $f(b_1, c_2) = i, \ldots, f(b_3, c_3) = 2$. The matrix of f in terms of ordered bases B and C for $\mathbb{C}^3$ is the following:

$$\begin{bmatrix} 1 & i & 2+i \\ -i & 1 & 1+i \\ 2-i & 1-i & 2 \end{bmatrix}.$$

Easy computations show that f is Hermitian (as well as its associated matrix).

In light of the above comments regarding the matrix of a sesquilinear form, it is clear that any Hermitian form on a complex vector space V is represented by a complex

Hermitian matrix, depending on the choice of the basis for V. Conversely, for any Hermitian matrix A, the sesquilinear form $f : V \times V \to \mathbb{C}$ associated with A is Hermitian.

Theorem 8.9 *Let $f : V \times V \to \mathbb{C}$ be a sesquilinear form on the complex vector space V. The function f is a Hermitian form if and only if $f(v, v) \in \mathbb{R}$, for any $v \in V$.*

Proof If we assume that f is Hermitian, then it is easy to see how the fact $f(v, v) = \overline{f(v, v)}$, for any $v \in V$, implies that $f(v, v) \in \mathbb{R}$.

Conversely, suppose that $f(v, v) \in \mathbb{R}$, for any $v \in V$. Thus, for any $u, v \in V$ we also have

$$f(u + v, u + v) = f(u, u) + f(u, v) + f(v, u) + f(v, v) \in \mathbb{R}$$

implying that

$$f(u, v) + f(v, u) = \alpha \in \mathbb{R}. \tag{8.9}$$

In particular,

$$f(iu, v) + f(v, iu) = \beta \in \mathbb{R}$$

that is

$$-if(u, v) + if(v, u) = \beta$$

and multiplying by i,

$$f(u, v) - f(v, u) = i\beta. \tag{8.10}$$

By comparing relations (8.9) and (8.10), we get both

$$f(u, v) = \frac{\alpha + i\beta}{2}$$

and

$$f(v, u) = \frac{\alpha - i\beta}{2},$$

which means $f(u, v) = \overline{f(v, u)}$, for any $u, v \in V$, as required.

Remark 8.10 In what follows, we will substantially go over the outlines of the presentation given for bilinear and symmetric forms. The most part of arguments and proofs also apply to Hermitian forms, so we omit them, unless when they need to be deeply altered.

Definition 8.11 Let $f : V \times V \to \mathbb{C}$ be a Hermitian form. The quadratic form associated with f is called *h-quadratic form*.

Notice that, in light of Theorem 8.9, if f is Hermitian then the corresponding quadratic function Q is real valued. For this reason, we may also catalog any *h*-quadratic form in relation to whether it assumes positive or negative values.

Definition 8.12 Let $f : V \times V \to \mathbb{C}$ be a Hermitian form and Q its associated h-quadratic form. The h-quadratic form Q is called:

(1) *Positive definite* if $Q(X) > 0$ for all nonzero vectors $X \in V$.
(2) *Negative definite* if $Q(X) < 0$ for all nonzero vectors $X \in V$.
(3) *Indefinite* if it is neither positive-definite nor negative-definite, in the sense that Q takes on V both positive and negative values.
(4) *Positive semi-definite* if $Q(X) \geq 0$ for all $X \in V$, but there is some nonzero vector $X_0 \in V$ so that $Q(X_0) = 0$.
(5) *Negative semi-definite* if $Q(X) \leq 0$ for all $X \in V$, but there is some nonzero vector $X_0 \in V$ so that $Q(X_0) = 0$.

8.2   The Effect of the Change of Bases

If $B = \{b_1, \ldots, b_n\}$ and $B' = \{b'_1, \ldots, b'_n\}$ are two different ordered bases for V, and $C = \{c_1, \ldots, c_m\}$, $C' = \{c'_1, \ldots, c'_m\}$ are ordered bases for W, we know that the sesquilinear form $f : V \times W \to \mathbb{C}$ can be represented by different matrices, according to the choice of a basis for V and W. Let A be the matrix of f in terms of the bases B for V and C for W, and A' the matrix of f in terms of the bases B' for V and C' for W.

By following the same procedure as in the case of bilinear forms, we may establish the following relationship between A and A':

$$A' = P^*AQ$$

where $P \in M_n(\mathbb{C})$ is the transition matrix of B' relative to B, and $Q \in M_m(\mathbb{C})$ is the transition matrix of C' relative to C. In particular:

Definition 8.13 Let A, A' be two $n \times n$ complex matrices. A, A' are called *H-congruent matrices* if there exists an invertible $n \times n$ complex matrix P, such that $A' = P^*AP$. The relationship $A' = P^*AP$ is usually called *H-congruence* and it is an equivalence relation.

Just as in the case of bilinear forms, the following holds:

Theorem 8.14 *Let $f : V \times V \to \mathbb{C}$ be a sesquilinear form on V. Two matrices A, A' represent f, in terms of two different bases for V, if and only if they are H-congruent.*

Proof The proof is unchanged with respect to the one of Theorem 7.7.

8.3 Orthogonality

The concepts of f-orthogonality, f-orthogonal, f-orthonormal vectors and f-orthogonal complement, as well as the definitions of isotropic vectors, nonisotropic vectors, degenerate and non-degenerate forms are unchanged from the previously studied case of bilinear and symmetric forms.

Of course, if $f : V \times V \to \mathbb{C}$ is a Hermitian form on V, the orthogonality relation is symmetric, that is $f(u, v) = 0$ if and only if $f(v, u) = 0$, for all $u, v \in V$.

We would like to focus our attention just on the proof of the following:

Lemma 8.15 *Let $f : V \times V \to \mathbb{C}$ be a Hermitian form on V. If $f \neq 0$ (in the sense that it is not identically zero on $V \times V$), then there exists a vector $v \in V$ such that $f(v, v) \neq 0$.*

Proof Since we assume that $f \neq 0$, there are $x_0, y_0 \in V$ such that $0 \neq f(x_0, y_0) = \alpha \in \mathbb{C}$. Let $0 \neq \beta = \overline{\alpha}$, so that $0 \neq \beta f(x_0, y_0) = \beta\alpha \in \mathbb{R}$. Thus

$$0 \neq \beta f(x_0, y_0) = f(x_0, \beta y_0) \in \mathbb{R}.$$

This means that, for $z_0 = \beta y_0$,

$$0 \neq f(x_0, z_0) = \overline{f(z_0, x_0)} = f(z_0, x_0).$$

To prove our result, we suppose on the contrary $f(v, v) = 0$ for all $v \in V$. Hence, it follows that $f(x_0, x_0) = 0$, $f(z_0, z_0) = 0$ and $f(x_0 + z_0, x_0 + z_0) = 0$. Therefore, we get the contradiction

$$0 = f(x_0 + z_0, x_0 + z_0) = f(x_0, z_0) + f(z_0, x_0) = 2f(x_0, z_0).$$

Theorem 8.16 *Let $f : V \times V \to \mathbb{C}$ be a Hermitian form on V. Then there is an f-orthogonal basis for V.*

Proof An inspection of the proof of Theorem 7.32 reveals that it applies unchanged.

Here we just apply Lemma 8.15 in place of Lemma 7.31. As above mentioned, the rest of the proof is unchanged.

Remark 8.17 The Gram-Schmidt orthonormalization process, introduced in order to construct an orthonormal basis for a symmetric space, also applies to the case of a complex space equipped with a Hermitian form f. No change is needed in the procedure previously described for symmetric spaces, if not the replacement of the transpose matrix U^t by the conjugate-transpose (adjoint) U^*, where U is the transition matrix.

Moreover, we also recall that, in the case of a complex space V equipped with a Hermitian form f, if we change to a new orthonormal basis for V, the transition matrix U will be unitary. In fact, since the matrix A associated with f is Hermitian, it is unitarily similar to a diagonal matrix. This means that there is an f-orthogonal basis for V in terms of which the real diagonal matrix A' representing f is obtained by $A' = U^*AU = U^{-1}AU$, where U is an unitary matrix.

Theorem 8.16 and Remark 8.17 will allow us to state the 'Hermitian' version of Theorem 7.38:

Theorem 8.18 *Let $f : V \times V \to \mathbb{C}$ be a Hermitian form on V. Then there is an f-orthogonal basis B for V such that the matrix of f in terms of B has the following form:*

$$\begin{bmatrix} 1 & & & & & & \\ & \ddots & & & & & \\ & & 1 & & & & \\ & & & -1 & & & \\ & & & & \ddots & & \\ & & & & & -1 & \\ & & & & & & 0 \\ & & & & & & & \ddots \\ & & & & & & & & 0 \end{bmatrix}$$

Example 8.19 Let $f : \mathbb{C}^3 \times \mathbb{C}^3 \to \mathbb{C}$ be the Hermitian form defined by the matrix

$$A = \begin{bmatrix} 2 & 1+2i & i \\ 1-2i & 0 & 2-i \\ -i & 2+i & 1 \end{bmatrix}$$

in terms of the canonical basis $B = \{e_1, e_2, e_3\}$ for $\mathbb{C}^3$. In order to apply the orthonormalization process, we firstly introduce the following new basis $B' = \{e'_1, e'_2, e'_3\}$ for $\mathbb{C}^3$:

$$e'_1 = e_1, \quad e'_2 = e_2 - \frac{f(e_1, e_2)}{f(e_1, e_1)}e_1, \quad e'_3 = e_3 - \frac{f(e_1, e_3)}{f(e_1, e_1)}e_1$$

that is

$$e'_1 = e_1, \quad e'_2 = e_2 - \frac{1+2i}{2}e_1, \quad e'_3 = e_3 - \frac{i}{2}e_1$$

and the corresponding transition matrix is

$$C_1 = \begin{bmatrix} 1 & -\frac{1+2i}{2} & -\frac{i}{2} \\ 0 & 1 & 0 \\ 0 & 0 & 1 \end{bmatrix}.$$

Hence, the new expression of the matrix of f is

$$A' = C_1^* A C_1 = \begin{bmatrix} 2 & 0 & 0 \\ 0 & -\frac{5}{2} & \frac{2-3i}{2} \\ 0 & \frac{2+3i}{2} & \frac{1}{2} \end{bmatrix}.$$

The next change of basis will be the following

$$e_1'' = e_1', \quad e_2'' = e_2', \quad e_3'' = e_3' - \frac{f(e_2', e_3')}{f(e_2', e_2')} e_2'.$$

We introduce the basis $B'' = \{e_1'', e_2'', e_3''\}$ such that

$$e_1'' = e_1', \quad e_2'' = e_2', \quad e_3'' = e_3' + \frac{2-3i}{5} e_2'$$

and the corresponding transition matrix is

$$C_2 = \begin{bmatrix} 1 & 0 & 0 \\ 0 & 1 & \frac{2-3i}{5} \\ 0 & 0 & 1 \end{bmatrix}.$$

Thus, the matrix of f assumes the following form:

$$A'' = C_2^* A' C_2 = \begin{bmatrix} 2 & 0 & 0 \\ 0 & -\frac{5}{2} & 0 \\ 0 & 0 & \frac{9}{5} \end{bmatrix}.$$

The basis $B'' = \{e_1'', e_2'', e_3''\}$ in terms of which the matrix of f is precisely A'' is obtained by the computation

$$C = C_1 C_2 = \begin{bmatrix} 1 & -\frac{1+2i}{2} & -\frac{4+3i}{5} \\ 0 & 1 & \frac{2-3i}{5} \\ 0 & 0 & 1 \end{bmatrix}$$

where C is the transition matrix of B'' relative to the starting basis B. Looking at the columns of C and reordering the vectors, we have the following basis for $\mathbb{C}^3$

$$B''' = \left\{ (1, 0, 0), \left(-\frac{4+3i}{5}, \frac{2-3i}{5}, 1\right), \left(-\frac{1+2i}{2}, 1, 0\right) \right\}$$

in terms of which the matrix of f is

$$A''' = \begin{bmatrix} 2 & 0 & 0 \\ 0 & \frac{9}{5} & 0 \\ 0 & 0 & -\frac{5}{2} \end{bmatrix}.$$

Finally, we construct the basis for $\mathbb{C}^3$, with respect to which the matrix of f has all nonzero entries equal to ± 1. To do this, we replace any vector

$$e_1''' = (1, 0, 0), \quad e_2''' = (-\frac{4+3i}{5}, \frac{2-3i}{5}, 1), \quad e_3''' = (-\frac{1+2i}{2}, 1, 0)$$

by the corresponding $\dfrac{e_i'''}{\sqrt{|f(e_i''', e_i''')|}}$. In particular, we have

$$\sqrt{|f(e_1''', e_1''')|} = \sqrt{2}, \quad \sqrt{|f(e_2''', e_2''')|} = \frac{3}{\sqrt{5}}, \quad \sqrt{|f(e_3''', e_3''')|} = \sqrt{\frac{5}{2}}.$$

By all replacements, we obtain the new basis

$$\tilde{B} = \left\{ (\frac{1}{\sqrt{2}}, 0, 0), (-\frac{4+3i}{3\sqrt{5}}, \frac{2-3i}{3\sqrt{5}}, \frac{5}{3\sqrt{5}}), (-\frac{1+2i}{\sqrt{10}}, \frac{2}{\sqrt{10}}, 0) \right\}.$$

The transition matrix of $\tilde{B}$ relative to the starting basis B is

$$\tilde{C} = \begin{bmatrix} \frac{1}{\sqrt{2}} & -\frac{4+3i}{3\sqrt{5}} & -\frac{1+2i}{\sqrt{10}} \\ 0 & \frac{2-3i}{3\sqrt{5}} & \frac{2}{\sqrt{10}} \\ 0 & \frac{5}{3\sqrt{5}} & 0 \end{bmatrix}$$

and the matrix of f in terms of $\tilde{B}$ is

$$\tilde{A} = \tilde{C}^* A \tilde{C} = \begin{bmatrix} 1 & 0 & 0 \\ 0 & 1 & 0 \\ 0 & 0 & -1 \end{bmatrix}.$$

Finally, we may extend the concept of signature to any Hermitian form:

Definition 8.20 Let $f : V \times V \to \mathbb{C}$ be a complex Hermitian form. The ordered pair $(p, r - p)$ such that, for any $X, Y \in V$ having coordinate vectors $[X] = [x_1, \ldots, x_n]^t$ and $[Y] = [y_1, \ldots, y_n]^t$, $f(X, Y) = \overline{x_1} y_1 + \cdots + \overline{x_p} y_p - \overline{x_{p+1}} y_{p+1} - \cdots - \overline{x_r} y_r$ with respect to an appropriate f-orthogonal basis for V, is called the *signature* of f.

Of course, in parallel, we give the following:

Definition 8.21 Let A be a complex $n \times n$ Hermitian matrix, U a complex $n \times n$ invertible matrix such that $U^* A U = D$ is a real diagonal matrix of the form

$$D = \begin{bmatrix} 1 & & & & & & & & \\ & \ddots & & & & & & & \\ & & 1 & & & & & & \\ & & & -1 & & & & & \\ & & & & \ddots & & & & \\ & & & & & -1 & & & \\ & & & & & & 0 & & \\ & & & & & & & \ddots & \\ & & & & & & & & 0 \end{bmatrix}$$

having p ones and $r - p$ negative ones on the main diagonal, where r is the rank of A (recall that A and D have the same rank, since they are congruent). The ordered pair $(p, r - p)$ is called the *signature* of A. Moreover, according to Definition 8.12, f is:

(1) *Positive definite* if its signature is $(n, 0)$.
(2) *Negative definite* if its signature is $(0, n)$.
(3) *Indefinite* if its signature is $(p, r - p)$, for $p \neq r$.
(4) *Positive semi-definite* if its signature is $(r, 0)$.
(5) *Negative semi-definite* if its signature is $(0, r)$.

Definition 8.22 Let V be a complex space equipped by the Hermitian form f. In general, one may refer to (V, f) as a *Hermitian space*. In particular, if f has the additional property $f(v, v) > 0$, for all $0 \neq v \in V$, then it is an inner product, and we may also refer to (V, f) as an *inner product complex space*.

We conclude the chapter by providing the 'Hermitian' versions of Theorems 7.58 and 7.59:

Theorem 8.23 *Let* A, A' *be two complex* $n \times n$ *Hermitian matrices.* A, A' *are H-congruent if and only if they have the same signature.*

Theorem 8.24 *Let* A *be a real* $n \times n$ *symmetric matrix having signature* $(p, r - p)$. *Then*

(i) *The number of positive eigenvalues of* A *is equal to* p.
(ii) *The number of negative eigenvalues of* A *is equal to* $r - p$.

To prove Theorems 8.23 and 8.24, it is sufficient to recall that the eigenvalues of any Hermitian matrix are real numbers. Therefore, we can use the argument contained in Theorems 7.58 and 7.59 without any change, but first replace the role of transpose matrices with one of adjoint matrices in all the proofs.

Exercises

1. Let $f : V \times V \to \mathbb{C}$ be a sesquilinear form. Prove that f is Hermitian if and only if the matrix A of f is Hermitian, whatever the choice of basis for V with respect to which A is related.
2. Let $f : \mathbb{C}^3 \times \mathbb{C}^3 \to \mathbb{C}$ be a Hermitian form associated with the matrix

$$
\begin{bmatrix}
2 & i & 2-i \\
-i & 1 & 1+i \\
2+i & 1-i & 1
\end{bmatrix}
$$

   in terms of the canonical basis for $\mathbb{C}^3$. Determine an orthogonal basis for $\mathbb{C}^3$ in terms of which the matrix of f is diagonal.
3. Let $f : \mathbb{C}^4 \times \mathbb{C}^4 \to \mathbb{C}$ be a Hermitian form associated with the matrix

$$
\begin{bmatrix}
1 & 2+i & 0 & i \\
2-i & 1 & i & -i \\
0 & -i & 2 & -i \\
-i & i & i & 2
\end{bmatrix}
$$

   in terms of the canonical basis for $\mathbb{C}^4$. Determine an orthogonal basis for $\mathbb{C}^4$ in terms of which the matrix of f is diagonal.
4. Let V be the vector space of all $n \times n$ matrices over $\mathbb{C}$ and $f : V \times V \to \mathbb{C}$ the map defined as $f(A, B) = trace(A^*B)$, for any $A, B \in V$. Prove that f is a positive definite Hermitian form on $V \times V$.
5. Let V be a complex vector space of dimension n, $f : V \times V \to \mathbb{C}$ a positive definite Hermitian form, H the $n \times n$ complex matrix of f in terms of a suitable basis B for V. Prove that the sesquilinear form associated with the matrix H^{-1} is a positive definite Hermitian form.
6. Let V be a complex vector space of dimension n, $f : V \times V \to \mathbb{C}$ a positive definite Hermitian form, H the $n \times n$ complex matrix of f in terms of a suitable basis B for V. Let N be the $n \times n$ complex matrix such that $N^2 = H$. Prove that the sesquilinear form associated with the matrix N is a positive definite Hermitian form on $V \times V$.
7. Let $H = \begin{bmatrix} 1 & 1+i & 2i \\ 1-i & 4 & 2-3i \\ -2i & 2+3i & 7 \end{bmatrix}$ be a Hermitian matrix. Find a nonsingular matrix P such that $D = P^*HP$ is diagonal. Also, find the signature of H.
8. Show that the relation "Hermitian congruent" is an equivalence relation.
9. Let f be a Hermitian form on V over $\mathbb{C}$. Then prove that there exists a basis of V in which f is represented by a diagonal matrix. Also, show that every other diagonal matrix representation of f has the same number of positive entries and negative entries.

Chapter 9
Tensors and Their Algebras

In Chap. 7 bilinear and quadratic forms with various ramifications have been discussed. In the present chapter we address an aesthetic concern raised by bilinear forms and, as a part of this study, the tensor product of vector spaces has been introduced. Further, besides the study of tensor product of linear transformations, in the subsequent sections, a tensor algebra will be developed, and the chapter concludes with the study of exterior algebra viz. Grassmann algebra.

9.1 The Tensor Product

Definition 9.1 Let U and V be vector spaces over the same field $\mathbb{F}$. Then tensor product of U and V is a pair (W, ψ) consisting of a vector space W over $\mathbb{F}$ together with a bilinear map $\psi : U \times V \to W$ satisfying the following universal property: for any vector space X over $\mathbb{F}$ and any bilinear map $f : U \times V \to X$, there exists a unique linear map $h : W \to X$ such that $h \circ \psi = f$, that is, the following diagram

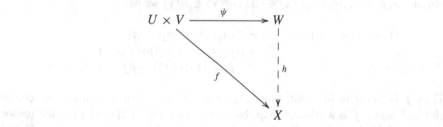

commutes.

Remark 9.2 (*i*) W is generally denoted by $U \otimes V$ and is called tensor product of U and V (if exists) and we shall see below that it is unique up to isomorphism.

(*ii*) Let $B_1 = \{u_i \mid i \in I\}$ and $B_2 = \{v_j \mid j \in J\}$ be bases of U and V, respectively. The bilinear mapping ψ on $U \times V$ can be uniquely determined by assigning arbitrary values to the pair (u_i, v_j). For each ordered pair (u_i, v_j), we can write a formal symbol $u_i \otimes v_j$ and define W to be a vector space with basis $B = \{u_i \otimes v_j \mid u_i \in B_1, v_j \in B_2\}$. Now define a mapping ψ on $U \times V$ by setting $\psi(u_i, v_j) = u_i \otimes v_j$. This uniquely defines a bilinear map ψ which is universal. If $f : U \times V \to X$ is bilinear, then the condition $f = h \circ \psi$ is equivalent to $h(u_i \otimes v_j) = f(u_i, v_j)$ which uniquely defines a linear map $h : W \to X$ and the pair (W, ψ) has the universal property for bilinearity. Any element w of W is a finite linear combination of $u_i \otimes v_j$, i.e., $w = \sum\limits_{i=1}^{m} \sum\limits_{j=1}^{n} \alpha_{ij}(u_{k_i} \otimes v_{k_j})$, and if $u = \sum\limits_{i=1}^{m} \alpha_i u_i$, $v = \sum\limits_{j=1}^{n} \beta_j v_j$, then

$$u \otimes v = \psi(u, v) = \psi\left(\sum_{i=1}^{m} \alpha_i u_i, \sum_{j=1}^{n} \beta_j v_j\right) = \sum_{i=1}^{m}\sum_{j=1}^{n} \alpha_i \beta_j u_i \otimes v_j.$$

(*iii*) As indicated above, for any $u \in U$ and $v \in V$, the image of (u, v) under the universal bilinear pairing into $U \otimes V$ shall be denoted $u \otimes v \in U \otimes V$. In view of the bilinearity of the pairing $(u, v) \mapsto u \otimes v$, we find the relations $(\alpha_1 u_1 + \alpha_2 u_2) \otimes v = \alpha_1 (u_1 \otimes v) + \alpha_2 (u_2 \otimes v)$ and $u \otimes (\beta_1 v_1 + \beta_2 v_2) = \beta_1 (u \otimes v_1) + \beta_2 (u \otimes v_2)$ in $U \otimes V$ for any $\alpha_1, \alpha_2, \beta_1, \beta_2 \in \mathbb{F}, u, u_1, u_2 \in U, v, v_1, v_2 \in V$.

(*iv*) Note that if $U = \{0\}$ or $V = \{0\}$, then $U \otimes V = \{0\}$. Indeed the only bilinear pairing $\psi : U \times V \to W$ is the zero pairing and hence the pairing $U \times V \to \{0\}$ yields the tensor product. Thus $\{0\} \otimes V = \{0\}$ and $U \otimes \{0\} = \{0\}$.

Example 9.3 Let $\mathbb{R}[x]$ and $\mathbb{R}[y]$ be vector spaces over $\mathbb{R}$. Then, $\mathbb{R}[x] \otimes \mathbb{R}[y] = \mathbb{R}[x, y]$.

Define a map $f : \mathbb{R}[x] \times \mathbb{R}[y] \to \mathbb{R}[x, y]$, given by, $f(p(x), q(y)) = p(x)q(y)$. For any $\alpha, \beta \in \mathbb{R}, p_1(x), p_2(x) \in \mathbb{R}[x], q(y) \in \mathbb{R}[y]$ we have,

$$\begin{aligned} f(\alpha p_1(x) + \beta p_2(x), q(y)) &= (\alpha p_1(x) + \beta p_2(x))q(y) \\ &= \alpha p_1(x)q(y) + \beta p_2(x)q(y) \\ &= \alpha f(p_1(x), q(y)) + \beta f(p_2(x), q(y)). \end{aligned}$$

Thus f is linear in first slot. Similarly, one can show that f is linear in second slot and hence f is a bilinear map. Now we claim that, $(\mathbb{R}[x, y], f)$ is the tensor product of $\mathbb{R}[x]$ and $\mathbb{R}[y]$. For this let X be any arbitrary vector space over $\mathbb{R}$ and $g : \mathbb{R}[x] \times \mathbb{R}[y] \to X$ be any arbitrary bilinear map. Then we have to construct a unique homomorphism $h : \mathbb{R}[x, y] \to X$ such that the following diagram commutes, i.e., $h \circ f = g$.

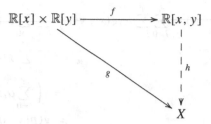

Define $h : \mathbb{R}[x, y] \to X$ such that,

$$h\left(\sum_{i,j\geq 0}^{finite} \alpha_{ij} x^i y^j \right) = \sum_{i,j\geq 0}^{finite} \alpha_{ij} g(x^i, y^j).$$

Obviously, h is well defined. Next we prove that h is a linear map.

$$h\left(\alpha \left(\sum_{i,j\geq 0}^{finite} \alpha_{ij} x^i y^j \right) + \beta \left(\sum_{i,j\geq 0}^{finite} \alpha_{ij} x^i y^j \right) \right)$$

$$= h\left(\sum_{i,j\geq 0}^{finite} (\alpha \alpha_{ij} + \beta \beta_{ij}) x^i y^j \right)$$

$$= \sum_{i,j\geq 0}^{finite} (\alpha \alpha_{ij} + \beta \beta_{ij}) g(x^i, y^j)$$

$$= \sum_{i,j\geq 0}^{finite} \alpha \alpha_{ij} g(x^i, y^j) + \sum_{i,j\geq 0}^{finite} \beta \beta_{ij} g(x^i, y^j)$$

$$= \alpha h\left(\sum_{i,j\geq 0}^{finite} \alpha_{ij} x^i y^j \right) + \beta h\left(\sum_{i,j\geq 0}^{finite} \beta_{ij} x^i y^j \right).$$

Hence, h is a linear map. Also

$$(h \circ f)(p(x), q(y)) = h(f(p(x), q(y))) = h(p(x)q(y)).$$

Let $p(x) = \sum_{i=0}^{m'} \alpha_i x^i$ and $q(y) = \sum_{j=0}^{n'} \beta_j y^j$. This implies that $p(x)q(y) = \sum_{i=0}^{m'} \sum_{j=0}^{n'} \delta_{ij} x^i y^j$ and $\delta_{ij} = \alpha_i \beta_j, 0 \leq i \leq m', 0 \leq j \leq n'$.

Therefore,

$$(h \circ f)(p(x), q(y)) = \sum_{i=0}^{m'} \sum_{j=0}^{n'} \delta_{ij} g(x^i, y^j) = \sum_{i=0}^{m'} \sum_{j=0}^{n'} \alpha_i \beta_j g(x^i, y^j)$$

$$= \sum_{i=0}^{m'} \sum_{j=0}^{n'} g(\alpha_i x^i, \beta_j y^j)$$

$$= g\left(\sum_{i=0}^{m'} \alpha_i x^i, \sum_{j=0}^{n'} \beta_j y^j \right)$$

$$= g(p(x), q(y)).$$

This implies that $h \circ f = g$. Finally, we show the uniqueness of h. If possible, suppose that $h_1 : \mathbb{R}[x, y] \to X$ be another linear map such that $h_1 \circ f = g$, then

$$h(r(x, y)) = h\left(\sum_{i,j \geq 0}^{finite} \gamma_{ij} x^i y^j \right) \quad \text{where } r(x, y) \in \mathbb{R}[x, y]$$

$$= \sum_{i,j \geq 0}^{finite} \gamma_{ij} g(x^i, y^j)$$

$$= \sum_{i,j \geq 0}^{finite} \gamma_{ij} h_1 f(x^i, y^j)$$

$$= h_1\left(\sum_{i,j \geq 0}^{finite} \gamma_{ij} f(x^i, y^j) \right)$$

$$= h_1\left(\sum_{i,j \geq 0}^{finite} \gamma_{ij} x^i y^j \right)$$

$$= h_1(r(x, y)).$$

This implies that, $h = h_1$. Therefore, h is unique and finally, we have $\mathbb{R}[x] \otimes \mathbb{R}[y] = \mathbb{R}[x, y]$. Here it is to be noted that x and y do not commute.

Existence and Uniqueness of Tensor Product of Two Vector Spaces

To prove the existence of tensor product of two vector spaces, we need the notion of "free vector space over a given set". Let S be any set and $\mathbb{F}$, a field. A pair (V, f) is called a free vector space over $\mathbb{F}$ on the set S, where V is a vector space over $\mathbb{F}$ and $f : S \to V$ is a function if for any arbitrary vector space X over $\mathbb{F}$ and for any arbitrary function $g : S \to X$, there exists a unique linear map $h : V \to X$ such that the following diagram commutes, i.e., $h \circ f = g$.

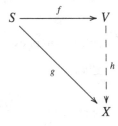

It can be easily shown that given any set S and $\mathbb{F}$, a field, there always exists a free vector space over $\mathbb{F}$ on S. If (V, f) is a free vector space on S, then f is one-to-one and $\langle f(S) \rangle = V$. Moreover, every vector space can be realized as a quotient space of a free vector space. If (V, f) is a free vector space over $\mathbb{F}$ on S, then as a convention, we say that V is a free vector space on S. The associated map f and the field $\mathbb{F}$ are assumed to be understood.

Theorem 9.4 *Let U and V be vector spaces over a field $\mathbb{F}$. Then their tensor product $U \otimes V$ exists and it is unique upto isomorphism.*

Proof Let (W, f) be a free vector space over the set $U \times V$. Now consider T, the subspace of W, generated by the elements $f(\alpha_1 u_1 + \alpha_2 u_2, v) - \alpha_1 f(u_1, v) - \alpha_2 f(u_2, v)$, $f(u, \beta_1 v_1 + \beta_2 v_2) - \beta_1 f(u, v_1) - \beta_2 f(u, v_2)$, for all elements $u, u_1,$ $u_2 \in U$, $v, v_1, v_2 \in V$ and $\alpha_1, \alpha_2, \beta_1, \beta_2 \in \mathbb{F}$. Now construct the quotient space $\frac{W}{T}$. It is obvious that there exists the quotient homomorphism $q : W \to \frac{W}{T}$. Clearly $f' = q \circ f : U \times V \to \frac{W}{T}$. We show that f' is a bilinear map from $U \times V$ to $\frac{W}{T}$;

$$f'(\alpha_1 u_1 + \alpha_2 u_2, v) - \alpha_1 f'(u_1, v) - \alpha_2 f'(u_2, v)$$
$$= q[f(\alpha_1 u_1 + \alpha_2 u_2, v)] - \alpha_1 q[f(u_1, v)] - \alpha_2 q[f(u_2, v)],$$
$$= q[f(\alpha_1 u_1 + \alpha_2 u_2, v) - \alpha_1 f(u_1, v) - \alpha_2 f(u_2, v)],$$
$$= 0.$$

Hence
$$f'(\alpha_1 u_1 + \alpha_2 u_2, v) = \alpha_1 f'(u_1, v) + \alpha_2 f'(u_2, v),$$

for all $\alpha_1, \alpha_2 \in \mathbb{F}$, $u_1, u_2 \in U$, $v \in V$.

This shows that f' is linear in the first coordinate. Similarly, it can be seen that f' is also linear in the second coordinate. Thus, f' is a bilinear map. Now, we claim that $(\frac{W}{T}, f')$ is a tensor product of U and V. For this let $g : U \times V \to X$ be any arbitrary bilinear map, where X is any vector space. Then we have to produce a unique linear map $h : \frac{W}{T} \to X$ such that the following diagram commutes, i.e., $h \circ f' = g$.

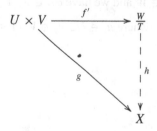

Since W is a free vector space over $U \times V$. Hence for the map $g : U \times V \to X$, there exists a unique linear map $h' : W \to X$ such that the following diagram commutes, i.e., $h' \circ f = g$.

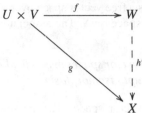

Now we show that $T \subseteq Kerh'$. For this, consider arbitrary elements $\alpha_1, \alpha_2 \in \mathbb{F}$, $u, u_1, u_2 \in U$, $v, v_1, v_2 \in V$. Then, we have

$$
\begin{aligned}
h'(f(\alpha_1 u_1 + \alpha_2 u_2, v) &- \alpha_1 f(u_1, v) - \alpha_2 f(u_2, v)) \\
&= g(\alpha_1 u_1 + \alpha_2 u_2, v) - \alpha_1 g(u_1, v) - \alpha_2 g(u_2, v) \\
&= 0.
\end{aligned}
$$

Hence, the element $f(\alpha_1 u_1 + \alpha_2 u_2, v) - \alpha_1 f(u_1, v) - \alpha_2 f(u_2, v) \in Kerh'$. Similarly, we can show that the element $f(u, \beta_1 v_1 + \beta_2 v_2) - \beta_1 f(u, v_1) - \beta_2 f(u, v_2) \in Kerh'$. But T is generated by these elements. Thus, we conclude that $T \subseteq Kerh'$. Now define a map $h : \frac{W}{T} \to X$ such that $h(w + T) = h'(w)$. Assume that $w_1 + T = w_2 + T$. This implies that $w_1 - w_2 \in T$, i.e., $w_1 - w_2 \in Ker\ h'$. As a result $h'(w_1 - w_2) = 0$, i.e., $h'(w_1) = h'(w_2)$. Finally, we have shown that $h(w_1 + T) = h(w_2 + T)$. Thus h is well defined. Obviously, h is a linear map. Next we show that $hf' = g$. Consider $(hf')(u, v) = h(f'(u, v)) = h(qf(u, v)) = h(q(f(u, v))) = h(f(u, v) + T) = h'(f(u, v)) = (h'f)(u, v) = g(u, v)$ for every element $(u, v) \in U \times V$. Hence $hf' = g$ stands proved.

To prove our claim, it only remains to be proved that h is unique. If possible, suppose that there exists another linear map $k : \frac{W}{T} \to X$ such that $k \circ f' = g$. As (W, f) is a free vector space over $U \times V$, we have $< f(U \times V) >\, = W$.

Let $w + T \in \frac{W}{T}$. Thus $w \in W$ and we have $\alpha_{ij} \in \mathbb{F}$, $1 \leq i \leq m$, $1 \leq j \leq n$ such that $w = \sum_{i=1}^{m} \sum_{j=1}^{n} \alpha_{ij} f(u_i, v_j)$; where $u_i \in U$, $v_j \in V$, $1 \leq i \leq m$, $1 \leq j \leq n$.

Hence,   •

$$h(w + T) = h\left[\left(\sum_{i=1}^{m}\sum_{j=1}^{n}\alpha_{ij}f(u_i, v_j)\right) + T\right]$$

$$= h'\left(\sum_{i=1}^{m}\sum_{j=1}^{n}\alpha_{ij}f(u_i, v_j)\right)$$

$$= \sum_{i=1}^{m}\sum_{j=1}^{n}\alpha_{ij}h'f(u_i, v_j)$$

$$= \sum_{i=1}^{m}\sum_{j=1}^{n}\alpha_{ij}g(u_i, v_j)$$

$$= \sum_{i=1}^{m}\sum_{j=1}^{n}\alpha_{ij}(k \circ f')(u_i, v_j)$$

$$= \sum_{i=1}^{m}\sum_{j=1}^{n}\alpha_{ij}k\left(f'(u_i, v_j)\right)$$

$$= k\left(\sum_{i=1}^{m}\sum_{j=1}^{n}\alpha_{ij}f'(u_i, v_j)\right)$$

$$= k\left(\sum_{i=1}^{m}\sum_{j=1}^{n}\alpha_{ij}(qf)(u_i, v_j)\right)$$

$$= k\left[q\left(\sum_{i=1}^{m}\sum_{j=1}^{n}\alpha_{ij}f(u_i, v_j)\right)\right]$$

$$= k\left[\left(\sum_{i=1}^{m}\sum_{j=1}^{n}\alpha_{ij}f(u_i, v_j)\right) + T\right]$$

$$= k(w + T).$$

This shows that $h = k$. Hence, h is unique linear map. Thus $(\frac{W}{T}, f')$ is a required tensor product of U and V. Thus, existence stands proved.

To prove the uniqueness of tensor product of U and V, let us assume that (W, ψ) and (W', ϕ) are two tensor products of U and V, then it can be easily seen that there exists a unique linear isomorphism $W \cong W'$ carrying ψ and ϕ (and vice-versa). In fact, if we factor ϕ through universal property of W,

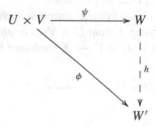

then $h \circ \psi = \phi$.

In a similar way, now factor ψ through universal property of W',

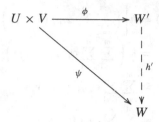

we arrive at $h' \circ \phi = \psi$. Now consider the identity map I_W on W and composition $h' \circ h$:

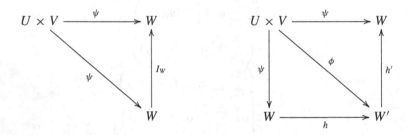

Both I_W and $h' \circ h$ factor ψ through ψ as in the universal property of W and hence by the uniqueness of the universal property, we arrive at $I_W = h' \circ h$. In a similar way, one can arrive at $I_{W'} = h \circ h'$ and hence we find that $W \cong W'$ which ensures that there is only one tensor product.

In this sense, the tensor product of U and V (equipped with its "universal" bilinear pairing from $U \times V$) is unique up to isomorphism, and hence we may speak of the tensor product of U and V and write $U \otimes V$ for the tensor product of U and V.

Theorem 9.5 *Let U, V be vector spaces over a field $\mathbb{F}$. The following statements are equivalent:*

(i) $(U \otimes V, \psi)$ *is the tensor product of U and V.*

(ii) *Let $\psi : U \times V \to U \otimes V$ be a bilinear function such that $\langle \psi(U \times V) \rangle = U \otimes V$. For any bilinear mapping f from $U \times V$ to another vector space X over $\mathbb{F}$ there exists a linear map $h : U \otimes V \to X$ such that $h \circ \psi = f$.*

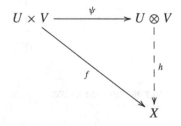

Proof *(i)* $\Rightarrow$ *(ii)* If $(U \otimes V, \psi)$ is the tensor product of U and V, $\psi : U \times V \to U \otimes V$ is a bilinear mapping and $\langle \psi(U \times V) \rangle$ is a subspace of $U \otimes V$. This shows that the inclusion homomorphism exists, i.e., $i : \langle \psi (U \times V) \rangle \to U \otimes V$ is a vector space homomorphism. Define a map $f : U \times V \to \langle \psi (U \times V) \rangle$ such that $f(u, v) = \psi(u, v)$. Obviously f is a bilinear mapping and since $U \otimes V$ is the tensor product, there exists a unique homomorphism $h : U \otimes V \to \langle \psi (U \times V) \rangle$ such that the following diagram

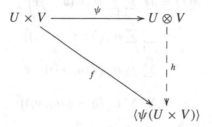

is commutative, i.e., $h \circ \psi = f$. Using the definition of tensor product again, there exists a unique homomorphism (the identity homomorphism) I on $U \otimes V$ such that the following diagram is commutative, i.e., $I \circ \psi = \psi$.

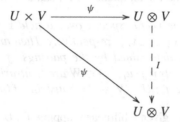

But we have
$$U \otimes V \xrightarrow{h} \langle \psi (U \times V) \rangle \xrightarrow{i} U \otimes V$$

Clearly $i \circ h$ is a homomorphism (the identity homomorphism I), i.e., $i \circ h : U \otimes V \to U \otimes V$. Since $f(u, v) = \psi(u, v)$, it is clear that

$$
\begin{aligned}
(i \circ h) \circ \psi &= i \circ (h \circ \psi) \\
&= i \circ f \\
&= f \\
&= \psi.
\end{aligned}
$$

This shows that homomorphism $i \circ h$ also makes the latter diagram commutative. Hence, the uniqueness of I in the same guarantees that $i \circ h = I$. On using ontoness of i and I is bijective, $i : \langle \psi (U \times V) \rangle \to U \otimes V$ becomes surjective, i.e., $i(\langle \psi (U \times V) \rangle) = U \otimes V$. But since i is the inclusion map, the later relation yields that $\langle \psi (U \times V) \rangle = U \otimes V$ and hence $\psi(U \times V)$ generates $U \otimes V$.

$(ii) \Rightarrow (i)$ Let $\psi : U \times V \to U \otimes V$ be a bilinear function such that $\langle \psi (U \times V) \rangle = U \otimes V$ and for any vector space X over $\mathbb{F}$ and any bilinear map $f : U \times V \to$

X there exists a homomorphism $h : U \otimes V \to X$ such that $h \circ \psi = f$. To prove the uniqueness of h, if possible suppose that $h' : U \otimes V \to X$ is another homomorphism which makes the above diagram commutative, i.e., $h' \circ \psi = f$. Since $U \otimes V$ is spanned by $\psi(U \times V)$, for any $x \in U \otimes V$ we find that

$$
\begin{aligned}
h(x) &= h\left(\sum_{i=1}^{m} \sum_{j=1}^{n} \alpha_{ij} \psi(u_i, v_j) \right) \\
&= \sum_{i=1}^{m} \sum_{j=1}^{n} \alpha_{ij} \left(h \circ \psi(u_i, v_j) \right) \\
&= \sum_{i=1}^{m} \sum_{j=1}^{n} \alpha_{ij} \left(f(u_i, v_j) \right) \\
&= \sum_{i=1}^{m} \sum_{j=1}^{n} \alpha_{ij} \left(h' \circ \psi(u_i, v_j) \right) \\
&= h'\left(\sum_{i=1}^{m} \sum_{j=1}^{n} \alpha_{ij} \psi(u_i, v_j) \right) \\
&= h'(x).
\end{aligned}
$$

This implies that $h = h'$ and the uniqueness of h stands proved.

Theorem 9.6 *Let U, V be vector spaces over a field $\mathbb{F}$ with the ordered bases $\{u_1, u_2, \ldots, u_m\}$ and $\{v_1, v_2, \ldots, v_n\}$, respectively. Then any bilinear mapping $f : U \times V \to W$ is uniquely determined by the pairings of $f(u_i, v_j) \in W$; $1 \leq i \leq m$; $1 \leq j \leq n$, and conversely if $f(u_i, v_j) \in W$ are arbitrarily given then there exists a unique bilinear mapping $f : U \times V \to W$ satisfying $f(u_i, v_j) = w_{ij}$.*

Proof Suppose that we are given a bilinear mapping $f : U \times V \to W$. Any $u \in U$ and $v \in V$ can be uniquely written as $u = \sum_{i=1}^{m} \alpha_i u_i$, $v = \sum_{j=1}^{n} \beta_j v_j$; $\alpha_i, \beta_j \in \mathbb{F}$. Hence using bilinearity of f, we arrive at

$$
f(u, v) = \sum_{i=1}^{m} \sum_{j=1}^{n} \alpha_i \beta_j f(u_i, v_j).
$$

This shows that f is uniquely determined by the pairing $f(u_i, v_i) \in W$.

Conversely, given any $w_{ij} \in W$ define the map $f : U \times V \to W$ such that $f(u, v) = \sum_{i=1}^{m} \sum_{j=1}^{n} \alpha_i \beta_j w_{ij}$; where $u = \sum_{i=1}^{m} \alpha_i u_i$, and $v = \sum_{j=1}^{n} \beta_j v_j$ are the unique representation of u and v with respect to the ordered bases of U and V respectively. The existence and uniqueness of such expressions for u and v ensure that f is well defined and obviously $f(u_i, v_j) = w_{ij}$. Hence, it remains only to show that f is bilinear. We check bilinearity on $u \in U$ for a fixed $v \in V$, and the other way around similarly. For any $u, u' \in U$, and $\alpha, \alpha' \in \mathbb{F}$ with $u = \sum_{i=1}^{m} \alpha_i u_i$, $u' = \sum_{i=1}^{m} \alpha'_i u_i$, and $v = \sum_{j=1}^{n} \beta_j v_j$, we find that $\alpha u + \alpha' u' = \sum_{i=1}^{m}(\alpha\alpha_i + \alpha'\alpha_i')u_i$ and

$$
\begin{aligned}
f(\alpha u + \alpha' u', v) &= f\left(\sum_{i=1}^{m}(\alpha\alpha_i + \alpha'\alpha_i')u_i, \sum_{j=1}^{n}\beta_j v_j\right) \\
&= \sum_{i=1}^{m}(\alpha\alpha_i + \alpha'\alpha_i') \sum_{j=1}^{n}\beta_j f(u_i, v_j) \\
&= \sum_{i=1}^{m}\sum_{j=1}^{n}(\alpha\alpha_i + \alpha'\alpha_i')\beta_j w_{ij} \\
&= \alpha\sum_{i=1}^{m}\sum_{j=1}^{n}\alpha_i\beta_j w_{ij} + \alpha'\sum_{i=1}^{m}\sum_{j=1}^{n}\alpha_i'\beta_j w_{ij} \\
&= \alpha f(u, v) + \alpha' f(u', v).
\end{aligned}
$$

This completes the proof.

Lemma 9.7 *Let U, V be vector spaces over a field $\mathbb{F}$, and let $\{u_1, u_2, \ldots, u_n\}$ be linearly independent subset of U. Then for arbitrary vectors $v_1, v_2, \ldots, v_n \in V$ $\sum_{i=1}^{n} u_i \otimes v_i = 0 \Rightarrow v_i = 0$ for all i. In particular, $u \otimes v = 0$ if and only if $u = 0$ or $v = 0$.*

Proof Let us suppose that $\sum_{i=1}^{n} u_i \otimes v_i = 0$ where it may be assumed that none of the vectors v_i are zero. The universal property of tensor product ensures that for any bilinear function $f : U \times V \rightarrow X$, there exists a unique linear map $h : U \otimes V \rightarrow X$ such that $h \circ \psi = f$. This implies that

$$
0 = h\left(\sum_{i=1}^{n} u_i \otimes v_i\right) = \sum_{i=1}^{n}(h \circ \psi)(u_i, v_i) = \sum_{i=1}^{n} f(u_i, v_i).
$$

The above relation holds for any bilinear function $f : U \times V \rightarrow X$. In particular one may choose two linear functionals $\phi \in U^*, \xi \in V^*$ such that $f(u, v) = \phi(u)\xi(v), u \in U, v \in V$. Since the set of vectors $\{u_1, u_2, \ldots, u_n\}$ is linearly independent, one can consider the dual vector $u_i^* \in U^*$ such that $u_i^*(u_j) = \delta_{ij}$. Now, if $\phi = u_k^*; 1 \leq k \leq n$, then we arrive at $0 = \sum_{i=1}^{n} u_k^*(u_i)\xi(v_i) = \xi(v_k)$ for all linear functionals $\xi \in V^*$. Hence, we easily obtain that $v_k = 0$.

Theorem 9.8 *Let U and V be vector spaces over a field $\mathbb{F}$ and let $B_1 = \{u_i \mid i \in I\}$ and $B_2 = \{v_j \mid j \in J\}$ be bases of U and V, respectively. Then the set $B = \{u_i \otimes v_j \mid i \in I, j \in J\}$ is a basis for $U \otimes V$.*

Proof To show that B is linearly independent, suppose that $\sum_{i=1}^{m}\sum_{j=1}^{n}\alpha_{ij}(u_i \otimes v_j) = 0$. This can be rewritten

$$\sum_{i=1}^{m} u_i \otimes \left(\sum_{j=1}^{n} \alpha_{ij} v_j \right) = 0$$

and hence application of Lemma 9.7 yields that $\sum_{j=1}^{n} \alpha_{ij} v_j = 0$ for all i and hence $\alpha_{ij} = 0$ for all i and j. Now to show that B spans $U \otimes V$, let $u \otimes v$ be an arbitrary element of $U \otimes V$. Then since $u = \sum_{i=1}^{m} \alpha_i u_i$ and $v = \sum_{j=1}^{n} \beta_j v_j$, we find that

$$
\begin{aligned}
u \otimes v &= \sum_{i=1}^{m} \alpha_i u_i \otimes \sum_{j=1}^{n} \beta_j v_j \\
&= \sum_{i=1}^{m} \alpha_i \left(u_i \otimes \sum_{j=1}^{n} \beta_j v_j \right) \\
&= \sum_{i=1}^{m} \alpha_i \left(\sum_{j=1}^{n} \beta_j (u_i \otimes v_j) \right) \\
&= \sum_{i=1}^{m} \sum_{j=1}^{n} \alpha_i \beta_j (u_i \otimes v_j).
\end{aligned}
$$

Hence, any sum of elements of the form $u \otimes v$ is a linear combination of the vectors $u_i \otimes v_j$, as desired.

Corollary 9.9 *If U and V are finite dimensional vector spaces over a field $\mathbb{F}$, then $dim(U \otimes V) = dim(U)dim(V)$.*

If U and V are finite dimensional vector spaces, then one can see the nature of linear functionals which exist on $U \otimes V$. The proof of the following theorem illustrates that any linear functional on the tensor product is nothing but a tensor product of linear functionals in isomorphic sense.

Theorem 9.10 *Let U and V be finite dimensional vector spaces. Then*

$$\widehat{U} \otimes \widehat{V} \cong \widehat{U \otimes V}$$

via the isomorphism $h : \widehat{U} \otimes \widehat{V} \to \widehat{U \otimes V}$ given by $h(\sigma \otimes \tau)(u \otimes v) = \sigma(u)\tau(v)$, for every $\sigma \otimes \tau \in \widehat{U} \otimes \widehat{V}$ and for every $u \otimes v \in U \otimes V$.

Proof Let us choose fixed elements $\sigma \in \widehat{U}$ and $\tau \in \widehat{V}$. Define a map $g' : U \times V \to \mathbb{F}$ such that $g'(u, v) = \sigma(u)\tau(v)$. It is obvious to observe that g' is a bilinear map. Let $(U \otimes V, f')$ be the tensor product of U and V. Hence, there exists a unique linear map $h_{\sigma,\tau} : U \otimes V \to \mathbb{F}$ such that the following diagram commutes:

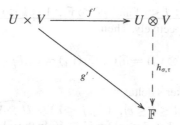

i.e., $h_{\sigma,\tau}(f'(u,v)) = g'(u,v)$ or $h_{\sigma,\tau}(u \otimes v) = g'(u,v)$, and hence $h_{\sigma,\tau}(u \otimes v) = \sigma(u)\tau(v)$. This shows that $h_{\sigma,\tau} \in \widehat{U \otimes V}$.

As for each fixed $\sigma \in \widehat{U}$ and for each fixed $\tau \in \widehat{V}$, we have unique linear functional $h_{\sigma,\tau}$ on $U \otimes V$. Thus, we define a map $g : \widehat{U} \times \widehat{V} \to \widehat{U \otimes V}$ such that $g(\sigma, \tau) = h_{\sigma,\tau}$. Noting the above arguments, g is well defined. Here, for any $\alpha, \beta \in \mathbb{F}, \sigma_1, \sigma_2 \in \widehat{U}$,

$$
\begin{aligned}
g(\alpha\sigma_1 + \beta\sigma_2, \tau)(u \otimes v) &= (h_{\alpha\sigma_1+\beta\sigma_2,\tau})(u \otimes v) \\
&= (\alpha\sigma_1 + \beta\sigma_2)(u)\tau(v) \\
&= (\alpha\sigma_1(u) + \beta\sigma_2(u))\tau(v) \\
&= \alpha\sigma_1(u)\tau(v) + \beta\sigma_2(u)\tau(v) \\
&= \alpha(h_{\sigma_1,\tau}(u \otimes v)) + \beta(h_{\sigma_2,\tau}(u \otimes v)) \\
&= (\alpha h_{\sigma_1,\tau})(u \otimes v) + (\beta h_{\sigma_2,\tau})(u \otimes v) \\
&= (\alpha h_{\sigma_1,\tau} + \beta h_{\sigma_2,\tau})(u \otimes v) \\
&= (\alpha g(\sigma_1, \tau) + \beta g(\sigma_2, \tau))(u \otimes v)
\end{aligned}
$$

implies that $g(\alpha\sigma_1 + \beta\sigma_2, \tau)(u \otimes v) = \alpha g(\sigma_1, \tau) + \beta g(\sigma_2, \tau)$.

Hence g is linear in the first coordinate. Similarly, it can be seen that g is also linear in the second coordinate. Thus, g is a bilinear map.

Let $(\widehat{U} \otimes \widehat{V}, f)$ be the tensor product of $\widehat{U}$ and $\widehat{V}$. Hence, corresponding to the above bilinear map g, there exists a unique linear map $h : \widehat{U} \otimes \widehat{V} \to \widehat{U \otimes V}$ such that the following diagram commutes:

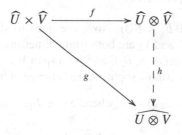

i.e, $h \circ f = g$, i.e., $h(f(\sigma, \tau)) = g(\sigma, \tau)$ or $h(\sigma \otimes \tau) = h_{\sigma,\tau}$. This implies that $h(\sigma \otimes \tau)(u \otimes v) = h_{\sigma,\tau}(u \otimes v)$, i.e., $h(\sigma \otimes \tau)(u \otimes v) = \sigma(u)\tau(v)$.

Next, we prove that the above homomorphism h is an isomorphism. For this, let $\{u_1, u_2, \ldots, u_m\}$ be a basis for U, with dual basis $\{\widehat{u_1}, \widehat{u_2}, \ldots, \widehat{u_m}\}$ and let $\{v_1, v_2, \ldots, v_n\}$ be a basis for V, with dual basis $\{\widehat{v_1}, \widehat{v_2}, \ldots, \widehat{v_n}\}$. We know that

$\{u_i \otimes v_j, \ 1 \le i \le m, \ 1 \le j \le n\}$ and $\{\widehat{u_i} \otimes \widehat{v_j}, \ 1 \le i \le m, \ 1 \le j \le n\}$ are bases of $U \otimes V$ and $\widehat{U} \otimes \widehat{V}$ respectively. Then

$$h(\widehat{u_i} \otimes \widehat{v_j})(u_\epsilon \otimes v_\mu) = \widehat{u_i}(u_\epsilon)\widehat{v_j}(v_\mu) = \delta_{i,\epsilon}\delta_{j,\mu} = \delta_{(i,j),(\epsilon,\mu)}.$$

Here $\delta_{i,\epsilon}$, etc. are Kronecker's deltas.

And, thus $\{h(\widehat{u_i} \otimes \widehat{v_j})|1 \le i \le m, \ 1 \le j \le n\} \subseteq \widehat{U \otimes V}$ is the dual basis to the basis $\{u_i \otimes v_j|1 \le i \le m, \ 1 \le j \le n\}$ for $U \otimes V$. In the other words we have $h(\widehat{u_i} \otimes \widehat{v_j}) = \widehat{u_i \otimes v_j}$.

In this way, we have proved the linear map h sends a basis of $\widehat{U} \otimes \widehat{V}$ to a basis of $\widehat{U \otimes V}$ and hence h is a bijective map. Thus h becomes an isomorphism and we have

$$\widehat{U} \otimes \widehat{V} \cong \widehat{U \otimes V}.$$

Theorem 9.11 *Let V_1, V_2 and W be vector spaces over a field $\mathbb{F}$. Let $\mu : V_1 \times V_2 \longrightarrow W$ be a bilinear map. Suppose, there exist bases B_1 and B_2 of V_1 and V_2 respectively such that; $\mu(B_1 \times B_2)$ is a basis for W. Then, for any choice of bases B_1' and B_2' for V_1 and V_2 respectively, $\mu(B_1' \times B_2')$ is a basis for W.*

Proof Let B_1' and B_2' be bases for V_1 and V_2, respectively. We first show that $\mu(B_1' \times B_2')$ spans W. Let $y \in W$. Since $\mu(B_1 \times B_2)$ spans W, we can write $y = \sum_{j=1}^{r} \sum_{k=1}^{s} a_{jk}\mu(z_{1j}, z_{2k})$, where $z_{1j} \in B_1, z_{2k} \in B_2$. But since B_1' is a basis for $V_1, z_{1j} = \sum_{\ell=1}^{t} b_{j\ell}x_{1\ell}$, where $x_{1\ell} \in B_1'$. Similarly, $z_{2k} = \sum_{m=1}^{p} c_{km}x_{2m}$, where $x_{2m} \in B_2'$. Thus,

$$y = \sum_{j=1}^{r} \sum_{k=1}^{s} a_{jk}\mu\left(\sum_{\ell=1}^{t} b_{j\ell}x_{1\ell}, \sum_{m=1}^{p} c_{km}x_{2m}\right)$$
$$= \sum_{j=1}^{r} \sum_{k=1}^{s} \sum_{\ell=1}^{t} \sum_{m=1}^{p} a_{jk}b_{j\ell}c_{km}\mu(x_{1\ell}, x_{2m}).$$

This implies that, $y \in \langle\mu(B_1' \times B_2')\rangle$. Now, we need to show that $\mu(B_1' \times B_2')$ is linearly independent. If V_1 and V_2 are both finite dimensional, this follows from the fact that $|\mu(B_1' \times B_2')| = |\mu(B_1 \times B_2)|$ and both span W.

For infinite dimensions, a more sophisticated change of basis argument is needed. Suppose $\sum_{\ell=1}^{t'} \sum_{m=1}^{p'} d_{\ell m}\mu(x_{1\ell}, x_{2m}) = 0$, where $x_{1\ell} \in B_1'$, $x_{2m} \in B_2'$. Then by change of basis, $x_{1\ell} = \sum_{j=1}^{r'} e_{\ell j}z_{1j}$, where $z_{1j} \in B_1$ and $x_{2m} = \sum_{k=1}^{s'} f_{mk}z_{2k}$, where $z_{2k} \in B_2$. Note that, the $e_{\ell j}$ form an inverse matrix to the matrix formed by, $b_{j\ell}$ above, then $\sum_{j=1}^{r'} e_{\ell j}b_{j\ell'} = \delta_{\ell\ell'}$ and similarly, $\sum_{k=1}^{s'} f_{mk}c_{km'} = \delta_{mm'}$ (where δ refers to the Kronecker's delta). Thus we have,

$$0 = \sum_{\ell=1}^{t'} \sum_{m=1}^{p'} d_{\ell m} \mu(x_{1\ell}, x_{2m})$$

$$= \sum_{\ell=1}^{t'} \sum_{m=1}^{p'} d_{\ell m} \mu\left(\sum_{j=1}^{r'} e_{\ell j} z_{1j}, \sum_{k=1}^{s'} f_{mk} z_{2k} \right)$$

$$= \sum_{j=1}^{r'} \sum_{k=1}^{s'} \sum_{\ell=1}^{t'} \sum_{m=1}^{p'} d_{\ell m} e_{\ell j} f_{mk} \mu(z_{1j}, z_{2k}).$$

Since $\mu(B_1 \times B_2)$ is linearly independent, $\sum_{\ell=1}^{t'} \sum_{m=1}^{p'} d_{\ell m} e_{\ell j} f_{mk} = 0$, for all j, k. But now,

$$d_{\ell' m'} = \sum_{\ell=1}^{t'} \sum_{m=1}^{p'} d_{\ell m} \delta_{\ell \ell'} \delta_{mm'}$$

$$= \sum_{\ell=1}^{t'} \sum_{m=1}^{p'} d_{\ell m} \left(\sum_{j=1}^{r'} b_{j\ell'} e_{\ell j} \right)\left(\sum_{k=1}^{s'} c_{km'} f_{mk} \right)$$

$$= \sum_{j=1}^{r'} \sum_{k=1}^{s'} b_{j\ell'} c_{km'} \left(\sum_{\ell=1}^{t'} \sum_{m=1}^{p'} d_{\ell m} e_{\ell j} f_{mk} \right)$$

$$= 0, \quad \text{for all } \ell', m'.$$

Thus, $\mu(B_1' \times B_2')$ is linearly independent. Hence $\mu(B_1' \times B_2')$ is a basis for W.

Theorem 9.12 *Let U, V, W be vector spaces over a field $\mathbb{F}$. Then*

(i) $U \otimes V \cong V \otimes U$,
(ii) $U \otimes \mathbb{F} \cong \mathbb{F} \otimes U \cong U$,
(iii) $(U \otimes V) \otimes W \cong U \otimes (V \otimes W)$.

Proof (i) Assume that $(U \otimes V, \psi)$ is a tensor product of U and V. Define a bilinear mapping $f : U \times V \to V \otimes U$ such that $f(u, v) = v \otimes u$. Using the definition of tensor product, we find that there exists a unique linear mapping $h : U \otimes V \to V \otimes U$ such that the diagram

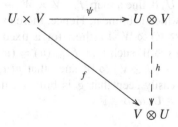

is commutative, i.e., $h \circ \psi = f$. Similarly, define $g : V \times U \to U \otimes V$ such that $g(v, u) = u \otimes v$. This leads to a unique linear mapping $h' : V \otimes U \to U \otimes V$ such that the diagram

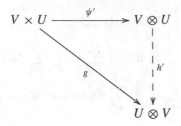

is commutative, i.e., $h' \circ \psi' = g$. Now for any $x \in U \otimes V$, we find that

$$
\begin{aligned}
h' \circ h(x) &= h' \circ h\left(\sum_{i=1}^{m} \sum_{j=1}^{n} \alpha_{ij}(u_i \otimes v_j) \right) \\
&= h'\left(\sum_{i=1}^{m} \sum_{j=1}^{n} \alpha_{ij} h(u_i \otimes v_j) \right) \\
&= h'\left(\sum_{i=1}^{m} \sum_{j=1}^{n} \alpha_{ij}(v_j \otimes u_i) \right) \\
&= \sum_{i=1}^{m} \sum_{j=1}^{n} \alpha_{ij} h'(v_j \otimes u_i) \\
&= \sum_{i=1}^{m} \sum_{j=1}^{n} \alpha_{ij}(u_i \otimes v_j) \\
&= x.
\end{aligned}
$$

Similarly, it can be seen that $h \circ h'(y) = y$ for all $y \in V \otimes U$. This shows that $h \circ h' = I$ on $V \otimes U$ and $h' \circ h = I$ on $U \otimes V$. Thus h is an isomorphism whose inverse is h' and hence $U \otimes V \cong V \otimes U$.

(ii) The above proof shows that the correspondence $u \otimes v \longleftrightarrow v \otimes u$ has been used to establish the isomorphism $U \otimes V \cong V \otimes U$. Similarly, using the correspondence $u \otimes \alpha \longleftrightarrow \alpha \otimes u \longleftrightarrow \alpha u$, where $\alpha \in \mathbb{F}, u \in U$ it can be seen that $U \otimes \mathbb{F} \cong \mathbb{F} \otimes U \cong U$.

(iii) For any fixed $u \in U$, define a map $f_u : V \times W \to (U \otimes V) \otimes W$ such that $f_u(v, w) = (u \otimes v) \otimes w$, which is bilinear. Hence, there exists a unique linear mapping $L_u : V \otimes W \to (U \otimes V) \otimes W$. Further, for a fixed $v \in V$ and $w \in W$, the map $f_{(v,w)} : U \to (U \otimes V) \otimes W$ such that $f_{(v,w)}(u) = (u \otimes v) \otimes w$ is linear. Now define $g : U \times (V \otimes W) \to (U \otimes V) \otimes W$ such that $g(u, v \otimes w) = L_u(v \otimes w) = f_{(v,w)}(u)$. Then it can be easily seen that g is bilinear, that is, linear in both the arguments. For any $u_1, u_2 \in U, \alpha, \beta \in \mathbb{F}$

$$
\begin{aligned}
g(\alpha u_1 + \beta u_2, v \otimes w) &= f_{(v,w)}(\alpha u_1 + \beta u_2) \\
&= \alpha f_{(v,w)}(u_1) + \beta f_{(v,w)}(u_2) \\
&= \alpha g(u_1, v \otimes w) + \beta g(u_2, v \otimes w)
\end{aligned}
$$

and

$$g\big(u, \alpha(v_1 \otimes w_1) + \beta(v_2 \otimes w_2)\big) = L_u\big(\alpha(v_1 \otimes w_1) + \beta(v_2 \otimes w_2)\big)$$
$$= \alpha L_u(v_1 \otimes w_1) + \beta L_u(v_2 \otimes w_2)$$
$$= \alpha g(u, v_1 \otimes w_1) + \beta g(u, v_2 \otimes w_2).$$

Thus, there exists a linear mapping $L_g : U \otimes (V \otimes W) \to (U \otimes V) \otimes W$ such that $L_g(u \otimes (v \otimes w)) = (u \otimes v) \otimes w$. Using similar arguments a linear map $L_{g'} : (U \otimes V) \otimes W \to U \otimes (V \otimes W)$ can be obtained and it is easy to see that $L_{g'} L_g = I$ on $U \otimes (V \otimes W)$ and $L_g L_{g'} = I$ on $(U \otimes V) \otimes W$, i.e., L_g and $L_{g'}$ are inverses of each other and hence L_g is an isomorphism, i.e., $(U \otimes V) \otimes W \cong U \otimes (V \otimes W)$.

Using the isomorphism given in (iii) as identification, define the tensor product of three vector spaces as follows:

$$U \otimes V \otimes W = (U \otimes V) \otimes W = U \otimes (V \otimes W).$$

More generally, when an ordered set of vector spaces $\{V_1, V_2, \ldots, V_n\}$ is given, it can be seen that their tensor product does not depend on the associativity and is uniquely determined up to a (canonical) isomorphism and in general by dropping the parenthesis it can be simply written as

$$\bigotimes_{i=1}^{n} V_i = V_1 \otimes V_2 \otimes \cdots \otimes V_n.$$

By definition, there exists a canonical map

$$(v_1, v_2 \cdots, v_n) \mapsto v_1 \otimes v_2 \otimes \cdots \otimes v_n \in \bigotimes_{i=1}^{n} V_i$$

and the following induction formula holds:

$$V_1 \otimes V_2 \otimes, \ldots \otimes V_k = (V_1 \otimes V_2 \otimes \cdots \otimes V_{k-1}) \otimes V_k$$

with correspondence $v_1 \otimes v_2 \otimes \cdots \otimes v_k = (v_1 \otimes v_2 \otimes \cdots \otimes v_{k-1}) \otimes v_k$.

If $V_1, V_2, \ldots, V_n, W$ are vector spaces over a field $\mathbb{F}$, a function $\psi : V_1 \times V_2 \times \cdots \times V_n \to W$ is said to be *multilinear*(n-linear) if it is linear in each argument (when the others are kept fixed), that is,

$$\psi(v_1, v_2, \ldots, \alpha v_i + \beta v_i', \ldots, v_n) = \alpha \psi(v_1, v_2, \ldots, v_i, \ldots, v_n)$$
$$+ \beta \psi(v_1, v_2, \ldots, v_i', \ldots, v_n)$$

for all $v_1 \in V_1, v_2 \in V_2, \ldots, v_i, v_i' \in V_i, \ldots, v_n \in V_n$ for all $1 \leq i \leq n$, $\alpha, \beta \in \mathbb{F}$. The collection of all such functions is a vector space over $\mathbb{F}$.

A bilinear mapping is a special case of n-linear mapping for $n = 2$. The above defined tensor product $v_1 \otimes v_2 \otimes \cdots \otimes v_n$ is n-linear and let us state the universal property for n-linear mappings.

In particular, a n-linear map $f : V \times V \times \cdots \times V \to W$ is called symmetric if interchanging any two coordinate positions, nothing changes in f, i.e.,

$$f(v_1, v_2, \ldots, v_i, \ldots, v_j, \ldots, v_n) = f(v_1, v_2, \ldots, v_j, \ldots, v_i, \ldots, v_n)$$

for any $i \neq j$.

A n-linear map $f : V \times V \times \cdots \times V \to W$ is called antisymmetric or skew symmetric if interchanging any two coordinate positions introduces a factor of (-1), i.e.,

$$f(v_1, v_2, \ldots, v_i, \ldots, v_j, \ldots, v_n) = -f(v_1, v_2, \ldots, v_j, \ldots, v_i, \ldots, v_n)$$

for any $i \neq j$.

A n-linear map $f : V \times V \times \cdots \times V \to W$ is called alternate or alternating if

$$f(v_1, v_2, \ldots, v_n) = 0$$

whenever any two of the vectors v_i are equal.

It is to be noted that if $char(\mathbb{F}) = 2$, then alternate $\Longrightarrow$ symmetric $\Longleftrightarrow$ skew symmetric. But if $char(\mathbb{F}) \neq 2$, then alternate $\Longleftrightarrow$ skew symmetric.

The pair $(V_1 \otimes V_2 \otimes \cdots \otimes V_n, \psi)$, where $\psi : V_1 \times V_2 \times \cdots \times V_n \to V_1 \otimes V_2 \otimes \cdots \otimes V_n$ is multilinear mapping defined by $\psi(v_1, v_2, \ldots, v_n) = v_1 \otimes v_2 \otimes \cdots \otimes v_n$ has the following universal property: If $f : V_1 \times V_2 \times \cdots \times V_n \to X$ is any multilinear function from $V_1 \times V_2 \times \cdots \times V_n$ to a vector space X over $\mathbb{F}$ then there exists a unique linear transformation $h : V_1 \otimes V_2 \otimes \cdots \otimes V_n \to X$ such that $h \circ \psi = f$, that is, the following diagram commutes:

$$
\begin{array}{ccc}
V_1 \times V_2 \times \cdots \times V_n & \xrightarrow{\ \psi\ } & V_1 \otimes V_2 \otimes \cdots \otimes V_n \\
 & {\scriptstyle f} \searrow & \downarrow {\scriptstyle h} \\
 & & X
\end{array}
$$

Exercises

1. If U and V are finite dimensional vector spaces over the same field, then show that $U \otimes V$ can be represented as the dual of the space of bilinear forms on $U \oplus V$.
2. Let U and V be vector spaces over the same field. If $V = V_1 \oplus V_2$, then show that $U \otimes V = (U \otimes V_1) \oplus (U \otimes V_2)$.

3. Show that the tensor product of $\mathbb{P}_3$ and $\mathbb{R}^2$ is $M_{2\times4}(\mathbb{R})$.
4. If $u = (1, i)$ and $v = (i, 1, -1)$ are vectors in $\mathbb{C}^2$ and $\mathbb{C}^3$, respectively, then find the representation of $u \otimes v$ in terms of the basis $e_i \otimes e_j$ for $\mathbb{C}^2 \otimes \mathbb{C}^3$.
5. Let U_1, U_2 be subspaces of a vector space U. Then show that

$$(U_1 \otimes U) \cap (U_2 \otimes U) \cong (U_1 \cap U_2) \otimes U.$$

6. Let U_1 and V_1 be subspaces of vector space U and V, respectively. Show that

$$(U_1 \otimes V) \cap (U \otimes V_1) \cong U_1 \otimes V_1.$$

7. Let $U_1, U_2 \subseteq U$ and $V_1, V_2 \subseteq V$ be subspaces of vector spaces U and V respectively. Show that

$$(U_1 \otimes V_1) \cap (U_2 \otimes V_2) \cong (U_1 \cap U_2) \otimes (V_1 \cap V_2).$$

8. Find examples of two vector spaces U and V and a nonzero vector $x \in U \otimes V$ such that at least two distinct (not including order of the terms) representation of the form $x = \sum_{i=1}^{n} u_i \otimes v_i$, where the u_i's are linearly independent and so are v_i's.
9. Let $\mathbb{R}^3$ and $\mathbb{R}^2$ be vector spaces over $\mathbb{R}$. Show that $\mathbb{P}_5 = \{\alpha_0 + \alpha_1 x + \alpha_2 x^2 + \alpha_3 x^3 + \alpha_4 x^4 + \alpha_5 x^5 | \alpha_0, \alpha_1, \alpha_2, \alpha_3, \alpha_4, \alpha_5 \in \mathbb{R}\}$ is a tensor product of $\mathbb{R}^3$ and $\mathbb{R}^2$.
(Hint: Consider associated bilinear map $f : \mathbb{R}^3 \times \mathbb{R}^2 \to \mathbb{P}_5$ such that

$$f\big((\alpha_1, \alpha_2, \alpha_3), (\beta_1, \beta_2)\big) = \alpha_1\beta_1 + \alpha_1\beta_2 x + \alpha_2\beta_1 x^2 + \alpha_2\beta_2 x^3 \\ + \alpha_3\beta_1 x^4 + \alpha_3\beta_2 x^5,$$

for every $\big((\alpha_1, \alpha_2, \alpha_3), (\beta_1, \beta_2)\big) \in \mathbb{R}^3 \times \mathbb{R}^2)$.

9.2 Tensor Product of Linear Transformations

Let $T_1 : U_1 \to V_1$ and $T_2 : U_2 \to V_2$ be any two arbitrarily given linear maps and consider the tensor products $U_1 \otimes U_2$ and $V_1 \otimes V_2$ together with their tensor maps τ_1 and τ_2. Define a map $g : U_1 \times U_2 \to V_1 \times V_2$ such that $g(u_1, u_2) = (T_1(u_1), T_2(u_2))$ for every $(u_1, u_2) \in U_1 \times U_2$. As $\tau_2 : V_1 \times V_2 \to V_1 \otimes V_2$ is a bilinear map, it follows that $\tau_2 \circ g$ is obviously a bilinear map. By definition of tensor product $U_1 \otimes U_2$, there exists a unique linear map $f : U_1 \otimes U_2 \to V_1 \otimes V_2$ such that $f \circ \tau_1 = \tau_2 \circ g$ holds in the following rectangle.

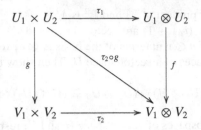

The unique map f is called the tensor product of linear maps T_1 and T_2 and usually f is represented by $f = T_1 \otimes T_2$. Here, we have $(f \circ \tau_1)(u_1, u_2) = (\tau_2 \circ g)(u_1, u_2)$, i.e., $f(u_1 \otimes u_2) = T_1(u_1) \otimes T_2(u_2)$. We also write as $(T_1 \otimes T_2)(u_1 \otimes u_2) = T_1(u_1) \otimes T_2(u_2)$.

Remark 9.13 (i) Let $I_U : U \to U$ and $I_V : V \to V$ be the identity linear maps on U and V respectively. Then $I_U \otimes I_V = I_{U \otimes V}$. Since $(I_U \otimes I_V)(u \otimes v) = I_U(u) \otimes I_V(v) = u \otimes v = I_{U \otimes V}(u \otimes v)$.

(ii) Let $\mathbf{0} : U_1 \to V_1$ be the zero linear map and $T : U_2 \to V_2$ be any arbitrary linear map then $\mathbf{0}' = \mathbf{0} \otimes T$, $T \otimes \mathbf{0} = \mathbf{0}''$, where $\mathbf{0}' : U_1 \otimes U_2 \to V_1 \otimes V_2$ is the zero linear map and $\mathbf{0}'' : U_2 \otimes U_1 \to V_2 \otimes V_1$ is the zero linear map. $(\mathbf{0} \otimes T)(u_1 \otimes u_2) = \mathbf{0}(u_1) \otimes T(u_2) = 0 \otimes T(u_2) = 0 = \mathbf{0}'(u_1 \otimes u_2)$ gives the clue.

Example 9.14 Let $D : \mathbb{R}[x] \to \mathbb{R}[x]$ be the linear map, which is called derivation and $\int : \mathbb{R}[y] \to \mathbb{R}[y]$ be the linear map, which is known as integration operator. We shall determine $D \otimes \int$.

Since, we know that $\mathbb{R}[x] \otimes \mathbb{R}[y] = \mathbb{R}[x, y]$, $D \otimes \int : \mathbb{R}[x] \otimes \mathbb{R}[y] \to \mathbb{R}[x] \otimes \mathbb{R}[y]$, i.e., $D \otimes \int : \mathbb{R}[x, y] \to \mathbb{R}[x, y]$, given by $(D \otimes \int)(p(x) \otimes q(y)) = (Dp(x)) \otimes (\int q(y))$.
  In other words,

$$(D \otimes \int)\left(\sum_{i \geq 0, j \geq 0}^{finite} \alpha_{ij} x^i y^j \right) = \sum_{i \geq 0, j \geq 0}^{finite} \alpha_{ij} D(x^i) \int y^j$$

$$= \sum_{i \geq 0, j \geq 0}^{finite} \alpha_{ij} i x^{i-1} \frac{y^{j+1}}{j+1}$$

$$= \sum_{i \geq 0, j \geq 0}^{finite} \frac{i}{j+1} \alpha_{ij} x^{i-1} y^{j+1}.$$

Here, it is to be noted that x and y do not commute.

Proposition 9.15 *Let* $T_1 : U_1 \to V_1$, $T_2 : V_1 \to W_1$, $S_1 : U_2 \to V_2$ *and* $S_2 : V_2 \to W_2$ *be linear maps. Then* $(T_2 \circ T_1) \otimes (S_2 \circ S_1) = (T_2 \otimes S_2) \circ (T_1 \otimes S_1)$.

Proof Clearly, $T_2 \circ T_1 : U_1 \to W_1$ and $S_2 \circ S_1 : U_2 \to W_2$ are linear maps. Hence by definition of tensor product of linear maps, we have $(T_2 \circ T_1) \otimes (S_2 \circ S_1) : U_1 \otimes$

$U_2 \to W_1 \otimes W_2$. On the other hand, we also have $T_1 \otimes S_1 : U_1 \otimes U_2 \to V_1 \otimes V_2$ and $T_2 \otimes S_2 : V_1 \otimes V_2 \to W_1 \otimes W_2$ as linear maps. As a result $(T_2 \otimes S_2) \circ (T_1 \otimes S_1) :$ $U_1 \otimes U_2 \to W_1 \otimes W_2$ is also a linear map. Consider, for any $u_1 \in U_1$, $u_2 \in U_2$,

$$\Big((T_2 \circ T_1) \otimes (S_2 \circ S_1)\Big)(u_1 \otimes u_2) = (T_2 \circ T_1)(u_1) \otimes (S_2 \circ S_1)(u_2)$$
$$= T_2(T_1(u_1)) \otimes S_2(S_1(u_2))$$
$$= (T_2 \otimes S_2)\big(T_1(u_1) \otimes S_1(u_2)\big)$$
$$= (T_2 \otimes S_2)\big((T_1 \otimes S_1)(u_1 \otimes u_2)\big)$$
$$= \big((T_2 \otimes S_2) \circ (T_1 \otimes S_1)\big)(u_1 \otimes u_2).$$

This implies that

$$(T_2 \circ T_1) \otimes (S_2 \circ S_1) = (T_2 \otimes S_2) \circ (T_1 \otimes S_1).$$

Theorem 9.16 *Let U_1, U_2, V_1, V_2 be vector spaces over the same field $\mathbb{F}$. If $U_1 \cong U_2$ and $V_1 \cong V_2$, then $U_1 \otimes V_1 \cong U_2 \otimes V_2$.*

Proof Given that $U_1 \cong U_2$ and $V_1 \cong V_2$. This confirms that there exist isomorphisms $T_1 : U_1 \to U_2$ and $T_2 : V_1 \to V_2$. Using the definition of tensor product of linear maps, we have $T_1 \otimes T_2 : U_1 \otimes V_1 \to U_2 \otimes V_2$, which is a linear map given as $(T_1 \otimes T_2)(u_1 \otimes v_1) = T_1(u_1) \otimes T_2(v_1)$. We claim that $T_1 \otimes T_2$ is a bijective map. For ontoness of $T_1 \otimes T_2$, let $u_2 \otimes v_2 \in U_2 \otimes V_2$. Since T_1 and T_2 are onto, there exist $u_1' \in U_1$ and $v_1' \in V_1$ such that $T_1(u_1') = u_2$ and $T_2(v_1') = v_2$. Obviously $(T_1 \otimes T_2)(u_1' \otimes v_1') = T_1(u_1') \otimes T_2(v_1') = u_2 \otimes v_2$. This shows that $T_1 \otimes T_2$ is onto. Before proving that $T_1 \otimes T_2$ is injective, we will prove a fact, i.e., kernel of $T_1 \otimes T_2$ is the subspace K of $U_1 \otimes V_1$ generated by the elements $x \otimes y$ of $U_1 \otimes V_1$ with $x \in Ker(T_1)$ or $y \in Ker(T_2)$. In otherwords,

$$Ker(T_1 \otimes T_2) = K = \langle\{x \otimes y \in U_1 \otimes V_1 | x \in Ker(T_1) \text{ or } y \in Ker(T_2)\}\rangle.$$

Let $L = \{x \otimes y \in U_1 \otimes V_1 | x \in Ker(T_1) \text{ or } y \in Ker(T_2)\}$. For each $x \otimes y \in L$, we have $(T_1 \otimes T_2)(x \otimes y) = T_1(x) \otimes T_2(y)$ but as $x \in Ker(T_1)$ or $y \in Ker(T_2)$, we conclude that $(T_1 \otimes T_2)(x \otimes y) = 0$. This implies that $x \otimes y \in Ker(T_1 \otimes T_2)$, i.e., $L \subseteq Ker(T_1 \otimes T_2)$. Since K is the smallest subspace of $U_1 \otimes V_1$, containing L, we arrive at $K \subseteq Ker(T_1 \otimes T_2)$. Next our target is to show that $Ker(T_1 \otimes T_2) \subseteq K$. Consider the quotient homomorphism $q : U_1 \otimes V_1 \to \frac{U_1 \otimes V_1}{K}$. This homomorphism q induces another homomorphism $q^* : \frac{U_1 \otimes V_1}{K} \to U_2 \otimes V_2$, such that $q^*\big((u \otimes v) + K\big) = T_1(u) \otimes T_2(v)$. Here, we have the following commutative diagram, i.e., $q^* q = T_1 \otimes T_2$.

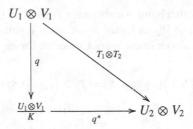

Now, we show that q^* is an injective homomorphism. For this purpose, we construct a function $g : U_2 \times V_2 \to \frac{U_1 \otimes V_1}{K}$ such that $g(x_2, y_2) = (x_1 \otimes y_1) + K$; where $T_1(x_1) = x_2$, $T_2(y_1) = y_2$. The existence of such elements $x_1 \in U_1, y_1 \in V_1$ is guaranteed because both T_1 and T_2 are onto. We claim that this association is a map. We show that choice of $x_1 \in U_1$, $y_1 \in V_1$ does not effect the value of $(x_1 \otimes y_1) + K$. For this let, we have another $x_1' \in U_1$, $y_1' \in V_1$ such that $T_1(x_1') = x_2$ and $T_2(y_1') = y_2$. This shows that $T_1(x_1 - x_1') = 0$ and $T_2(y_1 - y_1') = 0$, i.e., $x_1 - x_1' \in Ker T_1$ and $y_1 - y_1' \in Ker T_2$. Suppose that $x_1 - x_1' = \eta$ and $y_1 - y_1' = \delta$, for some $\eta \in Ker(T_1)$ and $\delta \in Ker(T_2)$. This implies that $x_1 = x_1' + \eta$ and $y_1 = y_1' + \delta$. In this case, since, $\eta \in Ker(T_1)$ and $\delta \in Ker(T_2)$

$$
\begin{aligned}
x_1 \otimes y_1 - x_1' \otimes y_1' &= (x_1' + \eta) \otimes (y_1' + \delta) - x_1' \otimes y_1' \\
&= x_1' \otimes y_1' + x_1' \otimes \delta + \eta \otimes y_1' + \eta \otimes \delta - x_1' \otimes y_1' \\
&= x_1' \otimes \delta + \eta \otimes y_1' + \eta \otimes \delta.
\end{aligned}
$$

Hence, we conclude that $x_1' \otimes \delta + \eta \otimes y_1' + \eta \otimes \delta \in K$ and hence $x_1 \otimes y_1 - x_1' \otimes y_1' \in K$. This forces us to conclude that $(x_1 \otimes y_1 + K) = (x_1' \otimes y_1') + K$. Thus above arguments are sufficient to say that g is a well-defined map. Next we prove that g is a bilinear map. For this, let $\alpha_1, \ \alpha_2 \in \mathbb{F}, \ x_3, \ x_4 \in U_2, \ y_3 \in V_2$. Since, both T_1 and T_2 are onto, there exists $x_3', \ x_4' \in U_1, \ y_3' \in V_1$ such that $T_1(x_3') = x_3, \ T_1(x_4') = x_4, \ T_2(y_3') = y_3$. Also, we have $T_1(\alpha_1 x_3' + \alpha_2 x_4') = \alpha_1 x_3 + \alpha_2 x_4$. Thus,

$$
\begin{aligned}
g(\alpha_1 x_3 + \alpha_2 x_4, y_3) &= \big((\alpha_1 x_3' + \alpha_2 x_4') \otimes y_3'\big) + K \\
&= \big(\alpha_1(x_3' \otimes y_3') + \alpha_2(x_4' \otimes y_3')\big) + K \\
&= \alpha_1\big((x_3' \otimes y_3') + K\big) + \alpha_2\big((x_4' \otimes y_3') + K\big) \\
&= \alpha_1 g(x_3, y_3) + \alpha_2 g(x_4, y_3).
\end{aligned}
$$

Hence, g is linear in the first coordinate. Similarly, we can show that g is linear in the second coordinate also. Thus g is a bilinear map. By the definition of $U_2 \otimes V_2$, there exists a unique linear map $h : U_2 \otimes V_2 \to \frac{U_1 \otimes V_1}{K}$ such that following diagram commutes, i.e., $h(u_2 \otimes v_2) = (u_1'' \otimes v_1'') + K$, where $T_1(u_1'') = u_2$ and $T_2(v_1'') = v_2$.

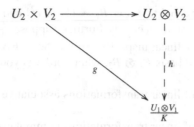

Here, we observe that $h \circ q^* : \frac{U_1 \otimes V_1}{K} \to \frac{U_1 \otimes V_1}{K}$ is a homomorphism such that

$$
\begin{aligned}
(h \circ q^*)(u_1 \otimes v_1 + K) &= h\big(q^*(u_1 \otimes v_1 + K)\big) \\
&= h\big(T_1(u_1) \otimes T_2(v_1)\big) \\
&= (u_1 \otimes v_1) + K.
\end{aligned}
$$

This shows that $h \circ q^*$ is the identity map. Hence, it is bijective map as a result q^* is one-to-one. Thus $Ker q^* = \{K\}$. Let us suppose that $z \otimes t \in Ker(T_1 \otimes T_2)$. This implies that $(T_1 \otimes T_2)(z \otimes t) = 0 \implies T_1(z) \otimes T_2(t) = 0$. Hence, $q^*\big((z \otimes t) + K\big) = 0$, i.e., $(z \otimes t) + K \in Ker(q^*)$. But since $Ker(q^*) = \{K\}$, this implies that $(z \otimes t) + K = K$, i.e., $z \otimes t \in K$. Thus, we have proved that $Ker(T_1 \otimes T_2) \subseteq K$. Finally, we have shown that $Ker(T_1 \otimes T_2) = K$. Given that T_1 and T_2 are injective, i.e., $Ker T_1 = \{0\}$ and $Ker T_2 = \{0\}$. As a result

$$
Ker(T_1 \otimes T_2) = \langle x \otimes y \in U_1 \otimes V_1 | x \in \{0\} \text{ or } y \in \{0\}\rangle.
$$

Therefore, $Ker(T_1 \otimes T_2) = \{0\}$. This implies that $T_1 \otimes T_2$ is injective. Finally, we have shown that $T_1 \otimes T_2$ is an isomorphism and thus $U_1 \otimes V_1 \cong U_2 \otimes V_2$.

Exercises

1. Let U_1, V_1, U_2, V_2 be vector spaces of finite dimensions. If $T_1 : U_1 \to V_1$ and $T_2 : U_2 \to V_2$ be linear maps, then prove that $rank(T_1 \otimes T_2) = rank T_1 rank T_2$.

2. Let U_1, V_1, U_2, V_2 be vector spaces of finite dimension n. Let $T_1 : U_1 \to V_1$ and $T_2 : U_2 \to V_2$ be nonsingular linear maps. Prove that $T_1 \otimes T_2$ is also nonsingular and $(T_1 \otimes T_2)^{-1} = T_1^{-1} \otimes T_2^{-1}$.

3. Let $\frac{d}{dx} : \mathbb{R}[x] \to \mathbb{R}[x]$ and $\frac{d}{dy} : \mathbb{R}[y] \to \mathbb{R}[y]$ be linear maps, known as derivations. Prove that $\frac{d}{dx} \otimes \frac{d}{dy} \equiv \frac{\partial^2}{\partial x \partial y}$. Here, $\mathbb{R}[x]$ and $\mathbb{R}[y]$ represent the vector spaces of all real polynomials in x and y, respectively. Here, x and y do not commute.

4. Let $\mathbb{P}_2(x)$ be the vector space of all polynomials over $\mathbb{R}$ in x of degree less than or equal to 2 and $\mathbb{P}_3(y)$ be the vector space of all polynomials over $\mathbb{R}$ in y

of degree less than or equal to 3. Let $B_1 = \{1, x, x^2\}$ and $B_2 = \{1, y, y^2, y^3\}$ be bases of $\mathbb{P}_2(x)$ and $\mathbb{P}_3(y)$, respectively. Further, suppose $\frac{d}{dx} : P_2(x) \to P_2(x)$ and $\frac{d}{dy} : P_2(y) \to P_2(y)$ be linear maps. Determine the matrix of linear map $\frac{d}{dx} \otimes \frac{d}{dy}$ with respect to ordered basis $B_1 \otimes B_2$, where order is your choice.

5. Is the tensor product of linear transformations associative? Justify your claim.

6. Is the tensor product of linear transformations commutative? Justify your claim.

7. Let U and V be vector spaces over $\mathbb{F}$, having dimensions m and n, respectively, where $\mathbb{F}$ is an algebraically closed field. Let $T_1 : U \to U$ be a linear map, having eigen values $\lambda_1, \lambda_2, \ldots, \lambda_m$ and $T_2 : V \to V$ be another linear map, having eigen values $\mu_1, \mu_2, \ldots, \mu_n$. Show that all the eigen values of $T_1 \otimes T_2$ will be given by $\lambda_i \mu_j$; $1 \le i \le m$, $1 \le j \le n$.

9.3 Tensor Algebra

In the previous section, we studied how to obtain the tensor product of finite number of vector spaces over the same field $\mathbb{F}$. Given a vector space V, we want to construct an algebra over $\mathbb{F}$. Before, doing that we want to give a brief idea of external direct product and external direct sum of an arbitrarily set of vector spaces over the same field $\mathbb{F}$. Let $\mathfrak{F} = \{V_i | i \in I\}$ be an arbitrarily given set of vector spaces V_i, where I is an indexing set. Let P denotes the set of all maps that can be defined from I to the union M of the sets V_i such that $f(i) \in V_i$ holds for every $i \in I$, i.e., $P = \{f : I \longrightarrow M = \cup_{i \in I} V_i$ and $f(i) \in V_i$ holds for every $i \in I\}$. Define addition and scalar multiplication in P as: for any $f, g \in P, \alpha \in \mathbb{F}$, the functions $f + g : I \longrightarrow M$ and $\alpha f : I \longrightarrow M$ are defined by $(f + g)(i) = f(i) + g(i)$ and $(\alpha f)(i) = \alpha(f(i))$ for each $i \in I$. One can easily verify that P is a vector space over F with regard to these operations. P is known as the external direct product of vector spaces $V_i, i \in I$. It is usually denoted by $\prod$, i.e., $P = \prod_{i \in I} V_i$.

Now, we construct a special type of subspace of the vector space P. Let us consider a subset S of P, which consists of all $f \in P$ such that $f(i) = 0$ holds for all except finite number of indices $i \in I$. It is obvious to observe that S is a subspace of P. This subspace is known as the external direct sum of given set $\mathfrak{F}$ of vector spaces. Usually, we denote it by $\bigoplus^{\text{ext}}$, i.e., $S = \bigoplus_{i \in I}^{\text{ext}} V_i$. It is to be remarked here that if the indexing set I is finite, then $P = S$. And, in this case, each of them can be called either the external direct product or the external direct sum of the given set $\mathfrak{F}$ of vector spaces.

For each index $j \in I$, we define a map $f_j : V_j \longrightarrow S$ such that $(f_j)(v) \in S$ for every $v \in V_j$, where $(f_j)(v) : I \longrightarrow M$ is a map defined by $((f_j)(v))(i) = v$ if $i = j$ and 0, otherwise. It can be easily shown that f_j is an injective linear transformation. Thus, for each index $j \in I$, we can identify V_j with its image $f_j(V_j)$ in S

and in this sense, we can say that V_j is a subspace of S for each index $j \in I$. Let $f \in S$ and suppose that $0 \neq f(i_1) = v_1 \in V_{i_1}, 0 \neq f(i_2) = v_2 \in V_{i_2}, \ldots, 0 \neq f(i_r) = v_r \in V_{i_r}$ but $f(i) = 0$ for each $i \in I - \{i_1, i_2, \ldots, i_r\}$, where r is any non-negative integer. Now, we can write $f = (f_{i_1})(v_1) + (f_{i_2})(v_2) + \cdots + (f_{i_r})(v_r)$. Here, the vectors $(f_{i_1})(v_1), (f_{i_2})(v_2), \ldots, (f_{i_r})(v_r)$ can be identified by the vectors $v_1, v_2, \ldots, v_r$ respectively. Thus, finally we can write $f \in S$ as $f = v_1 + v_2 + \cdots + v_r$, i.e., $f = \sum_{i=1}^{r} v_i$ in identified sense. Now, we start constructing an algebra over $\mathbb{F}$.

Let V be a vector space over $\mathbb{F}$. Define the symbol V_q^p by $V_q^p = V_1 \otimes V_2 \otimes \cdots \otimes V_p \otimes \widehat{V}_1 \otimes \widehat{V}_2 \otimes \cdots \otimes \widehat{V}_q$ where $V_i = V, i = 1, 2, \ldots, p$ and $\widehat{V}_j = \widehat{V}, j = 1, 2, \ldots, q$. Also, define $V_0^0 = \mathbb{F}$, $V_0^p = V_1 \otimes V_2 \otimes \cdots \otimes V_p$, $V_q^0 = \widehat{V}_1 \otimes \widehat{V}_2 \otimes \cdots \otimes \widehat{V}_q$, where $V_i = V, i = 1, 2, \ldots, p$, $\widehat{V}_j = \widehat{V}, j = 1, 2, \ldots, q$. Thus, the symbols V_q^p have been defined for each nonnegative integers p and q. Obviously, $\mathfrak{F} = \{V_q^p,$ where $(p, q) \in \mathbb{N} \cup \{0\} \times \mathbb{N} \cup \{0\}\}$ is a set of vector spaces, where $\mathbb{N} \cup \{0\} \times \mathbb{N} \cup \{0\}$ is an indexing set. Let $T(V)$ be the external direct sum of this set of vector spaces, i.e., $T(V) = \bigoplus_{(p,q) \in I}^{ext} V_q^p$, where $I = \mathbb{N} \cup \{0\} \times \mathbb{N} \cup \{0\}$. The vector space $T(V)$ is known as the tensor space of V and an element of $T(V)$ is called a tensor. By the previous arguments, it is clear that V_q^p are subspaces of $T(V)$ for each non-negative integers p and q. If p and q are positive integers, then an element of V_0^p is called a contravariant tensor, an element of V_q^0 is called a covariant tensor and an element of V_q^p is called as p times contravariant and q times covariant tensor or a mixed tensor of the type (p, q).

Define a multiplication in $T(V)$ as: let $x, y \in T(V)$, clearly $x = \sum_{i=1}^{r} \sum_{j=1}^{s} v_{q_j}^{p_i}, y = \sum_{i=1}^{m} \sum_{j=1}^{n} v_{q_j'}^{p_i'}$, where $v_{q_j}^{p_i} \in V_{q_j}^{p_i}, v_{q_j'}^{p_i'} \in V_{q_j'}^{p_i'}$ then

$$xy = \sum_{a=0}^{p_r+p_m'} \sum_{b=0}^{q_s+q_n'} \left(\sum_{p_i+p_i'=a} \sum_{q_j+q_j'=b} v_{q_j}^{p_i} \otimes v_{q_j'}^{p_i'} \right),$$

where $v_{q_j}^{p_i} \otimes v_{q_j'}^{p_i'} \in V_{q_j}^{p_i} \otimes V_{q_j'}^{p_i'} \cong V_{q_j+q_j'}^{p_i+p_i'}$. This can be easily seen that this multiplication is a binary operation in $T(V)$. Actually, this multiplication is a bilinear map from $T(V) \times T(V)$ to $T(V)$. It can be verified that the vector space $T(V)$ forms an algebra over $\mathbb{F}$ with regard to the multiplication defined above, which is in general noncommutative, infinite dimensional and with the identity V_0^0. This algebra constructed above is known as the tensor algebra of V and is usually denoted by $T(V)$. We will denote the set of all contravariant and covariant tensors in $T(V)$ by $T_0(V)$ and $T^0(V)$, respectively.

Theorem 9.17 $T_0(V)$ and $T^0(V)$ form subalgebras of the tensor algebra $T(V)$.

Proof Clearly $T_0(V) = \{ \sum_{finite} v_0^p | v_0^p \in V_0^p, p = 1, 2, 3, \ldots \}$. Let $x, y \in T_0(V)$. Thus $x = \sum_{i=1}^{r} v_0^{p_i}$ and $y = \sum_{i=1}^{m} v_0^{p_i'}$. For any $\alpha, \beta \in \mathbb{F}$, we have $\alpha x + \beta y = \sum_{i=1}^{r} \alpha v_0^{p_i} +$

$\sum_{i=1}^{m} \beta v_0^{p_i'}$. As $V_0^{p_i}$ and $V_0^{p_i'}$ are subspaces of $T(V)$, we have, say $w_0^{p_i} = \alpha v_0^{p_i} \in V_0^{p_i}$ and $w_0^{p_i'} = \beta v_0^{p_i'} \in V_0^{p_i'}$. Thus $\alpha x + \beta y = \sum_{i=1}^{r} w_0^{p_i} + \sum_{i=1}^{m} w_0^{p_i'}$. Which shows that $\alpha x + \beta y \in T_0(V)$, as a result $T_0(V)$ is a subspace of $T(V)$. Taking $\alpha = 1, \beta = -1$, we have $x - y \in T_0(V)$. Here $xy = \sum_{a=0}^{p_r + p_m'} (\sum_{p_i + p_i' = a} v_0^{p_i} \otimes v_0^{p_i'})$, where $v_0^{p_i} \otimes v_0^{p_i'} \in V_0^{p_i} \otimes V_0^{p_i'} \cong V_0^{p_i + p_i'}$. In turn, we conclude that $xy \in T_0(V)$ and hence $T_0(V)$ is a subring of $T(V)$. Only little observation shows that $T_0(V)$ is a subalgebra of $T(V)$. Using the similar arguments, it can be proved that $T^0(V)$ is also a subalgebra of $T(V)$.

Noting the definition of multiplication in $T(V)$, it is clear that the product of any two arbitrary tensors is completely determined by the product of tensors $x \in V_q^p$ and $y \in V_{q'}^{p'}$. Now, we see the calculus and the properties involved in the product of such tensors x and y. Let V be a finite dimensional vector space having dimension n. Let $\{e_i\}_{i=1}^{n}$ be a basis for V and $\{e^i\}_{i=1}^{n}$ be a dual basis for $\hat{V}$. By the previous section, we have $e_{i_1} \otimes e_{i_2} \otimes \cdots e_{i_p} \otimes e^{j_1} \otimes e^{j_2} \otimes \cdots \otimes e^{j_q}$, where i_k and j_m run independently over the integer 1 through n, is a basis for V_q^p. For the sake of conciseness, we write these tensors used in the basis as: $e_{i_1 i_2 \cdots i_p}^{j_1 j_2 \cdots j_q}$. In general, any element $x \in V_q^p$ is of the form

$$x = x_1 \otimes x_2 \otimes \cdots \otimes x_p \otimes y_1 \otimes y_2 \otimes \cdots \otimes y_q, \tag{9.1}$$

where $x_k = \sum_{i_k=1}^{n} x_k^{i_k} e_{i_k}, k = 1, 2, \ldots, p$; $y_m = \sum_{j_m=1}^{n} y_{j_m}^m e^{j_m}$, $m = 1, 2, \ldots, q$. If we put the values of x_k and y_m in the above expression for x, then a vast set of summations occurs. Thus to avoid all these summation signs, we separate these signs but it is understood. Now using the multilinearity, we obtain

$$x = x_1^{i_1} x_2^{i_2} \cdots x_p^{i_p} y_{j_1}^1 y_{j_2}^2 \cdots y_{j_q}^q e_{i_1 i_2 \cdots i_p}^{j_1 j_2 \cdots j_q}.$$

Using short notation, we can write the above expression as $x = \alpha_{j_1 j_2 \cdots j_q}^{i_1 i_2 \cdots i_p} e_{i_1 i_2 \cdots i_p}^{j_1 j_2 \cdots j_q}$. Here, it is understood that summation is carried out on each index. Thus, each tensor in V_q^p can be written in this form. Let $y = \beta_{h_1 h_2 \cdots h_{q'}}^{k_1 k_2 \cdots k_{p'}} e_{k_1 k_2 \cdots k_{p'}}^{h_1 h_2 \cdots h_{q'}}$. As a result

$$x \otimes y = \alpha_{j_1 j_2 \cdots j_q}^{i_1 i_2 \cdots i_p} \beta_{h_1 h_2 \cdots h_{q'}}^{k_1 k_2 \cdots k_{p'}} e_{i_1 i_2 \cdots i_p k_1 k_2 \cdots k_{p'}}^{j_1 j_2 \cdots j_q h_1 h_2 \cdots h_{q'}}.$$

A tensor x that can be expressed in the form (9.1) is known as decomposable tensor. Otherwise, tensor is known as indecomposable. Recalling the definition of a tensor, we conclude that, actually, each tensor is a finite sum of decomposable tensors. For example, the tensors

$$3e_1 \otimes e^1 \otimes e^2 \in V_2^1, -3e_1 \otimes e^1 + 10e_2 \otimes e^1 - 6e_1 \otimes e^2 + 20e_2 \otimes e^2$$

$$= (-3e_1 + 10e_2) \otimes (e^1 + 2e^2) \in V_1^1$$

etc. are decomposable tensors but tensors $5e_1 \otimes e^1 \otimes e^2 + e_1 \otimes e^1$, $e_1 \otimes e^1 + e^1 \otimes e^2$ etc. are indecomposable. Here, we notice that the product of any two tensors in $T(V)$ becomes a tensor of higher order, i.e., multiplication in $T(V)$ is an order enhancing phenomenon in tensors. Now, we introduce another phenomenon in $T(V)$, which reduces the orders of the tensors, that is known as contraction.

Contraction

Let V be any vector space and $T(V)$ be the tensor algebra of V. Define a mapping $f : V_1 \times V_2 \times \cdots \times V_p \times \widehat{V}_1 \times \widehat{V}_2 \times \cdots \times \widehat{V}_q \longrightarrow V_{q-1}^{p-1}$, where $V_i = V, i = 1, 2, \ldots, p$, $\widehat{V}_j = \widehat{V}, j = 1, 2, \ldots, q$ by

$$f(x_1, x_2, \ldots, x_p, y^1, y^2, \ldots, y^q)$$
$$= y^k(x_h)x_1 \otimes x_2 \otimes \cdots \otimes \widehat{x}_h \otimes x_p \otimes y^1 \otimes y^2 \otimes \cdots \otimes \widehat{y^k} \otimes \cdots \otimes y^q, \text{ where}$$

$\widehat{x}_h$ and $\widehat{y^k}$ indicate that these vectors are deleted from the tensor. Here, $y^k(x_h)$ represents the image of $x_h \in V$ under the linear functional $y^k \in \widehat{V}$. It can be verified that f is a $(p + q)$-linear mapping. Thus, using the definition of tensor product of finite number of vector spaces, there exists a unique linear map say $C_k^h : V_q^p \longrightarrow V_{q-1}^{p-1}$ such that

$$C_k^h(x_1 \otimes x_2 \otimes \cdots \otimes x_p \otimes y^1 \otimes y^2 \otimes \cdots \otimes y^q)$$
$$= y^k(x_h)x_1 \otimes x_2 \otimes \cdots \otimes \widehat{x}_h \otimes x_p \otimes y^1 \otimes y^2 \otimes \cdots \otimes \widehat{y^k} \otimes \cdots \otimes y^q.$$

Clearly, under this map, the orders of each tensor in V_q^p are being reduced. The mapping C_k^h is called a contraction of the hth contravariant index and the kth covariant index. Now, we examine some behaviors of C_k^h on the tensors belonging to V_q^p. For this, let V be of finite dimension n and further we will use some notations related with V and $T(V)$, which have the same meaning as they have in previous paragraphs. Consider the tensor $e_{i_1 i_2 \cdots i_p}^{j_1 j_2 \cdots j_q} \in V_q^p$, then $C_k^h(e_{i_1 i_2 \cdots i_p}^{j_1 j_2 \cdots j_q}) = e^{j_k}(e_{i_h})e_{i_1 i_2 \cdots \widehat{i}_h \cdots i_p}^{j_1 j_2 \cdots \widehat{j}_k \cdots j_q} = \delta_{i_h j_k} e_{i_1 i_2 \cdots \widehat{i}_h \cdots i_p}^{j_1 j_2 \cdots \widehat{j}_k \cdots j_q}$, where $\delta_{i_h j_k}$ stands for Kronecker's delta. If $x = \alpha_{j_1 j_2 \cdots j_q}^{i_1 i_2 \cdots i_p} e_{i_1 i_2 \cdots i_p}^{j_1 j_2 \cdots j_q}$, then using the linearity of C_k^h, we have $C_k^h(x) = \alpha_{j_1 j_2 \cdots j_{k-1} i_h j_{k+1} \cdots j_q}^{i_1 i_2 \cdots i_h \cdots i_p} e_{i_1 i_2 \cdots \widehat{i}_h \cdots i_p}^{j_1 j_2 \cdots \widehat{j}_k \cdots j_q}$. We consider some examples to clear these complicated symbols.

Let $x \in T(V)$ such that $x = e_1 \otimes e^2 - 5e_2 \otimes e^2 + e_1 \otimes e^2 - 7e_2 \otimes e^1$. Then $x \in V_1^1$ and $C_1^1(x) = -5$. Let $y \in T(V)$ such that

$$y = 5e_1 \otimes e_2 \otimes e^1 \otimes e^1 \otimes e^3 - 7e_1 \otimes e_2 \otimes e^1 \otimes e^1 \otimes e^3 + e_1 \otimes e_1 \otimes e^2 \otimes e^3 \otimes e^2.$$

Clearly $y \in V_3^2$ and $C_1^1(y) = 5e_2 \otimes e^1 \otimes e^3 - 7e_1 \otimes e^2 \otimes e^3$, $C_2^1(y) = 5e_2 \otimes e^1 \otimes e^3 - 7e_2 \otimes e^1 \otimes e^3$, $C_3^2(y) = 0$ but $C_2^3(y), C_4^1(y)$ etc. are undefined.

Example 9.18 Let V be a finite dimensional vector space with dimension n. Then V_1^1 is isomorphic to $\mathscr{A}(V)$ via the map $f, x \otimes y \longrightarrow f(x \otimes y)$ such that $f(x \otimes y)(v) = y(v)x$ for all $v \in V$, where $y(v)$ is the image of v under the linear functional y. Moreover, the contraction C_1^1 of any $(x \otimes y) \in V_1^1$ is precisely the trace of the linear operator $f(x \otimes y)$, where trace of a linear operator is defined as the trace of any matrix of the linear operator with regard to any ordered basis B of V.

Define a map $\eta : V \times \widehat{V} \longrightarrow \mathscr{A}(V)$ such that $\eta(x, y) = g(x, y)$, where $g(x, y)$ $(v) = y(v)x$ for all $v \in V$. Obviously, for any fixed x and y, $g(x, y)$ is a linear operator on V. It can be also easily verified that η is a bilinear map. Using the definition of tensor product, there exists a unique linear map $f : V_1^1$ to $\mathscr{A}(V)$ such that $f(x \otimes y)(v) = g(x, y)(v) = y(v)x$. Here, we observe that V_1^1 and $\mathscr{A}(V)$ have the same finite dimension. To prove that f is an isomorphism, it is only left to show that f is one-to-one. For this, let $z \in Kerf$, i.e., $z = \sum_{i=1}^{n} \sum_{j=1}^{n} \alpha_i^j (e_i \otimes e^j)$ because $e_i \otimes e^j$, $1 \leq i \leq n, 1 \leq j \leq n$ is a basis for V_1^1, where $\alpha_i^j \in \mathbb{F}$. This implies that $f(\sum_{i=1}^{n} \sum_{j=1}^{n} \alpha_i^j (e_i \otimes e^j) = 0$. Due to linearity of f, we have $\sum_{i=1}^{n} \sum_{j=1}^{n} \alpha_i^j f(e_i \otimes e^j) = 0$. It follows that $[\sum_{i=1}^{n} \sum_{j=1}^{n} \alpha_i^j f(e_i \otimes e^j)](v) = 0$ for all $v \in V$, i.e., $\sum_{i=1}^{n} \sum_{j=1}^{n} \alpha_i^j ((e^j(v))(e_i)) = 0$ for all $v \in V$. Using the fact that $\{e_1, e_2, \ldots, e_n\}$ is linearly independent set in V and varying v through all the vectors in the set $\{e_1, e_2, \ldots, e_n\}$, one concludes that $\alpha_i^j = 0$ for all i, j. In turn, we get $z = 0$ and hence f becomes injective. Hence V_1^1 is isomorphic to Hom (V, V). This shows that tensors in V_1^1 can be regarded as elements of $\mathscr{A}(V)$. In a similar fashion, one can show that tensors in V_2^1 can be regarded as elements of $\mathscr{A}(\widehat{V})$.

The contraction of $x \otimes y \in V_1^1$ is given by $C_1^1(x \otimes y) = y(x)$.

Case I: If at least one out of x and y is zero, then $C_1^1(x \otimes y) = 0$ and the operator $f(x \otimes y)$ will be the zero operator. As the trace of the zero operator is zero, result is obvious for this case.

Case II: Suppose that $x \neq 0$ and $y \neq 0$. As y is a nonzero linear functional, Ker $y \neq V$ and rank $y = 1$. Using rank nullity theorem, we have Nullity $y = n - 1$. Thus $V = $ Ker $y \bigoplus <u>$ for some nonzero $u \in V$. Now using the previous conclusion, one can show that the matrix of linear operator $f(x \otimes y)$ with regard to any ordered basis will have each of its diagonal entries equal to 0 except one that equals $y(x)$. Thus, trace of the linear operator equals $C_1^1(x \otimes y)$. This proves our result.

Using the same idea as in the previous example, each tensor in V_3^1 can be regarded as a linear transformation from V_0^3 into V. It can be easily seen that contraction need not be commutative. For example, let $x = \alpha_{j_1 j_2 j_3}^{i_1 i_2 i_3} e_{i_1 i_2 i_3}^{j_1 j_2 j_3}$ be an element of V_3^3. Then $C_1^2(x) = \alpha_{i_1 j_2 j_3}^{i_1 i_2 i_3} e_{i_1 i_3}^{j_2 j_3}$, $C_2^1(x) = \alpha_{j_1 i_1 j_3}^{i_1 i_2 i_3} e_{i_2 i_3}^{j_1 j_3}$, $C_1^2 \circ C_2^1(x) = \alpha_{i_2 i_1 j_3}^{i_1 i_2 i_3} e_{i_3}^{j_1}$, $C_2^1 \circ C_1^2(x) = \alpha_{i_1 i_1 j_3}^{i_1 i_2 i_3} e_{i_3}^{j_2}$. Clearly, this shows that $C_1^2 \circ C_2^1 \neq C_2^1 \circ C_1^2$. It is to be noted that product of contractions in one order may be defined but it may not be defined in the reverse order. The contractions C_1^1 are special in nature. Notice that if we take $x = \alpha_{j_1 j_2 j_3}^{i_1 i_2 i_3} e_{i_1 i_2 i_3}^{j_1 j_2 j_3}$, then $C_1^1(x) = \alpha_{i_1 j_2 j_3}^{i_1 i_2 i_3} e_{i_2 i_3}^{j_2 j_3}$, $C_1^1 \circ C_1^1(x) = \alpha_{i_1 i_2 j_3}^{i_1 i_2 i_3} e_{i_3}^{j_3}$, $C_1^1 \circ C_1^1 \circ C_1^1(x) = \alpha_{i_1 i_2 i_3}^{i_1 i_2 i_3}$. Thus $C_1^1 \circ C_1^1 \circ C_1^1$ maps V_3^3 into the scalars. Similarly, it follows that p copies of C_1^1 composed with each other map V_p^p into $\mathbb{F}$.

Exercises

1. If dimension of V is greater than 1, then prove that $T(V)$ is a ring, which is free from zero divisors.
2. Using only properties of tensor product, show that $V_q^p \otimes V_s^r \cong V_{q+s}^{p+r}$.
3. Find an isomorphism to show that V_2^1 is isomorphic to Hom $(\widehat{V}, V)$. Also determine the matrix of the elements of Hom $(\widehat{V}, \widehat{V})$.
4. Let V be an n-dimensional vector space. Let $X = \{x_i\}_{i=1}^n$ be a basis of V together with $X' = \{x^i\}_{i=1}^n$ as its dual basis. If $B = \{e_i\}_{i=1}^n$ is another basis of V together with $B' = \{e^i\}_{i=1}^n$ as its dual basis such that $e_i = \alpha_i^h$ and $e^j = \beta_k^j x^k$, then obtain the coefficients of an element of V_q^p for its expansion in the basis $x_{i_1 i_2 \cdots i_p}^{j_1 j_2 \cdots j_q}$.
5. Let $x, y \in V_1^1$ be decomposable. Find the conditions such that $x + y$ is also decomposable.
6. Prove that tensor algebra $T(V)$ is a graded algebra.

9.4 Exterior Algebra or Grassmann Algebra

Throughout this section, V represents a finite dimensional vector space over a field $\mathbb{F}$, where char $(\mathbb{F}) = 0$ and dim $(V) = n$. In this section, we define symmetric and antisymmetric or alternating tensors in V_0^p. Later, the constructions of symmetric tensor algebras $ST(V)$ and antisymmetric tensor algebras $AT(V)$ of the vector space V with their properties are interpreted. We have concluded this section by introducing the notions of exterior product or wedge product $\wedge$ and exterior algebra or Grassmann algebra $\bigwedge V$ of V.

Let S_p be the permutation group of p integers $1, 2, \ldots, p$. Let $\sigma \in S_p$. Define a map $g : V_1 \times V_2 \times \cdots \times V_p \longrightarrow V_0^p$, where $V_i = V$ for each i such that $1 \leq i \leq p$ and $g(x_1, x_2, \ldots, x_p) = x_{\sigma(1)} \otimes x_{\sigma(2)} \otimes \cdots \otimes x_{\sigma(p)}$. It can be shown that g is a p-linear mapping. Thus using the definition of tensor product, there exists a unique linear map, say $S_\sigma : V_0^p \longrightarrow V_0^p$, such that $S_\sigma(x_1 \otimes x_2 \otimes \cdots \otimes x_p) = x_{\sigma(1)} \otimes x_{\sigma(2)} \otimes \cdots \otimes x_{\sigma(p)}$. With the help of S_σ, we define symmetric and antisymmetric tensors in V_0^p. A tensor $x \in V_0^p$ is said to be symmetric if $S_\sigma(x) = x$ for every σ. If $S_\sigma(x) = x$ for some particular σ, then x is said to be symmetric with regard to σ. On the other hand, a tensor $x \in V_0^p$ is said to be antisymmetric or alternating if $S_\sigma(x) = (\text{Sign }\sigma)x$ for every σ. If $S_\sigma(x) = (\text{Sign }\sigma) x$ for some particular σ, then x is said to be antisymmetric or alternating with regard to σ, where Sign σ is 1 if σ is even and Sign σ is -1 if σ is odd.

Theorem 9.19 *Let $x \in V_0^p$. Then x is symmetric if and only if $S_\tau(x) = x$ for every transposition τ. The tensor x is antisymmetric if and only if $S_\tau(x) = -x$ for every transposition τ.*

Proof Let x be symmetric. Thus $S_\sigma(x) = x$ for every σ. But any transposition is also a particular permutation, hence $S_\tau(x) = x$. Conversely, suppose that $S_\tau(x) = x$ for every transposition τ. Let $\sigma \in S_p$. We know that each permutation can be written as product of transpositions. Thus $\sigma = \tau_1 \tau_2 \cdots \tau_m$ and hence $S_\sigma(x) = S_{(\tau_1 \tau_2 \cdots \tau_m)}(x) = x$, using the fact that $S_\sigma = S_{(\tau_1 \tau_2 \cdots \tau_m)} = S_{\tau_1} \circ S_{\tau_2} \circ \cdots \circ S_{\tau_m}$, we get as desired.

Let x be antisymmetric. Thus $S_\sigma(x) = (\text{Sign } \sigma)x$ for every σ and Sign σ is 1 if σ is even and Sign σ is -1 if σ is odd. If $\sigma = \tau$, then Sign $\sigma = -1$. We arrive at $S_\tau(x) = -x$. Conversely, let $S_\tau(x) = -x$ for every transposition τ. Let $\sigma \in S_p$. By using the above arguments we have $S_\sigma = S_{(\tau_1 \tau_2 \cdots \tau_m)} = S_{\tau_1} \circ S_{\tau_2} \circ \cdots \circ S_{\tau_m}$ and hence $S_\sigma(x) = (-1)^m x$. If σ is an odd permutation, then m will be an odd integer and hence $S_\sigma(x) = -x$. On the other hand if m is an even permutation, then m will be an even integer and hence $S_\sigma(x) = x$. Thus our result stands proved.

The set of all symmetric tensors of V_0^p is usually denoted by $ST^p(V)$, i.e., $ST^p(V) = \{x \in V_0^p | S_\sigma(x) = x, \text{ for all } \sigma \in S_p\}$. Let $x, y \in ST^p(V), \alpha, \beta \in \mathbb{F}$. Due to linearity of S_σ, we have $S_\sigma(\alpha x + \beta y) = \alpha S_\sigma(x) + \beta S_\sigma(y)$. The symmetry of x and y forces us to conclude that $\alpha x + \beta y \in ST^p(V)$. Thus $ST^p(V)$ is a subspace of V_0^p. The set of all antisymmetric tensors of V_0^p is usually denoted by $AT^p(V)$. The set of all antisymmetric tensors of V_0^p is usually denoted by $AT^p(V)$, i.e., $AT^p(V) = \{x \in V_0^p | S_\sigma(x) = (\text{Sign } \sigma)\, x, \text{ for all } \sigma \in S_p\}$, where Sign σ is 1 if σ is even and Sign σ is -1 if σ is odd. On the similar lines, one can show that $AT^p(V)$ is a subspace of V_0^p. The vector space $ST^p(V)$ is called the symmetric tensor space of V. On the other hand, the vector space $AT^p(V)$ is called the antisymmetric tensor space or exterior product space of V.

Theorem 9.20 *Let S and A be linear operators defined on V_0^p such that $S = \frac{1}{p!} \sum S_\sigma$, and $A = \frac{1}{p!} \sum (\text{Sign } \sigma)S_\sigma$, where the sum in both the operators is over all the elements of S_p. Then both S and A are projections. S is a projection on $ST^p(V)$ and A is a projection on $AT^p(V)$.*

Proof Let $x \in V_0^p$. Then for any $\xi \in S_p$, $S_\xi S(x) = \frac{1}{p!} \sum S_{\xi\sigma}(x)$. But S_p is a group. As σ runs over all the elements of S_p, so does $\xi\sigma$. This implies that $\sum S_{\xi\sigma}(x) = \sum S_\sigma(x)$. Hence $S_\xi S(x) = S(x)$. This implies that for any $x \in V_0^p$, $S(x)$ is symmetric. Furthermore, if x is symmetric, then $S_\sigma(x) = x$ for all $\sigma \in S_p$ and $S(x) = \frac{1}{p!} \sum S_\sigma(x) = \frac{1}{p!} \sum x = x$, because order of $S_p = p!$. Finally, we have shown that $S^2(x) = S(x)$ for every $x \in V_0^p$. Thus S is projection and it is projection on $ST^p(V)$.

Let τ be a transposition being an element of S_p. Then $\tau\sigma$ is odd if σ is even and $\tau\sigma$ is even if σ is odd. Thus sign $(\tau\sigma) = -$ sign (σ). Thus $S_\tau(A(x)) = \frac{1}{p!} \sum (\text{Sign } \sigma)S_{\tau\sigma}(x) = -\frac{1}{p!} \sum (\text{Sign } \tau\sigma)S_{\tau\sigma}(x) = -A(x)$. Hence, for any $x \in V_0^p$, $A(x)$ is an alternating tensor. Also, if x is alternating, then $(\text{sign } \sigma)S_\sigma(x) = (\text{sign } \sigma)^2 x = x$. Hence $A(x) = x$. Thus, since $A(x)$ is alternating, for any $x \in V_0^p$, $A^2(x) = A(x)$. Now we have proved that A is projection and it is projection on $AT^p(V)$.

Theorem 9.21 *If I_p denotes the linear span of all tensors in V_0^p, that are symmetric with respect to a transposition, then kernel of A is I_p.*

Proof Let $x \in V_0^p$, which is symmetric with regard to a transposition τ. Thus $S_\tau(x) = x$, and $A(S_\tau(x)) = \frac{1}{p!} \sum (\text{Sign } \sigma)S_{\sigma\tau}(x) = -\frac{1}{p!} \sum (\text{Sign } \sigma\tau)S_{\sigma\tau}(x) =$

$-A(x)$. Thus, we get $A(x) = -A(x)$, which implies that $A(x) = 0$ because char $(\mathbb{F}) = 0$. Hence $x \in$ Kernel (A). This implies that $I_p \subseteq$ Kernel A. Let $x \in$ Kernel A. Thus $A(x) = 0$. As we have $A(x) = \frac{1}{p!} \sum (\text{Sign } \sigma) S_\sigma(x)$. Now adding x on both sides of the preceding relation, we arrive at $x = \frac{1}{p!} \sum (x - \text{Sign } \sigma S_\sigma(x))$. To show that $x \in I_p$, it is sufficient to prove that $x - \text{Sign } \sigma S_\sigma(x) \in I_p$ for any $\sigma \in S_p$. We prove this statement by applying induction on the number of transpositions in which σ is expressible. The minimum number of transpositions in which σ is expressible will be 1 and this is the case, when σ is itself transposition. Thus let $\sigma = (ij)$. But $x \in V_0^p$, we have $x = x_1 \otimes x_2 \otimes \cdots \otimes x_p$. This shows that $y = x - \text{Sign } (ij) S_{(ij)}(x) = x_1 \otimes x_2 \otimes \cdots \otimes x_i \otimes \cdots \otimes x_j \otimes \cdots \otimes x_p + x_1 \otimes x_2 \otimes \cdots \otimes x_j \otimes \cdots \otimes x_i \otimes \cdots \otimes x_p = x_1 \otimes x_2 \otimes \cdots \otimes (x_i + x_j) \otimes \cdots \otimes (x_i + x_j) \otimes \cdots \otimes x_p - x_1 \otimes x_2 \otimes \cdots \otimes x_i \otimes \cdots \otimes x_i \otimes \cdots \otimes x_p - x_1 \otimes x_2 \otimes \cdots \otimes x_j \otimes \cdots \otimes x_j \otimes \cdots \otimes x_p$. Obviously $y \in I_p$. In fact, we have shown that if τ is any transposition then $x + S_\tau(x) \in I_p$. Assume the induction hypothesis, i.e., the statement is true for all permutations σ which have been expressible as product of r transpositions. Let $\sigma_1 \in S_p$, which is expressible as product of $r + 1$ transpositions. We can write $\sigma_1 = \sigma' \tau'$, where σ' is a permutation expressible as a product of r transpositions and τ' is a transposition. Now $x - \text{Sign } \sigma_1 S_{\sigma_1}(x) = x - \text{Sign } (\sigma' \tau') S_{\sigma' \tau'}(x) = x + \text{Sign }(\sigma') S_{\sigma' \tau'}(x) - S_{\tau'}(x) + S_{\tau'}(x) = x + S_{\tau'}(x) + \text{Sign }(\sigma') S_{\sigma' \tau'}(x) - S_{\tau'}(x) = x - \text{Sign } \tau' S_{\tau'}(x) - [S_{\tau'}(x) - \text{Sign }(\sigma') S_{\sigma'}(S_{\tau'}(x))]$. Using the fact that $A(\tau'(x)) = \frac{1}{p!} \sum (\text{Sign } \sigma) S_{\sigma \tau'}(x) = -\frac{1}{p!} \sum (\text{Sign } \sigma \tau') S_{\sigma \tau'}(x) = -A(x) = 0$ for any $\sigma \in S_p$ and the induction hypothesis we conclude that $x - \text{Sign } \sigma_1 S_{\sigma_1}(x) \in I_p$. Thus result follows by induction.

Let $\mathfrak{F} = \{ST^p(V) | p \in I = \{0, 1, 2, \ldots\}\}$ be a set of symmetric tensor spaces, where I is an indexing set. Let $ST(V)$ be the external direct sum of this set of vector spaces, i.e., $ST(V) = \bigoplus_{p \in I}^{ext} ST^p(V)$. $ST(V)$ is an infinite dimensional subspace of $T_0(V)$. $ST(V)$ is not closed with regard to $\otimes$. Now we define a new multiplication in $ST(V)$ in the following way. Let $x, y \in ST(V)$, clearly $x = \sum_{i=1}^{r} v_i$, $y = \sum_{j=1}^{s} v_j'$, where $v_i \in ST^i(V)$ and $v_j' \in ST^j(V)$ then

$$xy = \sum_{a=0}^{r+s} \left(\sum_{i+j=a} S(v_i \otimes v_j') \right).$$

It is easy to check that $ST(V)$ forms a commutative algebra with identity with regard to the above multiplication defined. This algebra is known as symmetric tensor algebra of V.

Theorem 9.22 *Let V be any vector space. Then the symmetric tensor algebra $ST(V)$ is isomorphic to the algebra of polynomials $\mathbb{F}[e_1, e_2, \ldots, e_n]$, where $\{e_1, e_2, \ldots, e_n\}$ is a basis of V.*

Proof We know that each element u of symmetric tensor algebra $ST(V)$ will be identified as $u = \sum_{i=1}^{r} v_i$, where $v_i \in ST^i(V)$. Thus let $x \in ST^i(V)$, as x is a symmetric element of V_0^i. Hence $x = \sum_{j_1,j_2,\ldots,j_i=1}^{n} \alpha_{j_1,j_2,\ldots,j_i} e_{j_1} \otimes e_{j_2} \otimes \cdots \otimes e_{j_i}$, where $\alpha_{j_1,j_2,\ldots,j_i} \in \mathbb{F}$. Define a map $f : ST(V) \longrightarrow \mathbb{F}[e_1, e_2, \ldots, e_n]$, such that

$$f\left(\sum_{j_1,j_2,\ldots,j_i=1}^{n} \alpha_{j_1,j_2,\ldots,j_i} e_{j_1} \otimes e_{j_2} \otimes \cdots \otimes e_{j_i} \right)$$

$$= \sum_{j_1,j_2,\ldots,j_i=1}^{n} \alpha_{j_1,j_2,\ldots,j_i} (e_{j_1} \bigvee e_{j_2} \bigvee \cdots \bigvee e_{j_i}),$$

where $\bigvee$ represents the multiplication of algebra $\mathbb{F}[e_1, e_2, \ldots, e_n]$. It can be easily verified that f is an algebra isomorphism. Thus the symmetric tensor algebra $ST(V)$ is isomorphic to the algebra of polynomials $\mathbb{F}[e_1, e_2, \ldots, e_n]$.

Theorem 9.23 *Let V be any vector space. Then the vector space $ST^p(V)$ of symmetric tensors is isomorphic to the vector space $\mathbb{F}_p[e_1, e_2, \ldots, e_n]$, of homogeneous polynomials of degree p for $p \geq 1$.*

Proof Let $x \in ST^p(V)$, as x is a symmetric element of V_0^p. Hence $x = \sum_{j_1,j_2,\ldots,j_i=1}^{n} \alpha_{j_1,j_2,\ldots,j_p} e_{j_1} \otimes e_{j_2} \otimes \cdots \otimes e_{j_p}$, where $\alpha_{j_1,j_2,\ldots,j_p} \in \mathbb{F}$. Define a map $f : ST^p(V) \longrightarrow \mathbb{F}_p[e_1, e_2, \ldots, e_n]$, such that

$$f\left(\sum_{j_1,j_2,\ldots,j_p=1}^{n} \alpha_{j_1,j_2,\ldots,j_p} e_{j_1} \otimes e_{j_2} \otimes \cdots \otimes e_{j_p} \right)$$

$$= \sum_{j_1,j_2,\ldots,j_i=1}^{n} \alpha_{j_1,j_2,\ldots,j_p} (e_{j_1} \bigvee e_{j_2} \bigvee \cdots \bigvee e_{j_p}),$$

where $\bigvee$ represents the multiplication of algebra $\mathbb{F}[e_1, e_2, \ldots, e_n]$. It can be easily verified that f is a vector space isomorphism. Thus the vector space $ST^p(V)$ of symmetric tensors is isomorphic to the vector space $\mathbb{F}_p[e_1, e_2, \ldots, e_n]$.

Theorem 9.24 *Let V be any vector space. Then, the vector space $\mathbb{F}_p[e_1, e_2, \ldots, e_n]$, of homogeneous polynomials of degree p is isomorphic to a quotient space of the vector space V_0^p for $p \geq 1$.*

Proof Define a map $f : V_0^p \longrightarrow \mathbb{F}_p[e_1, e_2, \ldots, e_n]$ such that

$$f\left(\sum_{j_1,j_2,\ldots,j_p=1}^{n} \alpha_{j_1,j_2,\ldots,j_p} e_{j_1} \otimes e_{j_2} \otimes \cdots \otimes e_{j_p} \right)$$

$$= \sum_{j_1,j_2,\ldots,j_i=1}^{n} \alpha_{j_1,j_2,\ldots,j_p} (e_{j_1} \bigvee e_{j_2} \bigvee \cdots \bigvee e_{j_p}).$$

Obviously, f is a surjective linear map. Kernel of f is given as $I_p = \langle \{S_\sigma(x) - x \mid x \in V_0^p, \sigma \in S_p\} \rangle$. By fundamental theorem of vector space homomorphism, we have $\frac{V_0^p}{I_p} \cong \mathbb{F}_p[e_1, e_2, \ldots, e_n]$, which is our required result. Using Theorem 9.23, we deduce that $\frac{V_0^p}{I_p} \cong STP(V)$. Due to this isomorphism the vector space $\frac{V_0^p}{I_p}$ is also referred as the symmetric tensor space of V.

Finally, in the end of this section, we want to explore the notion of exterior product and exterior algebra of a vector space V. Consider the quotient vector space $\frac{V_0^p}{I_p}$. Let $q : V_0^p \longrightarrow \frac{V_0^p}{I_p}$ be the quotient homomorphism, i.e., for any $x \in V_0^p$, we have $q(x) = x + I_p$. As the map $f : V_1 \times V_2 \times \cdots \times V_p \longrightarrow V_0^p$ where $V_i = V, i = 1, 2, \ldots, p$, such that $f(x_1, x_2, \ldots, x_p) = x_1 \otimes x_2 \otimes \cdots \otimes x_p$ is a p-linear map. Thus, the product map $q \circ f : V_1 \times V_2 \times \cdots \times V_p \longrightarrow \frac{V_0^p}{I_p}$ is also a p-linear map. The image of any element $(x_1, x_2, \ldots, x_n)$ under $q \circ f$ is denoted by $x_1 \wedge x_2 \wedge \cdots \wedge x_p$. The quotient space $\frac{V_0^p}{I_p}$ is denoted by $\bigwedge^p V$. The elements of $\bigwedge^p V$ are called p vectors over V. A p vector is called decomposable or a pure if it is of the form $x_1 \wedge x_2 \wedge \cdots \wedge x_p$. We define $\bigwedge^1 V = V$ and $\bigwedge^0 V = \mathbb{F}$. Now, we are interested in dealing with these p-vectors where $p = 0, 1, 2, \ldots$, which will give rise to an algebra. For this we prove the following.

Theorem 9.25 *Let U be any arbitrary vector space. If $h : \bigwedge^p V \longrightarrow U$ is a linear mapping, then $h \circ q \circ f : V_1 \times V_2 \times \cdots \times V_p \longrightarrow U$ where $V_i = V, i = 1, 2, \ldots, p$, is an alternating p-linear mapping, where q and f have been described in above paragraph. Conversely, if $g : V_1 \times V_2 \times \cdots \times V_p \longrightarrow U$ where $V_i = V, i = 1, 2, \ldots, p$, is an alternating p-linear mapping, then there exists a unique linear mapping $h : \bigwedge^p V \longrightarrow U$ such that $g(x_1, x_2, \ldots, x_p) = h(x_1 \wedge x_2 \wedge \cdots \wedge x_p)$.*

Proof Obviously, $h \circ q \circ f$ is a p-linear map. Let $(x_1, x_2, \ldots, x_p) \in V_1 \times V_2 \times \cdots \times V_p$ where $x_i = x_j$ for $i \neq j$, clearly $x_1 \otimes x_2 \otimes \cdots \otimes x_p$ is a symmetric tensor with regard to transposition (ij). By Theorem 9.21, $x_1 \otimes x_2 \otimes \cdots \otimes x_p \in I_p$. As a result $q(x_1 \otimes x_2 \otimes \cdots \otimes x_p) = I_p$. Finally, we have $h \circ q \circ f(x_1, x_2, \ldots, x_p) = hoq(x_1 \otimes x_2 \otimes \cdots \otimes x_p) = h(I_p) = 0$, because h is linear. This proves the first statement.

Conversely, suppose that $g : V_1 \times V_2 \times \cdots \times V_p \longrightarrow U$ is an alternating p-linear mapping. Thus $g(x_1, x_2, \ldots, x_p) = 0$ if $x_i = x_j$, where $i \neq j$. Also, using the above arguments if $x_i = x_j$ with $i \neq j$, then $x_1 \wedge x_2 \wedge \cdots \wedge x_p = 0$, i.e., $h(x_1 \wedge x_2 \wedge \cdots \wedge x_p) = 0$. Using these facts, we obtain that the association $(y_1 \wedge y_2 \wedge \cdots \wedge y_p) \longrightarrow g(y_1, y_2, \ldots, y_p)$ is a map. Now extending this map by linearity on $\bigwedge^p V$, define a map $h : \bigwedge^p V \longrightarrow U$ because the p-vectors $y_1 \wedge y_2 \wedge \cdots \wedge y_p$ span $\bigwedge^p(V)$. Further, for uniqueness of h, let $h_1 : \bigwedge^p V \longrightarrow U$ be any other linear mapping such that $h_1(x_1 \wedge x_2 \wedge \cdots \wedge x_p) = g(x_1, x_2, \ldots, x_p)$. This shows that h and h_1 agree on a spanning set for $\bigwedge^p(V)$. Hence $h = h_1$.

Now, we define a product of p-vectors and q-vectors. Consider vectors $x_1 \wedge x_2 \wedge \cdots \wedge x_p \wedge y_1 \wedge y_2 \wedge \cdots \wedge y_q \in \bigwedge^{p+q}(V)$. Now define a map $g_1 : V_1 \times V_2 \times$

$\cdots \times V_p \longrightarrow \bigwedge^{p+q}(V)$ such that $g_1(x_1, x_2, \ldots, x_p) = x_1 \wedge x_2 \wedge \cdots \wedge x_p \wedge y_1 \wedge y_2 \wedge \cdots \wedge y_q$. Clearly g_1 is alternating p-linear map. Hence by Theorem 9.25, there exists unique linear map $h_1 : \bigwedge^p(V) \longrightarrow \bigwedge^{p+q}(V)$ such that $g(x_1, x_2, \ldots, x_p) = h_1(x_1 \wedge x_2 \wedge \cdots \wedge x_p) = x_1 \wedge x_2 \wedge \cdots \wedge x_p \wedge y_1 \wedge y_2 \wedge \cdots \wedge y_q$. Similarly define a map $g_2 : V_1 \times V_2 \times \cdots \times V_q \longrightarrow \bigwedge^{p+q}(V)$ such that $g_2(y_1, y_2, \ldots, y_q) = x_1 \wedge x_2 \wedge \cdots \wedge x_p \wedge y_1 \wedge y_2 \wedge \cdots \wedge y_q$. Clearly g_1 is alternating p-linear map. Similarly, there exists unique linear map $h_2 : \bigwedge^q(V) \longrightarrow \bigwedge^{p+q}(V)$ such that $g(y_1, y_2, \ldots, y_q) = h_1(y_1 \wedge y_2 \wedge \cdots \wedge y_q) = x_1 \wedge x_2 \wedge \cdots \wedge x_p \wedge y_1 \wedge y_2 \wedge \cdots \wedge y_q$. Now define a map $f : \bigwedge^p(V) \times \bigwedge^q(V) \longrightarrow \bigwedge^{p+q}(V)$ such that $f(x_1 \wedge x_2 \wedge \cdots \wedge x_p, y_1 \wedge y_2 \wedge \cdots \wedge y_q) = x_1 \wedge x_2 \wedge \cdots \wedge x_p \wedge y_1 \wedge y_2 \wedge \cdots \wedge y_q$. As h_1 and h_2 are linear maps, therefore f is a bilinear map. Let $x = x_1 \wedge x_2 \wedge \cdots \wedge x_p$ and $y = y_1 \wedge y_2 \wedge \cdots \wedge y_q$. Then the image of (x, y) under f, i.e., $f(x, y) = x \wedge y = x_1 \wedge x_2 \wedge \cdots \wedge x_p \wedge y_1 \wedge y_2 \wedge \cdots \wedge y_q$ is called the exterior product or wedge product or the Grassmann product. If $p = 0$ or $q = 0$, then we define $\alpha \wedge y$ to be αy and $x \wedge \alpha$ to be αx. Since above map f is bilinear. Hence following are evident. $(x + y) \wedge (u + v) = x \wedge u + x \wedge v + y \wedge u + y \wedge v$, $(\alpha x) \wedge y = x \wedge (\alpha y) = \alpha(x \wedge y)$. For pure p-vectors x, q-vectors y, and r-vectors z, we have $(x \wedge y) \wedge z = x \wedge (y \wedge z)$.

Let $\mathfrak{F} = \{\bigwedge^p(V) | p \in I = \{0, 1, 2, \ldots\}\}$ be a set of vector spaces of p-vectors, where I is an indexing set. Let $\bigwedge(V)$ be the external direct sum of this set of vector spaces, i.e., $\bigwedge(V) = \bigoplus_{p \in I}^{ext} \bigwedge^p(V)$. Now we define a new multiplication in $\bigwedge(V)$ in the following way. Let $x, y \in \bigwedge(V)$, clearly $x = \sum_{i=1}^{r} v_i$, $y = \sum_{j=1}^{s} v'_j$, where $v_i \in \bigwedge^i(V)$ and $v'_j \in \bigwedge^j(V)$ then

$$xy = \sum_{a=0}^{r+s} \left(\sum_{i+j=a} (v_i \wedge v'_j) \right).$$

It is easy to check that $\bigwedge(V)$ forms a noncommutative algebra with identity with regard to the exterior product or wedge product defined above. This algebra is known as antisymmetric tensor algebra of V, or exterior algebra of V or Grassmann algebra of V. Usually it is represented by $\bigwedge(V)$ or $AT(V)$.

Exercises

1. Find the dimension of subspaces $ST^p(V)$ and $AT^p(V)$ of V_0^p.
2. Prove that the symmetric tensor algebra $ST(V)$ and exterior algebra $\bigwedge(V)$ are graded algebras.
3. Let the dimension of V be n. Prove that $\dim \bigwedge^p(V) = \binom{n}{p}$.
4. If $x \in \bigwedge^p(V)$ and $y \in \bigwedge^q(V)$, then show that $x \wedge y = (-1)^{pq} y \wedge x$.
5. Let $x_1 \wedge x_2 \wedge \cdots \wedge x_p \in \bigwedge^p(V)$. Then prove that $x_1 \wedge x_2 \wedge \cdots \wedge x_p = (\text{Sign}\sigma) x_{\sigma(1)} \wedge x_{\sigma(2)} \wedge \cdots \wedge x_{\sigma(p)}$, for any $\sigma \in S_p$.

6. Show that $x_1 \wedge x_2 \wedge \cdots \wedge x_m = 0$, $m \leq n$, if and only if $\{x_i\}_{i=1}^m$ is linearly dependent, where dim $V = n$.

7. Let $\{e_i\}_{i=1}^n$ be a basis for V. Prove that the set of $\binom{n}{p}$ vectors of the form $e_{i_1} \wedge e_{i_2} \wedge \cdots \wedge e_{i_p}$, where $i_1 i_2 \cdots i_p$ ranges over the combinations of p-distinct integers taken from the set $\{1, 2, \ldots, n\}$, is a basis for $\bigwedge^p(V)$.

Chapter 10
Applications of Linear Algebra to Numerical Methods

In this chapter, we shall study common problems in numerical linear algebra which includes LU and PLU decompositions together with their applications in solving a linear system of equations. Further, we shall briefly discuss the power method which gives an approximation to the eigenvalue of the greatest absolute value and corresponding eigenvectors. Finally, singular value decomposition (SVD) of matrices together with its properties and applications in diverse fields of studies are included.

10.1 LU Decomposition of Matrices

The Gauss elimination method reduces the system of equations to an equivalent upper triangular system which can be solved by back solution. A different approach can be used to solve the system of equations $AX = B$ by decomposing (or factoring) the coefficient matrix A into a product of lower and upper triangular matrices. The motivation for a triangular decomposition is based on the observation that system of equations involving triangular coefficient matrices is easier to deal with. Using the same arguments as used in Remark $1.101(ii)$, of Chap. 1, the coefficient matrix A can be factored in such a way.

Let A be an $m \times n$ matrix which can be reduced in row echelon form U by using sequence of elementary row operations (III)(see Note 1.86 which involves adding a constant multiple of a row R_j (namely αR_j) to another row R_i, in such a way that $R_i + \alpha R_j$ replaces R_i in the new set of arrays, usually denoted by $R_i \rightarrow R_i + \alpha R_j$, $i > j$ without interchanging any two rows. We zero out the entries below the pivot entry in the first column and then the next and so on while scanning from the left. By Remark $1.101(i)$, each of these operations can be accomplished by multiplying on the left by an appropriate elementary matrix. Hence, we can find elementary matrices $E_1, E_2, \ldots, E_r$ such that

© The Author(s), under exclusive license to Springer Nature Singapore Pte Ltd. 2022
M. Ashraf et al., *Advanced Linear Algebra with Applications*,
https://doi.org/10.1007/978-981-16-2167-3_10

$$E_r \cdots E_2 E_1 A = U.$$

Since the reduction of A to row echelon form can be achieved without interchanging any two rows, we can assume that the required elementary matrices E_k represent operations of the form $R_i \rightarrow R_i + \alpha R_j$ designed to add multiples αR_j of row R_j to the row R_i below. This means $i > j$ in all the cases. Each E_k is, therefore, a lower triangular elementary matrix with $1's$ on the diagonal. The inverse of operation $R_i \rightarrow R_i + \alpha R_j$ is $R_i \rightarrow R_i - \alpha R_j$, again with $i > j$. Hence, the inverse E_k^{-1} is also a lower triangular matrix with $1's$ on its main diagonal. Since elementary matrices $E_1, E_2, \ldots, E_r$ are nonsingular, multiplying both the sides of the above relation on the left successively by $E_r^{-1}, \ldots, E_2^{-1}, E_1^{-1}$, we get

$$A = E_1^{-1} E_2^{-1} \cdots E_r^{-1} U.$$

The matrix $L = E_1^{-1} E_2^{-1} \cdots E_r^{-1}$ is a lower triangular matrix with $1's$ on the main diagonal provided that no two rows are interchanged in reducing A to U, and the above yields that $A = LU$.

The following theorem summarizes the above result:

Theorem 10.1 (*LU* decomposition theorem) *Suppose that A is an $m \times n$ matrix that can be reduced to echelon form without interchanging any two rows. Then there exist an $m \times m$ lower triangular matrix L with $1's$ on the main diagonal and an $m \times n$ row echelon matrix U such that $A = LU$.*

Definition 10.2 A factorization of a matrix A as $A = LU$, where L is a lower triangular and U is an upper triangular matrix, is called an *LU* decomposition of A.

There is a convenient procedure for finding *LU* decomposition. In fact, it is only necessary to keep the track of the *multipliers* which are used to reduce the matrix in row reduced echelon form. This procedure is described in the following example and is called the multiplier method or Dolittle's method.

Example 10.3 In order to find the *LU* decomposition of the matrix $A = \begin{bmatrix} 1 & 2 & 4 \\ 3 & 8 & 14 \\ 2 & 6 & 13 \end{bmatrix}$,

write the identity matrix in the left, i.e.,

$$\begin{bmatrix} 1 & 0 & 0 \\ 0 & 1 & 0 \\ 0 & 0 & 1 \end{bmatrix} \begin{bmatrix} 1 & 2 & 4 \\ 3 & 8 & 14 \\ 2 & 6 & 13 \end{bmatrix}.$$

The procedure involves doing row operations to the matrix on the right while simultaneously updating the column of the matrix on the left. First, we perform the row operations $R_2 \rightarrow R_2 - 3R_1$ to make zero below 1 in the first column and second row. Note that 3 is added in the second row of the first column because -3 times of the first row of A is added to the second row.

$$\begin{bmatrix} 1 & 0 & 0 \\ 3 & 1 & 0 \\ 0 & 0 & 1 \end{bmatrix} \begin{bmatrix} 1 & 2 & 4 \\ 0 & 2 & 2 \\ 2 & 6 & 13 \end{bmatrix}.$$

We see that $E_1 = \begin{bmatrix} 1 & 0 & 0 \\ -3 & 1 & 0 \\ 0 & 0 & 1 \end{bmatrix}$ and hence $E_1^{-1} = \begin{bmatrix} 1 & 0 & 0 \\ 3 & 1 & 0 \\ 0 & 0 & 1 \end{bmatrix}$. We carry out the similar procedure for the third row and find that

$$\begin{bmatrix} 1 & 0 & 0 \\ 3 & 1 & 0 \\ 2 & 0 & 1 \end{bmatrix} \begin{bmatrix} 1 & 2 & 4 \\ 0 & 2 & 2 \\ 0 & 2 & 5 \end{bmatrix}.$$

Note that $E_2 = \begin{bmatrix} 1 & 0 & 0 \\ 0 & 1 & 0 \\ -2 & 0 & 1 \end{bmatrix}$ and hence $E_2^{-1} = \begin{bmatrix} 1 & 0 & 0 \\ 0 & 1 & 0 \\ 2 & 0 & 1 \end{bmatrix}$. Finally, similar arguments for the second column and third row yield that

$$A = \begin{bmatrix} 1 & 0 & 0 \\ 3 & 1 & 0 \\ 2 & 1 & 1 \end{bmatrix} \begin{bmatrix} 1 & 2 & 4 \\ 0 & 2 & 2 \\ 0 & 0 & 3 \end{bmatrix}.$$

Thus, we find LU decomposition of the matrix A. We see that $E_3 = \begin{bmatrix} 1 & 0 & 0 \\ 0 & 1 & 0 \\ 0 & -1 & 1 \end{bmatrix}$ and

hence $E_3^{-1} = \begin{bmatrix} 1 & 0 & 0 \\ 0 & 1 & 0 \\ 0 & 1 & 1 \end{bmatrix}$. It can be seen that $U = E_3 E_2 E_1 A$ and $L = E_1^{-1} E_2^{-1} E_3^{-1}$.
Notice that in each position below the main diagonal of L, the entry is the negative of the multiplier in the operation that introduced the zero in that position of U.

Remark 10.4 (*i*) This is natural to ask whether every matrix has LU decomposition. Sometimes it is impossible to write given matrix in this form. In fact, if a square matrix A can be reduced to row echelon form without using row interchanges, then A has LU decomposition. More generally, an invertible matrix A has LU decomposition provided that all its *leading submatrices* have nonzero determinants. The kth leading submatrix of A, denoted as A_k, is the $k \times k$ matrix obtained by retaining only the top k rows and left most k columns. For example, if

$$A = \begin{bmatrix} a_{11} & a_{12} & \cdots & a_{1n} \\ a_{21} & a_{22} & \cdots & a_{2n} \\ \vdots & \vdots & \vdots & \vdots \\ a_{n1} & a_{n2} & \cdots & a_{nn} \end{bmatrix},$$

then $A_1 = a_{11}$, $A_2 = \begin{bmatrix} a_{11} & a_{12} \\ a_{21} & a_{22} \end{bmatrix}$, $\ldots A_k = \begin{bmatrix} a_{11} & a_{12} & \cdots & a_{1k} \\ a_{21} & a_{22} & \cdots & a_{2k} \\ \vdots & \vdots & \vdots & \vdots \\ a_{k1} & a_{k2} & \cdots & a_{kk} \end{bmatrix}$, and A has LU

decompositions if $|A_i| \neq 0$; for all $1 \leq i \leq n$.

(ii) It is also interesting to ask whether a square matrix has more than one decomposition. In the absence of additional restrictions, it can be easily seen that LU decompositions are not unique. For example, if

$$A = LU = \begin{bmatrix} \ell_{11} & 0 & \cdots & 0 \\ \ell_{21} & \ell_{22} & \cdots & 0 \\ \vdots & \vdots & \vdots & \vdots \\ \ell_{n1} & \ell_{n2} & \cdots & \ell_{nn} \end{bmatrix} \begin{bmatrix} 1 & u_{12} & \cdots & u_{1n} \\ 0 & 1 & \cdots & u_{2n} \\ \vdots & \vdots & \cdots & \vdots \\ 0 & 0 & \cdots & 1 \end{bmatrix}$$

and L has nonzero entries on the main diagonal, then shift the diagonal entries from the left factor to the right factor as follows:

$$A = \begin{bmatrix} 1 & 0 & \cdots & 0 \\ \frac{\ell_{21}}{\ell_{11}} & 1 & \cdots & 0 \\ \vdots & \vdots & \cdots & \vdots \\ \frac{\ell_{n1}}{\ell_{11}} & \frac{\ell_{n2}}{\ell_{22}} & \cdots & 1 \end{bmatrix} \begin{bmatrix} \ell_{11} & 0 & \cdots & 0 \\ 0 & \ell_{22} & \cdots & 0 \\ \vdots & \vdots & \cdots & \vdots \\ 0 & 0 & \cdots & \ell_{nn} \end{bmatrix} \begin{bmatrix} 1 & u_{12} & \cdots & u_{1n} \\ 0 & 1 & \cdots & u_{2n} \\ \vdots & \vdots & \cdots & \vdots \\ 0 & 0 & \cdots & 1 \end{bmatrix}$$

$$= \begin{bmatrix} 1 & 0 & \cdots & 0 \\ \frac{\ell_{21}}{\ell_{11}} & 1 & \cdots & 0 \\ \vdots & \vdots & \cdots & \vdots \\ \frac{\ell_{n1}}{\ell_{11}} & \frac{\ell_{n2}}{\ell_{22}} & \cdots & 1 \end{bmatrix} \begin{bmatrix} \ell_{11} & \ell_{11}u_{12} & \cdots & \ell_{11}u_{1n} \\ 0 & \ell_{22} & \cdots & \ell_{22}u_{2n} \\ \vdots & \vdots & \cdots & \vdots \\ 0 & 0 & \cdots & \ell_{nn} \end{bmatrix} .$$

This is another triangular decomposition of A.

(iii) If A is an invertible matrix of order n that can be reduced to row echelon form without interchanging any two rows, then it can be seen that LU decomposition of A is unique. Suppose A has two decompositions say $A = L_1 U_1$ and $A = L_2 U_2$. Then since L_1, L_2, U_1, U_2 are nonsingular and

$$L_2^{-1} L_1 = L_2^{-1} I_n L_1 = L_2^{-1}(AA^{-1})L_1 = L_2^{-1}(L_2 U_2)(L_1 U_1)^{-1} L_1$$
$$= L_2^{-1} L_2 (U_2 U_1^{-1}) L_1^{-1} L_1$$
$$= U_2 U_1^{-1},$$

L_2^{-1} is also a lower triangular matrix and has $1's$ on its main diagonal, and $L_2^{-1}L_1$ is also a lower triangular matrix with $1's$ on its main diagonal. By the same process, one can conclude that $U_2 U_1^{-1}$ is an upper triangular matrix. Since $L_2^{-1}L_1 = U_2 U_1^{-1}$, their common value is a matrix that is both lower triangular and upper triangular with $1's$ on its main diagonal. The only matrix that meets these requirements is the identity matrix and hence $L_2^{-1}L_1 = U_2 U_1^{-1} = I_n$, which implies that $L_1 = L_2$ and $U_1 = U_2$. This ensures the uniqueness of decomposition of A.

The procedure adopted in Example 10.3 for finding LU decomposition of a matrix is also useful for obtaining an LU decomposition of a rectangular matrix, which can be seen by the following example.

Example 10.5 In order to find LU decomposition of $A = \begin{bmatrix} 1 & 2 & 1 & 2 & 1 \\ 2 & 0 & 2 & 1 & 1 \\ 2 & 3 & 1 & 3 & 2 \\ 1 & 0 & 1 & 1 & 2 \end{bmatrix}$, we write

the identity matrix in the left, i.e.,

$$A = \begin{bmatrix} 1 & 0 & 0 & 0 \\ 0 & 1 & 0 & 0 \\ 0 & 0 & 1 & 0 \\ 0 & 0 & 0 & 1 \end{bmatrix} \begin{bmatrix} 1 & 2 & 1 & 2 & 1 \\ 2 & 0 & 2 & 1 & 1 \\ 2 & 3 & 1 & 3 & 2 \\ 1 & 0 & 1 & 1 & 2 \end{bmatrix}.$$

Now applying the row operations $R_4 \to R_4 - R_1$, $R_2 \to R_2 - 2R_1$ and $R_3 \to R_3 - 2R_1$ on the matrix in the right and updating column of the identity matrix in the left, we find that

$$\begin{bmatrix} 1 & 0 & 0 & 0 \\ 2 & 1 & 0 & 0 \\ 2 & 0 & 1 & 0 \\ 1 & 0 & 0 & 1 \end{bmatrix} \begin{bmatrix} 1 & 2 & 1 & 2 & 1 \\ 0 & -4 & 0 & -3 & -1 \\ 0 & -1 & -1 & -1 & 0 \\ 0 & -2 & 0 & -1 & 1 \end{bmatrix}.$$

Further, performing the row operations $R_3 \to R_3 - \frac{1}{4}R_2$ followed by $R_4 \to R_4 - \frac{1}{2}R_2$, we get

$$\begin{bmatrix} 1 & 0 & 0 & 0 \\ 2 & 1 & 0 & 0 \\ 2 & \frac{1}{4} & 1 & 0 \\ 1 & \frac{1}{2} & 0 & 1 \end{bmatrix} \begin{bmatrix} 1 & 2 & 1 & 2 & 1 \\ 0 & -4 & 0 & -3 & -1 \\ 0 & 0 & -1 & -\frac{1}{4} & \frac{1}{4} \\ 0 & 0 & 0 & \frac{1}{2} & \frac{3}{2} \end{bmatrix}.$$

This is an LU decomposition of A.

It can be seen that $E_1 = \begin{bmatrix} 1 & 0 & 0 & 0 \\ -2 & 1 & 0 & 0 \\ 0 & 0 & 1 & 0 \\ 0 & 0 & 0 & 1 \end{bmatrix}$, $E_2 = \begin{bmatrix} 1 & 0 & 0 & 0 \\ 0 & 1 & 0 & 0 \\ -2 & 0 & 1 & 0 \\ 0 & 0 & 0 & 1 \end{bmatrix}$,

$$E_3 = \begin{bmatrix} 1 & 0 & 0 & 0 \\ 0 & 1 & 0 & 0 \\ 0 & 0 & 1 & 0 \\ -1 & 0 & 0 & 1 \end{bmatrix}, \quad E_4 = \begin{bmatrix} 1 & 0 & 0 & 0 \\ 0 & 1 & 0 & 0 \\ 0 & -\frac{1}{4} & 1 & 0 \\ 0 & 0 & 0 & 1 \end{bmatrix}, \quad E_5 = \begin{bmatrix} 1 & 0 & 0 & 0 \\ 0 & 1 & 0 & 0 \\ 0 & 0 & 1 & 0 \\ 0 & -\frac{1}{2} & 0 & 1 \end{bmatrix}.$$

with $U = E_5 E_4 E_3 E_2 E_1 A = \begin{bmatrix} 1 & 2 & 1 & 2 & 1 \\ 0 & -4 & 0 & -3 & -1 \\ 0 & 0 & -1 & -\frac{1}{4} & \frac{1}{4} \\ 0 & 0 & 0 & \frac{1}{2} & \frac{3}{2} \end{bmatrix}$ and

$$L = E_1^{-1} E_2^{-1} E_3^{-1} E_4^{-1} E_5^{-1} = \begin{bmatrix} 1 & 0 & 0 & 0 \\ 2 & 1 & 0 & 0 \\ 2 & \frac{1}{4} & 1 & 0 \\ 1 & \frac{1}{2} & 0 & 1 \end{bmatrix}.$$

Solution of the System of Equations Using LU Decomposition

The LU decomposition is an approach designed to exploit the triangular system and useful to solve the system of equations. In fact, this method is well suited for digital computers and is the basis of many practical computer programs.

Any given system can be solved in two stages. First, we shall show that how a linear system $AX = B$ can be readily solved once A is decomposed into a product of lower triangular and upper triangular matrices. Once A is factored, the system $AX = B$ can be solved using the following steps:

(1) Write the system $AX = B$ as $LUX = B$.
(2) Define an $n \times 1$ matrix Y by $UX = Y$.
(3) Rewrite the system in (1) as $LY = B$, where $L = \{(\ell_{ij}) \mid \ell_{ij} = 0, i < j; 1 \le$

$i, j \le n\}, Y = \begin{bmatrix} y_1 \\ y_2 \\ \vdots \\ y_n \end{bmatrix}, B = \begin{bmatrix} b_1 \\ b_2 \\ \vdots \\ b_n \end{bmatrix}$ and solve this system for Y.

$$\begin{aligned} \ell_{11} y_1 &= b_1 \\ \ell_{21} y_1 + \ell_{22} y_2 &= b_2 \\ \ell_{31} y_1 + \ell_{32} y_2 + \ell_{33} y_3 &= b_3 \\ \vdots \quad\quad &= \vdots \\ \ell_{n1} y_1 + \ell_{n2} y_2 + \ell_{n3} y_3 + \cdots + \ell_{nn} y_n &= b_n \end{aligned}$$

This yields value of y_1, and further using the successive equations, one can find $y_2, y_3, \ldots, y_n$.

(4) Once Y has been determined, solve the upper triangular system $UX = Y$ to find the solution X of the system.

Remark 10.6 (i) If any of the diagonal element ℓ_{ii} is zero, then the system is singular and can't be solved.

(ii) If all the diagonal elements ℓ_{ii} are nonzero, then the system has unique solution.

Although the above procedure replaces the problem of solving the single system into two different systems $LY = B$ and $UX = Y$, but because of the involvement of triangular matrices, the latter systems are easy to solve.

Example 10.7 Solve the system of equations involving five variables and four equations:

$$\begin{bmatrix} 1 & 2 & 1 & 2 & 1 \\ 2 & 0 & 2 & 1 & 1 \\ 2 & 3 & 1 & 3 & 2 \\ 1 & 0 & 1 & 1 & 2 \end{bmatrix} \begin{bmatrix} x \\ y \\ z \\ w \\ t \end{bmatrix} = \begin{bmatrix} 1 \\ 2 \\ 3 \\ 4 \end{bmatrix}.$$

By the above example note that we have the following *LU* decomposition of the coefficient matrix:

$$\begin{bmatrix} 1 & 2 & 1 & 2 & 1 \\ 2 & 0 & 2 & 1 & 1 \\ 2 & 3 & 1 & 3 & 2 \\ 1 & 0 & 1 & 1 & 2 \end{bmatrix} = \begin{bmatrix} 1 & 0 & 0 & 0 \\ 2 & 1 & 0 & 0 \\ 2 & \frac{1}{4} & 1 & 0 \\ 1 & \frac{1}{2} & 0 & 1 \end{bmatrix} \begin{bmatrix} 1 & 2 & 1 & 2 & 1 \\ 0 & -4 & 0 & -3 & -1 \\ 0 & 0 & -1 & -\frac{1}{4} & \frac{1}{4} \\ 0 & 0 & 0 & \frac{1}{2} & \frac{3}{2} \end{bmatrix}.$$

Represent the system of equations as $AX = B$ and let $UX = Y$. Consider $LY = B$, i.e.,

$$\begin{bmatrix} 1 & 0 & 0 & 0 \\ 2 & 1 & 0 & 0 \\ 2 & \frac{1}{4} & 1 & 0 \\ 1 & \frac{1}{2} & 0 & 1 \end{bmatrix} \begin{bmatrix} y_1 \\ y_2 \\ y_3 \\ y_4 \end{bmatrix} = \begin{bmatrix} 1 \\ 2 \\ 3 \\ 4 \end{bmatrix}.$$

The above yields that $Y = \begin{bmatrix} y_1 \\ y_2 \\ y_3 \\ y_4 \end{bmatrix} = \begin{bmatrix} 1 \\ 0 \\ 1 \\ 3 \end{bmatrix}$. Now by solving $UX = Y$, we find that

$$\begin{bmatrix} 1 & 2 & 1 & 2 & 1 \\ 0 & -4 & 0 & -3 & -1 \\ 0 & 0 & -1 & -\frac{1}{4} & \frac{1}{4} \\ 0 & 0 & 0 & \frac{1}{2} & \frac{3}{2} \end{bmatrix} \begin{bmatrix} x \\ y \\ z \\ w \\ t \end{bmatrix} = \begin{bmatrix} 1 \\ 0 \\ 1 \\ 3 \end{bmatrix}$$ and hence this reduces to $X = \begin{bmatrix} -\frac{1}{2} \\ \frac{4t-9}{2} \\ \frac{3t-1}{2} \\ 6 - 3t \\ t \end{bmatrix}$,

$t \in \mathbb{R}$.

This looks like trivial operations, but it is advantageous because it reduces the number of operations involved in finding a solution to a system of equations and makes a difference for a large system.

10.2 The *PLU* Decomposition

The *LU* decomposition is a useful tool to find the solution of a system of equations, but this does not work for every matrix. For example, if we consider $A = \begin{bmatrix} 0 & 1 \\ 1 & 0 \end{bmatrix}$, then it can be seen that there do not exist lower triangular matrix L and upper triangular matrix U such that $A = LU$. Motivation behind PLU decomposition, where P is a permutation matrix, is Gaussian elimination while reducing a matrix in echelon form where we need to perform elimination of type (III)(see Note 1.86), i.e., replacing a row R_k by $R_k + \alpha R_\ell$ and operation of type (I), i.e., interchanging two rows $R_k \leftrightarrow R_\ell$. Instead of using only type (III) operation, we shall permute two rows (operation (I)) first and then use operation of type (III) later.

Definition 10.8 Let $\sigma \in S_n$ be a permutation defined on n-symbols $\{1, 2, \ldots, n\}$. Then a permutation matrix P_σ associated with the permutation σ is defined as

$$P_\sigma = (a_{ij}), \quad \text{where } a_{ij} = \begin{cases} 1, & j = \sigma(i) \\ 0, & \text{otherwise} \end{cases}.$$

For example, if $\sigma = \begin{pmatrix} 1 & 2 & 3 & 4 \\ 2 & 4 & 3 & 1 \end{pmatrix}$ is a permutation defined on four symbols, then the

permutation matrix associated with σ is $P_\sigma = \begin{bmatrix} 0 & 1 & 0 & 0 \\ 0 & 0 & 0 & 1 \\ 0 & 0 & 1 & 0 \\ 1 & 0 & 0 & 0 \end{bmatrix}$.

Since the matrix P_σ is full of zeros, it is easy to deal with such matrix. It is called *permuting matrix* because it would equal to the identity matrix if we could permute with its rows.

Remark 10.9 (*i*) Let $\sigma, \tau \in S_n$. Then their composition $\sigma \circ \tau \in S_n$.

(*ii*) If P_σ and P_τ are permutation matrices associated with σ and τ, respectively, then $P_\sigma P_\tau = P_{\tau \circ \sigma}$. In fact, if a_{ij}, b_{ij} are (i, j)th entry of P_σ and P_τ, respectively, and c_{ij} is the (i, j)th entry of $P_{\tau \circ \sigma}$, then for each $i, j \in \{1, 2, \ldots, n\}$,

$c_{ij} = \sum_{k=1}^{n} a_{ik} b_{kj}$. Obviously, $a_{ik} = 1$ if and only if $k = \sigma(i)$ and $b_{kj} = 1$ if and only if $j = \tau(k)$. Therefore, $a_{ik} b_{kj} = 1$ if and only if $j = \tau(k) = \tau(\sigma(i))$. Hence, the product $P_\sigma P_\tau$ is the matrix of permutation $\tau \circ \sigma$ and $P_\sigma P_\tau = P_{\tau \circ \sigma}$.

(*iii*) The elementary matrix associated with the elementary operation of switching rows is a permutation matrix. Therefore, performing a series of row switches may be represented as a permutation matrix, since it is a product of permutation matrices.

(*iv*) If $E \in M_n(\mathbb{F})$ is an elementary matrix that represents the action $R_k \to R_k + \alpha R_\ell$ and if P_σ be the permutation matrix for $\sigma \in \{1, 2, \ldots, n\}$, then $E P_\sigma = P_\sigma E'$, where E' is the elementary matrix that represents the action $R_\sigma(k) \to R_\sigma(k) + \alpha R_\sigma(\ell)$.

Applying procedure of LU decomposition, we can say that when no interchanges are needed, we can factor a matrix $A \in M_n(\mathbb{C})$ as $A = LU$, where L is a lower triangular while U is upper triangular. When row interchanges are needed let P be the permutation matrix that creates these row interchanges, then the LU decomposition can be carried out for the matrix PA, i.e., $PA = LU$. This decomposition is known as PLU *decomposition*.

Theorem 10.10 (*PLU* decomposition theorem) *For any* $m \times n$ *matrix A, there exist a permutation matrix P, an* $m \times m$ *lower triangular matrix L and an* $m \times n$ *row echelon matrix U such that* $PA = LU$.

Proof It is clear that permutation of two rows has no bearing on the use of elementary row operation $R_i \rightarrow R_i + \alpha R_j$ in the reduction of A to row echelon form. Thus, for any matrix A, there exists a permutation matrix P such that PA can be reduced to row echelon form without requiring further permutation of rows. Hence, Theorem 10.1 guarantees that there exists a suitable matrix L and U such that $PA = LU$.

Example 10.11 *PLU* decomposition of the matrix $A = \begin{bmatrix} 2 & 1 & 0 & 1 \\ 2 & 1 & 2 & 3 \\ 0 & 0 & 1 & 2 \\ -4 & -1 & 0 & -2 \end{bmatrix}$. Since

LU decomposition of A is not possible, let $P = \begin{bmatrix} 1 & 0 & 0 & 0 \\ 0 & 0 & 0 & 1 \\ 0 & 0 & 1 & 0 \\ 0 & 1 & 0 & 0 \end{bmatrix}$ be a permutation matrix

and let

$$A' = PA = \begin{bmatrix} 1 & 0 & 0 & 0 \\ 0 & 0 & 0 & 1 \\ 0 & 0 & 1 & 0 \\ 0 & 1 & 0 & 0 \end{bmatrix} \begin{bmatrix} 2 & 1 & 0 & 1 \\ 2 & 1 & 2 & 3 \\ 0 & 0 & 1 & 2 \\ -4 & -1 & 0 & -2 \end{bmatrix} = \begin{bmatrix} 2 & 1 & 0 & 1 \\ -4 & -1 & 0 & -2 \\ 0 & 0 & 1 & 2 \\ 2 & 1 & 2 & 2 \end{bmatrix}.$$

Now we shall find *LU* decomposition of A'.

$$\begin{bmatrix} 2 & 1 & 0 & 1 \\ -4 & -1 & 0 & -2 \\ 0 & 0 & 1 & 2 \\ 2 & 1 & 2 & 2 \end{bmatrix} = \begin{bmatrix} 1 & 0 & 0 & 0 \\ 0 & 1 & 0 & 0 \\ 0 & 0 & 1 & 0 \\ 0 & 0 & 0 & 1 \end{bmatrix} \begin{bmatrix} 2 & 1 & 0 & 1 \\ -4 & -1 & 0 & -2 \\ 0 & 0 & 1 & 2 \\ 2 & 1 & 2 & 2 \end{bmatrix}.$$

Applying the row operations $R_2 \rightarrow R_2 + 2R_1$ and $R_4 \rightarrow R_4 - R_1$ on the matrix at the right side and updating column of the identity matrix in left, we get

$$\begin{bmatrix} 1 & 0 & 0 & 0 \\ -2 & 1 & 0 & 0 \\ 0 & 0 & 1 & 0 \\ 1 & 0 & 0 & 1 \end{bmatrix} \begin{bmatrix} 2 & 1 & 0 & 1 \\ 0 & 1 & 0 & 0 \\ 0 & 0 & 1 & 2 \\ 0 & 0 & 2 & 1 \end{bmatrix}.$$

Further, using the row operation $R_4 \rightarrow R_4 - 2R_3$, we find that

$$\begin{bmatrix} 1 & 0 & 0 & 0 \\ -2 & 1 & 0 & 0 \\ 0 & 0 & 1 & 0 \\ 1 & 0 & 2 & 1 \end{bmatrix} \begin{bmatrix} 2 & 1 & 0 & 1 \\ 0 & 1 & 0 & 0 \\ 0 & 0 & 1 & 2 \\ 0 & 0 & 0 & -3 \end{bmatrix}.$$

Thus, $A' = PA = LU$, where $L = \begin{bmatrix} 1 & 0 & 0 & 0 \\ -2 & 1 & 0 & 0 \\ 0 & 0 & 1 & 0 \\ 1 & 0 & 2 & 1 \end{bmatrix}$, $U = \begin{bmatrix} 2 & 1 & 0 & 1 \\ 0 & 1 & 0 & 0 \\ 0 & 0 & 1 & 2 \\ 0 & 0 & 0 & -2 \end{bmatrix}$.

Hence, $A = P^2 A = P(PA) = PLU$ which yields PLU decomposition of A, i.e.,

$$\begin{bmatrix} 2 & 1 & 0 & 1 \\ 2 & 1 & 2 & 3 \\ 0 & 0 & 1 & 2 \\ -4 & -1 & 0 & -2 \end{bmatrix} = \begin{bmatrix} 1 & 0 & 0 & 0 \\ 0 & 0 & 0 & 1 \\ 0 & 0 & 1 & 0 \\ 0 & 1 & 0 & 0 \end{bmatrix} \begin{bmatrix} 1 & 0 & 0 & 0 \\ -2 & 1 & 0 & 0 \\ 0 & 0 & 1 & 0 \\ 1 & 0 & 2 & 1 \end{bmatrix} \begin{bmatrix} 2 & 1 & 0 & 1 \\ 0 & 1 & 0 & 0 \\ 0 & 0 & 1 & 2 \\ 0 & 0 & 0 & -2 \end{bmatrix}.$$

Solution of the System of Equations Using PLU Decomposition

PLU decomposition of a matrix can be used to solve the system of equations. Let $AX = B$, where $A \in M_n(\mathbb{C})$ be a system of equations. Find a permutation matrix P such that $PA = LU$, where $L = \{(\ell_{ij}) \mid \ell_{ij} = 0, i < j\}$ is lower triangular and $U = \{(u_{ij}) \mid u_{ij} = 0, i > j\}$ is upper triangular. Therefore, we have $PAX = PB$ and hence $LUX = PB$. Now solve the systems

$$LY = PB \quad and \quad UX = Y.$$

Then $LUX = LY = PB$ and hence X is a solution of this system. This has an advantage over the direct Gaussian elimination because the systems $LY = PB$ and $UX = Y$ are triangular and are easy to solve.

For the first of the systems $LY = PB$, let $PB = \begin{bmatrix} b_1 \\ b_2 \\ \vdots \\ b_n \end{bmatrix}$. Then it can be easily

seen that back solution can be used to determine Y, i.e., we have recursive relations

$$y_1 = \frac{b_1}{\ell_{11}}$$
$$y_2 = \frac{b_2 - \ell_{21} y_1}{\ell_{22}}$$
$$\vdots$$
$$y_n = \frac{\left(b_n - \sum_{i=1}^{n-1} \ell_{ni} y_i\right)}{\ell_{nn}}$$

A similar procedure can be adopted to solve $UX = Y$. The upper triangular system $UX = Y$ can be written as the set of linear equations:

$$\begin{aligned} u_{11}x_1 + u_{12}x_2 + \cdots + u_{1n}x_n &= y_1 \\ u_{22}x_2 + \cdots + u_{2n}x_n &= y_2 \\ \vdots \quad\quad &= \vdots \\ u_{nn}x_n &= y_n \end{aligned}$$

The back solution is $x_n = \frac{y_n}{u_{nn}}$, $x_{n-1} = \frac{y_{n-1} - u_{n-1n}x_n}{u_{n-1n-1}}$, $x_1 = \frac{y_1 - \sum\limits_{k=1}^{n} u_{1k}x_k}{u_{11}}$.

Remark 10.12 (*i*) In practice, the step of determining and then multiplying by the permutation matrix is not actually carried out. Rather, an indent array is generated, while the elimination step is accomplished that effectively inter-changes a pointer to the row interchanges. This saves considerable time in solving potentially very large systems.

(*ii*) If any of the diagonal element u_{ii} is zero, then the system is singular and cannot be solved.

(*iii*) If all diagonal elements of U are nonzero, then the system has unique solution.

Example 10.13 Use *PLU* factorization of $A = \begin{bmatrix} 1 & 2 & 3 & 4 \\ 1 & 2 & 3 & 0 \\ 5 & 3 & 1 & 1 \end{bmatrix}$ and solve the system

of equations $AX = B$, where $B = \begin{bmatrix} 1 \\ 2 \\ 3 \end{bmatrix}$.

We proceed to find row reduced echelon form of the matrix A. First add -1 times of the first row to the second row and then add -5 times the first row to the third row of A to get

$$\begin{bmatrix} 1 & 0 & 0 \\ 1 & 1 & 0 \\ 5 & 0 & 1 \end{bmatrix} \begin{bmatrix} 1 & 2 & 3 & 4 \\ 0 & 0 & 0 & -4 \\ 0 & -7 & -14 & -19 \end{bmatrix}.$$

Now there is no way to obtain upper triangular matrix by using row operation of replacing a row with itself added to a multiple of another row to the second matrix (without interchanging any two rows). Now consider the matrix A' by switching the last two rows of A

$$A' = \begin{bmatrix} 1 & 2 & 3 & 4 \\ 5 & 3 & 1 & 1 \\ 1 & 2 & 3 & 0 \end{bmatrix} = \begin{bmatrix} 1 & 0 & 0 \\ 0 & 1 & 0 \\ 0 & 0 & 1 \end{bmatrix} \begin{bmatrix} 1 & 2 & 3 & 4 \\ 5 & 3 & 1 & 1 \\ 1 & 2 & 3 & 0 \end{bmatrix}.$$

Now add -1 times of the first row to the third row and then add -5 times the first row to the second row to get

$$\begin{bmatrix} 1 & 0 & 0 \\ 5 & 1 & 0 \\ 1 & 0 & 1 \end{bmatrix} \begin{bmatrix} 1 & 2 & 3 & 4 \\ 0 & -7 & -14 & -19 \\ 0 & 0 & 0 & -4 \end{bmatrix}.$$

The first matrix is lower triangular while the second matrix is upper triangular and hence A' has LU decomposition.

Thus, $A' = PA = LU$, where L and U are given above and $P = \begin{bmatrix} 1 & 0 & 0 \\ 0 & 0 & 1 \\ 0 & 1 & 0 \end{bmatrix}$.

Hence, $A = P^2 A = P(PA) = PLU$ and

$$\begin{bmatrix} 1\,2\,3\,4 \\ 1\,2\,3\,0 \\ 5\,3\,1\,1 \end{bmatrix} = \begin{bmatrix} 1\,0\,0 \\ 0\,0\,1 \\ 0\,1\,0 \end{bmatrix} \begin{bmatrix} 1\,0\,0 \\ 5\,1\,0 \\ 1\,0\,1 \end{bmatrix} \begin{bmatrix} 1 & 2 & 3 & 4 \\ 0 & -7 & -14 & -19 \\ 0 & 0 & 0 & -4 \end{bmatrix}.$$

To solve the system of equations $AX = B$, let $UX = Y$ and consider $PLY = B$. In other words, solve

$$\begin{bmatrix} 1\,0\,0 \\ 0\,0\,1 \\ 0\,1\,0 \end{bmatrix} \begin{bmatrix} 1\,0\,0 \\ 5\,1\,0 \\ 1\,0\,1 \end{bmatrix} \begin{bmatrix} y_1 \\ y_2 \\ y_3 \end{bmatrix} = \begin{bmatrix} 1 \\ 2 \\ 3 \end{bmatrix}.$$

Multiplying both sides by P, we find that $\begin{bmatrix} 1\,0\,0 \\ 5\,1\,0 \\ 1\,0\,1 \end{bmatrix} \begin{bmatrix} y_1 \\ y_2 \\ y_3 \end{bmatrix} = \begin{bmatrix} 1 \\ 3 \\ 2 \end{bmatrix}$ and hence $Y =$

$\begin{bmatrix} y_1 \\ y_3 \\ y_2 \end{bmatrix} = \begin{bmatrix} 1 \\ -2 \\ 1 \end{bmatrix}$. Now solve the system $UX = Y$, i.e.,

$$\begin{bmatrix} 1 & 2 & 3 & 4 \\ 0 & -7 & -14 & -19 \\ 0 & 0 & 0 & -4 \end{bmatrix} \begin{bmatrix} x_1 \\ x_2 \\ x_3 \\ x_4 \end{bmatrix} = \begin{bmatrix} 1 \\ -2 \\ 1 \end{bmatrix}.$$

This yields that

$$\begin{bmatrix} x_1 \\ x_2 \\ x_3 \\ x_4 \end{bmatrix} = \begin{bmatrix} \frac{49t+53}{7} \\ \frac{-14t-93}{7} \\ t \\ -\frac{1}{4} \end{bmatrix}, \ t \in \mathbb{R}.$$

Exercises

1. If A is any $n \times n$ matrix, then show that A can be factored as $A = PLU$, where L is lower triangular, U is upper triangular and P is a permutation matrix which can be obtained by the interchanging rows of I_n appropriately.
2. Show that the product of many finitely lower triangular matrices is a lower triangular matrix, and apply this result to show that the product of many finitely upper triangular matrices is upper triangular.
3. Let $A = \begin{bmatrix} p & q \\ r & s \end{bmatrix}$. Then

   (a) prove that if $p \neq 0$, then A has unique LU decomposition with $1's$ along the main diagonal of L;
   (b) find the LU decomposition described in (a).

4. Show that $A = \begin{bmatrix} 1 & 6 & 2 \\ 2 & 12 & 5 \\ -1 & -3 & -1 \end{bmatrix}$ does not have a LU decomposition. Moreover,

reorder the rows of A and find a LU decomposition of new matrix, and hence solve the system of equations:

$$\begin{aligned} x_1 + 6x_2 + 2x_3 &= 9, \\ 2x_1 + 12x_2 + 5x_3 &= 7, \\ -x_1 - 3x_2 - x_3 &= 17. \end{aligned}$$

5. Let A be an $n \times n$ matrix with triangular factorization LU. Show that $det(A) = u_{11} + u_{22} + \cdots + u_{nn}$.

6. Find LU factorization of $A = \begin{bmatrix} 1 & 1 & 1 \\ 2 & 4 & 1 \\ -3 & 1 & -2 \end{bmatrix}$, where L is a lower triangular matrix

with $1's$ along the main diagonal and U is an upper triangular matrix. Moreover, solve the system $AX = B$ for $B =$

$$(a) \quad \begin{bmatrix} 4 \\ 3 \\ -13 \end{bmatrix}; \qquad\qquad (b) \quad \begin{bmatrix} 7 \\ 23 \\ 0 \end{bmatrix}.$$

7. Use LU decomposition and forward and back solution to solve the system:

$$\begin{bmatrix} 1 & -3 & 2 & -2 \\ 3 & -2 & 0 & -1 \\ 2 & 36 & -28 & 27 \\ 1 & -3 & 22 & 5 \end{bmatrix} \begin{bmatrix} x_1 \\ x_2 \\ x_3 \\ x_4 \end{bmatrix} = \begin{bmatrix} -11 \\ -4 \\ 155 \\ 10 \end{bmatrix}.$$

8. Factor $A = \begin{bmatrix} 3 & -1 & 0 \\ 3 & -1 & 1 \\ 0 & 2 & 1 \end{bmatrix}$ as $A = PLU$, where P is obtained from I_3 by inter-

changing rows appropriately, L is lower triangular and U is upper triangular matrix.

10.3 Eigenvalue Approximations

The eigenvalues of a square matrix can be obtained by solving its characteristic equation. In practical problems, especially those involving matrices of large orders, that method for calculating eigenvalues has many computational difficulties; therefore, other methods for finding eigenvalues are needed. In this section, we will study a simple algorithm, called the power method, that gives an approximation to the eigenvalue of the greatest absolute value and a corresponding eigenvector. In many applications, only the dominant eigenvalue and eigenvector of a matrix are needed, there power method can be tried. However, if additional eigenvalues and eigenvectors are needed, then other methods are required. These methods will not involve the characteristic polynomial. To observe that there are some advantages to work directly with the matrix, we must determine the effect that minor changes in the entries of A have upon the eigenvalues. We have proved a result related to this.

Definition 10.14 Let A be a square matrix. An eigenvalue of A is called the *dominant eigenvalue* of A if its absolute value is larger than the absolute values of the remaining eigenvalues. An eigenvector corresponding to the dominant eigenvalue is called a *dominant eigenvector* of A.

Theorem 10.15 *Let A be a matrix of order $n \times n$ with a complete set of eigenvectors and let X be a matrix that diagonalizes A, i.e.,*

$$X^{-1}AX = D = \begin{bmatrix} \lambda_1 & & & \\ & \lambda_2 & & \\ & & \ddots & \\ & & & \lambda_n \end{bmatrix}.$$

If $A' = A + E$ and λ' is an eigenvalue of A', then $\displaystyle\min_{1 \le i \le n} |\lambda' - \lambda_i| \le cond_2(X)||E||_2$.

Proof If λ' is equal to any of the $\lambda_i's$, then nothing to do. Now suppose that λ' is unequal to any of the $\lambda_i's$. Thus, if we set $D_1 = D - \lambda'I$, then D_1 is a nonsingular matrix. As λ' is an eigenvalue of A', it is also an eigenvalue of $X^{-1}A'X$. Therefore, $X^{-1}A'X - \lambda'I$ is singular and hence $D_1^{-1}(X^{-1}A'X - \lambda'I)$ is also singular. On the other hand, $D_1^{-1}(X^{-1}A'X - \lambda'I) = D_1^{-1}X^{-1}(A + E - \lambda'I)X = D_1^{-1}X^{-1}EX + D_1^{-1}X^{-1}(A - \lambda'I)X = D_1^{-1}X^{-1}EX + D_1^{-1}X^{-1}(XDX^{-1} - \lambda'I)X = D_1^{-1}X^{-1}EX + D_1^{-1}(D - \lambda'I)$. Now using the fact that $D_1 = D - \lambda'I$, we conclude that $D_1^{-1}(X^{-1}A'X - \lambda'I) = D_1^{-1}X^{-1}EX + I$. This implies that $|D_1^{-1}X^{-1}EX - (-1)I| = 0$, i.e., (-1) is an eigenvalue of $D_1^{-1}X^{-1}EX$. It follows that $|-1| \le ||D_1^{-1}X^{-1}EX||_2 \le ||D_1^{-1}||_2 cond_2(X)||E||_2$. The two-norm of D_1^{-1} is given by $D_1^{-1} = \displaystyle\max_{1 \le i \le n} |\lambda' - \lambda_i|^{-1}$. The index i that maximizes $|\lambda' - \lambda_i|^{-1}$ is the same index that minimizes $|\lambda' - \lambda_i|$. Thus, $\displaystyle\min_{1 \le i \le n} |\lambda' - \lambda_i|^{-1} \le cond_2(X)||E||_2$.

Note 10.16 If the matrix A is symmetric, we can choose an orthogonal diagonalizing matrix. In general, if Q is any orthogonal matrix, then $cond_2(Q) = ||Q||_2||Q^{-1}||_2 = 1$ and hence the conclusion of Theorem 10.15 simplifies to $\displaystyle\min_{1 \le i \le n} |\lambda' - \lambda_i|^{-1} \le ||E||_2$.

Thus, if A is symmetric and $||E_2||$ is small, the eigenvalues of A' will be close to the eigenvalues of A.

Power Method

Let A be an $n \times n$ matrix with eigenvalues $\lambda_1, \lambda_2, \ldots, \lambda_n$ such that $|\lambda_1| > |\lambda_2| \ge \cdots \ge |\lambda_n|$. These eigenvalues are unknown to us. This method determines the dominant eigenvalue, i.e., λ_1 and a corresponding eigenvector. To see the idea behind the method, let us assume that A has n linearly independent eigenvectors $X_1, X_2, \ldots, X_n$ corresponding to the eigenvalues $\lambda_1, \lambda_2, \ldots, \lambda_n$, respectively. Given an arbitrary vector X_0 in $\mathbb{R}^n$, we can write $X_0 = \alpha_1 X_1 + \alpha_2 X_2 + \cdots + \alpha_n X_n$. This implies that $AX_0 = \alpha_1 \lambda_1 X_1 + \alpha_2 \lambda_2 X_2 + \cdots + \alpha_n \lambda_n X_n$. This further implies that $A^2 X_0 = \alpha_1 \lambda_1^2 X_1 + \alpha_2 \lambda_2^2 X_2 + \cdots + \alpha_n \lambda_n^2 X_n$. Moving ahead in this way at kth step, we arrive at $A^k X_0 = \alpha_1 \lambda_1^k X_1 + \alpha_2 \lambda_2^k X_2 + \cdots + \alpha_n \lambda_n^k X_n$. If we define $A^k X_0 = X_k$; $k = 1, 2, \ldots$, then we have $\frac{1}{\lambda_1^k} X_k = \alpha_1 X_1 + \alpha_2 (\frac{\lambda_2}{\lambda_1})^k X_2 + \cdots + \alpha_n (\frac{\lambda_n}{\lambda_1})^k X_n$. Since $|\frac{\lambda_i}{\lambda_1}| < 1$ for $i = 2, 3, \ldots, n$, it follows that $\frac{1}{\lambda_1^k} X_k \to \alpha_1 X_1$ as $k \to \infty$. Thus, if $\alpha_1 \neq 0$, the sequence $\{(\frac{1}{\lambda_1^k}) X_k\}$ converges to an eigenvector $\alpha_1 X_1$ of A. Since $\alpha_1 X_1$ is a dominant eigenvector of A, this shows that the limiting value of the sequence $\{(\frac{1}{\lambda_1^k}) X_k\}$ gives us a dominant eigenvector of A. Let us denote this limiting value, i.e., the dominant eigenvector by a vector Y. Thus, till this stage we have determined a dominant eigenvector Y of A. We now show how to approximate the dominant eigenvalue once an approximation Y to a dominant eigenvector is known. Let λ be an eigenvalue of A and X a corresponding eigenvector. If $\langle, \rangle$ denotes the Euclidean inner product, then

$$\frac{\langle X, AX \rangle}{\langle X, X \rangle} = \frac{\langle X, \lambda X \rangle}{\langle X, X \rangle} = \frac{\lambda \langle X, X \rangle}{\langle X, X \rangle} = \lambda.$$

Thus, if Y is an approximation to a dominant eigenvector, the dominant eigenvalue λ_1 can be approximated by

$$\lambda_1 \approx \frac{\langle Y, AY \rangle}{\langle Y, Y \rangle}.$$

It is to be noted that we have not scaled the sequence $\{(\frac{1}{\lambda_1^k}) X_k\}$ in the process. On the other hand, if we scale the sequence $\{X_k\}$, then one gets unit vectors at each step and the sequence converges to a unit vector in the direction of X_1. The eigenvalue λ_1 can be computed at the same time.

We now summarize the steps in the power method with scaling as following:

(1) Pick an arbitrary nonzero vector X_0.
(2) Compute AX_0 and scale down to obtain the first approximation to a dominant eigenvector. Say it X_1.
(3) Compute AX_1 and scale down to obtain the second approximation X_2.
(4) Compute AX_2 and scale down to obtain the third approximation X_3.

Continuing in this way, a succession $X_0, X_1, X_2, \ldots$ of better and better approximations to a dominant eigenvector will be obtained, and in each step, dominant eigenvalue λ_1 is approximated by $\frac{\langle X_i, AX_i \rangle}{\langle X_i, X_i \rangle}$, where $i : 1, 2, \ldots$.

Example 10.17 Approximate a dominant eigenvector and the dominant eigenvalue of the matrix A by using power method with scaling, where $A = \begin{bmatrix} 1 & 1 \\ 1 & 3 \end{bmatrix}$.

We arbitrarily choose

$$X_0 = \begin{bmatrix} 1 \\ 1 \end{bmatrix}$$

as an initial approximation of a dominant vector. Multiplying X_0 by A and scaling down yield

$$AX_0 = \begin{bmatrix} 1 & 1 \\ 1 & 3 \end{bmatrix} \begin{bmatrix} 1 \\ 1 \end{bmatrix} = \begin{bmatrix} 2 \\ 4 \end{bmatrix}, \quad X_1 = \frac{1}{4} \begin{bmatrix} 2 \\ 4 \end{bmatrix} = \begin{bmatrix} .5 \\ 1 \end{bmatrix}.$$

Multiplying X_1 by A and scaling down yield

$$AX_1 = \begin{bmatrix} 1 & 1 \\ 1 & 3 \end{bmatrix} \begin{bmatrix} .5 \\ 1 \end{bmatrix} = \begin{bmatrix} 1.5 \\ 3.5 \end{bmatrix}, \quad X_2 = \frac{1}{3.5} \begin{bmatrix} 1.5 \\ 3.5 \end{bmatrix} = \begin{bmatrix} .429 \\ 1 \end{bmatrix}.$$

Hence, the first approximation of the dominant eigenvalue is

$$\lambda_1 \approx \frac{\langle X_1, AX_1 \rangle}{\langle X_1, X_1 \rangle} = \frac{(.5)(1.5) + 3.5}{(.5)(.5) + 1} = 3.4.$$

Multiplying X_2 by A and scaling down yield

$$AX_2 = \begin{bmatrix} 1 & 1 \\ 1 & 3 \end{bmatrix} \begin{bmatrix} .429 \\ 1 \end{bmatrix} = \begin{bmatrix} 1.429 \\ 3.429 \end{bmatrix}, \quad X_3 = \frac{1}{3.429} \begin{bmatrix} 1.429 \\ 3.429 \end{bmatrix} = \begin{bmatrix} .417 \\ 1 \end{bmatrix}.$$

Hence, the second approximation of the dominant eigenvalue is

$$\lambda_1 \approx \frac{\langle X_2, AX_2 \rangle}{\langle X_2, X_2 \rangle} = \frac{(.429)(1.429) + 3.429}{(.429)(.429) + 1} = 3.414.$$

Multiplying X_3 by A and scaling down yield

$$AX_3 = \begin{bmatrix} 1 & 1 \\ 1 & 3 \end{bmatrix} \begin{bmatrix} .417 \\ 1 \end{bmatrix} = \begin{bmatrix} 1.417 \\ 3.417 \end{bmatrix}, \quad X_4 = \frac{1}{3.417} \begin{bmatrix} 1.417 \\ 3.417 \end{bmatrix} = \begin{bmatrix} .4147 \\ 1 \end{bmatrix}.$$

Hence, the third approximation of the dominant eigenvalue is

$$\lambda_1 \approx \frac{\langle X_3, AX_3 \rangle}{\langle X_3, X_3 \rangle} = \frac{(.417)(1.417) + 3.417}{(.417)(.417) + 1} = 3.414.$$

Continuing in this way, we generate a succession of approximations to a dominant eigenvector and the dominant eigenvalues. Thus, the approximate dominant eigenvalue is 3.414 and the corresponding dominant eigenvector is $\begin{bmatrix} .414 \\ 1 \end{bmatrix}$.

Exercises

1. Let $A = \begin{bmatrix} 2 & 1 \\ 1 & 2 \end{bmatrix}$. Apply three iterations of the power method with any nonzero starting vector and obtain the approximate value of dominant eigenvalue and dominant eigenvector of A. Determine the exact eigenvalues of A by solving characteristic equation and determine the eigenspace corresponding to the largest eigenvalue. Compare the answers you obtained in these two ways.

2. Find the dominant eigenvalue and dominant eigenvector if exist in the following matrices:

$$\begin{bmatrix} -1 & 4 \\ 1 & -1 \end{bmatrix}, \begin{bmatrix} 0 & 1 \\ 4 & 0 \end{bmatrix}, \begin{bmatrix} 4 & 2 & 1 \\ 0 & -5 & 3 \\ 0 & 0 & 6 \end{bmatrix}, \begin{bmatrix} 1 & -12 & 0 \\ 1 & 0 & 0 \\ 0 & 0 & 2 \end{bmatrix}.$$

3. Let $A = \begin{bmatrix} 1 & 2 \\ -1 & -1 \end{bmatrix}$ and $X_0 = \begin{bmatrix} 1 \\ 1 \end{bmatrix}$. Compute X_1, X_2, X_3 and X_4, using power method. Explain why power method will fail to converge in this case.

4. Let $A = \begin{bmatrix} 18 & 17 \\ 2 & 3 \end{bmatrix}$. Use the power method with scaling to approximate the dominant eigenvalue and the dominant eigenvector of A. Start with $X_0 = \begin{bmatrix} 1 \\ 1 \end{bmatrix}$. Round off all computations to three significant digits and stop after three iterations. Also, find the exact value for the dominant eigenvalue and eigenvector.

5. Let $A = \begin{bmatrix} 2 & 1 & 0 \\ 1 & 2 & 0 \\ 0 & 0 & 10 \end{bmatrix}$. Use the power method with scaling to approximate the dominant eigenvalue and the dominant eigenvector of A. Start with $X_0 = \begin{bmatrix} 1 \\ 1 \\ 1 \end{bmatrix}$. Round off all computations to three significant digits and stop after three iterations. Also, find the exact value for the dominant eigenvalue and eigenvector.

6. Let $X = (x_1, x_2, \ldots, x_n)^t$ be an eigenvector of A corresponding to eigenvalue λ. If $|x_i| = ||X||_\infty$, then show that $\sum_{j=1}^{n} a_{ij} = \lambda x_i$ and $|\lambda - a_{ii}| \leq \sum_{j=1, j \neq i}^{n} |a_{ij}|$.

7. Let A be a matrix with eigenvalues $\lambda_1, \lambda_2, \ldots, \lambda_n$ and let λ be an eigenvalue of $A + E$. Let X be a matrix that diagonalizes A, and let $C = X^{-1}EX$. Prove that for some i, $|\lambda - \lambda_i| \leq \sum_{j=1}^{n} |c_{ij}|$ and $\min_{1 \leq j \leq n} |\lambda - \lambda_j| \leq cond_\infty(X)||E||_\infty$.

10.4 Singular Value Decompositions

First two sections of this chapter deal with the LU and PLU decompositions of matrices. In this section, we shall discuss decomposition for rectangular matrices rather than square matrices. This decomposition is known as *Singular Value Decomposition* (SVD). This decomposition is fundamental to numerical analysis and linear algebra. We will describe here its properties and discuss its applications, which are many and growing. Throughout the section, A is an $m \times n$ matrix, where we have assumed $m \geq n$. This assumption is made for convenience only. All the results will also hold if $m < n$.

Definition 10.18 Let A be a matrix of order $m \times n$. The real numbers $\sigma_1, \sigma_2, \ldots, \sigma_n$ are called *singular values* of A if $\sigma_i = \sqrt{\lambda_i}$ for each $i = 1, 2, 3, \ldots, n$; where $\lambda_i's$ are eigenvalues of $A^t A$. The corresponding eigenvectors are called as *singular vectors* of A. It is to be noted that $A^t A$ is positive semi-definite. As a result, all the eigenvalues of $A^t A$ will be always nonnegative, i.e., $\lambda_i \geq 0$ for each $i = 1, 2, 3, \ldots, n$. Also, thus, all the singular values of A are nonnegative.

Definition 10.19 Let A be any matrix of order $m \times n$. Then A can be factored as $A = UDV^t$, where U is an $m \times m$ orthogonal matrix, V is an $n \times n$ orthogonal matrix and D is an $m \times n$ matrix whose off diagonal entries are all $0's$ and whose diagonal elements are shown as follows:

$$
D = \begin{bmatrix}
\sigma_1 & & & & & \\
 & \ddots & & & & \\
 & & \sigma_r & & & \\
 & & & 0 & & \\
 & & & & \ddots & \\
 & & & & & 0
\end{bmatrix},
$$

where $\sigma_1, \sigma_2, \ldots, \sigma_r$ are positive singular values of A such that $\sigma_1 \geq \sigma_2 \geq \cdots \geq \sigma_r$. Such type of factorization of A is known as *singular value decomposition* of A.

We initiate by showing that such a decomposition is always possible. Before moving ahead, we give a remark in which we describe different types of matrix norms which have been used throughout this section.

Remark 10.20 (*i*) Let A be any matrix of order $m \times n$ with real or complex entries. Then the Frobenius norm, sometimes also called the Euclidean norm of an $m \times n$ matrix A, is defined as the square root of the sum of the absolute squares of its elements, i.e., $\|A\|_F = \sqrt{\sum_{i=1}^{m} \sum_{j=1}^{n} |a_{ij}|^2}$ or equivalently $\|A\|_F = \sqrt{\text{trace}(AA^*)}$, where A^* is the tranjugate of A.

(*ii*) Let A be any matrix of order $m \times n$ with real or complex entries. Then

$\|A\|_1 = \max_{1 \leq j \leq n} \sum_{i=1}^{m} |a_{ij}|$, which is simply the maximum absolute column sum of the matrix A;

$\|A\|_\infty = \max_{1 \leq i \leq m} \sum_{j=1}^{n} |a_{ij}|$, which is simply the maximum absolute row sum of the matrix A;

$||A||_2 = [$ Dominant eigenvalue of $(A^*A)]^{\frac{1}{2}}$, where the dominant eigenvalue means the eigenvalue with greatest absolute value.

(iii) We have the following relations among different types of matrix norms. Let A be any matrix of order $m \times n$ with real or complex entries. Then $||A||_2 = \sqrt{||A||_1 ||A||_\infty}$. Also, we have $||A||_2 \leq ||A||_F$, equality holds if and only if A is of rank 1 or the zero matrix.

Theorem 10.21 *Every matrix of order $m \times n$ has a singular value decomposition.*

Proof Let A be any matrix of order $m \times n$. We have $X^t A^t A X = ||AX||_2^2 \geq 0$ for all $X \in \mathbb{R}^n$. This shows that the matrix $A^t A$ is positive semi-definite. As a result, all the eigenvalues of the matrix $(A^t A)$ will be nonnegative. We order these eigenvalues, so that $\lambda_1 \geq \lambda_2 \geq \cdots \geq \lambda_n \geq 0$. Define $\sigma_i = \sqrt{\lambda_i}$. Without loss of generality, one may suppose that exactly r of the $\sigma_i's$ are nonzero, so that $\sigma_1 \geq \sigma_2 \geq \cdots \geq \sigma_r > 0$ and $\sigma_{r+1} = \sigma_{r+2} = \cdots = \sigma_n = 0$.

Set

$$D = \begin{bmatrix} \sigma_1 & & & & & \\ & \ddots & & & & \\ & & \sigma_r & & & \\ & & & 0 & & \\ & & & & \ddots & \\ & & & & & 0 \end{bmatrix} = \begin{bmatrix} D_r & O \\ O & O \end{bmatrix},$$

where D_r is an $r \times r$ diagonal matrix whose diagonal entries are $\sigma_1, \ldots, \sigma_r$. Since $A^t A$ is symmetric, there is an orthogonal matrix V that diagonalizes $(A^t A)$. This implies that $V^t A^t A V = D^t D$. Let $V_1, V_2, \ldots, V_n$ be the column vectors of V. Hence, V_i is an eigenvector corresponding to eigenvalue σ_i^2 for $i = 1, \ldots, r$ and $V_{r+1}, \ldots, V_n$ are eigenvectors corresponding to eigenvalue 0. Now using the column vectors $V_1, V_2, \ldots, V_r, V_{r+1}, \ldots, V_n$, we construct two matrices of orders $n \times r$ and $n \times (n-r)$ denoted by P and Q, respectively, such that $P = [V_1, V_2, \ldots, V_r]$ and $Q = [V_{r+1}, V_{r+2}, \ldots, V_n]$. Since $A^t A V_i = 0$ for $i = r+1, \ldots, n$, it follows that $(AQ)^t AQ = Q^t A^t AQ = O$ and hence $AQ = O$. As $A^t A P = P D_r^2$, it follows that $D_r^{-1} P^t A^t A P D_r^{-1} = D_r^{-1} P^t P D_r^2 D_r^{-1} = D_r^{-1} P^t P D_r = I_r$, where I_r denotes the identity matrix of order $r \times r$. If we put $S = A P D_r^{-1}$, then $S^t S = I_r$ and, therefore, S is an $m \times r$ matrix with orthonormal column vectors $U_1, \ldots, U_r$. Now the set $\{U_1, \ldots, U_r\}$ can be extended to form an orthonormal basis for $\mathbb{R}^m$. Let the extended basis be $\{U_1, \ldots, U_r, U_{r+1}, \ldots, U_m\}$. Now set $T = [U_{r+1}, \ldots, U_m]$ and $U = [U_1, \ldots, U_r, U_{r+1}, \ldots, U_m] = [S \ T]$. It is obvious to observe that T and U are matrices of orders $m \times (m-r)$ and $m \times m$, respectively. This implies that

$$U^t A V = \begin{bmatrix} S^t \\ T^t \end{bmatrix} A [P \quad Q]$$

$$= \begin{bmatrix} S^t \\ T^t \end{bmatrix} [AP \quad O]$$

$$= \begin{bmatrix} S^t AP & O \\ T^t AP & O \end{bmatrix}.$$

But we get $S^t AP = D_r^{-1} P^t A^t AP = D_r$ and $T^t AP = T^t S D_r = O$. Therefore, $U^t A V = D$ and thus $A = U D V^t$. Here U and V are orthogonal matrices and D is a matrix in block form.

It is to be noted that A and D are equivalent matrices because U and V are nonsingular matrices. But we know that equivalent matrices have the same rank. Thus, we conclude that $rank\ A = rank\ D_r = r$, the number of nonzero singular values of A (provided counting should be done according to multiplicity).

In this proof, we have supposed that $m \geq n$. But if $m < n$, then using this proof we can also find the singular value decomposition described as follows: Let A be a matrix of order $m \times n$, where $m < n$. Then A^t will be a matrix of order $n \times m$, where $n > m$, now using the above proof we can get a singular value decomposition of A^t. This implies that there exist orthogonal matrices U and V such that $A^t = U D V^t$, where U is of order $n \times n$, V is of order $m \times m$ and D of order $n \times m$, respectively. Finally, we get $A = V D^t U^t = V D' U^t$, where $D^t = D'$. Clearly, A has a singular value decomposition and here D' is a block matrix whose off diagonal entries are zero and whose diagonal elements are given as follows:

$$D' = \begin{bmatrix} \sigma_1 & & & & & & \\ & \sigma_2 & & & & & \\ & & \ddots & & & & \\ & & & \sigma_r & & & \\ & & & & 0 & & \\ & & & & & \ddots & \\ & & & & & & 0 \end{bmatrix},$$

where $\sigma_1, \sigma_2, \ldots, \sigma_r$ are positive singular values of A^t. But we know that A and A^t have the same set of positive singular values. Therefore, A possesses a singular value decomposition.

Remark 10.22 Let A be an $m \times n$ matrix having a singular value decomposition $U D V^t$. Then

(i) Since $A = U D V^t$, this shows that $A^t A = V D^t U^t U D V^t$, i.e., $A^t A = V D^t D V^t$ and

$$D = \begin{bmatrix} \sigma_1^2 & & & & \\ & \sigma_2^2 & & & \\ & & \ddots & & \\ & & & \ddots & \\ & & & & \sigma_n^2 \end{bmatrix}.$$

It follows that the eigenvalues of $A^t A$ are $\sigma_1^2, \sigma_2^2, \ldots, \sigma_n^2$. As σ_i^s are the non-negative square roots of the eigenvalues of $A^t A$, they are unique.

(ii) Singular value decomposition of A is not unique. If we notice in the proof of the above theorem, then we come across the extension of the set $\{U_1, U_2, \ldots, U_r\}$ to an orthogonal basis of $\mathbb{R}^m$. With the help of this basis, we obtain U. But we know that this extension is not unique in general. As a result, U is also not unique. Also ,it is obvious to observe that V is also not unique. Thus, we have shown that singular value decomposition of A is not unique.

We can also justify this fact by giving a counter example as follows: Let $A = I_n$. If we set $D = I_n$, and take $U = V$ to be any arbitrary $n \times n$ orthogonal matrix, then $A = UDV^t$ will hold. Thus, uniqueness of decomposition stands disproved.

(iii) Since V diagonalizes $A^t A$, it shows that the $V_i^s : i = 1, 2, \ldots, n$ are eigenvectors of $A^t A$. Also, as $AA^t = UDV^t V D^t U^t = UDD^t U^t$, it follows that U diagonalizes AA^t and the $U_i^s : i = 1, 2, \ldots, m$ are eigenvectors of AA^t.

(iv) It is also easy to observe that $AV = UD$. Now comparing ith columns of each side of the previous equation, we have $AV_i = \sigma_i U_i; i = 1, 2, \ldots, n$. Similarly, we also have $A^t U = VD^t$ and we conclude that $A^t U_i = \sigma_i V_i$; for $i = 1, 2, \ldots, n$; and $A^t U_i = O$; for $i = n+1, n+2, \ldots, m$. The V_i^s are called the *right singular vectors* of A and the U_i^s are called the *left singular vectors* of A.

Note 10.23 Every complex matrix has a singular value decomposition.

Example 10.24 Find the singular values and a singular value decomposition of A, where $A = \begin{bmatrix} 0 & 2 \\ 2 & -1 \\ 1 & 0 \end{bmatrix}$.

The matrix

$$A^t A = \begin{bmatrix} 5 & -2 \\ -2 & 5 \end{bmatrix}$$

has eigenvalues $\lambda_1 = 7$ and $\lambda_2 = 3$. As a result, the singular values of A are $\sigma_1 = \sqrt{7}$ and $\sigma_2 = \sqrt{3}$. Thus, $D = \begin{bmatrix} \sqrt{7} & 0 \\ 0 & \sqrt{3} \\ 0 & 0 \end{bmatrix}$. The eigenvectors corresponding to the eigenvalues λ_1 and λ_2 will be of the form $\alpha \begin{bmatrix} 1 \\ 1 \end{bmatrix}$ and $\beta \begin{bmatrix} 1 \\ -1 \end{bmatrix}$, respectively, where α, β

are nonzero real numbers. Therefore, the orthogonal matrix $V = \frac{1}{\sqrt{2}} \begin{bmatrix} 1 & 1 \\ 1 & -1 \end{bmatrix}$ diag-

onalizes $A^t A$. As σ_1 and σ_2 both are nonzero, using (iv) of the above remark, U_1 and U_2 wil be given by

$$U_1 = \frac{1}{\sigma_1} A V_1 = \frac{1}{\sqrt{7}} \begin{bmatrix} 0 & 2 \\ 2 & -1 \\ 1 & 0 \end{bmatrix} \begin{bmatrix} \frac{1}{\sqrt{2}} \\ \frac{1}{\sqrt{2}} \end{bmatrix} = \begin{bmatrix} \frac{2}{\sqrt{14}} \\ \frac{1}{\sqrt{14}} \\ \frac{1}{\sqrt{14}} \end{bmatrix};$$

$$U_2 = \frac{1}{\sigma_2} A V_2 = \frac{1}{\sqrt{3}} \begin{bmatrix} 0 & 2 \\ 2 & -1 \\ 1 & 0 \end{bmatrix} \begin{bmatrix} \frac{1}{\sqrt{2}} \\ \frac{-1}{\sqrt{2}} \end{bmatrix} = \begin{bmatrix} \frac{-2}{\sqrt{6}} \\ \frac{3}{\sqrt{6}} \\ \frac{1}{\sqrt{6}} \end{bmatrix}.$$

The set $\{U_1, U_2\}$ can be extended to an orthonormal basis $\{U_1, U_2, U_3\}$ of $\mathbb{R}^3$, where

$$U_3 = \begin{bmatrix} \frac{1}{\sqrt{21}} \\ \frac{2}{\sqrt{21}} \\ \frac{-4}{\sqrt{21}} \end{bmatrix}.$$ It gives us

$$A = UDV^t = \begin{bmatrix} \frac{2}{\sqrt{14}} & \frac{-2}{\sqrt{6}} & \frac{1}{\sqrt{21}} \\ \frac{1}{\sqrt{14}} & \frac{3}{\sqrt{6}} & \frac{2}{\sqrt{21}} \\ \frac{1}{\sqrt{14}} & \frac{1}{\sqrt{6}} & \frac{-4}{\sqrt{21}} \end{bmatrix} \begin{bmatrix} \sqrt{7} & 0 \\ 0 & \sqrt{3} \\ 0 & 0 \end{bmatrix} \begin{bmatrix} \frac{1}{\sqrt{2}} & \frac{1}{\sqrt{2}} \\ \frac{1}{\sqrt{2}} & \frac{-1}{\sqrt{2}} \end{bmatrix},$$

which is a singular value decomposition of A.

Note 10.25 Let A be a matrix of order $m \times n$. Let r be the rank of A and $0 < k < r$. Let Λ be the set of $m \times n$ matrices of rank k or less. It can be proved that there exists a matrix $X \in \Lambda$ such that $||A - X|| = min_{S \in \Lambda} ||A - S||_F$, i.e., minimum is achieved in Λ with regard to the Frobenius norm.

Let A be a matrix of order $m \times n$. Let r be the rank of A and $0 < k < r$. Singular value decomposition of A can be used to find a matrix of order $m \times n$ of rank k which is closest to A with regard to the Frobenius norm. Assuming the above remark, we will discuss how such a matrix X can be derived from the singular value decomposition of A.

Lemma 10.26 *Let A be a matrix of order $m \times n$. If Q is an orthogonal matrix of order $m \times m$, then $||QA||_F = ||A||_F$.*

Proof Let $Q = [q_{ij}]_{m \times m}$ and $A = [a_{jk}]_{m \times n}$. Suppose that $QA = [\delta_{ik}]_{m \times n}$, where $\delta_{ik} = \sum_{j=1}^{m} q_{ij} a_{jk}$. As Q is an orthogonal matrix, we have $\sum_{i=1}^{m} q_{ij} q_{il} = 1$, if $j = l$ and 0 otherwise. Hence,

$$||QA||_F^2 = ||[\sum_{j=1}^{m} q_{ij} a_{jk}]||_F = \sum_{i=1}^{m} \sum_{k=1}^{n} [\sum_{j=1}^{m} q_{ij} a_{jk}]^2.$$

Using the relation just given above, we obtain that

$$||QA||_F^2 = \sum_{i=1}^{m} \sum_{k=1}^{n} [a_{ik}^2] = ||A||_F^2.$$

Note 10.27 If A has singular value decomposition UDV^t, then using the above lemma, we have $||A||_F = ||UDV^t||_F = ||DV^t||_F = ||(DV^t)^t||_F = ||VD^t||_F = ||D^t||_F = ||D||_F$. This follows that $||A||_F = (\sigma_1^2 + \sigma_2^2 + \cdots + \sigma_n^2)^{\frac{1}{2}}$, where $\sigma_1, \sigma_2, \ldots, \sigma_n$ are the singular values of A.

Theorem 10.28 *Let A be a matrix of order $m \times n$ of rank r and $0 < k < r$. Let $A = UDV^t$ be a singular value decomposition and let Λ denotes the set of all matrices of order $m \times n$ of rank k or less. Assuming that minimum is achieved in Λ, i.e., there exists a matrix $X \in \Lambda$ such that $||A - X||_F = min_{S \in \Lambda}||A - S||_F$, then $||A - X||_F = (\sigma_{k+1}^2 + \sigma_{k+2}^2 + \cdots + \sigma_n^2)^{\frac{1}{2}}$. In particular, if $A' = UD'V^t$, where*

$$D' = \begin{bmatrix} \sigma_1 & & & & & & \\ & \sigma_2 & & & & & \\ & & \ddots & & & & \\ & & & \sigma_k & & & \\ & & & & 0 & & \\ & & & & & \ddots & \\ & & & & & & 0 \end{bmatrix} = \begin{bmatrix} D_k' & O \\ O & O \end{bmatrix},$$

then $||A - A'||_F = (\sigma_{k+1}^2 + \sigma_{k+2}^2 + \cdots + \sigma_n^2)^{\frac{1}{2}} = min_{S \in \Lambda}||A - S||_F$.

Proof Let X be a matrix in Λ satisfying $||A - X||_F = min_{S \in \Lambda}||A - S||_F$. Since A' has its rank k, therefore, $A' \in \Lambda$. This implies that $||A - X||_F \leq ||A - A'||_F = ||U(D - D')V^t||_F = ||(D - D')||_F = (\sigma_{k+1}^2 + \sigma_{k+2}^2 + \cdots + \sigma_n^2)^{\frac{1}{2}}$. Next, we will prove that $||A - X||_F \geq (\sigma_{k+1}^2 + \sigma_{k+2}^2 + \cdots + \sigma_n^2)^{\frac{1}{2}}$. Let $X = U_1 D_1 V_1^t$ be a singular value decomposition, where

$$D_1 = \begin{bmatrix} \omega_1 & & & & & & \\ & \omega_2 & & & & & \\ & & \ddots & & & & \\ & & & \omega_k & & & \\ & & & & 0 & & \\ & & & & & \ddots & \\ & & & & & & 0 \end{bmatrix} = \begin{bmatrix} D_{1k} & O \\ O & O \end{bmatrix},$$

where $\omega_1, \omega_2, \ldots, \omega_k$ are at most nonzero singular values of X. If we put $B = U_1^t A V_1$, then $A = U_1 B V_1^t$, and it gives us

$$||A - X||_F = ||U_1(B - D_1)V_1^t||_F = ||(B - D_1)||_F.$$

Let us partition B as D_1, i.e., $B = \begin{bmatrix} B_{11} & B_{12} \\ B_{21} & B_{22} \end{bmatrix}$, where B_{11}, B_{12}, B_{21} and B_{22} are matrices of orders $k \times k$, $k \times (n - k)$, $(n - k) \times k$ and $(m - k) \times (m - k)$, respectively. This implies that

$$||A - X||_F^2 = ||B_{11} - D_{1k}||_F^2 + ||B_{12}||_F^2 + ||B_{21}||_F^2 + ||B_{22}||_F^2.$$

We claim that $B_{12} = O$. For otherwise, define $Y = V_1 \begin{bmatrix} B_{11} & B_{12} \\ O & O \end{bmatrix} U_1$. Obviously, $Y \in \Lambda$ and

$$||A - Y||_F^2 = ||B_{21}||_F^2 + ||B_{22}||_F^2 < ||A - X||_F^2,$$

which contradicts the definition of X. Hence, $B_{12} = O$. Similarly, it can be proved that $B_{21} = 0$. If we put $Z = V_1 \begin{bmatrix} B_{11} & O \\ O & O \end{bmatrix} U_1$, then $Z \in \Lambda$ and we have

$$||A - Z||_F^2 = ||B_{22}||_F^2 < ||B_{11} - D_k'||_F^2 + ||B_{22}||_F^2 = ||A - X||_F^2.$$

By using the definition of X, we conclude that $B_{11} = D_k'$. If B_{22} has a singular value decomposition $U_2^t D_2 V_2$, then

$$||A - X||_F = ||B_{22}||_F = ||D_2||_F.$$

Let $U_3 = \begin{bmatrix} I_k & O \\ O & U_2 \end{bmatrix}$ and $V_3 = \begin{bmatrix} I_k & O \\ O & V_2 \end{bmatrix}$. Now $U_3^t U_2^t A V_2 V_3 = \begin{bmatrix} D_{k2}' & O \\ O & D_2 \end{bmatrix}$. This implies that $A = U_2 U_3 \begin{bmatrix} D_{k2}' & O \\ O & D_2 \end{bmatrix} (V_2 V_3)^t$ and here it is clear that the diagonal elements of D_2 are singular values of A. Thus, it follows that

$$||A - X||_F = ||D_2||_F \geq (\sigma_{k+1}^2 + \sigma_{k+2}^2 + \cdots + \sigma_n^2)^{\frac{1}{2}}.$$

Finally, we conclude that

$$||A - X||_F = ||D_2||_2 = (\sigma_{k+1}^2 + \sigma_{k+2}^2 + \cdots + \sigma_n^2)^{\frac{1}{2}} = ||A - A'||_F.$$

Remark 10.29 (i) Let A be an $m \times n$ matrix having a singular value decomposition UDV^t. We define $E_j = U_j V_j^t, j = 1, 2, \ldots, n$; where U_j and V_j are column vectors as discussed in Theorem 10.21. Then each E_j is of rank 1 and we claim that $A = \sigma_1 E_1 + \sigma_2 E_2 + \cdots + \sigma_n E_n$. To prove the claim, notice that $(V_1 V_1^t + \cdots + V_n V_n^t)X = (V_1^t X)V_1 + \cdots + (V_n^t X)V_n = X$ holds for every $X \in \mathbb{R}^n$ due to the fact that the column vectors of V form an orthonormal basis for $\mathbb{R}^n$. This implies that null space $N(V_1 V_1^t + \cdots + V_n V_n^t - I) = $

$\mathbb{R}^n$. Thus, we conclude that $V_1 V_1^t + \cdots + V_n V_n^t = I$. This implies that

$$A = AI$$

$$= A(V_1 V_1^t + \cdots + V_n V_n^t)$$

$$= \sigma_1 U_1 V_1^t + \cdots + \sigma_n U_n V_n^t$$

$$= \sigma_1 E_1 + \cdots + \sigma_n E_n.$$

If A is of rank n, then

$$A' = U \begin{bmatrix} \sigma_1 & & & & & & \\ & \sigma_2 & & & & & \\ & & \ddots & & & & \\ & & & \sigma_{n-1} & & & \\ & & & & 0 & & \\ & & & & & \ddots & \\ & & & & & & 0 \end{bmatrix} V^t$$

$$= \sigma_1 E_1 + \cdots + \sigma_{n-1} E_{n-1}$$

will be a matrix of rank $(n-1)$ which is nearest to A with respect to the Frobenious norm. Similarly, $A'' = \sigma_1 E_1 + \cdots + \sigma_{n-2} E_{n-2}$ will be the nearest matrix of rank $(n-2)$ and so on. Particularly, if A is a nonsingular $n \times n$ matrix, then A' is singular and $||A - A'||_F = \sigma_n$. Thus, σ_n can be taken as a measure of how close a matrix is to being singular.

(ii) We should not use the value of $det\ A$ as a measure of how close A is to being singular. If, for example, A is the matrix of order 100×100 diagonal matrix whose diagonal entries are all $\frac{1}{3}$, then $det\ A = 3^{-100}$, however $\sigma_{100} = \frac{1}{3}$. On the other hand, the following example is very close to being singular even though its determinant is 1 and all of its eigenvalues are equal to 1. Let A be an upper triangular matrix whose diagonal entries are all 1 and whose entries above the main diagonal are all -1, i.e.,

$$A = \begin{bmatrix} 1 & -1 & -1 & \cdots & -1 & -1 \\ 0 & 1 & -1 & \cdots & -1 & -1 \\ 0 & 0 & 1 & \cdots & -1 & -1 \\ \cdots & \cdots & \cdots & \cdots & \cdots & \cdots \\ 0 & 0 & 0 & \cdots & 1 & -1 \\ 0 & 0 & 0 & \cdots & 0 & 1 \end{bmatrix}.$$

It is to be noted that $det\ A = det\ (A^{-1}) = 1$ and all the eigenvalues of A are 1. However, if n is large, then A is close to being singular. To observe this, let

$$
B = \begin{bmatrix}
1 & -1 & -1 & \cdots & -1 & -1 \\
0 & 1 & -1 & \cdots & -1 & -1 \\
0 & 0 & 1 & \cdots & -1 & -1 \\
\cdots & \cdots & \cdots & \cdots & \cdots & \cdots \\
0 & 0 & 0 & \cdots & 1 & -1 \\
\frac{-1}{3^{n-2}} & 0 & 0 & \cdots & 0 & 1
\end{bmatrix}.
$$

Clearly, B is singular, since the system $BX = O$ has a nontrivial solution $X = [3^{n-2}, 3^{n-3}, \ldots, 3^0, 1]^t$. The matrices A and B differ only at nth row and first column. Hence, we have $||A - B||_F = \frac{1}{3^{n-2}}$. Using Theorem 10.28, $\sigma_n = min_{X singular}||A - X||_F \leq ||A - B||_F = \frac{1}{3^{n-2}}$. Thus, if $n = 100$, then $\sigma_n \leq \frac{1}{3^{98}}$, and as a result, A is very close to singular.

Theorem 10.30 *Let A be a matrix of order $m \times n$ with singular value decomposition UDV^t. Then $||A||_2 = \sigma_1$, where σ_1 is the largest eigenvalue A.*

Proof Since U and V are orthogonal. We have $||A||_2 = ||UDV^t||_2 = ||D||_2$. Now

$$
||D||_2 = \max_{X \neq 0} \frac{||DX||_2}{||X||_2}
$$

$$
= \frac{\sqrt{\sum_{i=1}^{n}(\sigma_i x_i)^2}}{\sqrt{\sum_{i=1}^{n} x_i^2}}
$$

$$
\leq \sigma_1.
$$

In particular, if we choose $X = (1, 0, \ldots 0)$, then $\frac{||DX||_2}{||X||_2} = \sigma_1$. Therefore, it follows that $||A||_2 = ||D||_2 = \sigma_1$.

Corollary 10.31 *Let A be a nonsingular matrix with singular value decomposition UDV^t. Then $cond_2(A) = \frac{\sigma_1}{\sigma_n}$.*

Proof The singular values of $A^{-1} = VD^{-1}U^t$ arranged in decreasing order are $\frac{1}{\sigma_n} \geq \frac{1}{\sigma_{n-1}} \geq \cdots \geq \frac{1}{\sigma_1}$. Therefore, $||A^{-1}||_2 = \frac{1}{\sigma_n}$ and $cond_2(A) = \frac{\sigma_1}{\sigma_n}$.

10.5   Applications of Singular Value Decomposition

The singular value decomposition (SVD) is not only a classical theory in matrix computation and analysis but also is a powerful tool in machine learning and modern data analysis. Today, singular value decomposition has spread through many branches of sciences, in particular psychology and sociology, climate and atmospheric sciences, astronomy and descriptive and predictive statistics. This is also used in some important topics such as digital image processing, spectral decomposition, polar factorization to matrices, compression algorithm, ranking documents, discrete optimization, clustering a mixture of spherical Gaussians, principal component analysis and low-rank approximations of matrices. Finally, we have demonstrated an idea how SVD compression works on images.

SVD in Spectral Decomposition

Suppose B is a square matrix. If the vector X and scalar λ are such that $BX = \lambda X$, then X is an eigenvector of the matrix B and λ is the corresponding eigenvalue. We present here a spectral decomposition theorem for the special case where B is of the form $B = AA^t$ for some matrix A which may be possibly rectangular. If A is a real-valued matrix, then B is symmetric and positive definite. That is, $X^t BX > 0$ for all nonzero vectors X. The spectral decomposition theorem holds more generally. The theorem runs as follows:

If $B = AA^t$, then $B = \sum_i \sigma_i^2 u_i u_i^t$, where $A = \sum_i \sigma_i u_i v_i^t$ is the singular value decomposition of A. The proof of this theorem is described as follows:

$$B = AA^t = (\sum_i \sigma_i u_i v_i^t)(\sum_j \sigma_j u_j v_j^t)^t = \sum_i \sum_j \sigma_i \sigma_j u_i v_i^t v_j u_j^t = \sum_i \sigma_i^2 u_i u_i^t.$$

When the σ_i^s are all distinct, the u_i^s are the eigenvectors of B and the σ_i^2 are the corresponding eigenvalues. If the σ_i^s are not distinct, then any vector that is a linear combination of those u_i with the same eigenvalue is an eigenvector of B.

SVD in Principal Component Analysis

SVD is used in Principal Component Analysis (PCA). PCA is illustrated by an example: customer-product data where there are n customers buying d products. Let matrix A with elements a_{ij} represent the probability of customer purchasing product j. One hypothesizes that there are really only k underlying basic factors like age, income, family size, etc. that determine a customer's purchase behavior. An individual customer's behavior is determined by some weighted combination of these underlying factors. This implies that a customer's purchase behavior can be characterized by a k-dimensional vector where k is much smaller than n and d. The components of the vector are weights for each of the basic factors. Associated with each basic factor is a vector of probabilities, each component of which is the probability of purchasing a given product by someone whose nature depends only on that factor. More abstractly, A is an $n \times d$ matrix that can be expressed as the product of two matrices U and V, where U is an $n \times k$ matrix expressing the factor weights for each customer and V is a $k \times d$ matrix expressing the purchase probabilities of products that correspond to that factor. It is possible that A may not be exactly equal to UV, but close to it since there may be noise or random perturbations.

As discussed in the previous section, we take the best rank k approximation A_k from SVD, as a result, we get such a U, V. In this usual setting, one assumed that A was available completely and we wished to find U and V to identify the basic factors or in some applications to denoise A if we think of $A - UV$ as noise. Now suppose that n and d are very large, on the order of thousands or even millions, there is probably little one could do to estimate or even store A. In this setting, we may assume that we are given just a few elements of A and wish to estimate A. If A was an arbitrary matrix of size $n \times d$, this would require $\Omega(nd)$ pieces of information and cannot be done with a few entries. But again we hypothesize that A was a small rank matrix with added noise. If we also assume that the given entries are randomly drawn according to some known distribution, then there is a possibility that SVD can be used to estimate the whole of A. This area is called collaborative filtering and one of its uses is to target an ad to a customer based on one or two purchases.

SVD in Solving Discrete Optimization Problems

The use of SVD in solving discrete optimization problems is a relatively new subject with many applications. We start with an important Non-deterministic polynomial time hard problem, the Maximum Cut Problem for directed graph $G(V, E)$.

The Maximum Cut Problem is to partition the node set V of a directed graph into two subsets S and $\bar{S}$ so that the number of edges from S to $\bar{S}$ is maximized. Let A be the adjacency matrix of the graph. With each vertex i, associate an indicator variable x_i. The variable x_i will be set 1 for $i \in S$ and 0 for $i \in \bar{S}$. The vector $X = \{x_1, x_2, \ldots, x_n\}$ is unknown and we are trying to find it (or equivalently the cut), so as to maximize the number of edges across the cut. The number of edges across the cut is precisely $\sum_{ij}(1 - x_j)a_{ij}$. Thus, the Maximum Cut Problem can be posed as the optimization problem

$$\text{Maximize } \sum_{ij}(1 - x_j)a_{ij} \text{ subject to } x_i \in \{0, 1\}.$$

This can also be written in matrix notation as

$$\sum_{ij}(1 - x_j)a_{ij} = X^t A(I - X),$$

where I denotes the vector of all 1^s. So the problem can be restated as

$$\text{Maximize } X^t A(I - X) \text{ subject to } x_i \in \{0, 1\}. \tag{10.1}$$

SVD is used to solve this problem approximately by computing the SVD of A and replacing A by $A_k = \sum_{i=1}^{k} \sigma_i U_i V_i^t$ in the previous relation to get

$$\text{Maximize } X^t A_k(I - X) \text{ subject to } x_i \in \{0, 1\}. \tag{10.2}$$

It is to be noted that the matrix A_k is no longer a 0-1 adjacency matrix. We will prove that (i) for each 0-1 vector X, $X^t A_k(I - X)$ and $X^t A(I - X)$ differ by at most $\frac{n^2}{\sqrt{k+1}}$. Thus, maxima in (10.1) and (10.2) differ by at most this amount. Also, we will show that (ii) a near optimal X for (10.2) can be found by exploiting the low rank of A_k, which by (i) is near optimal for (10.1) where near optimal means with additive error of at most $\frac{n^2}{\sqrt{k+1}}$.

First, we prove (i), since X and $I - X$ are 0-1 n-vectors, each has a length at most $\sqrt{n}$. Using the definition of two-norm of a matrix , $|(A - A_k)(I - X)| \leq$

$\sqrt{n}\|A - A_k\|_2$. Now since $X^t(A - A_k)(I - X)$ is the dot product of the vector X with the vector $(A - A_k)(I - X)$, we get $X^t(A - A_k)(I - X) \le n\|A - A_k\|_2$. But we know that $\|A - A_k\|_2 = \sigma_{k+1}$, $(k + 1)^{st}$ singular value of A. The inequalities $(k + 1)\sigma_{k+1}^2 \le \sigma_1^2 + \sigma_2^2 + \cdots + \sigma_{k+1}^2 \le \|A\|_F^2 = \sum_{ij} a_{ij}^2 \le n^2$ imply that $\sigma_{k+1}^2 \le \frac{n^2}{k+1}$ and hence $\|A - A_k\|_2 \le \frac{n}{\sqrt{k+1}}$ showing (i).

Now we prove (ii) as given above. Let us look at the special case when $k = 1$ and A is approximated by the rank 1 matrix A_1. An even more special case when the left and the right singular vectors U and V are required to be identical is already hard to solve exactly because it subsumes the problem of whether for a set of n vectors $\{a_1, a_2, \ldots, a_n\}$, there is a partition into two subsets whose sums are equal. So, we look for algorithms that solve the Maximum Cut Problem approximately.

For (ii), we have to maximize $\sum_{i=1}^{k} \sigma_i(X^t U_i)(V_i^t(I - X))$ over 0-1 vectors X. For any $S \subseteq \{1, 2, \ldots, n\}$, write $U_i(S)$ for the sum of coordinates of the vector U_i corresponding to elements in the set S and also for V_i. That is, $U_i(S) = \sum_{j \in S} u_{ij}$. We will maximize $\sum_{i=1}^{k} \sigma_i U_i(S) V_i(\bar{S})$ using dynamic programming.

For a subset S of $\{1, 2, \ldots, n\}$, define the $2k$-dimensional vector $W(S) = (U_1(S), V_1(\bar{S}), U_2(S), V_2(\bar{S}), \ldots, U_k(S), V_k(\bar{S}))$. If we had the list of all such vectors, we could find $\sum_{i=1}^{k} \sigma_i U_i(S) V_i(\bar{S})$ for each of them and take the maximum. There are 2^n subsets S, but several S could have the same $W(S)$ and in that case it suffices to list just one of them. Round each coordinate of each U_i to the nearest integer multiple of $\frac{1}{n^2}$. Call the rounded vector $\bar{U}_i$. Similarly obtain $\bar{V}_i$. Let $\overline{W}(S)$ denote the vector $(\overline{U}_1(S), \overline{V}_1(\bar{S}), \overline{U}_2(S), \overline{V}_2(\bar{S}), \ldots, \overline{U}_k(S), \overline{V}_k(\bar{S}))$. We will construct a list of all possible values of the vector $\overline{W}(S)$. Again, if several different S^s lead to the same vector $\overline{W}(S)$, we will keep only one copy on the list. The list will be constructed by dynamic programming. For the recursive step of Dynamic Programming, assume we already have a list of all such vectors for $S \subseteq \{1, 2, \ldots, i\}$ and wish to construct the list for $S \subseteq \{1, 2, \ldots, i + 1\}$. Each $S \subseteq \{1, 2, \ldots, i\}$ leads to two possible $S' \subseteq \{1, 2, \ldots, i + 1\}$, namely S and $S \cup \{i + 1\}$. In the first case, the vector $\overline{W}(S') = (\overline{U}_1(S) + u_{1,i+1}, \overline{V}_1(\bar{S}), \overline{U}_2(S) + u_{2,i+1}, \overline{V}_2(\bar{S}), \ldots)$. In the second case, $\overline{W}(S') = (\overline{U}_1(S), \overline{V}_1(\bar{S}) + v_{1,i+1}, \overline{U}_2(S), \overline{V}_2(\bar{S}) + v_{2,i+1}, \ldots)$. We put these two vectors for each vector in the previous list. Then, crucially eliminate duplicates. Assume that k is constant. Now, we show that the error is at most $\frac{n^2}{\sqrt{k+1}}$ as claimed. Since U_i and V_i are unit length vectors, $|U_i(S)|, |(V_i(S)| \le \sqrt{n}$. Also $|\overline{U}_i(S) - U_i(S)| \le \frac{n}{nk^2} = \frac{1}{k^2}$ and similarly for V_i. To bound the error, we use an elementary fact: if a, b are reals with $|a|, |b| \le M$ and we estimate a by a' and b by b' so that $|a - a'|, |b - b'| \le \delta \le M$, then $|ab - a'b'| = |a(b - b') + b'(a - a')| \le |a||(b - b')| + |b'||(a - a')| \le 3M\delta$. Using this, we get that

$$\left| \sum_{i=1}^{k} \sigma_i \overline{U}_i(S) \overline{V}_i(\bar{S}) - \sum_{i=1}^{k} \sigma_i U_i(S) V_i(\bar{S}) \right| \le 3k\sigma_1 \frac{\sqrt{n}}{k^2} \le 3\frac{n^{\frac{3}{2}}}{k}$$

and this meets the claimed error bound.

Next, we prove that the running time is polynomially bounded. $|\overline{U}_i(S)|, |\overline{V}_i(S)| \leq 2\sqrt{n}$. Since $\overline{U}_i(S), \overline{V}_i(S)$ are all integer multiples of $\frac{1}{nk^2}$, there are at most $\frac{2}{\sqrt{n}k^2}$ possible values of $\overline{U}_i(S), \overline{V}_i(S)$ from which it follows that the list of $\overline{W}(S)$ never gets larger than $(\frac{1}{\sqrt{n}k^2})^{2k}$ which for fixed k is polynomially bounded. Finally, we have following conclusion:

"Given a directed graph $G(V, E)$, a cut of size at least the maximum cut minus $O(\frac{n^2}{\sqrt{k}})$ can be computed in polynomial time n for any fixed k".

SVD in Image Processing

Suppose A is the pixel intensity matrix of a large image. The entry a_{ij} gives the intensity of the ijth pixel. If A is $n \times n$, the transmission of A requires transmitting $O(n^2)$ real numbers. Instead, one could send A_k, that is, the top k singular values $\sigma_1, \sigma_2, \ldots, \sigma_k$ along with the left and right singular vectors $U_1, U_2, \ldots, U_k$ and $V_1, V_2, \ldots, V_k$. This would require sending $O(kn)$ real numbers instead of $O(n^2)$ real numbers. If k is much smaller than n, this results in saving. For many images, a k much smaller than n can be used to reconstruct the image provided that a very low-resolution version of the image is sufficient. Thus, one can use SVD as a compression method.

For an illustration, suppose a satellite takes a picture and wants to send it to Earth. The picture may contain 1000×1000 "pixels", a million little squares, each with a definite color. We can code the colors and send back 1000000 numbers. It is better to find the essential information inside the 1000×1000 matrix and send only that.

Suppose we know the SVD. The key is in the singular values (in D used in the previous section). Typically, some $\sigma's$ are significant and others are extremely small. If we keep 20 and throw away 980, then we send only the corresponding 20 columns of U and V (if $A = UDV^t$, as in the previous section). The other 980 columns are multiplied in UDV^t by the small $\sigma's$ that are being ignored. We can do the matrix multiplication as columns times rows:

$$A = UDV^t = U_1\sigma_1 V_1^t + U_2\sigma_2 V_2^t + \cdots + U_r\sigma_r V_r^t.$$

Any matrix is the sum of r matrices of rank 1. If only 20 terms are kept, we send 20 times 2000 numbers instead of a million (25 to 1).

The pictures are really striking, as more and more singular values are included. At first, you see nothing and suddenly you recognize everything. The cost is in computing the SVD, and this has become much more efficient, but it is expensive for a big matrix.

An example of SVD compression in image is demonstrated in Fig. 10.1, where Fig. 10.1(a) is the original image of Airplane of size $3 \times (512 \times 512)$ which is considered as the test image. In order to apply SVD compression on color image, firstly the color image is decomposed into three channels, i.e., Red, Green and Blue. After that, each color channel passes through SVD compression. Finally, all the three compressed channels are combined to generate the SVD compressed image. Figure 10.1(b) shows the recovered image using the first 10 singular values having Peak Signal-to-Noise Ratio (PSNR) value 22.62746 dB and Fig. 10.1(c) shows the

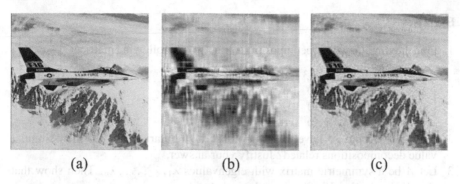

Fig. 10.1 SVD-based compression of Airplane image having size $3 \times (512 \times 512)$: (**a**) Original image; (**b**) Compressed image using the first 10 singular values; (**c**) Compressed image using the first 30 singular values

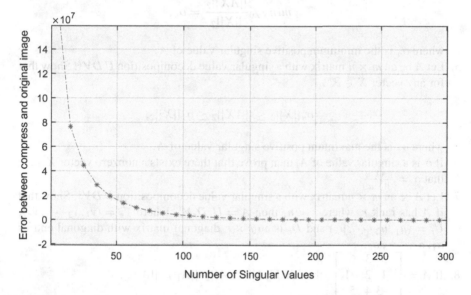

Fig. 10.2 Error of recovered image using different numbers of singular values

recovered image using the first 30 singular values having PSNR value 27.202 dB, where the visual quality of the recovered image is evaluated by PSNR, and a high value of PSNR shows that the reconstructed image has high visual quality. Moreover, in Fig. 10.2, we can analyze using approximately the first 200 singular values yields to approximately zero error of the generated image.

Exercises

1. Find the singular value decomposition of each of the following matrices. $\begin{bmatrix} 1 & -1 \\ 0 & 2 \end{bmatrix}$,

   $\begin{bmatrix} 2 & -1 \\ 1 & 2 \end{bmatrix}$, $\begin{bmatrix} 2 & 3 \\ 3 & 1 \\ 0 & 0 \\ 0 & 0 \end{bmatrix}$, $\begin{bmatrix} -3 & 0 & 0 \\ 0 & 1 & 1 \\ 0 & 0 & 2 \\ 0 & 0 & 0 \end{bmatrix}$.

2. Show that A and A^t have the same nonzero singular values. Do their singular value decompositions related? Justify your answer.

3. Let A be a symmetric matrix with eigenvalues $\lambda_1, \lambda_2, \ldots, \lambda_n$. Then show that $|\lambda_1|, |\lambda_2|, \ldots, |\lambda_n|$ are singular values of A.

4. Let A be an $m \times n$ matrix with a singular value decomposition $U D V^t$. Prove that

$$min_{X \neq 0} \frac{||AX||_2}{||X||_2} = \sigma_n,$$

   where σ_n is the minimum positive singular value of A.

5. Let A be an $m \times n$ matrix with a singular value decomposition $U D V^t$. Show that for any vector $X \in \mathbb{R}^n$,

$$\sigma_n ||X||_2 \leq ||AX||_2 \leq \sigma_1 ||X||_2,$$

   where σ_1 is the maximum positive singular value of A.

6. If σ is a singular value of A, then prove that there exists a nonzero vector X such that $\sigma = \frac{||AX||_2}{||X||_2}$.

7. Let A be an $m \times n$ matrix with a singular value decomposition $U D V^t$. Show that if A has rank r, where $r < n$, then $A = U_r D_r V_r^t$, where $V_r = (v_1, v_2 \cdots, v_r)$, $U_r = (u_1, u_2 \cdots, u_r)$ and D_r is an $r \times r$ diagonal matrix with diagonal entries $\sigma_1, \sigma_2, \cdots, \sigma_r$.

8. If $A = \begin{bmatrix} 5 & 3 & 6 \\ 1 & 2 & -1 \\ -3 & 4 & 5 \end{bmatrix}$, then find $||A||_1, ||A||_2, ||A||_F, ||A||_\infty$.

9. (a) If A changes to αA, what is the change in the SVD?
   (b) What is the SVD for A^t and for A^{-1}?

Chapter 11
Affine and Euclidean Spaces and Applications of Linear Algebra to Geometry

This chapter is initially devoted to the study of subspaces of an affine space, by applying the theory of vector spaces, matrices and system of linear equations. By using methods involved in the theory of inner product spaces, we then stress practical computation of distances between points, lines and planes, as well as angles between lines and planes in Euclidean spaces. Finally, we apply the development of eigenvalues, eigenvectors, diagonalization and allied concepts to the study and classification of the conic curves and quadric surfaces in the affine, Euclidean and projective spaces.

11.1 Affine and Euclidean Spaces

Initially, we shall describe briefly the framework within which we are about to set to work.

Definition 11.1 Let $\mathbb{F}$ be a field, V a vector space of dimension n over $\mathbb{F}$ and S a nonempty set. The pair (S, V) is said to be an *affine space* if there exists an application

$$\varphi : S \times S \to V$$

defined whereby two elements $A, B \in S$ are associated with a vector of V, which is denoted by $\varphi(A, B) = \mathbf{AB}$, such that the following conditions are satisfied:

(1) For any fixed element $O \in S$ and for any vector $\mathbf{v} \in V$, there is a unique element $P \in S$ such that $\mathbf{OP} = \mathbf{v}$.
(2) For any $P_1, P_2, P_3 \in S$, $\mathbf{P_1P_3} = \mathbf{P_1P_2} + \mathbf{P_2P_3}$.

For brevity, we shall denote the affine space (S, V) by $\mathbb{A}$. Any element of S is called a *point* of $\mathbb{A}$. The definition of an affine space $\mathbb{A} = (S, V)$ induces naturally a bijection

© The Author(s), under exclusive license to Springer Nature Singapore Pte Ltd. 2022   365
M. Ashraf et al., *Advanced Linear Algebra with Applications*,
https://doi.org/10.1007/978-981-16-2167-3_11

between the elements of S and vectors of V. In fact, after the choice of the origin $O \in S$, any vector $\mathbf{v} \in V$ can be represented as $\mathbf{v} = \mathbf{OP}$, where $P \in S$ is uniquely determined. This bijection $S \to V$ is then depending only by the choice of the origin point. Hence, if $dim_{\mathbb{F}} V = n$ then $\mathbb{A} \cong V \cong \mathbb{F}^n$. In this sense, we have the following:

Definition 11.2 An *affine coordinate system* of the affine space $\mathbb{A}$ consists of a fixed point $O \in S$, called *origin,* and a basis $\{\mathbf{e_1}, \ldots, \mathbf{e_n}\}$ of V. It is usually denoted by $\{O, \mathbf{e_1}, \ldots, \mathbf{e_n}\}$. In terms of this system, the coordinates of any point $P \in \mathbb{A}$ are defined as the coordinates of the vector $\varphi(O, P) = \mathbf{OP}$ with respect to the basis $\{\mathbf{e_1}, \ldots, \mathbf{e_n}\}$ of V. Hence, if

$$\mathbf{OP} = x_1\mathbf{e_1} + x_2\mathbf{e_2} + \cdots + x_n\mathbf{e_n},$$

we say that $x_1, \ldots, x_n$ are the coordinates of P and write $P \equiv (x_1, \ldots, x_n)$. The coordinate system $\{O, \mathbf{e_1}, \ldots, \mathbf{e_n}\}$ is also called *frame of reference* in the space $\mathbb{A}$.

If $dim_{\mathbb{F}}(V) = n \geq 1$, then we say that the affine space $\mathbb{A}$ has dimension n. Moreover, if $P, Q \in \mathbb{A}$ have respectively coordinates $(x_1, \ldots, x_n)$ and $(y_1, \ldots, y_n)$ with respect to $\{\mathbf{e_1}, \ldots, \mathbf{e_n}\}$, then the vector $\mathbf{PQ}$ has coordinates $(y_1 - x_1, \ldots, y_n - x_n)$ in terms of the frame of reference $\{O, \mathbf{e_1}, \ldots, \mathbf{e_n}\}$.

The introduction of an affine space (S, V) can be interpreted intuitively as saying that S and V are two different ways of looking at the same object. In some sense, $\mathbb{A}$ is the way of defining a vector space structure on the set S of points. However, let us recall that the addition of points does not make sense: points are not vectors. Nevertheless, fixed an origin point $O \in \mathbb{A}$, the coordinates of any point $P \in \mathbb{A}$ in terms of a frame of reference in $\mathbb{A}$ could be treated as coordinates of vector $\mathbf{OP}$ with respect to the basis of V, and so can be used in vector operations.

Remark 11.3 Let V be a vector space, n-dimensional over the field $\mathbb{F}$. If we define the map $\varphi : V \times V \to V$ by $\varphi(\mathbf{v}, \mathbf{w}) = \mathbf{w} - \mathbf{v}$, for any vectors $\mathbf{v}, \mathbf{w} \in V$, then the pair (V, V) is an affine space of dimension n. Indeed, the conditions introduced in Definition 11.1 are satisfied:

(*i*) For any fixed vector $\mathbf{u} \in V$ and for any vector $\mathbf{v} \in V$, there is a unique vector $\mathbf{w} \in V$ such that

$$\varphi(\mathbf{u}, \mathbf{v}) = \mathbf{v} - \mathbf{u} = \mathbf{w}.$$

(*ii*) For any vectors $\mathbf{v_1}, \mathbf{v_2}, \mathbf{v_3} \in V$, $\varphi(\mathbf{v_1}, \mathbf{v_3}) = \varphi(\mathbf{v_1}, \mathbf{v_2}) + \varphi(\mathbf{v_2}, \mathbf{v_3})$, because $\mathbf{v_3} - \mathbf{v_1} = (\mathbf{v_2} - \mathbf{v_1}) + (\mathbf{v_3} - \mathbf{v_2})$.

As a particular case, we consider $V = \mathbb{F}^n$. The affine space $(\mathbb{F}^n, \mathbb{F}^n)$ is usually called *n-dimensional affine space over* $\mathbb{F}$ and denoted by $\mathbb{F}\mathbb{A}^n$.

Definition 11.4 Let $\mathbb{A} = (S, V)$ be an affine space of dimension n, associated with the n-dimensional real vector space V. We say that $\mathbb{A}$ is an *affine Euclidean space* of dimension n if there is a fixed symmetric bilinear form $f : V \times V \to \mathbb{R}$ whose associated quadratic form is positive definite. For brevity, we shall denote the affine Euclidean space (S, V) by $\mathbb{E}$.

Hence, an affine Euclidean space $\mathbb{E}$ is associated with a real vector space V in which, for any vectors $\mathbf{v}_1, \mathbf{v}_2 \in V$, there corresponds a real nonnegative number $f(\mathbf{v}_1, \mathbf{v}_2)$ such that the following conditions are satisfied:

(1) For any vectors $\mathbf{v}_1, \mathbf{v}_2, \mathbf{w} \in V$, $f(\mathbf{v}_1 + \mathbf{v}_2, \mathbf{w}) = f(\mathbf{v}_1, \mathbf{w}) + f(\mathbf{v}_2, \mathbf{w})$.
(2) For any vectors $\mathbf{v}_1, \mathbf{v}_2 \in V$, $f(\mathbf{v}_1, \mathbf{v}_2) = f(\mathbf{v}_2, \mathbf{v}_1)$.
(3) For any vectors $\mathbf{v}_1, \mathbf{v}_2 \in V$ and scalar $\lambda \in \mathbb{R}$, $f(\lambda \mathbf{v}_1, \mathbf{v}_2) = f(\mathbf{v}_1, \lambda \mathbf{v}_2) = \lambda f(\mathbf{v}_1, \mathbf{v}_2)$.
(4) For any vector $0 \neq \mathbf{v} \in V$, $f(\mathbf{v}, \mathbf{v}) > 0$.

A *Euclidean coordinate system* of $\mathbb{E}$ (also called *Euclidean frame of reference* in $\mathbb{E}$) is an affine coordinate system $\{O, \mathbf{e}_1, \ldots, \mathbf{e}_n\}$ of $\mathbb{E}$, where vectors $\mathbf{e}_1, \ldots, \mathbf{e}_n$ are pairwise f-orthonormal. For $V = \mathbb{R}^n$, the affine Euclidean space $(\mathbb{R}^n, \mathbb{R}^n)$ is usually called *n-dimensional affine Euclidean space over* $\mathbb{R}$ and denoted by $\mathbb{R}\mathbb{E}^n$.

Example 11.5 The real affine Euclidean plane $\mathbb{R}\mathbb{E}^2$ in which $f(\mathbf{v}_1, \mathbf{v}_2) = \mathbf{v}_1 \cdot \mathbf{v}_2$ (the usual dot product) is a Euclidean space of dimension 2 associated with the vector space $\mathbb{R}^2$. It is well known how to introduce a coordinate system in the 2-dimensional Euclidean plane $\mathbb{R}\mathbb{E}^2$ (it is usually the 2-dimensional OXY coordinate system adopted for the Euclidean geometry in the plane). Choose an origin point O in the plane and draw two perpendicular axes through it, one horizontal and one vertical, respectively, called X and Y. Any point P can be uniquely represented by the pair of real numbers (x, y), called the *coordinates* of the point P in terms of the coordinate system OXY. If $\mathbf{i}, \mathbf{j}$ are the unit vectors of X and Y, respectively, the point P has coordinates (x, y) if and only if

$$OP = x\mathbf{i} + y\mathbf{j}.$$

Hence, if $\mathbf{v}_1 = \alpha_1 \mathbf{i} + \alpha_2 \mathbf{j}$ and $\mathbf{v}_2 = \beta_1 \mathbf{i} + \beta_2 \mathbf{j}$ are two vectors in the real Euclidean plane, then $\mathbf{v}_1 \cdot \mathbf{v}_2 = \alpha_1 \beta_1 + \alpha_2 \beta_2$.

Example 11.6 Analogously, one may introduce a coordinate system in the 3-dimensional affine Euclidean space $\mathbb{R}\mathbb{E}^3$ (it is usually the 3-dimensional $OXYZ$ coordinate system adopted for the Euclidean geometry in the space). One may fix an origin point O in the space and draw three pairwise perpendicular axes through it, respectively, called X, Y and Z. Any point P can be uniquely represented by the triplet of real numbers (x, y, z), called the *coordinates* of the point P in terms of the coordinate system $OXYZ$. If $\mathbf{i}, \mathbf{j}, \mathbf{k}$ are the unit vectors of X, Y and Z, respectively, the point P has coordinates (x, y, z) if and only if

$$OP = x\mathbf{i} + y\mathbf{j} + z\mathbf{k}.$$

Hence, if $\mathbf{v}_1 = \alpha_1 \mathbf{i} + \alpha_2 \mathbf{j} + \alpha_3 \mathbf{k}$ and $\mathbf{v}_2 = \beta_1 \mathbf{i} + \beta_2 \mathbf{j} + \beta_3 \mathbf{k}$ are two vectors in the real Euclidean space, then $\mathbf{v}_1 \cdot \mathbf{v}_2 = \alpha_1 \beta_1 + \alpha_2 \beta_2 + \alpha_3 \beta_3$.

Example 11.7 More generally, if $\mathbb{R}\mathbb{E}^n$ is the affine Euclidean space of dimension n over $\mathbb{R}$, then there exists an orthonormal basis B of $\mathbb{R}^n$ in terms of which the inner product of vectors is defined by

$$\mathbf{v}_1 \cdot \mathbf{v}_2 = \alpha_1 \beta_1 + \cdots + \alpha_n \beta_n$$

where $(\alpha_1, \ldots, \alpha_n)$ and $(\beta_1, \ldots, \beta_n)$ are the coordinates of $\mathbf{v}_1$ and $\mathbf{v}_2$, respectively, with respect to the basis B.

Definition 11.8 Let $\mathbb{A}$ be an affine space over the vector space V. A subset $\mathbb{A}' \subset \mathbb{A}$ is called an *affine subspace* of $\mathbb{A}$ if the set of all vectors $\mathbf{PQ}$, for any points $P, Q \in \mathbb{A}'$, forms a vector subspace V' of V. The dimension of V' as vector space is called *dimension* of $\mathbb{A}'$ as affine subspace of $\mathbb{A}$.

In particular, we recall that

(1) If $dim(\mathbb{A}') = 1$, $\mathbb{A}'$ is called *line* in $\mathbb{A}$.
(2) If $dim(\mathbb{A}') = 2$, $\mathbb{A}'$ is called *plane* in $\mathbb{A}$.
(3) If $dim(\mathbb{A}') = dim(\mathbb{A}) - 1$, $\mathbb{A}'$ is called *hyperplane* in $\mathbb{A}$.

The vector subspace V' of V associated with an affine subspace $\mathbb{A}'$ of $\mathbb{A}$ is called *direction* of $\mathbb{A}'$.

Example 11.9 Let $\mathbb{A} = \mathbb{R}\mathbb{A}^2$ be the real 2-dimensional affine plane. The affine subspaces of $\mathbb{A}$ are

(1) The points in $\mathbb{A}$: any point is an affine subspace of dimension 0.
(2) The lines in $\mathbb{A}$: any line is an affine subspace of dimension 1.

Example 11.10 Let $\mathbb{A} = \mathbb{R}\mathbb{A}^3$ be the real 3-dimensional affine space. The affine subspaces of $\mathbb{A}$ are

(1) The points in $\mathbb{A}$: any point is an affine subspace of dimension 0.
(2) The lines in $\mathbb{A}$: any line is an affine subspace of dimension 1.
(3) The planes in $\mathbb{A}$: any plane is an affine subspace of dimension 2.

Example 11.11 Let $\mathbb{A} = \mathbb{F}^n$ be the affine space over the vector space $V = \mathbb{F}^n$, for $\mathbb{F}$ a field. Consider the following system of linear equations in n unknowns:

$$
\begin{aligned}
a_{11}x_1 + a_{12}x_2 + \cdots + a_{1n}x_n &= b_1 \\
a_{21}x_1 + a_{22}x_2 + \cdots + a_{2n}x_n &= b_2 \\
&\cdots\cdots\cdots \\
a_{m1}x_1 + a_{m2}x_2 + \cdots + a_{mn}x_n &= b_m
\end{aligned}
\tag{11.1}
$$

where coefficients a_{ij} and constants b_i lie in $\mathbb{F}$. We recall that if we set

$$
A = \begin{bmatrix} a_{11} & a_{12} & \cdots & a_{1n} \\ a_{21} & a_{22} & \cdots & a_{2n} \\ \cdots & \cdots & \cdots & \cdots \\ a_{m1} & a_{m2} & \cdots & a_{mn} \end{bmatrix}, X = \begin{bmatrix} x_1 \\ x_2 \\ \cdots \\ \cdots \\ x_n \end{bmatrix}, B = \begin{bmatrix} b_1 \\ b_2 \\ \cdots \\ \cdots \\ b_m \end{bmatrix},
$$

the system (11.1) can be written in the compact form $AX = B$. The set of solutions $\mathbb{A}'$ is a subset of $\mathbb{A} = \mathbb{F}^n$; more precisely, it is an affine subspace of $\mathbb{A}$. To prove this fact, we show that if $Y_1, Y_2 \in \mathbb{A}'$ are solutions of (11.1), then the vector $\mathbf{Y_1Y_2}$ lies in the solution space of homogeneous system associated with (11.1), which is a vector subspace of $V = \mathbb{F}^n$. For any $P, Q \in \mathbb{A}$, we set $\mathbf{PQ} = Q - P$. By Remark 11.3, we know that this definition induces a structure of affine space. From $AY_1 = B$ and $AY_2 = B$, it follows $A(Y_2 - Y_1) = AY_2 - AY_1 = 0$, that is, $\mathbf{Y_1Y_2} = Y_2 - Y_1$ lies in the solution subspace of the homogeneous system associated with (11.1), as asserted.

Every affine subspace $\mathbb{A}'$ of $\mathbb{A} = (S, \mathbb{F}^n)$ can be defined by a system of linear equations. In fact, the direction V' of $\mathbb{A}'$ is a vector subspace of $\mathbb{F}^n$, so that it can be described by a system of homogeneous linear equations (r equations, if $dim(V') = n - r$):

$$f_1(x_1, \ldots, x_n) = 0$$
$$\vdots$$
$$\vdots \qquad (11.2)$$
$$\vdots$$
$$f_r(x_1, \ldots, x_n) = 0.$$

Now, fix an arbitrary point $P \in \mathbb{A}'$ having coordinates $(\eta_1, \ldots, \eta_n)$ in terms of a frame of reference in $\mathbb{A}$, and set

$$f_1(\eta_1, \ldots, \eta_n) = b_1, \ldots, f_r(\eta_1, \ldots, \eta_n) = b_r.$$

Let $Q \in \mathbb{A}'$ be any other arbitrary point whose coordinate vector is equal to $(\vartheta_1, \ldots, \vartheta_n)$. By the fact that $\mathbf{PQ} \in V'$, it follows that the vector $(\vartheta_1 - \eta_1, \ldots, \vartheta_n - \eta_n)$ is a solution of system (11.2), that is,

$$f_1(\eta_1, \ldots, \eta_n) = f_1(\vartheta_1, \ldots, \vartheta_n), \ldots, f_r(\eta_1, \ldots, \eta_n) = f_r(\vartheta_1, \ldots, \vartheta_n).$$

This means that the coordinate vector of an arbitrary point $Q \in \mathbb{A}'$ is a solution of the linear system

$$f_1(x_1, \ldots, x_n) = b_1$$
$$\vdots$$
$$\vdots$$
$$f_r(x_1, \ldots, x_n) = b_r.$$

Remark 11.12 Let $\mathbb{A} = \mathbb{F}^n$ be the affine space over the vector space $V = \mathbb{F}^n$, for $\mathbb{F}$ a field. Fix a point $p \in \mathbb{A}$ and let W be a vector subspace of V. The set

$$p + W = \{p + w | w \in W\}$$

is an affine subspace of $\mathbb{A}$ associated with vector space W. In fact, for $q_1 = p + \mathbf{w_1} \in p + W$ and $q_2 = p + \mathbf{w_2} \in p + W$, we have $\mathbf{q_1q_2} = q_2 - q_1 = \mathbf{w_2} - \mathbf{w_1} \in W$.

Let $\mathbb{A} = \mathbb{F}^n$ be the affine space over the vector space $V = \mathbb{F}^n$, for $\mathbb{F}$ a field. Under the assumption that $\mathbf{PQ} = Q - P$, for any $P, Q \in \mathbb{A}$, any affine subspace of $\mathbb{A}$ has the form $Q + W$, for some fixed point $Q \in \mathbb{A}$ and vector subspace W of V. Moreover, any affine subspace $\mathbb{A}'$ of $\mathbb{A}$ can be represented by the associated vector subspace W and by any of its points Q. To prove this, assume $\mathbb{A}' = Q + W$, let $P \in \mathbb{A}'$ be any other point of $\mathbb{A}'$ and set $\mathbb{A}'' = P + W$. If $R \in \mathbb{A}'$, then

$$\mathbf{PR} = \mathbf{PQ} + \mathbf{QR} = -\mathbf{QP} + \mathbf{QR} \in W$$

implying that $R \in \mathbb{A}''$. Conversely, if $S \in \mathbb{A}''$, then

$$\mathbf{QS} = \mathbf{QP} + \mathbf{PS} = -\mathbf{PQ} + \mathbf{PS} \in W,$$

that is, $S \in \mathbb{A}'$. Thus we have proved that $\mathbb{A}' = \mathbb{A}''$.

Starting from the above remark, let us now show how any affine subspace of $\mathbb{F}\mathbb{A}^n$ can be represented. Let V be the vector space associated with $\mathbb{F}\mathbb{A}^n$ and $B = \{\mathbf{e_1}, \ldots, \mathbf{e_n}\}$ be a basis of V. Set $\mathbb{A}' = P + W$, where $P \equiv (b_1, \ldots, b_n) \in \mathbb{F}\mathbb{A}^n$ and W is a subspace of V. Moreover, let $\{\mathbf{w_1}, \ldots, \mathbf{w_k}\}$ be a basis of W, and $\mathbf{w}_i \equiv (w_{1i}, \ldots, w_{ni})$ in terms of the basis B.

For any point $Q \in \mathbb{A}'$, $\mathbf{PQ} \in W$ implies that there exist suitable $t_1, \ldots, t_k \in \mathbb{F}$ such that

$$\mathbf{PQ} = t_1 \mathbf{w_1} + \cdots + t_k \mathbf{w_k}.$$

If we define $\mathbf{PQ}$ by its coordinate vector $(x_1, \ldots, x_n)$ with respect to basis B, we have that

$$x_1 = b_1 + t_1 w_{11} + \cdots + t_k w_{1k}$$
$$x_2 = b_2 + t_1 w_{21} + \cdots + t_k w_{2k}$$
$$\vdots$$
$$\vdots$$
$$x_n = b_n + t_1 w_{n1} + \cdots + t_k w_{nk}$$

(11.3)

called the *parametric equations* of $\mathbb{A}'$.

Example 11.13 A straight line r in $\mathbb{R}\mathbb{A}^3$ is an affine subspace of dimension 1. Hence, it can be represented by the coordinates of one of its point $P_0 \equiv (x_0, y_0, z_0)$ and the coordinates of a vector $\mathbf{v} \equiv (l, m, n)$ which is parallel to the line. In other words, if we denote by $P \equiv (x, y, z)$ an arbitrary point of r, then the vector $\mathbf{P_0P}$ is parallel to $\mathbf{v}$, that is, there exists some $t \in \mathbb{R}$ such that $\mathbf{P_0P} = t\mathbf{v}$, where t is depending on the choice of $P \in r$. Therefore, one may describe the coordinates of any point of the straight line r, regardless of the variation of the parameter value $t \in \mathbb{R}$. Hence, r can be represented by the following relations:

$$x = x_0 + tl$$
$$y = y_0 + tm \qquad\qquad (11.4)$$
$$z = z_0 + tn$$

which are called *parametric equations* of the straight line r. Thus, for any point $P \in r$, if we denote by $\mathbf{X}$ and $\mathbf{X_0}$ the coordinate vectors of P and P_0, respectively, we may write $\mathbf{X} = \mathbf{X_0} + t\mathbf{v}$.

The real numbers (l, m, n) are called *direction ratios* of the straight line r. Therefore, two straight lines r and r' are parallel if and only if their direction ratios are, respectively, (l, m, n) and $\rho(l, m, n)$, for a suitable $\rho \in \mathbb{R}$.

Moreover, if we consider two straight lines r and r' in $\mathbb{RE}^3$, having respectively direction ratios (l, m, n) and (l', m', n'), we see that they are perpendicular if and only if the vectors $\mathbf{v} \equiv (l, m, n)$ and $\mathbf{v'} \equiv (l', m', n')$ are orthogonal, that is, the inner product $\mathbf{v} \cdot \mathbf{v'}$ is zero. This means that $ll' + mm' + nn' = 0$.

Example 11.14 Let now π be a plane of $\mathbb{RA}^3$. It is an affine subspace of dimension 2 and could be uniquely represented by any of its points $P_0 \equiv (x_0, y_0, z_0)$ and by a pair of vectors $\mathbf{v_1}, \mathbf{v_2}$ that are parallel to the plane. Suppose that $\mathbf{v_1} \equiv (l_1, m_1, n_1)$ and $\mathbf{v_2} \equiv (l_2, m_2, n_2)$. The fact that π is a variety of dimension 2 implies that we may consider the plane as the set of all points $P \equiv (x, y, z)$ in $\mathbb{RA}^3$ such that the displacement vector from P_0 to P is a linear combination of $\mathbf{v_1}, \mathbf{v_2}$. Thus, $\mathbf{P_0P} = t\mathbf{v_1} + t'\mathbf{v_2}$, for $t, t' \in \mathbb{R}$, and any point of π is determined by an appropriate pair (t, t') of real parameters. On the other hand, $\mathbf{P_0P} \equiv (x - x_0, y - y_0, z - z_0)$, that is,

$$x - x_0 = tl_1 + t'l_2$$
$$y - y_0 = tm_1 + t'm_2$$
$$z - z_0 = tn_1 + t'n_2$$

so that the set of all points (x, y, z) belonging to the plane is described by

$$x = x_0 + tl_1 + t'l_2$$
$$y = y_0 + tm_1 + t'm_2$$
$$z = z_0 + tn_1 + t'n_2$$

which are called *parametric equations of the plane*.

Moreover, since the vectors of coordinates $(x - x_0, y - y_0, z - z_0)$, (l_1, m_1, n_1) and (l_2, m_2, n_2) are linearly dependent, it follows that the matrix

$$\begin{bmatrix} x - x_0 & y - y_0 & z - z_0 \\ l_1 & m_1 & n_1 \\ l_2 & m_2 & n_2 \end{bmatrix}$$

has rank ≤ 2, in particular, its determinant is zero. By the easy computation of this determinant, it follows that there exist suitable scalar coefficients a, b, c, d such that the coordinates (x, y, z) of any point of the plane should satisfy the following relation $ax + by + cz + d = 0$. It is called *Cartesian equation of the plane*.

Definition 11.15 Let $\mathbb{A}$ be an affine space over the vector space V and $\mathbb{A}'$, $\mathbb{A}''$ two affine subspaces of $\mathbb{A}$, having directions V' and V'', respectively. $\mathbb{A}'$ and $\mathbb{A}''$ are said to be *parallel* if either $V' \subseteq V''$ or $V'' \subseteq V'$. In particular, if $\mathbb{A}'$ and $\mathbb{A}''$ have the same dimension, we say that they are parallel if $V' = V''$.

Example 11.16 Consider the following affine subspaces in $\mathbb{RA}^3$:

$$\mathbb{A}' \text{ having direction } V' = \langle (1, -2, 2) \rangle,$$

$$\mathbb{A}'' \text{ having direction } V'' = \langle (0, 1, -1), (1, 1, -1) \rangle$$

and notice $V' \subset V''$ as vector spaces. Then $\mathbb{A}'$ and $\mathbb{A}''$ are parallel subspaces in $\mathbb{RA}^3$. In particular, we may look at that from the geometrical point of view, saying that any line having direction ratios $(1, -2, 2)$ and any plane containing vectors whose coordinates are of the form $\alpha(0, 1, -1) + \beta(1, 1, -1)$, for any $\alpha, \beta \in \mathbb{R}$, are parallel in the classical affine 3-dimensional space.

Example 11.17 Consider the following two linear systems of equations in 4 unknowns:

$$\begin{aligned} x_1 - x_2 + 2x_3 - x_4 &= 1, \\ 2x_1 + x_2 + x_3 + x_4 &= 1, \\ 2x_1 + x_2 - x_3 - x_4 &= 1, \end{aligned} \tag{11.5}$$

$$\begin{aligned} x_2 - x_3 + x_4 &= 1, \\ x_3 + x_4 &= 3. \end{aligned} \tag{11.6}$$

The set $\mathbb{A}'$ of solutions of (11.5) and the set $\mathbb{A}''$ of solutions of (11.6) are parallel affine subspaces in $\mathbb{RA}^4$. To prove it, firstly we compute the direction V' of $\mathbb{A}'$. It consists of the solutions of the homogeneous linear system

$$\begin{aligned} x_1 - x_2 + 2x_3 - x_4 &= 0, \\ 2x_1 + x_2 + x_3 + x_4 &= 0, \\ 2x_1 + x_2 - x_3 - x_4 &= 0, \end{aligned}$$

that is, $V' = \langle (1, -2, -1, 1) \rangle$. We may also notice that the point $P' = (\frac{2}{3}, -\frac{1}{3}, 0, 0)$ is a solution of system (11.5). Hence, $\mathbb{A}' = P' + V'$ can be represented by relations:

$$\begin{aligned} x_1 &= \tfrac{2}{3} + \alpha \\ x_2 &= -\tfrac{1}{3} - 2\alpha \\ x_3 &= -\alpha \\ x_4 &= \alpha \end{aligned}, \quad \alpha \in \mathbb{R}.$$

On the other hand, direction V'' of $\mathbb{A}''$ consists of the solutions of the homogeneous linear system

$$\begin{aligned} x_2 - x_3 + x_4 &= 0, \\ x_3 + x_4 &= 0, \end{aligned}$$

that is, $V'' = \langle (1, 0, 0, 0), (0, -2, -1, 1) \rangle$. Since the point $P'' = (0, 4, 3, 0)$ is solution of system (11.6), we write $\mathbb{A}'' = P'' + V''$ and it can be represented by

$$\begin{aligned} x_1 &= \beta \\ x_2 &= 4 - 2\gamma \\ x_3 &= 3 - \gamma \\ x_4 &= \gamma \end{aligned}, \quad \beta, \gamma \in \mathbb{R}.$$

Then $V' \subset V''$ and the assertion is proved.

Proposition 11.18 *Every hyperplane H in $\mathbb{RE}^n$ has the form*

$$H_{\mathbf{u},\lambda} = \{ P \in \mathbb{RE}^n \ \text{having coordinate vector } \mathbf{Y} \in \mathbb{R}^n : \mathbf{Y} \cdot \mathbf{u} = \lambda \} \qquad (11.7)$$

for some nonzero vector $\mathbf{u} \in \mathbb{R}^n$ that is orthogonal to the hyperplane and some $\lambda \in \mathbb{R}$.

Proof Fix any point $Q \in H$; let $\mathbf{X}$ be its coordinate vector. Then $H - Q$ is a hyperplane containing the origin of the frame of reference and parallel to H. If $\mathbf{u} \in \mathbb{R}^n$ is orthogonal to H, then a point P, having coordinate vector $\mathbf{Y}$, lies in H if and only if the vector $\mathbf{Y} - \mathbf{X}$ is orthogonal to $\mathbf{u}$, that is, $\mathbf{Y} \cdot \mathbf{u} = \mathbf{X} \cdot \mathbf{u}$. Thus H has the form (11.7) for $\lambda = \mathbf{X} \cdot \mathbf{u}$. More precisely, if $\mathbf{u} = [a_1, \ldots, a_n]^t \in \mathbb{R}^n$, then H is represented by the linear equation

$$a_1 x_1 + \cdots + a_n x_n - \lambda = 0.$$

Notice that, following the argument presented in the above proposition, the hyperplane H contains the origin point of the frame of reference if and only if $\lambda = 0$.

Example 11.19 Let $\mathbf{u} = [1, 2, -1]^t \in \mathbb{R}^3$. Then by the symbol $H_{\mathbf{u},\lambda}$ we mean all hyperplanes in $\mathbb{RE}^3$ that are orthogonal to vector $\mathbf{u}$. The different hyperplanes $H_{\mathbf{u},\lambda}$ are parallel to each other, as λ varies. Thus, for instance,

(1) for $\lambda = 0$, $H_{\mathbf{u},0}$ is represented by equation $x + 2y - z = 0$;
(2) for $\lambda = -1$, $H_{\mathbf{u},-1}$ is represented by equation $x + 2y - z + 1 = 0$;
(3) for $\lambda = 2$, $H_{\mathbf{u},2}$ is represented by equation $x + 2y - z - 2 = 0$.

By using the dot product of inner product vector space $\mathbb{R}^3$, we are in a position to compile an overview of the main formulae and techniques enabling the computation of angles and distances in the 3-dimensional real Euclidean space.

Projection of Vectors in $\mathbb{RE}^3$

One important use of dot product is in projections. Here, we describe two different kinds of projections of vectors: onto a line and onto a plane in the affine Euclidean space $\mathbb{RE}^3$.

Of course, the concept of orthogonal projection covers the more general situation related to inner product n-dimensional spaces. Here, we just would like to show how those arguments may be applied to the classical 3-dimensional analytic geometry. More precisely, given a vector $\mathbf{v} \in \mathbb{R}^3$ and a subspace W of $\mathbb{R}^3$ generated by vectors $\{\mathbf{e}_1, \dots, \mathbf{e}_k\}$ $(k = 1, 2)$, we say that $\mathbf{v}'$ is the orthogonal projection vector of $\mathbf{v}$ onto W if there exists $\mathbf{w} \in \mathbb{R}^3$ that is orthogonal to any vector in W, such that $\mathbf{v} = \mathbf{v}' + \mathbf{w}$. We know that in case $\{\mathbf{e}_1, \dots, \mathbf{e}_k\}$ is an orthogonal basis for W, then we may obtain the orthogonal projection as

$$\mathbf{v}' = \sum_{i=1}^{k} \frac{\mathbf{v} \cdot \mathbf{e}_i}{\mathbf{e}_i \cdot \mathbf{e}_i} \mathbf{e}_i. \tag{11.8}$$

Example 11.20 Let r be the line in $\mathbb{RE}^3$ defined by equations

$$\begin{aligned} x &= 1 + 2t \\ y &= 2 - t , \quad t \in \mathbb{R} \\ z &= \quad 3t \end{aligned}$$

and $\mathbf{v}$ be the vector of coordinates $(2, -1, 1)$ in terms of the standard frame of reference in $OXYZ$. In order to get the projection of $\mathbf{v}$ onto r, we may compute the projection onto r of a vector which is equipollent to $\mathbf{v}$ (that is, it has the same length, direction and sense of $\mathbf{v}$) but has its tail on r. We obtain this vector by a simple translation of $\mathbf{v}$. Actually, without loss of generality, we may assume that $\mathbf{v}$ is precisely applied to r. Notice that the direction of r is represented by the vector $\mathbf{e} = (2, -1, 3)$. At this point, performing the formula (11.8), we get

$$\mathbf{v}' = \frac{\mathbf{v} \cdot \mathbf{e}}{\mathbf{e} \cdot \mathbf{e}} \mathbf{e} = \left(\frac{8}{7}, -\frac{4}{7}, \frac{12}{7} \right).$$

Example 11.21 Let π be the plane in $\mathbb{RE}^3$ defined by equations

$$\begin{aligned} x &= 1 + t - t' \\ y &= 2 + 2t - t', \quad t, t' \in \mathbb{R} \\ z &= 1 - t + t' \end{aligned}$$

and $\mathbf{v}$ be the vector of coordinates $(1, 1, 2)$ in terms of the standard frame of reference in $OXYZ$. As above, we may assume that $\mathbf{v}$ is precisely applied to π. Starting from the definition of π and using the classical orthogonalization process, we arrive at the conclusion that the vector space representing the direction of π is generated by orthogonal vectors $\mathbf{e}_1 = (1, 2, -1)$, $\mathbf{e}_2 = (-1, 1, 1)$. At this point, performing the formula (11.8), we get

$$\mathbf{v}' = \frac{\mathbf{v} \cdot \mathbf{e}_1}{\mathbf{e}_1 \cdot \mathbf{e}_1} \mathbf{e}_1 + \frac{\mathbf{v} \cdot \mathbf{e}_2}{\mathbf{e}_2 \cdot \mathbf{e}_2} \mathbf{e}_2 = \left(-\frac{1}{2}, 1, \frac{1}{2} \right).$$

The Angle Enclosed Between Two Straight Lines

Let (l, m, n) be the direction ratios of a straight line r. The *direction cosines* of r are the cosines of the angles between the unit vector of r having coordinates

$$\left(\frac{l}{\sqrt{l^2 + m^2 + n^2}}, \frac{m}{\sqrt{l^2 + m^2 + n^2}}, \frac{n}{\sqrt{l^2 + m^2 + n^2}} \right)$$

and the positive X, Y and Z axes, respectively. Therefore, the direction cosines of r can be represented by

$$\alpha = \cos\widehat{(r, X)} \in \left\{ +\frac{l}{\sqrt{l^2 + m^2 + n^2}}, -\frac{l}{\sqrt{l^2 + m^2 + n^2}} \right\},$$

$$\beta = \cos\widehat{(r, Y)} \in \left\{ +\frac{m}{\sqrt{l^2 + m^2 + n^2}}, -\frac{m}{\sqrt{l^2 + m^2 + n^2}} \right\},$$

$$\gamma = \cos\widehat{(r, Z)} \in \left\{ +\frac{n}{\sqrt{l^2 + m^2 + n^2}}, -\frac{n}{\sqrt{l^2 + m^2 + n^2}} \right\}.$$

Let now (l, m, n) and (l', m', n') be the direction ratios of the straight lines r and r', respectively. The cosine of the angle enclosed between r and r' is equal to the cosine of the angle enclosed between the unit vectors of r and r', having coordinates

$$\left(\frac{l}{\sqrt{l^2 + m^2 + n^2}}, \frac{m}{\sqrt{l^2 + m^2 + n^2}}, \frac{n}{\sqrt{l^2 + m^2 + n^2}} \right)$$

and

$$\left(\frac{l'}{\sqrt{l'^2 + m'^2 + n'^2}}, \frac{m'}{\sqrt{l'^2 + m'^2 + n'^2}}, \frac{n'}{\sqrt{l'^2 + m'^2 + n'^2}} \right),$$

respectively, that is,

$$\cos\widehat{(rr')} = \pm \frac{ll' + mm' + nn'}{\sqrt{l^2 + m^2 + n^2}\sqrt{l'^2 + m'^2 + n'^2}}.$$

Example 11.22 Consider the lines

$$x = 1 + 2t \qquad x = 2 + 3t'$$
$$r : y = 1 - t \,, \quad r' : y = 1 + 2t', \quad t, t' \in \mathbb{R}.$$
$$z = 1 + t \qquad z = 1 + t'$$

The directions of r and r' are $(2, -1, 1)_r$ and $(3, 2, 1)_{r'}$, respectively. Their direction cosines are

$$\cos(\widehat{rX}) = \pm\frac{2}{\sqrt{6}}, \quad \cos(\widehat{rY}) = \mp\frac{1}{\sqrt{6}}, \quad \cos(\widehat{rZ}) = \pm\frac{1}{\sqrt{6}},$$

$$\cos(\widehat{r'X}) = \pm\frac{3}{\sqrt{14}}, \quad \cos(\widehat{r'Y}) = \pm\frac{2}{\sqrt{14}}, \quad \cos(\widehat{r'Z}) = \pm\frac{1}{\sqrt{14}}.$$

The cosine of the angle enclosed between r and r' is equal to

$$\cos(\widehat{rr'}) = \pm\frac{5}{\sqrt{6}\sqrt{14}}.$$

Distance Between a Point and a Hyperplane

Let H be a hyperplane in $\mathbb{RE}^n$, V the direction of H and $\mathbf{u} = (\alpha_1, \ldots, \alpha_n)$ an orthogonal vector to H. As remarked above, there exists $\lambda \in \mathbb{R}$ such that H is represented by equation

$$f(x_1, \ldots, x_n) = 0 \quad \text{where} \quad f(x_1, \ldots, x_n) = \sum_{i=1}^{n} \alpha_i x_i - \lambda. \tag{11.9}$$

Let now P_0 be a point of $\mathbb{RE}^n$, having coordinate vector $\mathbf{X}_0 = (\beta_1, \ldots, \beta_n)$ in terms of a fixed frame of reference in $\mathbb{RE}^n$. The *distance* from P_0 to H is defined as the length of vector $\mathbf{P_0Q_0}$, where Q_0 is the orthogonal projection of P_0 onto H. To obtain Q_0, we have to construct the affine subspace $H^{\perp} = P_0 + V^{\perp}$, which is orthogonal to H and containing P_0. Then compute $Q_0 = H \cap H^{\perp}$. If we describe $H^{\perp}$ in its parametric form

$$H^{\perp} : \begin{matrix} x_1 = \beta_1 + \alpha_1 t \\ \vdots \\ x_n = \beta_n + \alpha_n t \end{matrix} \quad , \quad t \in \mathbb{R} \tag{11.10}$$

and substitute (11.10) in (11.9), we get

$$\sum_{i=1}^{n} (\alpha_i \beta_i + \alpha_i^2 t) - \lambda = 0,$$

that is,

$$f(\mathbf{X_0}) + t\,\|\mathbf{u}\|^2 = 0.$$

Thus, point Q_0 is obtained by (11.10) for $t = -\frac{f(\mathbf{X_0})}{\|\mathbf{u}\|^2}$. The coordinate vector representing $\mathbf{P_0 Q_0}$ is then

$$\left(-\alpha_1 \frac{f(\mathbf{X_0})}{\|\mathbf{u}\|^2}, -\alpha_2 \frac{f(\mathbf{X_0})}{\|\mathbf{u}\|^2}, \ldots, -\alpha_n \frac{f(\mathbf{X_0})}{\|\mathbf{u}\|^2}\right)$$

whose length is

$$\sqrt{(\alpha_1^2 + \cdots + \alpha_n^2) \frac{f(\mathbf{X_0})^2}{\|\mathbf{u}\|^4}} = \frac{|f(\mathbf{X_0})|}{\|\mathbf{u}\|}.$$

Example 11.23 Let now $P_0 \equiv (x_0, y_0, z_0)$ be a point of $\mathbb{RE}^3$ and $ax + by + cz + d = 0$ the equation of a plane π. The distance from P_0 to π is equal to the length of the vector $\mathbf{P_0 Q_0}$, where Q_0 is the orthogonal projection point of P_0 on π.

Since

$$\mathbf{u} \equiv \left(\frac{a}{\sqrt{a^2 + b^2 + c^2}}, \frac{b}{\sqrt{a^2 + b^2 + c^2}}, \frac{c}{\sqrt{a^2 + b^2 + c^2}}\right)$$

represents the normal direction to π, then

$$\delta(P_0, H) = \frac{|ax_0 + by_0 + cz_0 + d|}{\sqrt{a^2 + b^2 + c^2}}.$$

Example 11.24 As a consequence, we may also obtain the distance between two parallel planes. To do this, without loss of generality, we consider the planes π and π' having equations

$$\pi : ax + by + cz + d = 0, \quad \pi' : ax + by + cz + d' = 0, \quad d \neq d'.$$

If $P_0 \equiv (x_0, y_0, z_0) \in \pi$, then the distance $\delta(\pi, \pi')$ between π and π' is equal to the distance from P_0 to π', that is,

$$\delta(\pi, \pi') = \delta(P_0, \pi') = \frac{|ax_0 + by_0 + cz_0 + d'|}{\sqrt{a^2 + b^2 + c^2}} = \frac{|d' - d|}{\sqrt{a^2 + b^2 + c^2}}$$

because $ax_0 + by_0 + cz_0 = -d$.

Distance Between Two Skew Lines

Consider now two skew lines (that is, non-parallel non-intersecting lines) r and r'. This means that there exist two parallel planes π and π', containing r and r', respectively. One of the more important problems in Geometry is finding the minimum distance between the two lines, which is the distance between π and π'. It is naturally the shortest distance between the lines, i.e., the length of the orthogonal line segment to both lines. The solution of the problem is very simple through the use of both scalar product $\cdot$ and cross product $\wedge$ of vectors. In detail, let $\mathbf{v} = (l, m, n)$ and $\mathbf{v}' = (l', m', n')$ be two vectors representing the direction ratios of r and r', respectively. The vector $\mathbf{v} \wedge \mathbf{v}'$ is orthogonal to $\mathbf{v}$ and $\mathbf{v}'$, that is, to r and r'. Hence, for any choice of two points $P \in r$ and $Q \in r'$, the absolute value of the scalar projection of $\mathbf{PQ}$ in the direction of $\mathbf{v} \wedge \mathbf{v}'$ is the minimum distance $\delta(r, r')$ between r and r':

$$\delta(r, r') = \left| \mathbf{PQ} \cdot \frac{\mathbf{v} \wedge \mathbf{v}'}{\|\mathbf{v} \wedge \mathbf{v}'\|} \right|.$$

Example 11.25 Consider the following lines in $\mathbb{RE}^3$:

$$r: \begin{array}{l} x = 0 \\ y = t \\ z = 1 + t \end{array}, \quad r': \begin{array}{l} x = -2 + t' \\ y = -1 - 3t' \\ z = -2t' \end{array}, \quad t, t' \in \mathbb{R}.$$

We may describe r as an affine subspace of the form $P + V$, where P is the point of r having coordinates $(0, 0, 1)$ and $V = \langle (0, 1, 1) \rangle$ is its direction. Analogously, r' is an affine subspace of the form $Q + W$, where Q is the point of r' having coordinates $(-2, -1, 0)$ and $W = \langle (1, -3, -2) \rangle$ is its direction. The conclusion that r and r' are skew lines stems from the fact that vectors $\mathbf{PQ} \equiv (-2, -1, -1)$, $(0, 1, 1)$ and $(1, -3, -2)$ are linearly independent. The cross product of directions $(0, 1, 1)$ and $(1, -3, -2)$ is equal to the vector

$$\begin{vmatrix} \mathbf{i} & \mathbf{j} & \mathbf{k} \\ 0 & 1 & 1 \\ 1 & -3 & -2 \end{vmatrix} = \mathbf{i} + \mathbf{j} - \mathbf{k}$$

and its length is equal to $\sqrt{3}$. Then, by performing the above-discussed formula, we obtain the minimum distance between r and r' as follows:

$$\delta(r, r') = \left| \mathbf{PQ} \cdot \frac{\mathbf{i} + \mathbf{j} - \mathbf{k}}{\sqrt{3}} \right| = \frac{2}{\sqrt{3}}.$$

Distance Between a Point and a Line

Let r be a line having direction ratios (l, m, n), P_1 a point of coordinates (x_1, y_1, z_1), which does not belong to r. To obtain the minimum distance $\delta(P_1, r)$ from P_1 to r, we may construct the plane π, containing P_1 and orthogonal to r. If $Q_1 = \pi \cap r$ is the intersection point of π and r, then the length of $\mathbf{P_1 Q_1}$ is the minimum distance between P_1 and r. In order to get the vector $\mathbf{P_1 Q_1}$, one may also proceed as follows. Let $Q_0 \equiv (x_0, y_0, z_0)$ be any point of r and consider the parallelogram $\mathscr{P}$ having height $\mathbf{P_1 Q_1}$ and base $\mathbf{Q_1 Q_0}$. Hence, the area of $\mathscr{P}$ is equal to

$$\|\mathbf{P_1Q_1}\|\,\|\mathbf{Q_1Q_0}\|\,.$$

On the other hand, since $\mathbf{Q_1Q_0}$ and $\mathbf{Q_0P_1}$ are the sides of $\mathscr{P}$, it can be also obtained as the norm of $\mathbf{Q_1Q_0} \wedge \mathbf{Q_0P_1}$. Therefore,

$$\|\mathbf{P_1Q_1}\|\,\|\mathbf{Q_1Q_0}\| = \|\mathbf{Q_1Q_0} \wedge \mathbf{Q_0P_1}\|\,.$$

Moreover, $\mathbf{Q_1Q_0}$ is a line segment of r, thus there exists a suitable $\alpha \in \mathbb{R}$ such that $\mathbf{Q_1Q_0} \equiv \alpha(l, m, n)$. From this we get

$$\mathbf{Q_1Q_0} \wedge \mathbf{Q_0P_1} = \begin{vmatrix} i & j & k \\ \alpha l & \alpha m & \alpha n \\ (x_1 - x_0) & (y_1 - y_0) & (z_1 - z_0) \end{vmatrix}$$

and the norm of $\mathbf{Q_1Q_0} \wedge \mathbf{Q_0P_1}$ is equal to

$$\alpha \sqrt{\left| \begin{matrix} x_1 - x_0 & y_1 - y_0 \\ l & m \end{matrix} \right|^2 + \left| \begin{matrix} x_1 - x_0 & z_1 - z_0 \\ l & n \end{matrix} \right|^2 + \left| \begin{matrix} y_1 - y_0 & z_1 - z_0 \\ m & n \end{matrix} \right|^2}.$$

Since $\|\mathbf{Q_1Q_0}\|$ is precisely $\alpha\sqrt{l^2 + m^2 + n^2}$, we conclude that $\delta(P_1, r)$

$$= \|\mathbf{P_1Q_1}\|$$

$$= \frac{\|\mathbf{Q_1Q_0} \wedge \mathbf{Q_0P_1}\|}{\|\mathbf{Q_1Q_0}\|}$$

$$= \sqrt{\frac{\left| \begin{matrix} x_1 - x_0 & y_1 - y_0 \\ l & m \end{matrix} \right|^2 + \left| \begin{matrix} x_1 - x_0 & z_1 - z_0 \\ l & n \end{matrix} \right|^2 + \left| \begin{matrix} y_1 - y_0 & z_1 - z_0 \\ m & n \end{matrix} \right|^2}{l^2 + m^2 + n^2}}.$$

Example 11.26 Let $P_0 \in \mathbb{RE}^3$ be the point of coordinates $(1, -1, 2)$ and r the line represented by the parametric form

$$r : \begin{aligned} x &= 1 + 2t \\ y &= 1 - t, \quad t \in \mathbb{R}. \\ z &= 1 + t \end{aligned}$$

In order to obtain the minimum distance from P_0 to r, we firstly compute the coordinate of vector $\mathbf{P_0Q}$, where Q is any point of r. For instance, if we choose Q as the point of coordinates $(1, 1, 1)$, it follows that $\mathbf{P_0Q} \equiv (0, 2, -1)$. Then the requested distance is

$$\delta(P_0, r) = \sqrt{\frac{\left| \begin{matrix} -1 & 1 \\ 2 & -1 \end{matrix} \right|^2 + \left| \begin{matrix} 2 & 1 \\ 0 & -1 \end{matrix} \right|^2 + \left| \begin{matrix} 2 & -1 \\ 0 & 2 \end{matrix} \right|^2}{4 + 1 + 1}}$$

$$= \frac{\sqrt{7}}{\sqrt{2}}.$$

Exercises

1. Let $\mathbb{A} = \mathbb{R}\mathbb{A}^5$ be equipped with the standard frame of reference. Determine parametric and linear equations representing the affine subspace of $\mathbb{A}$ having direction $V = \langle (0, 1, 1, 0, 0), (0, 0, 0, 1, 0), (1, 1, 0, 0, -1) \rangle$ and containing the point $P \equiv (1, -1, 0, 0, 0)$.

2. In the affine space $\mathbb{R}\mathbb{A}^6$ equipped with the standard frame of reference, represent the affine subspace containing the following points:

$$P_1 \equiv (2, 2, 0, 0, 0, 0),\ P_2 \equiv (1, 0, 1, 0, 0, 0),$$

$$P_3 \equiv (0, -2, 2, 0, 0, 0),\ P_4 \equiv (2, 1, -1, 1, 0, 0).$$

3. In the Euclidean space $\mathbb{R}\mathbb{E}^3$ equipped with the standard frame of reference, determine the projection of vector $\mathbf{v} \equiv (1, 2, -1)$ onto the hyperplane containing the origin point and having direction $V = \langle (2, 1, 0), (-1, 1, 1) \rangle$.

4. In the Euclidean space $\mathbb{R}\mathbb{E}^3$, consider the following lines: r_1 containing the point $P \equiv (1, 1, 2)$ and having direction $V = \langle (1, 1, 0) \rangle$, r_2 as intersection of planes having equations $x + y - z + 1 = 0$ and $2x + y + z - 1 = 0$. Prove that r_1 and r_2 are skew lines and determine their minimum distance.

5. Let P_1, P_2, P_3 be three points in the affine space $\mathbb{R}\mathbb{A}^2$ equipped with the standard frame of reference. Letting (α_1, α_2), (β_1, β_2) and (γ_1, γ_2) be their coordinates, prove that P_1, P_2, P_3 are collinear if and only if the matrix

$$\begin{bmatrix} \alpha_1 & \alpha_2 & 1 \\ \beta_1 & \beta_2 & 1 \\ \gamma_1 & \gamma_2 & 1 \end{bmatrix}$$

has determinant equal to zero.

6. Let P_1, P_2, P_3, P_4 be four points in the affine space $\mathbb{R}\mathbb{A}^3$ equipped by the standard frame of reference. Letting $(\alpha_1, \alpha_2, \alpha_3)$, $(\beta_1, \beta_2, \beta_3,)$ $(\gamma_1, \gamma_2, \gamma_3)$ and $(\delta_1, \delta_2, \delta_3)$ be their coordinates, prove that P_1, P_2, P_3, P_4 are coplanar if and only if the matrix

$$\begin{bmatrix} \alpha_1 & \alpha_2 & \alpha_3 & 1 \\ \beta_1 & \beta_2 & \beta_3 & 1 \\ \gamma_1 & \gamma_2 & \gamma_3 & 1 \\ \delta_1 & \delta_2 & \delta_3 & 1 \end{bmatrix}$$

has determinant equal to zero.

11.2 Affine Transformations

Definition 11.27 Let $\mathbb{F}$ be a field, V, V' two vector spaces over $\mathbb{F}$, S, S' two nonempty sets and $\mathbb{A} = (S, V)$, $\mathbb{A}' = (S', V')$ the corresponding affine spaces. An *affine transformation* of $\mathbb{A}$ into $\mathbb{A}'$ is a map $f : \mathbb{A} \to \mathbb{A}'$ such that there exists a linear transformation $\varphi : V \to V'$ satisfying the following condition:

$$\mathbf{f(P)f(Q)} = \varphi(\mathbf{PQ}), \quad \text{for all} \quad P, Q \in \mathbb{A}.$$

The homomorphism $\varphi : V \to V'$ is called the *linear part* of f (or also the *associated homomorphism* with f). An *affine transformation* of $\mathbb{A}$ is an isomorphism of $\mathbb{A}$ into itself and its linear part is an isomorphism $\varphi : V \to V$.

The set of all affine transformations of $\mathbb{A}$ is usually denoted by $Aff(\mathbb{A})$. The fact that it is a group can be easily checked.

Remark 11.28 In fact,

(i) the identity map $\eta : \mathbb{A} \to \mathbb{A}$ is the affine transformation having the identity map on V as associated automorphism,

(ii) the composition of two affine transformations $g \circ f$ of $\mathbb{A}$ (having associated automorphisms $\chi, \varphi : V \to V$, respectively) is the affine transformation associated with the automorphism $\chi \circ \varphi$ of V,

(iii) the inverse of the affine transformation $f : \mathbb{A} \to \mathbb{A}$, having associated automorphism $\varphi : V \to V$, is the affine transformation $f^{-1} : \mathbb{A} \to \mathbb{A}$ having $\varphi^{-1} : V \to V$ as associated automorphism.

Example 11.29 Let $\mathbb{A} = (S, V)$ be an affine space, where V is a vector space over the field $\mathbb{F}$, and let $\mathbf{v} \in V$ be a fixed vector. Consider the map $f_{\mathbf{v}} : \mathbb{A} \to \mathbb{A}$ defined as follows: for any $P \in \mathbb{A}$,

$$f_{\mathbf{v}}(P) = Q \in \mathbb{A} \iff \mathbf{PQ} = \mathbf{v}.$$

First, we notice that $f_{\mathbf{v}}(P_1) = f_{\mathbf{v}}(P_2) = Q \in \mathbb{A}$ if and only if $\mathbf{P_1 Q} = \mathbf{P_2 Q}$, i.e., if and only if $P_1 = P_2 \in \mathbb{A}$. Hence $f_{\mathbf{v}}$ is injective. Moreover, for any $Q \in \mathbb{A}$ and taking $P = f_{-\mathbf{v}}(Q)$, one has that $\mathbf{QP} = -\mathbf{v}$, that is, $\mathbf{PQ} = \mathbf{v}$. Thus $Q = f_{\mathbf{v}}(P)$, i.e., $f_{\mathbf{v}}$ is surjective. Let now $P, Q \in \mathbb{A}$, such that $P_1 = f_{\mathbf{v}}(P)$ and $P_2 = f_{\mathbf{v}}(Q)$, that is, $\mathbf{PP_1} = \mathbf{QP_2} = \mathbf{v}$. Thus, we have

$$\begin{aligned}
\mathbf{f_{\mathbf{v}}(P)f_{\mathbf{v}}(Q)} &= \mathbf{P_1 P_2} \\
&= \mathbf{P_1 P} + \mathbf{PQ} + \mathbf{QP_2} \\
&= -\mathbf{v} + \mathbf{PQ} + \mathbf{v} \\
&= \mathbf{PQ}.
\end{aligned}$$

Therefore, $f_{\mathbf{v}}$ is an affine transformation of $\mathbb{A}$ having the identity map of V as associated automorphism. The above- described map is usually called *translation of $\mathbb{A}$ defined by* $\mathbf{v}$.

Remark 11.30 Let $\mathbb{A} = (S, V)$ be an affine space and $f : \mathbb{A} \to \mathbb{A}$ any affine transformation associated with the identity map of V. It follows that, for any $P, Q \in \mathbb{A}$, both the following hold:

$$\mathbf{f(P)f(Q) = PQ}$$

and

$$\begin{aligned}\mathbf{Pf(P)} &= \mathbf{PQ + Qf(Q) + f(Q)f(P)} \\ &= \mathbf{PQ + Qf(Q) - PQ} \\ &= \mathbf{Qf(Q)}.\end{aligned}$$

Therefore, the vector $\mathbf{v} = \mathbf{Pf(P)}$ is not depending on the choice of point $P \in \mathbb{A}$. Thus, $f = f_{\mathbf{v}}$ is a translation in the sense of previous definition.

In other words, translations of $\mathbb{A}$ cover completely the set of affine transformations having the identity map of V as associated automorphism.

Remark 11.31 One can easily see that

(i) the identity map $\eta : \mathbb{A} \to \mathbb{A}$ is the translation f_0 of $\mathbb{A}$ defined by the zero vector of V.

(ii) the composition of two translations $f_{\mathbf{w}} \circ f_{\mathbf{v}}$ (defined by vectors $\mathbf{w}, \mathbf{v} \in V$) of $\mathbb{A}$ is the translation of $\mathbb{A}$ defined by the vector $\mathbf{w} + \mathbf{v}$.

(iii) the inverse of the translation $f_{\mathbf{v}}$ (defined by vector $\mathbf{v} \in V$) of $\mathbb{A}$ is the translation of $\mathbb{A}$ defined by vector $-\mathbf{v}$.

Hence, the set $T_{\mathbb{A}}$ of all translations of $\mathbb{A}$ is a group. Moreover, the map $\chi : T_{\mathbb{A}} \to V$ defined by $\chi(f_{\mathbf{v}}) = \mathbf{v} \in V$, for any $f_{\mathbf{v}} \in T_{\mathbb{A}}$, is bijective and

$$\chi(f_{\mathbf{w}} \circ f_{\mathbf{v}}) = \mathbf{w} + \mathbf{v} = \chi(f_{\mathbf{w}}) + \chi(f_{\mathbf{v}}),$$

that is, χ is an isomorphism.

Example 11.32 Let $\mathbb{A} = (S, V)$ be an affine space, where V is a vector space over the field $\mathbb{F}$. Let $P_0 \in \mathbb{A}$ be a fixed point and $0 \neq \lambda \in \mathbb{F}$. Consider the map $f_{P_0,\lambda} : \mathbb{A} \to \mathbb{A}$ defined as follows: for any $P \in \mathbb{A}$,

$$f_{P_0,\lambda}(P) = Q \in \mathbb{A} \iff \mathbf{P_0Q} = \lambda\mathbf{P_0P}.$$

The point P_0 is called the *center* of $f_{P_0,\lambda}$ and the nonzero scalar λ is called its *ratio*. If we assume $f_{P_0,\lambda}(P_1) = f_{P_0,\lambda}(P_2) = Q \in \mathbb{A}$, then $\lambda\mathbf{P_0P_1} = \lambda\mathbf{P_0P_2}$, implying $P_1 = P_2 \in \mathbb{A}$. Hence, $f_{P_0,\lambda}$ is injective. Moreover, for any $Q \in \mathbb{A}$ and taking $P = f_{P_0,\lambda^{-1}}(Q)$, one has that $\mathbf{P_0P} = \lambda^{-1}\mathbf{P_0Q}$, that is, $\mathbf{P_0Q} = \lambda\mathbf{P_0P}$. Thus $Q = f_{P_0,\lambda}(P)$, i.e., f_v is surjective.

Let now $P, Q \in \mathbb{A}$, such that $P_1 = f_{P_0,\lambda}(P)$ and $P_2 = f_{P_0,\lambda}(Q)$, that is, $\mathbf{P_0P_1} = \lambda\mathbf{P_0P}$ and $\mathbf{P_0P_2} = \lambda\mathbf{P_0Q}$. It follows that

$$\mathbf{f}_{P_0,\lambda}(P)\mathbf{f}_{P_0,\lambda}(Q) = \mathbf{P_1P_2}$$
$$= \mathbf{P_1P_0} + \mathbf{P_0P_2}$$
$$= -\lambda\mathbf{P_0P} + \lambda\mathbf{P_0Q}$$
$$= \lambda\mathbf{PQ}.$$

Therefore, $f_{P_0,\lambda}$ is an affine transformation of $\mathbb{A}$. The associated automorphism to $f_{P_0,\lambda}$ is the map $\varphi : V \to V$ defined by $\varphi(\mathbf{v}) = \lambda\mathbf{v}$, for any $\mathbf{v} \in V$. The above-described affine transformation is called *homothety of $\mathbb{A}$ centered at P_0* with *scale factor* (or *ratio*) λ.

Remark 11.33 Fix any point $P_0 \in \mathbb{A}$, the set of all homotheties of $\mathbb{A}$ centered at P_0 is usually denoted by $Hom(\mathbb{A})_{P_0}$. It is easy to see that $Hom(\mathbb{A})_{P_0}$ is a group. In fact

- (i)   the identity map $\eta : \mathbb{A} \to \mathbb{A}$ is the homothety $f_{P_0,1}$ of $\mathbb{A}$.
- (ii) the composition of two homotheties $f_{P_0,\mu} \circ f_{P_0,\lambda}$ (having ratios μ, λ, respectively) of $\mathbb{A}$ is the homothety $f_{P_0,\mu\lambda}$ of $\mathbb{A}$.
- (iii) the inverse of the homothety $f_{P_0,\mu}$ of $\mathbb{A}$ is the homothety $f_{P_0,\mu^{-1}}$ of $\mathbb{A}$.

Example 11.34 Let $\mathbb{A} = (S, V)$ be an affine space, where V is a vector space of dimension n over the field $\mathbb{F}$, $O \in \mathbb{A}$ be a fixed point and consider the map $f_O : \mathbb{A} \to \mathbb{A}$ defined by

$$P \in \mathbb{A}, \quad f_O(P) = P' \quad \Longleftrightarrow \quad \mathbf{OP'} = -\mathbf{OP}.$$

It is easy to see that f_O is an affine transformation. It is usually called *central symmetry* with respect to the point O. The linear part of the central symmetry is represented by the matrix $-I_n$, where I_n is the identity matrix in $M_n(\mathbb{F})$. A subset $S \subseteq \mathbb{A}$ is called *centrally symmetric* with respect to the point O, if $f_O(S) = S$. In this case, O is usually called the *center* of S.

We remark that any affine subspace S of an affine space $\mathbb{A}$ is symmetric with respect to any of its points. In fact, let $P_0 \in S$ be any point of S and W the vector subspace associated with S. Then, for any point $P \in S$, $\mathbf{P_0P} \in W$. The central symmetry $f_{P_0} : \mathbb{A} \to \mathbb{A}$ with respect to P_0 induces the identity $\mathbf{P_0f_{P_0}}(\mathbf{P}) = -\mathbf{P_0P}$. Thus, $\mathbf{P_0f_{P_0}}(\mathbf{P}) \in W$ and $f_{P_0}(P) \in S$. By the definition of central symmetry, it follows that

(1) $f : \mathbb{A} \to \mathbb{A}$ is a central symmetry with respect to the point P if and only if $f(P) = P$.

(2) Let $f : \mathbb{A} \to \mathbb{A}$ be a central symmetry of the n-dimensional affine space $\mathbb{A}$ with respect to the point P having coordinates $(\gamma_1, \dots, \gamma_n)$ in terms of a fixed frame of reference. If $Q \in \mathbb{A}$ has coordinates $(\alpha_1, \dots, \alpha_n)$, then its image $f(Q)$ has coordinates $(\beta_1, \dots, \beta_n)$ if and only if $\beta_i = 2\gamma_i - \alpha_i$, for any $i = 1, \dots, n$.

Definition 11.35 Let $\mathbb{A}' \subset \mathbb{A}$ be two affine spaces. An affine transformation $f : \mathbb{A} \to \mathbb{A}'$ is called *projection* of $\mathbb{A}$ onto $\mathbb{A}'$ if $f(\mathbb{A}) = \mathbb{A}'$.

Proposition 11.36 *Let $f : \mathbb{A} \to \mathbb{A}'$ be a projection of $\mathbb{A}$ onto its subspace $\mathbb{A}'$. Then, for any point $P' \in \mathbb{A}'$, the pre-image $f^{-1}(P') \subseteq \mathbb{A}$ is an affine subspace of $\mathbb{A}$. In particular, if $f : \mathbb{A} \to \mathbb{A}'$ is a projection of $\mathbb{A}$ onto its subspace $\mathbb{A}'$, then for distinct points P', $P'' \in \mathbb{A}'$, the affine subspaces $f^{-1}(P')$ and $f^{-1}(P'')$ are parallel.*

Proof Let V and V' be the respective vector spaces of $\mathbb{A}$ and $\mathbb{A}'$, and $\varphi : V \to V'$ the linear transformation associated with f. For any points $Q_1, Q_2 \in f^{-1}(P')$, we see that $\varphi(\mathbf{Q_1 Q_2}) = \mathbf{f(Q_1)f(Q_2)} = \mathbf{0}$ (since $f(Q_1) = f(Q_2) = P'$). Hence, the vector $\mathbf{Q_1 Q_2}$ lies in the null space of φ (which is a subspace of V).

Consider now $P \in f^{-1}(P') \subseteq \mathbb{A}$ and any vector $\mathbf{u} \in Ker(\varphi)$, the null space of φ. Then, there exists $Q \in \mathbb{A}$ such that $\mathbf{PQ} = \mathbf{u}$, that is, $\mathbf{0} = \varphi(\mathbf{PQ}) = \mathbf{f(P)f(Q)}$, and $f(P) = f(Q)$ follows. Thus, $Q \in f^{-1}(P')$ and $Ker(\varphi)$ is precisely the vector space associated with the affine subspace $f^{-1}(P')$.

Definition 11.37 Let $f : \mathbb{A} \to \mathbb{A}'$ be a projection of $\mathbb{A}$ onto its subspace $\mathbb{A}'$ and $S \subseteq \mathbb{A}'$ (not necessarily a subspace of $\mathbb{A}'$). The set

$$f^{-1}(S) = \bigcup_{P \in S} f^{-1}(P)$$

is called a *cylinder* in $\mathbb{A}$.

Let us now write down the action of affine transformations of affine spaces in coordinate form. To do this, let $\mathbb{A}$ be an affine n-dimensional space associated with the vector space V over the field $\mathbb{F}$, $B = \{O, \mathbf{e_1}, \dots, \mathbf{e_n}\}$ a frame of reference of $\mathbb{A}$. We prove the following.

Theorem 11.38 *The map $f : \mathbb{A} \to \mathbb{A}$ is an affine transformation if and only if there exist a nonsingular matrix $A \in M_n(\mathbb{F})$ and a fixed point $c \in \mathbb{A}$ such that, for any point $P \in \mathbb{A}$ having coordinate vector $X = [x_1, \dots, x_n]^t$ in terms of B, the point $f(P) \in \mathbb{A}$ has coordinate vector $Y = [y_1, \dots, y_n]^t$ in terms of B, satisfying the following relation:*

$$Y = AX + c. \tag{11.11}$$

Proof Assume that $f : \mathbb{A} \to \mathbb{A}$ is an affine transformation associated with the automorphism $\varphi : V \to V$, and let $A \in M_n(\mathbb{F})$ be the matrix of φ with respect to the basis $\{\mathbf{e_1}, \dots, \mathbf{e_n}\}$ of V. Set $c = f(O)$ and let $[c_1, \dots, c_n]^t$ be the coordinate vector of $f(O)$ in terms of B. Hence, we have that

$$\mathbf{OP} = x_1 \mathbf{e_1} + \cdots + x_n \mathbf{e_n}$$
$$\mathbf{Of(P)} = y_1 \mathbf{e_1} + \cdots + y_n \mathbf{e_n}$$
$$\mathbf{Of(O)} = c_1 \mathbf{e_1} + \cdots + c_n \mathbf{e_n}$$

and $\varphi(\mathbf{e_i}) = A\mathbf{e_i}$, for each $i = 1, \dots, n$, treating $\varphi(\mathbf{e_i})$ and $\mathbf{e_i}$ as column vectors. Thus,

$$\varphi(\mathbf{OP}) = \varphi(x_1 \mathbf{e_1} + \cdots + x_n \mathbf{e_n}) = A(x_1 \mathbf{e_1} + \cdots + x_n \mathbf{e_n})$$

and

$$\mathbf{f(O)f(P)} = (y_1 - c_1)\mathbf{e_1} + \cdots + (y_n - c_n)\mathbf{e_n}.$$

On the other hand, $\varphi(\mathbf{OP}) = \mathbf{f(O)f(P)}$, hence

$$A(x_1\mathbf{e_1} + \cdots + x_n\mathbf{e_n}) = (y_1 - c_1)\mathbf{e_1} + \cdots + (y_n - c_n)\mathbf{e_n},$$

that is,

$$A\begin{bmatrix} x_1 \\ \vdots \\ \vdots \\ x_n \end{bmatrix} = \begin{bmatrix} y_1 \\ \vdots \\ \vdots \\ y_n \end{bmatrix} - \begin{bmatrix} c_1 \\ \vdots \\ \vdots \\ c_n \end{bmatrix}$$

as required.

Conversely, for any invertible matrix $A \in M_n(\mathbb{F})$ associated with an automorphism of V with respect to the basis $\{\mathbf{e_1}, \ldots, \mathbf{e_n}\}$, and for any point $c \in \mathbb{A}$ having coordinate vector $[c_1, \ldots, c_n]^t$ in terms of B, let $f_{A,c} : \mathbb{A} \to \mathbb{A}$ be the map defined by relation (11.11). This map is an affine transformation. In fact, for any $P, Q \in \mathbb{A}$, having respectively coordinate vectors $X = [x_1, \ldots, x_n]^t$ and $Y = [y_1, \ldots, y_n]^t$ in terms of B, the following holds:

$$\begin{aligned} \mathbf{f_{A,c}(P)f_{A,c}(Q)} &= f_{A,c}(P) - f_{A,c}(Q) \\ &= AX + c - (AY + c) \\ &= A(X - Y) \\ &= \varphi(\mathbf{PQ}). \end{aligned}$$

Remark 11.39 Translations of an n-dimensional affine space $\mathbb{A}$ are precisely all the affine transformations of the form $f_{I_n,c}$, where $c \in \mathbb{A}$ is a fixed point and $I_n \in M_n(\mathbb{F})$ is the identity matrix.

One of the most relevant aspects of the affine transformations is that some properties are *invariant* under the action of such transformations. If a subset $S \subset \mathbb{A}$ possesses a property that is invariant under the action of f, then $f(S) \subset \mathbb{A}$ is a subset having the same property. Later, we describe in detail the application to the case of geometric figures having properties that are invariant under the action of affine transformations. Here, we firstly would like to fix some useful results:

Theorem 11.40 *Let $\mathbb{A}$ be the affine space $\mathbb{F}\mathbb{A}^n$ and $f : \mathbb{A} \to \mathbb{A}$ an affine transformation. Then*

(i) *f maps an affine subspace $\mathbb{A}'$ to an affine subspace having the same dimension of $\mathbb{A}'$.*

(ii) *f preserves the property of parallelism among affine subspaces.*

In particular, if $n = 2$, let $\mathbb{A}$ be the affine plane $\mathbb{F}\mathbb{A}^2$ or the affine 3-dimensional space $\mathbb{F}\mathbb{A}^3$ and $f : \mathbb{A} \to \mathbb{A}$ an affine transformation. Then, it is obvious to observe that f maps a line to a line and it also preserves the property of parallelism among lines.

Further, for $n = 3$ if $\mathbb{A}$ is the affine 3-dimensional space $\mathbb{F}\mathbb{A}^3$ and $f : \mathbb{A} \to \mathbb{A}$ an affine transformation, then f maps a plane to a plane; it preserves the property of parallelism among planes and preserves the property of parallelism among a plane and a line.

Proof (i) To show these properties, without loss of generality, we may consider $\mathbb{A}' = P_0 + W$, for a fixed point $P_0 \in \mathbb{A}'$ and a vector subspace W of $\mathbb{F}^n$. Here, we have identified V by $\mathbb{F}^n$ because $V \simeq \mathbb{F}^n$, where V is the vector space associated with the affine space $\mathbb{A}$. Assume that $\{w_1, \ldots, w_k\}$ is a basis of W. Thus, for any point $P \in \mathbb{A}'$,

$$P_0P = t_1 w_1 + \cdots + t_k w_k$$

for suitable $t_1, \ldots, t_k \in \mathbb{F}$. If we denote by X and X_0 the coordinate vector of P and P_0, respectively, we may write

$$X = X_0 + t_1 w_1 + \cdots + t_k w_k.$$

The coordinates of the point $f(P)$ may be computed using the expression $f(X) = AX + c$, where A is an invertible matrix of $M_n(\mathbb{F})$ and c is a fixed point of $\mathbb{A}'$. Thus

$$
\begin{aligned}
f(P) &= f(X_0 + t_1 w_1 + \cdots + t_k w_k) \\
&= A(X_0 + t_1 w_1 + \cdots + t_k w_k) + c \\
&= (AX_0 + c) + t_1 A w_1 + \cdots + t_k A w_k \\
&= P_1 + w'
\end{aligned}
$$

where P_1 is the point having coordinate vector $AX_0 + c$ and $w' = t_1 A w_1 + \cdots + t_k A w_k$. Hence, all points of the form $f(P)$ describe an affine subspace whose associated vector space W' has dimension k and is generated by vectors $\{A w_1, \ldots, A w_k.\}$ Moreover, since A represents a nonsingular operator of $\mathbb{F}^n$, $\{A w_1, \ldots, A w_k\}$ is an independent set of vectors, so it is a basis of $\varphi(W)$, where φ is the linear operator of $\mathbb{F}^n$ defined by A. Hence, we may write $f(\mathbb{A}') = f(P_0) + \varphi(W)$. As required, the dimension of $f(\mathbb{A}')$ coincides with the one of $\mathbb{A}'$.

(ii) Consider now $\mathbb{A}''$ a subspace which is parallel to $\mathbb{A}'$. As above, $\mathbb{A}'' = P_1 + U$, for a fixed point $P_1 \in \mathbb{A}''$ and a vector subspace U of $\mathbb{F}^n$. Without loss of generality, we may assume that U is a vector subspace of W. Hence, if $\{u_1, \ldots, u_h\}$ is a basis of U, we may complete a basis of W by adding appropriate $k - h$ vectors $w_1, \ldots, w_{k-h}$. By the above argument, $\{A u_1, \ldots, A u_h\}$ is a basis of $\varphi(U)$, $\{A u_1, \ldots, A u_h, A w_1, \ldots, A w_{k-h}\}$ is a basis of $\varphi(W)$ and we may represent the images of the affine subspaces as

$$f(\mathbb{A}') = f(P_0) + \varphi(W), \quad f(\mathbb{A}'') = f(P_1) + \varphi(U).$$

Since $\varphi(U)$ is a vector subspace of $\varphi(W)$, we conclude that $f(\mathbb{A}')$ and $f(\mathbb{A}'')$ are parallel.

As a consequence of Theorem 11.40, we have the following.

Corollary 11.41 *Let $\mathbb{A}$ be the affine space $\mathbb{F}\mathbb{A}^n$ and $f : \mathbb{A} \to \mathbb{A}$ a translation. Then f maps an affine subspace $\mathbb{A}'$ to an affine subspace that is parallel to $\mathbb{A}'$ and has the same dimension of $\mathbb{A}'$.*

Now, we spend a few lines in order to introduce the first step of the study of conics in affine spaces. We'll come back to a deep analysis of conics in the sequel; here, we just would like to remark some properties which are strictly related to the previous results.

A conic section or a conic is the locus of a point which moves in a plane so that its distance from a fixed point is in a constant ratio to its perpendicular distance from a fixed straight line. The fixed point is called *the focus*, the fixed straight line is called *the directrix* and the constant ratio is called *eccentricity* usually denoted by *e*. The line passing through the focus and perpendicular to the directrix is called *axis*, and the point of intersection of a conic with its axis is called a *vertex*.

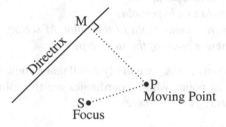

If S is the point (p, q) and the directrix is $\ell x + my + n = 0$, then $PS = \sqrt{(x - p)^2 + (y - q)^2}$; $PM = \frac{|\ell x + my + n|}{\sqrt{\ell^2 + m^2}}$. Then $\frac{PS}{PM} = e$ implies that

$$(\ell^2 + m^2)\{(x - p)^2 + (y - q)^2\} = e^2(\ell x + my + n)^2,$$

which is a particular form of

$$ax^2 + 2hxy + by^2 + 2gx + 2fy + c = 0, \tag{11.12}$$

the general equation of the second degree. Conversely, it can be proved that the general equation of the second degree (11.12) also represents a conic section. Hence, a conic section can be defined in the following way also:

A conic section is the set of all points in a plane, whose coordinates satisfy the general equation of the second degree given by (11.12). Now let $\triangle = \begin{vmatrix} a & h & g \\ h & b & f \\ g & f & c \end{vmatrix}$. Then the conic section represented by (11.12) is called

- a parabola if $\triangle \neq 0$, $h^2 = ab$;
- an ellipse if $\triangle \neq 0$, $h^2 < ab$, (either $a \neq b$ or $h \neq 0$);
- a circle if $\triangle \neq 0$, $h^2 = ab$, $a = b$, $h = 0$;
- a hyperbola if $\triangle \neq 0$, $h^2 > ab$.

Remark 11.42 (*i*) If $\triangle \neq 0$, then corresponding conic section is known as non-degenerate.

(*ii*) If $\triangle = 0$, then corresponding conic section is known as degenerate.
- If $\triangle = 0$, $h^2 > ab$, then (11.12) represents a pair of distinct real lines.

- If $\triangle = 0$, $h^2 = ab$, (either $g^2 = ac$ or $f^2 = bc$), then (11.12) represents a pair of coincident lines.
- If $\triangle = 0$, $h^2 < ab$, then (11.12) represents two imaginary lines/point.

Motivated by the above definition of a conic section, we have given a more general definition of a conic in the forthcoming Sect. 11.5.

Theorem 11.43 *Let* $\mathbb{A}$ *be the affine plane* $\mathbb{F}\mathbb{A}^2$ *and* $f : \mathbb{A} \rightarrow \mathbb{A}$ *an affine transformation. Then* f *maps a conic curve to a conic curve and in particular*

- (i) *f maps an ellipse to an ellipse.*
- (ii) *f maps a parabola to a parabola.*
- (iii) *f maps a hyperbola to a hyperbola.*
- (iv) *f maps a degenerate conic (split in the union of secant, parallel or merged lines) to a degenerate conic of the same type.*

Proof To represent a conic in $\mathbb{F}\mathbb{A}^2$, we firstly recall that a conic is a geometric locus in $\mathbb{F}\mathbb{A}^2$ consisting of all points whose coordinates are the solution of a quadratic equation of the form $g(x, y) = 0$, where

$$g(x, y) = a_{11}x^2 + 2a_{12}xy + a_{22}y^2 + 2a_{13}x + 2a_{23}y + a_{33} \qquad (11.13)$$

and $a_{ij} \in \mathbb{F}$, for any $i, j = 1, 2, 3$. It is the standard representing equation of a conic. The quadratic part of the polynomial $g(x, y)$ is the quadratic form

$$q(x, y) = a_{11}x^2 + 2a_{12}xy + a_{22}y^2 \qquad (11.14)$$

and can be represented by the symmetric matrix

$$A = \begin{bmatrix} a_{11} & a_{12} \\ a_{12} & a_{22} \end{bmatrix}$$

so that $q(x, y)$ can be written as $q(x, y) = X^t A X$, where $X = [x, y]^t$. Hence, if we denote $B = [2a_{13}, 2a_{23}]^t$, then the equation $g(x, y) = 0$ can be written in the following matrix notation:

$$X^t A X + B^t X + a_{33} = 0 \qquad (11.15)$$

or, also in the following very compact form:

$$Y^t C Y = 0, \quad \text{where} \quad Y = \begin{bmatrix} x \\ y \\ 1 \end{bmatrix} \quad \text{and} \quad C = \begin{bmatrix} a_{11} & a_{12} & a_{13} \\ a_{12} & a_{22} & a_{23} \\ a_{13} & a_{23} & a_{33} \end{bmatrix}. \qquad (11.16)$$

The general classification of conics is well known.

Case I: If $|C| \neq 0$ and $|A| > 0$, the conic is a non-degenerate (real or imaginary) ellipse.

Case II: If $|C| \neq 0$ and $|A| < 0$, the conic is a non-degenerate hyperbola.
Case III: If $|C| \neq 0$ and $|A| = 0$, the conic is a non-degenerate parabola.
Case IV: If $|C| = 0$, the conic is degenerate, that is, it consists of either the union of two (secant, parallel or merged) real lines or the union of two conjugate (secant or parallel) imaginary lines. In all these subcases, the rank of C is equal to 2 if the conic is a union of distinct lines; it is equal to 1 in either case (merged lines).

We now consider the affine transformation by its expression $f(X) = MX + c$, where M is an invertible matrix of $M_2(\mathbb{F})$ and c is a fixed point of $\mathbb{F}A^2$. Thus, the image of the conic $\Gamma : X^t A X + B^t X + a_{33} = 0$ is computed by the substitution

$$X = M^{-1}\left(f(X) - c\right) = HX' + d \qquad (11.17)$$

where

$$H = M^{-1} = \begin{bmatrix} \beta_{11} & \beta_{12} \\ \beta_{21} & \beta_{22} \end{bmatrix}, \quad X' = f(X) = [x', y']^t, \quad d = -M^{-1}c = \begin{bmatrix} \gamma_1 \\ \gamma_2 \end{bmatrix}.$$

Hence, by using relation (11.17) in (11.15), we have that the geometrical locus $f(\Gamma)$ consisting of all points whose coordinates X' are the solution of

$$(HX' + d)^t A(HX' + d) + B^t(HX' + d) + a_{33} = 0,$$

that is,

$$X'^t(H^t A H)X' + X'^t H^t A d + d^t A H X' + d^t A d + B^t H X' + B^t d + a_{33} = 0. \qquad (11.18)$$

Moreover, by the facts $X'^t H^t A d = d^t A^t H X'$ and $A^t = A$, (11.18) reduces to

$$X'^t(H^t A H)X' + (d^t A H + d^t A H + B^t H)X' + (d^t A d + B^t d + a_{33}) = 0. \qquad (11.19)$$

Therefore, under the notation $A' = H^t A H$, $B' = (d^t A H + d^t A H + B^t H)^t$ and $a'_{33} = d^t A d + B^t d + a_{33}$ in (11.19), $f(\Gamma)$ is represented by

$$X'^t A' X' + B'^t X' + a'_{33} = 0,$$

that is, a quadratic equation $h(x', y') = 0$. Then $f(\Gamma)$ is a conic curve. Notice that relation (11.17) also induces the following one:

$$\begin{bmatrix} x \\ y \\ 1 \end{bmatrix} = \begin{bmatrix} \beta_{11} & \beta_{12} & \gamma_1 \\ \beta_{21} & \beta_{22} & \gamma_2 \\ 0 & 0 & 1 \end{bmatrix} \begin{bmatrix} x' \\ y' \\ 1 \end{bmatrix}. \qquad (11.20)$$

Thus, by using (11.20) in (11.16), it follows that the conic $f(\Gamma)$ is represented by the equation

$$\left(\begin{bmatrix} \beta_{11} & \beta_{12} & \gamma_1 \\ \beta_{21} & \beta_{22} & \gamma_2 \\ 0 & 0 & 1 \end{bmatrix} \begin{bmatrix} x' \\ y' \\ 1 \end{bmatrix} \right)^t C \left(\begin{bmatrix} \beta_{11} & \beta_{12} & \gamma_1 \\ \beta_{21} & \beta_{22} & \gamma_2 \\ 0 & 0 & 1 \end{bmatrix} \begin{bmatrix} x' \\ y' \\ 1 \end{bmatrix} \right) = 0,$$

that is,

$$Y'' \begin{bmatrix} \beta_{11} & \beta_{21} & 0 \\ \beta_{12} & \beta_{22} & 0 \\ \gamma_1 & \gamma_2 & 1 \end{bmatrix} C \begin{bmatrix} \beta_{11} & \beta_{12} & \gamma_1 \\ \beta_{21} & \beta_{22} & \gamma_2 \\ 0 & 0 & 1 \end{bmatrix} Y' = 0. \tag{11.21}$$

So, the matrix associated with $f(\Gamma)$ is

$$\begin{aligned} C' &= \begin{bmatrix} \beta_{11} & \beta_{21} & 0 \\ \beta_{12} & \beta_{22} & 0 \\ \gamma_1 & \gamma_2 & 1 \end{bmatrix} C \begin{bmatrix} \beta_{11} & \beta_{12} & \gamma_1 \\ \beta_{21} & \beta_{22} & \gamma_2 \\ 0 & 0 & 1 \end{bmatrix} \\[2mm] &= \begin{bmatrix} \beta_{11} & \beta_{12} & \gamma_1 \\ \beta_{21} & \beta_{22} & \gamma_2 \\ 0 & 0 & 1 \end{bmatrix}^t C \begin{bmatrix} \beta_{11} & \beta_{12} & \gamma_1 \\ \beta_{21} & \beta_{22} & \gamma_2 \\ 0 & 0 & 1 \end{bmatrix}. \end{aligned}$$

Since C and C' are congruent, they have the same rank. Then we may admit that Γ is non-degenerate (respectively degenerate) if and only if $f(\Gamma)$ is. Moreover, if Γ were a degenerate conic, then it would be a pair of secant, parallel or merged lines. In this case, since f maps a line to a line and preserves the property of parallelism among lines; the image $f(\Gamma)$ would be again a pair of secant, parallel or merged lines, respectively.

Finally, we fix our attention to the case of a non-degenerate conics Γ (ellipse, hyperbola and parabola). The matrix A' represents the quadratic part of the polynomial $h(x', y')$, which describes the image $f(\Gamma)$. Since $A' = H^t A H$, then $|A'| = |H|^2|A|$, that is, the determinant of A' is zero, positive or negative, according with the determinant of A is zero, positive or negative. Thus, the type of the conic is unchanged.

Let us describe now one class of affine transformations of particular interest. Let us assume that $f : \mathbb{A} \to \mathbb{A}$ has a fixed point, that is, $O \in \mathbb{A}$ such that $f(O) = O$. In light of Theorem 11.38, f can be represented, with respect to a frame of reference $\{O, \mathbf{e}_1, \dots, \mathbf{e}_n\}$ of $\mathbb{A}$, by the action

$$f_{A,O}(X) = AX,$$

where $A \in M_n(\mathbb{F})$ is the invertible matrix of the automorphism φ associated with $f_{A,O}$ in terms of the basis $\{\mathbf{e}_1, \dots, \mathbf{e}_n\}$ of V. Then any affine transformation $f_{A,O}$, fixing the point O, can be identified with its linear part. Let us denote by $Aff(\mathbb{A})_O$, the set of all affine transformations of the form $f_{A,O}$. In this sense, there is a one-to-one correspondence between $Aff(\mathbb{A})_O$ and the matrices $A \in GL_n(\mathbb{F})$. For this reason, such affine transformations are usually called *linear*.

Consider then an affine transformation $f : \mathbb{A} \to \mathbb{A}$ and let O be any point of $\mathbb{A}$. If we denote $\mathbf{v} = \mathbf{Of}(O) \in V$ and $f_\mathbf{v}$ is the translation of $\mathbb{A}$ defined by $\mathbf{v}$, then the composition $g = f_\mathbf{v}^{-1} \circ f$ (that is, $g = f_{-\mathbf{v}} \circ f$) is clearly a linear affine transformation of $\mathbb{A}$ because g fixes the point O as discussed below. Hence, $f = f_\mathbf{v} \circ g$ is a representation of f precisely in terms of composition of one translation and one linear affine transformation.

Notice that if $\mathbf{v}' = -\mathbf{Of}^{-1}(O) \in V$ and $h = f \circ f_{\mathbf{v}'}^{-1}$, where $f_{\mathbf{v}'}$ is the translation of $\mathbb{A}$ defined by $\mathbf{v}'$, then $f = h \circ f_{\mathbf{v}'}$ is again a representation of f. Moreover, since

$$f_\mathbf{v}^{-1} \circ f(O) = T \iff \mathbf{f}(O)\mathbf{T} = -\mathbf{v}$$
$$\iff \mathbf{Tf}(O) = \mathbf{v}$$
$$\iff T = O$$

and

$$f_{\mathbf{v}'}^{-1}(O) = T \iff f_{-\mathbf{v}'}(O) = T$$
$$\iff \mathbf{OT} = -\mathbf{v}'$$
$$\iff \mathbf{OT} = \mathbf{Of}^{-1}(O)$$
$$\iff T = f^{-1}(O),$$

one has that

$$g(O) = f_\mathbf{v}^{-1} \circ f(O) = O$$

and

$$h(O) = f \circ f_{\mathbf{v}'}^{-1}(O) = f \circ f^{-1}(O) = O.$$

Therefore, g and h are linear affine transformations fixing the same point $O \in \mathbb{A}$ and associated with the same automorphism of V. Thus, in light of the previously mentioned bijection between $Aff(\mathbb{A})_O$ and $GL_n(\mathbb{F})$, g and h are represented by the same matrix $A \in GL_n(\mathbb{F})$ (by the choice of a basis of V), that is, $g = h$. As a conclusion, we have that

$$f = f_\mathbf{v} \circ g = g \circ f_{\mathbf{v}'}. \tag{11.22}$$

The meaning of this result is summarized in the following.

Theorem 11.44 *Let $\mathbb{A}$ be an affine space over the vector space V. For any affine transformation $f : \mathbb{A} \to \mathbb{A}$ and for any point $O \in \mathbb{A}$, there exist unique $\mathbf{v}, \mathbf{v}' \in V$ and a linear affine transformation $g : \mathbb{A} \to \mathbb{A}$ fixing O, such that relation (11.22) holds.*

11.3   Isometries

Definition 11.45 Let $\mathbb{E}$ be an affine Euclidean space over the vector space V. An affine transformation $f : \mathbb{E} \to \mathbb{E}$ is called an *isometry* of $\mathbb{E}$ if the associated automorphism $\varphi : V \to V$ is an orthogonal operator (an isometry of V).

It is easy to check that

(1) The identity map $1_{\mathbb{E}} : \mathbb{E} \to \mathbb{E}$ is an isometry and here the associated orthogonal operator is the identity map on V.
(2) The composition of two isometries f and g is an isometry and here the associated orthogonal operator is the composition of the orthogonal operators associated with f and g, respectively.
(3) The inverse of an isometry g is yet an isometry and here the associated orthogonal operator is the inverse of the orthogonal operator associated with g.

Thus, the set of all isometries of an affine Euclidean space $\mathbb{E}$ is a group, usually denoted by $Iso(\mathbb{E})$. A subgroup of $Iso(\mathbb{E})$ of particular interest is the one consisting of all isometries of $\mathbb{E}$ associated with automorphisms which are represented by matrices having determinant precisely equal to 1. These isometries are called *direct isometries* and the mentioned subgroup of $Iso(\mathbb{E})$ is called *group of direct isometries* and is denoted by $Iso^{+}(\mathbb{E})$.

Example 11.46 Any translation of $\mathbb{E}$ is a direct isometry of $\mathbb{E}$.

Example 11.47 Let H be a hyperplane in the n-dimensional affine Euclidean space $\mathbb{E}$, V the direction of H and $f : \mathbb{E} \to \mathbb{E}$ the map satisfying the following rules:

(1) for any $P \in \mathbb{E}$, $\mathbf{Pf(P)} \in V^{\perp}$;
(2) for any $P \in \mathbb{E}$, the distance from P to H is equal to the distance from $f(P)$ to H.

So, f fixes all points in H and interchanges the position of any other point in $\mathbb{E}$ along the orthogonal line to H, at equal distance from H. In the literature, such a map is usually called *reflection* across H.

In order to investigate the behavior of this kind of transformation, we firstly consider the case H that contains the origin point of the frame of reference in $\mathbb{E}$. If $\mathbb{V}$ is the associated vector space of the affine Euclidean space $\mathbb{E}$, then we identify $\mathbb{V}$ by $\mathbb{R}^{n}$. Similarly, we also identify the direction V of H by the vector space $\mathbb{R}^{n-1}$ and $V^{\perp}$ by the vector space $\mathbb{R}$. Since $\mathbb{R}^{n} = V \oplus V^{\perp}$, any vector $\mathbf{v} \in \mathbb{R}^{n}$ can be uniquely expressed as $\mathbf{w} + \mathbf{w}'$, where $\mathbf{w} \in V$ and $\mathbf{w}' \in V^{\perp}$. More precisely, since $dim(V) = n - 1$ and $dim(V^{\perp}) = 1$, if we fix a vector $0 \neq \mathbf{u} \in V^{\perp}$, we may write

$$\mathbf{v} = \mathbf{w} + \alpha \mathbf{u} \tag{11.23}$$

for $\alpha \in \mathbb{R}$ such that $\mathbf{w}' = \alpha \mathbf{u}$. By the fact that $\mathbf{w} \cdot \mathbf{u} = 0$ and performing the dot product by $\mathbf{u}$ in the identity (11.23), it follows $\mathbf{v} \cdot \mathbf{u} = \alpha \mathbf{u} \cdot \mathbf{u}$.

Let now $P \in \mathbb{E}$ having coordinate vector $\mathbf{v} \in \mathbb{R}^n$. Since f fixes $\mathbf{w} \in H$ and acts like -1 on $\alpha\mathbf{u}$, one has that

$$
\begin{aligned}
f(P) = f(\mathbf{w} + \alpha\mathbf{u}) \\
= \mathbf{w} - \alpha\mathbf{u} \\
= \mathbf{w} - \frac{\mathbf{v} \cdot \mathbf{u}}{\mathbf{u} \cdot \mathbf{u}} \mathbf{u} \\
= \mathbf{v} - \alpha\mathbf{u} - \frac{\mathbf{v} \cdot \mathbf{u}}{\mathbf{u} \cdot \mathbf{u}} \mathbf{u} \\
= \mathbf{v} - 2\frac{\mathbf{v} \cdot \mathbf{u}}{\mathbf{u} \cdot \mathbf{u}} \mathbf{u}
\end{aligned}
$$

which is the coordinate vector of the image $f(P)$. In particular, since $\mathbf{w}$ and $\alpha\mathbf{u}$ are orthogonal

$$\| f(\mathbf{v}) \| = \|\mathbf{w}\| + \|\alpha\mathbf{u}\| = \|\mathbf{v}\| .$$

Therefore, if $P, Q \in \mathbb{E}$ are represented respectively by the coordinate vectors $\mathbf{v}$ and $\mathbf{v}'$, it follows that

$$
\begin{aligned}
\|\mathbf{f}(Q)\mathbf{f}(P)\| = \left\| f(\mathbf{v}') - f(\mathbf{v}) \right\| \\
= \left\| f(\mathbf{v}' - \mathbf{v}) \right\| \\
= \left\| \mathbf{v}' - \mathbf{v} \right\| \\
= \|QP\|
\end{aligned}
$$

which proves that f is an isometry. To write the matrix representing f, pick an orthogonal basis $\{\mathbf{e}_1, \dots, \mathbf{e}_{n-1}\}$ for V and complete to a basis of $\mathbb{R}^n$ by adding a vector $\mathbf{e}_n \in V^\perp$. Hence, for any vector $\mathbf{v} \in \mathbb{R}^n$, there exist suitable $\alpha_1, \dots, \alpha_n \in \mathbb{R}$ such that $\mathbf{v} = \sum_{i=1}^{n} \alpha_i \mathbf{e}_i$ and $f(\mathbf{v}) = \sum_{i=1}^{n-1} \alpha_i \mathbf{e}_i - \alpha_n \mathbf{e}_n$, that is

$$
\begin{bmatrix} \alpha_1 \\ \vdots \\ \vdots \\ \vdots \\ \alpha_{n-1} \\ -\alpha_n \end{bmatrix}
=
\begin{bmatrix} 1 & & & & \\ & 1 & & & \\ & & \ddots & & \\ & & & 1 & \\ & & & & -1 \end{bmatrix}
\begin{bmatrix} \alpha_1 \\ \vdots \\ \vdots \\ \vdots \\ \alpha_{n-1} \\ \alpha_n \end{bmatrix} .
$$

This shows that every reflection across a hyperplane containing the origin consists only of its linear part, moreover, it is represented by a matrix having determinant equal to -1.

To get a formula for reflections across any hyperplane, not necessarily containing the origin, we describe the hyperplane in terms of coordinate vectors of its points:

$$H = \{P \in \mathbb{RE}^n \ \text{ having coordinate vector } \ \mathbf{Y} \in \mathbb{R}^n \ : \ \mathbf{Y} \cdot \mathbf{u} = \lambda\}$$

for some nonzero vector $\mathbf{u} \in \mathbb{R}^n$ that is orthogonal to the hyperplane and some $\lambda \in \mathbb{R}$.

Fix a point $Q \in \mathbb{E}$ having coordinate vector $\mathbf{w} \in \mathbb{R}^n$. To reflect a point P across H, whose coordinate vector is $\mathbf{v} \in \mathbb{R}^n$, we firstly perform a translation by subtracting the coordinates of Q; in this sense, we obtain a point P' having coordinate vector $\mathbf{v} - \mathbf{w}$ and a hyperplane $(-Q) + H$ which is parallel to H and contains the origin of the frame of reference. At this point, following the previous argument, we compute the reflection of the point P' across $(-Q) + H$. Finally, adding the coordinates of $\mathbf{w}$ back, we have the reflection of P across H.

More in detail, if we denote by g the reflection across the hyperplane $(-Q) + H$, we have

$$
\begin{aligned}
f(P) &= g(\mathbf{v} - \mathbf{w}) + \mathbf{w} \\
&= (\mathbf{v} - \mathbf{w}) - 2\frac{(\mathbf{v}-\mathbf{w})\cdot\mathbf{u}}{\mathbf{u}\cdot\mathbf{u}}\mathbf{u} + \mathbf{w}
\end{aligned}
\tag{11.24}
$$

for any vector $\mathbf{u}$ orthogonal to the direction of H. Moreover, since $Q \in H$ then $\mathbf{w} \cdot \mathbf{u} = \lambda$ and (11.24) can be written as

$$
f(v) = \mathbf{v} - 2\frac{\mathbf{v}\cdot\mathbf{u} - \lambda}{\mathbf{u}\cdot\mathbf{u}}\mathbf{u}
$$

which represents the coordinate vector of the reflected point from P across H.

To clarify much more the previous example, we illustrate some specific cases.

Example 11.48 Let r be the straight line in $\mathbb{RE}^2$ defined by equation $x - 2y + 1 = 0$. Its direction is $V = \langle(2, 1)\rangle$, whose orthogonal complement is $\langle\mathbf{u}\rangle$, where $\mathbf{u} \equiv (1, -2)$. Any point of r possesses coordinate vector (y_1, y_2) satisfying the relation $y_1 - 2y_2 = -1$ $(\lambda = -1)$. Hence, for any vector (x_1, x_2) representing a point of $\mathbb{RE}^2$, its reflection across r is obtained by

$$
\begin{aligned}
f((x_1, x_2)) &= \left[x_1 - \frac{2x_1 - 4x_2 + 2}{5}, \; x_2 + \frac{4x_1 - 8x_2 + 4}{5}\right] \\
&= \left[\frac{3x_1 + 4x_2 - 2}{5}, \; \frac{4x_1 - 3x_2 + 4}{5}\right] \\
&= \begin{bmatrix} \frac{3}{5} & \frac{4}{5} \\ \frac{4}{5} & -\frac{3}{5} \end{bmatrix} \begin{bmatrix} x_1 \\ x_2 \end{bmatrix} + \begin{bmatrix} -\frac{2}{5} \\ \frac{4}{5} \end{bmatrix}.
\end{aligned}
\tag{11.25}
$$

Example 11.49 Let π be the plane in $\mathbb{RE}^3$ defined by equation $x - 2y + 2z - 2 = 0$. Its direction is $V = \langle(2, 0, -1), (2, 1, 0)\rangle$, whose orthogonal complement is $V^\perp = \langle\mathbf{u}\rangle$, where $\mathbf{u} \equiv (1, -2, 2)$. The coordinates (y_1, y_2, y_3) of any point of π satisfy the relation $y_1 - 2y_2 + 2y_3 = 2$ $(\lambda = 2)$. Hence, for any vector (x_1, x_2, x_3) representing a point of $\mathbb{RE}^3$, its reflection across π is obtained by $f((x_1, x_2, x_3))$

$$= \left[x_1 - \frac{2x_1 - 4x_2 + 4x_3 - 4}{9}, \; x_2 - \frac{-4x_1 + 8x_2 - 8x_3 + 8}{9}, \; x_3 - \frac{4x_1 - 8x_2 + 8x_3 - 8}{9}\right]$$

$$= \left[\frac{7x_1 + 4x_2 - 4x_3 + 4}{9}, \; \frac{4x_1 + x_2 + 8x_3 - 8}{9}, \; \frac{-4x_1 + 8x_2 + x_3 + 8}{9}\right]$$

$$= \begin{bmatrix} \frac{7}{9} & \frac{4}{9} & -\frac{4}{9} \\ \frac{4}{9} & \frac{1}{9} & \frac{8}{9} \\ -\frac{4}{9} & \frac{8}{9} & \frac{1}{9} \end{bmatrix} \begin{bmatrix} x_1 \\ x_2 \\ x_3 \end{bmatrix} + \begin{bmatrix} \frac{4}{9} \\ -\frac{8}{9} \\ \frac{8}{9} \end{bmatrix}.$$

(11.26)

Definition 11.50 A direct isometry $f : \mathbb{E} \to \mathbb{E}$ of the affine Euclidean space $\mathbb{E}$, for which a point $P \in \mathbb{E}$ is a fixed point, i.e., $f(P) = P$, is called *rotation centered at* P.

By Theorem 11.38 and since any orthogonal operator of a vector space is represented by an orthogonal matrix, the following directly follows.

Theorem 11.51 *The map $f : \mathbb{E} \to \mathbb{E}$ is an isometry of the affine Euclidean space $\mathbb{E}$ if and only if there exist an orthogonal matrix $A \in M_n(\mathbb{F})$ and a fixed point $c \in \mathbb{A}$ such that, for any point $P \in \mathbb{A}$ having coordinate vector $X = [x_1, \ldots, x_n]^t$ in terms of B, the point $f(P) \in \mathbb{A}$ has coordinate vector $Y = [y_1, \ldots, y_n]^t$ in terms of B, satisfying the following relation:*

$$Y = AX + c. \tag{11.27}$$

Let us note that we are now able to reformulate the result contained in Theorem 11.44.

Theorem 11.52 *Let $\mathbb{E}$ be an affine Euclidean space over the vector space V. For any isometry $f : \mathbb{E} \to \mathbb{E}$ and for any point $O \in \mathbb{E}$, there exist unique $\mathbf{v} \in V$ and an isometry $g : \mathbb{A} \to \mathbb{A}$ fixing O, such that*

$$f = f_{\mathbf{v}} \circ g \tag{11.28}$$

where $f_{\mathbf{v}}$ is the translation of $\mathbb{E}$ defined by $\mathbf{v}$.

Remark 11.53 We know that, for any points P, Q of an affine Euclidean space $\mathbb{E}$ over the Euclidean vector space V, the vector $\mathbf{PQ}$ is determined. Hence, we may define the map

$$\delta : \mathbb{E} \times \mathbb{E} \to \mathbb{R}$$

such that $\delta(P, Q) = \|\mathbf{PQ}\|$, usually called *distance between P and Q*. As a metric map on a vector space, one may see that the distance δ satisfies the following properties:

(i) $\delta(P, Q) > 0$, for any $Q \neq P$ points of $\mathbb{E}$.
(ii) $\delta(P, P) = 0$, for any $P \in \mathbb{E}$.
(iii) $\delta(P, Q) = \delta(Q, P)$, for any $P, Q \in \mathbb{E}$.
(iv) $\delta(P, Q) = \delta(P, H) + \delta(H, Q)$, for any $P, H, Q \in \mathbb{E}$.

In other words, a distance satisfies the same condition of a *metric* of a vector space, thus any affine Euclidean space is a metric space.

Theorem 11.54 *Let* $\mathbb{E}$ *be an affine Euclidean space over the vector space* V *(over the field* $\mathbb{R}$*). The map* $f : \mathbb{E} \to \mathbb{E}$ *is an isometry if and only if there exists a distance* $\delta : \mathbb{E} \times \mathbb{E} \to \mathbb{R}$ *such that*

$$\delta\left(f(P), f(Q) \right) = \delta(P, Q) \tag{11.29}$$

for any points $P, Q \in \mathbb{E}$.

Proof If $f : \mathbb{E} \to \mathbb{E}$ is an isometry and $\varphi : V \to V$ is its associated automorphism (an isometry of V), then

$$\delta\left(f(P), f(Q) \right) = \|\mathbf{f(P)f(Q)}\|$$

$$= \|\varphi(\mathbf{PQ})\|$$

$$= \|\mathbf{PQ}\|$$

$$= \delta(P, Q).$$

Conversely, assume that condition (11.29) holds. We now fix a point $O \in \mathbb{E}$ and introduce the map $\sigma : \mathbb{E} \to V$ defined by $\sigma(P) = \mathbf{OP}$, for any point $P \in \mathbb{E}$. In light of the definition of affine spaces, it is clear that σ is a bijection between $\mathbb{E}$ and V, i.e., for any point $P \in \mathbb{E}$, there is a unique vector $\mathbf{v} \in V$ such that $\mathbf{v} = \mathbf{OP}$.

Consider the following function $\varphi : V \to V$ defined by $\varphi(v) = \mathbf{f(O)f(P)}$, for any $\mathbf{v} = \mathbf{OP} \in V$. Notice that

$$\varphi(\mathbf{0}) = \varphi(\mathbf{OO}) = \mathbf{f(O)f(O)} = \mathbf{0}.$$

Moreover, for any $\mathbf{v} = \mathbf{OP}$ and $\mathbf{w} = \mathbf{OQ}$, we have

$$\|\varphi(v) - \varphi(w)\| = \|\varphi(\mathbf{OP}) - \varphi(\mathbf{OQ})\|$$

$$= \|\mathbf{f(O)f(P)} - \mathbf{f(O)f(Q)}\|$$

$$= \|\mathbf{f(Q)f(P)}\|$$

$$= \|\mathbf{QP}\|$$

$$= \|v - w\|,$$

that is, φ is an isometry of V. Since f is an affine transformation of $\mathbb{E}$ associated with φ, f is an isometry of $\mathbb{E}$.

Thus, an isometry of an affine Euclidean space preserves distances between points. In the literature, if the mapping $f : \mathbb{E} \to \mathbb{E}$, of an affine Euclidean space $\mathbb{E}$ into itself, is an isometry of $\mathbb{E}$ as a metric space, then it is called *motion*.

11.4 A Natural Application: Coordinate Transformation in $\mathbb{R}\mathbb{E}^2$

A coordinate transformation in the affine Euclidean space $\mathbb{R}\mathbb{E}^2$ is the transition from a frame of reference of $\mathbb{R}\mathbb{E}^2$ to another one, corresponding to a transition from a basis of $\mathbb{R}^2$ to another one, so that any point of $\mathbb{R}\mathbb{E}^2$ (and so any vector of $\mathbb{R}^2$) is represented by different coordinates with respect to different systems.

We fix our attention to translations and rotations of the unit vectors $\mathbf{i}$ and $\mathbf{j}$ of a Cartesian coordinate system OXY. More precisely,

Translation: Introduce a second coordinate system $O'X'Y'$, having the origin O' and unit vectors $\mathbf{i}'$ and $\mathbf{j}'$, such that the coordinates of O' with respect to the system OXY are (x_0, y_0) and the base vectors $\mathbf{i}'$, $\mathbf{j}'$ are parallel to $\mathbf{i}$ and $\mathbf{j}$, respectively.

Rotation: Introduce a second coordinate system $OX'Y'$, having the same origin and unit vectors $\mathbf{i}'$ and $\mathbf{j}'$, such that the new coordinate system is obtained from the first one by a rotation of a certain angle ϑ of the base vectors. We recall that the sign convention of rotations is positive counterclockwise.

Let (x, y) be the coordinates of the point P in terms of OXY. In order to determine the coordinates (x', y') of P with respect to the new coordinate system, we firstly remark that they are the coordinates of the vector $\mathbf{OP}$, where O is the tail and P is the head. We express these coordinates both in terms of $\{\mathbf{i}, \mathbf{j}\}$ and in terms of $\{\mathbf{i}', \mathbf{j}'\}$:

$$\mathbf{OP} = x\mathbf{i} + y\mathbf{j} \qquad \mathbf{OP} = x'\mathbf{i}' + y'\mathbf{j}'.$$

In particular, in the case of the above-mentioned translation from the system OXY to $O'X'Y'$, the coordinate vector of $\mathbf{i}'$ and $\mathbf{j}'$ relative to the basis $\{\mathbf{i}, \mathbf{j}\}$ are $(1, 0)$ and $(0, 1)$, respectively. Hence, the following relation:

$$\begin{bmatrix} x - x_0 \\ y - y_0 \end{bmatrix} = \begin{bmatrix} 1 & 0 \\ 0 & 1 \end{bmatrix} \begin{bmatrix} x' \\ y' \end{bmatrix}$$

represents the translation, that is,

$$x' = x - x_0$$
$$y' = y - y_0$$

and conversely

$$x = x' + x_0$$
$$y = y' + y_0.$$

Moreover, one can see that rotations are nothing more than changes of basis in the vector space $\mathbb{R}^2$, then they can be represented by a transition matrix of the ordered basis $\{i', j'\}$ relative to the ordered basis $\{i, j\}$. Therefore, the coordinate vector of i' relative to the basis $\{i, j\}$ is the first column vector in the transition matrix relative to $\{i, j\}$; analogously, the coordinate vector of j' relative to the basis $\{i, j\}$ is the second column vector in the transition matrix relative to $\{i, j\}$.

Hence, in the case the new coordinate system is obtained from the first one by a counterclockwise rotation of a certain angle ϑ, the coordinate vector of i' and j' relative to the basis $\{i, j\}$ are $(cos(\vartheta), sin(\vartheta))$ and $(-sin(\vartheta), cos(\vartheta))$, respectively. Thus, the rotation is represented by

$$\begin{bmatrix} x \\ y \end{bmatrix} = \begin{bmatrix} cos(\vartheta) & -sin(\vartheta) \\ sin(\vartheta) & cos(\vartheta) \end{bmatrix} \begin{bmatrix} x' \\ y' \end{bmatrix}.$$

Conversely, since the previous transition matrix is orthogonal, we also have

$$\begin{bmatrix} x' \\ y' \end{bmatrix} = \begin{bmatrix} cos(\vartheta) & sin(\vartheta) \\ -sin(\vartheta) & cos(\vartheta) \end{bmatrix} \begin{bmatrix} x \\ y \end{bmatrix}.$$

Hence, the equations representing a rotation are

$$x = x'cos(\vartheta) - y'sin(\vartheta)$$
$$y = x'sin(\vartheta) + y'cos(\vartheta)$$

and

$$x' = xcos(\vartheta) + ysin(\vartheta)$$
$$y' = -xsin(\vartheta) + ycos(\vartheta).$$

We finally assume that the new coordinate system $O'X'Y'$ has been counterclockwise rotated of a certain angle ϑ with respect to OXY and then translated. As above, we denote as $O' \equiv (x_0, y_0)$ the coordinates of the translated origin. In this case, the roto-translation is represented by the relation

$$\begin{bmatrix} x - x_0 \\ y - y_0 \end{bmatrix} = \begin{bmatrix} cos(\vartheta) & -sin(\vartheta) \\ sin(\vartheta) & cos(\vartheta) \end{bmatrix} \begin{bmatrix} x' \\ y' \end{bmatrix}$$

and conversely

$$\begin{bmatrix} x' \\ y' \end{bmatrix} = \begin{bmatrix} cos(\vartheta) & sin(\vartheta) \\ -sin(\vartheta) & cos(\vartheta) \end{bmatrix} \begin{bmatrix} x - x_0 \\ y - y_0 \end{bmatrix}. \tag{11.30}$$

Thus, we have

$$x = x'\cos(\vartheta) - y'\sin(\vartheta) + x_0$$
$$y = x'\sin(\vartheta) + y'\cos(\vartheta) + y_0$$

(11.31)

and

$$x' = (x - x_0)\cos(\vartheta) + (y - y_0)\sin(\vartheta)$$
$$y' = -(x - x_0)\sin(\vartheta) + (y - y_0)\cos(\vartheta).$$

Example 11.55 Let $P \equiv (2, -1)$ and $r : x + y - 2 = 0$ be, respectively, the coordinates of the point P and the equation of the line r with respect to the system OXY. Introduce a new coordinate system $O'X'Y'$, where $O' \equiv (1, 4)$ and the base vectors $\{i', j'\}$ are counterclockwise rotated by $\frac{\pi}{4}$.

We now determine the coordinates of P and the equation of r in terms of the new coordinate system $O'X'Y'$. By using relation (11.30), we obtain the new coordinates of P:

$$\begin{bmatrix} x' \\ y' \end{bmatrix} = \begin{bmatrix} \frac{\sqrt{2}}{2} & \frac{\sqrt{2}}{2} \\ -\frac{\sqrt{2}}{2} & \frac{\sqrt{2}}{2} \end{bmatrix} \begin{bmatrix} 1 \\ -5 \end{bmatrix} = \begin{bmatrix} -2\sqrt{2} \\ -3\sqrt{2} \end{bmatrix}.$$

Moreover, by relation (11.31), we may replace indeterminates x, y in the equation of r, in order to get the equation of r with respect to the new system $O'X'Y'$:

$$\left(\frac{\sqrt{2}}{2}x' - \frac{\sqrt{2}}{2}y' + 1 \right) + \left(\frac{\sqrt{2}}{2}x' + \frac{\sqrt{2}}{2}y' + 4 \right) - 2 = 0$$

that is

$$\sqrt{2}x' + 3 = 0.$$

11.5 Affine and Metric Classification of Quadrics

We dedicate the present section to the study of quadrics in affine and Euclidean spaces. More precisely, we approach the study of the action of affine transformations and isometries on quadrics. After a general analysis and classification of all the potential cases, we will focus the attention on conic curves in $\mathbb{FA}^2$ and $\mathbb{FE}^2$ and quadric surfaces in $\mathbb{FA}^3$ and $\mathbb{FE}^3$ from a purely geometrical point of view.

In order to head in the right direction, the first step must therefore be to introduce the framework within the strategy for classification of geometric loci is usually implemented. Now, we start by some definitions:

Definition 11.56 Let S, S' be two subsets of an affine space $\mathbb{A}$. We say that S and S' are *affinely equivalent* if there exists a nonsingular affine transformation $f : \mathbb{A} \to \mathbb{A}$ such that $f(S) = S'$.

The significance of the previously defined equivalence (sometimes called *affinity*) lies in the fact that a frame of reference $\{O, \mathbf{e}_1, \ldots, \mathbf{e}_n\}$ for $\mathbb{A}$ is mapped in a new frame of reference $\{O', \mathbf{e}'_1, \ldots, \mathbf{e}'_n\}$. More in detail, if $\mathbf{e}_i = \mathbf{OP_i}$, then $O' = f(O)$, $P'_i = f(P_i)$ and $\mathbf{e}'_i = \mathbf{O'P'_i}$. Thus, any point $P \in S$ is mapped to a point $P' \in S'$ in such a way that P is represented in terms of $\{O, \mathbf{e}_1, \ldots, \mathbf{e}_n\}$ by the same coordinates which represent P' in terms of $\{O', \mathbf{e}'_1, \ldots, \mathbf{e}'_n\}$.

Typically, one is interested in geometric properties invariant under certain transformations. In the case of affine transformations, we know that parallelism between lines, planes and, more generally, between affine subspaces is preserved. Nevertheless, to ensure the invariance of distances (and angles) between objects, the transformation is required to be an isometry.

Definition 11.57 Let S, S' be two subsets of an affine Euclidean space $\mathbb{E}$. We say that S and S' are *metrically equivalent* or also *congruent* if there exists an isometry $f : \mathbb{E} \to \mathbb{E}$ such that $f(S) = S'$.

On the basis of the definitions we have just mentioned, we now investigate the question about how a subset $S \subset \mathbb{A}$ can be transformed into another $S' \subset \mathbb{A}$ by an affine transformation in such a way that S' could represent the simplest form among all the possible subsets of $\mathbb{A}$ that are affinely equivalent to S.

More specifically, here we examine the case when such a subset is a quadric.

Definition 11.58 Let $\mathbb{A}$ be an affine space associated with the vector space V over the field $\mathbb{F}$ and assume $dim_{\mathbb{F}} V = n$. A *quadric* in $\mathbb{A}$ is a nonempty set $\mathscr{Q}$ of points whose coordinates satisfy the equation $p(x_1, \ldots, x_n) = 0$, where p is a quadratic polynomial in the variables $x_1, \ldots, x_n$ and with coefficients in $\mathbb{F}$.

Note 11.59 In all that follows, we always assume that the characteristic of the mentioned field $\mathbb{F}$ is different from 2.

Starting from the definition of a quadric and collecting terms of the second, first and zeroth degrees, the polynomial p can be written as

$$p(x_1, \ldots, x_n) = \sum_{i=1}^{n} \sum_{j=1}^{n} a_{ij} x_i x_j + 2 \sum_{k=1}^{n} a_{k,n+1} x_k + a_{n+1,n+1} \tag{11.32}$$

for $a_{ij} = a_{ji} \in \mathbb{F}$, for any $i, j = 1, \ldots, n+1$. The quadratic part of the polynomial $p(x_1, \ldots, x_n)$ is the quadratic form

$$q(x_1, \ldots, x_n) = \sum_{i=1}^{n} \sum_{j=1}^{n} a_{ij} x_i x_j \tag{11.33}$$

and can be represented by the symmetric matrix

$$A = \begin{bmatrix} a_{11} & \cdots\cdots & a_{1n} \\ \vdots & & \vdots \\ \vdots & & \vdots \\ a_{1n} & \cdots\cdots & a_{nn} \end{bmatrix}.$$

Hence, $q(x_1, \ldots, x_n) = X^t A X$, where $X = [x_1, \ldots, x_n]^t$.

If $B = [2a_{1,n+1}, \ldots, 2a_{n,n+1}]^t$ then the equation $p(x_1, \ldots, x_n) = 0$ can be written in the following matrix notation:

$$X^t A X + B^t X + a_{n+1,n+1} = 0. \tag{11.34}$$

It is also easy to see that, if we set

$$Y = \begin{bmatrix} x_1 \\ \vdots \\ x_n \\ 1 \end{bmatrix}, \quad \mathbf{v} = \frac{1}{2} B \quad \text{and} \quad C = \begin{bmatrix} A & \mathbf{v} \\ \mathbf{v}^t & a_{n+1,n+1} \end{bmatrix},$$

we may express Eq. (11.34) in the compact form

$$Y^t C Y = 0. \tag{11.35}$$

A classical classification of quadrics is made in relationship with the determinant of matrix C:

Definition 11.60 A quadric of Eq. (11.35) is called *non-degenerate* if C is not singular, otherwise it is said to be *degenerate*.

However, the first step in order to observe quadrics from the right point of view is the distinction between quadrics having a center and those that don't. More precisely,

Definition 11.61 Let $P \in \mathbb{A}$ be a point and $\mathcal{Q} \subset \mathbb{A}$ a quadric. Then P is called *center* of the quadric if the set $\mathcal{Q}$ is invariant under the action of the central symmetry with respect to the point P (see Example 11.34).

In this sense, we have the following result.

Theorem 11.62 *Let $\mathcal{Q}$ be a quadric of Eq. (11.35), P a point of $\mathbb{F}\mathbb{A}^n$ having coordinate vector $\mathbf{u} \equiv [\gamma_1, \ldots, \gamma_n]$ in terms of a fixed coordinate system. The point P is a center of $\mathcal{Q}$ if and only if, for any $i = 1, \ldots, n$, $\sum_{j=1}^{n} a_{ij}\gamma_j + a_{i,n+1} = 0$, that is, $A\mathbf{u} + \mathbf{v} = \mathbf{0}$.*

Proof Let $f : \mathbb{F}\mathbb{A}^n \to \mathbb{F}\mathbb{A}^n$ be the central symmetry with respect to point P. By applying f to the polynomial (11.32) representing $\mathcal{Q}$, we obtain the polynomial representing the set of points $f(\mathcal{Q})$, that is,

$$f\left(p(x_1, \ldots, x_n)\right) = \sum_{i,j=1}^{n} a_{ij}(2\gamma_i - x_i)(2\gamma_j - x_j)$$
$$+2\sum_{i=1}^{n} a_{i,n+1}(2\gamma_i - x_i) + a_{n+1,n+1} \tag{11.36}$$
$$= 4\sum_{i,j=1}^{n} a_{ij}\gamma_i\gamma_j - 4\sum_{i,j=1}^{n} a_{ij}x_i\gamma_j + \sum_{i,j=1}^{n} a_{ij}x_ix_j$$
$$+4\sum_{i=1}^{n} a_{i,n+1}\gamma_i - 2\sum_{i=1}^{n} a_{i,n+1}x_i + a_{n+1,n+1}.$$

Since P is a center of $\mathcal{Q}$ if and only if $f(\mathcal{Q}) = \mathcal{Q}$, in an equivalent manner, P is a center of $\mathcal{Q}$ if and only if $f\left(p(x_1, \ldots, x_n)\right) = p(x_1, \ldots, x_n)$. On the other hand, by comparing relations (11.36) and (11.32) we see that $f\left(p(x_1, \ldots, x_n)\right) = p(x_1, \ldots, x_n)$ is true only if

$$\sum_{i=1}^{n}\left(\sum_{j=1}^{n} a_{ij}\gamma_j + a_{i,n+1}\right)(x_i - \gamma_i) = 0.$$

Since the last identity holds only if any coefficient is identically zero, we get the required conclusion.

Corollary 11.63 *A quadric $\mathcal{Q}$ of Eq. (11.35) has a center if and only if $rank(C) \leq rank(A) + 1$.*

Proof If we assume that $\mathcal{Q}$ has a center, then $\sum_{j=1}^{n} a_{ij}\gamma_j + a_{i,n+1} = 0$, where $[\gamma_1, \ldots, \gamma_n]$ is the coordinate vector of the center. Hence, the system of linear equations coming from the identity $AX = -\mathbf{v}$ has solutions, that is, $rank(A) = rank(A|\mathbf{v})$. Since $rank(C) \leq rank(A|\mathbf{v}) + 1$, the conclusion follows.

Conversely, if $rank(C) \leq rank(A) + 1$ but we assume that $\mathcal{Q}$ has no center, then we have to assert that $AX = -\mathbf{v}$ has no solution, that is, $rank(A|\mathbf{v}) = rank(A) + 1$. This should imply that the $(n + 1)$th column of C is linearly independent of the first n column vectors of C. On the other hand, since C is symmetric, the last assertion means that the $(n + 1)$th row of C is linearly independent of the first n row vectors of C. Thus $rank(C) = rank(A|\mathbf{v}) + 1 = rank(A) + 2$, which is a contradiction.

Theorem 11.64 *Let $\mathcal{Q}$ be a quadric of $\mathbb{A}$ represented by Eq. (11.34). If the matrix A is nonsingular then $\mathcal{Q}$ has a center. Moreover, it is unique.*

Proof By Theorem 11.62, a point $P \in \mathbb{A}$ should be a center of $\mathcal{Q}$ if and only if its coordinate vector $X_0 \equiv [\gamma_1, \ldots, \gamma_n]$ is a solution of the linear system $AX + a_{i,n+1} = 0$, for $i = 1, 2, \ldots, n$ where $X = [x_1, \ldots, x_n]^t$. However, since by our assumption the matrix A has rank equal to n, the system admits exactly one and only one solution, that is, the center exists and it is unique.

These fruitful discussions allow us to gain the first overview of the distinction between different categories of quadrics in relation to a possible existence of a center.

Let us now investigate the following question: into what simplest form can the equation of a quadric be written in terms of a suitable choice of a frame of reference of the n-dimensional affine space $\mathbb{A}^n$? The answer to this question is a consequence of the solution of the related problem regarding the establishment of appropriate conditions under which two quadrics can be transformed into each other by an affine transformation. Since a quadric is represented by a polynomial of degree 2, we may consider whether two quadratic polynomials are affinely (or metrically) equivalent, in the sense of the following definition:

Definition 11.65 Let $p_1(x_1, \ldots, x_n)$ and $p_2(x_1, \ldots, x_n)$ be two distinct polynomials with coefficients in a field $\mathbb{F}$. We say that p_1 and p_2 are affinely (or metrically) equivalent if there exists an affine transformation (respectively, an isometry) $f : \mathbb{F}\mathbb{A}^n \to \mathbb{F}\mathbb{A}^n$ such that $f(p_1(x_1, \ldots, x_n)) = p_2(x_1, \ldots, x_n)$.

The quadrics $\mathcal{Q}_1$ and $\mathcal{Q}_2$, represented by two affinely (or metrically) equivalent polynomials are said to be affine (or metrically) equivalent.

Here we extend the argument previously presented in Theorem 11.43 to the n-dimensional case. Hence, consider the affine transformation f of $\mathbb{A}$, described by $f(X) = MX + c$, where M is an invertible matrix of $M_n(\mathbb{F})$ and c is a fixed point of $\mathbb{A}$. The image of $\mathcal{Q}$ is computed by the substitution

$$X = M^{-1}\left(f(X) - c\right) = HX' + d \tag{11.37}$$

in the relation (11.34), where $H = M^{-1}$ and $d = -M^{-1}c = -Hc$. Therefore, $f(\mathcal{Q})$ consists of all points whose coordinates X' are the solution of the equation

$$(HX' + d)^t A(HX' + d) + B^t(HX' + d) + a_{n+1,n+1} = 0,$$

that is,

$$X'^t(H^t AH)X' + (d^t AH + d^t AH + B^t H)X' + (d^t Ad + B^t d + a_{n+1,n+1}) = 0. \tag{11.38}$$

Therefore, under the notation $A' = H^t AH$, $B' = (d^t AH + d^t AH + B^t H)^t$ and $a'_{n+1,n+1} = d^t Ad + B^t d + a_{n+1,n+1}$ in (11.38), $f(\mathcal{Q})$ is represented by the equation

$$X'^t A'X' + B'^t X' + a'_{n+1,n+1} = 0$$

implying that $f(\mathcal{Q})$ is yet a quadric of $\mathbb{A}$. Since the relation (11.37) can be also obtained by

$$\begin{bmatrix} x_1 \\ \vdots \\ x_n \\ 1 \end{bmatrix} = \begin{bmatrix} H & d \\ \mathbf{0}^t & 1 \end{bmatrix} \begin{bmatrix} x'_1 \\ \vdots \\ x'_n \\ 1 \end{bmatrix}, \quad \text{for } \mathbf{0} = \underbrace{[0, \ldots, 0]^t}_{n\text{-times}} \tag{11.39}$$

the substitution (11.39) in (11.35) gives

$$\left(\begin{bmatrix} H & d \\ \mathbf{0}^t & 1 \end{bmatrix} Y' \right)^t C \begin{bmatrix} H & d \\ \mathbf{0}^t & 1 \end{bmatrix} Y' = 0,$$

that is,

$$Y'^t \begin{bmatrix} H^t & \mathbf{0} \\ d^t & 1 \end{bmatrix} C \begin{bmatrix} H & d \\ \mathbf{0}^t & 1 \end{bmatrix} Y' = 0.$$

This means that the matrix representing $f(\mathcal{Q})$ is

$$C' = \begin{bmatrix} H^t & \mathbf{0} \\ d^t & 1 \end{bmatrix} C \begin{bmatrix} H & d \\ \mathbf{0}^t & 1 \end{bmatrix} = \begin{bmatrix} H & d \\ \mathbf{0}^t & 1 \end{bmatrix}^t C \begin{bmatrix} H & d \\ \mathbf{0}^t & 1 \end{bmatrix}.$$

Thus, using the terminology introduced in Definition 11.56, we may assert that

(1) The matrices A, A', representing the quadratic parts of polynomials defining two *affinely equivalent* quadrics, are congruent.
(2) The matrices C, C', representing the entire polynomials defining two *affinely equivalent* quadrics, are congruent.

One of the main properties of congruent matrices is that they have the same rank. For this reason, in the further discussions we refer to the ranks of A and C in order to indicate at the same time ranks of A' and C', respectively.

If we consider the case when $\mathcal{Q}$ is a quadric of an affine Euclidean space $\mathbb{E}$, then the affine transformation acting on the points of $\mathcal{Q}$ can be described by $f(X) = MX + c$, where M is an invertible orthogonal matrix of $M_n(\mathbb{F})$ and c is a fixed point of $\mathbb{E}$. So, following Definition 11.57, we say that

(1) The matrices A, A', representing the quadratic parts of polynomials defining two *metrically equivalent* quadrics, are congruent.
(2) The matrices C, C', representing the entire polynomials defining two *metrically equivalent* quadrics, are congruent.

Fixing a quadric $\mathcal{Q}$, we pay special attention to establishing what is a suitable choice of a frame of reference of the n-dimensional affine space $\mathbb{A}$, in terms of which the equation of $\mathcal{Q}$ can be written in the simplest form. We divide our argument into two main cases.

The Affine Classification for Quadrics with Center

We start from Eq. (11.34) for $\mathbf{v} = \frac{1}{2} B$, that is,

$$X^t A X + 2\mathbf{v}^t X + a_{n+1,n+1} = 0. \tag{11.40}$$

By assuming that $\mathcal{Q}$ has a center, and by Theorem 11.62, there exists a vector $\mathbf{u} \in \mathbb{F}^n$ such that $A\mathbf{u} + \mathbf{v} = \mathbf{0}$. Looking at relation (11.37), and since A is symmetric, we may choose $H \in M_n(\mathbb{F})$ such that $A' = H^t A H$ is a diagonal matrix. Hence, for $X = HX' + \mathbf{u}$ in (11.40), the polynomial representing $\mathcal{Q}$ in the new coordinate system is equal to

$$(HX' + \mathbf{u})^t A(HX' + \mathbf{u}) + 2\mathbf{v}^t(HX' + \mathbf{u}) + a_{n+1,n+1} = 0,$$

that is,

$$X'^t A' X' + e = 0 \tag{11.41}$$

where $e = a_{n+1,n+1} + \mathbf{v}^t \mathbf{u}$. The complete matrix associated with the quadric is

$$C' = \begin{bmatrix} A' & \mathbf{0} \\ \mathbf{0}^t & e \end{bmatrix}$$

and the polynomial p of $\mathcal{Q}$ may assume two different forms.

In case $e \neq 0$, then $r = rank(A) < rank(C)$ (so that $r = rank(A') < rank(C')$), meaning that $rank(C) = r + 1$. The polynomial is

$$p(x'_1, \ldots, x'_n) = \sum_{i=1}^r \alpha_i x_i'^2 + e \tag{11.42}$$

and, by the following final substitution (affine transformation)

$$x''_i = \begin{cases} \sqrt{\frac{|e|}{|\alpha_i|}} x'_i & i = 1, \ldots, r \\ x'_i & i = r+1, \ldots, n \end{cases}. \tag{11.43}$$

Equation (11.42) reduces to one of

$$p(x''_1, \ldots, x''_n) = \sum_{i=1}^h x_i''^2 - \sum_{j=h+1}^r x_j''^2 + 1 \tag{11.44}$$

or

$$p(x''_1, \ldots, x''_n) = \sum_{i=1}^h x_i''^2 - \sum_{j=h+1}^r x_j''^2 - 1. \tag{11.45}$$

If $e = 0$, then $r = rank(A) = rank(C)$ (so that $r = rank(A') = rank(C')$) and the polynomial is

$$p(x'_1, \ldots, x'_n) = \sum_{i=1}^r \alpha_i x_i'^2. \tag{11.46}$$

Here, we apply the substitution

$$x_i'' = \begin{cases} \frac{1}{\sqrt{|\alpha_i|}} x_i' & i = 1, \ldots, r \\ x_i' & i = r+1, \ldots, n \end{cases}$$

and (11.46) reduces to

$$p(x_1'', \ldots, x_n'') = \sum_{i=1}^{h} x_i''^2 - \sum_{j=h+1}^{r} x_j''^2. \tag{11.47}$$

Notice that we have renumbered the coordinates in the polynomial p in such a way that the first h are positive and the remaining $r - h$ are negative. In further discussions, we are going to do it again, but no mention of such action will be done. These replacements are just about substitutions of variables, that is, affine transformations.

We then conclude that any quadric with center in the n-dimensional affine space $\mathbb{A}$ is affinely equivalent to one of the form (11.44), (11.45) or (11.47). In detail, we have

(i) For $r = n$, polynomials of the form (11.44) or (11.45) represent a non-degenerate quadric. We say that $\mathcal{Q}$ is an *Ellipsoid* in the cases $h = 0$ or $h = r$; we say that $\mathcal{Q}$ is a *Hyperboloid* in case $1 \leq h \leq r - 1$.

(ii) For $r \leq n - 1$, polynomials of the form (11.44) or (11.45) represent a degenerate quadric. We say that $\mathcal{Q}$ is a *non-parabolic Cylinder*.

(iii) Polynomials of the form (11.47) represent a degenerate quadric. We say that $\mathcal{Q}$ is a *Cone*.

Remark 11.66 Some particular cases are the following:

(i) In case $\mathbb{F} = \mathbb{R}$ and for $h = 0$ in (11.45), $\mathcal{Q}$ is the empty set.

(ii) In case $\mathbb{F} = \mathbb{R}$ and for $h = r$ in (11.44), $\mathcal{Q}$ is the empty set.

(iii) For $h = 0$ or $h = r$ in (11.47), $\mathcal{Q}$ is an affine subspace of $\mathbb{A}^n$.

Example 11.67 In the 3-dimensional affine space $\mathbb{R}\mathbb{A}^3$, let

$$x^2 + 4xy + 2x + y^2 + 2yz + 2z + 1 = 0$$

be the equation representing a quadric $\mathcal{Q}$. The matrices associated with $\mathcal{Q}$ are

$$A = \begin{bmatrix} 1 & 2 & 0 \\ 2 & 1 & 1 \\ 0 & 1 & 0 \end{bmatrix} \quad C = \begin{bmatrix} 1 & 2 & 0 & 1 \\ 2 & 1 & 1 & 0 \\ 0 & 1 & 0 & 1 \\ 1 & 0 & 1 & 1 \end{bmatrix}.$$

Hence, $\mathcal{Q}$ can be represented in the following matrix notations:

$$\begin{bmatrix} x & y & z \end{bmatrix} \begin{bmatrix} 1 & 2 & 0 \\ 2 & 1 & 1 \\ 0 & 1 & 0 \end{bmatrix} \begin{bmatrix} x \\ y \\ z \end{bmatrix} + \begin{bmatrix} 1 & 0 & 1 \end{bmatrix} \begin{bmatrix} x \\ y \\ z \end{bmatrix} + 1 = 0$$

and

$$\begin{bmatrix} x & y & z & 1 \end{bmatrix} \begin{bmatrix} 1 & 2 & 0 & 1 \\ 2 & 1 & 1 & 0 \\ 0 & 1 & 0 & 1 \\ 1 & 0 & 1 & 1 \end{bmatrix} \begin{bmatrix} x \\ y \\ z \\ 1 \end{bmatrix} = 0.$$

Notice that $|C| \neq 0$ and $AX + \mathbf{v} = \mathbf{0}$ has a solution, thus $\mathcal{Q}$ is a non-degenerate quadric having exactly one center of symmetry. Solving the system

$$\begin{bmatrix} 1 & 2 & 0 \\ 2 & 1 & 1 \\ 0 & 1 & 0 \end{bmatrix} \begin{bmatrix} x \\ y \\ z \end{bmatrix} + \begin{bmatrix} 1 \\ 0 \\ 1 \end{bmatrix} = \begin{bmatrix} 0 \\ 0 \\ 0 \end{bmatrix},$$

we find the coordinates of center $(1, -1, -1)$. Looking at the symmetric matrix A, by the process of diagonalization of a bilinear symmetric form, we may obtain $H \in M_n(\mathbb{R})$ such that $A' = H^t A H$ is a diagonal matrix. The implementation of the standard process leads us to

$$H = \begin{bmatrix} 1 & -2 & -\frac{2}{3} \\ 0 & 1 & \frac{1}{3} \\ 0 & 0 & 1 \end{bmatrix}.$$

Performing the transformation

$$\begin{bmatrix} x \\ y \\ z \end{bmatrix} = \begin{bmatrix} 1 & -2 & -\frac{2}{3} \\ 0 & 1 & \frac{1}{3} \\ 0 & 0 & 1 \end{bmatrix} \begin{bmatrix} x' \\ y' \\ z' \end{bmatrix} + \begin{bmatrix} 1 \\ -1 \\ -1 \end{bmatrix},$$

we have that the equation representing $\mathcal{Q}$ in the new coordinate system is

$$x'^2 - 3y'^2 + \frac{1}{3}z'^2 + 1 = 0.$$

Finally, for

$$x' = X$$
$$y' = \frac{1}{\sqrt{3}}Y$$
$$z' = \sqrt{3}Z,$$

we obtain the following affine canonical form of $\mathcal{Q}$:

$$X^2 - Y^2 + Z^2 + 1 = 0,$$

which is a hyperboloid.

The Affine Classification for Quadrics Without Center

Starting again from Eq. (11.40), we now assume $rank(A) < rank(A|\mathbf{v})$, that is, $AX + \mathbf{v} = 0$ has no solution. Here, we denote by X the generic coordinate vector in terms of a fixed basis for $\mathbb{F}^n$. In this case, the quadric $\mathscr{Q}$ has no center. Consider the following subspace of $\mathbb{F}^n$

$$N = \{X \in \mathbb{F}^n : \mathbf{v}^t X = 0\}.$$

The dimension of N is equal to $n - 1$. Let $\{\mathbf{e}_1, \ldots, \mathbf{e}_{n-1}\}$ be a basis for N and $\mathbf{e}_n \in \mathbb{F}^n$ be a vector such that $\mathbf{e}_n^t A \mathbf{e}_i = 0$, for any $i = 1, \ldots, n - 1$. Let us extend $\{\mathbf{e}_1, \ldots, \mathbf{e}_{n-1}\}$ to the basis $\{\mathbf{e}_1, \ldots, \mathbf{e}_n\}$ for $\mathbb{F}^n$. In order to get the equation of $\mathscr{Q}$ in the frame of reference $\{O, \mathbf{e}_1, \ldots, \mathbf{e}_n\}$, we have to perform the affine transformation $X = EX'$, where E is the transition matrix of $\{\mathbf{e}_1, \ldots, \mathbf{e}_n\}$ relative to the fixed basis and obviously whose column vectors are precisely $\mathbf{e}_1, \ldots, \mathbf{e}_n$. Substitution of $X = EX'$ in (11.40) gives

$$X''^t E^t A E X' + 2\mathbf{v}^t E X' + a_{n+1,n+1} = 0 \tag{11.48}$$

where

(i) the nth column and nth row of matrix $E^t A E$ are zeros, except eventually the (n, n)-entry;
(ii) the product $\mathbf{v}^t E X'$ is equal to $\alpha_n x_n'$, for some $\alpha_n \in \mathbb{F}$.

Thus, the polynomial representing $\mathscr{Q}$ with respect to the basis $\{\mathbf{e}_1, \ldots, \mathbf{e}_n\}$ can be written as

$$p(x_1', \ldots, x_n') = X''^t A' X' + 2\alpha_n x_n' + a_{n+1,n+1} \tag{11.49}$$

where

$$A' = \begin{bmatrix} \tilde{A} & \mathbf{0} \\ \mathbf{0}^t & a_{nn}' \end{bmatrix}.$$

Notice that, since $\mathscr{Q}$ has no center, the coefficient α_n in (11.49) is not zero. Moreover $\tilde{A} \in M_{n-1}(\mathbb{F})$ is symmetric, so that there exists a suitable basis $\{\mathbf{c}_1, \ldots, \mathbf{c}_{n-1}\}$ for N in terms of which the bilinear symmetric form associated with $\tilde{A}$ is represented by a diagonal matrix; in other words, there exists a nonsingular matrix $D \in M_{n-1}(\mathbb{F})$ such that $A'' = D^t \tilde{A} D$ is a diagonal matrix, i.e.,

$$A'' = D^t \tilde{A} D = \begin{bmatrix} \alpha_1 & & & & & \\ & \ddots & & & & \\ & & \alpha_r & & & \\ & & & 0 & & \\ & & & & \ddots & \\ & & & & & 0 \end{bmatrix} \qquad 0 \neq \alpha_i \in \mathbb{F}, \quad r = rank(\tilde{A}).$$

We now observe that introducing the matrix

$$H = \begin{bmatrix} D & 0 \\ 0^t & 1 \end{bmatrix},$$

an easy computation gives

$$H^t A' H = \begin{bmatrix} A'' & 0 \\ 0^t & a'_{nn} \end{bmatrix}.$$

Hence, applying the affine transformation $X' = HX''$ to the polynomial (11.49), we get

$$p(x''_1, \ldots, x''_n) = X'''^t H^t A' H X'' + 2\alpha_n x''_n + a_{n+1,n+1}$$
$$= \sum_{i=1}^{r} \alpha_i x''^2_i + a'_{nn} x''^2_n + 2\alpha_n x''_n + a_{n+1,n+1}. \tag{11.50}$$

Looking at (11.50), assume firstly that $a'_{nn} \neq 0$. In this case, and since

$$a'_{nn} x''^2_n + 2\alpha_n x''_n + a_{n+1,n+1} = a'_{nn}\left(x''_n + \frac{\alpha_n}{a'_{nn}}\right)^2 + a_{n+1,n+1} - \frac{\alpha_n^2}{a'_{nn}}$$

we may write (11.50) as

$$p(x''_1, \ldots, x''_n) = \sum_{i=1}^{r} \alpha_i x''^2_i + a'_{nn}\left(x''_n + \frac{\alpha_n}{a'_{nn}}\right)^2 + a' \tag{11.51}$$

for $a' = a_{n+1,n+1} - \frac{\alpha_n^2}{a'_{nn}}$. If we now apply to (11.51) the translation

$$\begin{aligned} x''_i &= x'''_i \quad i = 1, \ldots, n-1 \\ x''_n &= \quad x'''_n - \frac{\alpha_n}{a'_{nn}}, \end{aligned}$$

the polynomial representing $\mathscr{Q}$ assumes the form

$$\sum_{i=1}^{r} \alpha_i x'''^2_i + a'_{nn} x'''^2_n + a'$$

which should give a quadric with center. This contradiction proves that the coefficient a'_{nn} of polynomial (11.50) must be zero. Hence,

$$p(x''_1, \ldots, x''_n) = \sum_{i=1}^{r} \alpha_i x''^2_i + 2\alpha_n x''_n + a_{n+1,n+1}. \tag{11.52}$$

Finally, starting from (11.52) and by performing the substitution

$$x_i'' = \frac{1}{\sqrt{|\alpha_i|}} x_i''' \quad i = 1, \ldots, n-1$$
$$x_n'' = \frac{x_n'''}{2\alpha_n} - \frac{a_{n+1,n+1}}{2\alpha_n},$$

the polynomial of $\mathcal{Q}$ is

$$p(x_1''', \ldots, x_n''') = \sum_{i=1}^{h} x_i'''^2 - \sum_{j=h+1}^{r} x_j'''^2 + x_n'''. \tag{11.53}$$

Hence, any quadric without a center in the n-dimensional affine space $\mathbb{A}$ is affinely equivalent to one of the form (11.53).

Then, for a quadric without center, we have that $rank(A) = r \leq n-1$ and $rank(C) = r+2$. In particular,

(i) for $r = n-1$, the quadric is non-degenerate and here we say that $\mathcal{Q}$ is a *Paraboloid*;

(ii) for $r \leq n-2$, the quadric is degenerate and here we say that $\mathcal{Q}$ is a *parabolic Cylinder*.

We can sum up what we have done saying that for any quadric $\mathcal{Q}$ in the n-dimensional affine space $\mathbb{A}^n$, over an arbitrary field $\mathbb{F}$ of characteristic different from 2, there is a suitable frame of reference for $\mathbb{A}^n$, in terms of which $\mathcal{Q}$ is specified by a particularly simple equation, called *canonical equation* or *canonical form* for $\mathcal{Q}$. Any possible canonical equation of $\mathcal{Q}$ is associated with a polynomial of the form (11.44), (11.45), (11.47) or (11.53). In particular,

(i) polynomials $\sum_{i=1}^{n} x_i^2 + 1$ and $\sum_{i=1}^{n} x_i^2 - 1$ represent the canonical form of an Ellipsoid (it is a non-degenerate quadric with exactly one center);

(ii) polynomials $\sum_{i=1}^{h} x_i^2 - \sum_{i=h+1}^{n} x_i^2 + 1$ and $\sum_{i=1}^{h} x_i^2 - \sum_{i=h+1}^{n} x_i^2 - 1$ $(h \neq 0, n)$ represent the canonical form of a Hyperboloid (once again, it is a non-degenerate quadric with exactly one center);

(iii) polynomials $\sum_{i=1}^{h} x_i^2 - \sum_{i=h+1}^{r} x_i^2 + 1$ and $\sum_{i=1}^{h} x_i^2 - \sum_{i=h+1}^{r} x_i^2 - 1$ $(r \leq n-1)$ represent the canonical form of a non-parabolic Cylinder (it is a degenerate quadric with an infinite number of centers);

(iv) polynomials $\sum_{i=1}^{h} x_i^2 - \sum_{i=h+1}^{r} x_i^2$ $(r \leq n)$ represent a Cone (it is a degenerate quadric having exactly one center);

(v) polynomial $\sum_{i=1}^{n-1} x_i^2 + x_n$ represents a Paraboloid (it is a non-degenerate quadric without center);

(vi) polynomials $\sum_{i=1}^{r} x_i^2 + x_n$ $(r \leq n-2)$ represent a parabolic Cylinder (it is a degenerate quadric without center).

Example 11.68 Let $\mathscr{Q}$ be the quadric in $\mathbb{R}A^3$ having equation

$$x^2 + 2xy + 2xz + 4x + 2yz + z^2 + 2z + 1 = 0.$$

The matrices associated with $\mathscr{Q}$ are

$$A = \begin{bmatrix} 1 & 1 & 1 \\ 1 & 0 & 1 \\ 1 & 1 & 1 \end{bmatrix} \quad C = \begin{bmatrix} 1 & 1 & 1 & 2 \\ 1 & 0 & 1 & 0 \\ 1 & 1 & 1 & 1 \\ 2 & 0 & 1 & 1 \end{bmatrix}.$$

Hence, the compact forms representing $\mathscr{Q}$ are

$$\begin{bmatrix} x & y & z \end{bmatrix} \begin{bmatrix} 1 & 1 & 1 \\ 1 & 0 & 1 \\ 1 & 1 & 1 \end{bmatrix} \begin{bmatrix} x \\ y \\ z \end{bmatrix} + \begin{bmatrix} 2 & 0 & 1 \end{bmatrix} \begin{bmatrix} x \\ y \\ z \end{bmatrix} + 1 = 0$$

and

$$\begin{bmatrix} x & y & z & 1 \end{bmatrix} \begin{bmatrix} 1 & 1 & 1 & 2 \\ 1 & 0 & 1 & 0 \\ 1 & 1 & 1 & 1 \\ 2 & 0 & 1 & 1 \end{bmatrix} \begin{bmatrix} x \\ y \\ z \\ 1 \end{bmatrix} = 0.$$

Notice that $|C| \neq 0$ and $AX + \mathbf{v} = \mathbf{0}$ has no solution. In particular, $rank(A) = 2$ and $rank(C) = 4$, so that $\mathscr{Q}$ is a non-degenerate quadric without any center of symmetry. Consider the subspace

$$N = \{X \in \mathbb{R}^3 : [2, 0, 1]X = 0\}.$$

The dimension of N is equal to 2 and $N = \langle (1, 0, -2), (0, 1, 0) \rangle$. We extend this basis of N to one of $\mathbb{R}^3$, by adding the vector $(1, 0, -1)$, which is A-orthogonal to both vectors generating N.

We now perform the affine transformation $X = EX'$, where E is the transition matrix of $\{(1, 0, -2), (0, 1, 0), (1, 0, -1)\}$ relative to the standard basis of $\mathbb{R}^3$ and obviously whose column vectors are precisely $(1, 0, -2), (0, 1, 0), (1, 0, -1)$. As a result, we have

$$\begin{aligned} x &= x' + z' \\ y &= y' \\ z &= -2x' - z'. \end{aligned}$$

Substitution of $X = EX'$ leads to the following transformation of the matrices associated with the quadric:

$$A' = E^t A E = \begin{bmatrix} 1 & 0 & -2 \\ 0 & 1 & 0 \\ 1 & 0 & -1 \end{bmatrix} A \begin{bmatrix} 1 & 0 & 1 \\ 0 & 1 & 0 \\ -2 & 0 & -1 \end{bmatrix} = \begin{bmatrix} 1 & -1 & 0 \\ -1 & 0 & 0 \\ 0 & 0 & 0 \end{bmatrix}$$

and

$$C' = \begin{bmatrix} E^t & 0 \\ 0 & 1 \end{bmatrix} C \begin{bmatrix} E & 0 \\ 0 & 1 \end{bmatrix} = \begin{bmatrix} 1 & -1 & 0 & 0 \\ -1 & 0 & 0 & 0 \\ 0 & 0 & 0 & 1 \\ 0 & 0 & 1 & 1 \end{bmatrix} = \begin{bmatrix} & & & 0 \\ & A' & & 0 \\ & & & 1 \\ 0 & 0 & 1 & 1 \end{bmatrix}.$$

By the standard process for diagonalization for bilinear forms, we find the basis $\{(1, 0, 0), (1, 1, 0), (0, 0, 1)\}$ in terms of which the bilinear symmetric form associated with A' is represented by a diagonal matrix. More precisely, if we denote by D, the transition matrix of $\{(1, 0, 0), (1, 1, 0), (0, 0, 1)\}$ relative to the standard basis of $\mathbb{R}^3$ which has column vectors $(1, 0, 0)$, $(1, 1, 0)$, $(0, 0, 1)$, then we get

$$A'' = D^t A' D = \begin{bmatrix} 1 & 0 & 0 \\ 1 & 1 & 0 \\ 0 & 0 & 1 \end{bmatrix} A' \begin{bmatrix} 1 & 1 & 0 \\ 0 & 1 & 0 \\ 0 & 0 & 1 \end{bmatrix} = \begin{bmatrix} 1 & 0 & 0 \\ 0 & -1 & 0 \\ 0 & 0 & 0 \end{bmatrix}.$$

Now, introducing the matrix

$$H = \begin{bmatrix} D & 0 \\ 0^t & 1 \end{bmatrix},$$

we get

$$H^t C' H = \begin{bmatrix} 1 & 0 & 0 & 0 \\ 0 & -1 & 0 & 0 \\ 0 & 0 & 0 & 1 \\ 0 & 0 & 1 & 1 \end{bmatrix}.$$

The quadric is then represented by the equation $x''^2 - y''^2 + 2z'' + 1 = 0$, which is a paraboloid. Finally, by the transformation

$$\begin{aligned} x'' &= X \\ y'' &= Y \\ z'' &= \tfrac{Z}{2} - \tfrac{1}{2}, \end{aligned}$$

we obtain the affine canonical form of the equation representing $\mathcal{Q}$, that is $X^2 - Y^2 + Z = 0$, which is a paraboloid.

At this point, we wonder what can we say in case $\mathbb{F}$ is an algebraically closed field (for instance, $\mathbb{F} = \mathbb{C}$). We are going to solve the same above problem by means of successive affine transformations; after each one, the equation of the quadric is formulated simplifying the original one. One may simply repeat arguments previously presented, acting with appropriate amendments in some steps. More precisely, for quadrics with center, we may start from the polynomial of the form (11.42):

$$p(x'_1, \ldots, x'_n) = \sum_{i=1}^{r} \alpha_i x_i^2 + e.$$

If $e \neq 0$, we perform the substitution

$$x''_i = \begin{cases} \sqrt{\frac{e}{\alpha_i}} x'_i & i = 1, \ldots, r \\ x'_i & i = r+1, \ldots, n \end{cases}$$

so that (11.42) reduces to

$$p(x''_1, \ldots, x''_n) = \sum_{i=1}^{r} x_i''^2 + 1. \tag{11.54}$$

If $e = 0$, we apply the following change of coordinates:

$$x''_i = \begin{cases} \frac{1}{\sqrt{\alpha_i}} x'_i & i = 1, \ldots, r \\ x'_i & i = r+1, \ldots, n \end{cases}$$

and the polynomial reduces to

$$p(x''_1, \ldots, x''_n) = \sum_{i=1}^{r} x_i''^2. \tag{11.55}$$

In the case of quadrics without center, we look at polynomial of the form (11.52):

$$p(x''_1, \ldots, x''_n) = \sum_{i=1}^{r} \alpha_i x_i''^2 + 2\alpha_n x''_n + a_{n+1,n+1},$$

and by performing the substitution

$$x''_i = \frac{1}{\sqrt{\alpha_i}} x'''_i \quad i = 1, \ldots, n-1$$
$$x''_n = \frac{x'''_n}{2\alpha_n} - \frac{a_{n+1,n+1}}{2\alpha_n}$$

the polynomial of $\mathcal{Q}$ is

$$p(x'''_1, \ldots, x'''_n) = \sum_{i=1}^{r} x_i'''^2 + x'''_n. \tag{11.56}$$

Therefore, we conclude that an arbitrary set in an n-dimensional affine space over an algebraically closed field, given by equating a second-degree polynomial in n

variables to zero, is affinely equivalent to one of the sets defined by equating to zero one of the polynomials (11.54), (11.55) and (11.56). In particular,

(*i*) polynomial $\sum_{i=1}^{n} x_i^2 + 1$ represents the canonical form of an Ellipsoid;

(*ii*) polynomials $\sum_{i=1}^{r} x_i^2 + 1$ ($r \leq n - 1$) represent the canonical form of a non-parabolic Cylinder;

(*iii*) polynomials $\sum_{i=1}^{r} x_i^2$ ($r \leq n$) represent a Cone;

(*iv*) polynomial $\sum_{i=1}^{n-1} x_i^2 + x_n$ represents a Paraboloid;

(*v*) polynomials $\sum_{i=1}^{r} x_i^2 + x_n$ ($r \leq n - 2$) represent a parabolic Cylinder.

Remark 11.69 We notice that the canonical equation of a degenerate quadric contains less than n variables.

We would like to dedicate a brief interlude to a considerable aspects of quadrics theory, due to the natural pathology of quadratic polynomials in n indeterminates: in some cases, they can be split in a product of two linear polynomials. When this event occurs, it is a good thing to understand what's going on to quadrics represented by such polynomials.

Definition 11.70 A quadric $\mathcal{Q} \subset \mathbb{A}^n$ is said to be *reducible* if its representing polynomial $p(x_1, \ldots, x_n)$ can be reduced as a product of two linear polynomials $h_1(x_1, \ldots, x_n)$ and $h_2(x_1, \ldots, x_n)$. Each of such polynomials defines a hyperplane in $\mathbb{A}^n$, namely

$$H_1 : h_1(x_1, \ldots, x_n) = 0 \quad H_2 : h_2(x_1, \ldots, x_n) = 0.$$

We write $\mathcal{Q} = H_1 \cup H_2$, meaning that the subset $\mathcal{Q}$ consists of all points from H_1 and all points from H_2.

In order to characterize reducible quadrics we fix the following.

Theorem 11.71 *Let $\mathcal{Q} \subset \mathbb{A}^n$ be a quadric represented by equation $Y^t C Y = 0$, as specified in (11.35). If $\mathcal{Q}$ is not the empty set, then it is reducible if and only if $rank(C) \leq 2$.*

Proof Assume firstly that $\mathcal{Q}$ is reducible and hyperplanes H_1 and H_2 are represented by the following equations:

$$H_1 : \alpha_1 x_1 + \cdots + \alpha_n x_n + \alpha_{n+1} = 0 \quad H_2 : \beta_1 x_1 + \cdots + \beta_n x_n + \beta_{n+1} = 0,$$

for some $\alpha_i, \beta_i \in \mathbb{F}$. By an appropriate affinity $f : \mathbb{A}^n \to \mathbb{A}^n$, we may map H_1 in such a way that the equation representing its image $f(H_1)$ is exactly $x_1 = 0$. After performing the same affinity on H_2, we denote by $\gamma_1 x_1 + \cdots + \gamma_n x_n = 0$ the equation representing $f(H_2)$.

The image $f(\mathcal{Q})$ consists precisely of all points from $f(H_1)$ and all points from $f(H_2)$. Then its representing polynomial comes from the product of polynomials representing hyperplanes $f(H_1)$ and $f(H_2)$, that is,

$$\gamma_1 x_1^2 + \gamma_2 x_2 + \cdots + \gamma_n x_n + \gamma_{n+1} = 0.$$

Hence, the matrix associated with $f(\mathcal{Q})$ is

$$C' = \begin{bmatrix} \gamma_1 & \frac{\gamma_2}{2} & \cdots\cdots\cdots & \frac{\gamma_{n+1}}{2} \\ \frac{\gamma_2}{2} & 0 & \cdots\cdots\cdots & 0 \\ \vdots & \vdots & & \vdots \\ \vdots & \vdots & & \vdots \\ \vdots & \vdots & & \vdots \\ \frac{\gamma_{n+1}}{2} & 0 & \cdots\cdots\cdots & 0 \end{bmatrix}$$

whose rank is ≤ 2, as well as the rank C, which is congruent with C'.

Conversely, suppose that $rank(C) \leq 2$ and consider now the canonical form of quadric $\mathcal{Q}$. In its simplest form, the quadric is described by a matrix C' having the same rank of C. Taking a look at all possible canonical forms of $\mathcal{Q}$, that is,

(*i*) Equations (11.44), (11.45), (11.47) and (11.53) in case $\mathbb{F}$ is not algebraically closed and
(*ii*) Equations (11.54), (11.55) and (11.56), in case $\mathbb{F}$ is algebraically closed,

it is clear that the only ones admitting rank ≤ 2 are

(*i*) polynomials $x_i^2 + 1$ and $x_i^2 - 1$, for some index $1 \leq i \leq n$;
(*ii*) polynomials $x_i^2 - x_j^2$ and $x_i^2 + x_j^2$, for some indices $i \neq j$ and $1 \leq i, j \leq n$;
(*iii*) polynomials x_i^2, for some index $1 \leq i \leq n$.

Initially, we consider polynomials $x_i^2 + 1$ and $x_i^2 + x_j^2$ when $\mathbb{F}$ is not algebraically closed. In both cases, $f(\mathcal{Q})$ should be the empty set, which contradicts the fact that $\mathcal{Q}$ is not.

In any other case, the corresponding polynomial is always reducible in the product of two linear polynomials, which are distinct, merged, real or also, in case $\mathbb{F}$ is algebraically closed, complex conjugate. These polynomials represent two hyperplanes K_1, K_2 in $\mathbb{A}^n$, whose union is $f(\mathcal{Q})$, namely $f(\mathcal{Q}) = K_1 \cup K_2$. Therefore, $f(\mathcal{Q})$ consists of all points from K_1 and all points from K_2. Since any affine transformation maps hyperplanes to hyperplanes (preserving the direction), we conclude that $\mathcal{Q}$ is the union of the hyperplanes $f^{-1}(K_1)$ and $f^{-1}(K_2)$.

All that has just been said can be applied to the case of quadrics in $\mathbb{A}^2$ and $\mathbb{A}^3$, in order to obtain the well known affine classification of conic curves in the affine plane and the affine classification of quadric surfaces in the 3-dimensional space. These are nothing more than reduced cases of the ones previously discussed.

Thus, in the case of conics in the plane, we may simply assert the following.

Theorem 11.72 *Let* $\mathbb{R}A^2$ *be the 2-dimensional affine space over the vector space* $\mathbb{R}^2$. *Any conic curve is affinely equivalent to one of the sets defined by the following equations:*

(i) $x^2 + y^2 - 1 = 0$, *non-degenerate real ellipse;*
(ii) $x^2 - y^2 - 1 = 0$, *non-degenerate hyperbola;*
(iii) $x^2 - y = 0$, *non-degenerate parabola;*
(iv) $x^2 + y^2 + 1 = 0$, *empty set (non-degenerate imaginary ellipse);*
(v) $x^2 - y^2 = 0$, *two secant real lines;*
(vi) $x^2 + y^2 = 0$, *one real point (two complex conjugate lines);*
(vii) $x^2 - 1 = 0$, *two real parallel lines;*
(viii) $x^2 + 1 = 0$, *empty set (two complex parallel lines);*
(ix) $x^2 = 0$, *two real merged lines.*

In the case of complex spaces, we have the following.

Theorem 11.73 *Let* $\mathbb{C}A^2$ *be the 2-dimensional affine space over the vector space* $\mathbb{C}^2$. *Any conic curve is affinely equivalent to one of the sets defined by the following equations:*

(i) $x^2 + y^2 + 1 = 0$, *non-degenerate ellipse;*
(ii) $x^2 - y = 0$, *non-degenerate parabola;*
(iii) $x^2 + y^2 = 0$, *two complex conjugate secant lines;*
(iv) $x^2 + 1 = 0$, *two complex parallel lines;*
(v) $x^2 = 0$, *two real merged lines.*

Shifting the focus to quadric surfaces in real affine space, we have the following.

Theorem 11.74 *Let* $\mathbb{R}A^3$ *be the 3-dimensional affine space over the vector space* $\mathbb{R}^3$. *Any quadric surface is affinely equivalent to one of the sets defined by the following equations:*

(i) $x^2 + y^2 + z^2 - 1 = 0$, *real ellipsoid;*
(ii) $x^2 + y^2 + z^2 + 1 = 0$, *empty set (non-degenerate imaginary ellipsoid);*
(iii) $x^2 + y^2 - z^2 + 1 = 0$, *elliptic hyperboloid;*
(iv) $x^2 + y^2 - z^2 - 1 = 0$, *hyperbolic hyperboloid;*
(v) $x^2 + y^2 - z = 0$, *elliptic paraboloid;*
(vi) $x^2 - y^2 - z = 0$, *hyperbolic paraboloid;*
(vii) $x^2 + y^2 + z^2 = 0$, *one real point (imaginary cone);*
(viii) $x^2 + y^2 - z^2 = 0$, *real cone;*
(ix) $x^2 + y^2 + 1 = 0$, *empty set (imaginary cylinder);*
(x) $x^2 + y^2 - 1 = 0$, *right circular cylinder;*
(xi) $x^2 - y = 0$, *parabolic cylinder;*
(xii) $x^2 - y^2 - 1 = 0$, *hyperbolic cylinder;*

$(xiii)$ $x^2 + y^2 = 0$, *a line (two secant complex planes)*;
(xiv) $x^2 - y^2 = 0$, *two secant real planes*;
 (xv) $x^2 + 1 = 0$, *empty set (two complex parallel planes)*;
(xvi) $x^2 - 1 = 0$, *two real parallel planes*;
$(xvii)$ $x^2 = 0$, *two merged real planes.*

And finally, in the case of complex spaces, we have the following.

Theorem 11.75 *Let* $\mathbb{C}A^3$ *be the 3-dimensional affine space over the vector space* $\mathbb{C}^3$. *Any quadric surface is affinely equivalent to one of the sets defined by the following equations:*

  (i) $x^2 + y^2 + z^2 - 1 = 0$, *ellipsoid*;
 (ii) $x^2 + y^2 - z = 0$, *paraboloid*;
(iii) $x^2 + y^2 + z^2 = 0$, *cone*;
 (iv) $x^2 + y^2 + 1 = 0$, *elliptic cylinder*;
  (v) $x^2 - y = 0$, *parabolic cylinder*;
 (vi) $x^2 + y^2 = 0$, *two secant planes*;
(vii) $x^2 + 1 = 0$, *two parallel planes*;
$(viii)$ $x^2 = 0$, *two merged planes.*

The classification of quadrics in an affine Euclidean space $\mathbb{R}E^n$ up to metric equivalence uses precisely the same arguments as those used in the case of the affine space $\mathbb{F}E^n$. Since isometries are affine transformations, we may apply the results obtained in the affine space to the Euclidean case. We start by considering quadrics given by Eq. (11.34) for $\mathbf{v} = \frac{1}{2}B$, that is,

$$X^t A X + 2\mathbf{v}^t X + a_{n+1,n+1} = 0. \tag{11.57}$$

The Metric Classification for Quadrics with Center

This is the case when $rank(A) = rank(A|\mathbf{v})$. Let $\mathbf{u} \in \mathbb{F}^n$ be such that $A\mathbf{u} + \mathbf{v} = \mathbf{0}$. We use the same transformation $X = HX' + \mathbf{u}$ as in the affine case. But in the present case $H \in M_n(\mathbb{F})$ is an orthogonal matrix, such that $A' = H^t A H$ is a diagonal matrix. Thus, we reduce to the equation

$$X'' A' X' + e = 0 \tag{11.58}$$

where A' has its eigenvalues on the main diagonal. The polynomial associated with the quadric is

$$p(x'_1, \ldots, x'_n) = \sum_{i=1}^{r} \alpha_i x_i'^2 + e \tag{11.59}$$

where $\alpha_1, \ldots, \alpha_r$ are the nonzero eigenvalues of A. In case $e \neq 0$, we divide $p(x_1', \ldots, x_n')$ by $\pm e$, according with the fact that $e > 0$ or $e < 0$, and obtain the form

$$p(x_1', \ldots, x_n') = \sum_{i=1}^{r} \alpha_i' x_i'^2 + 1 \tag{11.60}$$

where $\alpha_i' = \frac{\alpha_i}{e}$, for any $i = 1, \ldots, r$. Notice that the last affine transformation (11.43), previously used in the affine case, in the present Euclidean space doesn't make any sense, because it is not orthogonal.

We then conclude that any quadric with center in the n-dimensional affine Euclidean space $\mathbb{R}E^n$ is metrically equivalent to one of the following forms:

$$\sum_{i=1}^{r} \alpha_i' x_i'^2 + 1 \tag{11.61}$$

$$\sum_{i=1}^{r} \alpha_i' x_i'^2. \tag{11.62}$$

Example 11.76 Let $\mathcal{Q}$ be the quadric in $\mathbb{F}E^3$, where $\mathbb{F} = \mathbb{R}$ or $\mathbb{C}$ having equation

$$2x^2 + 2xy + 2x + 2y^2 + 4y + z^2 + 2 = 0.$$

The matrices associated with $\mathcal{Q}$ are

$$A = \begin{bmatrix} 2 & 1 & 0 \\ 1 & 2 & 0 \\ 0 & 0 & 1 \end{bmatrix} \quad C = \begin{bmatrix} 2 & 1 & 0 & 1 \\ 1 & 2 & 0 & 2 \\ 0 & 0 & 1 & 0 \\ 1 & 2 & 0 & 2 \end{bmatrix}.$$

Then $rank(A) = rank(C) = 3$ and $\mathcal{Q}$ is a degenerate quadric having exactly one center of symmetry and which is obviously a cone.

Solving the system

$$\begin{bmatrix} 2 & 1 & 0 \\ 1 & 2 & 0 \\ 0 & 0 & 1 \end{bmatrix} \begin{bmatrix} x_1 \\ x_2 \\ x_3 \end{bmatrix} + \begin{bmatrix} 1 \\ 2 \\ 0 \end{bmatrix} = \begin{bmatrix} 0 \\ 0 \\ 0 \end{bmatrix},$$

we find the coordinates $(0, -1, 0)$ of the center. To implement the transformation $X = HX' + \mathbf{u}$, where $H \in M_3(\mathbb{R})$ is an orthogonal matrix, we determine the eigenvalues $\lambda_1, \lambda_2, \lambda_3$ of A. We get

(1) $\lambda_1 = \lambda_2 = 1$ having associated eigenspace generated by orthogonal vectors $\left(\frac{1}{\sqrt{2}}, -\frac{1}{\sqrt{2}}, 0 \right)$ and $(0, 0, 1)$;

(2) $\lambda_3 = 3$, having associated eigenspace generated by vector $\left(\frac{1}{\sqrt{2}}, \frac{1}{\sqrt{2}}, 0\right)$.

Thus, the transformation of variables is

$$
\begin{bmatrix} x \\ y \\ z \end{bmatrix} = \begin{bmatrix} \frac{1}{\sqrt{2}} & 0 & \frac{1}{\sqrt{2}} \\ -\frac{1}{\sqrt{2}} & 0 & \frac{1}{\sqrt{2}} \\ 0 & 1 & 0 \end{bmatrix} \begin{bmatrix} X \\ Y \\ Z \end{bmatrix} + \begin{bmatrix} 0 \\ -1 \\ 0 \end{bmatrix}.
$$

Substitution of variables leads to the canonical equation

$$
X^2 + Y^2 + 3Z^2 = 0.
$$

Notice that, for $\mathbb{F} = \mathbb{R}$, the quadric consists of exactly one point. For $\mathbb{F} = \mathbb{C}$, the quadric is a degenerate non-reducible surface.

The Metric Classification for Quadrics Without Center

Starting again from Eq. (11.57), we now assume $rank(A) < rank(A|v)$, that is, $AX + \mathbf{v} = 0$ has no solution. Even in this case also if we could repeat the same argument as in the case of affine spaces, here we want to take the shortcut by taking advantage of the fact that we can use the resources provided by the Euclidean structure of the space $\mathbb{R}E^n$.

Denoted by r the rank of the matrix A, the null space $N(A)$ of A has dimension $n - r$. Moreover, since A is symmetric, $N(A)$ coincides with the orthogonal complement of the column space of A (that is $Im(A)$, the image of A). Hence, we may write $\mathbf{v} = \mathbf{v}_1 + \mathbf{v}_2$, where $\mathbf{v}_1 \in Im(A)$ and $\mathbf{v}_2 \in N(A)$. Hence, $A\mathbf{v}_2 = 0$ and there exists $\mathbf{u} \in \mathbb{R}^n$ such that $A\mathbf{u} = \mathbf{v}_1$. Of course, the first stage of the process is to guarantee the diagonalization of A. Once again, we introduce the transformation $X = HX' - \mathbf{u}$, where $H \in M_n(\mathbb{F})$ is an orthogonal matrix, whose column vectors are precisely the eigenvectors of A, that is, $A' = H^t A H = H^{-1} A H$ is a diagonal matrix, having its eigenvalues on the main diagonal.

The column vectors of H define an orthonormal basis $\{\mathbf{e}_1, \ldots, \mathbf{e}_n\}$ for $\mathbb{R}^n$, consisting of r eigenvectors corresponding to nonzero eigenvalues and $n - r$ eigenvectors corresponding to the eigenvalue zero. We may reorder the element in the basis and assume that $\{\mathbf{e}_1, \ldots, \mathbf{e}_r\}$ are precisely the eigenvectors associated with nonzero eigenvalues of A. Hence, since $\mathbf{v}_2$ is an eigenvector corresponding to eigenvalue zero, we may choose $\mathbf{e}_{r+1} = \frac{\mathbf{v}_2}{\|\mathbf{v}_2\|}$.

Thus, by substitution $X = HX' - \mathbf{u}$ in Eq. (11.57), we have

$$
\begin{aligned}
p(x_1', \ldots, x_n') &= X'' H^t A H X' + 2\mathbf{v}_2^t H X' + (-2\mathbf{v}_2^t \mathbf{u} - \mathbf{v}_1^t \mathbf{u} + a_{n+1,n+1}) \\
&= X'' H^t A H X' + 2\mathbf{v}_2^t H X' + e
\end{aligned} \tag{11.63}
$$

where $e = -2\mathbf{v}_2^t \mathbf{u} - \mathbf{v}_1^t \mathbf{u} + a_{n+1,n+1}$. Notice that $\mathbf{e}_{r+1}^t H$ is precisely the $(r + 1)$th column of $H^t H = I_n$, that is, $\frac{\mathbf{v}_2^t}{\|\mathbf{v}_2\|} H = \mathbf{e}_{r+1}$. Hence, $\mathbf{v}_2^t H = \|\mathbf{v}_2\| \mathbf{e}_{r+1}$. Therefore, the polynomial (11.63) reduces to

$$p(x'_1, \ldots, x'_n) = X'' H^t A H X' + 2 \|\mathbf{v}_2\| x'_{r+1} + e$$
$$= \sum_{i=1}^{r} \alpha_i x'^2_i + \beta_{r+1} x'_{r+1} + e \qquad (11.64)$$

where $\alpha_1, \ldots, \alpha_r$ are the nonzero eigenvalues of A and $\beta_{r+1} = 2 \|\mathbf{v}_2\|$. Finally, by performing the translation

$$x'_i = x''_i \ i = 1, \ldots, r$$
$$x'_{r+1} = x''_{r+1} - \frac{e}{\beta_{r+1}},$$

we may write the polynomial associated with the quadric as

$$\sum_{i=1}^{r} \alpha_i x''^2_i + \beta_{r+1} x''_{r+1}. \qquad (11.65)$$

Example 11.77 Let $\mathcal{Q}$ be the quadric in $\mathbb{RE}^4$ having equation

$$x_1^2 + 2x_1 x_2 + 4x_1 + x_2^2 + 2x_2 + x_3^2 + 2x_3 x_4 + x_4^2 + 3 = 0.$$

The matrices associated with $\mathcal{Q}$ are

$$A = \begin{bmatrix} 1 & 1 & 0 & 0 \\ 1 & 1 & 0 & 0 \\ 0 & 0 & 1 & 1 \\ 0 & 0 & 1 & 1 \end{bmatrix} \quad C = \begin{bmatrix} 1 & 1 & 0 & 0 & 2 \\ 1 & 1 & 0 & 0 & 1 \\ 0 & 0 & 1 & 1 & 0 \\ 0 & 0 & 1 & 1 & 0 \\ 2 & 1 & 0 & 0 & 3 \end{bmatrix}.$$

Then $rank(A) = 2$, $rank(C) = 4$ and $\mathcal{Q}$ is a degenerate quadric having no center of symmetry which is obviously a parabolic cylinder. The first step is finding the projection $\mathbf{v}_1$ of vector $\mathbf{v}$, having components $(2, 1, 0, 0)$, onto the subspace $Im(A)$. To do this, we easily see that an orthogonal basis for $Im(A)$ consists of vectors $\mathbf{c}_1 = (1, 1, 0, 0)$ and $\mathbf{c}_2 = (0, 0, 1, 1)$. Then,

$$\mathbf{v}_1 = \frac{\mathbf{v} \cdot \mathbf{c}_1}{\mathbf{c}_1 \cdot \mathbf{c}_1} \mathbf{c}_1 + \frac{\mathbf{v} \cdot \mathbf{c}_2}{\mathbf{c}_2 \cdot \mathbf{c}_2} \mathbf{c}_2 = \left(\frac{3}{2}, \frac{3}{2}, 0, 0 \right).$$

Therefore, since $\mathbf{v}_1 \in Im(A)$, the system

$$\begin{bmatrix} 1 & 1 & 0 & 0 \\ 1 & 1 & 0 & 0 \\ 0 & 0 & 1 & 1 \\ 0 & 0 & 1 & 1 \end{bmatrix} \begin{bmatrix} x_1 \\ x_2 \\ x_3 \\ x_4 \end{bmatrix} + \begin{bmatrix} \frac{3}{2} \\ \frac{3}{2} \\ 0 \\ 0 \end{bmatrix} = \begin{bmatrix} 0 \\ 0 \\ 0 \\ 0 \end{bmatrix}$$

has solutions and their general form is $(-\alpha - \frac{3}{2}, \alpha, \beta, -\beta)$, for any $\alpha, \beta \in \mathbb{R}$. We choose one of them, for instance, $\alpha = \beta = 0$, and denote it as $\mathbf{u} = (-\frac{3}{2}, 0, 0, 0)$. This is the vector we'll use for translation. We now determine the null space $N(A)$ of A: one of its orthonormal bases is

$$\left\{ \left(\frac{1}{\sqrt{2}}, -\frac{1}{\sqrt{2}}, 0, 0 \right), \left(0, 0, \frac{1}{\sqrt{2}}, -\frac{1}{\sqrt{2}} \right) \right\}.$$

Then compose an orthonormal basis for $\mathbb{R}^4$ by the union of bases from $Im(A)$ and $N(A)$, that is,

$$\mathscr{B} = \left\{ \left(\frac{1}{\sqrt{2}}, \frac{1}{\sqrt{2}}, 0, 0 \right), \left(0, 0, \frac{1}{\sqrt{2}}, \frac{1}{\sqrt{2}} \right), \left(\frac{1}{\sqrt{2}}, -\frac{1}{\sqrt{2}}, 0, 0 \right), \left(0, 0, \frac{1}{\sqrt{2}}, -\frac{1}{\sqrt{2}} \right) \right\}.$$

The transformation of variables is then the composition of

(1) translation by $\mathbf{u}$ and
(2) rotation by the matrix having elements from $\mathscr{B}$ as column vectors,

that is,

$$\begin{bmatrix} x_1 \\ x_2 \\ x_3 \\ x_4 \end{bmatrix} = \begin{bmatrix} \frac{1}{\sqrt{2}} & 0 & \frac{1}{\sqrt{2}} & 0 \\ \frac{1}{\sqrt{2}} & 0 & -\frac{1}{\sqrt{2}} & 0 \\ 0 & \frac{1}{\sqrt{2}} & 0 & \frac{1}{\sqrt{2}} \\ 0 & \frac{1}{\sqrt{2}} & 0 & -\frac{1}{\sqrt{2}} \end{bmatrix} \begin{bmatrix} x_1' \\ x_2' \\ x_3' \\ x_4' \end{bmatrix} + \begin{bmatrix} -\frac{3}{2} \\ 0 \\ 0 \\ 0 \end{bmatrix}.$$

Hence, we perform the following substitution of variables in the equation of the quadric:

$$\begin{aligned} x_1 &= \tfrac{1}{\sqrt{2}}(x_1' + x_3') - \tfrac{3}{2}, \\ x_2 &= \tfrac{1}{\sqrt{2}}(x_1' - x_3'), \\ x_3 &= \tfrac{1}{\sqrt{2}}(x_2' + x_4'), \\ x_4 &= \tfrac{1}{\sqrt{2}}(x_2' - x_4'). \end{aligned}$$

This leads us to arrive at the equation

$$2x_1'^2 + 2x_2'^2 + \sqrt{2}x_3' - \frac{3}{4} = 0.$$

Finally, for

$$\begin{aligned} x_1' &= X_1 \\ x_2' &= X_2 \\ x_3' &= X_3 + \tfrac{3}{4\sqrt{2}} \\ x_4' &= X_4, \end{aligned}$$

we have the metric canonical equation of the cylinder

$$2X_1^2 + 2X_2^2 + \sqrt{2}X_3 = 0.$$

Summarizing, we conclude that every quadric $\mathcal{Q}$ of the affine Euclidean space $\mathbb{RE}^n$ having equation

$$X^t A X + B^t X + a_{n+1,n+1} = 0$$

and in its more compact form

$$Y^t C Y = 0$$

can be given in some coordinate system by an equation of the form (11.61) in the case $\mathcal{Q}$ has a center, or by an equation of the form (11.65) if $\mathcal{Q}$ doesn't have any center. So, we have proved the following.

Theorem 11.78 *Every quadric of $\mathbb{RE}^n$ is metrically equivalent to one of the following:*

(i) $\sum\limits_{i=1}^{p} \alpha_i x_i^2 - \sum\limits_{j=p+1}^{r} \alpha_j x_j^2 = 0$ *if rank(C) = rank(A)*,

(ii) $\sum\limits_{i=1}^{p} \alpha_i x_i^2 - \sum\limits_{j=p+1}^{r} \alpha_j x_j^2 = 1$ *if rank(C) = rank(A)+1*,

(iii) $\sum\limits_{i=1}^{p} \alpha_i x_i^2 - \sum\limits_{j=p+1}^{r} \alpha_j x_j^2 - x_{r+1} = 0$ *if rank(C) = rank(A)+2*,

where $p \leq r \leq n$ and $0 \neq \alpha_k > 0$, for any $k = 1, \ldots, r$.

At first sight, we may remark that there exists an infinite number of classes of congruence for quadrics in $\mathbb{RE}^n$: any class is represented by one of the polynomials in Theorem 11.78 that are depending by the coefficients $\alpha_i \in \mathbb{R}$. On the contrary, the number of classes of affinity for quadrics in $\mathbb{RA}^n$ is finite, since it depends only on the number n of indeterminates and on the rank r of the matrix A.

Once again let us establish the natural connection between the results obtained earlier with the more familiar objects from analytic geometry: conic curves in the real plane and quadric surfaces in the real 3-dimensional space. In the light of Theorem 11.78, we have the following

Theorem 11.79 *Let $\mathbb{RE}^2$ be the 2-dimensional affine Euclidean space over the vector space $\mathbb{R}^2$. Any conic curve is metrically equivalent to one of the sets defined by the following equations:*

(i) $\frac{x^2}{a^2} + \frac{y^2}{b^2} - 1 = 0$, *real ellipse;*

(ii) $\frac{x^2}{a^2} - \frac{y^2}{b^2} - 1 = 0$, *hyperbola;*

(iii) $x^2 + ay = 0$, *parabola;*

(iv) $\frac{x^2}{a^2} + \frac{y^2}{b^2} + 1 = 0$, *empty set (imaginary ellipse);*

(v) $\frac{x^2}{a^2} - \frac{y^2}{b^2} = 0$, *two secant real lines;*

(vi) $\frac{x^2}{a^2} + \frac{y^2}{b^2} = 0$, *one real point (two complex conjugate lines);*

(vii) $\frac{x^2}{a^2} - 1 = 0$, *two real parallel lines;*

(viii) $\frac{x^2}{a^2} + 1 = 0$, *empty set (two complex parallel lines);*

 (ix) $x^2 = 0$, *two real merged lines.*

For quadric surfaces in real Euclidean space, we have the following.

Theorem 11.80 *Let* $\mathbb{RE}^3$ *be the 3-dimensional affine Euclidean space over the vector space* $\mathbb{R}^3$. *Any quadric surface is metrically equivalent to one of the sets defined by the following equations:*

   (i) $\frac{x^2}{a^2} + \frac{y^2}{b^2} + \frac{z^2}{c^2} - 1 = 0$, *real ellipsoid;*

  (ii) $\frac{x^2}{a^2} + \frac{y^2}{b^2} + \frac{z^2}{c^2} + 1 = 0$, *empty set (imaginary ellipsoid);*

 (iii) $\frac{x^2}{a^2} + \frac{y^2}{b^2} - \frac{z^2}{c^2} + 1 = 0$, *elliptic hyperboloid;*

 (iv) $\frac{x^2}{a^2} + \frac{y^2}{b^2} - \frac{z^2}{c^2} - 1 = 0$, *hyperbolic hyperboloid;*

  (v) $\frac{x^2}{a^2} + \frac{y^2}{b^2} - z = 0$, *elliptic paraboloid;*

 (vi) $\frac{x^2}{a^2} - \frac{y^2}{b^2} - z = 0$, *hyperbolic paraboloid;*

(vii) $\frac{x^2}{a^2} + \frac{y^2}{b^2} + \frac{z^2}{c^2} = 0$, *one real point (imaginary cone);*

(viii) $\frac{x^2}{a^2} + \frac{y^2}{b^2} - \frac{z^2}{c^2} = 0$, *real cone;*

 (ix) $\frac{x^2}{a^2} + \frac{y^2}{b^2} + 1 = 0$, *empty set (imaginary cylinder);*

  (x) $\frac{x^2}{a^2} + \frac{y^2}{b^2} - 1 = 0$, *elliptic cylinder;*

 (xi) $\frac{x^2}{a^2} - y = 0$, *parabolic cylinder;*

(xii) $\frac{x^2}{a^2} - \frac{y^2}{b^2} - 1 = 0$, *hyperbolic cylinder;*

(xiii) $\frac{x^2}{a^2} + \frac{y^2}{b^2} = 0$, *a line (two secant complex planes);*

(xiv) $\frac{x^2}{a^2} - \frac{y^2}{b^2} = 0$, *two secant real planes;*

 (xv) $\frac{x^2}{a^2} + 1 = 0$, *empty set (two complex parallel planes);*

(xvi) $\frac{x^2}{a^2} - 1 = 0$, *two real parallel planes;*

(xvii) $x^2 = 0$, *two merged real planes.*

11.6 Projective Classification of Conic Curves and Quadric Surfaces

We now aim at exploiting the algebraic properties of the representations of geometric entities, recalling the concept of homogeneous coordinates and projective spaces. As affine and Euclidean coordinates represent geometric entities in an Affine and a Euclidean space, respectively, homogeneous coordinates represent geometric elements in a projective space. Throughout any classical course of Geometry, the usefulness of homogeneous coordinates for constructions and transformations is usually well highlighted. The projective geometry substantially focuses on the study of properties which are invariant under the action of projective transformations: a characteristic feature of projective geometry is the symmetry of relationships between points and lines, called duality.

Nevertheless, we will restrict ourselves here to briefly mentioning the main definitions, which will contribute to achieving the primary objective of this chapter: the study of conic curves and quadric surfaces from a projective point of view.

Let $\mathbb{F}$ be a field and consider the vector space $\mathbb{F}^{n+1}$ of dimension $n + 1$. Let $0 \neq \mathbf{v} \in \mathbb{F}^{n+1}$ be a nonzero vector, then the set

$$\langle \mathbf{v} \rangle = \{ \lambda \mathbf{v} \mid \lambda \in \mathbb{F} \}$$

is called a *ray* of $\mathbb{F}^{n+1}$.

Definition 11.81 The *projective space* $\mathbb{FP}^n$, of dimension n, associated with $\mathbb{F}^{n+1}$, is the set of rays of $\mathbb{F}^{n+1}$. Any element of $\mathbb{FP}^n$ is called a *point*.

Therefore, a point of the projective space $\mathbb{FP}^n$ is represented by a vector of coordinates $X = [x_1, \ldots, x_{n+1}]^t \in \mathbb{F}^{n+1}$, where at least one of x_i's is not zero.

Definition 11.82 Let $X = [x_1, \ldots, x_{n+1}]^t \in \mathbb{F}^{n+1}$ be the coordinates representing a point $P \in \mathbb{FP}^n$. The scalars $\{x_1, \ldots, x_{n+1}\}$ are called *projective* (or *homogeneous*) *coordinates* of P.

Two vectors $X, Y \in \mathbb{F}^{n+1}$ represent the same point of $\mathbb{FP}^n$ when there exists $\lambda \in \mathbb{F} \setminus \{0\}$ such that $X = \lambda Y$. Hence, the projective coordinates of a point are defined up to a scale factor. This relates to the following.

Definition 11.83 Let $M_{n+1}(\mathbb{F})$ be the ring of $(n + 1) \times (n + 1)$ matrices over $\mathbb{F}$. Any nonsingular matrix $C \in M_{n+1}(\mathbb{F})$ defines a linear transformation $F : \mathbb{FP}^n \to \mathbb{FP}^n$ which is called *projective transformation* and transforms a point having projective coordinates X into a point having projective coordinates X' via $X' \approx CX$, where $\approx$ indicates equality up to a scale factor.

Hence, a projective transformation F of $\mathbb{FP}^n$ is an automorphism of $\mathbb{FP}^n$ induced by an automorphism of $\mathbb{F}^{n+1}$. The matrix representing this automorphism is not unique: two different matrices $A, B \in M_{n+1}(\mathbb{F})$ induce the same projective transformation if and only if there exists $\lambda \in \mathbb{F}$ such that $A = \lambda B$.

Definition 11.84 Let S and S' be two subsets of the projective space $\mathbb{FP}^n$. We say that S is *projectively equivalent* to S' if there exists a projective transformation $F : \mathbb{FP}^n \to \mathbb{FP}^n$ such that $F(S) = S'$.

Projective Coordinates in 2-Dimensional Projective Space

It is well known how to introduce a Cartesian coordinate system in the 2-dimensional affine plane $\mathbb{FA}^2$. Any point P of $\mathbb{FA}^2$ can be uniquely represented by the pair of scalars (x, y), called the *coordinates* of the point P in terms of the coordinate system OXY, we say $P \equiv (x, y)$. The origin itself has coordinates $O \equiv (0, 0)$.

If $P \equiv (x, y)$ is a point in the affine plane $\mathbb{F}A^2$, we may represent it in the Projective plane $\mathbb{F}P^2$ by its homogeneous coordinates. We simply add a third coordinate equal to 1. Thus, $P \equiv (x, y)$ is represented by its homogeneous coordinates $(x, y, 1)$.

More generally, the homogeneous coordinates of a point P are (x_1, x_2, x_3) iff $x = \frac{x_1}{x_3}$ and $y = \frac{x_2}{x_3}$ are its Euclidean coordinates. Thus, homogeneous coordinates are invariant up to scaling: (x_1, x_2, x_3) and $(\alpha x_1, \alpha x_2, \alpha x_3)$ represent the same point, for any $0 \neq \alpha \in \mathbb{F}$.

To represent a line in the projective plane, we start from the standard formula $ax + by + c = 0$ and introduce the homogeneous coordinates to arrive at the equation $ax_1 + bx_2 + cx_3 = 0$.

The coordinates $(x_1, x_2, 0)$ cannot represent any point of the form $(x, y, 1)$ as they don't share the same third coordinates. In fact, $(x_1, x_2, 0)$ is a point representing the slope of a line: parallel lines r and r' in $\mathbb{F}P^2$ meet at the line $x_3 = 0$ (line at infinity). More precisely, if $r : ax_1 + bx_2 + cx_3 = 0$ and $r' : ax_1 + bx_2 + c'x_3 = 0$, then their intersection point has homogeneous coordinates equal to $(b, -a, 0)$. This point is called *point at infinity* of r (and r') and represents the class of all parallel lines having the same direction.

Hence, the general idea is to let every couple of lines in the projective plane have an intersection point. By this approach, the projective plane can be defined as

$$\mathbb{F}P^2 = \mathbb{F}A^2 \cup \{\text{the set of all directions in } \mathbb{F}A^2\}.$$

Projective Classification of Conics in $\mathbb{F}P^2$

As above, to represent a conic in the projective plane, we start from the standard representing equation. Hence, we firstly recall that a conic is a locus in $\mathbb{F}A^2$ ($\mathbb{F} = \mathbb{R}$ or $\mathbb{F} = \mathbb{C}$) consisting of all points whose coordinates are solution of a quadratic equation of the form $f(x, y) = 0$, where

$$f(x, y) = a_{11}x^2 + 2a_{12}xy + a_{22}y^2 + 2a_{13}x + 2a_{23}y + a_{33} \qquad (11.66)$$

and $a_{ij} \in \mathbb{F}$, for any $i, j = 1, 2, 3$. Introducing the homogeneous coordinates, we define a conic Γ as a locus in the Projective plane $\mathbb{F}P^2$ consisting of all points whose homogeneous coordinates are solution of a quadratic equation of the form $f(x_1, x_2, x_3) = 0$, where

$$f(x_1, x_2, x_3) = a_{11}x_1^2 + 2a_{12}x_1x_2 + a_{22}x_2^2 + 2a_{13}x_1x_3 + 2a_{23}x_2x_3 + a_{33}x_3^2$$
$$(11.67)$$

and $a_{ij} \in \mathbb{F}$, for any $i, j = 1, 2, 3$.

The quadratic polynomial $f(x_1, x_2, x_3)$ can be represented by the symmetric matrix

$$A = \begin{bmatrix} a_{11} & a_{12} & a_{13} \\ a_{12} & a_{22} & a_{23} \\ a_{13} & a_{23} & a_{33} \end{bmatrix}$$

so that

$$f(x_1, x_2, x_3) = \begin{bmatrix} x_1 & x_2 & x_3 \end{bmatrix} \begin{bmatrix} a_{11} & a_{12} & a_{13} \\ a_{12} & a_{22} & a_{23} \\ a_{13} & a_{23} & a_{33} \end{bmatrix} \begin{bmatrix} x_1 \\ x_2 \\ x_3 \end{bmatrix}.$$

Hence, if we denote $X = \begin{bmatrix} x_1 & x_2 & x_3 \end{bmatrix}^t$, then the equation $f(x_1, x_2, x_3) = 0$ can be written in the following matrix notation: $X^t A X = 0$.

We say that A is the matrix associated with the conic Γ, or also that the conic Γ is determined by the matrix A.

Following Definition 11.84, we say that two conics Γ and Γ' of the projective space $\mathbb{FP}^2$ are *projectively equivalent* if there exists a projective transformation $F : \mathbb{FP}^2 \to \mathbb{FP}^2$ such that $F(\Gamma) = \Gamma'$.

Theorem 11.85 *Let Γ be a conic of the projective space $\mathbb{FP}^2$, determined by the 3×3 matrix A with coefficients in $\mathbb{F}$. If $F : \mathbb{FP}^2 \to \mathbb{FP}^2$ is a projective transformation and C is the matrix representing F, then $F(\Gamma)$ is a conic determined by the matrix $(C^{-1})^t A C^{-1}$.*

Proof By the definition of projective transformations, we have that $F(X) = CX$, for any $X \in \mathbb{FP}^2$. We now denote by Γ' the conic determined by the matrix $(C^{-1})^t A C^{-1}$ and prove that $\Gamma' = F(\Gamma)$.

Firstly, we notice that if $X \in \Gamma$, then obviously $C^{-1} C X \in \Gamma$. Hence,

$$(C^{-1} C X)^t A (C^{-1} C X) = 0,$$

that is,

$$(CX)^t \left((C^{-1})^t A C^{-1} \right) CX = 0,$$

therefore, $CX = F(X) \in \Gamma'$, so that $F(\Gamma) \subseteq \Gamma'$. Similarly, by using the fact that $F^{-1}(X) = C^{-1} X$, for any $X \in \mathbb{FP}^2$, one may prove that $F^{-1}(\Gamma') \subseteq \Gamma$, concluding that $\Gamma' = F(\Gamma)$, as required.

As a consequence of the previous theorem, we have the following.

Theorem 11.86 *Let Γ and Γ' be two projective conics, represented by the symmetric matrices $A \in M_3(\mathbb{F})$ and $A' \in M_3(\mathbb{F})$, respectively. Then Γ is projectively equivalent to Γ' if and only if there exists a nonsingular matrix $C \in M_3(\mathbb{F})$ such that $A' = C^t A C$.*

Proof Firstly, we assume that Γ is projectively equivalent to Γ', that is, there exists a projective transformation $F : \mathbb{FP}^2 \to \mathbb{FP}^2$ such that $F(\Gamma) = \Gamma'$. Let $C \in M_3(\mathbb{F})$ be the nonsingular matrix associated with F. Hence, $CX_0 \in \Gamma'$, for any point $X_0 \in \Gamma$. The replacement of X by CX in the matrix notation $X^t A X = 0$ of Γ leads to the relation $(CX)^t A (CX) = 0$, that is, $X^t (C^t A C) X = 0$, which represents Γ'. Therefore, $A' = C^t A C$ as required.

Conversely, let C be any nonsingular 3×3 matrix with coefficients in $\mathbb{F}$, and let $F : \mathbb{FP}^2 \to \mathbb{FP}^2$ be the projective transformation represented by the matrix C^{-1}. By Theorem 11.85, the conic $F(\Gamma)$ is determined by the matrix $C^t A C$. Hence, $F(\Gamma) = \Gamma'$, that is, Γ is projectively equivalent to Γ'.

In other words, two projective conics Γ and Γ' are projectively equivalent if and only if their associated matrices are congruent.

Definition 11.87 Let Γ be a conic of $\mathbb{FP}^2$, represented by the matrix notation $X^t A X = 0$, where $A \in M_3(\mathbb{F})$. The rank of the matrix A is also called the *rank* of Γ.

In light of Definition 11.87 and Lemma 7.50, we are now able to state the following.

Theorem 11.88 *If two conics Γ and Γ' of $\mathbb{FP}^2$ are projectively equivalent, then they have the same rank.*

In particular,

Theorem 11.89 *Let $\mathbb{F}$ be an algebraically closed field. Two conics Γ and Γ' of $\mathbb{FP}^2$ are projectively equivalent if and only if they have the same rank.*

The fact that the matrix associated with a conic is symmetric makes it possible for us to transform it to a diagonal matrix. In this way, we determine the number of congruence classes for matrices in $M_3(\mathbb{F})$, which correspond to congruence classes of conics. Any class is represented by a diagonal matrix notation, usually called *projective canonical form of a conic*, such that any conic of $\mathbb{FP}^2$ is projectively equivalent to one (and only one) of them. To do this, we refer to the results contained in Theorems 7.46 and 7.47, devoted to the description of the diagonal forms of quadratic functions. More precisely, in the case $\mathbb{F}$ is algebraically closed and using Theorem 7.46, we have the following.

Theorem 11.90 *Let $\mathbb{F}$ be an algebraically closed field. Any conic Γ of $\mathbb{FP}^2$ is projectively equivalent to one (and only one) of the following:*

(i) $x_1^2 + x_2^2 + x_3^2 = 0$ *(non-degenerate or ordinary conic);*

(ii) $x_1^2 + x_2^2 = 0$ *(degenerate conic of rank 2);*

(iii) $x_1^2 = 0$ *(degenerate conic of rank 1).*

Proof By Theorem 11.86, any projective transformation associated with a nonsingular matrix $C \in M_3(\mathbb{F})$ transforms Γ to the projectively equivalent Γ'. In particular, if A is the matrix representing Γ, then $C^t A C$ is the matrix representing Γ'. Hence, as an application of Theorem 7.46, it follows that there exists an appropriate $C \in M_3(\mathbb{F})$ such that $C^t A C$ has one of the following forms:

$$\begin{bmatrix} 1 & 0 & 0 \\ 0 & 1 & 0 \\ 0 & 0 & 1 \end{bmatrix}, \quad \begin{bmatrix} 1 & 0 & 0 \\ 0 & 1 & 0 \\ 0 & 0 & 0 \end{bmatrix}, \quad \begin{bmatrix} 1 & 0 & 0 \\ 0 & 0 & 0 \\ 0 & 0 & 0 \end{bmatrix}.$$

Therefore, in case $\mathbb{F}$ is algebraically closed, there exist precisely 3 classes of congruence, each of which is determined by the rank of the belonging conics.

If $\mathbb{F}$ is not algebraically closed, the congruence classes are 5.

Theorem 11.91 *Any conic Γ of $\mathbb{RP}^2$ is projectively equivalent to one (and only one) of the following:*

(i) $x_1^2 + x_2^2 - x_3^2 = 0$ (non-degenerate or ordinary conic);

(ii) $x_1^2 + x_2^2 + x_3^2 = 0$ (non-degenerate or ordinary conic, containing no real points);

(iii) $x_1^2 - x_2^2 = 0$ (degenerate conic of rank 2);

(iv) $x_1^2 + x_2^2 = 0$ (degenerate conic of rank 2);

(v) $x_1^2 = 0$ (degenerate conic of rank 1).

Proof By using the same argument as in Theorem 11.90 and applying Theorem 7.47, one has that any matrix representing a conic of $\mathbb{RP}^2$ is congruent to one of the following:

$$\begin{bmatrix} 1 & 0 & 0 \\ 0 & 1 & 0 \\ 0 & 0 & -1 \end{bmatrix}, \begin{bmatrix} 1 & 0 & 0 \\ 0 & 1 & 0 \\ 0 & 0 & 1 \end{bmatrix}, \begin{bmatrix} 1 & 0 & 0 \\ 0 & -1 & 0 \\ 0 & 0 & 0 \end{bmatrix}, \begin{bmatrix} 1 & 0 & 0 \\ 0 & 1 & 0 \\ 0 & 0 & 0 \end{bmatrix}, \begin{bmatrix} 1 & 0 & 0 \\ 0 & 0 & 0 \\ 0 & 0 & 0 \end{bmatrix}.$$

Remark 11.92 We notice that, in case $\mathbb{F}$ is algebraically closed, a degenerate conic of $\mathbb{FP}^2$ having rank 2 is projectively equivalent to the union of two distinct lines. In fact, the polynomial defining the conic factors into a product of linear polynomials.

If $\mathbb{F} = \mathbb{R}$ and Γ is a degenerate conic having rank 2, we have two different cases: the congruence class $x_1^2 - x_2^2 = 0$, that is again the union of two distinct lines; and the congruence class $x_1^2 + x_2^2 = 0$ which represents a real point (intersection of two conjugate complex lines).

When Γ is a degenerate conic having rank 1, there is no difference between the case $\mathbb{F} = \mathbb{R}$ and $\mathbb{F}$ is algebraically closed: in any case, the conic is projectively equivalent to two superposed lines, that is, the polynomial defining the curve is a square of a linear polynomial.

Projective Classification of Quadrics in $\mathbb{FP}^3$

Before proceeding with the description of the quadric surfaces in $\mathbb{FP}^3$, we need to recall the definition of the projective coordinates in 3-dimensional projective space.

As in the 2-dimensional Euclidean space, any point P of the 3-dimensional Euclidean Space $\mathbb{FE}^3$ can be uniquely represented by the triplet of scalars (x, y, z), called the *coordinates* of the point P with respect to the coordinate system $OXYZ$, we say $P \equiv (x, y, z)$. The origin itself has coordinates $O \equiv (0, 0, 0)$. As above we recall that if $P \equiv (x, y, z)$ is a point of $\mathbb{FE}^3$, we may represent it in the projective space $\mathbb{FP}^3$ by its homogeneous coordinates, adding a fourth coordinate equal to 1. Thus, $P \equiv (x, y, z)$ is represented by its homogeneous coordinates $(x, y, z, 1)$.

The homogeneous coordinates of a point P are (x_1, x_2, x_3, x_4) iff $x = \frac{x_1}{x_4}$, $y = \frac{x_2}{x_4}$ and $z = \frac{x_3}{x_4}$ are its Euclidean coordinates. Since homogeneous coordinates are invariant up to scaling, then (x_1, x_2, x_3, x_4) and $(\alpha x_1, \alpha x_2, \alpha x_3, \alpha x_4)$ represent the same point, for any $0 \neq \alpha \in \mathbb{R}$.

To represent a plane in the projective space $\mathbb{FP}^3$, we homogenize the general equation for planes $ax + by + cz + d = 0$ by introducing the homogeneous coordinates, so that each term has the same degree. Hence, the general equation of a plane in $\mathbb{FE}^3$ has the form $ax_1 + bx_2 + cx_3 + dx_4 = 0$. Therefore, in order to represent a line r in $\mathbb{FE}^3$, we just recall that it is the intersection of two non-parallel planes $\pi : ax_1 + bx_2 + cx_3 + dx_4 = 0$ and $\pi' : a'x_1 + b'x_2 + c'x_3 + d'x_4 = 0$, thus the line r can be described as

$$r : \quad ax_1 + bx_2 + cx_3 + dx_4 = 0, \quad a'x_1 + b'x_2 + c'x_3 + d'x_4 = 0.$$

Let now P be a point having projective coordinates $(l, m, n, 0)$. The first three coordinates (l, m, n) represent the direction vector of a line: parallel lines r and r' in $\mathbb{FP}^3$ meet at the plane $x_4 = 0$ (plane at infinity). More precisely, if r and r' are parallel lines in $\mathbb{FP}^3$, then their intersection point has homogeneous coordinates equal to $(l, m, n, 0)$, where (l, m, n) are the direction vector of r and r'. The point $(l, m, n, 0)$ is called *point at infinity* of r (and r') and represents the class of all parallel lines having the same direction vector.

To represent a quadric of $\mathbb{FP}^3$, we start from the standard representing equation. A quadric is a locus in $\mathbb{F}^3$ ($\mathbb{F} = \mathbb{R}$ or $\mathbb{F} = \mathbb{C}$) consisting of all points whose coordinates are a solution of a quadratic equation of the form $f(x, y, z) = 0$, where

$$f(x, y, z) = a_{11}x^2 + 2a_{12}xy + 2a_{13}xz + 2a_{14}x + a_{22}y^2$$

$$+2a_{23}yz + 2a_{24}y + a_{33}z^2 + 2a_{34}z + a_{44}$$

and $a_{ij} \in \mathbb{F}$, for any $i, j = 1, 2, 3, 4$. Introducing the homogeneous coordinates, we define a quadric Σ as a locus in $\mathbb{FP}^3$ consisting of all points whose homogeneous coordinates are solution of a quadratic equation of the form $f(x_1, x_2, x_3, x_4) = 0$, where

$$f(x_1, x_2, x_3, x_4) = a_{11}x_1^2 + 2a_{12}x_1x_2 + 2a_{13}x_1x_3 + 2a_{14}x_1x_4 + a_{22}x_2^2$$

$$+2a_{23}x_2x_3 + 2a_{24}x_2x_4 + a_{33}x_3^2 + 2a_{34}x_3x_4 + a_{44}x_4^2,$$

and $a_{ij} \in \mathbb{F}$, for any $i, j = 1, 2, 3, 4$.

Hence, the quadratic polynomial $f(x_1, x_2, x_3, x_4)$ can be determined by the symmetric matrix

$$A = \begin{bmatrix} a_{11} & a_{12} & a_{13} & a_{14} \\ a_{12} & a_{22} & a_{23} & a_{24} \\ a_{13} & a_{23} & a_{33} & a_{34} \\ a_{14} & a_{24} & a_{34} & a_{44} \end{bmatrix}.$$

Therefore, if we denote $X = \begin{bmatrix} x_1 & x_2 & x_3 & x_4 \end{bmatrix}^t$, we have that $f(x_1, x_2, x_3, x_4) = X^t A X$. In this way, we obtain the following matrix notation for the equation of a quadric: $X^t A X = 0$.

We say that A is the matrix associated with the quadric Σ, or also that the quadric Σ is determined by the matrix A.

Definition 11.93 Let Σ and Σ' be two quadrics of the projective space $\mathbb{FP}^3$. We say that Σ is *projectively equivalent* to Σ' if there exists a projective transformation $F : \mathbb{FP}^3 \to \mathbb{FP}^3$ such that $F(\Sigma) = \Sigma'$.

At this point, we may remind the arguments previously developed in the section devoted to the classification of conics in the projective space $\mathbb{FP}^2$. Following the same line we are able to state the following.

Theorem 11.94 *Let Σ be a quadric of the projective space $\mathbb{FP}^3$, determined by the 4×4 matrix A with coefficients in $\mathbb{F}$. If $F : \mathbb{FP}^3 \to \mathbb{FP}^3$ is a projective transformation and C is the matrix representing F, then $F(\Sigma)$ is a quadric determined by the matrix $(C^{-1})^t A C^{-1}$.*

Proof See the proof of Theorem 11.85.

Theorem 11.95 *Let Σ and Σ' be two projective quadrics, represented by the symmetric matrices $A \in M_4(\mathbb{F})$ and $A' \in M_4(\mathbb{F})$, respectively. Then Σ is projectively equivalent to Σ' if and only if there exists a nonsingular matrix $C \in M_4(\mathbb{F})$ such that $A' = C^t A C$.*

Proof See the proof of Theorem 11.86.

Hence, two projective quadrics Σ and Σ' are projectively equivalent if and only if their associated matrices are congruent.

Definition 11.96 Let Σ be a quadric of $\mathbb{FP}^3$, represented by the matrix notation $X^t A X = 0$, where $A \in M_4(\mathbb{F})$. The rank of the matrix A is also called the *rank* of Σ.

Moreover:

Theorem 11.97 *If two quadrics of $\mathbb{FP}^3$ are projectively equivalent, then they have the same rank. In particular, two quadrics of $\mathbb{CP}^3$ are projectively equivalent if and only if they have the same rank. ($\mathbb{C}$ is an algebraically closed field.)*

Therefore, also in the case of classification of quadrics, we may determine a number of congruence classes for matrices in $M_4(\mathbb{F})$, which correspond to congruence classes of quadrics. The classes are usually called *projective canonical forms of a quadric*, such that any quadric of $\mathbb{FP}^3$ is projectively equivalent to one (and only one) of them. By using again Theorems 7.46 and 7.47, we have the following.

Theorem 11.98 *Let $\mathbb{F}$ be an algebraically closed field. Any quadric of $\mathbb{FP}^3$ is projectively equivalent to one (and only one) of the following:*

(i) $x_1^2 + x_2^2 + x_3^2 + x_4^2 = 0$ *(non-degenerate or ordinary quadric);*
(ii) $x_1^2 + x_2^2 + x_3^2 = 0$ *(degenerate quadric of rank 3);*
(iii) $x_1^2 + x_2^2 = 0$ *(degenerate quadric of rank 2);*
(iv) $x_1^2 = 0$ *(degenerate quadric of rank 1).*

Proof By Theorem 11.95, any projective transformation associated with a nonsingular matrix $C \in M_4(\mathbb{F})$ transforms Σ to the projectively equivalent Σ'. In particular, if A is the matrix representing Σ, then $C^t A C$ is the matrix representing Σ'. By Theorem 7.46, it follows that there exists an appropriate $C \in M_4(\mathbb{F})$ such that $C^t A C$ has one of the following forms:

$$\begin{bmatrix} 1&0&0&0 \\ 0&1&0&0 \\ 0&0&1&0 \\ 0&0&0&1 \end{bmatrix}, \begin{bmatrix} 1&0&0&0 \\ 0&1&0&0 \\ 0&0&1&0 \\ 0&0&0&0 \end{bmatrix}, \begin{bmatrix} 1&0&0&0 \\ 0&1&0&0 \\ 0&0&0&0 \\ 0&0&0&0 \end{bmatrix}, \begin{bmatrix} 1&0&0&0 \\ 0&0&0&0 \\ 0&0&0&0 \\ 0&0&0&0 \end{bmatrix}.$$

Theorem 11.99 *Any quadric Σ of $\mathbb{RP}^3$ is projectively equivalent to one (and only one) of the following:*

(i) $x_1^2 + x_2^2 + x_3^2 + x_4^2 = 0$ *(non-degenerate or ordinary quadric, containing no real points);*
(ii) $x_1^2 + x_2^2 + x_3^2 - x_4^2 = 0$ *(non-degenerate or ordinary quadric);*
(iii) $x_1^2 + x_2^2 - x_3^2 - x_4^2 = 0$ *(non-degenerate or ordinary quadric);*
(iv) $x_1^2 + x_2^2 + x_3^2 = 0$ *(degenerate quadric of rank 3, containing only one real point);*
(v) $x_1^2 + x_2^2 - x_3^2 = 0$ *(degenerate quadric of rank 3);*
(vi) $x_1^2 + x_2^2 = 0$ *(degenerate quadric of rank 2);*
(vii) $x_1^2 - x_2^2 = 0$ *(degenerate quadric of rank 2);*
(viii) $x_1^2 = 0$ *(degenerate quadric of rank 1).*

Proof As in Theorem 11.98 and applying Theorem 7.47, one has that any matrix representing a quadric of $\mathbb{RP}^3$ is congruent to one of the following:

$$\begin{bmatrix} 1&0&0&0 \\ 0&1&0&0 \\ 0&0&1&0 \\ 0&0&0&1 \end{bmatrix}, \begin{bmatrix} 1&0&0&0 \\ 0&1&0&0 \\ 0&0&1&0 \\ 0&0&0&-1 \end{bmatrix}, \begin{bmatrix} 1&0&0&0 \\ 0&1&0&0 \\ 0&0&-1&0 \\ 0&0&0&-1 \end{bmatrix}, \begin{bmatrix} 1&0&0&0 \\ 0&1&0&0 \\ 0&0&1&0 \\ 0&0&0&0 \end{bmatrix},$$

$$\begin{bmatrix} 1&0&0&0 \\ 0&1&0&0 \\ 0&0&-1&0 \\ 0&0&0&0 \end{bmatrix}, \begin{bmatrix} 1&0&0&0 \\ 0&1&0&0 \\ 0&0&0&0 \\ 0&0&0&0 \end{bmatrix}, \begin{bmatrix} 1&0&0&0 \\ 0&-1&0&0 \\ 0&0&0&0 \\ 0&0&0&0 \end{bmatrix}, \begin{bmatrix} 1&0&0&0 \\ 0&0&0&0 \\ 0&0&0&0 \\ 0&0&0&0 \end{bmatrix}.$$

Example 11.100 Let $\mathcal{Q}$ be the projective quadric of $\mathbb{FP}^3$ having equation $x_1^2 + x_2^2 - x_3^2 + x_1 x_2 + x_1 x_4 - 3x_4^2 = 0$. The matrix associated with $\mathcal{Q}$ is

$$A = \begin{bmatrix} 1 & \frac{1}{2} & 0 & \frac{1}{2} \\ \frac{1}{2} & 1 & 0 & 0 \\ 0 & 0 & -1 & 0 \\ \frac{1}{2} & 0 & 0 & -3 \end{bmatrix}.$$

By performing the standard process for diagonalization of quadratic forms, we may transform the matrix A into a diagonal matrix, which is congruent with A. After computations, we see that two steps are needed for diagonalization of A.

In the first step, the transition matrix needed to annihilate the entries on the first row and column in A is

$$E_1 = \begin{bmatrix} 1 & -\frac{1}{2} & 0 & -\frac{1}{2} \\ 0 & 1 & 0 & 0 \\ 0 & 0 & 1 & 0 \\ 0 & 0 & 0 & 1 \end{bmatrix}$$

so that

$$A' = E_1^t A E_1 = \begin{bmatrix} 1 & 0 & 0 & 0 \\ 0 & \frac{3}{4} & 0 & -\frac{1}{4} \\ 0 & 0 & -1 & 0 \\ 0 & -\frac{1}{4} & 0 & -\frac{13}{4} \end{bmatrix}.$$

Starting from this last one, the second transition of basis is represented by the matrix

$$E_2 = \begin{bmatrix} 1 & 0 & 0 & 0 \\ 0 & 1 & 0 & \frac{1}{3} \\ 0 & 0 & 1 & 0 \\ 0 & 0 & 0 & 1 \end{bmatrix}$$

and we obtain

$$A'' = E_2^t A' E_2 = \begin{bmatrix} 1 & 0 & 0 & 0 \\ 0 & \frac{3}{4} & 0 & 0 \\ 0 & 0 & -1 & 0 \\ 0 & 0 & 0 & -\frac{10}{3} \end{bmatrix}.$$

In case $\mathbb{F} = \mathbb{R}$ and by substitutions

$$x_1 = X_1$$

$$x_2 = \frac{2}{\sqrt{3}} X_2$$

$$x_3 = X_3$$

$$x_4 = \frac{\sqrt{3}}{\sqrt{10}} X_4,$$

we get the projective canonical equation $X_1^2 + X_2^2 - X_3^2 - X_4^2 = 0$.
   If $\mathbb{F} = \mathbb{C}$, by substitutions

$$x_1 = X_1$$

$$x_2 = \frac{2}{\sqrt{3}} X_2$$

$$x_3 = i X_3$$

$$x_4 = \frac{i\sqrt{3}}{\sqrt{10}} X_4,$$

we get the projective canonical equation $X_1^2 + X_2^2 + X_3^2 + X_4^2 = 0$.

Exercises

1. In the affine space $\mathbb{R}A^2$, consider the transformation $f : \mathbb{R}A^2 \rightarrow \mathbb{R}A^2$ defined by

$$f(x, y) = \left(-\frac{1}{4}x - \frac{\sqrt{3}}{4}y + \frac{3}{4}, \frac{\sqrt{3}}{4}x - \frac{1}{4}y + \sqrt{34} \right).$$

   Prove that f is an affinity on $\mathbb{R}A^2$ and describe geometrically what its action is. (translation, rotation, reflection, a composition of some of them, etc.)
2. In the affine space $\mathbb{R}A^2$, consider the transformation $f : \mathbb{R}A^2 \rightarrow \mathbb{R}A^2$ such that

$$f(0, 0) = (2, 0) \quad f(2, 0) = (4, 0) \quad f(1, 1) = (3, -1)$$

   where coordinates of points are referred to the standard frame of reference. Prove that f is an affinity on $\mathbb{R}A^2$, determine its representation and describe geometrically what its action is.
3. Represent the reflection in $\mathbb{R}E^4$ across the hyperplane of equation $x_1 + 2x_2 - x_3 + x_4 + 1 = 0$.
4. Given a hyperbola Γ in space $\mathbb{R}E^2$, prove that there exists an affine transformation mapping Γ to the hyperbola represented by equation $xy = 1$.
5. Let S_1, S_2 be two ordered sets of $\mathbb{R}A^2$ consisting of 3 non-collinear points each. Prove that there exists a unique affine transformation $f : \mathbb{R}A^2 \rightarrow \mathbb{R}A^2$ such that $f(S_1) = S_2$.
6. Determine the affine and metric classification of the following quadrics and the transformations needed to obtain them, in both cases $\mathbb{F} = \mathbb{R}$ and $\mathbb{F} = \mathbb{C}$:

   (a) $2x^2 + y^2 + z^2 - x - 2y + 1 = 0$ in $\mathbb{F}A^3$ and $\mathbb{F}E^3$.
   (b) $x^2 + y^2 + 2xy + 2xz + 2yz - 2x + 2 = 0$ in $\mathbb{F}A^4$ and $\mathbb{F}E^4$.

(c) $2x^2 - 8xy + 2y^2 - 2x + 1 = 0$ in $\mathbb{FA}^2$ and $\mathbb{FE}^2$.

(d) $x^2 + y^2 + 2xy + 2yz - 2x + y + 1 = 0$ in $\mathbb{FA}^3$ and $\mathbb{FE}^3$.

(e) $x^2 + y^2 + 2xy - 2x + y + 1 = 0$ in $\mathbb{FA}^4$ and $\mathbb{FE}^4$.

7. Determine the projective classification of the following quadrics, in both cases $\mathbb{F} = \mathbb{R}$ and $\mathbb{F} = \mathbb{C}$:

(a) $2x_1^2 + x_2^2 + x_3^2 - x_1x_4 - 2x_2x_4 + x_4^2 = 0$ in $\mathbb{FP}^3$.

(b) $x_1^2 - 2x_1x_2 + x_2^2 - x_1x_3 = 0$ in $\mathbb{FP}^2$.

(c) $x_1^2 - x_2^2 + 2x_1x_2 + 2x_1x_3 + 4x_2x_3 - 2x_1x_4 + x_4^2 = 0$ in $\mathbb{FP}^3$.

(d) $x_1^2 + x_2^2 + x_1x_2 + 2x_2x_3 - 4x_1x_4 + 2x_2x_4 + x_4^2 = 0$ in $\mathbb{FP}^3$.

(e) $3x_1^2 - 2x_1x_2 - 2x_1x_4 + x_2x_4 + x_3^2 + 2x_4^2 = 0$ in $\mathbb{FP}^3$.

Chapter 12
Ordinary Differential Equations and Linear Systems of Ordinary Differential Equations

In this chapter, we provide a method for solving systems of linear ordinary differential equations by using techniques associated with the calculation of eigenvalues, eigenvectors and generalized eigenvectors of matrices. We learn in calculus how to solve differential equations and the system of differential equations. Here, we firstly show how to represent a system of differential equations in a matrix formulation. Then, using the Jordan canonical form and, whenever possible, the diagonal canonical form of matrices, we will describe a process aimed at solving systems of linear differential equations in a very efficient way. To do this, we also give a short description of the so-called vector-valued functions. Finally, as a further application, in the last part of the chapter, we show that the method linked with the solution of systems also supplies a way of dealing with the problem of resolution of differential equations of order n.

12.1 A Brief Overview of Basic Concepts of Ordinary Differential Equations

A differential equation is an equation which involves one or more independent variables, one or more dependent variables and derivatives of the dependent variables with respect to some or all of the independent variables. If there is just one independent variable, then the derivatives are all ordinary derivatives, and the equation is an *ordinary differential equation*. Hence, an *ordinary differential equation* is an equation relating to one unknown function $y(x)$ of one variable x and some of its derivatives. The domain of the unknown function is an interval of the real line. Such an equation is then of the form

$$F(x, y(x), y'(x), \ldots, y^{(n)}(x)) = g(x), \tag{12.1}$$

© The Author(s), under exclusive license to Springer Nature Singapore Pte Ltd. 2022
M. Ashraf et al., *Advanced Linear Algebra with Applications*,
https://doi.org/10.1007/978-981-16-2167-3_12

where x is a scalar parameter, $g(x)$ is a function of the variable x which is defined and continuous in an interval $I \subseteq \mathbb{R}$, $y(x)$ is the unknown function of x, $y'(x)$ is the first derivative of $y(x)$ and, for any i, $y^{(i)}$ is the ith derivative of $y(x)$. We recall that in literature, the functions $y(x), y'(x), \ldots, y^{(n)}(x)$ of the variable x, that are involved in an ordinary differential equation, are commonly replaced by the $y, y', \ldots, y^{(n)}$ so that Eq. (12.1) can be written as

$$F(x, y, y', \ldots, y^{(n)}) = g(x).$$

We say that the differential equation is *homogeneous* if the function $g(x)$ is just the constant function 0, that is, the differential equation is precisely

$$F(x, y(x), y'(x), \ldots, y^{(n)}(x)) = 0. \tag{12.2}$$

The *order* of a differential equation is the order of the highest derivative which appears in the equation. A *solution* (or *integral*) of an ordinary differential equation of order n of the form (12.1) is a function $y(x)$ defined in the interval $I \subseteq \mathbb{R}$ and satisfying the following:

(1) It is n-times differentiable in the domain of definition I.
(2) $F(x_0, y(x_0), y'(x_0), \ldots, y^{(n)}(x_0)) = g(x_0)$, for any $x_0 \in I$.

Example 12.1 Any solution of the first-order differential equation $y' = ay$, where $a \in \mathbb{R}$ is a fixed constant, has the form $y(x) = ce^{ax}$ ($c \in \mathbb{R}$). In fact, assuming that $y(x)$ is a solution, we notice that the product $y(x)e^{-ax}$ has derivative equal to zero; hence, it is a constant, say $y(x)e^{-ax} = c$, for some $c \in \mathbb{R}$. Moreover, the function ce^{ax} is differentiable in the whole $\mathbb{R}$.

Example 12.2 Any solution of the second-order differential equation $y'' = a^2 y$, where $a \in \mathbb{R}$ is a fixed constant, has the form $y(x) = c_1 e^{ax} + c_2 e^{-ax}$ ($c_1, c_2 \in \mathbb{R}$). In fact,

$$y'(x) = ac_1 e^{ax} - ac_2 e^{-ax} \quad \text{and} \quad y''(x) = a^2 c_1 e^{ax} + a^2 c_2 e^{-ax} = a^2 y(x).$$

Moreover, the function $c_1 e^{ax} + c_2 e^{-ax}$ is two times differentiable in the whole $\mathbb{R}$.

Example 12.3 Any solution of the first-order differential equation $y' = ay + be^{ax}$, where $a, b \in \mathbb{R}$ are fixed constants, has the form $y(x) = ce^{ax} + bxe^{ax}$ ($c \in \mathbb{R}$). More in general, any solution of the differential equation $y' = f(x)y + g(x)$ has the form

$$y(x) = e^{\int f(x)dx} \left\{ \int g(x)e^{-\int f(x)dx} dx + c \right\} \quad \text{for any constant} \quad c \in \mathbb{R}.$$

Note that the constants of integration are included in each of the previous examples. They are always included in the general solution of differential equations. This means that the solution of a differential identity represents a family of infinite curves (∞^1

and ∞^2 curves for Examples 12.1 and 12.2, respectively). More in general, the solution of a differential equation of order n is a function depending on n arbitrary constants and represents a family of ∞^n curves. The set of solutions

$$y = y(x, c_1, \ldots, c_n) \quad \text{for any} \quad c_1, \ldots, c_n \in \mathbb{R}$$

is called the *general solution* of the differential equation. In speaking of ordinary differential equations, we say that we have an *initial value problem* if all the specified values of the solution and its derivatives are given at one point of the domain of definition I. Any standard text on differential equations discusses the *initial value problem*, that is, whether we can construct the unique function $y(x)$, among those representing the general solution, such that

$$F(x, y, y', \ldots, y^{(n)}) = g(x) \quad \text{and} \quad \begin{cases} y(x_0) = y_0 \\ y'(x_0) = y_1 \\ \quad \cdots \\ \quad \cdots \\ y^{(n-1)}(x_0) = y_{n-1} \end{cases} \tag{12.3}$$

for a specific element $x_0 \in I$, the domain of definition of $y(x)$. This procedure is in line with the assignment of specific values to the arbitrary constants $c_1, \ldots, c_n$. The function satisfying (12.3) is usually said to be a *particular solution* of the differential equation.

Example 12.4 The general solution of the second-order differential equation $2yy' - xy'^2 - 4x = 0$ has the form $y(x) = \frac{x^2 + c^2}{c}$ ($0 \neq c \in \mathbb{R}$).
If we consider the following initial value problem:

$$2yy' - xy'^2 - 4x = 0 \quad \text{and} \quad y(1) = 2,$$

we get the particular solution correlating to the value of the constant c which satisfies

$$2 = y(1) \implies \frac{1^2 + c^2}{c} = 2 \implies (c - 1)^2 = 0 \implies c = 1, 1,$$

that is, $c = 1$ and $y(x) = x^2 + 1$.

However, sometimes, solutions of the differential equation that cannot be regarded as part of general solution exist. The characteristic of any such solution (usually called *singular solution*) is such that it is not derived from the general solution by assigning particular values to the constants.

Example 12.5 Let us return to Example 12.4. Notice that both $y = 2x$ and $y = -2x$ are solutions of the second-order differential equation $2yy' - xy'^2 - 4x = 0$. Nevertheless, none of them can be deducted by the general solution by assigning particular values to the constant c.

Example 12.6 The general solution of the first-order differential equation $2y' - (y - 1)^2 x = 0$ has the form $y(x) = -\frac{2}{x^2+c} + 1$ ($c \in \mathbb{R}$).
Even if $y = 1$ is a solution of the equation, it cannot be derived by the general one.

12.2 System of Linear Homogeneous Ordinary Differential Equations

An ordinary differential equation of the general form

$$F(x, y(x), y'(x), \ldots, y^{(n)}(x)) = g(x) \tag{12.4}$$

is *linear* if it is linear in the unknown function $y(x)$ and its derivatives.

If $g(x)$ is not the zero function, the equation is said to be *nonhomogeneous*. In this case, the equation

$$F(x, y(x), y'(x), \ldots, y^{(n)}(x)) = 0 \tag{12.5}$$

is called the *associated homogeneous equation*. It is known that, if $s(x)$ is the general solution of the associated homogeneous equation (12.5), ($s(x)$ is usually called a *complementary solution*) and $s_0(x)$ is any solution of the nonhomogeneous equation (12.4), then $y(x) = cs(x) + s_0(x)$ is a general solution of the nonhomogeneous equation, for an arbitrary constant c. In other words, the general solution to (12.4) is, thus, obtained by adding all possible homogeneous solutions to one fixed particular solution.

From this family, we can select one which satisfies the initial condition $y(x_0) = y_0$. In any course on ordinary differential equations, all of us probably encountered the general method known as *variation of parameters* for constructing particular solutions of nonhomogeneous ordinary differential equations with constant coefficients. Here, it is not our intention to discuss this aspect. Actually, we would like to reiterate that, from the point of view of the search for solutions, the study of the associated homogeneous equation plays a crucial role. Thus, in this section, we shall concentrate our attention exactly on the case of linear homogeneous first-order ordinary differential equations in which one or more than one unknown function occurs.

Let $y_1(x), \ldots, y_n(x)$ be differentiable functions of the scalar parameter x, a_{ij} given scalars (for $i, j = 1, \ldots, n$) and $f_1(x), \ldots, f_n(x)$ arbitrary functions of x. Assume that $f_1(x), \ldots, f_n(x)$ are defined and continuous in an interval $I \subseteq \mathbb{R}$.

A system of n linear first-order ordinary differential equations in n unknowns with constant coefficients (an $n \times n$ system of linear equations) has the general form

$$\begin{aligned}
y_1'(x) &= a_{11}y_1(x) + a_{12}y_2(x) + \cdots + a_{1n}y_n(x) + f_1(x) \\
y_2'(x) &= a_{21}y_1(x) + a_{22}y_2(x) + \cdots + a_{2n}y_n(x) + f_2(x) \\
&\quad \ldots\ldots\ldots \\
y_n'(x) &= a_{n1}y_1(x) + a_{n2}y_2(x) + \cdots + a_{nn}y_n(x) + f_n(x).
\end{aligned} \tag{12.6}$$

We may write the system as a *matrix differential equation*

$$\mathbf{y}'(x) = A\mathbf{y}(x) + \mathbf{f}(x), \tag{12.7}$$

where

$$\mathbf{y}'(x) = \begin{bmatrix} y_1'(x) \\ y_2'(x) \\ \cdots \\ \cdots \\ y_n'(x) \end{bmatrix}, \quad A = \begin{bmatrix} a_{11} & a_{12} & \cdots & a_{1n} \\ a_{21} & a_{22} & \cdots & a_{2n} \\ \cdots & \cdots & \cdots \\ a_{n1} & a_{n2} & \cdots & a_{nn} \end{bmatrix},$$

$$\mathbf{y}(x) = \begin{bmatrix} y_1(x) \\ y_2(x) \\ \cdots \\ \cdots \\ y_n(x) \end{bmatrix}, \quad \mathbf{f}(x) = \begin{bmatrix} f_1(x) \\ f_2(x) \\ \cdots \\ \cdots \\ f_n(x) \end{bmatrix}.$$

With the aim of developing our discussion of systems of differential equations, we need to introduce the following terminology:

Definition 12.7 A *vector-valued function* is a vector whose entries are functions of x. To be more specific, let $g_1(x), \ldots, g_n(x)$ be functions of the independent variable x. The vector-valued function having component $g_1(x), \ldots, g_n(x)$ is the vector

$$\mathbf{g}(x) = \begin{bmatrix} g_1(x) \\ g_2(x) \\ \cdots \\ \cdots \\ g_n(x) \end{bmatrix}$$

whose domain of definition is the largest possible interval for which all components are defined, that is, the intersection of their domains of definition.

The calculus processes of taking limits, differentiating and integrating are extended to vector-valued functions by evaluating the limit (derivative or integral, respectively) of each entry $g_i(x)$ separately, that is,

$$\lim_{x \to x_0} \mathbf{g}(x) = \begin{bmatrix} \lim_{x \to x_0} g_1(x) \\ \lim_{x \to x_0} g_2(x) \\ \cdots \\ \cdots \\ \lim_{x \to x_0} g_n(x) \end{bmatrix}, \quad \mathbf{g}'(x) = \begin{bmatrix} g_1'(x) \\ g_2'(x) \\ \cdots \\ \cdots \\ g_n'(x) \end{bmatrix}, \quad \int \mathbf{g}(x) = \begin{bmatrix} \int g_1(x) \\ \int g_2(x) \\ \cdots \\ \cdots \\ \int g_n(x) \end{bmatrix}.$$

More precisely, for any point x_0 in the domain of definition of $\mathbf{g}(x)$, the following hold:

(1) The limit $\lim\limits_{x \to x_0} \mathbf{g}(x)$ exists if and only if the limits of all components exist. If any of the limits $\lim\limits_{x \to x_0} g_i(x)$ fail to exist, then $\lim\limits_{x \to x_0} \mathbf{g}(x)$ does not exist.

(2) $\mathbf{g}(x)$ is continuous at $x = x_0$ if and only if all components are continuous at $x = x_0$.

(3) $\mathbf{g}(x)$ is differentiable at $x = x_0$ if and only if all components are differentiable at $x = x_0$.

The objects $\mathbf{y}(x)$, $\mathbf{y}'(x)$ and $\mathbf{f}(x)$ in relation (12.7) are examples of vector-valued functions. In this sense, we say that a *solution* of (12.7) is a vector-valued function $\mathbf{y}(x) = [y_1(x), \ldots, y_n(x)]^T$ satisfying the following conditions:

(1) Each function $y_i(x)$ is defined and differentiable in the domain of definition I.

(2) $\mathbf{y}(x_0)$ satisfies (12.7), for any $x_0 \in I$, that is, for any $x_0 \in I$,

$$y_1'(x_0) = Ay_1(x_0) + f_1(x_0)$$

$$\cdots \quad \cdots \quad \cdots$$

$$\cdots \quad \cdots \quad \cdots$$

$$y_n'(x_0) = Ay_n(x_0) + f_n(x_0).$$

Just as in the case of a single differential equation, we are usually interested in solving an initial value problem; that is, we seek the vector function $\mathbf{y}(x) = [y_1(x), y_2(x), \ldots, y_n(x)]^T$ that satisfy not only the differential equations given by (12.4) but also a set of initial conditions of the form

$$\mathbf{y_0} = \mathbf{y}(x_0) = \begin{bmatrix} y_1(x_0) \\ y_2(x_0) \\ \cdots \\ \cdots \\ y_n(x_0) \end{bmatrix},$$

where $x_0, y_1(x_0), \ldots, y_n(x_0)$ are known constants.

The system is said to be *homogeneous* if each of the functions $f_1(x), \ldots, f_n(x)$ is precisely equal to the constant function 0, that is,

$$\mathbf{y}'(x) = A\mathbf{y}(x). \tag{12.8}$$

Since linear homogeneous systems have linear structure, if $\mathbf{s_1}(x), \ldots, \mathbf{s_n}(x)$ are solutions of system (12.8), then

$$\mathbf{s}(x) = c_1 \mathbf{s_1}(x) + \cdots \cdots + c_n \mathbf{s_n}(x)$$

is also a solution of (12.8) for any $c_1, \ldots, c_n \in \mathbb{R}$. In fact,

$$\left(c_1 \mathbf{s_1}(x) + \cdots\cdots + c_n \mathbf{s_n}(x) \right)' = c_1 \mathbf{s_1}'(x) + \cdots\cdots + c_n \mathbf{s_n}'(x)$$

$$= c_1 A \cdot \mathbf{s_1}(x) + \cdots\cdots + c_n A \cdot \mathbf{s_n}(x)$$

$$= A \cdot \left(c_1 \mathbf{s_1}(x) + \cdots\cdots + c_n \mathbf{s_n}(x) \right).$$

In other words, given a homogeneous system, any linear combination of a finite number of its solutions is also a solution. Moreover, we notice that the vector function $\mathbf{0}$ is a solution for (12.8). Hence, the set $\mathbb{S}$ of all solutions for (12.8) has a natural structure of vector space.

As in every vector space, we need to know whether a set of vectors is linearly dependent or independent. This leads to the following definition:

Definition 12.8 Let $\mathbf{s_1}(x), \ldots, \mathbf{s_k}(x) \in \mathbb{S}$. We say that $\mathbf{s_1}(x), \ldots, \mathbf{s_k}(x)$ are *linearly dependent* vector-valued functions if there exist constants $c_1, \ldots, c_k$ not all zero such that $c_1 \mathbf{s_1}(x) + \cdots\cdots + c_k \mathbf{s_k}(x) = \mathbf{0}$, for any $x \in I$.

Practically speaking, $\mathbf{s_1}(x), \ldots, \mathbf{s_k}(x) \in \mathbb{S}$ are *linearly dependent vector-valued functions* if and only if $\mathbf{s_1}(x_0), \ldots, \mathbf{s_k}(x_0)$ are linearly dependent vectors of $\mathbb{R}^n$, for any $x_0 \in I$. In this sense, we say that $\mathbf{s_1}(x), \ldots, \mathbf{s_k}(x) \in \mathbb{S}$ are *linearly independent vector-valued functions* if there exists $x_0 \in I$ such that $\mathbf{s_1}(x_0), \ldots, \mathbf{s_k}(x_0)$ are linearly independent vectors of $\mathbb{R}^n$.

Now, if we fix a constant vector $\mathbf{c} \in \mathbb{R}^n$, there exists an unique solution $\mathbf{s}(x)$ of system (12.8) satisfying the condition $\mathbf{s}(0) = \mathbf{c}$. In this way, we may define the map $\Phi : \mathbb{R}^n \to \mathbb{S}$ and, for any $\mathbf{c} \in \mathbb{R}^n$, there exists an unique $\mathbf{s}(x) \in \mathbb{S}$ such that $\Phi(\mathbf{c}) = \mathbf{s}(x)$. The map Φ is linear and bijective, implying that $\mathbb{S} \cong \mathbb{R}^n$. This means that $\mathbb{S}$, viewed as a vector space, has dimension n. Therefore, any solution of (12.8) can be obtained as a linear combination of n linearly independent elements of $\mathbb{S}$, that is, n linearly independent solutions $\{\mathbf{s_1}, \ldots\ldots, \mathbf{s_n}\}$ of the system. Such a set is called a *fundamental system of solutions*. To formalize the previous definition, we say that $\{\mathbf{s_1}, \ldots\ldots, \mathbf{s_n}\}$ is a fundamental system of solutions if the following two conditions are satisfied:

(1) There exists $x_0 \in I$ such that $\mathbf{s_1}(x_0), \ldots\ldots, \mathbf{s_n}(x_0)$ are linearly independent vectors.
(2) Any vector function solution of the system is a linear combination of $\mathbf{s_1}(x), \ldots\ldots, \mathbf{s_n}(x)$.

A simple test for the linear independence of a set of vector functions $\mathbf{s_1}(x), \ldots\ldots, \mathbf{s_n}(x)$ that are real-valued and defined on an interval $I \subseteq \mathbb{R}$ can be formulated as follows:

(1) Put the components of these vector functions into the columns of a square matrix $\mathbf{M}(x)$ of order n:

$$\mathbf{M}(x) = \left[\mathbf{s_1}(x) | \cdots\cdots | \mathbf{s_n}(x) \right].$$

(2) Compute the determinant of the matrix $\mathbf{M}(x)$: it is called the *Wronskian* of the n vector functions $\mathbf{s}_1(x), \ldots\ldots, \mathbf{s}_n(x)$ and usually denoted by $W(x)$.

(3) If the Wronskian is different from zero for any $x \in I$, then $\mathbf{s}_1(x), \ldots\ldots, \mathbf{s}_n(x)$ are linearly independent.

As a consequence of well-known Liouville's theorem (also called Abel's formula), we observe that if $W(x_0) \neq 0$ for any given $x_0 \in I$, then $W(x) \neq 0$ for any $x \in I$.

Remark 12.9 In general, if $\{\mathbf{y}_i(x)\}$ is a linearly dependent set of functions, then the Wronskian must vanish. However, the converse is not necessarily true, as one can find examples in which the Wronskian vanishes without the functions being dependent. Nevertheless, if $\{\mathbf{y}_i(x)\}$ are solutions for a linear system of ordinary differential equations, then the converse does hold. In other words, if $\{\mathbf{y}_i(x)\}$ are solutions for a linear system of ordinary differential equations and the Wronskian of the $\{\mathbf{y}_i(x)\}$ vanishes, then $\{\mathbf{y}_i(x)\}$ is a linearly dependent set of functions.

Example 12.10 The vector functions

$$\mathbf{s}_1(x) = \begin{bmatrix} e^{3x} \\ -e^{3x} \end{bmatrix}, \quad \mathbf{s}_2(x) = \begin{bmatrix} e^{5x} \\ e^{5x} \end{bmatrix}$$

are solutions of the system

$$y_1' = 4y_1 + y_2,$$
$$y_2' = y_1 + 4y_2.$$

Those are independent solutions, in fact

$$W(x) = \begin{vmatrix} e^{3x} & e^{5x} \\ -e^{3x} & e^{5x} \end{vmatrix} \neq 0.$$

Thus, a general solution of the system is given by

$$\mathbf{y}(x) = c_1 \begin{bmatrix} 1 \\ -1 \end{bmatrix} e^{3x} + c_2 \begin{bmatrix} 1 \\ 1 \end{bmatrix} e^{5x}.$$

Example 12.11 Solve the system

$$y_1' = 4y_1,$$
$$y_2' = -2y_2,$$
$$y_3' = 3y_3,$$

and find a solution that satisfies the initial conditions

$$y_1(2) = 0, \quad y_2(2) = -1, \quad y_3(2) = 3.$$

Solving each equation separately, we easily find

$$y_1 = c_1 e^{4x},$$

$$y_2 = c_2 e^{-2x},$$

$$y_3 = c_3 e^{3x}.$$

Thus, the vector functions

$$\mathbf{s}_1(x) = \begin{bmatrix} 1 \\ 0 \\ 0 \end{bmatrix} e^{4x}, \quad \mathbf{s}_2(x) = \begin{bmatrix} 0 \\ 1 \\ 0 \end{bmatrix} e^{-2x}, \quad \mathbf{s}_3(x) = \begin{bmatrix} 0 \\ 0 \\ 1 \end{bmatrix} e^{3x}$$

are solutions of the system. Moreover, the determinant of the matrix

$$\begin{bmatrix} e^{4x} & 0 & 0 \\ 0 & e^{-2x} & 0 \\ 0 & 0 & e^{3x} \end{bmatrix},$$

whose columns are precisely the coordinates of solutions $\mathbf{s}_1, \mathbf{s}_2, \mathbf{s}_3$, is equal to e^{5x}. The determinant is then different from zero, for any $x \in \mathbb{R}$. Therefore, vectors $\mathbf{s}_1(x), \mathbf{s}_2(x), \mathbf{s}_3(x)$ are linearly independent (for any $x \in \mathbb{R}$) and form a fundamental system of solutions. Hence, a general solution for the system is the vector function

$$\mathbf{y}(x) = c_1 \begin{bmatrix} 1 \\ 0 \\ 0 \end{bmatrix} e^{4x} + c_2 \begin{bmatrix} 0 \\ 1 \\ 0 \end{bmatrix} e^{-2x} + c_3 \begin{bmatrix} 0 \\ 0 \\ 1 \end{bmatrix} e^{3x} = \begin{bmatrix} c_1 e^{4x} \\ c_2 e^{-2x} \\ c_3 e^{3x} \end{bmatrix}.$$

To achieve the same general solution, we now introduce a different approach. The system has the matrix form

$$\mathbf{y}'(x) = \begin{bmatrix} y_1'(x) \\ y_2'(x) \\ y_3'(x) \end{bmatrix} = \begin{bmatrix} 4 & 0 & 0 \\ 0 & -2 & 0 \\ 0 & 0 & 3 \end{bmatrix} \begin{bmatrix} y_1(x) \\ y_2(x) \\ y_3(x) \end{bmatrix}.$$

The coefficient matrix of the system

$$\begin{bmatrix} 4 & 0 & 0 \\ 0 & -2 & 0 \\ 0 & 0 & 3 \end{bmatrix}$$

is diagonal and has three distinct real eigenvalues $\{4, -2, 3\}$. The corresponding eigenvectors are as follows:

- $X_1 = [1, 0, 0]$ for $\lambda_1 = 4$.
- $X_2 = [0, 1, 0]$ for $\lambda_2 = -2$.
- $X_3 = [0, 0, 1]$ for $\lambda_3 = 3$.

At this point, we notice that $s_1(x) = e^{4x} X_1$, $s_2(x) = e^{-2x} X_2$, $s_3(x) = e^{3x} X_3$, where s_1, s_2, s_3 are precisely the solutions previously deducted from the system. Thus, the general solution is given by

$$\mathbf{y}(x) = \begin{bmatrix} c_1 e^{4x} \\ c_2 e^{-2x} \\ c_3 e^{3x} \end{bmatrix} = c_1 e^{4x} X_1 + c_2 e^{-2x} X_2 + c_3 e^{3x} X_3$$

for X_1, X_2, X_3 linearly independent eigenvectors associated with the coefficient matrix of the system.

From the given initial conditions, we have

$$\begin{aligned} 0 = y_1(2) &= c_1 e^8, \\ -1 = y_2(2) &= c_2 e^{-4}, \\ 3 = y_3(2) &= c_3 e^6. \end{aligned}$$

Thus, $c_1 = 0$, $c_2 = -e^4$, $c_3 = 3e^{-6}$ and the solution satisfying the initial conditions is

$$y_1 = 0,$$

$$y_2 = -e^4 e^{-2x},$$

$$y_3 = 3e^{-6} e^{3x}.$$

The previous example is easy to solve, thanks to the fact that each equation in the system involves one and only one unknown function and its derivative. More precisely, the ith equation involves precisely y_i, y_i', so that the matrix associated with the system is diagonal. This allows us to easily find the solution of any equation separately. It is clear that the simplest systems are those in which the associated matrix is diagonal.

Here, we'll give an answer to the question of how we might solve an homogeneous system

$$\mathbf{y}'(x) = A\mathbf{y}(x) \tag{12.9}$$

whose coefficient matrix A is not diagonal. The first case we analyze is related to systems whose associated coefficient matrix is diagonalizable. To work this out, we prove the following:

Theorem 12.12 *Let $y_1(x), \dots, y_n(x)$ be differentiable functions of the scalar parameter x, $A = (a_{ij})_{n \times n}$, a given diagonalizable scalar matrix such that*

$$\begin{bmatrix} y_1'(x) \\ y_2'(x) \\ \dots \\ \dots \\ y_n'(x) \end{bmatrix} = \begin{bmatrix} a_{11} & a_{12} & \dots & a_{1n} \\ a_{21} & a_{22} & \dots & a_{2n} \\ \dots & \dots & \dots & \dots \\ a_{n1} & a_{n2} & \dots & a_{nn} \end{bmatrix} \begin{bmatrix} y_1(x) \\ y_2(x) \\ \dots \\ \dots \\ y_n(x) \end{bmatrix}. \tag{12.10}$$

If $\{\lambda_1, \ldots, \lambda_n\}$ *is the spectrum of A (the eigenvalues are not necessarily distinct) and* $\{X_1, \ldots, X_n\}$ *is a linearly independent set of eigenvectors of A, such that* $AX_i = \lambda_i X_i$, *for any* $i = 1, \ldots, n$, *then* $\{X_1 e^{\lambda_1 x}, \ldots \ldots, X_n e^{\lambda_n x}\}$ *is a fundamental system of solutions and any solution of system (12.10) is a vector-valued function having the form*

$$\mathbf{y}(x) = c_1 X_1 e^{\lambda_1 x} + \cdots + c_n X_n e^{\lambda_n x} \tag{12.11}$$

for $c_1, \ldots, c_n$ *arbitrary constants.*

Proof Since A is a diagonalizable matrix, there exists an $n \times n$ invertible matrix $P = (p_{ij})_{n \times n}$ such that

$$P^{-1}AP = \begin{bmatrix} \lambda_1 & & & \\ & \ddots & & \\ & & \ddots & \\ & & & \lambda_m \end{bmatrix},$$

where the diagonal entries λ_i are not necessarily distinct and any eigenvalue λ_i repeatedly occurs on the main diagonal as many times as it occurs as a root of the characteristic polynomial of A. We recall that the columns of P coincide with the eigenvectors of A. In this regard, we may write

$$P = [X_1 \quad X_2 \quad \ldots \quad X_n]$$

with

$$X_1 = \begin{bmatrix} p_{11} \\ p_{21} \\ \ldots \\ p_{n1} \end{bmatrix} \quad X_2 = \begin{bmatrix} p_{12} \\ p_{22} \\ \ldots \\ p_{n2} \end{bmatrix} \quad \ldots \quad X_n = \begin{bmatrix} p_{1n} \\ p_{2n} \\ \ldots \\ p_{nn} \end{bmatrix}.$$

Introduce a new vector-valued function

$$\mathbf{g}(x) = \begin{bmatrix} g_1(x) \\ g_2(x) \\ \ldots \\ \ldots \\ g_n(x) \end{bmatrix},$$

where $g_1, \ldots, g_n$ are differentiable functions of the variable x such that $\mathbf{g}(x)$ is related to the unknown function $\mathbf{y}(x)$ by the equation $\mathbf{g}(x) = P^{-1} \cdot \mathbf{y}(x)$. Hence,

$$\mathbf{y}(x) = P\mathbf{g}(x) \tag{12.12}$$

which means that

$$\begin{bmatrix} y_1(x) \\ y_2(x) \\ \cdots \\ \cdots \\ y_n(x) \end{bmatrix} = \begin{bmatrix} p_{11} & p_{12} & \cdots & p_{1n} \\ p_{21} & p_{22} & \cdots & p_{2n} \\ \cdots & \cdots & \cdots & \cdots \\ p_{n1} & p_{n2} & \cdots & p_{nn} \end{bmatrix} \begin{bmatrix} g_1(x) \\ g_2(x) \\ \cdots \\ \cdots \\ g_n(x) \end{bmatrix}. \tag{12.13}$$

Since P is a constant coefficient matrix, differentiation of both sides of (12.13) leads to

$$\begin{bmatrix} y_1'(x) \\ y_2'(x) \\ \cdots \\ \cdots \\ y_n'(x) \end{bmatrix} = \begin{bmatrix} p_{11} & p_{12} & \cdots & p_{1n} \\ p_{21} & p_{22} & \cdots & p_{2n} \\ \cdots & \cdots & \cdots & \cdots \\ p_{n1} & p_{n2} & \cdots & p_{nn} \end{bmatrix} \begin{bmatrix} g_1'(x) \\ g_2'(x) \\ \cdots \\ \cdots \\ g_n'(x) \end{bmatrix}, \tag{12.14}$$

that is,

$$\mathbf{y}'(x) = P\mathbf{g}'(x). \tag{12.15}$$

Substitution of (12.12) and (12.15) in (12.10) leads us to $P\mathbf{g}'(x) = AP\mathbf{g}(x)$. Thus, $\mathbf{g}'(x) = P^{-1}AP\mathbf{g}(x)$, which means

$$\begin{bmatrix} g_1'(x) \\ g_2'(x) \\ \cdots \\ \cdots \\ g_n'(x) \end{bmatrix} = \begin{bmatrix} \lambda_1 & & & \\ & \ddots & & \\ & & \ddots & \\ & & & \lambda_n \end{bmatrix} \begin{bmatrix} g_1(x) \\ g_2(x) \\ \cdots \\ \cdots \\ g_n(x) \end{bmatrix} \tag{12.16}$$

so that

$$g_i'(x) = \lambda_i g_i(x) \quad \forall i = 1, \ldots, n. \tag{12.17}$$

We can solve equations (12.17) individually and obtain

$$g_i(x) = c_i e^{\lambda_i x} \quad \forall i = 1, \ldots, n. \tag{12.18}$$

From the given relation (12.13), for any $i = 1, \ldots, n$, we get

$$y_i(x) = \sum_{j=1}^{n} p_{ij} g_j(x)$$

$$= \sum_{j=1}^{n} p_{ij} c_j e^{\lambda_j x},$$

that is,

$$\mathbf{y}(x) = \begin{bmatrix} y_1(x) \\ y_2(x) \\ \cdots \\ \cdots \\ y_n(x) \end{bmatrix}$$

$$= \begin{bmatrix} p_{11}c_1 e^{\lambda_1 x} + \cdots + p_{1n}c_n e^{\lambda_n x} \\ p_{21}c_1 e^{\lambda_1 x} + \cdots + p_{2n}c_n e^{\lambda_n x} \\ \cdots \\ \cdots \\ p_{n1}c_1 e^{\lambda_1 x} + \cdots + p_{nn}c_n e^{\lambda_n x} \end{bmatrix}$$

$$= c_1 X_1 e^{\lambda_1 x} + \cdots + c_n X_n e^{\lambda_n x},$$

where the coefficients $c_1, \ldots, c_n$ are arbitrary and can be determined by assigning *initial conditions* in terms of the value of $y_1(x), \ldots, y_n(x)$ at some particular $x = x_0$. In particular, $\{X_1 e^{\lambda_1 x}, \ldots \ldots, X_n e^{\lambda_n x}\}$ is a set of n solutions. Moreover, the matrix whose columns are the coordinates of these solutions is

$$\begin{bmatrix} e^{\lambda_1 x} p_{11} & e^{\lambda_2 x} p_{12} & \cdots & e^{\lambda_n x} p_{1n} \\ e^{\lambda_1 x} p_{21} & e^{\lambda_2 x} p_{22} & \cdots & e^{\lambda_n x} p_{2n} \\ \cdots & \cdots & \cdots & \cdots \\ e^{\lambda_1 x} p_{n1} & e^{\lambda_2 x} p_{n2} & \cdots & e^{\lambda_n x} p_{nn} \end{bmatrix}.$$

Its determinant is equal to $e^{(\lambda_1 + \cdots + \lambda_n)x} \cdot det(P)$ and is different from zero, for any $x \in \mathbb{R}$. Therefore, we conclude that $\{X_1 e^{\lambda_1 x}, \ldots \ldots, X_n e^{\lambda_n x}\}$ is a fundamental system of solutions.

Example 12.13 Solve the system

$$\begin{aligned} y_1' &= 3y_1 - y_2 + y_3 - 2y_4, \\ y_2' &= -y_1 + 3y_2 - 2y_3 + y_4, \\ y_3' &= 2y_3 + 2y_4, \\ y_4' &= 8y_3 + 2y_4, \end{aligned}$$

and find a solution that satisfies the initial conditions

$$y_1(0) = 1, \quad y_2(0) = 1, \quad y_3(0) = 2, \quad y_4(0) = -1.$$

First we write the system in its matrix form

$$\mathbf{y}'(x) = \begin{bmatrix} y_1'(x) \\ y_2'(x) \\ y_3'(x) \\ y_4'(x) \end{bmatrix} = \begin{bmatrix} 3 & -1 & 1 & -2 \\ -1 & 3 & -2 & 1 \\ 0 & 0 & 2 & 2 \\ 0 & 0 & 8 & 2 \end{bmatrix} \cdot \begin{bmatrix} y_1(x) \\ y_2(x) \\ y_3(x) \\ y_4(x) \end{bmatrix}.$$

The coefficient matrix of the system

$$\begin{bmatrix} 3 & -1 & 1 & -2 \\ -1 & 3 & -2 & 1 \\ 0 & 0 & 2 & 2 \\ 0 & 0 & 8 & 2 \end{bmatrix}$$

has four distinct real eigenvalues $\{-2, 2, 4, 6\}$; thus, it is diagonalizable. The corresponding eigenvectors are as follows:

- $X_1 = [-7, 5, 8, -16]^T$ for $\lambda_1 = -2$.
- $X_2 = [1, 1, 0, 0]^T$ for $\lambda_2 = 2$.
- $X_3 = [1, -1, 0, 0]^T$ for $\lambda_3 = 4$.
- $X_4 = [-9, 3, 8, 16]^T$ for $\lambda_4 = 6$.

Hence,

$$\mathbf{y}(x) = c_1 \begin{bmatrix} -7 \\ 5 \\ 8 \\ -16 \end{bmatrix} e^{-2x} + c_2 \begin{bmatrix} 1 \\ 1 \\ 0 \\ 0 \end{bmatrix} e^{2x} + c_3 \begin{bmatrix} 1 \\ -1 \\ 0 \\ 0 \end{bmatrix} e^{4x} + c_4 \begin{bmatrix} -9 \\ 3 \\ 8 \\ 16 \end{bmatrix} e^{6x},$$

that is,

$$y_1 = -7c_1 e^{-2x} + c_2 e^{2x} + c_3 e^{4x} - 9c_4 e^{6x},$$

$$y_2 = 5c_1 e^{-2x} + c_2 e^{2x} - c_3 e^{4x} + 3c_4 e^{6x},$$

$$y_3 = 8c_1 e^{-2x} + 8c_4 e^{6x},$$

$$y_4 = -16c_1 e^{-2x} + 16c_4 e^{6x}.$$

From the given initial conditions, we have

$$1 = y_1(0) \quad = -7c_1 + c_2 + c_3 - 9c_4,$$

$$1 = y_2(0) \quad = 5c_1 + c_2 - c_3 + 3c_4,$$

$$2 = y_3(0) \quad = 8c_1 + 8c_4,$$

$$-1 = y_4(0) = -16c_1 + 16c_4.$$

Solving the above linear system, one has $c_1 = \frac{5}{32}$, $c_2 = \frac{23}{16}$, $c_3 = \frac{3}{2}$, $c_4 = \frac{3}{32}$. Hence, the solution satisfying the initial conditions is

$$y_1 = -\tfrac{35}{32}e^{-2x} + \tfrac{23}{16}e^{2x} + \tfrac{3}{2}e^{4x} - \tfrac{27}{32}e^{6x},$$

$$y_2 = \tfrac{25}{32}e^{-2x} + \tfrac{23}{16}e^{2x} - \tfrac{3}{2}e^{4x} + \tfrac{9}{32}e^{6x},$$

$$y_3 = \tfrac{5}{4}e^{-2x} + \tfrac{3}{4}e^{6x},$$

$$y_4 = -\tfrac{5}{2}e^{-2x} + \tfrac{3}{2}e^{6x}.$$

Example 12.14 Find the general solution of the system

$$y_1' = y_1 + 2y_2,$$
$$y_2' = 3y_1 + 2y_2,$$
$$y_3' = 3y_1 - 2y_2 + 4y_3.$$

We write the system in its matrix form

$$\mathbf{y}'(x) = \begin{bmatrix} y_1'(x) \\ y_2'(x) \\ y_3'(x) \end{bmatrix} = \begin{bmatrix} 1 & 2 & 0 \\ 3 & 2 & 0 \\ 3 & -2 & 4 \end{bmatrix} \begin{bmatrix} y_1(x) \\ y_2(x) \\ y_3(x) \end{bmatrix}.$$

The coefficient matrix of the system

$$\begin{bmatrix} 1 & 2 & 0 \\ 3 & 2 & 0 \\ 3 & -2 & 4 \end{bmatrix}$$

has two distinct real eigenvalues $\{-1, 4\}$, in particular the algebraic multiplicity of $\lambda = 4$ is equal to 2. The corresponding eigenvectors are as follows:

- $X_1 = [2, 3, 0]'$ and $X_2 = [0, 0, 1]'$ for $\lambda = 4$.
- $X_3 = [1, -1, -1]'$ for $\lambda_2 = -1$.

Thus, the geometric multiplicity of $\lambda = 4$ is equal to the algebraic one, that is, the matrix is diagonalizable. So, the general solution has the following form:

$$\mathbf{y}(x) = c_1 \begin{bmatrix} 2 \\ 3 \\ 0 \end{bmatrix} e^{4x} + c_2 \begin{bmatrix} 0 \\ 0 \\ 1 \end{bmatrix} e^{4x} + c_3 \begin{bmatrix} -1 \\ 1 \\ 1 \end{bmatrix} e^{-x},$$

that is,

$$y_1 = 2c_1 e^{4x} - c_3 e^{-x},$$

$$y_2 = 3c_1 e^{4x} + c_3 e^{-x},$$

$$y_3 = c_2 e^{4x} + c_3 e^{-x}.$$

Example 12.15 Find the general solution of the system

$$y_1' = 3y_1 + y_2 - 4y_3,$$
$$y_2' = 3y_1 + 4y_2 - 3y_3,$$
$$y_3' = -y_1 + 4y_2.$$

The coefficient matrix of the system is

$$\begin{bmatrix} 3 & 1 & -4 \\ 3 & 4 & -3 \\ -1 & 4 & 0 \end{bmatrix}.$$

It has one real eigenvalue $\lambda = -1$ and two complex conjugate eigenvalues $\mu = 4 + 3i$, $\overline{\mu} = 4 - 3i$. It is clear that the matrix is diagonalizable.

The eigenspace corresponding to $\lambda = -1$ is generated by the eigenvector $X_1 = [1, 0, 1]^t$.

To find the part of the general solution associated with the pair of complex conjugate eigenvalues, it is sufficient to take only one of them and find the associated eigenvector. For example, if we consider $\mu = 4 + 3i$ and solve the system

$$\begin{bmatrix} 3 - \mu & 1 & -4 \\ 3 & 4 - \mu & -3 \\ -1 & 4 & -\mu \end{bmatrix} \begin{bmatrix} x_1 \\ x_2 \\ x_3 \end{bmatrix} = \begin{bmatrix} 0 \\ 0 \\ 0 \end{bmatrix},$$

we get the complex eigenvector $X_2 = [i, 1 + i, 1]^t$. Thus, the complex eigenvector corresponding to $\overline{\mu} = 4 - 3i$ is $\overline{X_2} = [-i, 1 - i, 1]^t$. So, the general solution has the following form:

$$\mathbf{y}(x) = c_1 \begin{bmatrix} 1 \\ 0 \\ 1 \end{bmatrix} e^{-x} + c_2 \begin{bmatrix} i \\ 1 + i \\ 1 \end{bmatrix} e^{(4+3i)x} + c_3 \begin{bmatrix} -i \\ 1 - i \\ 1 \end{bmatrix} e^{(4-3i)x},$$

that is,

$$y_1 = c_1 e^{-x} + i c_2 e^{(4+3i)x} - i c_3 e^{(4-3i)x},$$

$$y_2 = (1 + i)c_2 e^{(4+3i)x} + (1 - i)c_3 e^{(4-3i)x},$$

$$y_3 = c_1 e^{-x} + c_2 e^{(4+3i)x} + c_3 e^{(4-3i)x}.$$

Recalling that, for $a, b \in \mathbb{R}$, $e^{a+ib} = e^a(\cos b + i \sin b)$, we may write the above solution as

$$y_1 = c_1 e^{-x} + i c_2 e^{4x}\left(\cos 3x + i \sin 3x\right) - i c_3 e^{4x}\left(\cos 3x - i \sin 3x\right)$$

$$= c_1 e^{-x} + i c_2 e^{4x} \cos 3x - c_2 e^{4x} \sin 3x - i c_3 e^{4x} \cos 3x - c_3 e^{4x} \sin 3x$$

$$= \left(c_1 e^{-x} - (c_2 + c_3)e^{4x} \sin 3x\right) + i\left((c_2 - c_3)e^{4x} \cos 3x\right),$$

$$y_2 = (1+i)c_2 e^{(4+3i)x} + (1-i)c_3 e^{(4-3i)x}$$

$$= (1+i)c_2 e^{4x}\left(\cos 3x + i \sin 3x\right) + (1-i)c_3 e^{4x}\left(\cos 3x - i \sin 3x\right)$$

$$= c_2 e^{4x} \cos 3x + i c_2 e^{4x} \cos 3x + i c_2 e^{4x} \sin 3x - c_2 e^{4x} \sin 3x +$$

$$c_3 e^{4x} \cos 3x - i c_3 e^{4x} \cos 3x - i c_3 e^{4x} \sin 3x - c_3 e^{4x} \sin 3x$$

$$= (c_2 + c_3)e^{4x}\left(\cos 3x - \sin 3x\right) + i(c_2 - c_3)e^{4x}\left(\cos 3x + \sin 3x\right),$$

$$y_3 = c_1 e^{-x} + c_2 e^{4x}\left(\cos 3x + i \sin 3x\right) + c_3 e^{4x}\left(\cos 3x - i \sin 3x\right)$$

$$= \left(c_1 e^{-x} + (c_2 + c_3)e^{4x} \cos 3x\right) + i\left((c_2 - c_3)e^{4x} \sin 3x\right).$$

They give us the complex solution

$$\mathbf{y}(x) = \begin{bmatrix} \left(c_1 e^{-x} - (c_2 + c_3)e^{4x} \sin 3x\right) + i\left((c_2 - c_3)e^{4x} \cos 3x\right) \\ (c_2 + c_3)e^{4x}\left(\cos 3x - \sin 3x\right) + i(c_2 - c_3)e^{4x}\left(\cos 3x + \sin 3x\right) \\ \left(c_1 e^{-x} + (c_2 + c_3)e^{4x} \cos 3x\right) + i\left((c_2 - c_3)e^{4x} \sin 3x\right) \end{bmatrix}.$$

Notice that every solution of the form (12.11) will be a linear combination of the special solutions

$$X_1 e^{\lambda_1 x}, \ldots, X_n e^{\lambda_n x}$$

each of which must be interpreted as the product of the variable scalar $e^{\lambda_i x}$ and the constant vector X_i, for any eigenvalue λ_i and corresponding eigenvector X_i. Actually, any particular solution $X_i e^{\lambda_i x}$ can be obtained from the general one, assigning the values $c_i = 1$ and $c_j = 0$ for any $j \neq i$.

The previously mentioned method allows us to solve the system (12.10) whenever the associated coefficient matrix A is diagonalizable. The difficulty arises when there exist some eigenvalues of A having its geometric multiplicity strictly lesser than the algebraic one. In this case, A has fewer than n linearly independent eigenvectors and it is not diagonalizable. Nevertheless, when the matrix A has real eigenvalues, the eigenvectors and generalized eigenvectors form a basis for $\mathbb{R}^n$. From this basis, we may construct the complete solution of system (12.10), by using a similar argument as in Theorem 12.12.

Theorem 12.16 Let $y_1(x), \ldots, y_n(x)$ be differentiable functions of the scalar parameter x, $A = (a_{ij})_{n \times n}$, a given jordanizable scalar matrix such that

$$\begin{bmatrix} y_1'(x) \\ y_2'(x) \\ \cdots \\ \cdots \\ y_n'(x) \end{bmatrix} = \begin{bmatrix} a_{11} & a_{12} & \ldots & a_{1n} \\ a_{21} & a_{22} & \ldots & a_{2n} \\ \cdots & \cdots & \cdots & \cdots \\ a_{n1} & a_{n2} & \ldots & a_{nn} \end{bmatrix} \begin{bmatrix} y_1(x) \\ y_2(x) \\ \cdots \\ \cdots \\ y_n(x) \end{bmatrix}. \tag{12.19}$$

Assume that

$$\begin{bmatrix} J_{n_1}(\lambda_1) & & \\ & \ddots & \\ & & J_{n_r}(\lambda_r) \end{bmatrix}$$

is the Jordan canonical form of matrix A, where eigenvalues $\lambda_1, \ldots, \lambda_r$ are not necessarily distinct and

- $J_{n_i}(\lambda_i)$ is the Jordan block associated with eigenvalue λ_i (for any $i = 1, \ldots, r$);
- $n_1, \ldots, n_r$ are the orders of Jordan blocks $J_{n_1}(\lambda_1), \ldots, J_{n_r}(\lambda_r)$, respectively, with $\sum_{i=1}^{r} n_i = n$;
- $\{X_{i,1}, \ldots, X_{i,n_i}\}$ is the chain of generalized eigenvectors generating the $n_i \times n_i$ Jordan block associated with λ_i (for any $i = 1, \ldots, r$).

Then, the set

$$\left\{ e^{\lambda_i x} \left(\sum_{j=1}^{k} \frac{x^{k-j}}{(k-j)!} X_{i,j} \right) \Big| i = 1, \ldots, r \quad k = 1, \ldots, n_i \right\}$$

is a fundamental system of solutions and any solution of system (12.19) is a vector-valued function having the form

$$\mathbf{y}(x) = \sum_{i=1}^{r} \sum_{k=1}^{n_i} c_{i,k} e^{\lambda_i x} \left(\sum_{j=1}^{k} \frac{x^{k-j}}{(k-j)!} X_{i,j} \right) \tag{12.20}$$

for n arbitrary constants $c_{i,k}$ ($i = 1, \ldots, r$ and $k = 1, \ldots, n_i$).

Proof By our main assumptions, there exists an $n \times n$ invertible matrix $P = (p_{ij})_{n \times n}$ such that

$$P^{-1}AP = \begin{bmatrix} J_{n_1}(\lambda_1) & & \\ & \ddots & \\ & & J_{n_r}(\lambda_r) \end{bmatrix}.$$

We recall that column vectors from P coincide with a Jordan basis for $\mathbb{R}^n$ and represent a set of n linearly independent generalized eigenvectors of A. Any Jordan block $J_{n_k}(\lambda_k)$ of size n_k is associated with a subset of n_k Jordan generators, that is, a chain of n_k linearly independent generalized eigenvectors corresponding to the eigenvalue λ_k. Moreover, the first vector in the chain is exactly an eigenvector of A corresponding to λ_r. We may write

$$P = [\ \underbrace{X_{1,1}, \ldots, X_{1,n_1}}_{\text{chain related to } J_{n_1}} \mid \cdots\cdots \mid \underbrace{X_{r,1}, \ldots, X_{r,n_r}}_{\text{chain related to } J_{n_r}}\ ].$$

Introducing the vector-valued function

$$\mathbf{g}(x) = \begin{bmatrix} g_{1,1}(x) \\ \cdots \\ g_{1,n_1}(x) \\ \cdots \\ \cdots \\ g_{r,1}(x) \\ \cdots \\ g_{r,n_r}(x) \end{bmatrix}$$

for $g_{i,j}$ differentiable functions of the variable x and defined by $\mathbf{g}(x) = P^{-1}\mathbf{y}(x)$, one can see that solution of (12.19) is given by

$$\mathbf{y}(x) = P\mathbf{g}(x)$$

$$= X_{1,1}g_{1,1}(x) + \cdots + X_{1,n_1}g_{1,n_1}(x)+$$

$$\cdots\cdots \tag{12.21}$$

$$+X_{r,1}g_{r,1}(x) + \cdots + X_{r,n_r}g_{r,n_r}(x)$$

$$= \sum_{i=1}^{r}\sum_{j=1}^{n_i} X_{i,j}g_{i,j}(x).$$

As in the proof of Theorem 12.12, we arrive at

$$\mathbf{g}'(x) = P^{-1}AP\mathbf{g}(x),$$

that is,

$$
\begin{bmatrix} g'_{1,1}(x) \\ \cdots \\ g'_{1,n_1}(x) \\ \cdots \\ g'_{r,1}(x) \\ \cdots \\ g'_{r,n_r}(x) \end{bmatrix} = \begin{bmatrix} J_{n_1}(\lambda_1) & & \\ & \ddots & \\ & & J_{n_r}(\lambda_r) \end{bmatrix} \begin{bmatrix} g_{1,1}(x) \\ \cdots \\ g_{1,n_1}(x) \\ \cdots \\ g_{r,1}(x) \\ \cdots \\ g_{r,n_r}(x) \end{bmatrix}. \tag{12.22}
$$

Relation (12.22) produces r different systems of linear first-order ordinary differential equations. More precisely, we get one and only one system for every diagonal Jordan block. Notice that, given any diagonal Jordan block $J_{n_k}(\lambda_k)$ of size n_k, the n_k unknown functions $g_{i,j}$ (and their first derivatives $g'_{i,j}$) that are involved in the associated linear system do not occur in any other system related to any other Jordan block. Thus, the above-mentioned systems can be solved independently of each other. By virtue of this independence, we'll now proceed to determine the solutions of any one of them. To simplify the exposition, we denote by

J_k any Jordan block of order k corresponding to eigenvalue λ;

$\{X_1, \ldots, X_k\}$ the chain of corresponding eigenvectors; (12.23)

$\{g_1, \ldots, g_k\}$ the unknown functions involved in the associated system.

Solving the system

$$
\begin{bmatrix} g'_1(x) \\ g'_2(x) \\ \cdots \\ g'_k(x) \end{bmatrix} = \begin{bmatrix} \lambda & 1 & & \\ & \ddots & & \\ & & \lambda & 1 \\ & & & \lambda \end{bmatrix} \begin{bmatrix} g_1(x) \\ g_2(x) \\ \cdots \\ g_k(x) \end{bmatrix}, \tag{12.24}
$$

we get

$$
\begin{aligned} g'_i &= \lambda g_i + g_{i+1} \quad \text{for} \quad i = 1, \ldots, k-1; \\ g'_k &= \lambda g_k. \end{aligned} \tag{12.25}
$$

We integrate starting from the last equation (see Example 12.1)

$$
g_k(x) = c_k e^{\lambda x} \quad \text{for arbitrary constant} \quad c_k.
$$

Then, substitution of g_k leads to

$$
g'_{k-1}(x) = \lambda g_{k-1}(x) + g_k(x) = \lambda g_{k-1}(x) + c_k e^{\lambda x}
$$

having solution (see Example 12.3)

$g_{k-1}(x) = \lambda g_{k-1}(x) + g_k(x) = c_{k-1}e^{\lambda x} + c_k xe^{\lambda x}$ for arbitrary constants c_{k-1}, c_k.

Analogously,

$$g'_{k-2}(x) = \lambda g_{k-2}(x) + g_{k-1}(x) = \lambda g_{k-2}(x) + c_{k-1}e^{\lambda x} + c_k xe^{\lambda x}$$

whose integral is (see again Example 12.3)

$$g_{k-2}(x) = c_{k-2}e^{\lambda x} + c_{k-1}xe^{\lambda x} + c_k\frac{x^2}{2}e^{\lambda x} \quad \text{for arbitrary constants} \quad c_{k-2}, c_{k-1}, c_k.$$

Continuing this backward substitutions process, we arrive at the solution of system (12.25), more precisely,

$$g_j(x) = \sum_{h=j}^{k} c_h \frac{x^{h-j}}{(h-j)!}e^{\lambda x}, \quad j = 1, \ldots, k \quad \text{for arbitrary constants} \quad c_1, \ldots, c_k.$$

$$(12.26)$$

Go back to the general case. The unknown functions $\{g_{i,1}, \ldots, g_{i,n_i}\}$, that are involved in the system associated with J_{n_i} and related to eigenvalue λ_i (for any $i = 1, \ldots, r$) , can be determined as follows:

$$g_{i,j}(x) = \sum_{h=j}^{n_i} c_{i,h} \frac{x^{h-j}}{(h-j)!}e^{\lambda_i x}, \quad j = 1, \ldots, n_i \quad \text{for arbitrary constants} \quad c_{i,h}.$$

$$(12.27)$$

Substitution of (12.27) in (12.21) gives

$$\mathbf{y}(x) = \sum_{i=1}^{r}\sum_{j=1}^{n_i} X_{i,j}g_{i,j}(x)$$

$$= \sum_{i=1}^{r}\sum_{j=1}^{n_i} X_{i,j}e^{\lambda_i x}\left\{\sum_{h=j}^{n_i} c_{i,h}\frac{x^{h-j}}{(h-j)!}\right\} \quad (12.28)$$

$$= \sum_{i=1}^{r}\sum_{h=1}^{n_i} c_{i,h}e^{\lambda_i x}\left(\sum_{j=1}^{h}\frac{x^{h-j}}{(h-j)!}X_{i,j}\right),$$

where

$$\left\{e^{\lambda_i x}\left(\sum_{j=1}^{h}\frac{x^{h-j}}{(h-j)!}X_{i,j}\right)|i = 1, \ldots, r, \quad h = 1, \ldots, n_i\right\} \quad (12.29)$$

is a set of n solutions to the initial system. We may describe this set by marking with an appropriate symbol and any of its element as follows:

$$s_{i,h}(x) = e^{\lambda_i x}\left(\sum_{j=1}^{h} \frac{x^{h-j}}{(h-j)!} X_{i,j}\right) \quad i = 1, \ldots, r \quad h = 1, \ldots, n_i$$

and denote by **S** the $n \times n$ matrix-valued function whose columns are precisely the coordinates of any $s_{i,h}(x)$:

$$\mathbf{S}(x) = \left[s_{1,1}(x) \ldots s_{1,n_1}(x) \ldots \ldots s_{r,1}(x) \ldots s_{r,n_r}(x)\right].$$

To show that (12.29) is a fundamental system of solutions, we now prove that it is a linearly independent set of solutions.

To do this, we see that $\mathbf{S}(x) = P\mathbf{B}(x)$, where $\mathbf{B}(x)$ is matrix-valued function having the following block diagonal form:

$$\mathbf{B}(x) = \begin{bmatrix} \mathbf{B}_{n_1}(x) & & \\ & \ddots & \\ & & \mathbf{B}_{n_r}(x) \end{bmatrix}$$

for

$$\mathbf{B}_{n_i}(x) = \begin{bmatrix} e^{\lambda_i x} & xe^{\lambda_i x} & \frac{x^2}{2}e^{\lambda_i x} & \frac{x^3}{3!}e^{\lambda_i x} & \frac{x^4}{4!}e^{\lambda_i x} & \cdots & \cdots & \frac{x^{n_i-1}}{(n_i-1)!}e^{\lambda_i x} \\ 0 & e^{\lambda_i x} & xe^{\lambda_i x} & \frac{x^2}{2}e^{\lambda_i x} & \frac{x^3}{3!}e^{\lambda_i x} & \cdots & \cdots & \frac{x^{n_i-2}}{(n_i-2)!}e^{\lambda_i x} \\ 0 & 0 & e^{\lambda_i x} & xe^{\lambda_i x} & \frac{x^2}{2}e^{\lambda_i x} & \frac{x^3}{3!}e^{\lambda_i x} & \cdots & \frac{x^{n_i-3}}{(n_i-3)!}e^{\lambda_i x} \\ 0 & 0 & 0 & e^{\lambda_i x} & xe^{\lambda_i x} & \frac{x^2}{2}e^{\lambda_i x} & \cdots & \frac{x^{n_i-4}}{(n_i-4)!}e^{\lambda_i x} \\ & & & & \ddots & & & \\ & & & & & \ddots & & \\ 0 & 0 & 0 & \cdots & \cdots & \cdots & e^{\lambda_i x} & xe^{\lambda_i x} \\ 0 & 0 & 0 & \cdots & \cdots & \cdots & 0 & e^{\lambda_i x} \end{bmatrix}.$$

By the fact that P is the constant invertible matrix whose columns are the coordinates of linearly independent generalized eigenvectors of A, and since the determinant of $\mathbf{B}(x)$ is a function of the variable x which is trivially nowhere zero, we arrive at the conclusion that $\mathbf{S}(x)$ is an invertible matrix-valued function, for any $x \in \mathbb{R}$. This means that its rank is equal to n and its columns are linearly independent, as required.

□

Remark 12.17 Returning to the simplified case (12.23), we can assert that, for any Jordan block J_k of length k and related to the eigenvalue λ, a set of k linearly independent solutions is given by

$$s_1(x) = e^{\lambda x} X_1; \quad s_2(x) = e^{\lambda x}\left(x X_1 + X_2\right);$$

$$s_3(x) = e^{\lambda x}\left(\frac{x^2}{2} X_1 + x X_2 + X_3\right); \quad s_4(x) = e^{\lambda x}\left(\frac{x^3}{3!} X_1 + \frac{x^2}{2} X_2 + x X_3 + X_4\right);$$

$$\cdots \quad \cdots \quad \cdots \quad \cdots$$

$$s_{k-1}(x) = e^{\lambda x}\left(\frac{x^{k-2}}{(k-2)!} X_1 + \cdots + x X_{k-2} + X_{k-1}\right);$$

$$s_k(x) = e^{\lambda x}\left(\frac{x^{k-1}}{(k-1)!} X_1 + \cdots + \frac{x^2}{2} X_{k-2} + x X_{k-1} + X_k\right);$$

$$\tag{12.30}$$

where $\{X_1, \ldots, X_k\}$ is the chain of generalized eigenvectors associated with eigenvalue λ and generating J_k. Thus, the contribution to the general solution (12.21) of system (12.19) has the form

$$\sum_{t=1}^{k} c_t s_t(x) = \sum_{t=1}^{k} c_t e^{\lambda x} \sum_{j=1}^{t} \frac{x^{t-j}}{(t-j)!} X_j, \tag{12.31}$$

where $c_1, \ldots, c_k$ are arbitrary constants. So, a fundamental system of solutions can be obtained by the union of sets of linearly independent solutions of the form (12.30), corresponding to all Jordan blocks. The general solution (12.21) is then given by the total sum of particular solutions of the form (12.31), each of which is obtained from a different Jordan block of the matrix $P^{-1}AP$.

Let us finally summarize the method for solving a system of the form (12.19):

- Find eigenvalues $\{\lambda_1, \ldots, \lambda_r\}$ of A.
- For any Jordan block $J_k(\lambda)$ of order k (related to some eigenvalue λ), find the corresponding chain of k generalized eigenvectors $\{X_1, \ldots, X_k\}$.
- Construct the particular solution (12.31) corresponding with J_k, that is,

$$c_1 e^{\lambda x} X_1 +$$

$$c_2 e^{\lambda x}\left(x X_1 + X_2\right) +$$

$$c_3 e^{\lambda x}\left(\frac{x^2}{2} X_1 + x X_2 + X_3\right) +$$

$$\cdots \quad \cdots \quad \cdots \quad + \tag{12.32}$$

$$c_{k-1} e^{\lambda x}\left(\frac{x^{k-2}}{(k-2)!} X_1 + \cdots + x X_{k-2} + X_{k-1}\right) +$$

$$c_k e^{\lambda x}\left(\frac{x^{k-1}}{(k-1)!} X_1 + \cdots + \frac{x^2}{2} X_{k-2} + x X_{k-1} + X_k\right).$$

- Finally, add up all the particular solutions of the form (12.32).

Example 12.18 Solve the system

$$
\begin{aligned}
y_1' &= 2y_1 - y_2 + y_3 + y_4,\\
y_2' &= y_1 + 4y_2 + y_4,\\
y_3' &= 4y_3 + y_4,\\
y_4' &= y_3 + 4y_4,
\end{aligned}
$$

and find a solution that satisfies the initial conditions

$$
y_1(0) = 1, \quad y_2(0) = 0, \quad y_3(0) = 0, \quad y_4(0) = 1.
$$

The coefficient matrix of the system

$$
A = \begin{bmatrix} 2 & -1 & 1 & 1 \\ 1 & 4 & 0 & 1 \\ 0 & 0 & 4 & 1 \\ 0 & 0 & 1 & 4 \end{bmatrix}
$$

has two distinct real eigenvalues $\lambda_1 = 3$ and $\lambda_2 = 5$. The eigenvalues $\lambda_1 = 3$ has algebraic multiplicity equal to 3, but its geometric multiplicity is equal to 1. Thus, the matrix is not diagonalizable. However, we may obtain its Jordan canonical form, that is,

$$
A' = \begin{bmatrix} 3 & 1 & 0 & 0 \\ 0 & 3 & 1 & 0 \\ 0 & 0 & 3 & 0 \\ 0 & 0 & 0 & 5 \end{bmatrix}.
$$

To arrive at the general solution of the system, we now find the eigenvectors and generalized eigenvectors of matrix A.

Since there exists only one Jordan block of order 3 with eigenvalue $\lambda_1 = 3$, we need to find generalized eigenvectors corresponding to λ_1 and having exponents 1, 2, 3. Hence, we must solve the homogeneous linear systems associated with matrices

$$
(A - 3I) = \begin{bmatrix} -1 & -1 & 1 & 1 \\ 1 & 1 & 0 & 1 \\ 0 & 0 & 1 & 1 \\ 0 & 0 & 1 & 1 \end{bmatrix},
$$

$$
(A - 3I)^2 = \begin{bmatrix} 0 & 0 & 1 & 0 \\ 0 & 0 & 2 & 3 \\ 0 & 0 & 2 & 2 \\ 0 & 0 & 2 & 2 \end{bmatrix}
$$

and

$$(A - 3I)^3 = \begin{bmatrix} 0 & 0 & 1 & 1 \\ 0 & 0 & 5 & 5 \\ 0 & 0 & 4 & 4 \\ 0 & 0 & 4 & 4 \end{bmatrix}.$$

Starting with $(A - 3I)$, we see that solutions of

$$\begin{bmatrix} -1 & -1 & 1 & 1 \\ 1 & 1 & 0 & 1 \\ 0 & 0 & 1 & 1 \\ 0 & 0 & 1 & 1 \end{bmatrix} \begin{bmatrix} x_1 \\ x_2 \\ x_3 \\ x_4 \end{bmatrix} = \begin{bmatrix} 0 \\ 0 \\ 0 \\ 0 \end{bmatrix}$$

are vectors from the space $N_{1,\lambda_1} = \langle (1, -1, 0, 0) \rangle$.
For the generalized eigenvectors of exponent 2, we solve the system

$$\begin{bmatrix} 0 & 0 & 1 & 0 \\ 0 & 0 & 2 & 3 \\ 0 & 0 & 2 & 2 \\ 0 & 0 & 2 & 2 \end{bmatrix} \begin{bmatrix} x_1 \\ x_2 \\ x_3 \\ x_4 \end{bmatrix} = \begin{bmatrix} 0 \\ 0 \\ 0 \\ 0 \end{bmatrix}$$

and find $N_{2,\lambda_1} = \langle (1, 0, 0, 0), (0, 1, 0, 0) \rangle$.
Then, for generalized eigenvectors of exponent 3, by the system

$$\begin{bmatrix} 0 & 0 & 1 & 1 \\ 0 & 0 & 5 & 5 \\ 0 & 0 & 4 & 4 \\ 0 & 0 & 4 & 4 \end{bmatrix} \begin{bmatrix} x_1 \\ x_2 \\ x_3 \\ x_4 \end{bmatrix} = \begin{bmatrix} 0 \\ 0 \\ 0 \\ 0 \end{bmatrix},$$

we get $N_{3,\lambda_1} = \langle (1, 0, 0, 0), (0, 1, 0, 0), (0, 0, 1, -1) \rangle$.
To construct the chain of generalized eigenvectors $\{X_1, X_2, X_3\}$, we start from $X_3 = (0, 0, 1, -1) \in N_{3,\lambda_1} \setminus N_{2,\lambda_1}$. Thus,

$$X_2 = (A - 3I)X_3 = \begin{bmatrix} 0 \\ -1 \\ 0 \\ 0 \end{bmatrix}$$

and

$$X_1 = (A - 3I)X_2 = \begin{bmatrix} 1 \\ -1 \\ 0 \\ 0 \end{bmatrix}.$$

Finally, we compute the eigenvector corresponding to $\lambda_2 = 5$, whose multiplicity is equal to one. The homogeneous system associated with $(A - 5I)$ is

$$\begin{bmatrix} -3 & -1 & 1 & 1 \\ 1 & -1 & 0 & 1 \\ 0 & 0 & -1 & 1 \\ 0 & 0 & 1 & -1 \end{bmatrix} \begin{bmatrix} x_1 \\ x_2 \\ x_3 \\ x_4 \end{bmatrix} = \begin{bmatrix} 0 \\ 0 \\ 0 \\ 0 \end{bmatrix}$$

having solution $Y_1 = \alpha(1, 5, 4, 4)$, for any $\alpha \in \mathbb{R}$. Therefore, the Jordan basis with respect of which we have the Jordan canonical form A' is $\{X_1, X_2, X_3, Y_1\}$ and the general solution of the original system of differential equations is

$$c_1 e^{3x} \begin{bmatrix} 1 \\ -1 \\ 0 \\ 0 \end{bmatrix} + c_2 e^{3x} \left\{ x \begin{bmatrix} 1 \\ -1 \\ 0 \\ 0 \end{bmatrix} + \begin{bmatrix} 0 \\ -1 \\ 0 \\ 0 \end{bmatrix} \right\} +$$

$$c_3 e^{3x} \left\{ \frac{x^2}{2} \begin{bmatrix} 1 \\ -1 \\ 0 \\ 0 \end{bmatrix} + x \begin{bmatrix} 0 \\ -1 \\ 0 \\ 0 \end{bmatrix} + \begin{bmatrix} 0 \\ 0 \\ 1 \\ -1 \end{bmatrix} \right\} + c_4 e^{5x} \begin{bmatrix} 1 \\ 5 \\ 4 \\ 4 \end{bmatrix},$$

that is,

$$y_1(x) = c_1 e^{3x} + c_2 x e^{3x} + c_3 \frac{x^2}{2} e^{3x} + c_4 e^{5x},$$

$$y_2(x) = -c_1 e^{3x} - c_2 (x e^{3x} + e^{3x}) - c_3 (\frac{x^2}{2} e^{3x} + x e^{3x}) + 5 c_4 e^{5x},$$

$$y_3(x) = c_3 e^{3x} + 4 c_4 e^{5x},$$

$$y_4(x) = -c_3 e^{3x} + 4 c_4 e^{5x}.$$

From the given initial conditions, we have

$$1 = y_1(0) = c_1 + c_4,$$

$$0 = y_2(0) = -c_1 - c_2 + 5 c_4,$$

$$0 = y_3(0) = c_3 + 4 c_4,$$

$$1 = y_4(0) = -c_3 + 4 c_4,$$

whose solutions are

$$c_1 = \frac{7}{8}, c_2 = -\frac{1}{4}, c_3 = -\frac{1}{2}, c_4 = \frac{1}{8}.$$

So the solution satisfying the initial conditions is

$$y_1(x) = \tfrac{7}{8}e^{3x} - \tfrac{1}{4}xe^{3x} - \tfrac{1}{2}\tfrac{x^2}{2}e^{3x} + \tfrac{1}{8}e^{5x},$$

$$y_2(x) = -\tfrac{7}{8}e^{3x} + \tfrac{1}{4}(xe^{3x} + e^{3x}) + \tfrac{1}{2}(\tfrac{x^2}{2}e^{3x} + xe^{3x}) + \tfrac{5}{8}e^{5x},$$

$$y_3(x) = -\tfrac{1}{2}e^{3x} + \tfrac{1}{2}e^{5x},$$

$$y_4(x) = \tfrac{1}{2}e^{3x} + \tfrac{1}{2}e^{5x}.$$

Example 12.19 Find the general solution of the system

$$y_1' = 3y_1 + 2y_2 - y_3 - y_4,$$
$$y_2' = 2y_2 + y_4,$$
$$y_3' = y_1 + y_2 + y_3 + y_4,$$
$$y_4' = 2y_4.$$

The coefficient matrix of the system

$$A = \begin{bmatrix} 3 & 2 & -1 & -1 \\ 0 & 2 & 0 & 1 \\ 1 & 1 & 1 & 1 \\ 0 & 0 & 0 & 2 \end{bmatrix}$$

has only one eigenvalue $\lambda = 2$, with algebraic multiplicity equal to 4, but its geometric multiplicity is equal to 1. Thus, the matrix has the following Jordan canonical form:

$$A' = \begin{bmatrix} 2 & 1 & 0 & 0 \\ 0 & 2 & 1 & 0 \\ 0 & 0 & 2 & 1 \\ 0 & 0 & 0 & 2 \end{bmatrix}.$$

To arrive at the general solution of the system, we now find the eigenvectors and generalized eigenvectors corresponding to λ and having exponents 1, 2, 3, 4. Starting with $(A - 2I)$, we must solve the system

$$\begin{bmatrix} 1 & 2 & -1 & -1 \\ 0 & 0 & 0 & 1 \\ 1 & 1 & -1 & 1 \\ 0 & 0 & 0 & 0 \end{bmatrix} \begin{bmatrix} x_1 \\ x_2 \\ x_3 \\ x_4 \end{bmatrix} = \begin{bmatrix} 0 \\ 0 \\ 0 \\ 0 \end{bmatrix}.$$

It is easy to see that $N_{1,\lambda} = \langle (1, 0, 1, 0) \rangle$.

For the generalized eigenvectors of exponent 2, we look at the homogeneous system associated with the matrix $(A - 2I)^2$, that is,

$$
\begin{bmatrix} 0 & 1 & 0 & 0 \\ 0 & 0 & 0 & 0 \\ 0 & 1 & 0 & -1 \\ 0 & 0 & 0 & 0 \end{bmatrix} \begin{bmatrix} x_1 \\ x_2 \\ x_3 \\ x_4 \end{bmatrix} = \begin{bmatrix} 0 \\ 0 \\ 0 \\ 0 \end{bmatrix}
$$

and find $N_{2,\lambda} = \langle (1, 0, 0, 0), (0, 0, 1, 0) \rangle$.

Then, for generalized eigenvectors of exponent 3, we solve the homogeneous system associated with $(A - 2I)^3$, that is,

$$
\begin{bmatrix} 0 & 0 & 0 & 1 \\ 0 & 0 & 0 & 0 \\ 0 & 0 & 0 & 1 \\ 0 & 0 & 0 & 0 \end{bmatrix} \begin{bmatrix} x_1 \\ x_2 \\ x_3 \\ x_4 \end{bmatrix} = \begin{bmatrix} 0 \\ 0 \\ 0 \\ 0 \end{bmatrix}
$$

and obtain $N_{3,\lambda} = \langle (1, 0, 0, 0), (0, 1, 0, 0), (0, 0, 1, 0) \rangle$.

Finally, since $(A - 2I)^4 = 0$, $N_{4,\lambda} = \mathbb{R}^4$. To complete a chain of generalized eigenvectors $\{X_1, X_2, X_3, X_4\}$, we start from $X_4 = (0, 0, 0, 1) \in N_{4,\lambda} \setminus N_{3,\lambda}$. Thus,

$$
X_3 = (A - 2I)X_4 = \begin{bmatrix} -1 \\ 1 \\ 1 \\ 0 \end{bmatrix},
$$

$$
X_2 = (A - 2I)X_3 = \begin{bmatrix} 0 \\ 0 \\ -1 \\ 0 \end{bmatrix}
$$

and

$$
X_1 = (A - 2I)X_2 = \begin{bmatrix} 1 \\ 0 \\ 1 \\ 0 \end{bmatrix}.
$$

Therefore, the Jordan basis with respect of which we have the Jordan canonical form A' is $\{X_1, X_2, X_3, X_4\}$ and the general solution of the original system of differential equations is

$$c_1 e^{2x} \begin{bmatrix} 1 \\ 0 \\ 1 \\ 0 \end{bmatrix} + c_2 e^{2x} \left\{ x \begin{bmatrix} 1 \\ 0 \\ 1 \\ 0 \end{bmatrix} + \begin{bmatrix} 0 \\ 0 \\ -1 \\ 0 \end{bmatrix} \right\} +$$

$$c_3 e^{2x} \left\{ \frac{x^2}{2} \begin{bmatrix} 1 \\ 0 \\ 1 \\ 0 \end{bmatrix} + x \begin{bmatrix} 0 \\ 0 \\ -1 \\ 0 \end{bmatrix} + \begin{bmatrix} -1 \\ 1 \\ 1 \\ 0 \end{bmatrix} \right\} +$$

$$c_4 e^{2x} \left\{ \frac{x^3}{6} \begin{bmatrix} 1 \\ 0 \\ 1 \\ 0 \end{bmatrix} + \frac{x^2}{2} \begin{bmatrix} 0 \\ 0 \\ -1 \\ 0 \end{bmatrix} + x \begin{bmatrix} -1 \\ 1 \\ 1 \\ 0 \end{bmatrix} + \begin{bmatrix} 0 \\ 0 \\ 0 \\ 1 \end{bmatrix} \right\},$$

that is,

$$y_1(x) = \left\{ c_1 + c_2 x + c_3 (\tfrac{x^2}{2} - 1) + c_4 (\tfrac{x^3}{6} - x) \right\} e^{2x},$$

$$y_2(x) = \left\{ c_3 + c_4 x \right\} e^{2x},$$

$$y_3(x) = \left\{ c_1 + c_2 (x - 1) + c_3 (\tfrac{x^2}{2} - x + 1) + c_4 (\tfrac{x^3}{6} - \tfrac{x^2}{2} + x) \right\} e^{2x},$$

$$y_4(x) = c_4 e^{2x}.$$

Example 12.20 Find the general solution of the system

$$\begin{aligned} y_1' &= 2y_1 + 4y_2 - 8y_3, \\ y_2' &= 4y_3, \\ y_3' &= -y_2 + 4y_3. \end{aligned}$$

The coefficient matrix of the system

$$A = \begin{bmatrix} 2 & 4 & -8 \\ 0 & 0 & 4 \\ 0 & -1 & 4 \end{bmatrix}$$

has only one eigenvalue $\lambda = 2$, with algebraic multiplicity equal to 3, but its geometric multiplicity is equal to 2. Thus, the matrix has the following Jordan canonical form:

$$A' = \begin{bmatrix} 2 & 1 & 0 \\ 0 & 2 & 0 \\ 0 & 0 & 2 \end{bmatrix}.$$

To arrive at the general solution of the system, we firstly find the generalized eigenvectors corresponding to generating the Jordan block of order 2. Starting with $(A - 2I)$, we must solve the system

$$
\begin{bmatrix} 0 & 4 & -8 \\ 0 & -2 & 4 \\ 0 & -1 & 2 \end{bmatrix} \begin{bmatrix} x_1 \\ x_2 \\ x_3 \end{bmatrix} = \begin{bmatrix} 0 \\ 0 \\ 0 \end{bmatrix}.
$$

Then $N_{1,\lambda} = \langle (1, 0, 0), (0, 2, 1) \rangle$.

Since $(A - 2I)^2 = 0$, we may choose $X_2 = (0, 0, 1) \in N_{2,\lambda} \setminus N_{1,\lambda}$ as generalized eigenvector of exponent 2. To get a chain of generalized eigenvectors $\{X_1, X_2\}$, we start from $X_2 = (0, 0, 1)$. Thus,

$$
X_1 = (A - 2I)X_2 = \begin{bmatrix} -8 \\ 4 \\ 2 \end{bmatrix}.
$$

For the chain of order 1, we can easy choose $Y_1 = (1, 0, 0) \in N_{1,\lambda}$. Therefore, the Jordan basis with respect of which we have the Jordan canonical form A' is $\{X_1, X_2, Y_1\}$ and the general solution of the original system of differential equations is

$$
c_1 e^{2x} \begin{bmatrix} -8 \\ 4 \\ 2 \end{bmatrix} + c_2 e^{2x} \left\{ x \begin{bmatrix} -8 \\ 4 \\ 2 \end{bmatrix} + \begin{bmatrix} 0 \\ 0 \\ 1 \end{bmatrix} \right\} + c_3 e^{2x} \begin{bmatrix} 1 \\ 0 \\ 0 \end{bmatrix},
$$

that is,

$$
y_1(x) = \{-8c_1 - 8c_2 x + c_3\} e^{2x},
$$

$$
y_2(x) = \{4c_1 + 4c_2 x\} e^{2x},
$$

$$
y_3(x) = \{2c_1 + c_2(2x + 1)\} e^{2x}.
$$

Example 12.21 Find the general solution of the system

$$
\begin{aligned}
y_1' &= -y_3 - y_4, \\
y_2' &= -y_4, \\
y_3' &= y_1 + y_2, \\
y_4' &= y_2.
\end{aligned}
$$

The coefficient matrix of the system

$$
A = \begin{bmatrix} 0 & 0 & -1 & -1 \\ 0 & 0 & 0 & -1 \\ 1 & 1 & 0 & 0 \\ 0 & 1 & 0 & 0 \end{bmatrix}
$$

has two complex conjugate eigenvalues $\lambda = \pm i$, each of which has algebraic multiplicity equal to 2, but geometric multiplicity equal to 1. Thus, the matrix has the following Jordan canonical form

$$A' = \begin{bmatrix} i & 1 & 0 & 0 \\ 0 & i & 0 & 0 \\ 0 & 0 & -i & 1 \\ 0 & 0 & 0 & -i \end{bmatrix}.$$

To determine eigenvectors and generalized eigenvectors corresponding to both $\lambda = \pm i$, it is sufficient to take just one of them. So, for $\lambda = -i$, we have that the associated eigenspace is generated by $X_1 = [-1, 0, -i, 0]^t$ and the chain of generalized eigenvectors is completed by $X_2 = [0, -i, 0, 1]^t$. Thus, for $\lambda = i$, we obtain the chain $\overline{X}_1 = [-1, 0, i, 0]^t$, $\overline{X}_2 = [0, i, 0, 1]^t$. Therefore, the general complex solution of the original system of differential equations is

$$c_1 e^{-ix} \begin{bmatrix} -1 \\ 0 \\ -i \\ 0 \end{bmatrix} + c_2 e^{-ix} \left\{ x \begin{bmatrix} -1 \\ 0 \\ -i \\ 0 \end{bmatrix} + \begin{bmatrix} 0 \\ -i \\ 0 \\ 1 \end{bmatrix} \right\} +$$

$$c_3 e^{ix} \begin{bmatrix} -1 \\ 0 \\ i \\ 0 \end{bmatrix} + c_4 e^{ix} \left\{ x \begin{bmatrix} -1 \\ 0 \\ i \\ 0 \end{bmatrix} + \begin{bmatrix} 0 \\ i \\ 0 \\ 1 \end{bmatrix} \right\},$$

that is,

$$c_1 (\cos x - i \sin x) \begin{bmatrix} -1 \\ 0 \\ -i \\ 0 \end{bmatrix} + c_2 (\cos x - i \sin x) \left\{ x \begin{bmatrix} -1 \\ 0 \\ -i \\ 0 \end{bmatrix} + \begin{bmatrix} 0 \\ -i \\ 0 \\ 1 \end{bmatrix} \right\} +$$

$$c_3 (\cos x + i \sin x) \begin{bmatrix} -1 \\ 0 \\ i \\ 0 \end{bmatrix} + c_4 (\cos x + i \sin x) \left\{ x \begin{bmatrix} -1 \\ 0 \\ i \\ 0 \end{bmatrix} + \begin{bmatrix} 0 \\ i \\ 0 \\ 1 \end{bmatrix} \right\}$$

and

$$y_1(x) = -(c_1 + c_3) \cos x - (c_2 + c_4) x \cos x + i(c_1 - c_3) \sin x + i(c_2 - c_4) x \sin x,$$

$$y_2(x) = -(c_2 + c_4) \sin x - i(c_2 - c_4) \cos x,$$

$$y_3(x) = -(c_1 + c_3) \sin x - (c_2 + c_4) x \sin x - i(c_1 - c_3) \cos x - i(c_2 - c_4) x \cos x,$$

$$y_4(x) = (c_2 + c_4) \cos x - i(c_2 - c_4) \sin x.$$

12.3 Real-Valued Solutions for Systems with Complex Eigenvalues

Summarizing what is seen in the above brief presentation, when solving

$$\mathbf{y}'(x) = A\mathbf{y}(x), \tag{12.33}$$

we know what has to be done. After determining eigenvalues, we have to compute the corresponding eigenvectors and generalized eigenvectors, then the solution is in the form of the linear combination (12.11) or (12.20) according to whether the coefficient matrix is diagonalizable or Jordanizable, respectively. In case the coefficient matrix A has some complex eigenvalues, we find a complex vector-valued function that is the general complex solution of system (see Examples 12.15 and 12.21). Even if complex functions find applications in several sectors, from engineering to telecommunications and surgery, real functions are more appropriate for many other purposes.

Now, assume that the real matrix A has two complex conjugate eigenvalues $\lambda, \overline{\lambda}$, with associated complex conjugate (generalized) eigenvectors $X_1, \ldots, X_k$ and $\overline{X}_1, \ldots, \overline{X}_k$, respectively. Corresponding to both the first sequence of vectors and to the second one, we may construct two solutions of system $\mathbf{s}_1(x)$ and $\mathbf{s}_2(x)$. In particular, these solutions come in a conjugate pair, that is, $\mathbf{s}_2(x) = \overline{\mathbf{s}_1(x)}$. Thus, the real and imaginary parts of $\mathbf{s}_1(x)$ are precisely

$$Re\big(\mathbf{s}_1(x)\big) = \frac{1}{2}\Big\{\mathbf{s}_1(x) + \overline{\mathbf{s}_1(x)}\Big\}, \quad Im\big(\mathbf{s}_1(x)\big) = \frac{1}{2i}\Big\{\mathbf{s}_1(x) - \overline{\mathbf{s}_1(x)}\Big\}.$$

Therefore, the real and imaginary parts of $\mathbf{s}_1(x)$ are real solutions of the system, because they are linear combinations of solutions. The arguments that were put forward are all we need for obtaining the real solutions for systems with complex eigenvalues.

Example 12.22 Let's get back to Example 12.15. We now would like to find the general real solution of the system

$$\begin{aligned}
y_1' &= 3y_1 + y_2 - 4y_3, \\
y_2' &= 3y_1 + 4y_2 - 3y_3, \\
y_3' &= -y_1 + 4y_2.
\end{aligned}$$

The coefficient matrix of the system has one real eigenvalue $\lambda = -1$ and two complex conjugate eigenvalues $\mu = 4 + 3i, \overline{\mu} = 4 - 3i$. It is diagonalizable. The eigenspace corresponding to $\lambda = -1$ is generated by the eigenvector $X_1 = [1, 0, 1]^t$. The particular solution associated with $\lambda = -1$ is then

$$\mathbf{s}_1(x) = \begin{bmatrix} 1 \\ 0 \\ 1 \end{bmatrix} e^{-x}.$$

To find the part of the general real solution associated with the pair of complex conjugate eigenvalues, it is sufficient to take only the real and imaginary parts of eigenvector corresponding to one of them. For example, the complex eigenvector corresponding to $\mu = 4 + 3i$ is $X_2 = [i, 1+i, 1]^t = [0, 1, 1]^t + i[1, 1, 0]^t$. Thus, the real and imaginary parts of X_2 are, respectively,

$$Re(X_2) = [0, 1, 1]^t, \quad Im(X_2) = [1, 1, 0]^t,$$

and the associated particular real solution is

$$\mathbf{s}_2(x) = e^{(4+3i)x} \left\{ \begin{bmatrix} 0 \\ 1 \\ 1 \end{bmatrix} + i \begin{bmatrix} 1 \\ 1 \\ 0 \end{bmatrix} \right\},$$

that is,

$$\mathbf{s}_2(x) = e^{4x}(\cos 3x + i \sin 3x) \left\{ \begin{bmatrix} 0 \\ 1 \\ 1 \end{bmatrix} + i \begin{bmatrix} 1 \\ 1 \\ 0 \end{bmatrix} \right\}$$

$$= e^{4x} \begin{bmatrix} -\sin 3x \\ \cos 3x - \sin 3x \\ \cos 3x \end{bmatrix} + i e^{4x} \begin{bmatrix} \cos 3x \\ \cos 3x + \sin 3x \\ \sin 3x \end{bmatrix}.$$

Hence, the linear combination of $\mathbf{s}_1(x)$, $Re(\mathbf{s}_2(x))$ and $Im(\mathbf{s}_2(x))$ gives us the general real solution of the system, that is,

$$\mathbf{y}(x) = c_1 e^{-x} \begin{bmatrix} 1 \\ 0 \\ 1 \end{bmatrix} + c_2 e^{4x} \begin{bmatrix} -\sin 3x \\ \cos 3x - \sin 3x \\ \cos 3x \end{bmatrix} + c_3 e^{4x} \begin{bmatrix} \cos 3x \\ \cos 3x + \sin 3x \\ \sin 3x \end{bmatrix}$$

for c_1, c_2, c_3 arbitrary constants.

Example 12.23 In conclusion, let us return to Example 12.21 and find the general real solution of the system

$$\begin{aligned} y_1' &= -y_3 - y_4, \\ y_2' &= -y_4, \\ y_3' &= y_1 + y_2, \\ y_4' &= y_2. \end{aligned}$$

The coefficient matrix of the system has two complex conjugate eigenvalues $\lambda = \pm i$, each of which has algebraic multiplicity equal to 2, but geometric multiplicity equal to 1. As pointed out above, to find the general real solution associated with the pair of complex conjugate eigenvalues, we take only the real and imaginary parts of generalized eigenvectors corresponding to one of them, more precisely we choose $\lambda = -i$. The chain of generalized eigenvector associated with $\lambda = -i$ is $X_1 = [-1, 0, -i, 0]^t$ and $X_2 = [0, -i, 0, 1]^t$. Therefore, the particular corresponding to $\lambda = -i$ is

$$\mathbf{s}(x) = e^{-ix} \begin{bmatrix} -1 \\ 0 \\ -i \\ 0 \end{bmatrix} + e^{-ix} \left\{ x \begin{bmatrix} -1 \\ 0 \\ -i \\ 0 \end{bmatrix} + \begin{bmatrix} 0 \\ -i \\ 0 \\ 1 \end{bmatrix} \right\},$$

that is,

$$\mathbf{s}(x) = (\cos x - i \sin x) \begin{bmatrix} -1 \\ 0 \\ -i \\ 0 \end{bmatrix} + (\cos x - i \sin x) \left\{ x \begin{bmatrix} -1 \\ 0 \\ -i \\ 0 \end{bmatrix} + \begin{bmatrix} 0 \\ -i \\ 0 \\ 1 \end{bmatrix} \right\}$$

$$= \begin{bmatrix} -\cos x - x \cos x \\ -\sin x \\ -\sin x - x \sin x \\ \cos x \end{bmatrix} + i \begin{bmatrix} \sin x + x \sin x \\ -\cos x \\ -\cos x - x \cos x \\ -\sin x \end{bmatrix}.$$

The linear combination of real and imaginary parts of $\mathbf{s}(x)$ is then the real solution

$$\mathbf{y}(x) = c_1 \begin{bmatrix} -\cos x - x \cos x \\ -\sin x \\ -\sin x - x \sin x \\ \cos x \end{bmatrix} + c_2 \begin{bmatrix} \sin x + x \sin x \\ -\cos x \\ -\cos x - x \cos x \\ -\sin x \end{bmatrix}$$

for c_1, c_2 arbitrary constants.

12.4   Homogeneous Differential Equations of nth Order

We now consider the linear, homogeneous differential equation with constant coefficients of order n

$$y^{(n)}(x) + \sum_{i=1}^{n} a_i y^{(n-i)}(x) = 0, \quad \text{where} \quad a_1, \dots, a_n \quad \text{are fixed real constants.}$$

$$(12.34)$$

Any function $z(x)$ defined in an interval $I \subseteq \mathbb{R}$ is a solution of (12.34) if

(1) $z(x)$ is at least n-times differentiable in I;

(2) $z^{(n)}(x) + \sum_{i=1}^{n} a_i z^{(n-i)}(x) = 0$, for any $x \in I$.

Let L be the differential operator of order n with constant coefficients defined by

$$Ly = y^{(n)} + \sum_{i=1}^{n} a_i y^{(n-i)} \qquad (12.35)$$

for any function $y = y(x)$ at least n-times differentiable over I. Thus, L defines a map $C^n(I) \to C^0(I)$ such that Ly is a continuous function, for any complex-valued function $y = y(x)$ at least n-times differentiable over I. We trivially note that, for any $c_1, c_2 \in \mathbb{C}$ and $y_1, y_2 \in C^n(I)$,

$$L(c_1 y_1 + c_2 y_2) = c_1 L y_1 + c_2 L y_2,$$

that is, L is a linear operator. In particular, this means that if $y_1, \ldots, y_k$ are k solutions of (12.34), then any linear combination of them is a solution of (12.34). Therefore, the set

$$V = \{y \in C^n(I) \; : \; Ly = 0\}$$

is a complex vector space.

Theorem 12.24 *The dimension of the vector space V is precisely equal to n.*

Proof Consider the following set of initial values problems:

(P_1) $Ly = 0, y(x_0) = 1, y'(x_0) = 0, \ldots, y^{(n-1)}(x_0) = 0$

(P_2) $Ly = 0, y(x_0) = 0, y'(x_0) = 1, \ldots, y^{(n-1)}(x_0) = 0$

$$\qquad (12.36)$$

$\cdots \cdots \cdots \cdots$

(P_n) $Ly = 0, y(x_0) = 0, y'(x_0) = 0, \ldots, y^{(n-1)}(x_0) = 1$

and assume that functions $y_1(x), \ldots, y_n(x)$ are solutions of the problems $(P_1), \ldots, (P_n)$, respectively. Thus, $y = c_1 y_1 + \cdots + c_n y_n$ $(c_i \in \mathbb{R})$ is solution of the initial value problem

$$Ly = 0, y(x_0) = c_1, y'(x_0) = c_2, \ldots, y^{(n-1)}(x_0) = c_n.$$

Hence, if we suppose that $c_1 y_1 + \cdots + c_n y_n = 0$, i.e., $y = 0$, it follows $c_1 = c_2 = \cdots = c_n = 0$, that is, $y_1, \ldots, y_n \in V$ are linearly independent functions.
Let now $w(x) \in V$ be any solution of (12.34) and define the following function:

$$y(x) = w(x_0) y_1(x) + w'(x_0) y_2(x) + \cdots + w^{(n-1)}(x_0) y_n(x),$$

where $x_0 \in I$ is precisely the previously fixed point in the initial values problems $(P_1), \ldots, (P_n)$. By the definition of $y_1, \ldots, y_n$, we observe that $Ly = 0$ and also

$$y(x_0) = w(x_0), \, y'(x_0) = w'(x_0), \ldots, y^{(n-1)(x_0)} = w^{(n-1)(x_0)}.$$

Therefore, y and w are solutions of the same initial value problem and this implies that $y(x) = w(x)$, for any $x \in I$. Therefore, the arbitrary solution $w(x) \in V$ is a linear combination of the linearly independent solutions $y_1, \ldots, y_n$. The arbitrariness of w allows us to conclude that $\{y_1, \ldots, y_n\}$ is a set of linear independent generators for the vector space V, that is, $dim_{\mathbb{R}} V = n$.

At this point, before proceeding with the determination of methods to obtain the solutions of (12.34), we recall how to estimate if a set of known solutions is linear independent or dependent. To do this, we suppose $\{y_1, \ldots, y_n\}$ is a set of solutions for (12.34).

Definition 12.25 The *Wronskian* of $\{y_1, \ldots, y_n\}$ is defined on the interval I to be the determinant

$$W(x) = \begin{vmatrix} y_1(x) & y_2(x) & \cdots & y_n(x) \\ y_1'(x) & y_2'(x) & \cdots & y_n'(x) \\ y_1''(x) & y_2''(x) & \cdots & y_n''(x) \\ \cdots & \cdots & \cdots & \cdots \\ y_1^{(n-1)}(x) & y_2^{(n-1)}(x) & \cdots & y_n^{(n-1)}(x) \end{vmatrix}.$$

Solutions $\{y_1, \ldots, y_n\}$ are linearly independent if and only if $W(x_0) \neq 0$ at some point $x_0 \in I$.

The Reduction to a Linear System
A first approach for solving (12.34) is to show how we can reduce it to a system of linear ordinary differential equations and solve it by using the above-discussed method.
In fact, setting

$$y_1 = y, \, y_2 = y', \, y_3 = y'', \ldots, y_n = y^{(n-1)}, \, y_n' = y^{(n)},$$

we obtain the system of the form

$$\begin{aligned} y_1' &= y_2 \\ y_2' &= y_3 \\ \cdots &\quad \cdots \cdots \\ y_{n-1}' &= y_n \\ y_n' &= -a_n y_1 - a_{n-1} y_2 - a_{n-2} y_3 - \cdots \cdots - a_1 y_n \end{aligned} \tag{12.37}$$

whose coefficient matrix is

$$A = \begin{bmatrix} 0 & 1 & 0 & \cdots & 0 \\ 0 & 0 & 1 & \cdots & 0 \\ \cdots & \cdots & \cdots & \cdots & \cdots \\ 0 & 0 & 0 & \cdots & 1 \\ -a_n & -a_{n-1} & \cdots & & -a_1 \end{bmatrix}.$$ (12.38)

Hence, if the vector-valued function $\mathbf{f}(x) = [y_1(x), \ldots, y_n(x)]^T$ is the solution of system (12.37), the solution of the original differential equation (12.34) is precisely the first component $y_1(x)$ of $\mathbf{f}(x)$.

The above matrix A is usually called *the Frobenius companion matrix of the monic polynomial*

$$p(x) = x^n + a_1 x^{n-1} + a_2 x^{n-2} + \cdots + a_{n-1} x + a_n = x^n + \sum_{i=1}^n a_i x^{n-i}.$$

Note that, if n is the degree of a monic polynomial, its companion matrix has order n. For example, the 4×4 companion matrix of $x^4 + 2x^3 - 3x^2 + x + 1$ is

$$\begin{bmatrix} 0 & 1 & 0 & 0 \\ 0 & 0 & 1 & 0 \\ 0 & 0 & 0 & 1 \\ -1 & -1 & 3 & -2 \end{bmatrix}.$$

The above discussions clearly show that it is important, from an analytic viewpoint, to describe some properties of companion matrices. In particular, we focus our attention in order to compute eigenvalues and eigenvectors of a companion matrix. To do this, we prove some easy known results.

Theorem 12.26 *Let A be the companion matrix of the polynomial*

$$p(x) = x^n + a_1 x^{n-1} + a_2 x^{n-2} + \cdots + a_{n-1} x + a_n = x^n + \sum_{i=1}^n a_i x^{n-i}.$$

Then the characteristic polynomial of A is $(-1)^n p(x)$. Moreover, the minimal polynomial of A coincides with the characteristic one.

Proof We prove the first part of the theorem by induction on the order n of the matrix (the degree of $p(x)$).

For $n = 1$, $p(x) = x + a$ and $A = [-a]$. Hence, the characteristic polynomial of A is $-a - \lambda = (-1)^1(\lambda + a) = (-1)^1 p(\lambda)$.

Suppose the assertion is true for any $(n-1) \times (n-1)$ companion matrix (with $n \geq 2$). Since A has order n, its characteristic polynomial is equal to the determinant

$$|A - \lambda I| = \begin{vmatrix} -\lambda & 1 & 0 & \cdots & & 0 \\ 0 & -\lambda & 1 & \cdots & & 0 \\ \cdots & \cdots & \cdots & \cdots & & \cdots \\ 0 & 0 & \cdots & -\lambda & & 1 \\ -a_n & -a_{n-1} & \cdots & \cdots & & -a_1 - \lambda \end{vmatrix}.$$

We compute $|A - \lambda I|$ by using the cofactor expansion with respect to the first column:

$$|A - \lambda I| = (-\lambda) \begin{vmatrix} -\lambda & 1 & 0 & \cdots & & 0 \\ 0 & -\lambda & 1 & \cdots & & 0 \\ \cdots & \cdots & \cdots & \cdots & & \cdots \\ 0 & 0 & \cdots & -\lambda & & 1 \\ -a_{n-1} & -a_{n-2} & \cdots & \cdots & & -a_1 - \lambda \end{vmatrix}$$

$$+ (-1)^{n+1}(-a_n) \begin{vmatrix} 1 & 0 & 0 & \cdots & 0 \\ -\lambda & 1 & 0 & \cdots & 0 \\ \cdots & \cdots & \cdots & \cdots & \cdots \\ 0 & \cdots & -\lambda & 1 & 0 \\ 0 & 0 & \cdots & -\lambda & 1 \end{vmatrix}.$$

Notice that the first determinant is equal to the characteristic polynomial of the companion matrix of the polynomial

$$x^{n-1} + a_1 x^{n-2} + \cdots + a_{n-2}x + a_{n-1}.$$

So, by induction hypothesis, the first determinant is

$$(-1)^{n-1} \left(\lambda^{n-1} + a_1 \lambda^{n-2} + \cdots + a_{n-2}\lambda + a_{n-1} \right).$$

Moreover, the second determinant is trivially equal to $(-1)^{n+1}(-a_n) = (-1)^n(a_n)$. Hence,

$$|A - \lambda I| = (-1)^{n-1}(-\lambda)\left(\lambda^{n-1} + a_1 \lambda^{n-2} + \cdots + a_{n-2}\lambda + a_{n-1} \right) + (-1)^n(a_n)$$

$$= (-1)^n \left(\lambda^n + a_1 \lambda^{n-1} + \cdots + a_{n-2}\lambda^2 + a_{n-1}\lambda + a_n \right)$$

as desired.

Let now λ_0 be any eigenvalue of A and consider the matrix $A - \lambda_0 I$. Since its determinant must be zero, its rank is less than or equal to $n - 1$. On the other hand, writing

$$A - \lambda_0 I = \begin{bmatrix} -\lambda_0 & 1 & 0 & \ldots & & 0 \\ 0 & -\lambda_0 & 1 & \ldots & & 0 \\ \ldots & \ldots & \ldots & \ldots & & \ldots \\ 0 & 0 & \ldots & -\lambda_0 & & 1 \\ -a_n & -a_{n-1} & \ldots & \ldots & & -a_1 - \lambda_0 \end{bmatrix}$$

and deleting the first column and the last row, we obtain the $(n-1) \times (n-1)$ lower triangular submatrix

$$\begin{bmatrix} 1 & 0 & \ldots & 0 \\ -\lambda_0 & 1 & \ldots & 0 \\ \ldots & \ldots & \ldots & \ldots \\ 0 & \ldots & -\lambda_0 & 1 \end{bmatrix}$$

whose determinant is equal to 1. Thus, the rank of $A - \lambda_0 I$ is precisely equal to $n - 1$. This is enough to conclude that the dimension of the eigenspace associated with λ_0 is equal to 1, that is, there will be only one Jordan block corresponding to λ_0 (when we consider the canonical form of the matrix), and its dimension is equal to the algebraic multiplicity of λ_0 as the root of the characteristic polynomial. Repeating this discussion for any eigenvalue of A, we can affirm that the minimal polynomial of A coincides with its characteristic one.

Theorem 12.27 *Let A be the companion matrix of the polynomial*

$$p(x) = x^n + a_1 x^{n-1} + a_2 x^{n-2} + \cdots + a_{n-1} x + a_n = x^n + \sum_{i=1}^{n} a_i x^{n-i}.$$

If λ is an eigenvalue of A, then its corresponding eigenspace is generated by the eigenvector $[1, \lambda, \lambda^2, \ldots, \lambda^{n-1}]^T$.

Proof Assume that $X = [x_1, \ldots, x_n]^t$ is the eigenvector corresponding to λ. Thus, X is the solution of the homogeneous linear system whose coefficient matrix is $A - \lambda I$:

$$\begin{bmatrix} -\lambda & 1 & 0 & \ldots & & 0 \\ 0 & -\lambda & 1 & \ldots & & 0 \\ \ldots & \ldots & \ldots & \ldots & & \ldots \\ 0 & 0 & \ldots & -\lambda & & 1 \\ -a_n & -a_{n-1} & \ldots & \ldots & & -a_1 - \lambda \end{bmatrix} \begin{bmatrix} x_1 \\ x_2 \\ \ldots \\ \ldots \\ x_n \end{bmatrix} = 0.$$

We have already proved that the rank of $A - \lambda I$ is equal to $n - 1$; more precisely, the first $n - 1$ rows of the matrix are linearly independent. Hence, we may construct the homogeneous linear system by using exactly these lines:

$$-\lambda x_1 + x_2 \quad = \ 0$$

$$-\lambda x_2 + x_3 \quad = \ 0$$

(12.39)

$$\cdots \qquad \cdots\cdots$$

$$-\lambda x_{n-1} + x_n = \ 0.$$

The easy solution of (12.39) gives the required eigenvector.

Example 12.28 Solve the following initial value problem:

$$y''' - 3y'' + 4y = 0,$$

$$y(0) = 1, \quad y'(0) = 0, \quad y''(0) = -1.$$

We firstly set

$$y_1 = y, y_2 = y', y_3 = y'', y_3' = y'''.$$

Then we have the system

$$y_1' = y_2,$$

$$y_2' = y_3,$$

(12.40)

$$y_3' = -4y_1 + 3y_3,$$

whose coefficient matrix is

$$A = \begin{bmatrix} 0 & 1 & 0 \\ 0 & 0 & 1 \\ -4 & 0 & 3 \end{bmatrix}.$$

The matrix A has two distinct real eigenvalues: $\lambda_1 = -1$ having algebraic multiplicity equal to 1; $\lambda_2 = 2$ having algebraic multiplicity equal to 2.
The eigenvector generating the null space of $A + I$ is $X_1 = (1, -1, 1)$ (that is, the eigenvector associated with $\lambda_1 = -1$).
Notice that, since $\lambda_2 = 2$ has geometric multiplicity equal to 1, there exists one Jordan block of order 2 generated by the generalized eigenvectors of λ_2. By easy computation, we get $N_{1,\lambda_2} = \langle (1, 2, 4) \rangle$ and $N_{2,\lambda_2} = \langle (1, 0, -4), (0, 1, 4) \rangle$. The corresponding chain of generalized eigenvectors of order 2 is $X_1 = (1, 2, 4)$, $X_2 = (0, 1, 4)$.
Hence, the Jordan canonical form of A is

$$A' = \begin{bmatrix} 2 & 1 & 0 \\ 0 & 2 & 0 \\ 0 & 0 & -1 \end{bmatrix}$$

relative to the Jordan basis $\{(1, 2, 4), (0, 1, 4), (1, -1, 1)\}$. Thus, a general solution for the system (12.40) is

$$c_1 e^{2x} \begin{bmatrix} 1 \\ 2 \\ 4 \end{bmatrix} + c_2 e^{2x} \left\{ x \begin{bmatrix} 1 \\ 2 \\ 4 \end{bmatrix} + \begin{bmatrix} 0 \\ 1 \\ 4 \end{bmatrix} \right\} + c_3 e^{-x} \begin{bmatrix} 1 \\ -1 \\ 1 \end{bmatrix},$$

that is,

$$y_1(x) = \{c_1 + c_2 x\} e^{2x} + c_3 e^{-x},$$

$$y_2(x) = \{2c_1 + c_2(2x + 1)\} e^{2x} - c_3 e^{-x},$$

$$y_3(x) = \{4c_1 + c_2(4x + 4)\} e^{2x} + c_3 e^{-x}.$$

In particular,

$$y(x) = y_1(x) = \{c_1 + c_2 x\} e^{2x} + c_3 e^{-x}$$

is the general solution of the original differential equation, so that

$$y'(x) = 2c_1 e^{2x} + 2c_2 x e^{2x} + c_2 e^{2x} - c_3 e^{-x},$$

$$y''(x) = 4c_1 e^{2x} + 4c_2 x e^{2x} + 4c_2 e^{2x} + c_3 e^{-x}.$$

From the given initial conditions, we have

$$1 = y(0) \quad = c_1 + c_3,$$

$$0 = y'(0) \quad = 2c_1 + c_2 - c_3,$$

$$-1 = y''(0) = 4c_1 + 4c_2 + c_3.$$

Solving the above linear system, one has $c_1 = \frac{2}{3}, c_2 = -1, c_3 = \frac{1}{3}$. So the solution satisfying the initial conditions is

$$y(x) = \frac{2}{3} e^{2x} - x e^{2x} + \frac{1}{3} e^{-x}.$$

The Characteristic Polynomial of a Differential Equation

Here we look at a different method for solving homogeneous differential equations of nth order. To do this, we observe that the exponential function $e^{\lambda x}$ ($\lambda \in \mathbb{C}$) has appeared several times in the solutions of differential equations of the form (12.34). We then consider the function $y(x) = e^{\lambda x}$ and ask ourselves when it is possible that this is a solution for (12.34).

Since $y^{(j)}(x) = \lambda^j e^{\lambda x}$ for any $j \geq 0$, it follows that $y(x)$ is a solution for (12.34) if and only if

$$e^{\lambda x}\{\lambda^n + \sum_{i=1}^{n} a_i \lambda^{n-i}\} = 0 \quad \forall x \in I.$$

This will be zero only if

$$\lambda^n + \sum_{i=1}^{n} a_i \lambda^{n-i} = 0 \quad \forall x \in I$$

since the exponential is never zero. We then observe that the complex exponential $e^{\lambda x}$ can be a solution for the given differential equation. More precisely, it is a solution for (12.34) if and only if λ is a root of the polynomial

$$p(t) = t^n + \sum_{i=1}^{n} a_i t^{n-i}.$$

This polynomial is typically called the *characteristic polynomial* of the differential equation. Factoring the polynomial over $\mathbb{C}$, we obtain the n roots of $p(t)$, namely $\lambda_1, \ldots, \lambda_n \in \mathbb{C}$. These roots are not necessarily all distinct. If each of them is counted with its multiplicity, then the polynomial factors as

$$p(t) = (t - \lambda_1)^{m_1} (t - \lambda_2)^{m_2} \cdots (t - \lambda_k)^{m_k},$$

where $\lambda_1, \ldots, \lambda_k$ are the distinct roots of $p(t)$.

Let L be the linear differential operator of order n with constant coefficients defined by (12.35). We notice that L can be written as

$$L = D^{(n)} + \sum_{1=1}^{n} a_i D^{(n-i)} = p(D), \tag{12.41}$$

where $p(t)$ is precisely the characteristic polynomial associated with the differential equation and $D^k = (\frac{d}{dx})^k$ ($k = 1, \ldots, n$), denoting by $\frac{d}{dx}$ the differentiation operator. Recalling that Eq. (12.34) can also be written $Ly = 0$, we see that any solution $y(x)$ of (12.34) actually is an element of the null space of L, that is, $y(x) \in Ker(L)$.

Remark 12.29 Suppose there exist $L_1, \ldots, L_k$ linear differential operators with constant coefficients mapping $C^n(I) \to C^0(I)$, such that $L = L_1 L_2 \cdots L_k$. Then $Ker(L_i) \subseteq Ker(L)$, for any $i = 1, \ldots, k$.

The idea is now to discuss the characteristic polynomial associated with the differential equation, in order to obtain the linearly independent functions generating the null space of L, that is, a basis for $Ker(L)$. The roots of $p(t)$ will have three possible forms:

(1) They are all real and distinct.
(2) They are all real but not all distinct (at least one of them has algebraic multiplicity greater than 1).
(3) There exist at least two complex roots $\lambda = \alpha + i\beta$ and $\bar{\lambda} = \alpha - i\beta$, for $\alpha, \beta \in \mathbb{R}$ and $\beta \neq 0$. Also in this case, each complex root must be counted with its multiplicity.

Hence, we look at each of these cases to get the general solution for Eq. (12.34). To do this, it is important to recall the following well-known fact:

Remark 12.30 Suppose there exist $L_1, \ldots, L_k$ linear differential operators with constant coefficients mapping $C^n(I) \to C^0(I)$, such that $L = L_1 L_2 \cdots L_k$. Then $Ker(L_i) \subseteq Ker(L)$, for any $i = 1, \ldots, k$.

Case (1): The roots of $p(t)$ are all real and distinct.
In this case, the characteristic polynomial factors as

$$p(t) = (t - \lambda_1)(t - \lambda_2) \cdots (t - \lambda_n),$$

where $\lambda_1, \ldots, \lambda_n$ are the distinct roots of $p(t)$, as well as the differential operator L similarly factors as

$$L = (D - \lambda_1)(D - \lambda_2) \cdots (D - \lambda_n),$$

where

$$(D - \lambda_i)y = y' - \lambda_i y \quad \forall i = 1, \ldots, n. \tag{12.42}$$

Solving (12.42) for any $i = 1, \ldots, n$, we have the solutions

$$y_1(x) = e^{\lambda_1 x}, \ldots, y_n(x) = e^{\lambda_n x}$$

and, by Remark 12.30, $y_1, \ldots, y_n \in Ker(L)$, that is, $\{y_1, \ldots, y_n\}$ is a set of n solutions for (12.34). Moreover, computing the Wronskian, we get

$$W(x) = \begin{vmatrix} e^{\lambda_1 x} & e^{\lambda_2 x} & \cdots & e^{\lambda_n x} \\ \lambda_1 e^{\lambda_1 x} & \lambda_2 e^{\lambda_2 x} & \cdots & \lambda_n e^{\lambda_n x} \\ \lambda_1^2 e^{\lambda_1 x} & \lambda_2^2 e^{\lambda_2 x} & \cdots & \lambda_n^2 e^{\lambda_n x} \\ \cdots & \cdots & \cdots & \cdots \\ \lambda_1^{n-1} e^{\lambda_1 x} & \lambda_2^{n-1} e^{\lambda_2 x} & \cdots & \lambda_n^{n-1} e^{\lambda_n x} \end{vmatrix}$$

$$= e^{(\lambda_1 + \cdots + \lambda_n)x} \begin{vmatrix} 1 & 1 & \cdots & 1 \\ \lambda_1 & \lambda_2 & \cdots & \lambda_n \\ \lambda_1^2 & \lambda_2^2 & \cdots & \lambda_n^2 \\ \cdots & \cdots & \cdots & \cdots \\ \lambda_1^{n-1} & \lambda_2^{n-1} & \cdots & \lambda_n^{n-1} \end{vmatrix}$$

$$= e^{(\lambda_1 + \cdots + \lambda_n)x} \prod_{i > j} (\lambda_i - \lambda_j) \neq 0 \quad \forall x \in I.$$

Hence, $\{y_1, \ldots, y_n\}$ is a linearly independent set, that is, a basis for the vector space of solutions. The general solution can be written as

$$y(x) = c_1 e^{\lambda_1 x} + \cdots + c_n e^{\lambda_n x} \quad c_1, \ldots, c_n \in \mathbb{R}.$$

Example 12.31 Solve the differential equation

$$y''' - \frac{9}{2} y'' + 5y' - \frac{3}{2} y = 0.$$

The characteristic polynomial associated with the equation is

$$p(t) = t^3 - \frac{9}{2} t^2 + 5t - \frac{3}{2}$$

having three distinct real roots: $\lambda_1 = 1$, $\lambda_2 = \frac{1}{2}$ and $\lambda_3 = 3$, all of them of algebraic multiplicity equal to 1.

Thus, the general solution of the differential equation is

$$y(x) = c_1 e^x + c_2 e^{\frac{1}{2} x} + c_3 e^{3x}.$$

Case (2): The roots of $p(t)$ are all real but not all distinct.
Here we assume

$$p(t) = (t - \lambda_1)^{m_1} (t - \lambda_2)^{m_2} \cdots (t - \lambda_k)^{m_k},$$

where $\lambda_1, \ldots, \lambda_k$ are the distinct roots of $p(t)$ and m_i is the algebraic multiplicity of λ_i, for $i = 1, \ldots, k$. In parallel, we can factor L as

$$L = (D - \lambda_1)^{m_1} (D - \lambda_2)^{m_2} \cdots (D - \lambda_k)^{m_k}.$$

To fully describe the present case, we need to premise some results. More precisely,

Proposition 12.32 *Let λ be a root of $p(t)$, having algebraic multiplicity equal to m. Then the functions*

$$e^{\lambda x}, x e^{\lambda x}, \ldots, x^{m-1} e^{\lambda x}$$

form a basis for the null space $Ker(D - \lambda)^m$.

Proof For $m = 1$, it is clear that $(D - \lambda) e^{\lambda x} = 0$. By induction, assume that

$$e^{\lambda x}, x e^{\lambda x}, \ldots, x^{m-2} e^{\lambda x} \in Ker(D - \lambda)^{m-1}.$$

Hence,

$$(D - \lambda)^m x^{m-1} e^{\lambda x} = (D - \lambda)^{m-1} (D - \lambda) x^{m-1} e^{\lambda x}$$

$$= (D - \lambda)^{m-1} (m - 1) x^{m-2} e^{\lambda x}$$

$$= 0 \quad \forall x \in I$$

proving that

$$e^{\lambda x}, x e^{\lambda x}, \ldots, x^{m-1} e^{\lambda x} \in Ker(D - \lambda)^m.$$

To complete the proof, we then prove that those functions are linearly independent. Let $c_1, \ldots, c_m \in \mathbb{R}$ be such that

$$c_1 e^{\lambda x} + c_2 x e^{\lambda x} + \cdots + c_m x^{m-1} e^{\lambda x} = 0 \tag{12.43}$$

and, by contradiction, assume there exists at least one index $i \in \{1, \ldots, m\}$ such that $c_i \neq 0$. Since $e^{\lambda x}$ is never zero, (12.43) says that

$$c_1 + c_2 x + \cdots + c_m x^{m-1} = 0.$$

On the other hand, functions $1, x, x^2, \ldots, x^{m-1}$ are clearly linearly independent, since their Wronskian is

$$\begin{vmatrix} 1 & x & x^2 & \cdots & x^{m-1} \\ 0 & 1 & 2x & \cdots & (m-1)x^{m-2} \\ 0 & 0 & 2 & \cdots & (m-2)(m-1)x^{m-3} \\ \cdots\cdots\cdots\cdots\cdots & & \cdots \\ 0 & 0 & 0 & \cdots & (m-1)! \end{vmatrix} \neq 0 \quad \forall x \in \mathbb{R}.$$

Of course, this is not possible, since $c_i \neq 0$.

Proposition 12.33 Let $\lambda_1, \ldots, \lambda_k$ be real distinct numbers and $f_1(x), \ldots, f_k(x)$ polynomials. If each $f_i(x)$ $(i = 1, \ldots, k)$ is not identically zero in $\mathbb{R}$, then the functions $f_1(x)e^{\lambda_1 x}, \ldots, f_k(x)e^{\lambda_k x}$ are linearly independent.

Proof We prove the result by induction. Assume firstly $k = 2$ and let $c_1, c_2 \in \mathbb{R}$ be such that

$$c_1 f_1(x)e^{\lambda_1 x} + c_2 f_2(x)e^{\lambda_2 x} = 0 \quad \forall x \in \mathbb{R}. \tag{12.44}$$

Since $\lambda_1 \neq \lambda_2$, without loss of generality, we may assume $\lambda_2 \neq 0$. Hence, by (12.44), we get

$$c_1 f_1(x)e^{(\lambda_1 - \lambda_2)x} + c_2 f_2(x) = 0 \quad \forall x \in \mathbb{R}. \tag{12.45}$$

If $c_2 \neq 0$, relation (12.45) implies

$$c_1 c_2^{-1} f_1(x)e^{(\lambda_1 - \lambda_2)x} = -f_2(x) \quad \forall x \in \mathbb{R}$$

which cannot occur, due to the fact that $\lambda_1 - \lambda_2 \neq 0$ and both $f_1(x)$ and $f_2(x)$ are not identically zero. Hence, we may assert that $c_2 = 0$ and, by (12.45), $c_1 = 0$ follows trivially.

Suppose now that the result is true for the $k - 1$ functions

$$f_1(x)e^{\lambda_1 x}, \ldots, f_{k-1}(x)e^{\lambda_{k-1} x}.$$

Our final aim is to show that it holds again for the k functions

$$f_1(x)e^{\lambda_1 x}, \ldots, f_k(x)e^{\lambda_k x}.$$

In this sense, we suppose there are $c_1, \ldots, c_k \in \mathbb{R}$ such that

$$\sum_{i=1}^{k} c_i f_i(x) e^{\lambda_i x} = 0 \quad \forall x \in \mathbb{R}. \tag{12.46}$$

As above, we may assume $\lambda_k \neq 0$ and reduce (12.46) to

$$\sum_{i=1}^{k-1} c_i f_i(x) e^{(\lambda_i - \lambda_k) x} = -c_k f_k(x) \quad \forall x \in \mathbb{R}. \tag{12.47}$$

If we denote $n = degree(f_k)$ and suppose $c_k \neq 0$, by (12.47), it follows

$$\left(\frac{d}{dx} \right)^{n+1} \left\{ \sum_{i=1}^{k-1} c_i f_i(x) e^{(\lambda_i - \lambda_k) x} \right\} = 0 \quad \forall x \in \mathbb{R}$$

which is a contradiction, since $\lambda_i - \lambda_k \neq 0$ and $f_i(x)$ is not identically zero, for any $i = 1, \ldots, k - 1$.

Thus, $c_k = 0$ and relation (12.47) reduces to

$$\sum_{i=1}^{k-1} c_i f_i(x) e^{\mu_i x} = 0 \quad \forall x \in \mathbb{R}, \tag{12.48}$$

where

$$\mu_1 = (\lambda_1 - \lambda_k), \mu_2 = (\lambda_2 - \lambda_k), \ldots, \mu_{k-1} = (\lambda_{k-1} - \lambda_k)$$

are all distinct. Then, by induction and relation (12.48), $c_1 = \cdots = c_{k-1} = 0$, as required.

Proposition 12.34 *Let*

$$p(t) = (t - \lambda_1)^{m_1} (t - \lambda_2)^{m_2} \cdots (t - \lambda_k)^{m_k}$$

be the characteristic polynomial for the differential equation (12.34), where $\lambda_1, \ldots,$ λ_k are the distinct roots of $p(t)$ and m_i is the algebraic multiplicity of λ_i, for $i = 1, \ldots, k$. Then functions

$$y_{is}(x) = x^{s-1} e^{\lambda_i x} \quad \forall 1 \le i \le k \quad \forall 1 \le s \le m_i \tag{12.49}$$

are linearly independent.

Proof For $i = 1, \ldots, k$ and $s = 1, \ldots, m_i$, let $c_{is} \in \mathbb{R}$ be such that

$$\sum_{i=1}^{k} \sum_{s=1}^{m_i} c_{is} x^{s-1} e^{\lambda_i x} = 0 \quad \forall x \in \mathbb{R}. \tag{12.50}$$

For any $i \in \{1, \ldots, k\}$, here we denote $f_i(x) = \sum_{s=1}^{m_i} c_{is} x^{s-1}$.

In case, any function $f_i(x)$ is identically zero in $\mathbb{R}$, it follows trivially $c_{is} = 0$, for any $i \in \{1, \ldots, k\}$ and for any $s \in \{1, \ldots, m_i\}$.
Then we may assume that

- there exist some $i_1, \ldots, i_h$ ($1 \le h \le k$) such that the polynomials $f_{i_1}, \ldots, f_{i_h}$ are not identically zero;
- $f_r(x) = 0$, for any $x \in \mathbb{R}$, if $r \notin \{i_1, \ldots, i_h\}$.

Hence, we reduce relation (12.50) to

$$\sum_{j=1}^{h} f_{i_j}(x) e^{\lambda_{i_j} x} = 0 \quad \forall x \in \mathbb{R}. \tag{12.51}$$

But, in light of Proposition 12.33 and since $f_{i_1}, \ldots, f_{i_h}$ are not identically zero, the relation (12.51) represents a contradiction.

At this point, we are ready to describe the general solution of Eq. (12.34) in the case its characteristic polynomial is

$$p(t) = (t - \lambda_1)(t - \lambda_2) \cdots (t - \lambda_n),$$

where $\lambda_1, \ldots, \lambda_n$ are the distinct roots of $p(t)$.
Using the results contained in Propositions 12.32, 12.33 and 12.34, we firstly list the n linearly independent solutions, that is, the functions of the basis for the null space $Ker(L)$. Those functions are

$$e^{\lambda_1 x}, x e^{\lambda_1 x}, \ldots, x^{m_1 - 1} e^{\lambda_1 x}$$

$$e^{\lambda_2 x}, xe^{\lambda_2 x}, \ldots, x^{m_2-1}e^{\lambda_2 x}$$

$$\ldots, \ldots, \ldots, \ldots$$

$$e^{\lambda_k x}, xe^{\lambda_k x}, \ldots, x^{m_k-1}e^{\lambda_k x}.$$

Then we write the general solution

$$y(x) = \sum_{i=1}^{k}\sum_{s=1}^{m_i} c_{is}x^{s-1}e^{\lambda_i x}. \qquad (12.52)$$

Example 12.35 Repeat Example 12.28 and solve the differential equation

$$y''' - 3y'' + 4y = 0.$$

The characteristic polynomial associated with the equation is

$$p(t) = t^3 - 3t^2 + 4$$

having two distinct real roots: $\lambda_1 = -1$ of algebraic multiplicity equal to 1; $\lambda_2 = 2$ of algebraic multiplicity equal to 2.
Thus, the general solution of the differential equation is

$$y(x) = c_1 e^{-x} + c_2 e^{2x} + c_3 x e^{2x}.$$

Example 12.36 Solve the differential equation

$$y^{iv} - 2y''' - 3y'' + 8y' - 4y = 0.$$

The characteristic polynomial associated with the equation is

$$p(t) = t^4 - 2t^3 - 3t^2 + 8t - 4$$

having two distinct real roots: $\lambda_1 = 1$ and $\lambda_2 = 2$, both of them of algebraic multiplicity equal to 2.
Thus, the general solution of the differential equation is

$$y(x) = c_1 e^x + c_2 x e^x + c_3 e^{2x} + c_4 x e^{2x}.$$

Example 12.37 Solve the differential equation

$$y^{iv} - 8y''' + 18y'' - 27y = 0.$$

The characteristic polynomial associated with the equation is

$$p(t) = t^4 - 8t^3 + 18t^2 - 27$$

having two distinct real roots: $\lambda_1 = -1$ of multiplicity equal to 1 and $\lambda_2 = 3$ of multiplicity equal to 3.

Thus, the general solution of the differential equation is

$$y(x) = c_1 e^{-x} + c_2 e^{3x} + c_3 x e^{3x} + c_4 x^2 e^{3x}.$$

Case (3): $p(t)$ has some complex roots.

We finally need to deal with complex roots. Assuming that $\lambda = \alpha + i\beta$ $(0 \neq \beta \in \mathbb{R})$ occurs m-times in the list of roots (i.e., λ has a multiplicity of m), we have that $\bar{\lambda} = \alpha - i\beta$ is again a root for $p(t)$, also having multiplicity equal to m. In this case, we can use the work from the repeated roots above to get the following set of $2m$ complex-valued solutions:

$$e^{\alpha x}(\cos \beta x + i \sin \beta x), \, x e^{\alpha x}(\cos \beta x + i \sin \beta x), \ldots, x^{m-1} e^{\alpha x}(\cos \beta x + i \sin \beta x)$$

which represent a basis for the null space $Ker(D - \lambda)^m$, and

$$e^{\alpha x}(\cos \beta x - i \sin \beta x), \, x e^{\alpha x}(\cos \beta x - i \sin \beta x), \ldots, x^{m-1} e^{\alpha x}(\cos \beta x - i \sin \beta x)$$

which represent a basis for the null space $Ker(D - \bar{\lambda})^m$. Exactly as seen in the previous cases, those functions give their contribution to the constitution of the whole basis for $Ker(L)$.

If we need to express the real-valued solutions of (12.34), we use Euler's formula on the first set of complex-valued solutions above. Then we split each complex solution into its real and imaginary parts to arrive at the following set of $2m$ real-valued solutions:

$$e^{\alpha x} \cos \beta x, \, x e^{\alpha x} \cos \beta x, \ldots, x^{m-1} e^{\alpha x} \cos \beta x,$$

$$e^{\alpha x} \sin \beta x, \, x e^{\alpha x} \sin \beta x, \ldots, x^{m-1} e^{\alpha x} \sin \beta x.$$

Those functions contribute to the general real solution, together with any other solutions corresponding to any real roots of $p(t)$.

Example 12.38 Solve the differential equation

$$y^{iv} - 16y = 0.$$

The characteristic polynomial associated with the equation is

$$p(t) = t^4 - 16$$

having two distinct real roots $\lambda_1 = 2$ and $\lambda_2 = -2$, both of which of multiplicity equal to 1, and two complex roots $\lambda_3 = 2i$ and $\overline{\lambda}_3 = -2i$.
Thus, the general solution of the differential equation is

$$y(x) = c_1 e^{2x} + c_2 e^{-2x} + c_3 e^{2ix} + c_4 e^{-2ix}$$

$$= c_1 e^{2x} + c_2 e^{-2x} + c_3(\cos 2x + i \sin 2x) + c_4(\cos 2x - i \sin 2x)$$

$$= c_1 e^{2x} + c_2 e^{-2x} + (c_3 + c_4) \cos 2x + i(c_3 - c_4) \sin 2x,$$

and the general real solution is

$$y(x) = c_1 e^{2x} + c_2 e^{-2x} + c_3 \cos 2x + c_4 \sin 2x.$$

Example 12.39 Find the real solution of the following initial value problem:

$$y''' - y'' + 4y' - 4y = 0,$$

$$y(0) = 2, \quad y'(0) = 1, \quad y''(0) = 1.$$

The characteristic polynomial associated with the equation is

$$p(t) = t^3 - t^2 + 4t - 4$$

having one real root $\lambda_1 = 1$ of multiplicity equal to 1, and two complex roots $\lambda_2 = 2i$ and $\overline{\lambda}_2 = -2i$.
Thus, the general real solution of the differential equation is

$$y(x) = c_1 e^x + c_2 \cos 2x + c_3 \sin 2x$$

so that the general real solution is

$$y'(x) = c_1 e^x - 2c_2 \sin 2x + 2c_3 \cos 2x,$$
$$y''(x) = c_1 e^x - 4c_2 \cos 2x - 4c_3 \sin 2x.$$

By the initial conditions
$$y(0) = c_1 + c_2 = 2,$$
$$y'(0) = c_1 + 2c_3 = 1,$$
$$y''(0) = c_1 - 4c_2 = 1$$

implying $c_1 = \frac{9}{5}, c_2 = \frac{1}{5}, c_3 = -\frac{2}{5}$. Hence, the solution for the initial values problem is the function
$$y(x) = \frac{9}{5}e^x + \frac{1}{5}\cos 2x - \frac{2}{5}\sin 2x.$$

Example 12.40 Solve the differential equation

$$y^{vi} + 4y^v + 4y^{iv} + 18y''' + 36y'' + 81y = 0.$$

The characteristic polynomial associated with the equation is

$$p(t) = t^6 + 4t^5 + 4t^4 + 18t^3 + 36t^2 + 81$$

having the following roots:

- $\lambda_1 = -3$ of multiplicity equal to 2.
- $\lambda_2 = \frac{1}{2} + i\frac{\sqrt{11}}{2}$ of multiplicity equal to 2.
- $\overline{\lambda_2} = \frac{1}{2} - i\frac{\sqrt{11}}{2}$ of multiplicity equal to 2.

Thus, the general solution of the differential equation is then

$$
\begin{aligned}
y(x) = \quad & c_1 e^{-3x} + c_2 x e^{-3x} \\[6pt]
& + c_3 e^{\frac{1}{2}x}\left(\cos \tfrac{\sqrt{11}}{2}x + i \sin \tfrac{\sqrt{11}}{2}x\right) \\[6pt]
& + c_4 x e^{\frac{1}{2}x}\left(\cos \tfrac{\sqrt{11}}{2}x + i \sin \tfrac{\sqrt{11}}{2}x\right) \\[6pt]
& + c_5 e^{\frac{1}{2}x}\left(\cos \tfrac{\sqrt{11}}{2}x - i \sin \tfrac{\sqrt{11}}{2}x\right) \\[6pt]
& + c_6 x e^{\frac{1}{2}x}\left(\cos \tfrac{\sqrt{11}}{2}x - i \sin \tfrac{\sqrt{11}}{2}x\right),
\end{aligned}
$$

and the general real solution is

$$
\begin{aligned}
y(x) = \quad & c_1 e^{-3x} + c_2 x e^{-3x} \\[6pt]
& + c_3 e^{\frac{1}{2}x} \cos \tfrac{\sqrt{11}}{2}x + c_4 x e^{\frac{1}{2}x} \cos \tfrac{\sqrt{11}}{2}x \\[6pt]
& + c_5 e^{\frac{1}{2}x} \sin \tfrac{\sqrt{11}}{2}x + c_6 x e^{\frac{1}{2}x} \sin \tfrac{\sqrt{11}}{2}x.
\end{aligned}
$$

Exercises

1. Solve the system

$$
\begin{aligned}
y_1' &= y_2, \\
y_2' &= 4y_1, \\
y_3' &= y_4, \\
y_4' &= 4y_3,
\end{aligned}
$$

and find a solution that satisfies the initial conditions

$$y_1(0) = 1, \quad y_2(0) = 0, \quad y_3(0) = 1, \quad y_4(0) = 0.$$

2. Find the general solution of the system

$$
\begin{aligned}
y_1' &= -5y_1 + 2y_2 + 2y_4, \\
y_2' &= 2y_1 - 2y_2 - y_4, \\
y_3' &= -5y_3 + 2y_4, \\
y_4' &= 2y_3 - 2y_4.
\end{aligned}
$$

3. Find the general solution of the system

$$
\begin{aligned}
y_1' &= y_1, \\
y_2' &= -y_2 - 4y_3 + 2y_4, \\
y_3' &= 3y_2 + y_3 - 2y_4, \\
y_4' &= y_2 - 4y_3 + y_4.
\end{aligned}
$$

4. Find the general solution of the system

$$
\begin{aligned}
y_1' &= 2y_1 + y_2 - y_3 + y_4, \\
y_2' &= 2y_2 + y_3 + 2y_4, \\
y_3' &= 2y_3 + y_4, \\
y_4' &= 2y_4.
\end{aligned}
$$

5. Find the general solution of the system

$$
\begin{aligned}
y_1' &= y_1 + 2y_2 - y_3 - y_4, \\
y_2' &= -2y_1 + y_2 + y_3 + y_4, \\
y_3' &= y_3 + 2y_4, \\
y_4' &= -2y_3 + y_4.
\end{aligned}
$$

6. Find the real-valued solution of the system

$$
\begin{aligned}
y_1' &= y_1 + 2y_2, \\
y_2' &= -2y_1 + y_2.
\end{aligned}
$$

7. Find the real-valued solution of the system

$$
\begin{aligned}
y_1' &= y_2, \\
y_2' &= -3y_1 - 2y_2.
\end{aligned}
$$

8. Find the real-valued solution of the system

$$
\begin{aligned}
y_1' &= y_1 + y_2 + y_3, \\
y_2' &= -y_1 + y_2 - y_4, \\
y_3' &= y_3 + y_4, \\
y_4' &= -y_3 + y_4.
\end{aligned}
$$

9. Find the real-valued solution of the system

$$
\begin{aligned}
y_1' &= 3y_2 - y_4, \\
y_2' &= -3y_1 + y_3, \\
y_3' &= 2y_4, \\
y_4' &= -2y_3.
\end{aligned}
$$

10. Find the real-valued solution of the system

$$
\begin{aligned}
y_1' &= 2y_1 - 5y_2 + y_4, \\
y_2' &= y_1 - 2y_2 - y_3 - y_4, \\
y_3' &= -y_3 - 6y_4, \\
y_4' &= 3y_3 + 5y_4.
\end{aligned}
$$

11. Find the solution of the following initial value problem:

$$
y'' - 2y' + 2y = 0,
$$

$$
y(0) = \frac{1}{3}, \quad y'(0) = 1.
$$

12. Solve the following initial value problem:

$$
y'' - 2y' - 8y = 0,
$$

$$
y(1) = 1, \quad y'(1) = -1.
$$

13. Solve the following initial value problem:

$$
y''' - 7y'' + 16y' - 12y = 0,
$$

$$
y(0) = 0, \quad y'(0) = 1, \quad y''(0) = 1.
$$

14. Solve the following initial value problem:

$$
y''' - y'' + 9y' - 9y = 0,
$$

$$
y(0) = 1, \quad y'(0) = 0, \quad y''(0) = 0.
$$

15. Find the general solution of the differential equation

$$y^v - 5y^{iv} + 14y''' - 22y'' + 17y' - 5y = 0.$$

16. Find the solution of the following equation:

$$y^{iv} - 2y''' + 2y'' - 2y' + y = 0.$$

17. Find the general solution of the differential equation

$$y^{iv} + y'' = 0.$$

18. Find the real solution of the differential equation

$$y^{iv} + 8y'' + 16y = 0.$$

19. Describe all possible solutions of the differential equation

$$y''' + (1 - k)y'' - ky' = 0$$

as parameter k varies in real numbers.
20. Describe all possible solutions of the differential equation

$$y''' - ky'' + k^2y' - k^3y = 0$$

as parameter $k > 0$ varies in real positive numbers.

References

1. Anton, H.: Elementary Linear Algebra. Wiley, New York (1987)
2. Artin, M.: Algebra. Pearson Education Limited, Inc. (2013)
3. Axler, S.: Linear Algebra Done Right. Springer International Publishing, Berlin (2015)
4. Clay, D.C., Lay, S.R., MacDonald, J.J.: Linear Algebra and its Applications. Pearson, London (2014)
5. Friedberg, S.H., Insel, A.J., Spence, L.E.: Linear Algebra. Pearson New International Edition (2013)
6. Garcia, S.R., Horn, R.A.: A Second Course in Linear Algebra. Cambridge University Press, Cambridge (2017)
7. Greub, W.H.: Linear Algebra. Springer, Berlin (1967)
8. Herstein, I.N.: Topics in Algebra. Wiley, New York (1975)
9. Hoffman, K., Kunze, R.: Linear Algebra. Prentice Hall of India Pvt. Ltd, New Delhi (2005)
10. Jacobson, N.: Lectures in Abstract Algebra 2: Linear Algebra. Springer, New York (1953)
11. Kostrikin, A., Manin, Y.: Linear Algebra and Geometry. Gordon and Breach Science Publishers, London (1997)
12. Lam, T.Y.: A First Course in Noncommutative Rings. Springer, New York (2001)
13. Leon, S.J.: Linear Algebra with Applications. Macmillan Publishing Company, New York (1980)
14. Marcus, M.: Finite Dimensional Multilinear Algebra Part II. Marcel Dekker, New York (1975)
15. Nobel, B., Daniel, J.: Applied Linear Algebra. Prentice Hall, Hoboken (1977)
16. Roger, B.: Linear Algebra. Rinton Press, Princeton (2001)
17. Roman, Steven: Advanced Linear Algebra. Graduate Texts in Mathematics. Springer Science Inc, New York (2005)
18. Satake, I.: Linear Algebra. Marcel Dekker Inc., New York (1975)
19. Scroggs, J.E.: Linear Algebra. Brooks/Cole Publishing Company, Belmount (1970)
20. Singh, S.: Linear Algebra. Vikas Publishing House Pvt, Ltd (1997)
21. Strang, G.: Linear Algebra and its Applications. Hardcourt Brace Jovanovich, New York (1980)
22. Strang, G.: Introduction to Linear Algebra. Wellesley-Cambridge Press, Wellesley (2016)

M. Ashraf et al., *Advanced Linear Algebra with Applications*,
https://doi.org/10.1007/978-981-16-2167-3

Index

© The Editor(s) (if applicable) and The Author(s), under exclusive license to Springer   491
Nature Singapore Pte Ltd. 2022
M. Ashraf et al., *Advanced Linear Algebra with Applications*,
https://doi.org/10.1007/978-981-16-2167-3

Printed in the United States
by Baker & Taylor Publisher Services

Printed in the United States
by Baker & Taylor Publisher Services